Positron Emission Particle Tracking

A comprehensive guide

Positron Emission Particle Tracking

A comprehensive guide

Kit Windows-Yule
School of Chemical Engineering, University of Birmingham, Edgbaston, Birmingham, UK

Leonard Nicuşan
School of Chemical Engineering, University of Birmingham, Edgbaston, Birmingham, UK

Matthew T Herald
School of Chemical Engineering, University of Birmingham, Edgbaston, Birmingham, UK

Samuel Manger
School of Chemical Engineering, University of Birmingham, Edgbaston, Birmingham, UK

David Parker
Positron Imaging Centre, School of Physics and Astronomy, University of Birmingham, Edgbaston, Birmingham, UK

IOP Publishing, Bristol, UK

ISBN 978-0-7503-3071-8 (ebook)
ISBN 978-0-7503-3069-5 (print)
ISBN 978-0-7503-3072-5 (myPrint)
ISBN 978-0-7503-3070-1 (mobi)

DOI 10.1088/978-0-7503-3071-8

Version: 20220601

IOP ebooks

British Library Cataloguing-in-Publication Data: A catalogue record for this book is available from the British Library.

Published by IOP Publishing, wholly owned by The Institute of Physics, London

IOP Publishing, Temple Circus, Temple Way, Bristol, BS1 6HG, UK

US Office: IOP Publishing, Inc., 190 North Independence Mall West, Suite 601, Philadelphia, PA 19106, USA

Multimedia content and notebook files are available for this book from https://iopscience.iop.org/book/978-0-7503-3071-8.

In memory of Dr Mike Hawkesworth, whose work helped to create PEPT, and whose generosity helped educate the next generation of PEPT researchers.

Contents

Preface

To see a World in a Grain of Sand,
And a Heaven in a Wild Flower,
Hold Infinity in the palm of your hand,
And Eternity in an hour

–William Blake

Positron emission particle tracking (PEPT) is a powerful experimental technique which allows the user to track the motion of one or more tracer particles with high spatial and temporal resolution, even through the interior of large, dense, optically-opaque systems. While PEPT may not quite allow us to see a world in a grain of sand, it nonetheless enables us to use a single, tiny particle to visualise the detailed dynamics of large, complex industrial and natural systems, and in many cases to unlock the mysteries of processes unobservable using the human eye alone.

As its title suggests, this book is intended to provide a comprehensive, 'one-stop' resource for all those who wish to learn about, use, develop, or expand their existing knowledge of, the PEPT technique. As we explain in detail in the opening chapter of the book, this guide is written so as to be accessible to those with no prior knowledge of PEPT, whilst simultaneously providing a valuable reference guide for experts in the field. In addition to text and images, it includes a host of carefully-designed interactive examples and 'sandboxes' in which the reader may use and experiment with real PEPT algorithms and data, which we hope will allow the reader to learn at a faster rate, and in greater depth, than a conventional tome.

We hope that you find this book both enlightening and enjoyable.

Acknowledgement

The authors would like to thank a number of colleagues who have provided information and material for this book, including:

Mr Daniel Weston, for providing a detailed introduction to computational fluid dynamics (CFD).

Mr Dominik Werner, for the development and inclusion of an effective and user-friendly jump-time algorithm.

Professor Jonathan Seville, for his expert insight into the field of fluidisation, and the application of PEPT in the study thereof.

Dr Aleksandar Vasiljevic, MBChB MRCS, for providing detailed information regarding the MRI technique.

Author biographies

Kit Windows-Yule

Kit Windows-Yule is a Turing Fellow and two-time Royal Academy of Engineering Industrial Fellow, currently working as a Lecturer based in the University of Birmingham's School of Chemical Engineering. He has used PEPT in diverse projects addressing various issues of contemporary importance, including waste plastic recycling, biofuels production, pharmaceutical secondary manufacture and the study of cardiovascular disease, in projects funded by EPSRC, the Royal Academy of Engineering, the Royal Society, the Henry Royce Institute, the British Academy, InnovateUK and the British Heart Foundation.

Leonard Nicuşan

Leonard Nicuşan published the PEPT-ML algorithm; the paper was selected as a featured article and scientific highlight by IOP Publishing. He is the lead developer of the open-source PEPT Python library, unifying computational PEPT research, including tracking, simulation, post-processing and visualisation tools.

Matthew T Herald

Matthew is a PhD student at the University of Birmingham whose work focuses on applications of simulating PEPT using Monte Carlo radiation transport. His research includes validating the performance of detector models, optimising experimental parameters, benchmarking PEPT algorithms, and contributing to the development of new PEPT algorithms.

Samuel Manger

Samuel Manger is a contributor to the construction of superPEPT and the Pept library at the University of Birmingham. He created a software implementation of the Pept-EM algorithm. Currently, Samuel is working at the University of Manchester studying the interplay of patient motion and spot-scanned proton therapy treatment.

David Parker

David Parker is the original co-inventor of the PEPT technique, winning the 2008 Institute of Physics Joule Medal 'for the creation of positron emission particle tracking as a practical tool in a wide variety of engineering applications'. He has over 30 years of experience developing the PEPT technique and applying it across an expansive range of disciplines and applications.

Chapter 0

Using the book

As the title of this book suggests, it is intended to provide a truly comprehensive guide to the positron emission particle tracking (PEPT) technique, including its methodology, its history, its associated equipment, its past (and potential future) application in academia and industry, and its place in the wider world of particle imaging techniques.

However, it is inevitable that different readers will possess varying levels of prior knowledge regarding PEPT, and further will have strongly differing motivations for reading this book. Some, for example, may possess limited knowledge of the technique, and simply wish to gain a minimal understanding of its operation such that they can successfully apply it to their own research. Others, meanwhile, may already possess an expert knowledge of the technique itself, yet wish to learn more about its possible wider applications, or even to gain greater knowledge of its underlying algorithms and detector systems, such that they may develop the technique further. This book has been carefully designed and structured so as to accommodate, as far as possible, the full range of potential readers.

In this prefatory chapter, we consider four main categories into which the majority of readers may be broadly grouped, and provide for each a guide outlining the manner in which the book may be read to most efficiently achieve their aims, whilst avoiding (if desired) any information unnecessary to their particular goal. They are as follows:

The User: This category describes the reader who wishes simply to learn how to use the PEPT technique effectively for their own specific research or application, without necessarily caring how it works.

The Researcher: Similar to 'the User', 'the Researcher' represents a reader who does not require a deep knowledge of the history or underlying physics of PEPT, but who may wish to gain a better understanding of the significant volume of research performed using PEPT, and perhaps also use the technique to add to this body of work themselves.

doi:10.1088/978-0-7503-3071-8ch0

The Developer: 'The Developer' represents those who are less interested in the application of PEPT, but rather are interested in the technique itself, and may wish either to actively develop the technique, its algorithms, detector systems and/or tracers, or to apply concepts from PEPT to other experimental systems or wider applications.

The Expert: The final category describes those who which to gain a deep, expert knowledge of the PEPT technique in its entirety.

For all of the above, we provide both 'beginner' and 'advanced' routes for, respectively, those with little or no prior knowledge of PEPT, and those familiar with the technique, but looking to expand and consolidate their knowledge.

0.1 The 'User'

Those simply wishing to gain an operational knowledge of PEPT can focus predominantly on chapters 6 and 7. Chapter 6 provides details regarding the manner in which data acquired from a PEPT experiment may be converted from raw line-of-response data (see chapter 5) into particle trajectory data, and gives insight into the best method through which this conversion may be performed for a given system. Chapter 7 then provides information on how this trajectory data may be manipulated to give meaningful physical quantities and fields. The entirely unfamiliar reader may additionally find chapters 1 and 2 useful in terms of providing a basic introduction to the PEPT technique (and indeed the wider field of particle imaging/tracking) and furnishing them with the terminology necessary to fully understand the later chapters. Finally, chapter 10 will provide the user with the information necessary to gain access to a PEPT facility and apply the knowledge gained (figure 0.1).

0.2 The 'Researcher'

Those wishing not only to *use* the PEPT technique but also to gain an overview of the research performed using the technique—and thus deeper insight into the type of

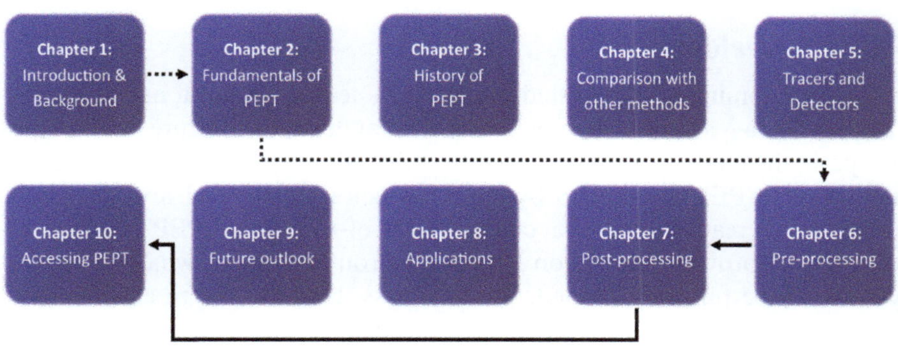

Figure 0.1. Schematic diagram illustrating the recommended path through the book for those wishing simply to acquire an operational knowledge of PEPT. The path depicted using solid arrows represents that suggested for readers already possessing a basic familiarity with PEPT, while those entirely new to the technique should read also the chapters connected via dotted lines.

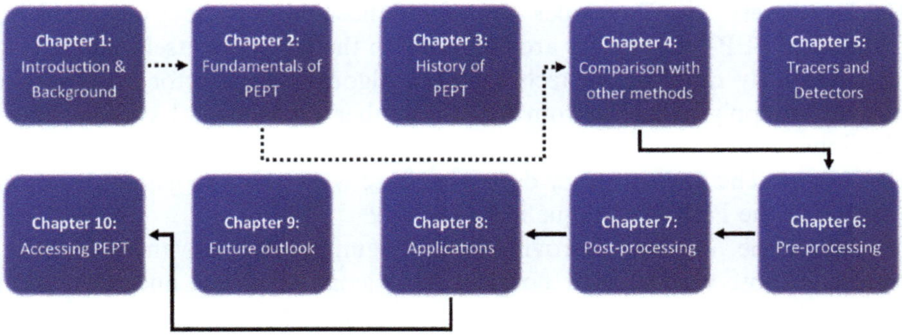

Figure 0.2. Schematic diagram illustrating the recommended path through the book for those wishing to understand and/or apply the PEPT technique for research purposes. The path depicted using solid arrows represents that suggested for readers already possessing a basic familiarity with PEPT, while those entirely new to the technique should read also the chapters connected via dotted lines.

work that can be performed effectively using PEPT, and current open questions in the literature that PEPT studies may help resolve—are recommended to follow the path outlined in figure 0.2. This path incorporates chapters 6 and 7 which—as described in section 0.1—provide an operational knowledge of the PEPT technique, as well as chapters 4 and 8. Chapter 4 contextualises the technique in terms of other contemporary experimental and numerical methods, providing insight into which applications and scientific problems are most suitable for PEPT, and conversely which may potentially be better addressed using other techniques, as well as discussing possible manners in which PEPT may be applied synergistically with other methods. Chapter 8 provides a synopsis of past research performed using PEPT, providing a background knowledge which may prove valuable to the design of future experiments.

As with the 'User' path, readers entirely unfamiliar with PEPT and/or its related concepts are advised to incorporate also chapters 1 and 2, while chapter 10 may be used to provide a route towards applying the knowledge gained in practice.

0.3 The 'Developer'

For those predominantly interested in the PEPT technique itself, as opposed to the use of PEPT as a research tool, a path such as that depicted in figure 0.3 is suggested. In this path, the reader is recommended to start with chapter 9 (for readers currently unfamiliar with PEPT, preceded by the introductory chapters 1 and 2), which will familiarise the reader with the current state-of-the-art in PEPT software and hardware, and provide inspiration for possible routes through which the technique may valuably be further extended. The reader is then directed to chapter 5, which provides detailed information regarding the detector hardware used to perform PEPT, and the manners in which PEPT tracers are produced. Having gained an understanding of PEPT hardware in this chapter, the reader may then gain insight into the relevant software, being finally directed to chapters 6 and 7, which detail the various pre- and post-processing algorithms currently developed for PEPT.

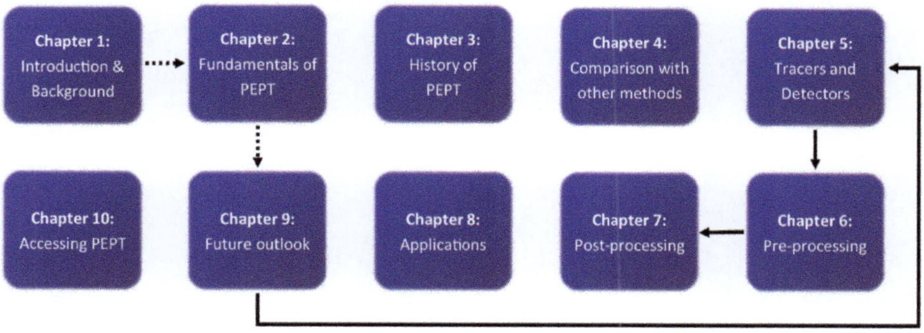

Figure 0.3. Schematic diagram illustrating the recommended path through the book for those wishing to actively develop PEPT and its associated systems and techniques, or to use the technique's algorithms and systems as inspiration for the development of other imaging methodologies. The path depicted using solid arrows represents that suggested for readers already possessing a basic familiarity with PEPT, while those entirely new to the technique should read also the chapters connected via dotted lines.

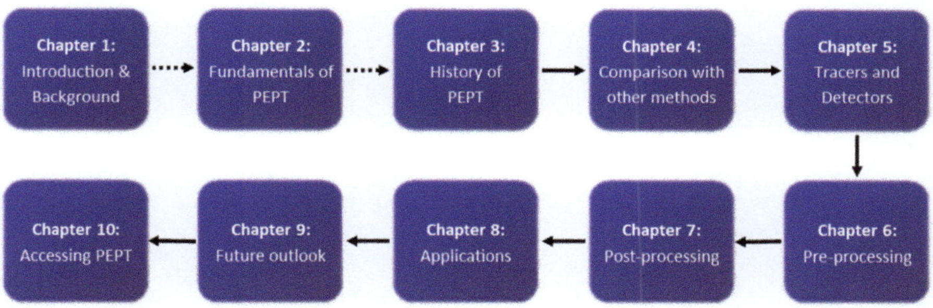

Figure 0.4. Schematic diagram illustrating the recommended path through the book for those wishing to gain a comprehensive, rounded knowledge of the PEPT technique. The path depicted using solid arrows represents that suggested for readers already possessing a basic familiarity with PEPT, while those entirely new to the technique should read also the chapters connected via dotted lines.

0.4 The 'Expert'

Finally, those wishing to gain the deepest possible insight into the PEPT technique may follow the simplest possible path through the book, as illustrated in figure 0.4. This route will arm the reader with a comprehensive knowledge of the PEPT technique, its detailed operation, its history and early development, its underlying software and hardware, and its applications in science and industry.

Part I

Introduction and background knowledge

IOP Publishing

Positron Emission Particle Tracking

A comprehensive guide
Kit Windows-Yule, Leonard Nicuşan, Matthew T Herald, Samuel Manger and David Parker

Chapter 1

Imaging particulate and multiphase systems

Particulate and multiphase media feature prominently in science, industry, and indeed our everyday lives. Understanding their behaviours is of paramount importance to a vast range of applications, from the manufacture of safe pharmaceutical products to the prediction of earthquakes and other catastrophic geophysical phenomena. However, unlike classical solids, liquids and gases, the dissipative nature of particulate and multiphase systems makes their behaviours highly complex, and thus difficult to model and predict. In order to gain a deeper understanding of these materials, we need to be able to study them carefully, rigorously and in detail. One of the main methods through which this is achieved is through experimental imaging, using methods such as particle imaging velocimetry (PIV), x-ray computed tomography (CT) and, of course, positron emission particle tracking (PEPT). In the present chapter, we provide a brief insight into the importance and applications of particulate and multiphase media across a multitude of disciplines, and an overview of the particle and flow imaging methodologies used to advance our knowledge thereof.

1.1 Particulate and multiphase systems: why do they matter?

If one were to ask the proverbial person on the street to name a common, 'everyday' type of matter, most would answer solid, liquid, gas or perhaps—if they were more scientifically-minded (and/or keen to show off)—plasma. It is unlikely, however, that many would give as their answer powder, granulate, slurry, or indeed any other form of particulate or particle-laden medium. Our lack of familiarity, as a society, with these materials is rather peculiar, as most of us will have encountered at least one granular material already today, be it when eating our breakfast cereal, walking along a gravel path, or searching through pocket-change to buy our morning coffee (itself made from a granular material, processed from another granular material!).

In fact, particulate media—which may be simply defined as any collection of multiple, discrete, macroscopic 'particles'—are, after only water, the second-most widely-handled material in the world [1]. They play crucial roles in a vast array of

doi:10.1088/978-0-7503-3071-8ch1

processes, spanning almost all industrial sectors. Consider as an example the pharmaceutical industry, where both initial components (e.g. active ingredients and excipients) and final products (e.g. capsules and tablets) are particulate media, or the mining industry, which necessitates the handling of fine powders, coarse granulates and slurries (liquid-immersed particulate media). Myriad other examples may be found in the chemical, petrochemical, food, agriculture and biomedical sectors, to name but a few [2–7]. Outside of industry, a knowledge of granular materials is also vital to our understanding of numerous significant natural phenomena, from avalanches and landslides and earthquakes to volcanic eruptions [8–14], and even our own biological processes [15, 16].

Particulate media are simultaneously extremely similar to, yet markedly distinct from, 'classical' media—i.e. those composed of microscopic atoms and molecules, as opposed to macroscopic particles or 'grains'. On the one hand, both can exist as solids, liquids and gases [3], can be thought of as possessing a temperature and exerting a pressure [17], and exhibit phenomena such as convection [18] and surface waves [19]. On the other hand, granular materials freely break many of the fundamental laws of physics as defined for classical, molecular materials: for example they violate the equipartition theorem [20, 21], display anisotropic temperatures [22] and highly heterogeneous stress distributions [23], and undergo spontaneous segregation due to differences in particle size, density, shape and various other properties [24].

The multitudinous complex, non-classical behaviours exhibited by granular media stem from the inherently dissipative nature of macroscopic particles [25]. When particles collide, kinetic energy is converted to heat and sound—i.e. transferred to the molecules which form the macroscopic granular particles and/or the surrounding fluid. As this 'microscopic kinetic energy' cannot be recovered by the macroscopic particles (if you put a grain of sand in an oven, no matter how much you heat it, it will never begin to float around!) it is effectively lost. To put it more scientifically, granular media are typically considered *athermal, non-equilibrium systems*.

The non-conservative, athermal nature of particulate media means that their behaviours cannot be modelled using classical frameworks such as Navier–Stokes or kinetic theory; indeed, there currently exists no accepted continuum theory to describe the dynamics of granular media. To put it bluntly, our understanding of these media is comparable to our understanding on classical fluids at the turn of the 19th century. As such, our current understanding of particulate systems is heavily reliant on experimentation. If we wish to develop a granular 'unified model', it is experimental results upon which such a model will be built, and experimental data by which they must be verified.

1.2 The importance of imaging

It is abundantly clear from the preceding section that if we wish to better understand, and thus predict, the complex dynamics of particulate and multiphase media, it is vital that we are actually able to experimentally *observe* these dynamics. Early works concerning the dynamics of particulate media were performed by some of history's most eminent scientists, including Chladni (the Father of Acoustics) [26], Coulomb [27], Faraday [28], and Reynolds [29]. The 'particle imaging' performed by these researchers was elegant yet rudimentary—in essence merely visually observing their systems of interest and sketching (or simply describing) their results. But before we mock these early scientists for their elementary methods it is worth noting that, even in the new millennium, results of studies involving no more than the researchers' eyes (and perhaps a stopwatch) are still being published in highly-respected scientific journals [30].

The ability to image the motion and interactions of particles and the larger particulate media which they form grants us insight into the fundamental physics of these systems, allowing us both to test the validity of theoretical models (as in the case of Coulomb [27]) and to develop empirical models (as in the case of Geldart [31] and many other famed chemical engineers). Since the advent of numerical modelling techniques such as computational fluid dynamics (CFD) [32] and the discrete element method (DEM)[1] [33], the ability to accurately and precisely image the dynamics of particulate and multiphase media has become more important still, due to the requirement of these models for rigorous calibration and validation [34]. While comparison of numerical results with experimental data offered by pressure sensors, temperature probes and other conventional monitoring devices can give some indication of the validity of a model, these offer only localised measurements of bulk (macroscopic) quantities. If we want a true, rigorous validation of our models, we need information pertaining to the whole-field, particle-level (microscopic[2]) behaviours of the system. It is just such information that particle imaging methods such as the PEPT technique, which forms the focus of this book, are adept at providing.

1.3 Particle and flow imaging: an overview

Before we embark on a detailed discussion of the relatively complex positron emission particle tracking technique, it is perhaps worth—for those less familiar with the field—providing a basic introduction to particle tracking techniques in general.

Since the times of Coulomb and Faraday, a multitude of more advanced imaging techniques have been developed, including optical techniques such as particle tracking velocimetry (PTV) [35] and refractive index matched scanning (RIMS) [36], magnetism-based techniques such as magnetic resonance imaging (MRI) [37] and magnetic particle tracking (MPT) [38], and radiation-based techniques such as x-ray CT [39], computer-aided radioactive particle tracking (CARPT) [40] and of course PEPT, all of which offer the ability to observe the microscopic dynamics of particulate systems[3].

[1] These numerical techniques are discussed in detail in section 4.5.

[2] A note on terminology: when discussing particulate materials, the terms 'microscopic' or 'micro-scale' do not refer to absolute sizes. Rather, they refer to the particle scale, with 'microscopic' referring to the motion of individual particles and 'macroscopic' to the bulk behaviour of the particulate system as a whole. Absolute size scales $\sim 10^{-6}$ are instead referred to as the *micron scale*.

[3] These experimental methods, and several others besides, are discussed in detail in chapter 4.

Despite their diversity in terms of applicability, temporal and spatial resolution, method of operation, and various other factors, all of the above-mentioned techniques share the ability to track and record the motion of particles, and operate in the same fundamental manner:

1. Using a given subset of the data available, locate one or more particles at a given point in time and record their positions.
2. Find and identify the same particle(s) at a subsequent point in time, linking the new position of each particle to its previous position.
3. Repeat the above steps to build up a picture of the particles' trajectories.

The above is perhaps more clearly illustrated using a specific example. To ensure maximal simplicity, let us consider the particle tracking velocimetry (PTV) imaging of a simple, dilute, monodisperse system of spheres in a quasi-two-dimensional, vibrated container (see figure 1.1 (a)). The system is captured using a high-speed camera. Each frame of the video recorded is then analysed using image processing techniques (described in detail in section 4.1) so that the positions of the dark-coloured particles can be located (figure 1.1(b)). By working through the video frame-by-frame, the trajectories of the dark-coloured 'tracer particles' can be constructed.

With the known trajectory of particles over an extended period of time, it is then possible to calculate a host of relevant information regarding the system's dynamics. For example, pseudo-instantaneous particle velocities, $\vec{v} = \frac{\Delta\vec{r}}{\Delta t}$, can be calculated simply by dividing the distance, $\Delta\vec{r}$, travelled by a particle between two consecutive frames by the frame rate, i.e. the time step, Δt, between said frames (see figure 1.1(c)). Many other, more complex quantities ranging from packing densities and granular temperatures to diffusion coefficients and shear rates may also be extracted from particle trajectories, as discussed in detail in chapter 7.

To observe the dynamics of a dilute, quasi-two-dimensional medium such as that in figure 1.1, all we require is a video camera with an adequately high resolution and frame rate. Unfortunately, the vast majority of 'real' industrial systems are neither dilute nor quasi-two-dimensional, and in most cases are housed in optically-opaque systems. To image systems more representative of these 'real' cases, we require a more powerful technique, capable of imaging the three-dimensional dynamics of

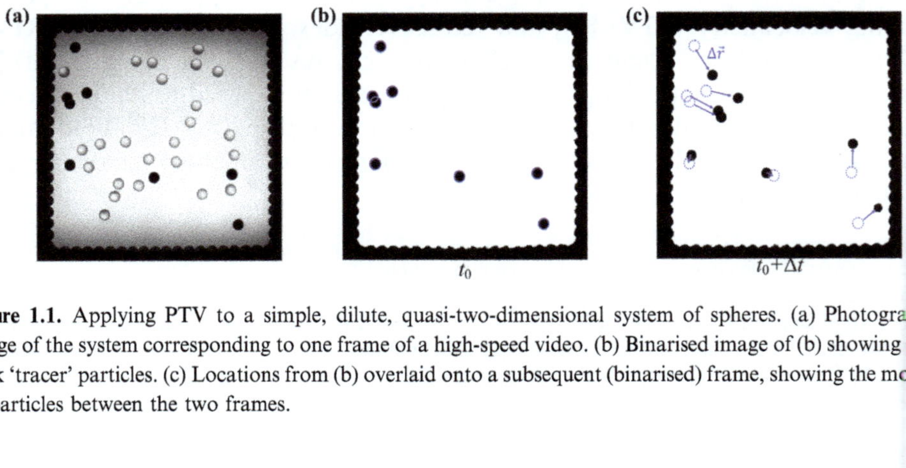

Figure 1.1. Applying PTV to a simple, dilute, quasi-two-dimensional system of spheres. (a) Photographic image of the system corresponding to one frame of a high-speed video. (b) Binarised image of (b) showing only dark 'tracer' particles. (c) Locations from (b) overlaid onto a subsequent (binarised) frame, showing the motion of particles between the two frames.

large, dense, optically opaque systems, ideally with high temporal and spatial resolution. It is such a technique—positron emission particle tracking—which forms the focus of this book.

References

[1] de Gennes P-G 1999 Granular matter: a tentative view *Rev. Mod. Phys.* **71** S374

[2] Williams J C and Shields G 1967 The segregation of granules in a vibrated bed *Powder Technol.* **1** 134–42

[3] Jaeger H M, Nagel S R and Behringer R P 1996 Granular solids, liquids, and gases *Rev. Mod. Phys.* **68** 1259

[4] Cleary P W 1998 Discrete element modelling of industrial granular flow applications *TASK. Q. Sci. Bull.* **2** 385–416

[5] Muzzio F J, Shinbrot T and Glasser B J 2002 Powder technology in the pharmaceutical industry: the need to catch up fast *Powder Technol.* **124** 1–7

[6] Coussot P 2005 *Rheometry of Pastes, Suspensions, and Granular Materials: Applications in Industry and Environment* (New York: Wiley)

[7] Merkus H G and Meesters G M H 2016 *Production, Handling and Characterization of Particulate Materials Particle Technology Series* (Berlin: Springer)

[8] Han Z, Ueda H, Matsuda K, Zhang R, Arao K, Kanai Y and Hasome H 2004 Model study on particle size segregation and deposition during Asian dust events in March 2002 *J. Geophys. Res. Atmos* **109**

[9] Mitani N K, Matuttis H-G and Kadono T 2004 Density and size segregation in deposits of pyroclastic flow *Geophys. Res. Lett.* **31** L15606

[10] Gray J M N T and Chugunov V A 2006 Particle-size segregation and diffusive remixing in shallow granular avalanches *J. Fluid Mech.* **569** 365–98

[11] Gray J M N T and Ancey C 2011 Multi-component particle-size segregation in shallow granular avalanches *J. Fluid Mech.* **678** 535–88

[12] Johnson C G, Kokelaar B P, Iverson R M, Logan M, LaHusen R G and Gray J M N T 2012 Grain size segregation and levee formation in geophysical mass flows *J. Geophys. Res. Earth Surf.* **117** F01032

[13] Shroder J F, Haeberli W and Whiteman C 2015 Snow and ice-related hazards *Risks, and Disasters* (Amsterdam: Elsevier)

[14] Köhler A, McElwaine J N and Sovilla B B 2018 Geodar data and the flow regimes of snow avalanches *J. Geophys. Res. Earth* **123** 1272–94

[15] Østergaard L, Smith D F, Vestergaard-Poulsen P, Hansen S B, Gee A D, Gjedde A and Gyldensted C 1998 Absolute cerebral blood flow and blood volume measured by magnetic resonance imaging bolus tracking: comparison with positron emission tomography values *J. Cereb. Blood Flow Metab.* **18** 425–32

[16] Stamatopoulos K, Karandikar S, Goldstein M, C O'farrell, Marciani L, Sulaiman S, Hoad C L, Simmons M J H and Batchelor H K 2020 Dynamic colon model (DCM): a cine-MRI informed biorelevant *in vitro* model of the human proximal large intestine characterized by positron imaging techniques *Pharmaceutics* **12** 659

[17] Falcon E, Fauve S and Laroche C 1999 Cluster formation, pressure and density measurements in a granular medium fluidized by vibrations *Eur. Phys. J.* **B9** 183–6

[18] Ehrichs E E, Jaeger H M, Karczmar G S, Knight J B, Kuperman V Y and Nagel S R 1995 Granular convection observed by magnetic resonance imaging *Science* **267** 1632–4

[19] Pak H K and Behringer R P 1993 Surface waves in vertically vibrated granular materials *Phys. Rev. Lett.* **71** 1832

[20] Feitosa K and Menon N 2002 Breakdown of energy equipartition in a 2D binary vibrated granular gas *Phys. Rev. Lett.* **88** 198301

[21] Wildman R D and Parker D J 2002 Coexistence of two granular temperatures in binary vibrofluidized beds *Phys. Rev. Lett.* **88** 064301

[22] van der Meer D and Reimann P 2006 Temperature anisotropy in a driven granular gas *Europhys. Lett.* **74** 384

[23] Majmudar T S and Behringer R P 2005 Contact force measurements and stress-induced anisotropy in granular materials *Nature* **435** 1079–82

[24] Rosato A D and Windows-Yule C 2020 *Segregation in Vibrated Granular Systems* (New York: Academic)

[25] Luding S 2002 Structures and non-equilibrium dynamics in granular media *C. R. Phys.* **3** 153–61

[26] Chladni E F F 2015 *Treatise on Acoustics: The First Comprehensive English Translation of EFF Chladni's Traité d'Acoustique* (Berlin: Springer)

[27] Coulomb C-A d 1773 In memories de mathematique et de physique *Acad. R. Sci. Divers Sans* **7** 343–82

[28] Faraday M 1831 XVII. On a peculiar class of acoustical figures; and on certain forms assumed by groups of particles upon vibrating elastic surfaces *Phil. Trans. R. Soc.* 299–340

[29] Reynolds O 1885 LVII. On the dilatancy of media composed of rigid particles in contact. With experimental illustrations *London Edinburgh Dublin Philos. Mag. J. Sci.* **20** 469–81

[30] Huerta D A and Ruiz-Suárez J C 2004 Vibration-induced granular segregation: a phenomenon driven by three mechanisms *Phys. Rev. Lett.* **92** 114301

[31] Geldart D 1973 Types of gas fluidization *Powder Technol.* **7** 285–92

[32] Versteeg H K and Malalasekera W 2007 *An Introduction to Computational Fluid Dynamics: The Finite Volume Method* (Harlow: Pearson Education)

[33] Guo Y and Curtis J S 2015 Discrete element method simulations for complex granular flows *Annu. Rev. Fluid Mech.* **47** 21–46

[34] Windows-Yule C R K, Tunuguntla D R and Parker D J 2016 Numerical modelling of granular flows: a reality check *Computat. Part. Mech.* **3** 311–32

[35] Grant I 1997 Particle image velocimetry: a review *Proc. Inst. Mech. Eng.* **C211** 55–76

[36] Dijksman J A, Rietz F, Lőrincz K A, van Hecke M and Losert W 2012 Invited article: refractive index matched scanning of dense granular materials *Rev. Sci. Instrum.* **83** 011301

[37] Kawaguchi T 2010 MRI measurement of granular flows and fluid-particle flows *Adv. Powder Technol.* **21** 235–41

[38] Buist K A, van der Gaag A C, Deen N G and Kuipers J A M 2014 Improved magnetic particle tracking technique in dense gas fluidized beds *AIChE J.* **60** 3133–42

[39] Wang L B, Frost J D and Lai J S 2004 Three-dimensional digital representation of granular material microstructure from x-ray tomography imaging *J. Comput. Civil Eng.* **18** 28–35

[40] Lin J S, Chen M M and Chao B T 1985 A novel radioactive particle tracking facility for measurement of solids motion in gas fluidized beds *AIChE J.* **31** 465–73

IOP Publishing

Positron Emission Particle Tracking

A comprehensive guide

Kit Windows-Yule, Leonard Nicuşan, Matthew T Herald, Samuel Manger and David Parker

Chapter 2

The fundamentals of PEPT

In this chapter, we provide a simple and accessible overview of the positron emission particle tracking (PEPT) technique and its associated physics. We explain how solid particles can be 'labelled' with positron-emitting materials, allowing them to be used as 'tracers' to non-invasively track the flow of particles and/or fluids, and describe the fundamental operation of the detector systems used to record their motion. We begin with a somewhat idealised picture involving 'perfect' detectors and a point-like source, before introducing complicating factors such as the limited spatial resolution and finite detection rate of real positron cameras, and the scattering of the gamma rays used to track particles. We include also interactive examples in which the user can directly observe the effect of varying key parameters such as the detector resolution, the prevalence of erroneous events, and the positron range[1], and thus gain a deeper understanding of the technique.

PEPT is a non-invasive experimental imaging technique capable of tracking and recording the dynamics of one or more radioactively-labelled 'tracer particles' through three-dimensional space with sub-millimetre spatial resolution and sub-millisecond temporal resolution [1]. Thanks to its use of highly-penetrating 511 keV gamma rays, PEPT is capable of tracking particle motion even deep within the interior of dense, optically-opaque systems.

The above is typical of the 'abridged' description that you are likely to find in articles utilising PEPT as a research tool. However, it gives little real indication of the physics or algorithms underlying the technique. In the remainder of this chapter, we give a more detailed explanation of the fundamental physics and basic operation of the PEPT technique. We then, in chapters 5 and 6, go on to provide further details regarding the types of tracers, detectors and algorithms which allow PEPT to be performed.

2.1 Positron emission…

Consider a dynamic granular system comprising a number of monodisperse (that is to say equally-sized) glass beads—a simple, canonical system commonly used by

[1] These, and other relevant terms, will be explained in detail later in this chapter.

doi:10.1088/978-0-7503-3071-8ch2

researchers to explore the fundamental dynamics of granular systems [2] (see figure 2.1). Suppose that we want to image the internal dynamics of said system using PEPT. How would we go about it?

In order to perform positron emission particle tracking, we firstly require a particle which emits positrons. Unlike techniques such as x-ray radiography or computed tomography which fire ionising radiation at the object to be imaged, the PEPT technique tracks the particle(s) of interest using gamma rays *emitted from* said particle(s).

A suitably radioactive 'tracer particle' may be created by 'labelling' a normal, non-radioactive material with a β^+-emitting radioisotope. There exist several manners in which this labelling process may be performed (these processes are described in detail in chapter 5) and an almost limitless range of materials which may be labelled, but for now we will focus on the simple case in hand. Glass—or, more specifically, silicon dioxide (SiO_2)—contains oxygen atoms, which can be bombarded with high-energy helium-3 (^{3}He) nuclei to produce positron-emitting fluorine-18 (^{18}F), via the reactions:

$$^{16}\mathrm{O}(^3\mathrm{He},\, \mathrm{p})^{18}\mathrm{F} \tag{2.1}$$

and

$$^{16}\mathrm{O}(^3\mathrm{He},\, \mathrm{n})^{18}\mathrm{Ne} \rightarrow {}^{18}\mathrm{F}. \tag{2.2}$$

As such, by simply placing one of our glass particles in the path of a suitably energetic helium-3 beam from a cyclotron for a suitable duration, we can induce significant β^+ activity in the particle (see figure 2.1). Importantly, this 'activation' process converts only a relatively small proportion of the material's ^{16}O to ^{18}F and does not cause any significant structural changes to the particle, meaning it remains effectively physically identical to all the other particles in the system—i.e. PEPT can be performed truly non-invasively.

The fluorine-18 atoms produced decay to oxygen-18 with a half-life of 109.7 minutes, producing a positron and an electron neutrino [3]:

$$^{18}_{9}\mathrm{F} \rightarrow {}^{18}_{8}\mathrm{O} + \beta^+ + \nu_e. \tag{2.3}$$

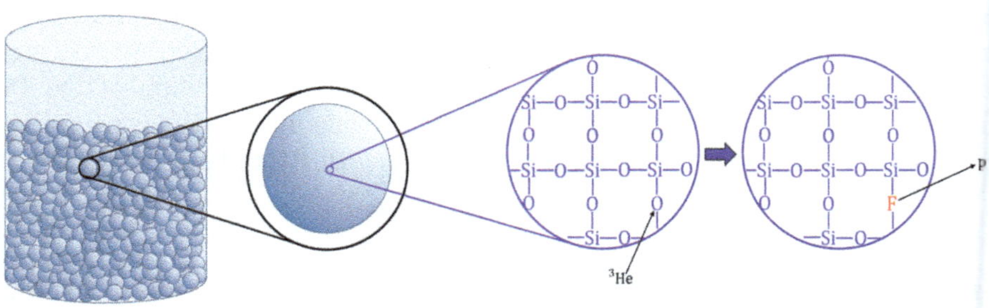

Figure 2.1. From left to right: A granular bed of glass particles, from which a single particle—identical to all others in the system—is removed. The particle is then bombarded with high-energy helium-3 nuclei, which convert oxygen atoms within the glass to fluorine-18.

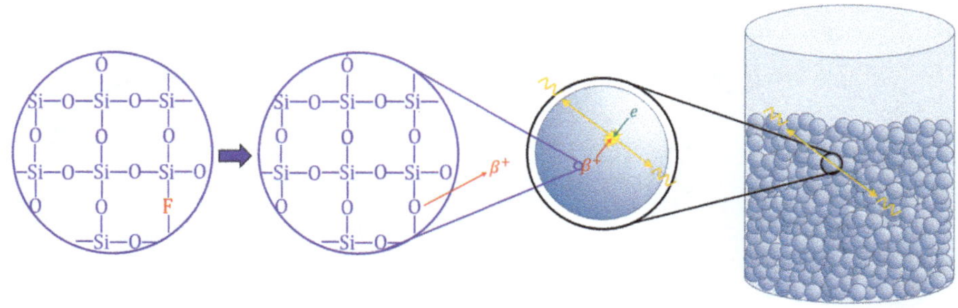

Figure 2.2. From left to right: a fluorine-18 atom decays to oxygen-18, releasing a positron (β^+). The positron then annihilates with an electron within the tracer medium, producing two 511 keV gamma photons, whose trajectories are separated by 180° ± 0.5°.

In a relatively dense medium such as glass, the emitted positron (also called an 'anti-electron') will only travel a very short distance before colliding with a (regular) electron [4, 5]. As with the collision of any particle and anti-particle pair, the electron and positron will annihilate [6], converting their masses into energy in the form of two 511 keV ($\approx 8.2 \times 10^{-14}$ J) gamma photons (see figure 2.2) whose trajectories are approximately collinear and antiparallel (i.e. separated by roughly[2] 180°). It is the collinear nature of these emitted gamma rays which makes them so useful for locating and tracking particles—as we will see in the next section.

2.2 …particle tracking

Once we have, as described in the preceding section, created from one of the particles in our system a non-invasive 'tracer particle', we can place this particle back in the system, and begin to track its motion (see figure 2.3). To do so, we must firstly place our system, with its newly-reintroduced tracer particle, within the field of view of a suitable detector system, often referred to as a 'gamma camera' or 'positron camera' (see figure 2.3(a)). Detailed descriptions of the various types of gamma camera which may be used to perform PEPT, as well as the physics underlying these systems, can be found in chapter 5. For now, it suffices for us to know that, if a gamma photon of a suitable energy interacts with one of the gamma camera's detectors, it is able to record the position of this interaction—i.e. where the photon 'hit' the camera. If two gamma photons hit the detector system within a very short timing window (typically of the order of nanoseconds), it is assumed that the two photons were produced simultaneously—i.e. they both originated from a positron annihilation event within our tracer particle. The registering of two consecutive photons within the aforementioned timing window is known as a 'coincidence event'.

From the previous section, we know that any pair of electrons produced by such an event follow a collinear, antiparallel trajectory—in other words, if both members of the pair interact with our detector system we must, by definition, be able to connect the two recorded points by a straight line (see figure 2.3(b)). Further, we know that this straight

[2] We say here 'roughly' as while the motion of the positron is largely arrested immediately before annihilation, the momentum associated with the electron's orbital state may nonetheless exert a small influence on the resultant photons' trajectories. As such, a slight acollinearity (of the order of 0.5 degrees) may in fact exist.

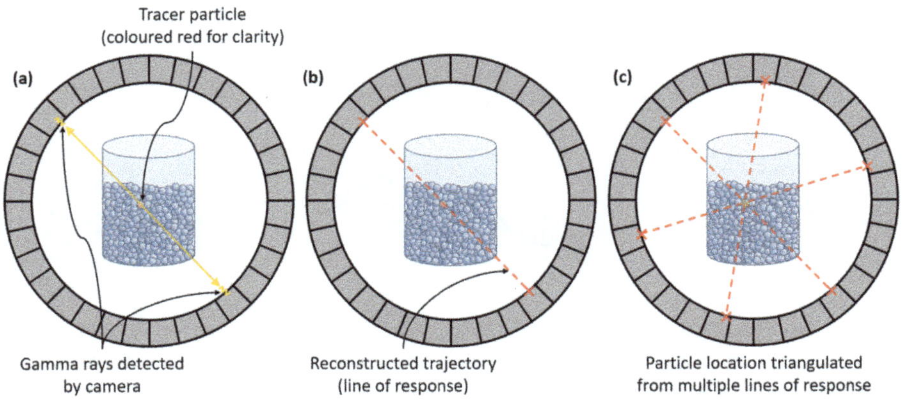

Figure 2.3. Schematic diagrams showing an idealised representation of the manner in which PEPT may be used to locate a static tracer. (a) The tracer emits a pair of back-to-back gamma rays produced by the annihilation of a positron. (b) The coordinates at which the gamma rays hit the detector are recorded, allowing their trajectory to be recorded. (c) Multiple recorded trajectories (lines of response) can be used to triangulate the position of the tracer.

line must pass through (or at least near) the source of these positron annihilation events—i.e. our tracer particle. Our gamma camera system will record the locations of all detected photon pairs, allowing it to store and later 'reconstruct' these straight line trajectories, which are typically referred to as 'lines of response' or 'LoRs'.

Let us consider first, for simplicity, the case of a static tracer. As the ^{18}F nuclei within the tracer decay, the particle will emit multiple pairs of gamma photons, of which a fraction will interact with the detector system, and have their trajectories recorded as LoRs. Clearly, if we know the point of intersection of these LoRs, then we can triangulate the position of our tracer (see figure 2.3(c)).

If the tracer particle is adequately active—i.e. produces a large enough number of positrons per second—it is also possible to track its motion through three-dimensional space by regularly triangulating its position.

2.2.1 Interactive example: PEPT—an idealised case

The notebook files accompanying this section are available in the supplementary data online at https://iopscience.iop.org/book/978-0-7503-3071-8.

PEPT: an idealised case

In this notebook we will generate some *ideal* lines of response (LoRs) emitted from a simple spherical tracer—*ideal* meaning that all LoRs are emitted within the tracer, there are no corrupted events and the PEPT detector has perfect spatio-temporal resolution. These LoRs will then be used to triangulate the tracer's location as simply the minimum distance point (MDP) to all LoRs. Finally, the spatial error in the tracer location found is computed for varying numbers of LoRs used when computing the MDP.

This is very much an **interactive** notebook, with many free parameters that the reader is invited to vary to intuitively discover many of the interesting properties of PEPT that will be discussed in detail in the later chapters of the book.

[1]:
```
# Importing necessary libraries for numerical array manipulation and plotting,
# plus the `pept` library for straightforward PEPT data processing
import numpy as np
import matplotlib.pyplot as plt
import pept

# Set Matplotlib theme (try "darkgrid" too for a more modern look)
import seaborn
seaborn.set_style("whitegrid")

# Set Plotly figure output (try "browser" too for interactive graphs)
import plotly
plotly.io.renderers.default = "png"
```

Monte Carlo line of response generation

In the code cell below we define subroutines for the Monte Carlo simulation of a PEPT experiment containing a single spherical tracer inside a parallel screens PEPT scanner; 'Monte Carlo' is the name of a family of algorithms which, in simple terms, use random numbers to simulate physical phenomena such as stochastic radioactive decay (section 4.5 provides a more in-depth explanation of the Monte Carlo technique, while section 5.3 describes its use for the scientific simulation of positron emission).

There are a number of steps involved in the acquisition of lines of response in a typical PEPT experiment:

- A single fluorine-18 radioisotope undergoes radioactive decay, producing a single positron. Note that there are other positron-emitting radioisotopes (and even direct gamma ray emitters), but F-18 generally yields the best spatial resolution.
- The positron travels some distance before annihilating with a nearby electron; this distance, or *positron range*, depends on the positron energy—which depends on the radioisotope used and its surrounding environment. **In this notebook it is assumed that the positron range is zero**, i.e. all gamma rays are emitted exactly at the radioactive decay point *inside the tracer*.
- The positron–electron annihilation event produces two back-to-back gamma rays. If both gamma rays hit two opposing detectors within a given *resolving time*, they are registered as a *coincidence event* and the two intersection* points are logged, producing a single *line of response*, or LoR.

The above is of course a simplified picture, not accounting for the various sources of error, but provides a useful illustration of the processes involved in recording LoRs; for a detailed discussion on the subject, the reader is referred to chapter 5.

As mentioned above, the code below contains multiple free parameters that the reader is invited to have a play with:

[2]:
```
#
# User-definable PEPT simulation parameters
#
screen_height = 500                              # Detector screen height, mm
screen_width = 500                               # Detector screen width, mm
screen_separation = 500                          # Detector heads separation, mm

tracer_radius = 5                                # Tracer radius, mm
tracer_location = [250, 250, 250]                # Tracer 3D location, mm

tracer_activity = 10e6                           # Tracer activity, Bq


#
# Functions simulating a PEPT experiment with a static spherical tracer. For
# simplicity, they have been defined as free functions which make use of the
# global parameters defined above.
#
# In a production codebase, such "global parameters" would be encapsulated
# within a class, and the functions below would be class methods making use of
# the aforementioned class variables.
#

# Instantiate a random number generator with a seed for deterministic
# pseudo-random numbers
rng = np.random.default_rng(0)


def generate_positron():
    '''Generate a radioactive decay / positron emission point *inside* a
    spherical tracer centred at (0, 0, 0) of radius `tracer_radius`.

    Returns an array with the X, Y, Z positions of the emission point.
    '''

    # Create random spherical coordinates
    r = rng.uniform(0, tracer_radius)
    theta = rng.uniform(0, np.pi)
    phi = rng.uniform(0, 2 * np.pi)

    # Transform spherical coordinates into cartesian coordinates
    x = r * np.sin(theta) * np.cos(phi)
    y = r * np.sin(theta) * np.sin(phi)
    z = r * np.cos(theta)

    return np.array([x, y, z])


def generate_lor(annihilation):
    '''Given an input positron annihilation 3D point, generate a 3D LoR and
    check it hits both detector screens; if the LoR hits both detector screens,
    it is acquired and returned as a list with the two LoR interception points
    ([x1, y1, z1, x2, y2, z2]). If at least one of the two LoR ends fall outside
    the detector heads, None is returned.
    '''
```

```
    # Generate LoR direction in spherical coordinates
    phi = rng.uniform(0, np.pi)
    theta = rng.uniform(0, np.pi)

    # If the LoR is perfectly vertical / horizontal (parallel to the detector
    # screens), it is not intercepted
    if phi == np.pi / 2 or theta == np.pi / 2:
        return None

    # Unpack positron annihilation point 3D coordinates
    xp, yp, zp = annihilation

    # Find the 2D points where the LoR intersects the two *planes* of the two
    # detector screens
    x1 = xp - zp * np.tan(phi)
    x2 = xp + (screen_separation - zp) * np.tan(phi)

    y1 = yp - zp / np.cos(phi) * np.tan(theta)
    y2 = yp + (screen_separation - zp) / np.cos(phi) * np.tan(theta)

    # Check the rays fell within the PEPT screens, otherwise return None
    if 0 < x1 < screen_width and 0 < x2 < screen_width:
        if 0 < y1 < screen_height and 0 < y2 < screen_height:
            return [x1, y1, 0, x2, y2, screen_separation]

    return None


def record_lors(num_lors):
  '''Run a Monte Carlo simulation generating `num_lors` LoRs emitted from
  a spherical tracer and return them as a NumPy array with `num_lors` rows
  and 7 columns formatted as [timestamp, x1, y1, z1, x2, y2, z2], where
  (x1, y1, z1) is the LoR interception point on the first detector screen and
  (x2, y2, z2) is the LoR interception point on the second detector screen.
  '''
  # Pre-allocate array for LoRs
  lors = np.zeros((num_lors, 7))

  # Keep track of generated LoRs' timestamps
  current_time = 0
  dt = 1 / tracer_activity

  # Generate enough LoRs until `num_lors` hit both detector screens and are
  # recorded.
  #
  # Because the LoRs generated may fall outside the detector screens, we
  # cannot know in advance how many LoRs we have to generate; simply keep
  # track of a current LoR index
  lor_index = 0
```

```
    while lor_index < num_lors:
        # Create radioactive decay / positron emission point *inside* the
        # tracer. In reality, the positron travels before annihilating with a
        # nearby electron (producing the two gamma rays); here, however,
        # assume the LoR is emitted right at the emission point
        emission = tracer_location + generate_positron()

        # Generate LoR / pair of back-to-back gamma rays and check it hit both
        # detector screens
        lor = generate_lor(emission)

        # If the LoR was invalid, `None` was returned; simply advance the
        # simulation time and generate a new LoR at the next loop iteration
        if lor is None:
            current_time += dt
            continue

        # If the LoR hits both detector screens, add it to the `lors` array and
        # record the current timestamp
        lors[lor_index, 1:8] = lor
        lors[lor_index, 0] = current_time

        # Increment current LoR index and advance simulation time
        lor_index += 1
        current_time += dt

    return lors
```

The PEPT-oriented `PlotlyGrapher` plotting interface provided by the `pept` library can be used for straighforward plotting of PEPT data; while not required *per se* for illustrating PEPT algorithms, it significantly reduces the amount of repetitive code and visual noise, allowing the reader to focus on the main concepts.

In the code cell below we generate `num_lors` 'ideal' LoRs and create a 3D visualisation of them:

```
[3]: # Generate `num_lors` LoRs and plot them using the `pept` library.
from pept.plots import PlotlyGrapher

num_lors = 200
lors = record_lors(num_lors)

grapher = PlotlyGrapher()
grapher.add_lines(lors)

# Set default zoom level in `camera_eye`
grapher.fig.update_scenes(camera_eye=dict(x=1.5, y=1.5, z=1.5))
grapher.show()
```

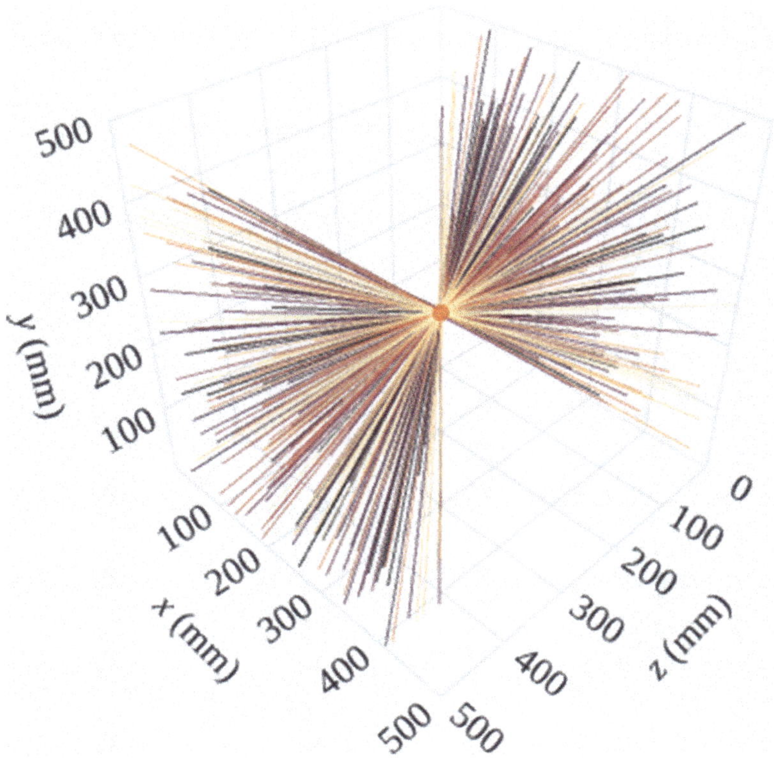

Note the PEPT 3D axes convention, differing from the common mathematical coordinates with *Y* and *Z* being switched, such that the *Y* axis points upwards, while the *Z* axis is perpendicular to the two detector heads.

Triangulate tracer's location

Now that we can simulate a PEPT experiment and record the LoRs emitted from a spherical tracer, we can try to backcompute its location from those LoRs. The simplest method of doing so is to compute the *minimum distance point* (MDP), or the 3D point minimising the sum of distances to all LoRs.

In the code cell below, we compute the MDP and print the *spatial error*, or distance between the MDP and the true tracer location (as defined in the `tracer_location` variable above):

```
[4]: # Find the minimum distance point (MDP) to all LoRs by minimising the sum of
     # distances to all lines.
     #
     # Note that the MDP can also be determined analytically (as is done in the C
     # implementation from the `pept` library), but the derivation is rather tedious
     # and takes away from the didactic purposes of this notebook
     from scipy.optimize import minimize
```

```
def distances(point, lines):
    '''Return an array of distances from a `point` to an array of `lines`
    formatted as [t, x1, y1, z1, x2, y2, z2].'''

    p1 = lines[:, 1:4]
    p2 = lines[:, 4:7]

    d1 = np.linalg.norm(np.cross(point - p1, point - p2), axis=1)
    d2 = np.linalg.norm(p2 - p1, axis=1)

    return d1 / d2

def sum_distances(point, lines):
    '''Return the sum of distances from a `point` to an array of `lines`.'''
    return np.sum(distances(point, lines))

# Find the minimum distance point via an optimisation approach - the MDP is
# simply the point minimising the sum of distances to all LoRs
p0 = np.array([200, 200, 200])                # initial guess
mdp = minimize(sum_distances, p0, lors).x

# Insert MDP timestamp at column 0
mdp = np.insert(mdp, 0, lors[:, 0].mean())

# The spatial error between the MDP and the tracer position
error = np.linalg.norm(mdp[1:4] - tracer_location)

print(f"The minimum distance point (MDP) found:\n{mdp}\n")
print(f"The spatial error is {error:4.4f} mm")
```

```
The minimum distance point (MDP) found:
[4.53280000e-05 2.50024324e+02 2.50069425e+02 2.50001383e+02]

The spatial error is 0.0736 mm
```

[5]:
```
# Plot all LoRs and the minimum distance point (MDP)
grapher = PlotlyGrapher()

grapher.add_lines(lors)
grapher.add_points([mdp], size=5)

grapher.fig.update_scenes(camera_eye=dict(x=1.5, y=1.5, z=1.5))
grapher.show()
```

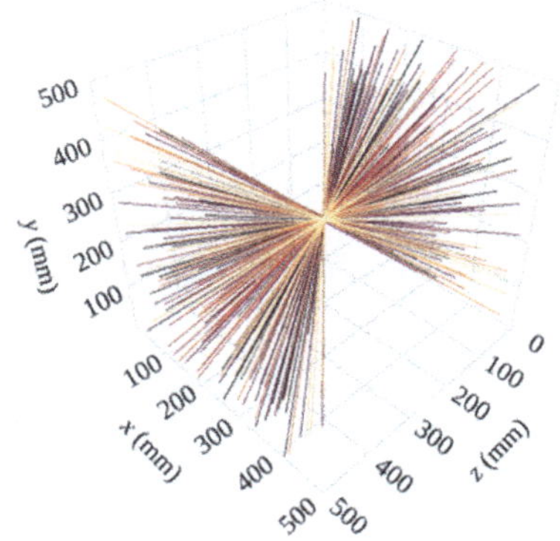

Spatial error versus number of LoRs used

How does the spatial error change when different numbers of LoRs are used to compute the minimum distance point (MDP)? The code cell below generates a 'large' dataset of LoRs, selects *samples* of different numbers of LoRs at a time, and computes the spatial error of the MDP—then plots the spatial errors recorded:

[6]:
```
# Generate a "large" number of LoRs from which we will choose a handful of
lors = record_lors(10_000)

# Generate `sample_sizes` for each sample that will be used to compute an MDP
#
# Note that `np.arange(min, max, step)` generates consecutive numbers between
# `min` and `max` in steps of `step`.
min_lors = 10
max_lors = 1000
sample_sizes = np.arange(min_lors, max_lors, 10)

# For each sample size in `sample_sizes`, select the first `num_lors` LoRs from
# the `lors` dataset, compute the MDP, then calculate the spatial error and
# append it to the `errors` list
errors = []

for num_lors in sample_sizes:
    sample = lors[:num_lors]
    mdp = minimize(sum_distances, p0, sample).x
    error = np.linalg.norm(mdp - tracer_location)

    errors.append(error)

# Plot the spatial errors found
fig, ax = plt.subplots()
ax.scatter(sample_sizes, errors, s=10)
ax.set(xlabel="Number of LoRs used", ylabel="Spatial Error (mm)");
```

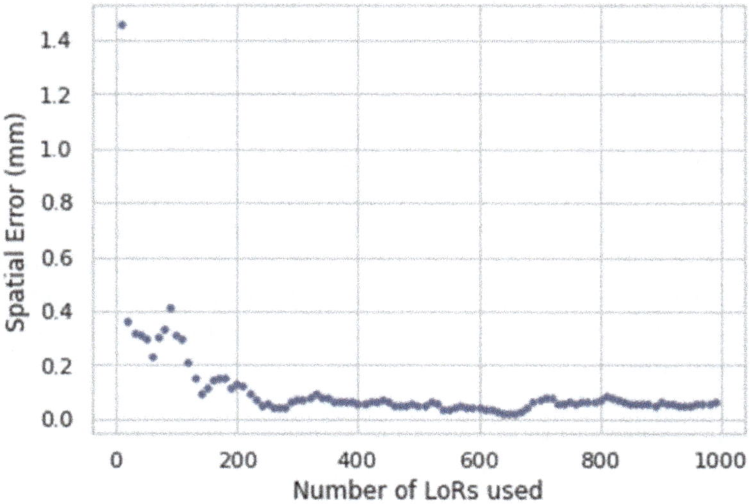

Note that this is a relatively naive assessment of how the spatial error varies with the number of LoRs used to compute the MDP—why are there such large fluctuations in the spatial error recorded? There are some segments where using *more* LoRs yields *less* accurate tracer locations! As you will see in the Monte Carlo section (chapter 4, section 4.5), when random numbers are used to simulate a stochastic physical process (such as generating LoRs in random directions), it is essential that a computation is run *multiple times* for different samples of random data to get statistically accurate measurements. In the above example, for each *sample size*—or numbers of LoRs used for a single data point—multiple samples of LoRs can be generated and the *mean* spatial error can be recorded to minimise *random fluctuations*—we invite the reader to 'hack' the code above and make it statistically sound!

2.3 A more realistic picture

In the preceding sections we have presented a highly idealised version of the PEPT procedure, in order to explain its basic principles in a simple and easily-comprehensible manner. Of course, in reality, the situation is significantly more complex. In this section, we attempt to present a more nuanced and realistic picture, illustrating some of the major issues which must be overcome in order to accurately locate and track particles using PEPT[3].

2.3.1 Issue 1: false coincidences

In the preceding section, it was implicitly assumed that any two particles detected within the chosen coincidence time window correspond to a photon pair produced via a β^+ annihilation event—a 'true coincidence'. In reality, of course, there are numerous manners in which two unrelated—or otherwise non-collinear—photons may happen to be detected within a given time window. For example, if a high-activity source is used, it is possible (indeed likely) that two annihilation events may occur within the coincidence window[4]. If only one photon from each pair interacts with the detector, the camera will reconstruct a false trajectory connecting these two points (see figure 2.4(a)). Similarly, if one (or both) of the photons released by an annihilation event undergoes Compton scattering [7]—meaning that the particles' trajectories are no longer collinear—the detector's algorithms will still fit a straight line to the detected points, again resulting in a false LoR (see figure 2.4(b)). It is also possible that two photons entirely unrelated to the experiment at hand may simply happen to interact with the detector within the coincidence time window, once again creating a false trajectory (see figure 2.4(c)).

[3] Without wishing to dismay the reader, it should be noted that the below still does not reveal the *full* complexity of the matter. Additional complicating issues are discussed in detail in chapter 5. Those outlined here nonetheless represent the most significant issues with which a typical PEPT practitioner will generally concern themselves.

[4] This may also occur due to the presence of *secondary radiation*—i.e. additional gamma rays produced by the tracer that are not due to β^+ annihilation.

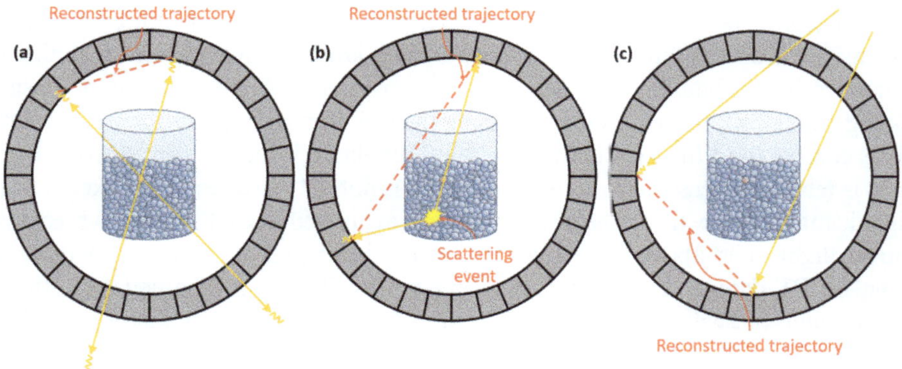

Figure 2.4. Schematic diagrams illustrating three possible causes of 'false coincidences'. (a) A single photon from two separate β^+ annihilation events is detected by the camera. (b) One photon in a pair undergoes a scattering event, meaning that gamma photons' trajectories are no longer collinear or antiparallel. (c) Two photons unrelated to the experiment happen to be detected within the coincidence time window.

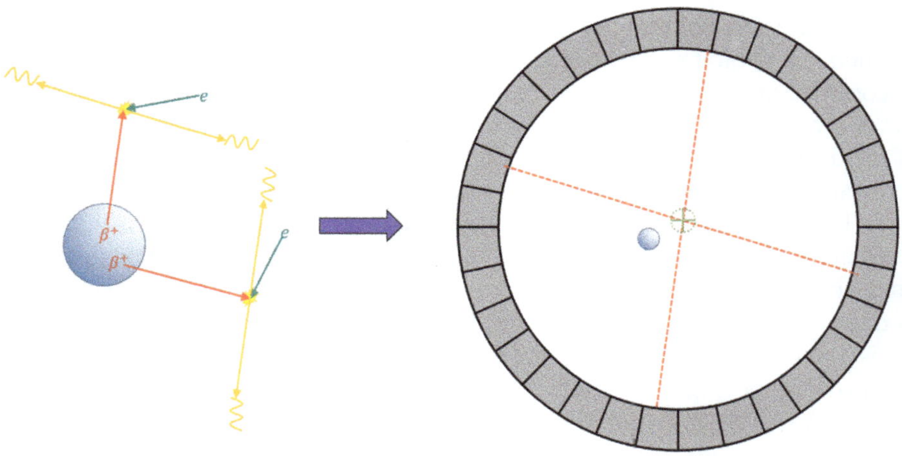

Figure 2.5. Illustration of the manner in which the non-negligible distance of flight of positrons may create false locations when using rudimentary triangulation techniques.

2.3.2 Issue 2: positron flight

A second problematic assumption from the picture presented in section 2.1 is that all gamma-ray pairs originate from the centre of the tracer, meaning that hypothetically the intersection point of any two LoRs could be used to give the particle location. In reality, however, positrons will travel non-negligible distances even in relatively dense media, and have a range several millimetres in air [8]. As illustrated schematically in figure 2.5, this means that the gamma rays emitted will not necessarily pass through the tracer itself, and thus that the intersection of a given pair of rays does not necessarily correspond to the tracer's centre, or indeed to the tracer at all!

The effects of positron flight are particularly pronounced when using small and/or low-density tracers in dilute interstitial media. With larger, denser tracers (assuming they are uniformly irradiated), there is clearly a significantly higher chance of positrons colliding with electrons in, and thus releasing their photon pairs from, the tracer medium itself as compared to the surrounding medium. Similarly, if the medium surrounding the tracers is relatively dense, even if the tracer itself is not, β^+ particles are unlikely to travel a significant distance from the tracer surface, again reducing the negative effects of positron flight. The above creates an interesting balancing act if we wish to acquire optimal PEPT data—a denser interstitial medium will mitigate the negative effects of positron flight, but will also increase the amount of Compton scatter, leading to a higher incidence of false LoRs. A more dilute medium will reduce the proportion of false coincidences, but means that the true coincidences obtained are less likely to pass directly through the tracer. Generally speaking, optimal results may be obtained using a large, dense, internally-irradiated tracer in a comparatively dilute medium.

2.3.3 Issue 3: imperfect detectors

The detectors used in positron imaging, like any experimental apparatus, possess inherent errors. All detectors, for example, possess a limited spatial resolution. This means that the x-position of a photon which hits the face of a given detector at a position X may actually be recorded at any position $X - \sigma_{\text{det.}} \leqslant x \leqslant X + \sigma_{\text{det.}}$, where $\sigma_{\text{det.}}$ characterises the spatial error of the detectors. The same holds for the y-position. Let us assume, for simplicity, that the detector's resolution in the x- and y-directions is equal (i.e. $\sigma^x_{\text{det.}} = \sigma^y_{\text{det.}} = \sigma_{\text{det.}}$). This means that a gamma photon interacting with a detector at coordinates (X, Y) may have its location recorded as falling anywhere within a region of approximate size $\sim 2\backslash\sigma_{\text{det.}}$ centred at (X, Y). Our lines of response, therefore, have some significant width, and may perhaps be considered better as 'cylinders of response'! As illustrated in figure 2.6, this introduces a significant potential error in the location of our tracer.

It is important to note that in the above we are still considering a relatively simplified case in which we assume that all photons are detected at the surface of the detector. In reality, different photons will penetrate the detector crystal to differing extents, leading to an additional source of error. As illustrated in section 2.3.6, and discussed further in chapter 5 this error can be considered as simply further expanding the radius of the 'cylinder of response'.

2.3.4 Issue 4: finite detection rate

In the preceding section, we discussed limitations on the *spatial* resolution of the gamma cameras used to perform PEPT. The PEPT technique also possesses certain limitations on its *temporal* resolution—i.e. the number of gamma ray pairs that can be detected per unit time, and thus the rate at which a given particle may be located. This 'event rate' is dependent on two main factors: the activity of the particle used, and the efficiency of the detectors used.

Let us consider first the matter of tracer activity. First and foremost, as should be clear from our description in section 2.1, the emission of positrons (and hence back-to-back gamma rays) is a discrete process, thus inherently implying a finite temporal resolution for the PEPT process (i.e. particles cannot be tracked truly continuously). For a given system,

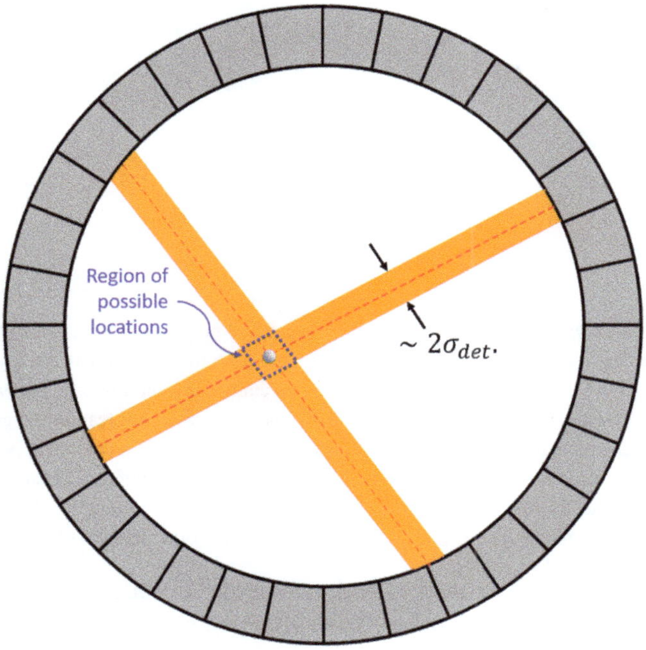

Figure 2.6. In reality, due to the detector error $\sigma_{\text{det.}}$, lines of response should more realistically be considered as 'cylinders of response'. Note that, due to the curved nature of the detector pictured, the radius of the cylinder does not correspond exactly to the detector error.

under given conditions, and with a perfect detector, one can naturally expect the temporal resolution of PEPT to increase with the activity of the tracer—i.e. the number of β^+ particles emitted per unit time. It is important to note here that positron emission is a stochastic process, meaning positrons are not emitted at a constant, fixed rate, but rather at random intervals in time. Nonetheless, a given activity can be associated with a characteristic timescale (e.g. the mean time between successive β^+ decays). As the activity of a tracer halves, this characteristic timescale—and correspondingly the temporal error of our PEPT measurements—can thus be expected to double. Clearly, then, for our hypothetical perfect detector system, it is desirable in all cases to maximise tracer activity[5].

In reality, there exist considerable limitations on the amount of activity one may induce in a tracer particle. The most significant impediment in most cases is the *size* of the tracer: clearly, a tracer with a finite size can only be labelled with so much fluorine-18 (or any other isotope for that matter). The smaller the tracer, the less activity can be held.

Even in the deeply unphysical situation of an infinitely active particle, however, we could not extract a continuous particle trajectory from PEPT data. For reasons which will be explained in detail in chapter 5, there exists a limit to the number of photons a given detector system can register within a given time period. This limitation is due to a variety of factors, including the 'dead time' of the detectors themselves and the maximal data rate of the electronics which process and store the resultant information.

[5] We will see in chapter 5 that with real detectors, more activity is not always better, as too many gamma rays may begin to 'blind' the detector, and indeed cause other issues.

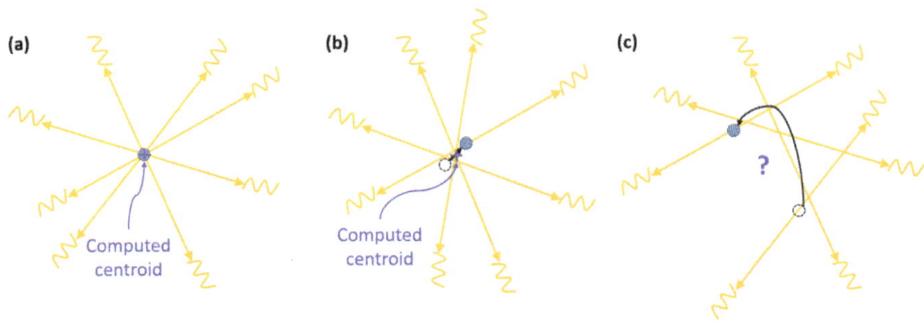

Figure 2.7. Schematic diagrams showing the most likely computed positions of (a) a static tracer, (b) a tracer of similar activity travelling a short distance at a low velocity and (c) a tracer of similar activity travelling a longer route at a larger velocity. In (a), as the tracer remains static, all emitted gamma-ray pairs intersect within the tracer volume, allowing its position to be precisely located. In (b), there exists some spread in the gamma rays, but the majority of intersection points lie within a relatively small volume, resulting in a computed centroid lying close to the particle's mean position during its motion. In (c), the particle travels a significant distance in between gamma ray emissions, meaning that the intersections between LoRs do not yield any meaningful information. Note that for the sake of clear demonstration in this example, we are neglecting the other sources of error discussed above.

So why is this limited temporal resolution an issue? For a static particle, in fact, it is not—we need merely to wait until we have amassed enough LoRs to reconstruct an accurate location.[6] For a *moving* tracer, however, things become more problematic.

As we will see in chapter 6, most PEPT algorithms work on the assumption that all gamma rays are emanating from a point-like source (corresponding to the centre of the tracer). If our tracer is moving, this assumption will no longer hold true as the LoRs required to compute a location will be spread along a line. If our data rate is suitably high, and our tracer suitably slow, the length of this line will be relatively short, and the trajectory effectively straight, meaning that the midpoint of this line represents a good approximation to the mean position of the particle during the relevant time window. If the path is long and meandering, however, and/or the activity too low, determining the central point of the gamma rays produced will be largely meaningless (see figure 2.7).

2.3.5 The real picture

In the previous sections, we have used simple schematics to individually illustrate the main sources of error in the PEPT location of tracer particles. Of course, in reality, all of these individual effects occur simultaneously, creating a much more complex and messy picture. Figure 2.8 shows a number of LoRs taken from real PEPT data, and clearly illustrates the effects of these sources of error in a real-world data set. As is clear from this figure, determining the precise location of a particle is not an easy job. Luckily, there exist a number of carefully-designed algorithms capable of accurately extracting the positions of particles even from this noisy data. These algorithms are described in detail in chapter 6.

In the interactive example below, we are presented with a 'virtual tracer' in a simulated gamma camera. In the simulation, we are able to explore individually the

[6] Note that this does not necessarily hold for the case in which the level of background radiation is significant compared to that emitted by the tracer particle.

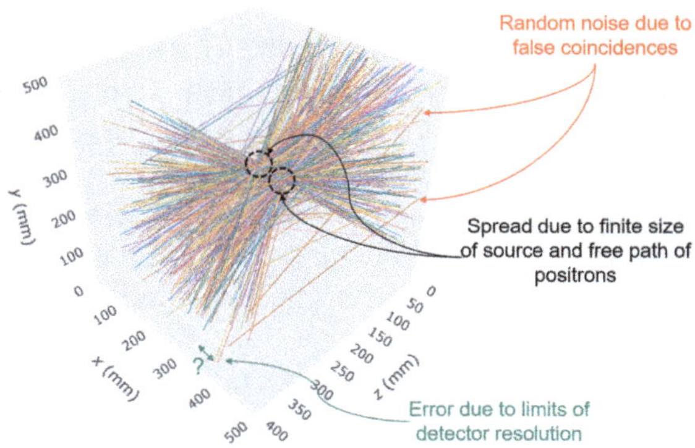

Figure 2.8. Image showing a sample of LoRs extracted from real, experimental PEPT data. The data corresponds to a pair of 1 mm tracer particles separated by a distance of 32 mm.

effects of false coincidences, positron flight and detector error, as well as the combined influence of these effects.

2.3.6 Interactive example: sources of error in PEPT

The notebook files accompanying this section are available in the supplementary data online at https://iopscience.iop.org/book/978-0-7503-3071-8.

Sources of error in PEPT

In this notebook we will generate some more realistic lines of response (LoRs) emitted from a simple spherical tracer—that is to say that all LoRs will be subject to the multiple sources of error discussed above, such as positron range, the finite spatial resolution of detectors, plus the existence of corrupted LoRs. Even though accounting for all sources of errors in a PEPT experiment is a research area in its own right—as detailed in chapter 5—this interactive tutorial will serve to illustrate the very non-trivial task of finding tracer locations from 'real' PEPT data. The generated 'real' LoRs will then be used to triangulate the tracer's location as simply the minimum distance point (MDP) to all LoRs. Finally, the spatial error in the tracer location found is computed for varying numbers of LoRs used when computing the MDP.

This is very much an **interactive** notebook, with many free parameters that the reader is invited to vary to intuitively discover many of the interesting properties of PEPT that will be discussed in detail in the later chapters in the book.

[1]:
```
# Importing necessary libraries for numerical array manipulation and plotting,
# plus the `pept` library for straightforward PEPT data processing
import numpy as np
import matplotlib.pyplot as plt
import pept

# Set Matplotlib theme (try "darkgrid" too for a more modern look)
import seaborn
seaborn.set_style("whitegrid")

# Set Plotly figure output (try "browser" too for interactive graphs)
import plotly
plotly.io.renderers.default = "png"
```

Monte Carlo line of response generation

In the code cell below we define subroutines for the Monte Carlo simulation of a PEPT experiment containing a single spherical tracer inside a parallel screens PEPT scanner; 'Monte Carlo' is the name of a family of algorithms which, in simple terms, use random numbers to simulate physical phenomena such as stochastic radioactive decay (section 4.5 provides a more in-depth explanation of the Monte Carlo technique, while section 5.3 describes its use for the scientific simulation of positron emission).

There are a number of steps involved in the acquisition of lines of response in a typical PEPT experiment:

- A single fluorine-18 radioisotope undergoes radioactive decay, producing a single positron. Note that there are other positron-emitting radioisotopes (and even direct gamma ray emitters), but F-18 generally yields the best spatial resolution.
- The positron travels some distance before annihilating with a nearby electron; this distance, called *positron range*, depends on the positron energy—which in turn depends on the radioisotope used and its surrounding environment.
- The positron–electron annihilation event produces two back-to-back gamma rays. If both gamma rays hit two opposing detectors within a given *resolving time*, they are registered as a *coincidence event* and the two intersection points are logged, producing a single *line of response*, or LoR.

This is of course a simplified picture, which nontheless provides a useful illustration of the processes involved in recording LoRs; for a detailed discussion on the subject, the reader is referred to chapter 5.

As mentioned above, the code below contains multiple free parameters that the reader is invited to have a play with:

[2]:
```
#
# User-definable PEPT simulation parameters
#
screen_height = 500                            # Detector screen height, mm
screen_width = 500                             # Detector screen width, mm
screen_separation = 500                        # Detector heads separation, mm

tracer_radius = 5                              # Tracer radius, mm
tracer_location = [250, 250, 250]              # Tracer 3D location, mm

tracer_activity = 10e6                         # Tracer activity, Bq


# Once a positron is created, it travels through its environment until it
# annihilates with a nearby electron. The distance it travels before
# annihilation is called the "positron range"; it can be approximated as a
# Gaussian distribution (Palmer MR, Brownell GL. Annihilation density
# distribution calculations for medically important positron emitters.
# IEEE transactions on medical imaging. 1992 Sep) - for simplicity, define
# here the standard deviation of the positron range Gaussian distribution
positron_range_std = 5


# Instantiate a random number generator with a seed for deterministic
# pseudo-random numbers
rng = np.random.default_rng(0)
```

```
#
# Functions simulating a PEPT experiment with a static spherical tracer. For
# simplicity, they have been defined as free functions which make use of the
# global parameters defined above.
#
# In a production codebase, such "global parameters" would be encapsulated
# within a class, and the functions below would be class methods making use of
# the aforementioned class variables.
#
def generate_positron():
    '''Generate a radioactive decay / positron emission point *inside* a
    spherical tracer centred at (0, 0, 0) of radius `tracer_radius`.

    Returns an array with the X, Y, Z positions of the emission point.
    '''

    # Create random spherical coordinates
    r = rng.uniform(0, tracer_radius)
    theta = rng.uniform(0, np.pi)
    phi = rng.uniform(0, 2 * np.pi)

    # Transform spherical coordinates into cartesian coordinates
    x = r * np.sin(theta) * np.cos(phi)
    y = r * np.sin(theta) * np.sin(phi)
    z = r * np.cos(theta)

    return np.array([x, y, z])

def generate_lor(annihilation):
    '''Given an input positron annihilation 3D point, generate a 3D LoR and
    check it hits both detector screens; if the LoR hits both detector screens,
    it is acquired and returned as a list with the two LoR interception points
    ([x1, y1, z1, x2, y2, z2]). If at least one of the two LoR ends fall outside
    the detector heads, None is returned.
    '''
    # Generate LoR direction in spherical coordinates
    phi = rng.uniform(0, np.pi)
    theta = rng.uniform(0, np.pi)

    # If the LoR is perfectly vertical / horizontal (parallel to the detector
    # screens), it is not intercepted
    if phi == np.pi / 2 or theta == np.pi / 2:
        return None

    # Unpack positron annihilation point 3D coordinates
    xp, yp, zp = annihilation

    # Find the 2D points where the LoR intersects the two *planes* of the two
    # detector screens
    x1 = xp - zp * np.tan(phi)
    x2 = xp + (screen_separation - zp) * np.tan(phi)

    y1 = yp - zp / np.cos(phi) * np.tan(theta)
    y2 = yp + (screen_separation - zp) / np.cos(phi) * np.tan(theta)
```

```
    # Check the rays fell within the PEPT screens, otherwise return None
    if 0 < x1 < screen_width and 0 < x2 < screen_width:
        if 0 < y1 < screen_height and 0 < y2 < screen_height:
            return [x1, y1, 0, x2, y2, screen_separation]

    return None


def record_lors(num_lors):
    '''Run a Monte Carlo simulation generating `num_lors` LoRs emitted from
    a spherical tracer and return them as a NumPy array with `num_lors` rows
    and 7 columns formatted as [timestamp, x1, y1, z1, x2, y2, z2], where
    (x1, y1, z1) is the LoR interception point on the first detector screen and
    (x2, y2, z2) is the LoR interception point on the second detector screen.
    '''
    # Pre-allocate array for LoRs
    lors = np.zeros((num_lors, 7))

    # Keep track of generated LoRs' timestamps
    current_time = 0
    dt = 1 / tracer_activity


    # Generate enough LoRs until `num_lors` hit both detector screens and are
    # recorded.
    #
    # Because the LoRs generated may fall outside the detector screens, we
    # cannot know in advance how many LoRs we have to generate; simply keep
    # track of a current LoR index
    lor_index = 0


    while lor_index < num_lors:
        # Create radioactive decay / positron emission point *inside* the tracer
        emission = tracer_location + generate_positron()

        # Create positron-electron annihilation point according to a Gaussian
        # distribution
        annihilation = emission + rng.normal(0, positron_range_std, size=3)

        # Generate LoR / pair of back-to-back gamma rays and check it hit both
        # detector screens
        lor = generate_lor(annihilation)

        # If the LoR was invalid, `None` was returned; simply advance the
        # simulation time and generate a new LoR at the next loop iteration
        if lor is None:
            current_time += dt
        continue
```

```
            # If the LoR hit both detector screens, add it to the `lors` array and
            # record the current timestamp
            lors[lor_index, 1:8] = lor
            lors[lor_index, 0] = current_time

            # Increment current LoR index and advance simulation time
            lor_index += 1
            current_time += dt

    return lors
```

The PEPT-oriented `PlotlyGrapher` plotting interface provided by the `pept` library can be used for straighforward plotting of PEPT data; while not required *per se* for illustrating PEPT algorithms, it significantly reduces the amount of repetitive code and visual noise, allowing the reader to focus on the main conceptual procedures.

In the code cell below we generate `num_lors` 'ideal' LoRs—though still being subject to positron range—and create a 3D visualisation of them:

```
[3]: # Generate `num_lors` LoRs and plot them using the `pept` library.
     from pept.plots import PlotlyGrapher

     num_lors = 200
     lors = record_lors(num_lors)

     grapher = PlotlyGrapher()
     grapher.add_lines(lors)

     # Set default zoom level in `camera_eye`
     grapher.fig.update_scenes(camera_eye=dict(x=1.5, y=1.5, z=1.5))
     grapher.show()
```

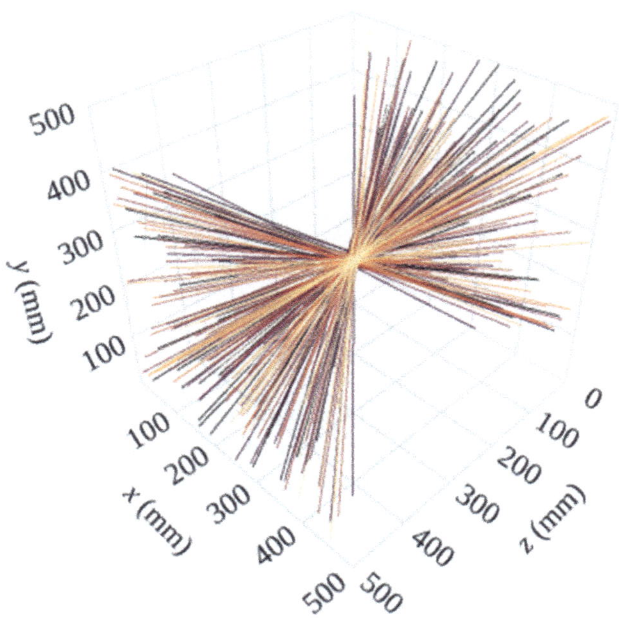

Note the PEPT 3D axes convention, differing from the common mathematical coordinates with Y and Z being switched, such that the Y axis points upwards, while the Z axis is perpendicular to the two detector heads.

Adding noise: scattered events and spread

Even in a simplified picture of a PEPT experiment, there are number of sources of error:

- The *positron range* mentioned above serves to generate a spread in the triangulated tracer locations due to LoRs possibly being emitted from points *outside* the tracer.
- If one gamma ray hits another subatomic particle and is *scattered*, it will yield a *corrupted event* whose LoR may be very far from the actual tracer. Due to the finite interval of the *resolving time*, other gamma rays may be intercepted from different electron–positron annihilation events, again yielding a corrupted event. Due to the existence of 'corrupted events', **a given fraction of LoRs recorded will simply be noise**–this fraction varies with the density of the environment the tracer is in, its activity, and the specific gamma detector used.
- The gamma rays travel a certain unknown distance *inside* the detector before being captured, such that the two ends of an LoR will always have some spatial error on the detector screens.

The positron range is accounted for in the LoR generation stage within the `record_lors` function above. In the code cell below we define subroutines for adding noise (i.e. scattered LoRs) and spread (on the detector screens).

[4]:
```
#
# Functions adding noise and spread to generated LoRs
#


def add_noise(lors, noise_fraction):
    '''Given an array of generated LoRs, insert scattered events and return an
    array of "real" LoRs in which a fraction of `noise_fraction` (between 0 and
    1) LoRs will be erroneous. The scattered LoRs will be inserted randomly
    in the `lors` array such that the timestamps will still be sorted.
    '''

    # Number of scattered events required such that the final array of "real"
    # LoRs will have `noise_fraction` scattered LoRs
    num_scattered = int(len(lors) * noise_fraction / (1 - noise_fraction))

    # Pre-allocate array of scattered events
    noise = np.zeros((num_scattered, 7))

    # Generate uniformly-distributed XY points on the two detector screens
    noise[:, [1, 4]] = rng.uniform(0, screen_width, size=(num_scattered, 2))
    noise[:, [2, 5]] = rng.uniform(0, screen_height, size=(num_scattered, 2))

    # The Z coordinates are fixed, 0 and screen_separation
    noise[:, 0] = 0
    noise[:, 6] = screen_separation

    # Generate random timestamps for the scattered LoRs between the first and
    # last timestamps in the input `lors`
    noise[:, 0] = rng.uniform(lors[0, 0], lors[-1, 0], size=num_scattered)

    # Insert the noise in the input `lors` such that the timestamps are still
    # sorted (and therefore the noise will be randomly distributed)
    insertion_indices = np.searchsorted(lors[:, 0], noise[:, 0])
    lors_noise = np.insert(lors, insertion_indices, noise, axis=0)

    return lors_noise


def add_spread(lors, max_spread):
    '''Add detector spread to a given array of generated LoRs; essentially, the
    XY positions on the two detector heads (columns 1, 2, 4, 5 in `lors`) will
    have a random number added between (-max_spread, max_spread). A typical
    value for max_spread is 0.5 mm.
    '''

    # Create a copy of the LoRs array to avoid mutating it in-place
    lors = lors.copy()

    # Generate random erroroneous values that will be added to the XY LoR
    # positions on the two detector heads
    spread = rng.uniform(-max_spread, max_spread, size=(len(lors), 4))
    lors[:, [1, 2, 4, 5]] += spread

    return lors
```

Now let's generate some LoRs, add noise and spread to make them 'real', and plot them:

[5]:
```
# Generate "real" LoRs and plot them; do vary the variables below and see their
# effects!

# Number of "ideal" LoRs to generate
num_lors = 200

# Ratio of scattered events to all LoRs
noise_ratio = 0.3

# Maximum error added to the gamma ray XY interception points on the detector
# screens to simulate finite spatial resolution of the scanner
max_spread = 1.0

# Generate `num_lors` ideal LoRs, then add noise and spread to them
ideal_lors = record_lors(num_lors)
real_lors = add_noise(ideal_lors, noise_ratio)
real_lors = add_spread(real_lors, max_spread)

# Plot the "real" LoRs
grapher = PlotlyGrapher()
grapher.add_lines(real_lors)

# Set default zoom level in `camera_eye`
grapher.fig.update_scenes(camera_eye=dict(x=1.5, y=1.5, z=1.5))
grapher.show()
```

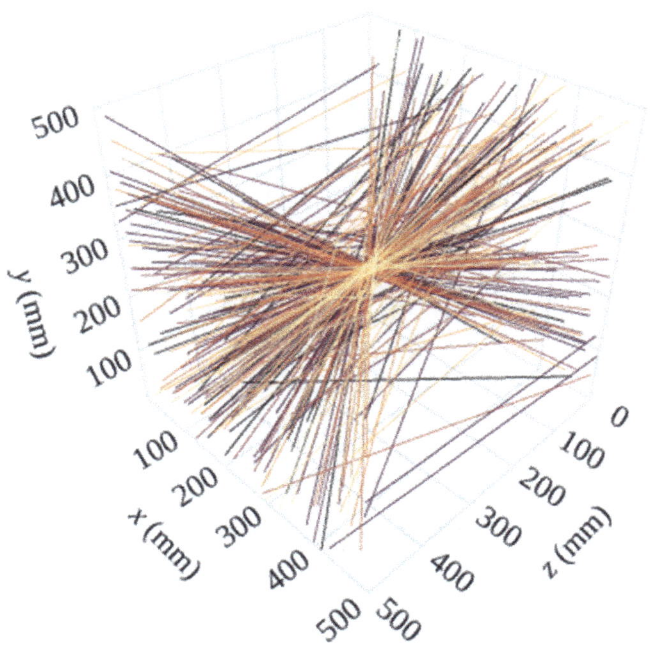

Triangulate tracer's location

Now that we can simulate a PEPT experiment and record the LoRs emitted from a spherical tracer, we can try to backcompute its location from those LoRs. The simplest method of doing so is to compute the *minimum distance point* (MDP), or the 3D point minimising the sum of distances to all LoRs.

In the code cell below, we compute the MDP and print the *spatial error*, or distance between the MDP and the true tracer location (as defined in the `tracer_location` variable above):

[6]:
```
# Find the minimum distance point (MDP) to all LoRs by minimising the sum of
# distances to all lines.
#
# Note that the MDP can also be determined analytically (as is done in the C
# implementation from the `pept` library), but the derivation is rather tedious
# and takes away from the didactic purposes of this notebook
from scipy.optimize import minimize


def distances(point, lines):
    '''Return an array of distances from a `point` to an array of `lines`
    formatted as [t, x1, y1, z1, x2, y2, z2].'''

    p1 = lines[:, 1:4]
    p2 = lines[:, 4:7]

    d1 = np.linalg.norm(np.cross(point - p1, point - p2), axis=1)
    d2 = np.linalg.norm(p2 - p1, axis=1)

    return d1 / d2


def sum_distances(point, lines):
    '''Return the sum of distances from a `point` to an array of `lines`.'''
    return np.sum(distances(point, lines))


# Find the minimum distance point via an optimisation approach - the MDP is
# simply the point minimising the sum of distances to all LoRs
p0 = np.array([200, 200, 200])                    # initial guess
mdp = minimize(sum_distances, p0, real_lors).x

# Insert MDP timestamp at column 0
mdp = np.insert(mdp, 0, real_lors[:, 0].mean())

# The spatial error between the MDP and the tracer position
error = np.linalg.norm(mdp[1:4] - tracer_location)

print(f"The minimum distance point (MDP) found:\n{mdp}\n")
print(f"The spatial error is {error:4.4f} mm")
```

```
The minimum distance point (MDP) found:
[4.72150033e-05 2.49498231e+02 2.49880891e+02 2.50438135e+02]

The spatial error is 0.6767 mm
```

Notice that (using the default parameters in the notebook) the spatial error is 9 times larger than what we calculated in the previous notebook, where we simulated 'ideal' LoRs! In this case, the tracer location we found is completely outside the actual particle.

[7]:
```
# Plot all LoRs and the minimum distance point (MDP)
grapher = PlotlyGrapher()

grapher.add_lines(real_lors)
grapher.add_points([mdp], size=5)

# Set default zoom level in `camera_eye`
grapher.fig.update_scenes(camera_eye=dict(x=1.5, y=1.5, z=1.5))
grapher.show()
```

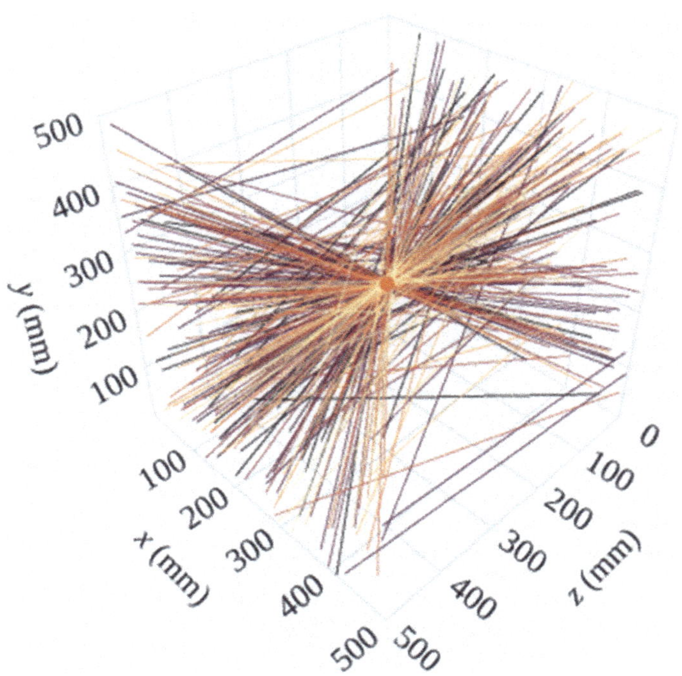

Spatial error versus noise ratio

How does the spatial error change when different numbers of LoRs are used to compute the minimum distance point (MDP)? The code cell below generates a 'large' dataset of LoRs, selects *samples* of different numbers of LoRs at a time, and computes the spatial error of the MDP—then plots the spatial errors recorded:

```
[8]: # Compute spatial error at varying levels of noise in the set of LoRs. Do
     # multiple repeats of the measurement for accurate statistics

     # Number of LoRs used to compute MDP for a single measurement
     sample_size = 200

     # Number of samples of LoRs to compute MDP for, for a single error value
     repeats = 10

     # Maximum error added to the gamma ray XY interception points on the detector
     # screens to simulate finite spatial resolution of the scanner
     spread = 1.0

     # Fractions of scattered events to number of LoRs recorded for which to compute
     # the spatial error
     noise_fractions = [0.0, 0.1, 0.2, 0.3, 0.4, 0.5, 0.6, 0.7, 0.8, 0.9]

# Generate a "large" batch of "ideal" LoRs which will be used when adding noise,
# spread and recording the spatial error
ideal_lors = record_lors(sample_size * repeats)


def spatial_error(noise_ratio):
    '''Measure the spatial error in the MDP from `repeats` samples of LoRs with
    `noise_ratio` fraction of scattered LoRs. Return the mean error and standard
    deviation.
    '''
    # Add noise and spread to the "ideal" LoRs to simulate "real" LoRs
    real_lors = add_noise(ideal_lors, noise_ratio)
    real_lors = add_spread(real_lors, spread)

    # Pre-allocate array of errors
    errors = np.zeros(repeats)

    # For `repeats` consecutive samples of "real" LoRs, compute the MDP and
    # save the spatial error in the `errors` array
    for i in range(repeats):
        # Select a sample of LoRs
        sample = real_lors[i * sample_size:(i + 1) * sample_size]

        # Find the minimum distance point via an optimisation approach - the MDP
        # is simply the point minimising the sum of distances to all LoRs
        p0 = np.array([200, 200, 200])                    # initial guess
        mdp = minimize(sum_distances, p0, sample).x

        # The spatial error between the MDP and the tracer position
        errors[i] = np.linalg.norm(mdp - tracer_location)

    # Return the mean error and standard deviation
    return errors.mean(), errors.std()
```

```
# Measure the spatial error for each level of noise in `noise_fractions`.
# `errors` will be a 2D array, where the first column is the mean spatial error
# measured and the second column is the standard deviation
errors = np.array([spatial_error(fraction) for fraction in noise_fractions])

# Plot the spatial errors found with the standard deviations as error bars
fig, ax = plt.subplots()
ax.errorbar(noise_fractions, errors[:, 0], yerr=errors[:, 1], capsize=5)
ax.set(xlabel="Fraction Scattered Events", ylabel="Spatial Error (mm)");
```

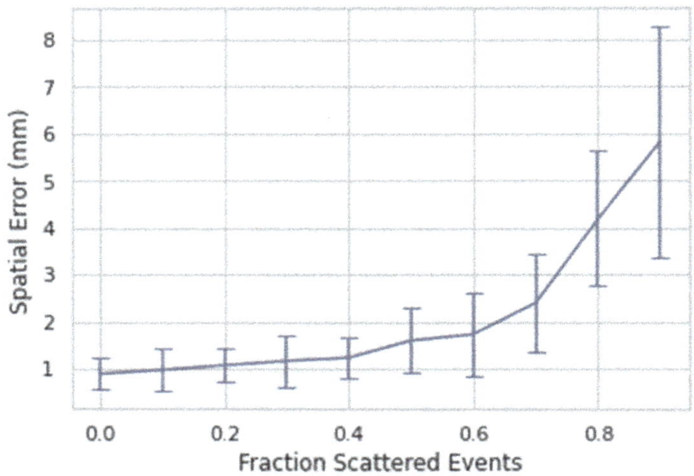

Note that even at zero noise fraction, the spatial error is larger than the actual tracer radius due to the positron range and finite detector spatial resolution; increasing fractions of scattered events then yield exponentially worse tracer locations. The usefulness of PEPT is directly tied to the *tracer finding algorithm* used, which must robustly discriminate between real LoRs and scattered events. **Any tracer finding algorithm must be able to robustly handle noise, even in large amounts!**

2.4 Not *just* particle tracking

In the preceding sections we have explained, in simple terms, how PEPT may be used to locate and record the trajectory of a radioactively-labelled tracer particle. We must now ask ourselves—what's the point? Surely a single-particle trajectory alone offers us only a very limited insight into the dynamics of a particulate system!

In fact, one of the great strengths of PEPT is that these high-resolution, three-dimensional trajectories can—despite only corresponding to a single particle's motion—be manipulated and post-processed to provide a wealth of information regarding the dynamics of a whole system. This is most easily done for the case of systems in an ergodic steady state. The principle of ergodicity states that in such a system, the long-term average of the behaviour of a single particle is equivalent to that of the system as a whole [9]. Consider, as an example, a dynamic system of 1000 particles moving freely within a container, including 1 designated tracer particle. Let us then notionally divide our container into a number of individual sections. Suppose that after measuring our system for a long time (specifically, long enough for all particles to fully explore the relevant phase space), we find that our single tracer has spent 15% of its time in a given section. If our system is *ergodic*, we can assume that at any given point in time, an average of 150 particles (i.e. 15% of the total number) will be found in the aforementioned section. This concept—and the manner in which it can be valuably applied in PEPT imaging—is explained in much greater detail in chapter 7.

Using the principle of ergodicity, we can extract a wide variety of properties using PEPT, including density fields, velocity fields and velocity vector fields, velocity probability density functions, mean squared displacements, dispersion and diffusion coefficients, granular temperatures and a host of other important physical quantities which help us explore, characterise and understand particulate systems. The various physical quantities and fields which may be extracted from PEPT data, and the manners in which they may be accurately computed, are described in detail in chapter 7.

References

[1] Nicuşan A L and Windows-Yule C R K 2020 Positron emission particle tracking using machine learning *Rev. Sci. Instrum.* **91** 013329

[2] Rosato A D and Windows-Yule C 2020 *Segregation in Vibrated Granular Systems* (New York: Academic)

[3] Kónya J and Nagy N M 2018 *Nuclear and Radiochemistry* (Amsterdam: Elsevier)

[4] Sanchez-Crespo A, Andreo P and Larsson S A 2004 Positron flight in human tissues and its influence on PET image spatial resolution *Eur. J. Nucl. Med. Mol. Imaging* **31** 44–51

[5] Davila H O, Serrato H F C and Ovalle S A M 2017 Monte Carlo simulation of different positron emitting radionuclides incorporated in a soft tissue volume *Ciencia en Desarrollo* **8** 35–40

[6] Peskin M E 2019 *Concepts of Elementary Particle Physics* vol 26 (Oxford: Oxford University Press)

[7] Cooper M J 1985 Compton scattering and electron momentum determination *Rep. Prog. Phys.* **48** 415

[8] Bai B, Ruangma A, Laforest R, Tai Y-C and Leahy R M 2003 Positron range modeling for statistical PET image reconstruction *Nuclear Science Symp. Conf. Record (IEEE Cat. No. 03CH37515)* vol 4 (Piscataway, NJ: IEEE) p 2501–052003

[9] Wildman R D, Huntley J M, Hansen J-P, Parker D J and Allen D A 2000 Single-particle motion in three-dimensional vibrofluidized granular beds *Phys. Rev.* **E62** 3826

Chapter 3

A history of PEPT

This chapter presents an overview of the history of the positron emission particle tracking technique, presented from the vantage point of one of its original pioneers, Professor David Parker. It provides a unique, first-hand account of PEPT's evolution from a niche offshoot of positron emission tomography to a powerful and widely-used tool, detailing key moments in its development, and introducing the people and places involved therein, as well as recounting some amusing anecdotes, including a now-infamous incident involving radioactive crisps.

Note: Though a variety of PEPT researchers and facilities around the world—notably the groups in Bergen, Cape Town, KwaZulu-Natal, Tennessee and, more recently, Virginia, Stanford, and King's College London—have contributed significantly to the development of the PEPT technique, this brief history focuses in particular on developments at Birmingham. This is in part because this is where the technique was first developed, and in part because these are best known to the authors, and thus allow us to tell the most compelling (and most accurate) story. The work of these, and other, PEPT researchers is nonetheless clearly flagged throughout the book, in particular in chapter 5, where important work on detector development performed at Cape Town is detailed, and chapter 8, where key research performed by all of the above-named institutions is discussed.

3.1 Adding the 'P': from PET to PEPT (origins to 1989)

The technique of positron emission particle tracking (PEPT) arose as a development from positron emission tomography (PET), and in particular the University of Birmingham's use of PET as a tool for engineering studies (in contrast to its use elsewhere as a branch of nuclear medicine). The development of PET as a major tool for imaging human physiology is often attributed to Ter-Pogossian, Phelps and Hoffman [1, 2] at Washington University in the mid-1970s. A breakthrough in technology was the development in 1985 by Casey and Nutt of CTI, Knoxville, of the scintillation block detector in which four photomultiplier tubes are used to read out an array of eight (later 8 × 8) crystal elements [3]. The first clinical PET

doi:10.1088/978-0-7503-3071-8ch3

scanner based on this principle was delivered to Hammersmith Hospital, UK, in 1987 [4].

The use of these scintillator blocks rapidly became the standard for PET scanners, but in the 1970s and 80s various groups had been pursuing alternative detector technologies for PET. In the UK, at the Rutherford Appleton Laboratory (RAL), Eddie Bateman led a group developing gas-filled multiwire proportional chambers (MWPCs) for high energy physics. As a spin-off from this work, they proposed a 'positron camera' based on a pair of MWPCs, which they hoped would offer the nuclear medicine community a simplified system for clinical PET [5]. Unfortunately, the medical community was largely unenthusiastic about this approach, although one system ('MUPPET') was eventually installed at the Royal Marsden Hospital. Meanwhile, Bateman's concept had come to the attention of Mike Hawkesworth at the University of Birmingham, who was exploring the possible use of PET for studying flows in engineering—specifically the behaviour of fuels and lubricants in aero-engines.

Hawkesworth's background was in neutron radiography, and he had been working with Peter Stewart, Roger Witcomb and colleagues at Rolls Royce on the use of this technique to study aero-engines. Unlike x-ray radiography, which highlights dense, high-Z material, neutron radiography penetrates metal casings and reveals the presence of hydrogenous material, such as the fuel or lubricant inside [6]. It is a powerful technique, able to show, for example, blockages in cooling channels, but with the major limitation that studies must be performed at a research reactor. As an alternative way of visualising fuel or lubricant inside an operating aero engine, Hawkesworth proposed the use of radioisotope tracers. In 1980 Stewart, Witcomb and Hawkesworth obtained a patent for the application of PET to oil flow [7].

Hawkesworth realised that Bateman's proposed positron camera could be used for this work. A collaboration was formed between the University of Birmingham, Rolls Royce, Burmah Castrol (lubricant specialists) and RAL. In 1981 the UK Science and Engineering Research Council (SERC—predecessor of the later EPSRC) awarded this consortium its largest grant to date under its co-operative research grants scheme to explore 'fluid flow tracing using radioisotopes' (supposedly the title was deliberately vague so as not to reveal too much to industrial competitors).

The RAL MWPC camera was completed late 1983 [8]. Over the next few years, it was used at Birmingham to perform PET studies on bearing test rigs. This work culminated in trials on a Viper gas turbine, operated at full power on a testbed at the Staverton engine test facility in early 1987, by means of which a real problem encountered by the Brazilian Air Force was identified as a lubrication failure and successfully resolved. Before this, the project was the subject of a half-page feature in the Financial Times on 8th October 1985.

Radioisotope tracers (radiotracers) for all the above-mentioned work were produced by Peter Fowles of Birmingham's School of Chemistry, who developed techniques for incorporating ^{68}Ga (eluted from a generator containing the parent

^{68}Ge), or ^{18}F (manufactured using one of the cyclotrons at the University of Birmingham) into an oil-soluble form[1].

While the project was technically successful, its impact on aero-engine development was questionable, perhaps partly due to the long lead times required for approving any change in aero-engine design. In parallel, Hawkesworth, Fowles and colleagues at Birmingham investigated other opportunities for exploiting the new positron camera. As well as applications in mechanical engineering they explored its use in earth sciences (imaging the passage of a bolus of labelled water or oil through rock in laboratory samples) but the most promising applications were found in chemical engineering. John Bridgwater and John Botterill of Birmingham's School of Chemical Engineering obtained funding for feasibility studies of the use of PET to study solids motion, and recruited as research assistants Colin Bemrose to supervise the chemical engineering and Mark O'Dwyer to perform the data analysis. For studying the mixing of particulate material, two approaches were tried: imaging the dispersion of a clump of labelled particles, or tracking a single labelled particle, and from this second approach came what would eventually be called PEPT.

The researchers soon realised that backprojected images were not the best way to visualise the motion of a particle. O'Dwyer solved the equations to derive a probable particle location from a set of back-to-back gamma ray lines, developing the 'minimum distance' method [9], which later became known simply as the 'Birmingham Method'. The technique was validated by tracking a ^{22}Na source rotating on a turntable, and then demonstrated by tracking the motion inside a small fluidised bed of a 6 mm glass bead inside which ^{22}Na had been encapsulated [9]. Thus by 1988 the technique that would later be called PEPT had already been demonstrated, but funders were slow to appreciate its potential. Bemrose's and O'Dwyer's contracts were not renewed, and the future seemed unclear.

3.2 MWPC PEPT (1989–1999)

In 1989 David Parker joined Birmingham's School of Physics as Lecturer in Environmental and Biomedical Physics in Derek Beynon's Applied Physics Group. Parker's first degree was in mathematics but he had worked for the previous 10 years as an experimental nuclear physicist at the UKAEA's Harwell Laboratory. Casting around for a new area of research, he teamed up with Hawkesworth, who needed help with the computer processing of positron camera data due to O'Dwyer's recent departure.

Parker worked on 2D and 3D PET imaging as well as particle tracking. Early tracking studies were performed on a horizontal axis bladed mixer, tracking a steel-encapsulated ^{22}Na source: the mixer was treated as three compartments and in each the data were analysed as 2D plots showing the occupancy and time-averaged velocity field, as well as to determine residence times in each compartment and locations of transfers between compartments [10][2].

[1] The development and production of radiotracers is discussed in detail chapter 5.

[2] This and other early PEPT studies are discussed further in chapter 8. Details regarding occupancy plots, velocity fields and other quantities which may be extracted from PEPT—including the methods by which they may be computed—are discussed in chapter 7.

By now the term 'positron emission particle tracking' had been adopted, to distinguish the new approach from conventional PET imaging. A full description of the technique was published in 1993 [11]; with hindsight, this first publication was perhaps unnecessarily complicated by describing an automated method for varying slice size during tracking.

As well as Bridgwater (originally at Birmingham, later at Cambridge), Hawkesworth, Fowles and Parker collaborated with Jonathan Seville, then at the University of Surrey. To mark the partnership between the Schools of Physics and Chemical Engineering, the University of Birmingham Positron Imaging Centre was established in 1992. Unfortunately research councils initially seemed reluctant to fund this new activity: several applications to SERC for funding for a new camera (an improved RAL model) and new chemical engineering programmes were unsuccessful—reviewers persistently compared the capabilities of PET and PEPT unfavourably with those of MRI, in a way that seemed ill informed. To try to break this impasse, the Birmingham group invited SERC to set up a review panel. This panel visited Birmingham in 1993; in their report they wrote 'the key feature was the development of the particle tracking facility. This opens up the possibility of markedly improved understanding of the fundamentals of particle dynamics. Any improved understanding would readily feed through into applications to practical problems.' The possibility of formally establishing a national PEPT facility was aired, but considered premature.

In the years following this review, SERC funding was obtained for three research associates, and applications of PEPT proliferated. Writing in 1999, Parker noted 'over a recent 21 month period measurements were made on 315 days (74% of the days available) on behalf of over 20 distinct research programmes employing 14 PhD students. Collaborations currently include 5 other UK universities as well as a number of overseas universities'.

The studies to which he referred (and of course countless studies since!) relied on the availability of more realistic tracer particles than the large glass beads initially tracked. As well as direct activation of glass beads (down to 1 mm in diameter), Fowles introduced a technique for indirectly activating small ion-exchange resin beads with $^{18}F^{-}$, produced in aqueous solution. A weakness of these particles was that if they were exposed to water the activity rapidly leached out, so crude techniques were developed for sealing the activity inside (the first method used nail varnish). Mike Hawkesworth retired from the University of Birmingham in 1994, having never held a permanent post. He remained in close touch with Parker until his death from pancreatic cancer in 2005.

3.3 Out with the old, in with the new (1999–2002)

The major weakness of the MWPC positron camera was its low sensitivity and consequent low data rates[3]. Each head had an efficiency of only 7% for detecting an

[3] Sensitivity, efficiency, data rates and other related topics as concerns positron cameras are introduced and discussed in detail in sections 5.2 and 5.3.

incident 511 keV photon, so that the overall coincidence efficiency was at best less than 0.5%. Compensating for this low efficiency by increasing the source activity rapidly resulted in unacceptable rates of random coincidences, so in practice the useful data rate from the camera was limited to around 3000 events per second.

State of the art PET scanners based on bismuth germanate (BGO) blocks offered much higher sensitivity and data rates, but were extremely expensive and even if funding could have been obtained the field of view of these ring systems was not ideal for most engineering studies. In 1988 the nuclear medicine department of Birmingham's Queen Elizabeth Hospital acquired a Vertex 'gamma camera PET' system; this system, manufactured by the Californian company ADAC, consisted of a pair of sodium iodide gamma camera heads (with thicker than usual crystals) which could be operated in coincidence and without collimation for PET imaging. Parker recognised that a system like this would be an ideal successor to the MWPC camera, having a very similar geometry but much higher sensitivity.

A case to purchase one of these systems was made under HEFC's JREI scheme, which required applicants to obtain matching contributions from other sources, especially industry. Parker called in favours and obtained promises of support from industry in the form of Amoco Polymers, BNFL, Merck Sharpe and Dohme, Unilever, and Weetabix, and academia in the form of the Universities of Cambridge, Exeter and Loughborough; together with a contribution from the University of Birmingham, just over 50% of the required total of £360k was raised, with HEFCE providing the balance.

The ADAC Forté camera (successor model to the aforementioned Vertex) was delivered in 1999. Compared to the MWPC camera it has better spatial resolution, and (for the first time) energy resolution, as well as significantly higher efficiency. Most importantly, it offers a useful data rate of up to 100k events/s [12]. The camera was mounted on rails, and subsequently motorised so that slow-moving tracers could be tracked over an extended field of view.

Over 20 years later, at the time of writing the Forté is still in daily use for PEPT studies. Although there have been occasional problems with the gantry, the detectors and electronics have proved remarkably reliable. An unfortunate accident in early 2021 led to cracking of both NaI crystals, but these were rapidly replaced.

Soon after installation of the new camera, Parker inadvertently became briefly involved with the UK media. Asked to write about the Centre's work for the monthly *Physics World*, he included an account of how the process of flavouring potato crisps was being studied by labelling a crisp with ^{18}F and tracking its motion inside a rotating drum. This story about 'radioactive crisps' was picked up by a number of newspapers and by the BBC, and Parker found himself pinned between the media asking 'what's the story here?' and the sponsors of the work, who had not approved its disclosure and were terrified of a public scare about radioactivity in crisps.

In 2000 EPSRC awarded the Positron Imaging Centre one of its first Platform Grants. Part of the case made for this grant was the need to replace Fowles, who had retired in 1999 but continued to work part-time on tracer labelling pending the recruitment of a successor. Xianfeng Fan was appointed to this role, and was responsible for producing PEPT tracers for the next eight years.

Another appointment around this time was Andy Ingram, who arrived from industry and following a period as research fellow in the Centre was appointed to a lectureship in chemical engineering.

The next major development was Parker's acquisition of a much newer cyclotron to replace the Radial Ridge Cyclotron which had become increasingly unreliable as a source of PEPT tracers. In early 2002 the Scanditronix MC40 Cyclotron was purchased second hand from the VA Hospital in Minneapolis. It was dismantled, moved to Birmingham and recommissioned, becoming operational in 2004.

The MC40 is the most versatile cyclotron in the UK, accelerating protons, deuterons, ^{3}He and ^{4}He. Although the justification for its purchase was to produce tracers for the expanding range of PEPT studies, the cyclotron has also been used extensively [13] for nuclear physics studies [14, 15], detector testing and studies of radiation hardness and radiation damage [16], the development of novel methods for proton therapy [17, 18], as well as for radioisotope production for diverse other applications. On five days per week the cyclotron produces ^{81}Rb for hospitals across the UK.

3.4 Positron cameras of all shapes and sizes (2002–present)

The next phase of Birmingham positron cameras came about almost by chance. While visiting Hammersmith (home of another MC40 cyclotron) to discuss aspects of the cyclotron transfer, Parker was asked if he had any interest in acquiring their old ECAT930 PET scanner. As mentioned above, this system had been installed in 1987 and was then the first modern PEPT scanner in a clinical environment. 25 years later it was due for disposal. Parker leapt at this opportunity and the scanner was dismantled and moved to Birmingham in 2002.

It was stored for a time while Parker concentrated on the cyclotron, then recommissioned. A PhD student, Alirezar Sadrmomtaz, found that many of the 32 detector buckets were faulty, so instead of the original double ring just a single ring of 16 buckets was reassembled and PET imaging was demonstrated. In order to extend the scanner to PEPT it was necessary to restore the capability of recording data event-by-event in listmode, an original feature of the scanner which had been lost at some stage in its development. Fortunately, expertise in the data acquisition system was available in the form of Larry Byars, a former employee of CTI, now working as a self-employed consultant. In 2004 Byars reinstalled listmode acquisition, and the first PEPT measurements were made with the 930 scanner.

Since the 1 m diameter single ring of detectors offered too narrow a field of view to be useful for most applications, the decision was made to reconfigure the buckets from the 930 into a more flexible geometry. The modular nature of PET scanners (discussed further in section 5.2) makes this in many ways a natural evolution, though at the time it represented an entirely novel idea. The 930 scanner comprised a total of 128 detector blocks grouped into 'buckets' each consisting of four blocks plus their associated electronics, controlled by a 'bucket controller'. As initially reconfigured, the four detector blocks of each bucket were spread out and mounted within a box with their controller. A modular positron camera was thus produced, comprising 16 boxes which could be arranged in a variety of geometries [19][4].

Because of faults in many of the bucket controllers, initial trials were performed with just 14 buckets. This modular PEPT system was first used seriously in a series of studies of a large (75 cm diameter) pressurised fluidised bed at BP's Hull site performed in 2006 [20]. Rather than trying to track over the full height of the bed, modules were mounted on opposite sides of the bed at two heights, one set near the base and the other near the top, in order to observe circulation of the tracer particle around a central baffle. This remains the only time, to date, that the Birmingham group has performed PEPT studies away from the Birmingham campus, though a recently-awarded grant from the Royal Academy of Engineering aims to expand the use of '*in situ*' PEPT.

Subsequently, more bucket controllers were purchased from Turku, where an ECAT930 scanner had recently been decommissioned, and the modular camera was used for further fluidisation studies, as well as for tracking the movement of inclusions during casting of molten metal[5].

Meanwhile Parker had moved onto the next generation of ECAT scanners. The trigger for this was a proposal from John Clark, doyen of PET radiochemistry, to develop a PET scanner for equine veterinary medicine. After Hammersmith, the next PET centres to be established in England had been at London's Guys and St Thomas' Hospitals, each of which acquired an ECAT951 scanner in the 1990s. These two scanners were due to be replaced in 2004, and Clark had the idea of combining them into a single double-sized scanner, large enough to accommodate a horse, enlisting Parker to construct the scanner.

An initial application for funding by the Wellcome Trust was unsuccessful, so to avoid the St Thomas scanner being sent for scrap Parker used his own funds to acquire it. With a student Monjur Ahasan he reconfigured the 1 m diameter double ring of detector blocks into a single ring of 2 m diameter to demonstrate the feasibility of imaging on this scale [21, 22]. Once again, Larry Byars provided invaluable input in modernising the data acquisition system.

In 2005 Parker also acquired the Guys scanner, but the idea of combining the two scanners never took off. Instead, Parker's latest PhD student, Tom Leadbeater, incorporated ECAT951 components into the existing modular camera, and developed a dedicated PC-based listmode acquisition system for this [23]. After completing his PhD [24] Leadbeater remained at Birmingham until 2016, overseeing a wide variety of PEPT studies using both the ADAC and modular cameras.

Parker further supplemented his stock of second-hand scanners, acquiring an EXACT31 from Aberdeen in 2006, an ART from Hammersmith in 2009 and finally a HR+ from Hammersmith in 2011. Components from all these three systems are being combined in a new '*SuperPEPT*' instrument (discussed further below).

In 2003 EPSRC awarded Parker a Senior Fellowship, allowing him to focus exclusively on research for 5 years while other colleagues undertook his teaching

[4] Modular PEPT cameras are discussed in further detail in section 5.2.4.

[5] This series of experiments is discussed in detail in section 7.4.

duties. In 2008 the Institute of Physics awarded him its first Joule Medal ‘for the creation of positron emission particle tracking as a practical tool in a wide variety of engineering applications’—whether, in view of all the other people involved, this citation was justified, the reader can judge.

On a lighter note, the Times newspaper picked up a paper [25] about the use of PEPT to study flow inside a domestic dishwasher, and its leader for 17th Jan 2015 began ‘Phew! Thank heavens for positron emission particle tracking. As most domestic sous-chefs will know by now, this is a technique devised in Britain for following the progress of tiny particles through sealed machines to measure accurately what’s happening inside them’.

3.5 PEPT elsewhere

Outside Birmingham, the first group to develop a PEPT system was at the Delft University of Technology. In 1996 the Delft group coupled together a pair of gamma cameras (operating without collimators and in coincidence) and used them to track particles in a four-compartment interconnected fluidised bed. As tracers, copper-loaded glass beads were activated in the Delft research reactor to generate the positron emitting nuclide ^{64}Cu [26]. This system was in use until 2003.

In 2003 Alex Hoffmann and co-workers used a medical PET scanner (ECAT HR+) in a hospital at Groningen to perform PEPT studies of a small fluidised bed, tracking a 470 μm ion exchange resin bead labelled with ^{18}F [27].

After moving to the University of Bergen, Hoffmann has used a newer scanner (Siemens Truepoint) in a Bergen hospital to perform detailed studies of particle motion in the vortex of a hydrocyclone [28, 29].

At the University of Tennessee-Knoxville, Art Ruggles’ group performed PEPT using a pre-clinical PET scanner (Concorde MicroPET P4) [30]. In a separate development, they used a newer pre-clinical scanner (Siemens Inveon) to track simultaneously a large number of weakly labelled yeast cells suspended in water [31]—somewhat of a groundbreaking study in terms of the size and biological nature of the materials tracked. Rather than employing the Birmingham algorithm, most of the studies mentioned above used a location method based on generating the cutpoint for each pair of lines of response and then analysing the resulting cloud of cutpoints[6].

Currently the most extensive use of PEPT outside Birmingham is at the University of Cape Town. Having previously performed PEPT studies at Birmingham, in 2009 the Cape Town group obtained their own second-hand PET scanner (ECAT HR++, previously at Hammersmith) and established PEPT Cape Town at iThemba Labs, with particular interest in minerals processing. Subsequently they also obtained the ADAC Vertex previously used at the Queen Elizabeth Hospital Birmingham. In 2016 Tom Leadbeater moved from Birmingham to Cape Town, strengthening the links between the two groups. To date, PEPT at Cape Town has predominantly used tracers labelled with ^{68}Ga eluted from a generator containing the parent ^{68}Ge [32].

[6] The various algorithms used to perform PEPT are discussed at length in chapter 6.

3.6 The future—SuperPEPT, MicroPEPT and beyond

At the time of writing, the Birmingham group is commissioning two new instruments, dubbed '*SuperPEPT*' and '*MicroPEPT*'.

SuperPEPT is constructed out of components from ECAT EXACT31, ART and HR+ scanners, combined to form a high sensitivity positron camera. It has a cylindrical geometry with internal diameter 40 cm and axial length 50 cm, and is mounted on a versatile gantry so that it can operate with its axis either horizontal (for studying pipeline flow) or vertical (for studying stirred tanks and other vertically-oriented process equipment) [33].

MicroPEPT is essentially a reconfiguration of a Concorde MicroPET P4 pre-clinical scanner into a more compact geometry (13 cm diameter × 16 cm axial length) to create a high-sensitivity system for studying small-scale systems.

For the future, perhaps the most exciting development—discussed further in chapter 9—would be incorporating time of flight (ToF) information into a PEPT system. ToF is already available on a number of commercially-available clinical and pre-clinical PET systems, and it must be only a matter of time before the information on difference in arrival times of the two annihilation photons is used to enhance PEPT tracking.

References

[1] Ter-Pogossian M M, Phelps M E, Hoffman E J and Mullani N A 1975 A positron-emission transaxial tomograph for nuclear imaging (PETT) *Radiology* **114** 89–98

[2] Phelps M E, Hoffman E J, Mullani N A and Ter-Pogossian M M 1975 Application of annihilation coincidence detection to transaxial reconstruction tomography *J. Nucl. Med.* **16** 210–24

[3] Hoffman E J, Nutt R, Casey M, Carroll L R and Dahlbom M 1985 A new multi-crystal two dimensional detector block for PET *J. Nucl. Med.* **26** P28

[4] Spinks T J, Jones T, Gilardi M C and Heather J D 1988 Physical performance of the latest generation of commercial positron scanner *IEEE Trans. Nucl. Sci.* **35** 721–5

[5] Bateman J E, Locke C E R and Ferrari C A 1986 A powerful, low-cost histogramming memory for digital radiography with multi-wire proportional counters *Technical report* Rutherford Appleton Lab

[6] Hawkesworth M R 1977 Neutron radiography. equipment and methods *At. Energy Rev.* **15** 169–220

[7] Stewart P A E, Hawkesworth M R and Witcomb R C 1987 Analyzing fluid flows within hollow bodies *US Patent* 4,697,079

[8] Hawkesworth M R, O'Dwyer M A, Walker J, Fowles P, Heritage J, Stewart P A E, Witcomb R C, Bateman J E, Connolly J F and Stephenson R 1986 A positron camera for industrial application *Nucl. Instrum. Methods Phys. Res.* A **253** 145–57

[9] Bemrose C R, Fowles P, Hawkesworth M R and O'dwyer M A 1988 Application of positron emission tomography to particulate flow measurement in chemical engineering processes *Nucl. Instrum. Methods Phys. Res.* A **273** 874–80

[10] Broadbent C J, Bridgwater J, Parker D J, Keningley S T and Knight P 1993 A phenomenological study of a batch mixer using a positron camera *Powder Technol.* **76** 317–29

[11] Parker D J, Broadbent C J, Fowles P, Hawkesworth M R and McNeil P 1993 Positron emission particle tracking—a technique for studying flow within engineering equipment *Nucl. Instrum. Methods Phys. Res.* A **326** 592–607

[12] Parker D J, Forster R N, Fowles P and Takhar P S 2002 Positron emission particle tracking using the new Birmingham positron camera *Nucl. Instrum. Methods Phys. Res.* A **477** 540–5

[13] Parker D and Wheldon C 2018 The Birmingham MC40 cyclotron facility *Nucl. Phys. News* **28** 15–20

[14] Wheldon C, Tz Kokalova, Freer M, Glenn A, Parker D J, Roberts T and Walmsley I 2014 States at high excitation in ^{12}C from the ^{12}C (^{3}He, ^{3}He) 3α reaction *Phys. Rev.* C **90** 014319

[15] Smith R, Tz Kokalova, Wheldon C, Bishop J E, Freer M, Curtis N and Parker D J 2017 New measurement of the direct 3α decay from the ^{12}C Hoyle state *Phys. Rev. Lett.* **119** 132502

[16] Allport P *et al* 2017 Recent results and experience with the Birmingham MC40 irradiation facility *J. Instrum.* **12** C03075

[17] Price T and PRaVDA Consortium *et al* 2016 PRaVDA: high energy physics towards proton computed tomography *Nucl. Instrum. Methods Phys. Res.* A **824** 226–7

[18] Price T *et al* 2015 Expected proton signal sizes in the PRaVDA range telescope for proton computed tomography *J. Instrum.* **10** P05013

[19] Sadrmomtaz A, Parker D J and Byars L G 2007 Modification of a medical PET scanner for PEPT studies *Nucl. Instrum. Methods Phys. Res.* **573** 91–4

[20] Ingram I, Hausard M, Fan X, Parker D J, Seville J P K, Finn N and Evans M 2007 Portable positron emission particle tracking (PEPT) for industrial use *12th Int Conf. on Fluidization—New Horizons in Fluidization Engineering (Engineering Confs. Int. Symp. Series)*

[21] Ahasan M M and Parker D J 2008 Initial results from a prototype large (2M) ring PET scanner *Proc. of the World Congress on Engineering* vol 1

[22] Ahasan M M and Parker D J 2009 Design and performance evaluation of a prototype large ring PET scanner *Advances in Electrical Engineering and Computational Science* (Berlin: Springer) pp 161–72

[23] Leadbeater T W and Parker D J 2009 A high speed pc-based data acquisition and control system for positron imaging *Nucl. Instrum. Methods Phys. Res.* **604** 355–8

[24] Leadbeater T W 2009 The development of positron imaging systems for applications in industrial process tomography *PhD Thesis* University of Birmingham

[25] Pérez-Mohedano R, Letzelter N, Amador C, VanderRoest C T and Bakalis S 2015 Positron emission particle tracking (PEPT) for the analysis of water motion in a domestic dishwasher *Chem. Eng. J.* **259** 724–36

[26] Stellema C S, Vlek J, Mudde R F, De Goeij J J M and Van den Bleek C M 1998 Development of an improved positron emission particle tracking system *Nucl. Instrum. Methods Phys. Res.* **404** 334–48

[27] Hoffmann A C, Dechsiri C, Van de Wiel F and Dehling H G 2005 Pet investigation of a fluidized particle: spatial and temporal resolution and short term motion *Meas. Sci. Technol.* **16** 851

[28] Chang Y-F, Ilea C G, Aasen Ø L and Hoffmann A C 2011 Particle flow in a hydrocyclone investigated by positron emission particle tracking *Chem. Eng. Sci.* **66** 4203–11

[29] Hoffmann A C, Skorpen Å and Chang Y-F 2019 Positron emission particle tracking and CFD investigation of hydrocyclones acting on liquids of varying viscosity *Chem. Eng. Sci.* **200** 310–9

[30] Langford S, Wiggins C, Tenpenny D and Ruggles A 2016 Positron emission particle tracking (PEPT) for fluid flow measurements *Nucl. Eng. Des.* **302** 81–9

[31] Langford S T, Wiggins C S, Santos R, Hauser M, Becker J M and Ruggles A E 2017 Three-dimensional spatiotemporal tracking of fluorine-18 radiolabeled yeast cells via positron emission particle tracking *PLoS One* **12** e0180503

[32] Volkwyn T S, Buffler A, Govender I, Franzidis J-P, Morrison A J, Odo A, Van Der Meulen N P and Vermeulen C 2011 Studies of the effect of tracer activity on time-averaged positron emission particle tracking measurements on tumbling mills at PEPT cape town *Miner. Eng* **24** 261–6

[33] Hampel D M *et al* 2021 SuperPEPT: A new tool for positron emission particle tracking; first results *Nucl. Instrum. Methods Phys. Res.* A **1028** 166254

IOP Publishing

Positron Emission Particle Tracking
A comprehensive guide
Kit Windows-Yule, Leonard Nicuşan, Matthew T Herald, Samuel Manger and David Parker

Chapter 4

Comparison with other techniques

There exists a diverse range of techniques used by scientists and those in industry to gain insight into the mechanics and dynamics of particulate and multiphase systems. These include both experimental methods, encompassing imaging techniques, such as particle imaging velocimetry x-ray CT or MRI, and characterisation techniques, such as powder rheometry or shear testing, as well as numerical methods such as the discrete element method (DEM) or continuum fluid dynamics (CFD). Each of these techniques, however, carries its own unique set of strengths and weaknesses, meaning that when planning a scientific study, one must carefully choose the tool—or combination of tools—most suited to the task at hand. In this chapter, we provide an overview of the experimental and numerical tools most commonly used in the study of particulate and multiphase systems, providing in each case an elementary introduction to the technique and its underlying physics, and a summary of its major strengths and weaknesses relative to the PEPT technique which forms the focus of this book. In each case, we also provide a range of key references for those wishing to gain a deeper insight into the technique.

Although PEPT forms the main focus of this book, it is important to remember that it is not the *only* technique which may be used to study particulate or multiphase systems and their dynamics. Indeed, there exists a plethora of techniques developed for this purpose. An immediate question that may arise in the reader's mind is why, in a book about PEPT, should we spend time discussing other experimental techniques? To this hypothetical reader one may respond that all experimental techniques—PEPT included—possess certain strengths, and certain weaknesses (see figure 4.1).

PEPT, while powerful, is not an academic panacea, and there exist certain situations and certain systems for which another experimental approach may be superior, or where the use of PEPT in tandem with another method may allow enhanced outcomes (as can be seen in our review of the PEPT literature presented in chapter 8). The main aims of this section are (a) to contextualise the PEPT technique, showing how it compares to other commonly-used particle imaging

doi:10.1088/978-0-7503-3071-8ch4

Figure 4.1. Table summarising the main strengths and limitations of the various experimental imaging modalities explored in this chapter. Green colouring indicates a technique is strong in a given area. Properties coloured in orange and red, respectively, indicate relatively minor and comparatively significant weaknesses—for example orange colouring in 'temporal resolution' indicates an inability to track fast-moving flows, whereas red indicates that only static or quasi-static systems can be imaged. Similarly, for 'orientation information', orange would indicate that particle orientation *can* be tracked, but doing so is comparatively difficult and/or only achievable under specific circumstances (e.g. with PEPT, where such tracking is possible, but only with relatively large particles). The specific reasoning underlying the rating of each feature for each technique is discussed in detail in the text below.

modalities, and (b) to allow the reader, when planning an experiment, to determine the optimal technique (or techniques) to apply.

In the sections below, we introduce the most commonly-used particle imaging techniques, providing for each an overview of its underlying principles, and a brief summary of its major strengths and weaknesses as compared to PEPT. We then go on to describe also *non-imaging* techniques which are widely used to investigate granular systems, as well as numerical techniques (computer simulations) which are increasingly used to complement PEPT and other experimental studies.

4.1 Quasi-two-dimensional techniques

We begin this chapter by discussing the (relatively) simple case of imaging techniques which record only quasi-two-dimensional information from a given system, as opposed to a full, three-dimensional reconstruction. Such techniques may provide either full information pertaining to a shallow and/or highly dilute system, or partial images or depth-averaged two-dimensional projections from a fully-three-dimensional system. As such, they cannot be easily applied to—or at least, cannot extract comprehensive information from—a majority of 'real' industrial or natural systems, which tend to be dense, three-dimensional and (in the former case) constrained by solid, optically-opaque walls. Nonetheless, these techniques can provide a great deal of fundamental insight into the behaviours of particulate and multiphase systems, and as such are extremely widely-used [1].

4.1.1 Particle tracking velocimetry (PTV)

Technique overview

Particle tracking velocimetry (PTV) and particle imaging velocimetry (PIV) are two of the most commonly-used approaches to the imaging of particulate systems. This is due largely to the fact that—compared to other techniques such as x-ray CT, MRI and indeed PEPT which require expensive equipment—they are (relatively) cheap[1] and simple to perform, requiring, in their simplest forms, only a video camera. The techniques can also provide data with impressively high temporal *and* spatial resolution [2, 3]. The main downside of PTV and PIV—and indeed all optical techniques—is that (unlike PEPT) they utilise electromagnetic radiation in the optical range, which is inherently unable to penetrate opaque objects. As such, these techniques are only suitable for imaging shallow (quasi-two-dimensional) systems or highly dilute three-dimensional systems with at least one optically-transparent wall [4, 5].

We will discuss first particle *tracking* velocimetry, which bears the greater similarity to the PEPT technique, and which we have already introduced briefly in chapter 2. In order to perform PTV, one in essence requires only a digital video of a particulate system with suitably high contrast, resolution and frame rate. For each frame of the video, the positions of the tracer particles used are extracted using image processing techniques. In the example shown in figure 4.2, we begin with a simple black and white (or, more accurately, greyscale) image of a system of particles of which a fraction, chosen as tracers, have been coloured black. The image is then *binarised* [6]—that is to say any pixel within the system whose grey value falls below a chosen threshold is set to black, and all those above set to white. By choosing a suitable threshold, it is possible to remove all objects other than our chosen tracers from the image (see figure 4.2(b)). From this simple binary image, and in our simple case of spherical particles, the centroids are easily determined using a circle finding algorithm (figure 4.2(c)), in this instance Matlab's implementation of a circular hough transform [7], `imfindcircles`, though ImageJ, Python and various other programmes and programming/scripting languages of course possess similar functions.

It is important to note that there are many different image processing methodologies through which the positions of particles may be extracted from both black-

[1] Note that this is cheap in the context of research budgets as opposed to personal or household budgets—a PTV set up with reasonable spec will still typically carry a five-figure price tag.

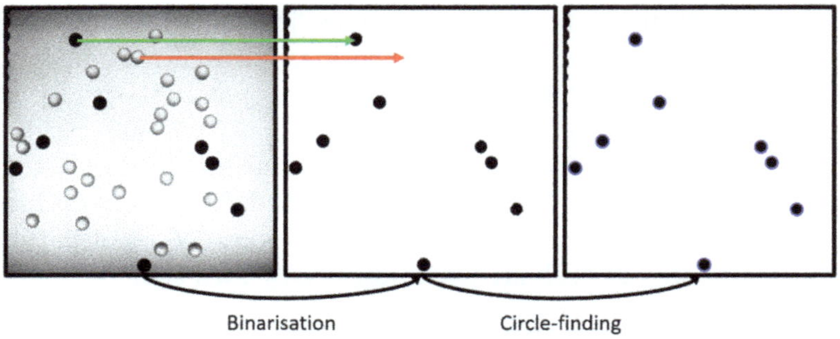

Figure 4.2. Example of an image processing method to extract the positions of tracer particles from a dilute granular system. The original greyscale image is binarised using a suitable threshold such that the all pixels corresponding to the light background and non-tracer particles are converted to white, leaving only those belonging to the dark tracers, which are converted to black. From this simple black-and-white image, the particle's edges can be easily detected, and thus their centroids determined.

and-white and RGB images [6], the above representing one of the simpler and more commonly-used approaches. The process can be repeated for all frames in the video allowing—for a suitably high frame rate—the trajectories of the tracer particles to be reconstructed.

Comparison with PEPT

Advantages:

- Comparatively cheap to set up and operate a basic system[2].
- Capable of imaging large numbers of particles in close proximity.

Disadvantages:

- Only able to image quasi-two-dimensional or highly dilute three-dimensional systems.
- Unable to penetrate opaque containers, e.g. metal-walled process equipment, or systems immersed in opaque fluids.

4.1.2 Particle imaging velocimetry (PIV)

While PTV, like PEPT, is a *Lagrangian* technique (i.e. we follow the motion of particles through space and time), PIV is an *Eulerian* technique (i.e. we consider a series of fixed regions in space and observe how particles flow through these regions)[3].

There exist several distinct manners in which PIV may be performed [8], but all share the same fundamental methodology, namely the use of spatial correlations to determine the mean velocity of small groups of particles in each given region of the system.

[2] Note, however, that this is not necessarily the case for systems using high-end CCD cameras and laser light sources.

[3] It is valuable to note that while Eulerian information can be easily extracted from data acquired using Lagrangian techniques such as PEPT and PTV, Lagrangian information cannot typically be recovered from data acquired using Eulerian techniques.

For the sake of simplicity and brevity, we provide here only a brief overview of the most commonly-applied method, such that the reader may gain a fundamental understanding of the technique, and how it compares with PEPT. The interested reader may find more detailed information in one of several texts dedicated to the technique, for example references [9–11].

The method we describe (single-exposure/double frame PIV) can be performed as follows:

1. A camera system captures two consecutive images of the system of interest, separated by a known interval Δt (see figure 4.3(a)–(c)). This can be achieved by firing two pulses of laser light at the system [12, 13] (more common for imaging flow tracers in fluids) or simply by taking two consecutive frames of a continuous video under normal conditions [14] (more common for imaging granular systems).
2. This pair of frames is divided into a series of 2D cells or *interrogation windows*, each capturing the images of a number of particles (see figure 4.3(b), (c)).
3. If the timestep Δt is small enough, and the window size large enough, the same group of particles—or at least the majority thereof—appearing in a given interrogation window in the first frame should also be found in the

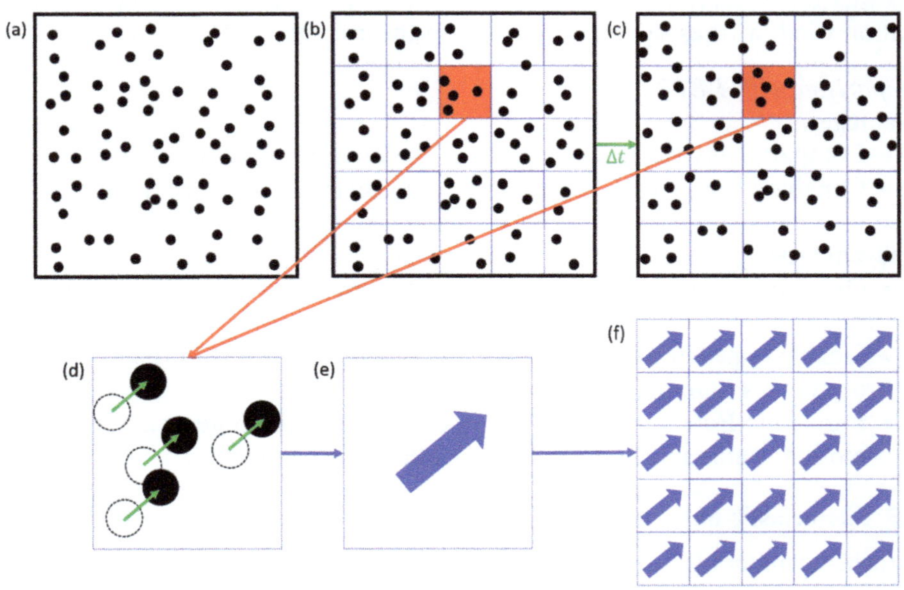

Figure 4.3. Schematic diagram showing an example workflow for performing particle imaging velocimetry. (a) We begin with a digital image of our system of interest, corresponding to a single frame from a video. (b) The image is subdivided into a series of 'interrogation windows'. (c) This process is also performed for the next frame in the video. (d) Individual interrogation windows from a given frame are compared to the equivalent window in the next frame. (e) Using a suitable correlation function, a mean velocity is determined for the current interrogation window. (f) By repeating this process for each interrogation window, the flow field of the system may be mapped.

equivalent window in the second image, though with a shifted position due to their motion (see figure 4.3(b)–(d)).

4. A spatial cross-correlation function is applied to the two pairs of images corresponding to each interrogation window. While we omit the mathematics here (the interested reader may find more detail in [10]), this can be simply thought of as determining the 'degree of match-up' between particle positions as we shift one image over the other (see figure 4.3(d)).
5. The maximum value of the above-described correlation function in essence corresponds to the shift in the x–y plane which gives the best match-up between the two images. Thus, by finding this shift, we provide an estimate of the average displacement vector, $\Delta\vec{r}$, for the group of particles in question, and thus their mean velocity, $\vec{v} = \frac{\Delta\vec{r}}{\Delta t}$ (see figure 4.3(e)).
6. By performing this operation for each of the interrogation windows in the image, one may build up a picture of the system's flow field (see figure 4.3(f)).

The above technique is highly useful and accurate when studying relatively laminar fluid flow (or at least flows for which the length scale of turbulence is large compared to the particle size), as well as relatively uniform granular flows (e.g. hopper flow [15, 16] or chute flow [17]). However, for systems (e.g. dilute vibrated beds [18]) where particle motion is more random and less correlated [19, 20], particles are likely to possess strongly varying velocity vectors even within a single, relatively small interrogation window, thus significantly reducing the accuracy of the technique.

Comparison with PEPT

Advantages:

- Comparatively cheap to set up and operate a basic system.
- Capable of imaging large numbers of particles in close proximity.

Disadvantages:

- Only able to image quasi-two-dimensional or highly dilute three-dimensional systems.
- Unable to penetrate opaque containers, e.g. metal-walled process equipment, or systems immersed in opaque fluids.
- Limited to Eulerian information → only suitable for systems in which particle motion is comparatively strongly spatially correlated.

4.1.3 Photoelastic imaging

Photoelastic materials have been used in scientific and engineering applications for a number of decades [21, 22], but their widespread use in granular systems is a comparatively modern occurrence, being popularised by the late Bob Behringer and colleagues in the 90s [23–25].

Photoelastic imaging in our particular field of interest (that is, particle imaging) uses photoelastic materials to qualitatively visualise and even—with suitable calibration—quantitatively measure stresses and stress fields within a granular system. In this regard, photoelastic imaging is very much unique in the field of experimental imaging.

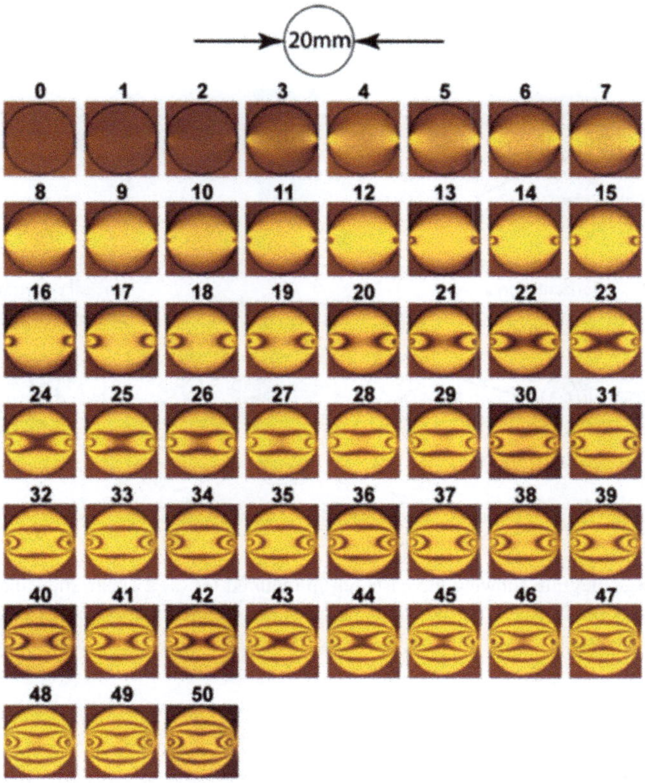

Figure 4.4. Images of a plane-glass photoelastic disk, viewed under circularly-polarised light, under diametric compression. The load applied to the disk increases from image 1 to image 50, creating at first two bright regions at the points at which the compression is applied, which expand to fill the disk as further compression is applied. As the load is increased further, a series of dark fringes appear, increasing in number with the increasing load. Image reprinted with permission from Springer (GmbH) from [28], (©) The Authors 2010.

A photoelastic material is one whose birefringence[4] depends on the local stress at each point in the material [26, 27]. This means that the intensity and pattern of (polarised) light transmitted through such a material varies with the stress applied (see figure 4.4).

In order to interpret the information transmitted by a photoelastic material, such as the disk pictured in figure 4.4, one may utilise an imaging device known as a *polariscope* [29]. A polariscope comprises simply a light source, a camera, and a pair of oppositely-oriented polarising filters placed on either side of the sample to be imaged; such a set-up is depicted schematically in figure 4.5. The pairings of oppositely-oriented filters may either be, for example, vertical and horizontal linear polarisers (figure 4.5(a)) or left-circular (LCP) and right-circular (RCP) polarisers (figure 4.5(b)). The use of circular polarisers typically produces better results [30].

[4] A birefringent material is a medium for which the refractive index varies with both the direction of travel of incident light waves and their polarisation.

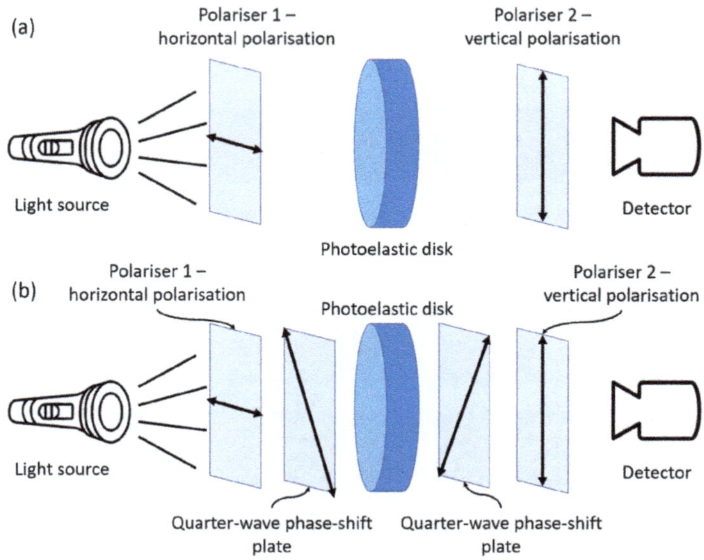

Figure 4.5. Schematic diagram showing the basic set-up required for (a) a linear and (b) a circular (darkfield transmission) polariscope. In panel (b), the combination of a horizontal/vertical polariser and a quarter-wave phase-shift plate (collectively termed a left/right circular polariser) act to induce circular, as opposed to linear polarisation.

Viewed through such a pair of filters, an entirely-unstressed photoelastic disk will block all light, thus providing an entirely dark image in the polariscope's detector (see figure 4.4(0)). Conversely, a stressed region of the photoelastic material will, due to its birefringent nature, convert the incident circularly polarised light to *elliptically polarised* light, meaning that light emitted from this region will not be completely blocked by the second, oppositely-oriented polariser. Thus, the stressed regions of the disc will appear as bright spots in the resultant image (see figure 4.4(1–8)). When the stress applied to the material is such that it propagates through the entire disk, one will instead see a pattern of fringes corresponding to the spatial variation in the local stresses across it, with each fringe corresponding to a 2π phase shift between the principle components of the light refracted by the birefringent material (see figure 4.4(9–50)).

Photoelastic imaging can be applied both semi-quantitatively and fully quantitatively. The semi-quantitative application of photoelastic imaging is extremely simple: we know from the above that as the stress applied to a given photoelastic particle increases, its image will become brighter and/or develop a larger number of fringes. As such one may, for example, assess the qualitative distribution of stresses across a system comprising multiple photoelastic particles simply by recording the measured brightness and/or fringe numbers of particles across different regions of the system. As the photoelastic response is rapid, one may also investigate the variation of these stresses in time, even measuring the stresses developed during individual particle collisions [31, 32].

Obtaining a *fully quantitative* measure of the stresses applied to one or more photoelastic particles can, however, be considerably more mathematically complex and computationally intensive, and as such we will not discuss it in detail in the brief overview provided here (the interested reader may find fuller, more detailed descriptions in references [30, 33, 34]). For the case of relatively small stresses (in the case of planar glass particles, $\lesssim 15\%$ of the disk strength [28]), the mean measured light intensity scales approximately linearly with the applied stress, meaning that measurements of the *scalar* force can be extracted via a relatively simple calibration. Above a certain threshold force, however, this linear relation no longer holds, and one must adopt more involved methods. For higher applied loads, the total, scalar force on a given particle or set of particles may be measured using not the overall intensity of the image, but the *gradient* of the two dimensional intensity distribution [33, 35, 36], which is expected to vary monotonically—though not linearly [24]—with the applied pressure. It is important to note that the calibration curves obtained using this method are sensitive not only to the specific particles used, but also to the specific lighting conditions and image resolution, making careful calibration a necessity [37].

Finally, it is also possible to obtain *vector* force measurements using the photoelastic imaging technique—though this requires still more complex methods. The specific pattern formed on a photoelastic particle is dependent not only on the total force applied to said object, but also the specific points at which this force is applied, and the relative strengths of the individual forces applied at these contact points. In other words, the pattern, and thus intensity map recorded by our hypothetical camera system, contains all information pertaining to the vector contact forces on each particle in a given system. As such, by solving the relevant inverse problem, it is (hypothetically[5]) possible to determine these forces from the images collected during photoelastic imaging. Details regarding the precise manner in which this may be performed, and the relevant underlying mathematics, can be found in references [34, 38–44].

Comparison with PEPT

Advantages:

- Provides force and stress information.

Disadvantages:

- Only able to image quasi-two-dimensional or highly dilute three-dimensional systems.
- Unable to penetrate opaque containers, e.g. metal-walled process equipment, or systems immersed in opaque fluids.
- Requires detailed calibration.
- Limitations in terms of usable material.

[5] It should be noted that the solutions to these inverse problems are not by definition unique, meaning that the true force distribution may not be recovered in all cases.

4.1.4 X-ray radiography

The basic concept of x-ray radiography will be very familiar to any of us who have broken a bone: an object (in our case a human) is placed between a source of relatively high-energy electromagnetic radiation (x-rays) and some form of a detector system, such as x-ray film. By emitting a short pulse of x-rays, much like the flash of a camera, from our source, we can form an image on our detector (film). Again just like a conventional camera, regions of the film exposed to more 'light' (i.e. x-ray radiation) will become darker, and those exposed to less will remain lighter. However, unlike an optical camera, which uses light reflected by an object to form an image, radiographic images instead rely on the attenuation of x-rays as they travel through an object.

To continue the above example, if a radiographer wishes to image our bones (and thus ignore the surrounding tissue), they will use x-rays with an energy suitably high that they can relatively easily penetrate comparatively low-density materials such as skin, muscle and other soft tissue, but adequately low that they are comparatively strongly attenuated by the denser material of bone[6]. As such, x-ray photons are comparatively unlikely to reach regions of our notional x-ray film behind bones, meaning that these regions will gain relatively little exposure, and remain bright (see figure 4.6). Conversely, regions lying behind only soft tissue will be bombarded by a significantly larger number of x-ray photons, causing the film to become darker, and regions entirely unobstructed by solid material darker still. Due to this clear 'contrast' between bone and the surrounding media, we are able to create a clear image thereof, as illustrated in figure 4.6.

In the current digital age, film detectors—both in optical cameras and x-ray imaging systems—are all but obsolete. Modern x-ray sensors instead use pixellated photosensitive detectors, much like those used by modern (optical) cameras. These digital systems carry the significant advantage of allowing many radiographs to be acquired in rapid succession, thus allowing the creation of moving images.

The principle described above can, naturally, be applied not just to the bones in a human body, but to the imaging of any system in which there exists a suitably large difference in densities to produce a clear '*radiographic contrast*'. Let us now turn to some examples more relevant to the focus of this book: the imaging of particulate and multiphase media.

In figure 4.7 we provide an extremely simple example in which a single, dense particle is exposed to an x-ray beam, creating an intensity pattern on a detector. In this elementary example, the centroid position of our spherical particle can be easily determined simply by determining the position of the minimum x-ray intensity received by the detector (i.e. the minimum exposure). Similarly, the size of our particle can be back-computed from the diameter of the less-exposed region and the known position of the source, system and detector[7].

[6] The energy of the electromagnetic radiation generally used for x-ray imaging is typically lower than that of the gamma-rays used in PEPT imaging, allowing PEPT to penetrate larger, denser systems. It also means that we humans are less likely to absorb the radiation emitted by PEPT tracers!

[7] For a set-up such as that illustrated in illustrated in figure 4.7 with a point-like source (a 'cone-beam geometry'), any object imaged will be magnified in the detector plane [45]—as illustrated in the figure. However, it is possible to determine the size of the object from the size of the image using simple geometrical arguments if we know the relative positions of the source, object and detector.

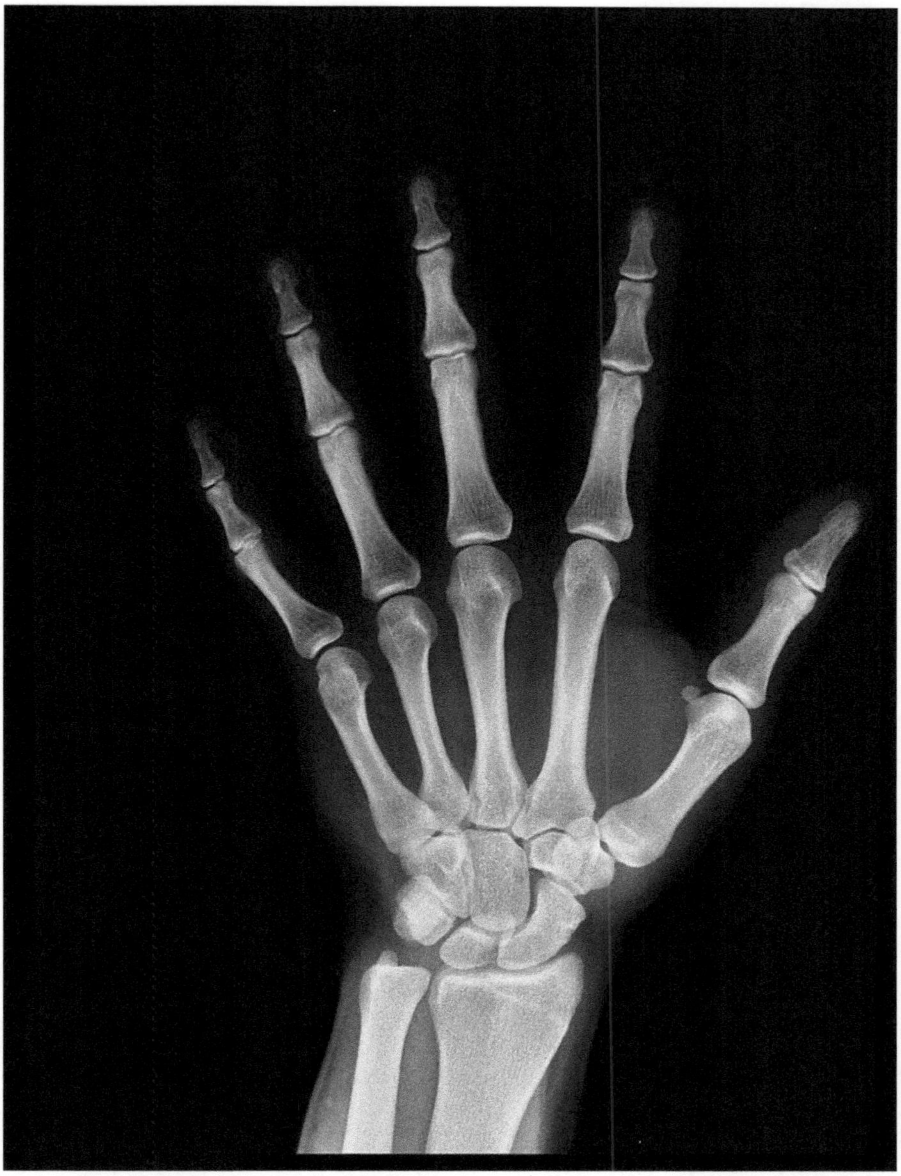

Figure 4.6. X-ray image of a human hand. Areas of the x-ray detector lying directly underneath bones, which absorb most incident x-rays, receive relatively little exposure, and thus remain bright, while other regions become darkened by the interacting x-ray photons. While soft tissue has a lower x-ray absorption cross-section (i.e. it is less likely to absorb a given photon) than bone, it is still non-negligible, and significantly greater than of the surrounding air, and thus provides some degree of contrast (i.e. we can see a ghostly shadow corresponding to the patient's skin and muscle). Note also that thicker bones (e.g. the wrist bones) cause greater attenuation than thinner bones (e.g. the finger tips), and thus appear brighter in the image.

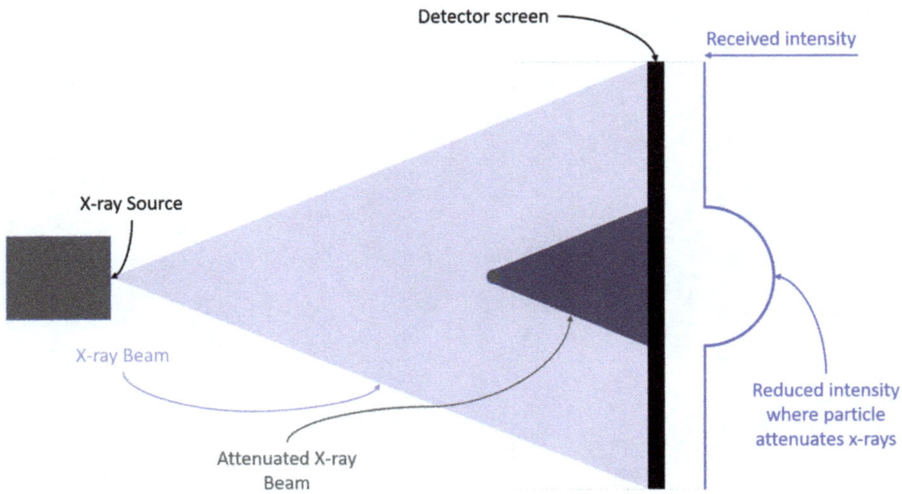

Figure 4.7. Simple schematic diagram illustrating the acquisition of a radiograph of a single dense particle.

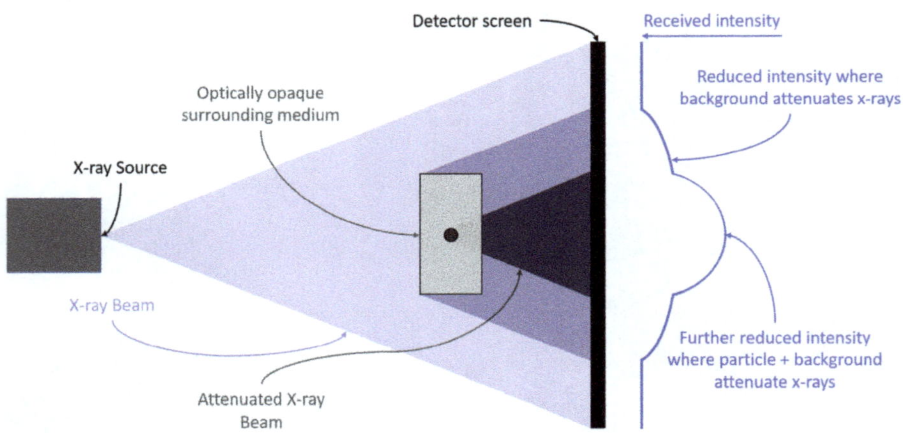

Figure 4.8. Simple schematic diagram illustrating the acquisition of a radiograph of a single dense particle surrounded by a comparatively less-dense 'background medium'.

Of course, a single particle floating in free space is neither a particularly interesting system to study, nor does it require an expensive x-ray set-up to image it! The value of x-ray radiography is made more apparent in our second schematic, figure 4.8, which depicts a single, dense particle, which can be used as a tracer, surrounded by a comparatively less-dense yet still optically-opaque medium. This '*background medium*' can be either a fluid, a granular medium or indeed a mixture of the two. Clearly, this dense particle's position and motion would be impossible to determine using the conventional, optical methods described in the preceding sections. However, if the density contrast between the particle and the background medium is adequate, and the x-ray energy suitably chosen, the position of this particle can be expected to show up in our x-ray image just as clearly as the skeletal fingers of our x-rayed hand in

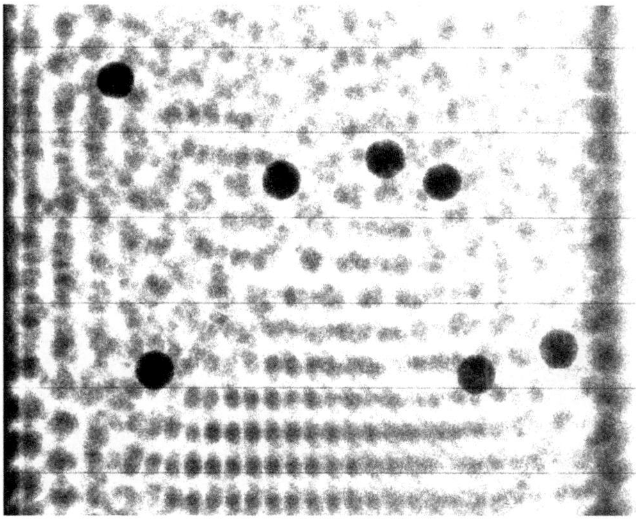

Figure 4.9. X-ray radiograph of a system of glass particles 'seeded' with a number of dense (steel) tracers. As an interesting aside, one may notice a series of thin parallel lines cutting across the image. These correspond to the gaps between the detector regions of the particular x-ray camera used—a common artefact in x-ray images.

figure 4.6. Figure 4.9 shows an example of a *real* x-ray image of a granular bed, acquired from a system comprising a number of comparatively dense, steel spheres (ρ = 7800 kg m^{-3}) mixed into a background material composed of glass spheres (ρ = 2500 kg m^{-3}), all housed within an aluminium container (ρ = 2700 kg m^{-3}). From this image, we can clearly see the locations of the steel spheres against the background, even though said background is significantly less uniform than in our schematic example (figure 4.8). Having produced such an image, or better still a series of such images, the positions and trajectories of the particles within the system may be extracted in the exact same manner as with the optical images discussed in section 4.1.

As those with a knowledge of granular dynamics will doubtless be aware, a heavy steel tracer is liable to undergo segregation in a dynamic system containing a significant proportion of lighter particles [46, 47, 48]. As such, in this instance, our tracer is only going to provide us with information regarding the motion of similarly dense particles, offering only limited insight into the behaviours of the less-dense particles which make up the majority of the bed. If we wish to make a truly non-invasive tracer for a system of such particles, we must attempt to match the densities (and indeed other properties e.g. friction, stiffness, restitution coefficient, moment of inertia, …) of the tracer and bulk particles. A commonly used method is to embed a small, highly-dense 'core' into a particle the same size as those which make up the majority of the bed, but fabricated from a slightly lighter material, such that the overall density and moment of inertia of the tracer approximately match those of the bed (bulk) medium.

As we touched upon earlier, in addition to ensuring a good density contrast between the materials used, one must also ensure the energy of the x-rays used—which determines how easily they may penetrate a given material—is well chosen. If

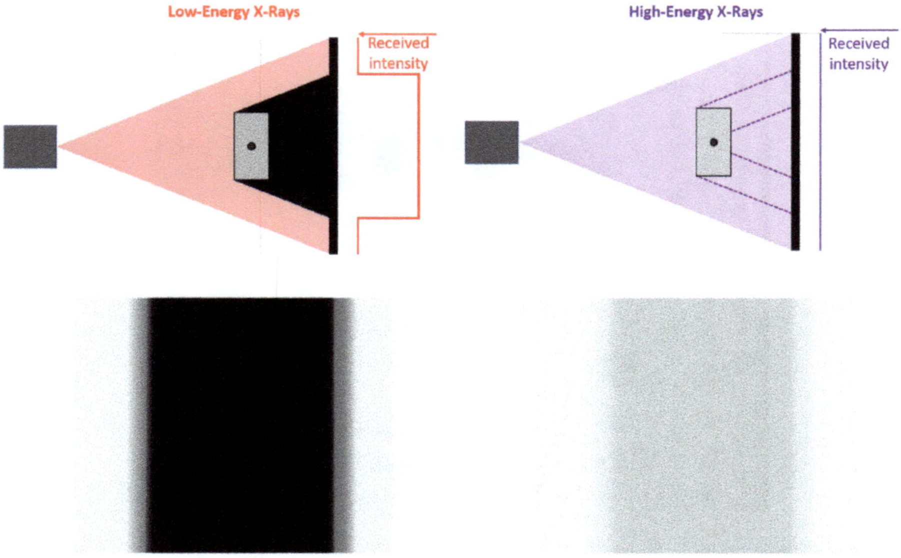

Figure 4.10. Above: simple schematics showing the intensity patterns received when operating at an unsuitably low x-ray energy (practically all photons are absorbed by sample) and an unsuitably high x-ray energy (practically all photons successfully penetrate the system), in both cases resulting in a negligible contrast between tracer and background in the resultat image. Below: real x-ray images showing both extremal cases.

the energy of the x-rays is too low, they will not be able to penetrate either the source *or* the tracer to an adequate extent, leading to an overly dark image with poor contrast (see figure 4.10, left-hand side). Conversely, if the energy is too high, neither tracer nor background material will be able to attenuate the incident x-rays to any significant extent, leading to a 'washed-out' image (see figure 4.10, right-hand side).

Comparison with PEPT

Advantages:

- Capable of imaging large numbers of particles in close proximity (if suitably oriented[8]).

Disadvantages:

- Provides only quasi-two-dimensional projections[9].
- Difficult to perform truly non-invasively due to requirement for density contrast.

---

[8] As x-ray radiography only provides a 2D projection, any two or more particles falling in a line perpendicular to the detector will 'mask' one another.

[9] Although, unlike PTV and other optical techniques, x-ray radiography can extract information from deep, dense, fully-three-dimensional systems, it still only provides quasi-two-dimensional information. In this instance, however, the information acquired corresponds in effect to an average through the entire system, whereas optical techniques provide only information pertaining to a thin outer layer of the system.

4.2 Tomographic techniques

The word 'tomography' stems from the Greek root *tomos* meaning 'slice' or 'section', and the suffix '-graphy', from the Greek *graphein*, meaning 'to draw'. Thus the name tomography describes very well its function—to provide an image of an object from a series of individual slices. Tomographic images may be acquired via a variety of methodologies, and using electromagnetic radiation in a variety of forms, from gamma rays and x-rays, to optical light, and even radio waves. In the sub-sections below, we discuss the methodologies most widely-used in the imaging of particulate and multiphase systems.

4.2.1 X-ray computed tomography

We begin this section with a discussion of x-ray CT for two main reasons: firstly, the information provided here builds directly upon that presented in the preceding subsection, which should still be fresh in the reader's mind. Secondly, x-ray CT is the best known tomographic technique, to the point where it is often referred to simply as 'CT', omitting the 'x-ray' which should, for correctness, be used to differentiate it from various other forms of computed tomography!

X-ray CT begins with a series of two-dimensional, digital radiographs, such as those discussed in detail in section 4.1.4, each of which can be considered as a 'slice' (*tomos*) through a full three-dimensional image. The goal of CT is to manipulate these individual slices in such a way that we can faithfully reconstruct this three-dimensional image. There exist various manners in which this may be performed, for example the Algebraic Reconstruction Technique (ART) [49], iterative reconstruction [50, 51], and Fourier domain reconstruction [52]. More recently, deep learning methods have also been applied to the problem of tomographic reconstruction [53]. However, as our aim in this section is merely to provide an understanding of the basic principles of each major imaging technique described, we choose as our example the most conceptually simple methodology: back-projection [54].

For the sake of clarity and simplicity, let us consider in our discussion not the reconstruction of a three-dimensional image from a series of two-dimensional projections, but instead the reconstruction of a two-dimensional image from a series of one-dimensional projections. A 1D $\rightarrow$ 2D reconstruction still follows the same major steps as 2D $\rightarrow$ 3D, yet is significantly easier to visually represent in a clear and meaningful way, making it a much more useful pedagogical example.

To further simplify our example, we return to the case of a single particle being imaged, and instead of the 'fan beam' projections illustrated in section 4.1.4, we consider instead 'parallel projections', where the incident x-rays are, as the name suggests, parallel, as opposed to diverging from a point source (see figures 4.10 and 4.11(a)).

Figure 4.11(b) shows the intensity map corresponding to a single 1D projection of our 2D object. Figure 4.11(c) shows this same intensity profile being 'back-projected' all the way through the image in the direction of the incident photons. By re-orienting our x-ray source and detector at a different angle, θ, and taking another image, we can providing a second (and third, and fourth, …) distinct projection of our object. By overlaying multiple back-projections taken at multiple angles,

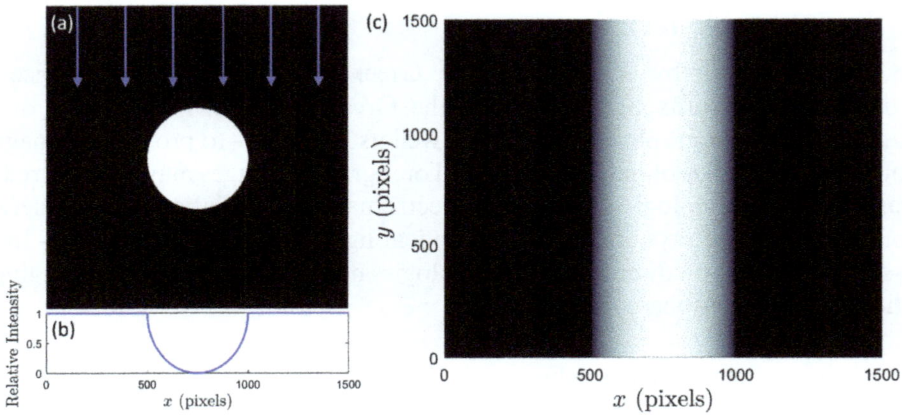

Figure 4.11. (a) Parallel projection imaging of a circular object in vacuum; blue arrows represent the directions of incident x-ray photons. (b) Normalised 1D intensity map corresponding to a single 1D projection (i.e. one radiograph). (c) Backprojection of the single radiograph from (b).

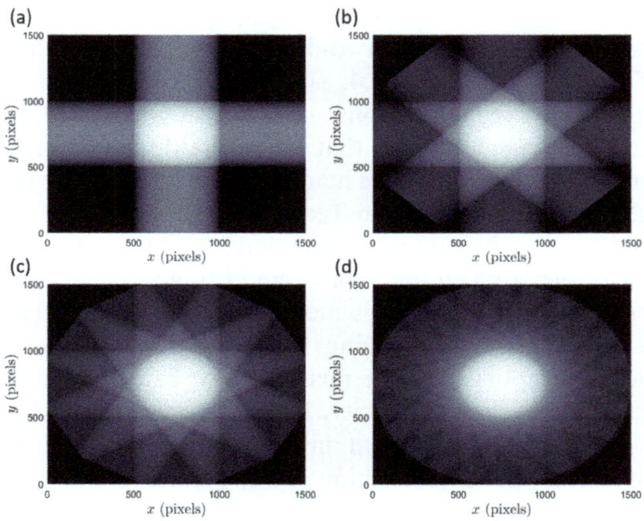

Figure 4.12. Images showing reconstructions of our 2D image from figure 4.11 produced from the summations of (a) 2 projections at $\theta = 0°, 90°$; (b) 4 projections at $\theta = 0°, 45°, 90°, 135°$; (c) 6 projections from $\theta = 0°$ to $\theta = 150°$ in steps of 30°; (d) 12 projections from $\theta = 0°$ to $\theta = 165°$ in steps of 15°. Note that the clarity of the image increases as the number of projections used increases.

summing the received (inverse) intensities, we are able to reconstruct an image of our 2D object from this series of one-dimensional projections (see figure 4.12). In these images, particularly panel (d), we can see a clear reconstruction of our original image, though there exists a significant amount of blurring due to the overlap of projections in other parts of the image. This can be overcome, to an extent, by using —for example—a high-pass filter to remove artefacts [55]. The complete process is then known as *filtered back-projection*.

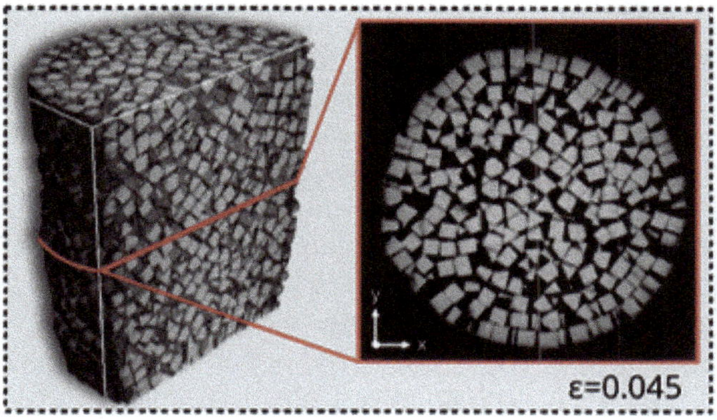

Figure 4.13. Example of a full, three-dimensional reconstruction of a complex granular system composed of aspherical particles. Reproduced from [56], with the permission of AIP Publishing.

The above, as is so often the case in this chapter, clearly represents a very simple example. However, this exact methodology can also be used to accurately map out the entire 3D volume of a complex system, even to the point of individually imaging all particles within a system (as demonstrated in figure 4.13).

The above process can also be described mathematically. For a parallel-beam set-up such as that discussed here, but extended now to two dimensions, a given projection can be expressed as a Radon transform, $\mathcal{R}$, which is simply the line integral (i.e. sum of intensities) along a (parallel) tomography beam at an angle θ, and at a distance $|s|$ from the origin:

$$\mathcal{R} = \int_{\infty}^{\infty} \int_{\infty}^{\infty} f(x, y)\delta(x\cos\theta + y\sin\theta - s)dxdy. \tag{4.1}$$

The back projection process can then be represented by the operator $\mathcal{B}$, defined as:

$$\mathcal{B}\mathcal{R} = \int_{0}^{\pi} \mathcal{R}(x\cos\theta + y\sin\theta)d\theta. \tag{4.2}$$

The above gives the sum of the projections passing through a point (x, y), thus allowing us to build up a full picture, as illustrated in figure 4.12. This image can then be 'un-blurred' using a suitable filter.

As is immediately clear both from the above discussion and from figure 4.12, the fidelity with which our tomographic image is created depends strongly on the number of projections used (i.e. the number of θ values explored), with an increase in the number of projections leading to an improved image. Hypothetically, an infinite number of images taken at infinitesimally varying angles could (with suitable data treatment) produce a perfect reconstruction. However, as each projection takes a finite time to acquire, it would unfortunately take infinite time to acquire such an image! Though this example is extreme, it highlights a very real problem relating to the imaging of dynamic granular systems: if we wish to obtain higher *spatial* resolution (i.e. more accurate images of our particles), we require a larger number of individual projections to be acquired. However, this will increase the time required to produce each full reconstruction, thus lowering the effective 'frame rate' (i.e. *temporal* resolution) with which data can be acquired. As such, when performing x-ray CT, one must perform a careful balancing act between obtaining enough information to accurately image particles, whilst keeping the acquisition process fast enough to accurately capture the dynamics of the system of interest. Due to this limitation, x-ray CT is only typically applied to quasi-static systems and those with relatively slow dynamics. Nonetheless, researchers are making significant strides in the development of 'fast x-ray techniques' [57, 58], which may in the future allow these methods to be applied to faster-moving systems.

Another balancing act which must be performed when performing x-ray tomography, as with x-ray radiography, is the determination of a suitable beam energy. This is an especially significant issue when attempting to image all particles within a system (as per figure 4.13) as opposed to simply a few density-contrast tracers: in the latter case, one can simply choose a high beam energy and 'ignore' the main bulk of the material, ensuring that a suitable number of x-rays reach the detector even in the case of comparatively large systems. In the former, however, it is of course imperative that all particles within the system attenuate the beam to some extent, such that they may be detected and reconstructed. As such, this inherently places limits on the size of system that can be imaged: if some finite fraction of the imaging beam must be attenuated by a particle in order for said particle to be imaged, then only a finite number of particles can be imaged before information no longer reaches the detector[10].

Comparison with PEPT

Advantages:

- Capable of imaging all particles in a system simultaneously.
- Capable of imaging transient processes in real-time for systems with suitably slow dynamics.

Disadvantages:

- Relatively low temporal resolution, as each image requires multiple individual projections to be acquired.
- Limitations in terms of system size as image reconstruction requires attenuation.
- When using tracers, as opposed to whole-system imaging, difficult to perform truly non-invasively due to requirement for density contrast.

4.2.2 Single-photon emission computed tomography (SPECT)

SPECT is, as we shall see in the coming pages, in many ways very similar to x-ray CT, in that it involves the acquisition of a number of two-dimensional intensity maps (projections), which are then combined to reconstruct a full three dimensional

[10] Note that for other radiation-based techniques such as PEPT, PET or SPECT (described in the next section) which do not rely on attenuation to produce images, the above is not an issue.

image. In fact, the reconstruction techniques used for SPECT are practically identical to those used in x-ray CT—and as such will not be repeated here. The difference between the two techniques lies instead in the manner in which the initial projections are acquired.

In x-ray CT, an *external* x-ray source is used to image an object, in essence using ‘shadows’ cast by the object to locate its position in a given 2D plane. In SPECT, the *object itself* is the source of the electromagnetic radiation (here gamma rays) used for imaging.

SPECT can be used for the imaging the spatial distributions of both liquids and gases that have been labelled with a radioactive isotope. It is also often used to image solid regions which said radioactive fluids have been absorbed into or deposited upon—allowing us, for example, to measure the uptake of tracers in the human brain (thus assessing its function) [59, 60] or the amount of radioactive carbon-11 absorbed by a carbon-capture process [61].

For reasons that will shortly become clear, SPECT—unlike PEPT—does not explicitly require positron-emitting isotopes—though these can be used if desired. Radioisotopes commonly used in humans include iodine (specifically ^{123}I and ^{131}I) [62, 63] and technetium (specifically the metastable isotope ^{99m}Tc) [64], all of which release gamma radiation with energies in the vicinity of ~150 KeV (comparable to those used in diagnostic x-ray CT, but higher than typically used in medical radiography). In granular media—which tend to be denser than human flesh and therefore attenuate low-energy gamma rays more strongly—isotopes producing higher energy radiation may be preferable.

SPECT is typically performed using a gamma camera such as that pictured in figure 4.14. When a sample of radioactive fluid is injected into a system in the

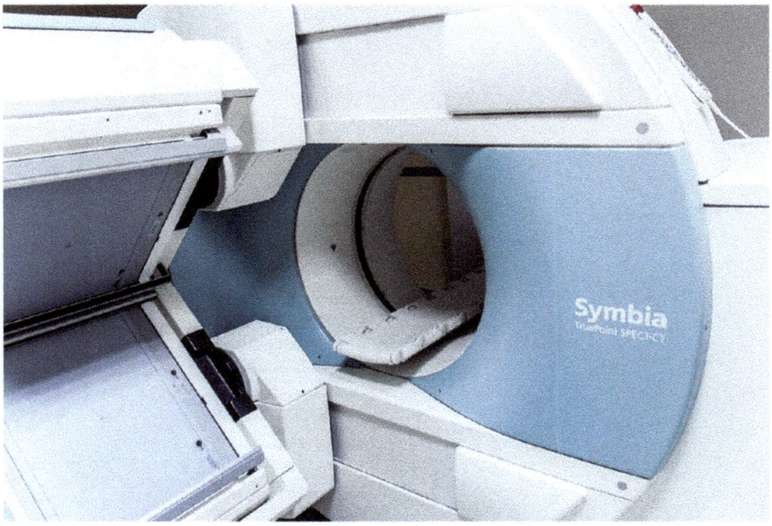

Figure 4.14. Image of a dual-headed SPECT scanner. Note that, unlike for PEPT and PET, the camera heads do not need to form opposing pairs, as only single photons (as opposed to photon pairs) are required. In fact, only a single camera head is required for SPECT, though in practice dual- or even triple-headed cameras are used to accelerate the imaging process.

camera's field of view, gamma rays emitted from said fluid, and indeed any solid objects which have absorbed some of the radioactive material (or upon which said material has otherwise been deposited) will be detected. Clearly, one may expect the intensity of the radiation received at a given point on the detector's 2D surface to be proportional to the amount of radioactive material lying perpendicular to this point. However, as we have discussed in chapter 2, the emission of electromagnetic radiation from radioactive decay is a random process, meaning that gamma photons will be emitted from our labelled materials at random times, from random positions, and in random directions. Thus, as illustrated in figure 4.15(a), the intensity map produced is likely to be highly noisy.

In order to overcome this issue, SPECT cameras are fitted with *collimators*. A collimator is in essence a grating fabricated from a highly dense—and hence highly attenuating—material (often lead). Collimators are designed such that, when an x-ray or gamma-ray beam passes through, any photons whose trajectories are not approximately parallel to a particular chosen direction (in this case, perpendicular to the face of the gamma camera) are

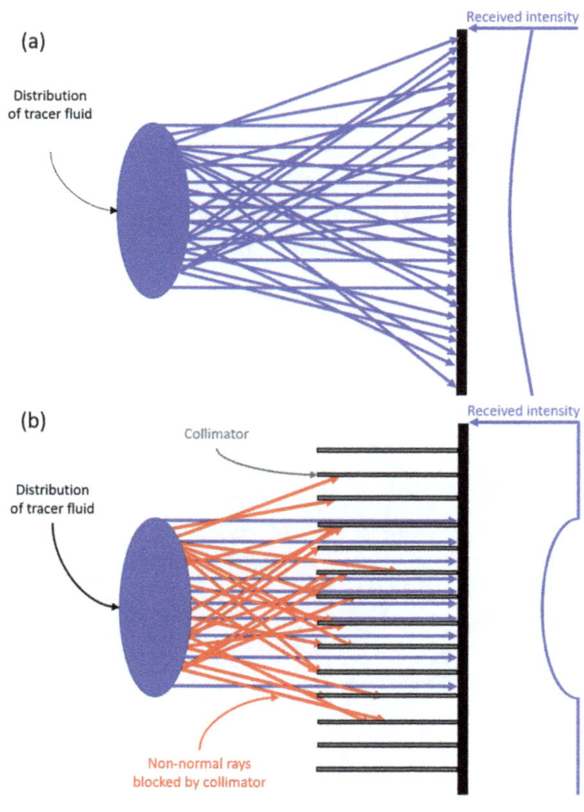

Figure 4.15. Simplified schematic illustrating the manner in which collimation improves SPECT's ability to precisely map tracer fluid distributions. Note that, for clarity, only gamma rays which are incident upon the detector (or would be in the absence of a collimator) are depicted in the schematic. In reality gamma rays would be emitted from all angles.

absorbed, leaving only those rays travelling in the desired orientation. A simplified schematic showing the basic operation of a collimator may be seen in figure 4.15(b).

If the collimator functions ideally, only photons lying within the 2D projection of the tracer region will be detected by the camera. The projection produced is thus similar to that acquired from x-ray radiography, except that the position of the tracer corresponds to a *peak* in intensity, rather than a *dip* in intensity as in figure 4.7. By rotating the detector array around the system of interest, a tomographic image may thus be reconstructed in a manner directly analogous to that described in section 4.2.1.

In reality, of course, physical collimators are not perfect—some degree of noise will inevitably still exist in the images acquired, most notably due to scattered photons (see figure 4.16(a)). Specifically, if a photon undergoes a scattering event in a region outside that occupied by the tracer fluid, there is inherently a chance that the scattered ray will possess a trajectory normal to the detector, and thus be absorbed by the detector in a region outside the 'true' 2D projection of the tracer region onto the detector.

If the size of the tracer used is comparable to the separation of the collimator slats used, this may also lead to a significant degree of blurring due to the potential acceptance of slightly non-parallel x-rays, as illustrated in figure 4.16(b).

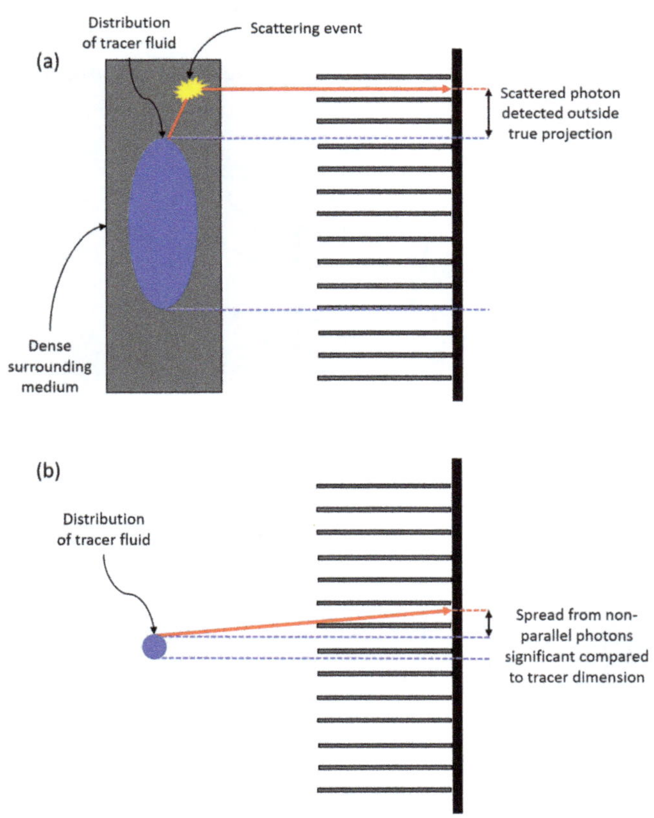

Figure 4.16. Schematic diagrams representing major error sources in SPECT imaging, namely those due to (a) the scattering of photons and (b) the case in which the tracer region is small compared to the collimator grating size.

The above highlights an additional issue raised by the use of collimators: in order for them to function correctly, the collimator slats must be both highly absorptive so as to ensure gamma-rays cannot simply penetrate them, and adequately densely packed to minimise blur from non-parallel photons. As such, collimators typically absorb over 99% of the gamma rays emitted by a source. This means that, when choosing a collimator, there exists an inherent trade-off between resolution and detection efficiency. In industrial and scientific applications, this issue can be overcome to an extent by simply increasing the activity of the tracer medium. In medical applications, however, where one must consider the dose received by the patient, this is not a desirable option. This issue is, however, gradually improving due to the development of more advanced collimator systems [65, 66].

A final complicating factor that must be considered when performing SPECT—and another manner in which SPECT diverges significantly from x-ray CT—is that attenuation by the surrounding medium is here a problem, as opposed to a prerequisite. For example, as mentioned above, one of the valuable applications of SPECT is that the intensity of detected gamma radiation can be used to measure the amount of activity absorbed in a given region of a system. However, the intensity measured by the detector is a function of both the amount of absorbed activity (i.e. the number of photons emitted per unit time) *and* the degree of attenuation along the path between the source and the camera (i.e. the number of the emitted photons which are blocked on their way). Figure 4.17 illustrates, using a highly simplified

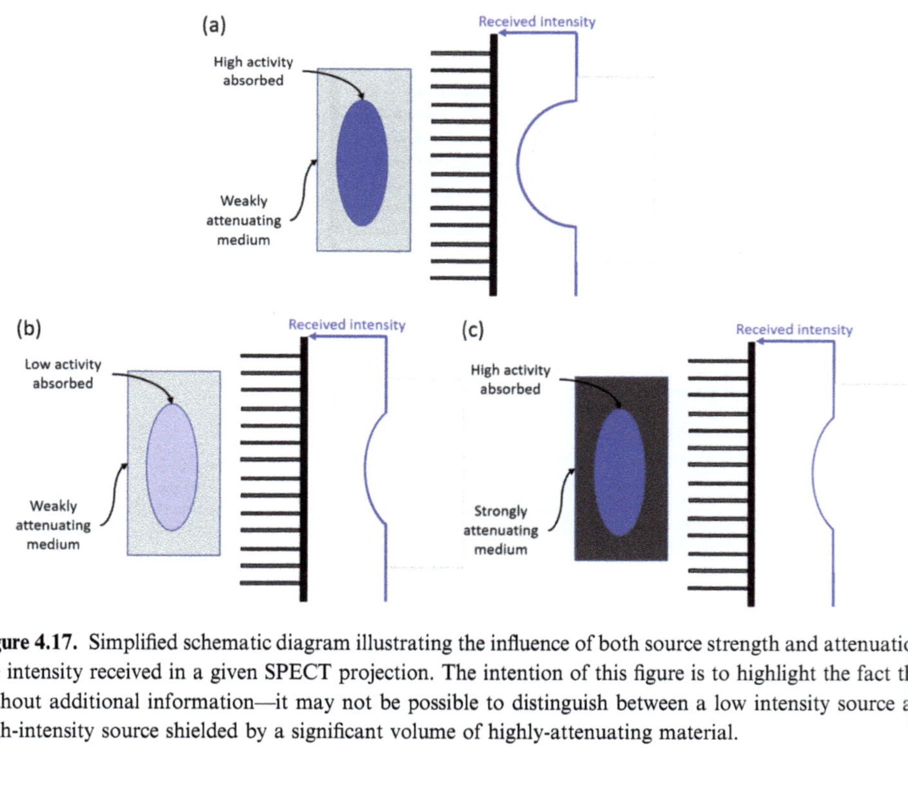

Figure 4.17. Simplified schematic diagram illustrating the influence of both source strength and attenuation on the intensity received in a given SPECT projection. The intention of this figure is to highlight the fact that—without additional information—it may not be possible to distinguish between a low intensity source and a high-intensity source shielded by a significant volume of highly-attenuating material.

example, the difficulty of differentiating between an intensity map produced by a weak source and that produced by a strong source suffering strong attenuation.

The above issue can, however, be overcome through the combined application of both SPECT and x-ray CT (commonly, albeit imprecisely, referred to as SPECT/CT) [67]. The structural information provided by the CT scan can be used to perform an *attenuation correction* [68, 69] on the projections recorded, thus allowing strong sources being strongly attenuated to be differentiated from weak sources being weakly attenuated.

Comparison with PEPT

Advantages:

- Capable of imaging transient processes in real-time for systems with suitably slow dynamics.
- Measures shape and intensity distribution of tracer as opposed to simply position.

Disadvantages:

- Relatively low temporal resolution, as each image requires multiple individual projections to be acquired.
- Low sensitivity—less than 1% of gamma radiation produced is typically detected.

4.2.3 Positron emission tomography (PET)

Positron emission tomography operates using the same fundamental principle as SPECT: a number of 2D projections created by a collimated beam of gamma-rays emitted by a radioactive tracer material are used to build up a three-dimensional image of said material and its spatial distribution within a system[11]. The main area in which the two techniques diverge is the manner in which collimation is achieved. In SPECT, as discussed in the preceding section, the alignment of gamma rays is achieved using a physical collimator (*absorptive collimation*). In PET, however, the same is achieved using *electronic collimation*; that is to say that all incident photons are detected by the camera as opposed to being blocked by the collimator. Photons with orientations lying non-perpendicular to a desired orientation are instead discarded electronically in post-processing. The use of an electronic collimator means that SPECT has a significantly higher sensitivity than SPECT [70].

The algorithms used to perform electronic collimation rely on the use of positron-emitting radioisotopes and the knowledge, discussed in detail in chapter 2, that pairs of gamma rays emitted by positron annihilation events follow approximately collinear, antiparallel trajectories. As per our discussion of PEPT in chapter 2, the lines of response (LoRs) pertaining to each pair of gamma rays detected are recorded and saved electronically. From these LoRs, the angles at which incident gamma-rays hit the camera can be determined, allowing only those possessing the desired incident angle to be considered when creating a projection.

[11] Also like SPECT, PET can additionally measure the relative intensity of active material within a given object.

For the sake of clarity, let us first use the case of a dual-headed, planar gamma camera to illustrate the relevant principles. It is important to note that to perform PET (as opposed to SPECT), one requires a dual-headed gamma camera such as that pictured in figure 4.18(a) or a ring scanner such as that in figure 4.18(b) (which we will discuss shortly). PET cannot be performed using single-headed gamma cameras, as these are inherently incapable of recording gamma-ray pairs produced by annihilation events. Similarly, even dual-headed cameras with geometries such as that pictured in figure 4.14, while not entirely incapable of imaging photon pairs, are certainly unsuitable for PET due to the small fraction of back-to-back photons that would be recorded by such a geometry, and the extremely limited region of space from which suitably-angled pairs could be emitted.

As discussed previously, our radioactively-labelled material will release photon pairs from random positions, and at random angles, as illustrated in the simplified image of figure 4.19(a). From the multiple pairs of photons detected, those whose detected interaction points with the detectors describe a line perpendicular to the camera faces are isolated (figure 4.19(b)). These LoRs alone are then used to construct an intensity map (figure 4.19(c)).

The scanner must then be rotated, and another projection obtained in the same manner at a different angle θ to the stationary object. The series of projections obtained may then be used to produce a tomographic image, as described in section 4.2.1. As with SPECT, this is a slow process, leading again to a somewhat disappointing temporal resolution. However, as PET—unlike SPECT—uses electronic collimation as opposed to absorptive collimation, LoRs not used in a given projection at a given angle, rather than simply being blocked, are *still detected and stored*. As such, if instead of a *planar* camera (such as that depicted in figures 4.18(a) and 4.19) we utilise a *ring* scanner (such as that depicted in figures 4.18(b) and 4.20) which completely surrounds the system and thus allows photons from all angles to be detected, we can in fact obtain data from *all projection angles simultaneously*. A simple illustration of how this may be achieved is presented in figure 4.20.

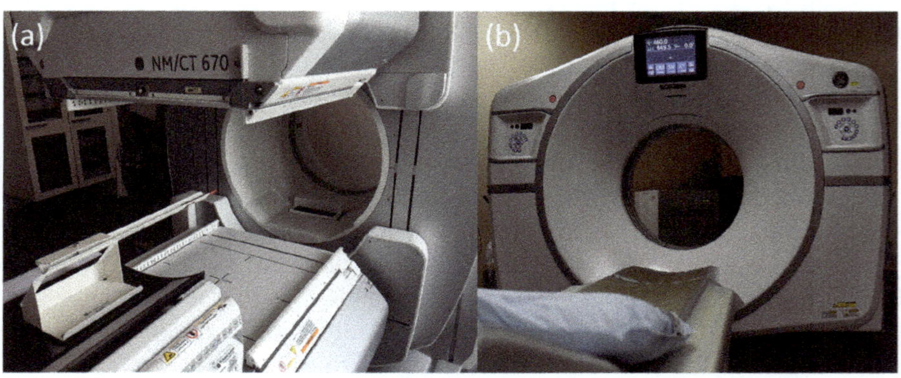

Figure 4.18. Examples of the two main types of positron camera which may be used to perform PET (or indeed PEPT). Image (a) shows a dual-headed planar camera, and image (b) a ring scanner.

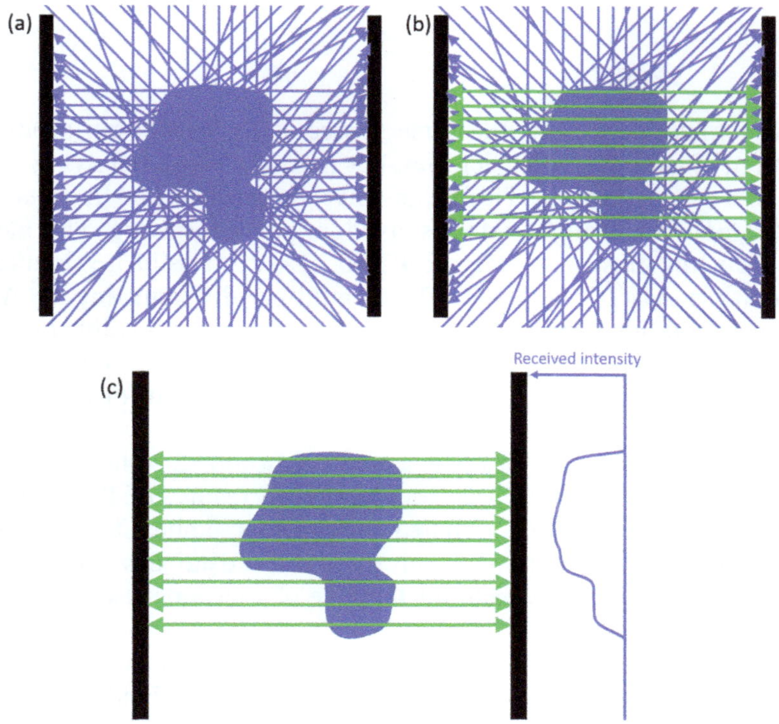

Figure 4.19. Simplified illustration of the manner in which a single projection may be acquired via PET using a planar, dual-headed gamma camera.

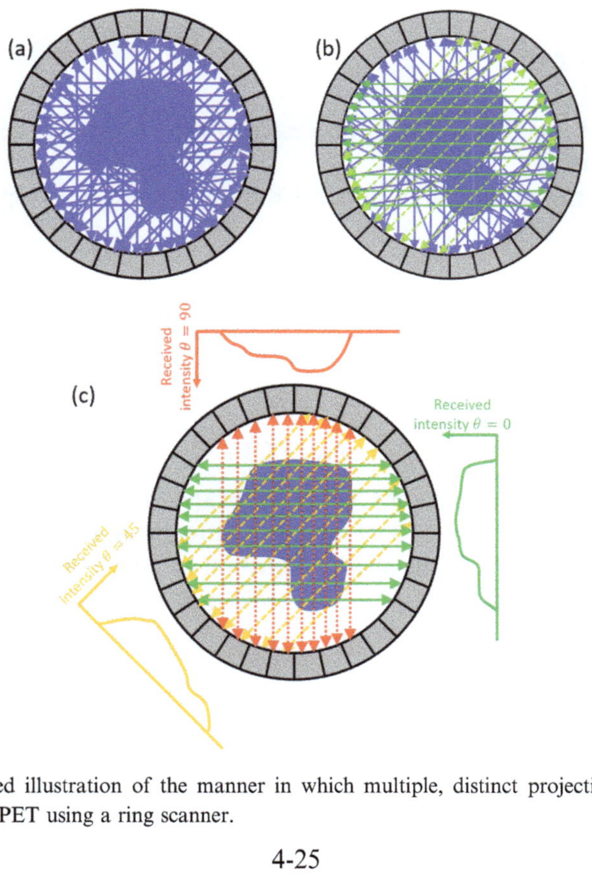

Figure 4.20. Simplified illustration of the manner in which multiple, distinct projections may be simultaneously acquired via PET using a ring scanner.

In figure 4.20(a), we show the same source as figure 4.19(a), producing the same pattern of gamma-ray trajectories, but this time housed in a ring scanner as opposed to a planar camera. This means that photons emitted at *all angles* may be detected, and thus their trajectories saved for post-processing. This means that —as simplistically illustrated in figure 4.20(b, c)—it is possible to simultaneously obtain multiple sets of parallel LoRs at multiple different angles. Assuming adequate computational resources, it is thus hypothetically possible to acquire all information necessary for a full 3D reconstruction in the time required for SPECT to obtain a single projection, and without the need for any mechanical motion of the detector.

It is important to note that, just like figure 2.3 in section 2.2, the picture provided in figure 4.20 presents a highly idealised picture. In reality, PET suffers from the same issues as PEPT (false and scattered coincidences, limited detector resolution, positron flight, etc) which we have already discussed in detail in section 2.3 (and will discuss further in chapter 5) which, for the sake of conciseness, we will not repeat here.

Images acquired using PET are typically noisy compared to x-ray CT due to the comparatively low flux of incident photons, meaning that individual projections in essence contain less information [71]. They are also, at a comparable acquisition rate, significantly noisier than PEPT due to the simple fact that for a given number of data points (i.e. lines of response) it is significantly easier to determine a single point in space than to map out a whole three-dimensional volume!

Comparison with PEPT

Advantages:

- Capable of imaging transient processes in real-time for systems with suitably slow dynamics.
- Measures shape and intensity distribution of tracer as opposed to simply position.

Disadvantages:

- Relatively low temporal resolution compared to PEPT[12].
- Relatively low spatial resolution compared to PEPT.

4.2.4 Electrical capacitance tomography (ECT)

Electrical capacitance tomography (ECT) operates on a principle in many ways similar to x-ray CT, but instead of utilising differences in materials' attenuation coefficients, ECT determines the spatial distributions and positions of particles and other objects based on differences in their *electrical permittivities*. ECT is typically

[12] As discussed in the main text, although PET can acquire multiple projections simultaneously, making it faster than other tomographic techniques such as SPECT, the larger amount of LoRs required to fully map out a region as opposed to simply determining its centroid means that PET is still significantly slower than PEPT.

performed using an array of electrodes located on the exterior of a system, or region of interest thereof. Figure 4.21 shows a schematic example of a typical set-up for the two-dimensional ECT measurement of a cylindrical system.

In a system such as that depicted in figure 4.21, each pair of electrodes can, in essence, be thought of as forming the two plates of a capacitor, with the material in between acting as the hypothetical capacitor's dielectric. The capacitance measured across each pair of electrodes can therefore be used to determine the permittivity of the material lying between them. In the simplest case, a linear relationship may be assumed, and indeed such assumptions can yield reasonable results; however, in order to obtain higher-quality reconstructions, more complex, non-linear relations must be used [72, 73]. With multiple pairs of electrodes, and thus the measurement of multiple inter-electrode capacitances (and hence permittivities), one may—for a system in which the various solid, liquid and/or gaseous components possess strongly-differing permittivities—back-compute the distribution of materials within the system.

Under the aforementioned simplifying assumption of a linear relation between capacitance and permittivity, reconstruction may be achieved using linear back projection (LBP), a method very similar to that discussed in section 4.2.1, which we illustrate schematically in figure 4.22.

Clearly, just as the quality of reconstruction for x-ray CT improves with the number of projections (see section 4.2.1, figure 4.12), the resolution of an ECT image

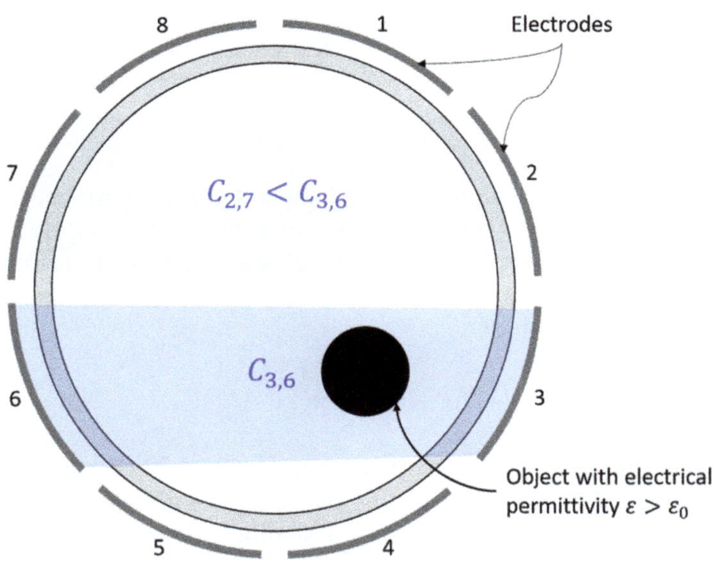

Figure 4.21. Simple schematic diagram showing an archetypal set-up for two-dimensional ECT. The set-up depicted utilises an array of eight electrodes, though more or fewer can easily be used. The region of interest encapsulated by the electrode array contains a single object with a permittivity considerably higher than that of the surrounding medium. As explained in the main text, one may measure the capacitance between any individual pair of detectors within the system. Clearly, in our example, the capacitance $C_{3,6}$ between electrodes 3 and 6 will be significantly higher than that between electrodes 2 and 7.

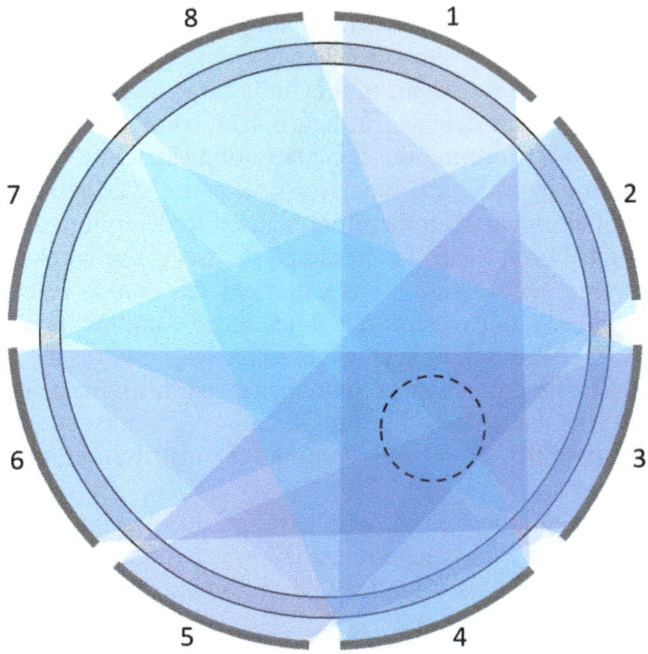

Figure 4.22. Simple schematic depicting the manner in which the position of the high-permittivity object from figure 4.21 may be determined using the data from our electrodes via LBP. Lighter (darker) shading of the path between a pair of electrodes represents a lower (higher) capacitance. For clarity, paths upon which our high-permittivity object does not fall remain entirely transparent. The region for which the sum of the backprojected permittivity values is greatest (in our figure the darkest blue-shaded area) corresponds to the approximate location of our object. In this example, as in real ECT, the presence of gaps between detectors can complicate the interpretation of results.

improves with the number of electrodes used. In fact, the improvement is even more marked as in ECT, individual sensors provide measurements not only with a single, diametrically opposite sensor, but with all other sensors in the system. Specifically, the number of independent capacitance measurements, (N_M) varies with the number of individual electrodes (n_e) as:

$$N_M = \frac{n_e(n_e - 1)}{2}. \tag{4.3}$$

Immediately notable from a comparison between figure 4.22 and our earlier, similar example for x-ray CT (figure 4.12) is that, even with a smaller number of 'projections', x-ray CT typically provides a significantly clearer representation of the position of our object. This is due to the fact that, for x-ray CT, each projection corresponds to a one-dimensional intensity map, whereas in ECT we have only a single scalar value. The issue of ECT's low spatial resolution is further compounded by the nonlinear nature of the inverse problems which must be solved to produce an image, as alluded to above. The combination of these factors means that ECT typically possesses poor resolution as compared to other tomographic techniques [74].

The image quality achievable with ECT has nonetheless improved somewhat since its inception with the introduction of improved reconstruction algorithms [75–77].

Finally, it is worth noting that in addition to electrical *capacitance* tomography, one may also perform electrical *resistance* tomography and electrical *impedance* tomography. The techniques operate largely on a similar principle to ECT though—as their names imply—measuring different electrical properties of the materials being imaged. Though, for brevity, we will not discuss ERT and EIT in detail here, the interested reader may find further information in our references [78] and [79], respectively.

Comparison with PEPT

Advantages:

- Comparatively cheap to set up and operate.
- Geometrically flexible as compared to other tomographic techniques.

Disadvantages:

- Typically poor spatial resolution.
- Requirement for strongly-differing permittivities limits choice of materials and combinations thereof.

4.2.5 Magnetic resonance imaging (MRI)

Magnetic resonance imaging (MRI), also referred to as nuclear magnetic resonance imaging (NMR or NMRI), operates by using carefully controlled magnetic fields to manipulate a 'spin system'—for example a system of hydrogen atoms. It is worth noting that the 'nuclear' in its full title refers not to any use of ionising (nuclear) radiation, as with PEPT, PET or RPT, but rather to the fact that it involves the *magnetic resonance of nuclei*. Nonetheless, due to the widespread negative connotations of the word 'nuclear', practitioners typically refer to the method simply as MRI.

An MRI scanner consists of a series of concentric magnets, alongside various other components, many of whose main role is simply to cool the (superconducting) magnets and thus allow them to operate optimally. A simple schematic illustrating these main components is provided in figure 4.23. The system we wish to image (be this a person or a piece of process equipment) is placed at the centre of these concentric coils.

In order to perform MRI, we require a suitable spin system—that is, the material whose motion we wish to track must contain a suitable number of atoms whose spins yield a (small) magnetic field, and thus can be successfully manipulated via the magnetic fields produced by our coils. While medical MRI typically works with hydrogen atoms (i.e. protons) due to the abundance of water in the human body, MRI can also be applied to a variety of other atoms, or even molecules. For the sake of simplicity, though, we will refer here simply to protons, though the description provided can just as easily be applied to other nuclei.

In the absence of a magnetic field, the spins of atoms will be randomly aligned (see figure 4.24(a)), yielding no net magnetic moment—i.e. all their randomly-oriented

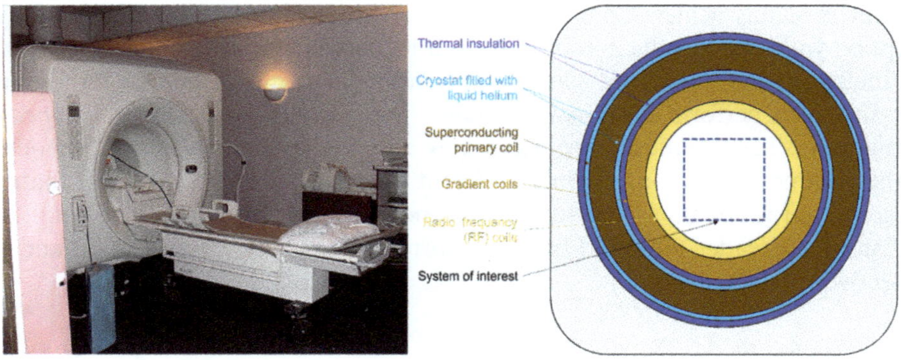

Figure 4.23. Left: Image of a real medical MRI scanner (copyright BrokenSphere/Wikimedia Commons). Right: Schematic showing a cross section of such a system, indicating the placement of the various coils discussed in the main text.

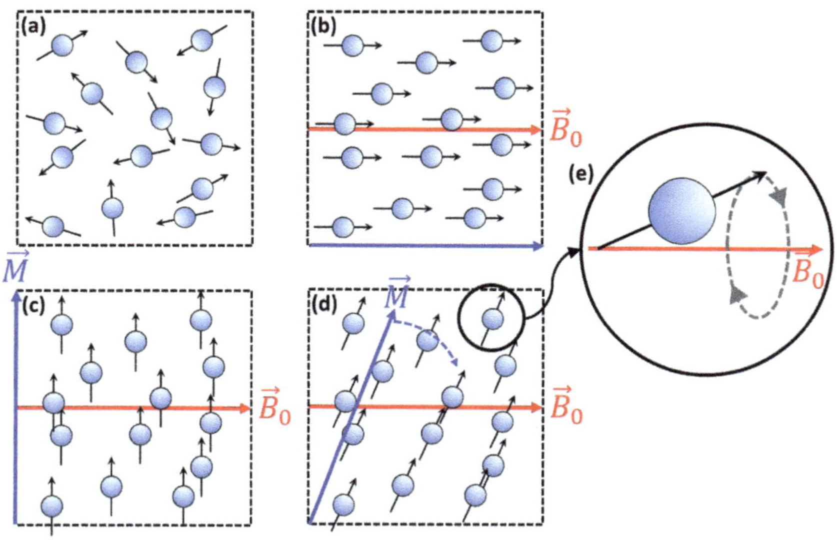

Figure 4.24. Highly simplified schematic depicting a spin system (a) in the absence of an external magnetic field; (b) in the presence of the strong, static magnetic field $\vec{B}_0$; (c) immediately following the application of a 90° RF pulse; (d) at a later time following the RF pulse, as the magnetic moments relax toward their equilibrium position in line with $\vec{B}_0$. Note that these images represent very much an oversimplification of the true picture. In reality, even a strong magnetic field will only cause a *preferential* alignment, as opposed to complete alignment, as shown in (b)–(d), due to the vibrations and collisions inherent to any molecular system.

magnetic moments simply cancel out. When placed in the bore of an MRI scanner, the superconducting primary coil (the outermost coil in figure 4.23) produces a strong, static magnetic field, $\vec{B}_0$, of typically 1.5–3 T—more than four orders of magnitude greater than the Earth's magnetic field. This strong field causes spin particles in the system to preferentially align with $\vec{B}_0$, in this case along the axis of the scanner bore (see figure 4.24(b)), producing a net magnetisation vector, $\vec{M}$, in the direction of $\vec{B}_0$.

The scanner's radiofrequency (RF) coils (the innermost coil in figure 4.23) are used to apply a 'pulse' of electromagnetic radiation at the resonant frequency of the protons, namely the frequency at which they precess [80, 81] (see figure 4.24(e)). This RF pulse provides the protons with energy and in essence 'knocks' them—and thus the net magnetisation vector $\vec{M}$—out of alignment with $\vec{B}_0$. RF pulses may be applied in a variety of manners [82], but as our intention here is just to provide a fundamental understanding of the MRI process, we will consider the simple case in which the duration of the applied pulse causes the net magnetisation vector to rotate 90°, such that it now lies perpendicular to $\vec{B}_0$.

Having been knocked out of equilibrium by the RF pulse, $\vec{B}_1$, the protons will naturally—in the absence of any continuing energy input—begin to revert to their equilibrium state, aligned with $\vec{B}_0$, releasing the energy provided by the RF pulse as they do so. In doing so, they allow a current, alternating at the precession frequency of the protons, to be detected by receiver coils placed around the system of interest see figure 4.25. The received current is known as the free induction decay (FID) signal, which can be analysed to deduce the density (i.e. local concentration) of the proton-containing material (our tracer).

Clearly, for the situation described thus far, our MRI scanner would provide only a single-valued signal for our entire system, which is of rather limited use! In reality, a third set of coils, known as gradient coils, are used—as their name implies—to add small, additional magnetic fields so as to induce a spatial variation, or gradient, in the magnetic field strength throughout the system. Let us consider, for simplicity, a one-dimensional gradient along the *z*-axis of the MRI system. In the presence of such a gradient, the magnetic field strength, and thus the precession frequency of protons, will be different at each given axial position within the system. Thus, the resonant frequency of protons in different positions within the systems will be different, meaning that an RF pulse at a given frequency *will only excite protons in a given axial slice* of the system. In this manner, we can calculate the *local* tracer density in the relevant slice; by repeating the process using RF pulses at different frequencies, we can thus build up a one-dimensional image of our system. By applying gradients also along the *x*- and *y*-axes, one can similarly build up a full, three-dimensional image.

A significant strength of magnetic resonance imaging is that (unlike, for example, x-ray CT) it does not require any physical movement to 'scan through' a system, simply the application of different electromagnetic pulses. This means that the method can achieve significantly higher spatial resolution, whilst still collecting (unlike PEPT) 'whole field' data, thus allowing the direct imaging of transient motion. Like PEPT, MRI can also acquire information in full 3D, from the interior of optically opaque systems, making it a very powerful technique. The main limitation of the method, however, is that one must be much more selective than with PEPT in terms of the materials to be used—ferromagnetic materials, for example, cannot be used either within the system to be imaged, or in its construction, making the direct imaging of real process equipment difficult in many cases. Further, the tracer to be used must contain nuclei with suitable spin properties, further limiting the range of (and combinations of) materials that can be successfully used for magnetic resonance

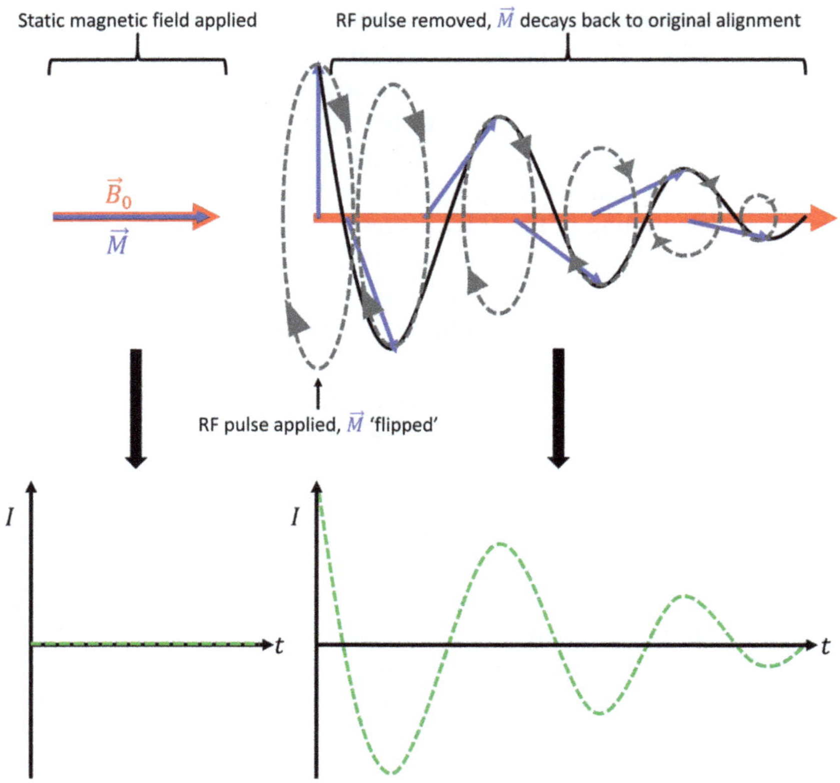

Figure 4.25. Simplified schematic illustrating the excitation/decay process of a single proton as it is excited by an RF pulse and subsequently relaxes toward its equilibrium state (above), and how this process maps to the observed free induction decay signal (below).

imaging. An additional issue is the relatively small bore size of the systems, though unlike the previous issue, this limitation is potentially surmountable.

As with other techniques described within this section, the above represents a significant simplification of the 'real' picture, intended solely to give the reader an elementary insight into the technique, such that (a) meaningful comparison can be drawn between PEPT and MRI, and indeed the other techniques mentioned in this section, and (b) references to MRI studies in chapter 8 will be more meaningful. The interested reader may find more complete details regarding MRI and the related physics in our references [80, 80, 82–85].

Comparison with PEPT

Advantages:

- Able to image transient behaviour, in 3D, in real time.

Disadvantages:

- Bore size limits size of system that can be imaged.
- Limitations on materials that can be used

4.2.6 Refractive index matched scanning (RIMS)

Of all techniques described in this section, RIMS is certainly the method which takes the concept of tomography—i.e. the reconstruction of an image from slices—most literally. Where other tomographic techniques take depth-averaged projections through the whole system and use relatively complex methods such as FBP, ART or iterative reconstruction to piece together a 3D image from these projections, RIMS uses a laser sheet to highlight and image individual 2D slices through a system, which are then simply 'stacked' together to form a 3D image. As ever, this one-line description is an over-simplification of the true process, but nonetheless captures the underlying concept of the RIMS technique.

Although refractive index matched imaging in general has been used in other fields for many decades [86], its use in tomographic techniques and application to dense, three-dimensional granular systems as RIMS is a relatively recent occurrence [87]. Indeed, RIMS is actually a more modern invention than many of the more conceptually challenging and technologically complex methods described above!

In order to perform RIMS, one requires a system of optically transparent particles, immersed in a fluid which (a) is index-matched to said particles[13] and (b) is labelled with a fluorescent material such that it produces light in the visible spectrum when exposed to a laser. A commonly-used combination is glass spheres immersed in a mixture of benzyl alcohol and ethanol and labelled with the fluorescent dye rhodamine [88]. A more comprehensive list of usable particle/fluid combinations can be found in [89, 90]. It is notable that the list compiled in the review of Dilksman *et al* [89] comprises a total of seven possible particle materials and 13 possible fluids—a far cry from the near-infinite range of tracers and fluids suitable for PEPT imaging. This is a significant limitation of the RIMS methodology. Most significantly, RIMS cannot be performed in air or vacuum, making it unusable for dry granular systems.

As with previous techniques, we present in figure 4.26 a simplified, lower-dimensionality representation of the RIMS process, though the description provided here is general to both 1D → 2D and 2D → 3D reconstruction. A sheet laser is shone on an index-matched system such as that described above in a direction perpendicular to the desired direction of imaging. As the 'labelled' fluid in which the particles are immersed fluoresces (i.e. emits photons) when stimulated by the laser light but the ('unlabelled') particles do not, images recorded by a camera whose field of view is perpendicular to the laser sheet will show particles as dark spots on a bright background. A schematic, one-dimensional illustration of such an image may be seen in figure 4.26(c), while figure 4.27 shows an equivalent 2D projection from a real RIMS experiment. The sheet laser, which is typically mounted on a translating stage, then scans across the system, moving in the direction perpendicular to the laser beam itself, creating a series of such projections across its entire length. A single projection can be acquired in approximately 10 ms, allowing scan speeds of roughly one volume per second [89]. This limits the flow rate of systems which can be successfully imaged to approximately one particle diameter per second.

[13] When the refractive indices of (transparent) particles and fluids are matched, the particles in essence become 'invisible' under normal lighting conditions.

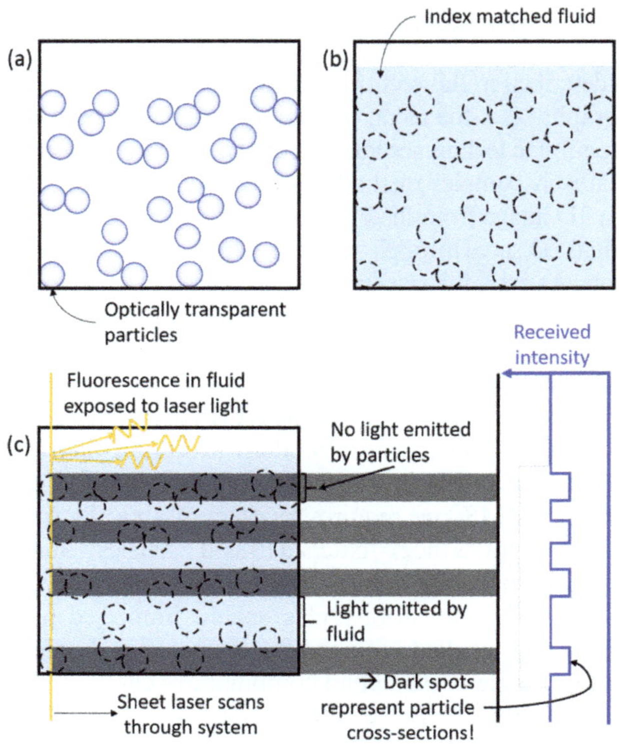

Figure 4.26. Simple schematic illustrating the fundamental principle of RIMS for a hypothetical 2D system.

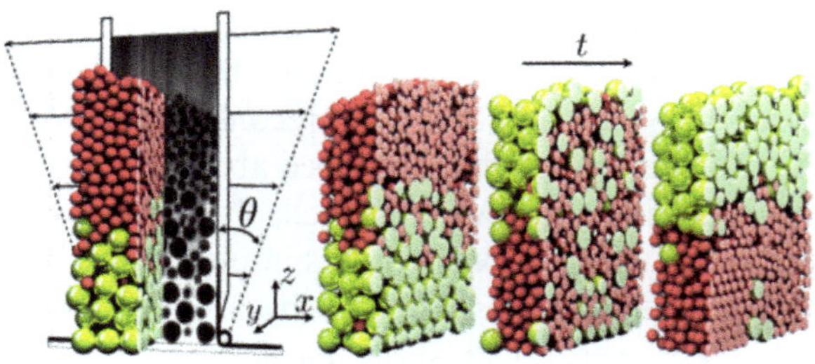

Figure 4.27. RIMS reconstructions of a binary granular system. Image reprinted with permission from [88], Copyright 2015 by the American Physical Society.

Once a suitable number of 2D slices have been produced across the full length of the system, these images can then be computationally 'stacked together' to form a full 3D image. For simple, spherical particles, this process typically involves, for each slice, applying relevant image processing techniques [91] to determine the centroids and diameters of the circles corresponding to the cross-sections of 3D particles. By matching the centres of these circular cross-sections in consecutive frames, it is possible to build up a three-dimensional image of the particles in the system [92–94], as illustrated in figure 4.27.

Comparison with PEPT

Advantages:

- Comparatively cheap to set up and operate.
- Capable of imaging large numbers of particles in close proximity.

Disadvantages:

- Low temporal resolution compared to PEPT.
- Unable to penetrate opaque containers, e.g. metal-walled process equipment, or systems immersed in opaque fluids.
- Highly limited range of usable materials due to requirements that particles and interstitial fluid are both transparent and index-matched.

4.3 Three-dimensional particle-tracking techniques

We have already, in our discussions regarding the fundamentals of Particle imaging PEPT and (chapters 1 and 2, respectively) and further in our introduction to particle tracking velocimetry (section 4.1.1), introduced the basic concept of particle tracking. In this section, we provide an introduction to other methods which, like PEPT, are capable of tracking the three-dimensional motion of one or more tracer particles through three-dimensional space. This class of techniques typically provide significantly better temporal resolution than tomographic techniques, at the expense of losing information regarding the shape of the material being tracked and contacts between particles. As such, when imaging slow or quasi-static flows, in particular those exhibiting transient dynamics, it is generally preferable to utilise tomographic techniques. Conversely, if we wish to investigate the behaviours of more dynamic systems, particle tracking methods are typically superior. As ever, this is somewhat of a generalisation and simplification of the true picture, and the relative strengths and weaknesses of 3D tracking techniques are somewhat more nuanced, as can be seen from the sections below.

4.3.1 Radioactive particle tracking (RPT)

RPT, often referred to as '*computer aided* radioactive particle tracking' (CARPT)[14], like PEPT, uses gamma radiation to track the motion of one or more [95, 96] particles. While in PEPT only positron-emitting radionuclides may be used, CARPT can be successfully performed using a wide range of gamma-emitting radioisotopes [97, 98]. Again unlike PEPT, which uses known gamma-ray trajectories to triangulate the position of a source, CARPT instead uses multiple detectors distributed around a system, and determines the particle position from the intensity (i.e. frequencies of recorded gamma rays) measured by each detector.

[14] The addition of 'computer aided' to the name is intended to ensure distinction is drawn between the technique described here and the rather primitive method occasionally employed in industry whereby a series of (analogue) Geiger counters are used to extremely roughly track the motion of a tracer through a piece of process equipment.

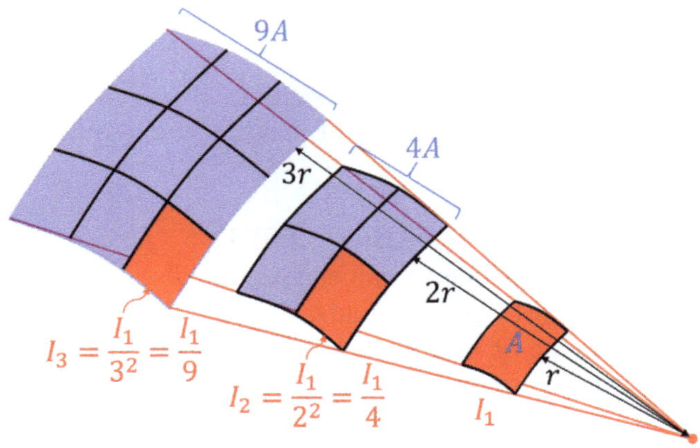

Figure 4.28. Visual illustration of the inverse square law. Suppose a detector, represented by a red-shaded square, of area A placed at a distance r from a source of electromagnetic radiation detects a number of photons per second corresponding to an intensity I_1. If the same detector is instead placed at a distance $2r$, the photons emitted by the source will be more spread out. Specifically, they will be spread over an area 4 (2^2) times larger, meaning that only a quarter ($\frac{1}{2^2}$) of the number of photons will be detected—i.e. $I_2 = \frac{I_1}{4}$. At a distance of $3r$, they will be spread over an area 9 (3^2) times larger, meaning that $I_3 = \frac{I_1}{9}$.

At its heart, CARPT works by leveraging the well-known inverse square law for radiation from a source, which states that, for a given point-like radiation source, the intensity, I, of radiation received by a detector of fixed surface area A will decrease as the square of the scalar distance, r, between the source and the detector, i.e.

$$I \propto \frac{1}{r^2}. \tag{4.4}$$

This concept is illustrated schematically in figure 4.28.

Clearly, if we know the intensity I_0 received by a detector at some given source-detector separation r_0 from a radioactive tracer particle, when said tracer is moved to a new location, the new source-detector separation can be determined from the new intensity received. This measured intensity, however, is only a scalar quantity, meaning that we do not know the precise location of the tracer, only that this position must fall on the surface of a sphere of radius r surrounding the detector (see figure 4.29(a)). Nonetheless, with multiple detectors, one may determine the points of intersection of the spheres corresponding to each (see figure 4.29(b)). With a suitably large number of detectors, one may determine a unique point of intersection between all circles, corresponding to the position of the tracer particle (see figure 4.29(c))[15]. For an (ideal) system lying in a 2D plane, this requires a minimum of three detectors. In reality, significantly larger numbers of detectors are typically used [99] to ensure a degree of redundancy—and to accommodate for the imperfect nature of real measurements, which we discuss in detail below.

[15] For a pair of spheres, as opposed to circles as depicted in figure 4.29, the initial intersection of any two spheres produces a circle, as opposed to a pair of points as depicted in the two-dimensional example of figure 4.29. However, this in essence simply necessitates an additional step in the process depicted.

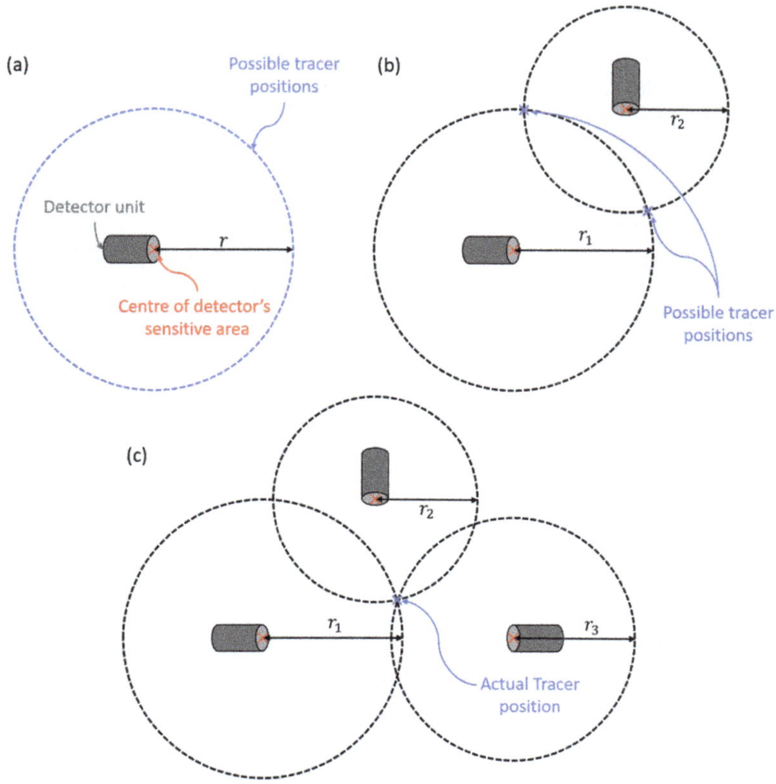

Figure 4.29. Based on the intensity received from a (calibrated) source–detector pair, it is possible to determine (assuming an ideal detector and no obstructions between source and detector) the scalar separation, r, of the two, meaning that the detector must lie at some point on a sphere of radius r. For visual clarity, we consider here only the simplified case of finding the position of a tracer in a two-dimensional plane, though the same principle is applicable also in 3D. By using multiple detectors and determining the points of intersection of their circles drawn in the plane of interest by their respective spheres, one can ultimately determine the singular point of shared intersection which must correspond to the position of the particle.

The above description, however, relies on a number of assumptions—many of which are decidedly dubious for the types of system commonly imaged with RPT [97, 99–101], and the types of detector used. The most critical of these are discussed below.

Firstly, it is assumed that the loss in intensity between source and detector occurs only due to distance, i.e. the inverse square law holds exactly. In reality, this is only going to hold perfectly true in vacuum, where there is no material to absorb or scatter gamma rays—and clearly such a system could be much more easily, cheaply and safely imaged using optical techniques! As the main advantage of RPT over such techniques is its ability to 'see through' opaque materials, it is naturally only applied to systems in which the above does not hold true—i.e. there exist significant volumes of material separating source from detector. However, the effects of attenuation—and thus deviations from the inverse square law—can be mitigated to an extent through the utilisation of tracers which produce particularly high-energy gamma-rays. Tracers such as scandium-46 and cobalt-60, for example, which produce gamma radiation in the MeV range (an order of magnitude higher than that typically used in x-ray imaging, for example), are commonly used [98, 102, 103]. At such energies, the mass absorption coefficient of this radiation (i.e. the rate at which it is absorbed by a given material) is relatively independent of the material, for a reasonable range of media [104]. Nonetheless, for lower-energy gamma rays and/or denser surrounding media, deviation from the inverse square law will become more significant, meaning that more complex, *in situ* calibration is required for each new system explored [99, 105].

Even with careful calibration, it must be remembered that dynamic granular systems are typically highly inhomogeneous, meaning that the bulk densities of material between the source and detector may vary significantly with time, often in an entirely random manner. Even a static tracer may register a strongly varying activity on a given detector in the presence of such fluctuations. The distinction of variations in intensity due to these fluctuations from those due to particle motion is not a trivial matter.

The above discussion also includes a number of implicit assumptions regarding the detectors used, most notably that they can be considered point-like, and thus that a single, specific centre for the above-discussed spheres may be determined. In reality, all detectors possess an active area of finite size, leading to a degree of uncertainty regarding the centre of said spheres, and thus their surfaces. Further, as the particle moves, it will take on differing relative orientations to this active area, which may affect the intensity received from the particle from different angles, even at the same separation distance r.

All of the above issues contribute to an uncertainty on various aspects of our RPT measurements which, for the sake of clarity and brevity, we will lump together into a single error, σ_r, on the source-detector separation. These errors, in turn, lead to a degree of uncertainty on the precise position of the tracer, as illustrated in figure 4.30. As such, locating a particle is not, in reality, a simple case of finding the intersection points of several circles (or, in 3D, spherical shells), but instead requires the application of more complex algorithms—some examples of which can be found in references [99, 106, 107].

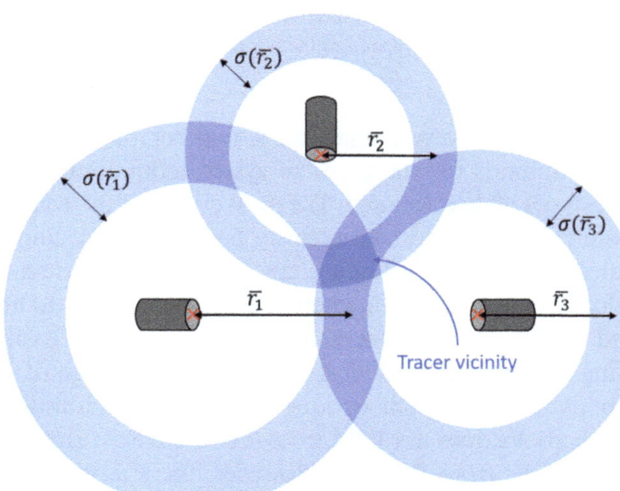

Figure 4.30. A more realistic depiction of the RPT process, incorporating the inherent errors in measurement, σ_r.

Comparison with PEPT

Advantages:

- Can be performed with any of a wide range of radioisotopes.
- Can utilise highly-penetrating ($E_\gamma = \mathcal{O}$(MBq)) gamma radiation, facilitating use in particularly large, dense systems.

Disadvantages:

- Relatively low spatial resolution compared to PEPT.
- System specific—requires significantly more calibration than PEPT.
- Using current algorithms, can track notably fewer individual tracers than PEPT [95, 108].

4.3.2 Magnetic particle tracking (MPT)

Magnetic particle tracking (MPT) works on a principle fundamentally similar to that of RPT, but relying on the measurement of the *magnetic field intensity*, $\vec{H}$, as opposed to the intensity of gamma radiation. Like RPT, MPT uses an array of sensors an known configuration (see, as an example, figure 4.31) to determine the relative intensity of the magnetic field produced by a tracer (or 'marker') particle which either is, or contains, a permanent magnet.

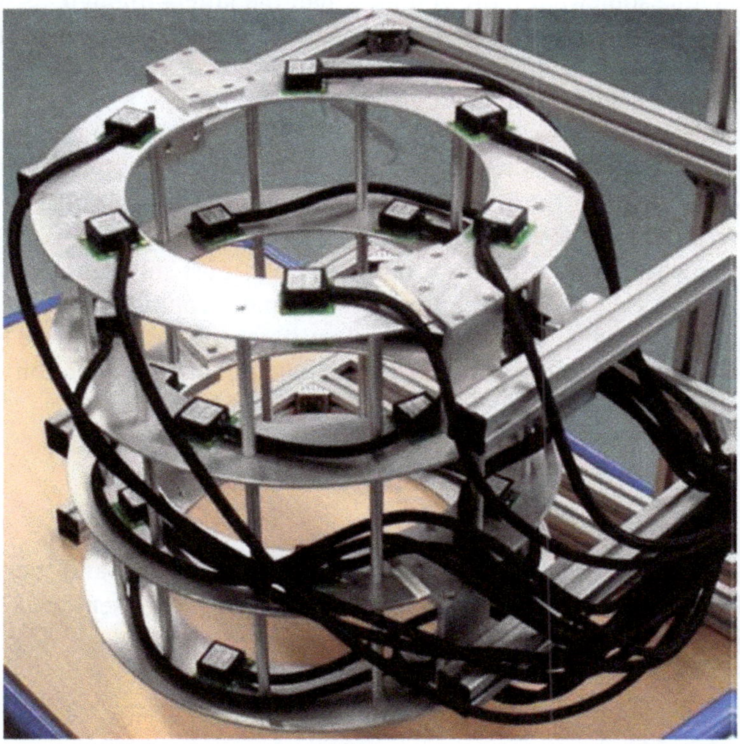

Figure 4.31. Image of a real 3D MPT sensor array, comprising 24 triaxis AMR sensors in a 4 × 6 cylindrical configuration. Image reproduced with permission from [109]. John Wiley & Sons. (©) 2014 American Institute of Chemical Engineers AIChE J.

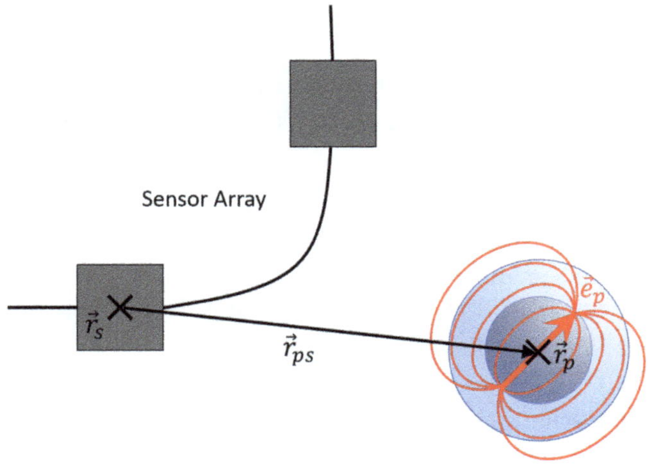

Figure 4.32. Simple schematic showing a tracer particle with a magnetic core and a section of a hypothetical detector system, showing two AMR sensors, demonstrating the meaning of the terms in equation (4.5).

The use by MPT of a magnetic field, as opposed to the comparatively simple case of high-energy electromagnetic radiation for CARPT, makes the application of the former technique somewhat more complex than the latter—though, as we will see, also carries some notable benefits. Firstly, in order to accurately measure the field produced by a magnetic tracer, one requires a known, highly consistent 'background' magnetic field: any fluctuations in such a field may—for reasons which will shortly become apparent—cause significant errors in the location and tracking of a magnetic tracer. Secondly, whereas the intensity of gamma photons represents only a simple, scalar quantity, the magnetic field intensity $\vec{H}$ is a vector quantity. As such, in order to estimate the position of a magnetic tracer—even for the case of a known magnetic moment and known sensor positions—we must solve an inverse problem possessing 5 degrees of freedom (3 translational and 2 rotational) [109].

The theoretical magnetic field strength at given sensor position $\vec{r}_s$ (see figure 4.32) arising from a tracer particle with magnetic moment μ_m and orientation unit vector $\vec{e}_p$ lying at a position $\vec{r}_p$ can be expected to follow the form:

$$\vec{H}\left(\vec{e}_p\right) = \frac{1}{4\pi}\left(-\frac{\mu_m \vec{e}_p}{|\vec{r}_{ps}|^3} + \frac{3\mu_m(\vec{e}_p \cdot \vec{r}_{ps})\vec{r}_{ps}}{|\vec{r}_{ps}|^5}\right), \tag{4.5}$$

where $\vec{r}_{ps} = \vec{r}_p - \vec{r}_s$ is the vector distance between particle and sensor[16]. The theoretical signal strength, S_t, that should be received by the sensor at $\vec{r}_s$ can then be determined as:

$$S_t = \vec{H}\left(\vec{e}_p\right) \cdot \vec{e}_s, \tag{4.6}$$

where $\vec{e}_s$ is the unit vector orientation of said sensor.

[16] Note that the above form holds only for a sufficiently large separation $|\vec{r}_{ps}|$.

The position—and indeed orientation—of our tracer can be determined by comparing the *measured* signal strength, $S_{M,i}$, at each sensors, i, to a series of possible theoretical signal strengths, $S_{t,i}$. In an ideal scenario, the vector position and orientation of our particle could be determined by finding the values of $\vec{r}_p$ and $\vec{e}_p$ which solve equation (4.6) for all N_s sensors. In reality, of course, the presence of noise, interference, and otherwise non-uniform and/or inconsistent background fields mean that such a perfect solution does not exist. Instead, a 'best fit' solution—i.e. that which minimises the total deviation between all pairs of measured and theoretical signal strengths across all sensors—may be determined using a quality function such as:

$$Q = \sum_{i=1}^{N_s} \frac{[(S_{m,i} - \langle S_m \rangle) - (S_{t,i} - \langle S_t \rangle)]^2}{\Delta S_{m,\, i}^2}. \tag{4.7}$$

In the above, the term $\langle S_m \rangle$ accounts for 'stray fields' which, clearly, may influence our results. These may be measured in the experimental system using a gradiometer. Each sensor, i, in the system will carry an innate sensitivity, and thus a min/max deviation, which we denote $\Delta S_{m,i}$. The term $\frac{1}{\Delta S_{m,\, i}^2}$ thus provides a correction factor, ensuring that less accurate sensors carry a smaller impact on the quality function.

The above provides a broad explanation of the fundamental principles upon which MPT operates. In reality, the data acquisition and analysis process typically involves a more refined optimisation methodology. Further details of this process, including specific algorithms applied, and the application of relevant initialisation and signal filtering processes may be found in references [109–112].

Comparison with PEPT

Advantages:

- Capable of tracking particle rotation for smaller tracers than currently possible with PEPT.
- Comparatively cheap to set up and operate.

Disadvantages:

- Limitations on systems which can be imaged—motors or other devices producing electromagnetic fields cannot be present in or near the system's field of view.
- Lowest usable particle size significantly larger than for PEPT imaging.
- Comparatively low spatial resolution (though sub-millimetre accuracy can be achieved for larger tracers).
- Requires non-identical tracers (though through careful density matching can achieve representative behaviour [109, 113]).

4.4 Non-imaging techniques

While the focus of this chapter—indeed of this book—lies on techniques facilitating the imaging (either direct or indirect) of particulate or multiphase systems, it is important to remember that this is not the only manner in which to extract information from such systems. There exist in fact a wide range of techniques through which one may gain

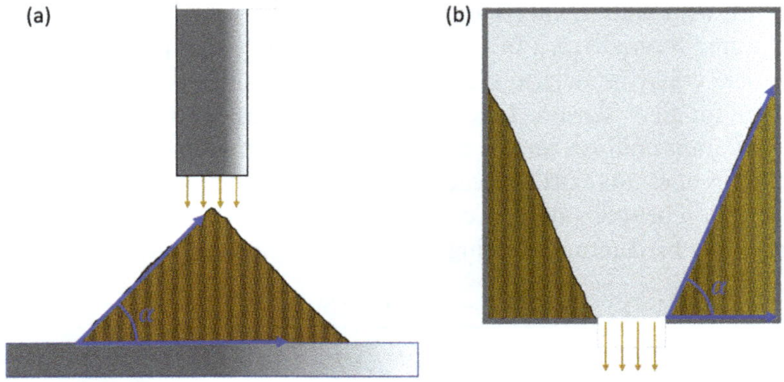

Figure 4.33. Simple schematic diagram illustrating the basic equipment and methodology through which one may measure (a) the external (poured) angle of friction and (b) the internal (drained) angle of friction.

insight into a particulate system's properties and dynamics *without* the need for imaging. Such techniques are not only useful in their own right, as they are typically more capable of measuring properties relating, for example, to interaction forces between particles and their surroundings than most of the imaging techniques discussed above, but they can also provide a valuable counterpart to such techniques.

The techniques we describe here are typically referred to as '*(powder) characterisation tests*', and are widely used both in industry and academia as a means to aid the design of systems, processes, products and experiments, as well as to gain a more fundamental insight into the particulate media used therein—i.e. to 'characterise' them. In the preceding sections, we have very much focused on the direct measurement of microscopic or 'particle-level' characteristics of our systems of interest—particle positions, velocities and orientations, inter-particle forces and other properties belonging to the individual particles that form a given particulate medium. The techniques described in this section, however, are largely empirical quantities, representing the bulk (macroscopic) behaviour of such a medium.

There exists a wide and diverse range of characterisation tests and testing devices available to researchers. In the subsections below, we discuss the main, most widely-used tests, which are generally recognised as industry standards.

Angle of repose (AOR) tests

The angle of repose of a powder or granular medium is simply defined as the angle formed between the horizontal and either the slope of a static, cone-like pile of powder (the *static angle of repose*) or the free-surface angle of a flowing powder (the *dynamic angle of repose*).

The static angle of repose may, in its simplest form, be measured simply by pouring a granular material through a funnel at a given height, and simply measuring the angle of the conical heap formed[17], as illustrated in figure 4.33(a), giving a value of the 'external' or 'poured' angle of repose [114]. An alternative

[17] Note that if a symmetric, conical heap cannot be formed by the material in question, then this measurement technique is unsuitable for said material.

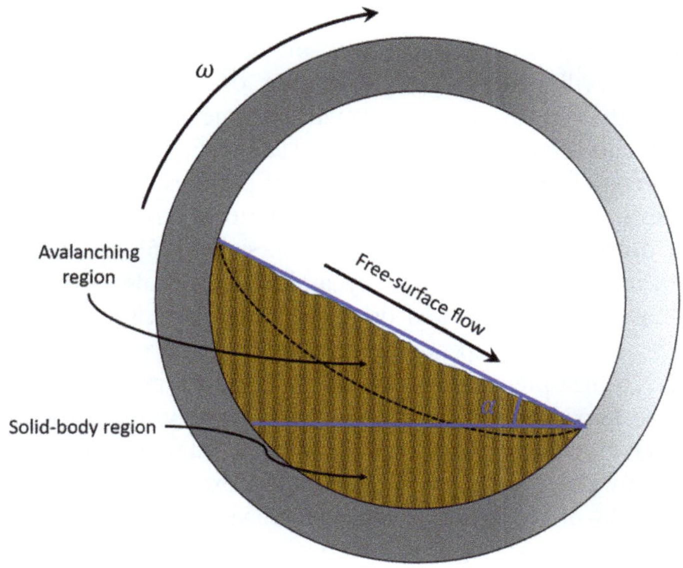

Figure 4.34. Simple schematic of a rotating drum system from which a value for the dynamic angle of repose may be extracted. The dashed line represents the delineation of the upper free-flowing region and the lower solid-body rotation region. The dynamic angle of repose is measured as the angle, α, between the horizontal and the average inclination of the free-surface flow.

method is to pour material out of a flat-bottomed hopper (see figure 4.33(b)) and measure the angle formed by the material remaining inside the hopper after the cessation of flow, providing a measure of the 'internal' or 'drained' angle of repose. [114]. It is notable that the internal angle of repose typically possesses a value significantly higher than the external AOR [115]. Indeed, these methods are highly sensitive to a number of factors relating to the precise methodology used, in addition to factors such as segregation in the case of non-uniform powders, as well as consolidation (particularly problematic in cohesive powders), or aeration, meaning that great care must be taken when performing such tests [116, 117].

The *dynamic* angle of repose is most commonly measured using a rotating cylinder or 'drum', as depicted in figure 4.34. A granular medium placed within a rotating cylinder will (assuming adequate friction at the wall [118] and a suitable rotation rate [119–121]) enter a dynamic steady state known as the *rolling* or *continuously avalanching* regime [119, 122]. In this regime, the granular bed within the cylinder is divided into two distinct regions. The first of these is the solid-body rotation region, in which particles exist (as the name implies) in a solid-like state, simply following the rotating motion of the drum. The remaining particles in the system—namely those at or near the bed's free surface—flow downward across the solid inclined slope formed by particles in the solid-body region. The typical delineation between these two regions is illustrated in figure 4.34. The dynamic angle of repose is taken simply as the angle between the horizontal and the mean angle of the free-flowing surface layer, as illustrated.

When measuring the dynamic angle of repose, one must be careful to operate between a specific range of rotation rates—or, more specifically Froude numbers: too low and one will observe either slipping or discrete, periodic avalanches [123, 124]; too high, and the free surface will become curved, eventually beginning to centrifuge around the edges of the system [122, 125].

The angle of repose is affected by the friction of particles (i.e. their micro-/nano-scale surface roughness) as well as by the shape and (to a lesser extent) size—or, more specifically, size distribution—of particles, their cohesivity [126], and the roughness of the base used for measurement [127]. In the dynamic case, it is also a function the Froude number, and thus rotation speed and drum diameter, as well as the fill height of the system [122, 128, 129]. This dependence on multiple parameters makes it difficult to extract from such a measurement any specific particle-level information. Nonetheless, it can provide some important bulk properties; most notably, for particles of a uniform shape and size, the measured (static) angle of repose is equal to the internal friction angle (an important and widely-used quantity relevant, for example, to the design of hoppers and silos [130–132])—of the material under investigation [133]. Recent work has also shown that, by in essence solving a suitable inverse problem, dynamic angle of repose testers can, in fact, under suitable circumstances be used to back-compute some microscopic quantities.

More generally, the angle of repose is used to assess the 'flowability' of a granular medium, using the classification developed by Carr [134–136], shown in table 4.1.

It is important to note that there are many other ways to measure both static and dynamic friction angles, and many nuances to the manner in which they should be performed which, for brevity, are not given in this brief overview. The interested reader may find further information in the recent review of Al-Hashemi *et al* [137] and the references therein.

Strengths:

- Simple to perform.
- Extremely low cost for basic set-up.

Limitations:

- Only limited, empirical information.
- Measured values sensitive to a number of external factors.

Table 4.1. Flow properties ('flowability') corresponding to different values of the angle of repose, as defined by Carr [134].

Angle of Repose	Flow character
25° – 30°	Excellent
31° – 35°	Good
36° – 40°	Fair—aid not needed
41° – 45°	Passable—may hang up
46° – 55°	Poor—must agitate/vibrate
56° – 65°	Very poor
⩾66°	Very, very poor

Tapped density tests—the Carr index and Hausner ratio

It is well known [138–140] that a 'tapped' granular bed (that is to say, one exposed to a series of short, discrete, vertical excitations or 'taps') will become compacted. That is to say the density of an initially loosely-packed system of particles with volume V_0 will, as it experiences an increasing number of taps, gradually pack more and more densely, and thus reduce in volume, eventually reaching a plateau volume, V_f, known as the final tapped volume (i.e. the volume at which the material cannot be compressed any further by tapping alone) (figure 4.35).

The amount of compaction experienced by a given material, often termed its *compressibility*, is a function of particle size and geometry, surface area, moisture content and cohesivity, as all of these factors may influence the degree to which a given granular material is able to pack. There exist two main parameters used to quantify the compressibility of a given material: the Carr compressibility index:

$$\mathrm{CI} = 100\left(\frac{V_0 - V_f}{V_0}\right), \tag{4.8}$$

and the Hausner ratio:

$$\mathrm{HR} = \frac{V_0}{V_f}. \tag{4.9}$$

The above metrics may also be calculated using density values, namely the material's bulk density (ρ_{bulk} and the final, asymptotic 'tapped density', ρ_{tapped}, possessed by the material when compressed to V_f. The above equations can then be rewritten as:

$$\mathrm{CI} = 100\left(\frac{\rho_{\mathrm{tapped}} - \rho_{\mathrm{bulk}}}{\rho_{\mathrm{tapped}}}\right), \tag{4.10}$$

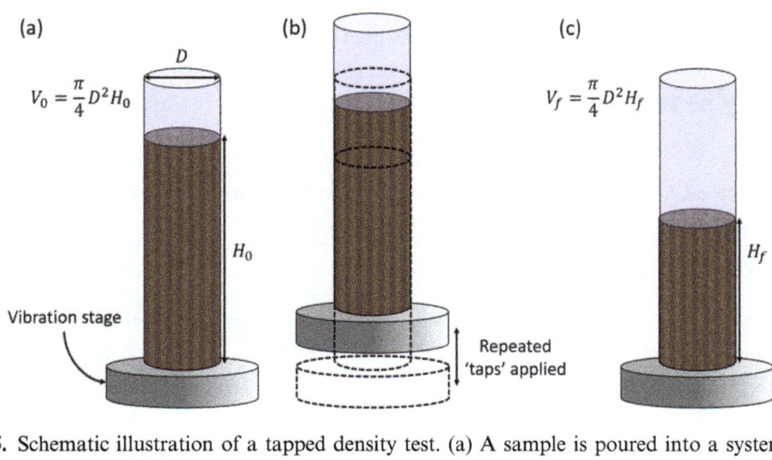

Figure 4.35. Schematic illustration of a tapped density test. (a) A sample is poured into a system to form a loosely-packed bed with initial volume V_0. (b) The sample is exposed to a series of discrete, vertical taps, causing its volume to decrease. (c) Eventually, the system's volume reaches an asymptotic minimum—the final tapped volume, V_f.

and

$$\mathrm{HR} = \frac{\rho_{\mathrm{tapped}}}{\rho_{\mathrm{bulk}}}. \tag{4.11}$$

As with the angle of repose discussed above, the large number of variables which may affect a granular medium's compressibility (and thus its Hausner ratio and compressibility index) mean that it is not a suitable tool for the direct measurement of any of these specific (particle-level) properties individually. Rather, these metrics are typically used—again as with the AoR—simply to assess the (bulk) flowability of a given material. Values of the two parameters corresponding to the same set of flow characteristics as used in table 4.1 can be seen in table 4.2.

As is often the case with these simple measures, the measured values of the above indices are very sensitive to the precise manner in which experiments are conducted, and indeed the details of the systems used. Examples of factors which may influence the measured CI and HR if not properly controlled include the size and geometry of the system used for tapping, the amount of material used, and even the details (e.g. frequency and amplitude) of the applied taps [141, 142].

Strengths:

- Simple to perform.
- Low cost for basic set-up.

Limitations:

- Only limited, empirical information.
- Measured values sensitive to a number of external factors.

Flow through an orifice

In many ways the most direct (and literal) manner in which to test a powder's flowability is simply to measure its flow through an orifice. There exist several approaches through which this may be achieved, but all share a common basic set-up, namely a hopper filled with the particulate material of interest, out of which material may be removed through a series of orifices of differing sizes.

Table 4.2. Flow properties ('flowability') corresponding to different values of the Carr index and Hausner ratio (defined in the main text).

Carr Index (%)	Flow character	Hausner ratio
⩽10	Excellent	1.00–1.11
11–15	Good	1.12–1.18
16–20	Fair—aid not needed	1.19–1.25
21–25	Passable—may hang up	1.26–1.34
26–31	Poor—must agitate/vibrate	1.35–1.45
32–37	Very poor	1.46–1.59
⩾38	Very, very poor	⩾1.60

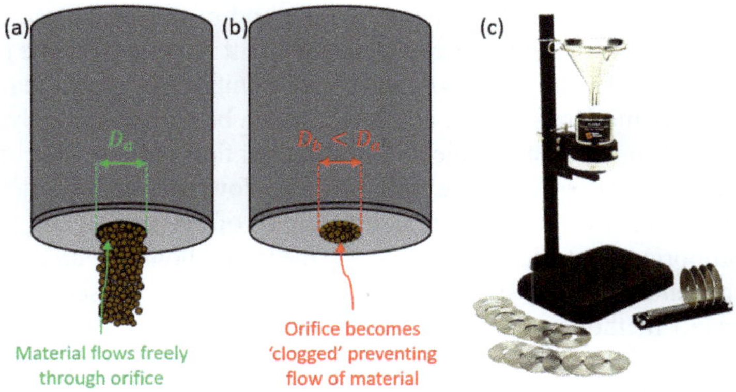

Figure 4.36. Illustration of the principle of operation of a *Flodex* powder flowability tester. Panel (a) shows the case of a disc possessing an orifice size D_a suitably large to allow the consistent flow of material. Panel (b) illustrates the case in which bridging, jamming or cohesion effects [143–145] cause the orifice to become clogged, preventing flow. If the testing system in question possesses no intermediary sizes between D_a and D_b, then a measure of flowability will be determined using D_a. Panel (c) shows an image of a real *Flodex* flow tester (reproduced with permission from Teledyne Hanson Research).

The simplest commonly-used approach, developed in the 80s by the Hanson Research Corporation, tests the ability of a powder to flow through a series of disks, each possessing an orifice of a differing diameter (see figure 4.36(c)). The flowability is characterised by determining the smallest-diameter orifice through which the particulate material of interest will successfully pass three times in three individual tests. Using Hanson's original *Flodex* tester, the flowability is taken as the inverse of this diameter, measured in centimetres [146]. It is also suggested [146] that the internal coefficient of friction, φ, may be determined as:

$$\varphi \leqq \frac{1}{2}\rho_{\text{bulk}} g r, \tag{4.12}$$

where $r = \frac{D}{2}$ is the orifice radius, g the gravitational acceleration and ρ_{bulk}, as above, the (un-tapped) bulk density of the powder being tested.

An alternative approach to determining powder properties from observations of their flow through an orifice is to measure the flow *rate* thereof. A simple and widely-used methodology is the Hall flow method. In this method, a known quantity (specifically 50 g) of the particulate material to be tested is loaded into a funnel with a 60° cone angle and a 2.5 mm diameter outlet[18]. The material is then released, and the time taken for the funnel to empty recorded. From the known mass of material used and the time taken for the material to exit the funnel one may estimate the average flow rate of the material, which can used as an alternative measure of flowability.

[18] Other geometries, such as the Carney funnel or the Gustavsson funnel [147] may also be used if the Hall geometry is unsuitable due, for instance, to the presence of 'ratholing' [148] or other flow issues.

A potential issue with the above-described method of determining a powder flow rate is that the flow rate of a powder can be inconsistent if the head of the powder (that is, the height of powder above the outlet) is not significantly larger than the funnel outlet size. In more modern systems, the flow rate can be more accurately acquired by logging data from an electronic scale as bed material flows into a vessel placed upon said scale. A comparatively precise value of the flow rate can be determined by extracting the gradient of the linear region of a plot of mass versus time.

Another important consideration is that the measured flow rate of a powder is also nonlinearly dependent on the specific size of the orifice used. As such, measurements such as the Hall method, which use only a single orifice size, may provide only a limited picture of a material's flow properties. More modern systems, such as the *GranuFlow* device (GranuTools, Belgium) take measurements of a sample's flow rate, F, through multiple different orifice sizes. Values of F acquired at different orifice diameters, D, are then fitted to the well-known Beverloo equation [149]:

$$F = C_b\sqrt{g}(D - D_{\min})^{\frac{5}{2}} \tag{4.13}$$

to obtain a value of the coefficient C_b, which can be used as another alternative flowability index. The method can also be used as an alternative means of determining the minimum aperture size, $D_{\min}$, discussed at the start of this section.

A general limitation common to all of the methodologies described above is that, as we have repeatedly alluded to, the measurements obtained are sensitive to a variety of factors relating both to the properties of the material studied and the system used to study it. Interestingly, this limitation can also be seen as a strength of these methods, in that one may perform the relevant tests in a cylinder, funnel or hopper whose design and operation are directly representative of the equipment in which the powder being tested will ultimately be used.

Strengths:

- Simple to perform.
- Low cost for basic set-up.
- Test system used can be adapted to suit specific systems/processes of interest.

Limitations:

- Only suitable for free-flowing (i.e. low-cohesion) materials.
- No single, generally-accepted measurement scale [150].
- Only limited, empirical information.
- Measured values sensitive to a number of external factors.

Shear testing

The methods discussed in the previous sub-sections, though simple, provide only a somewhat broad and non-specific characterisation of a given granular material. In this subsection, we discuss measurements conducted using a variety of different 'shear testers', which can provide a wider range of more detailed information regarding a material's properties. These include somewhat more specific (though still bulk) properties such as materials' confined and unconfined yield stresses, their yield loci [151], angles of internal friction and tensile strengths, as well as more abstract properties such as those discussed in the previous sections, including the Jenike flow function [152] and various flowability indices. The price paid for this increased depth of information is a more involved procedure, as will become apparent in the following pages. In this section, we will discuss two commonly-applied methods through which such information may be obtained.

Uniaxial compression test: The simplest form of shear test is the uniaxial compression test which, as its name implies, involves the application of stress (i.e. compression) of a sample of particulate material along a single axis (normally the vertical) (see figure 4.37).

Uniaxial compression testing begins by filling a hollow container (typically a cylinder) of cross-sectional area A with a sample of the material to be tested. The system is then compressed and consolidated by applying a downward vertical force F_1 (and thus normal stress $\sigma_1 = \frac{F_1}{A}$) to the top of the sample.

Following this initial consolidation phase, the container is removed, and a new (lower) normal stress is applied to the newly 'unconfined' system. The stress applied is gradually increased until the specimen fails (i.e. breaks or, more accurately for a particulate system, begins to flow). The stress at which this failure occurs is denoted σ_c, the *unconfined yield strength* of the system. This is an important quantity, as it corresponds to the required stress to achieve incipient flow in this particular material, under the initial consolidation conditions applied.

As is inherent in the above statement, σ_c is not an inherent property of a material, but is also a function of the degree of consolidation it has undergone. As such, uniaxial compression tests are typically performed for a range of different consolidation stress (σ_1) values. These values can then be plotted and joined with a curve to provide a *flow function* curve for the material.

From the above tests, one may also determine a measure of a material's flowability, known as the *Jenike flow factor*, ff_c, by taking the ratio of the consolidation stress σ_1 to the unconfined yield stress, σ_c, i.e.:

$$ff_c = \frac{\sigma_1}{\sigma_c}. \tag{4.14}$$

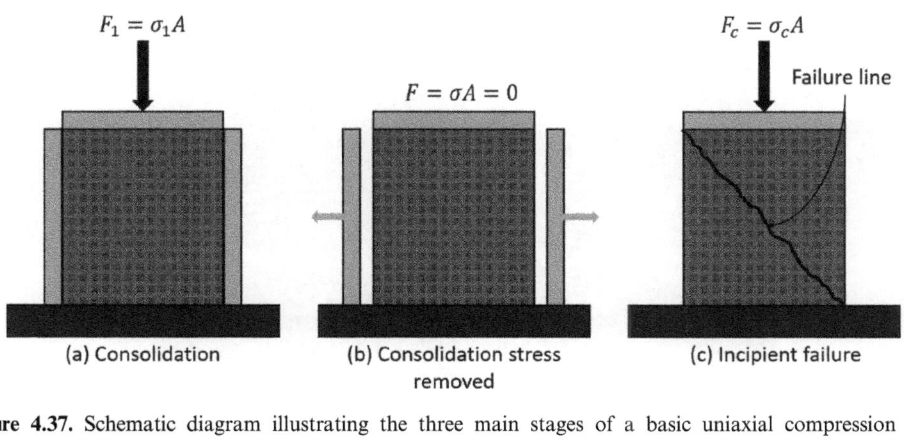

Figure 4.37. Schematic diagram illustrating the three main stages of a basic uniaxial compression test, described in detail in the main text.

The smaller the value of value of ff_c, the greater the unconfined yield stress for a given consolidation stress, i.e. the less freely-flowing (more cohesive) the powder is [132]. Table 4.3 shows how the value of ff_c may be used to classify the flowability of a material.

While simple uniaxial tests such as those described above may certainly provide some valuable information, they nonetheless possess certain significant limitations. Firstly, they are only suitable for relatively coarse granulates—for more fine-grained, cohesive media, measured values of a material's unconfined yield strength using such methods will often be too low [153]. Further, this simple test provides no information regarding, for example, internal friction or wall friction. This more detailed information may, however, be acquired in a similar manner using a slightly more advanced system, known as a shear cell.

Shear cells: There exist many types of shear cell available for powder testing, the first being the translational shear tester developed by Jenike in the 1960s [132, 154–156], and the current 'industry standard' widely considered to be the Schulze ring shear tester (or one of its variants) [156–160]. Both systems, however, operate using a similar methodology [161]; we describe below the comparatively simple case of the Jenike device, pictured schematically in figure 4.38.

Table 4.3. Flow properties ('flowability') corresponding to different values of the Jenike flow factor.

Jenike flow factor, ff_c	Flow character
<1	Not flowing
1–2	Very cohesive
2–4	Cohesive
4–10	Easy-flowing
>10	Free-flowing

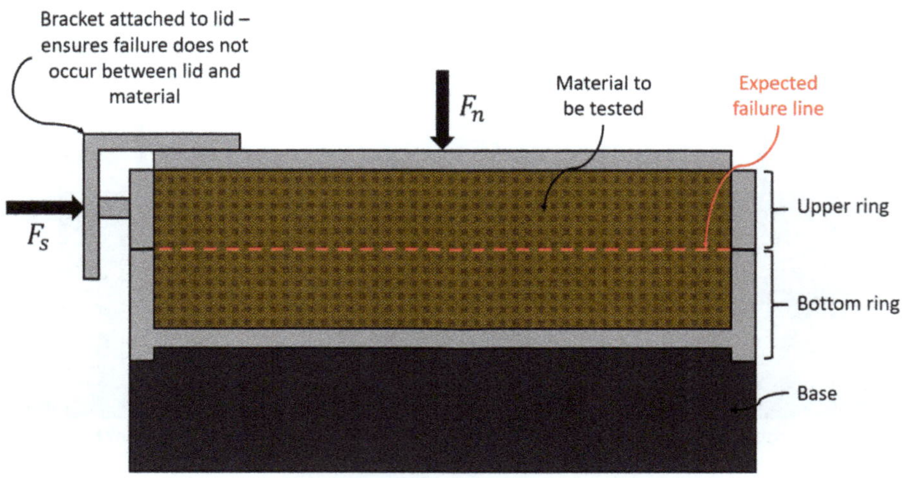

Figure 4.38. Simplified schematic showing the cross-section of a translational shear tester such as that developed by Jenike [132].

One of the great issues when dealing with any bulk solid is that their yield strength—i.e. the necessary stress that one must apply in order to achieve flow—is *strongly history dependent* [162]. For example, a powder previously 'loaded' by a relatively heavy weight will become more consolidated, and thus stronger, than an unloaded material, or indeed one compressed by a smaller weight. Further, a powder compressed by a given load for a long period will typically (though not always) become stronger than another system exposed to the *same load* for a shorter period of time; this is known as *time consolidation* [163, 164]. This being the case, if we wish to consistently measure the intrinsic properties of a material in a reproducible manner, it is vital that before any given test the system receives some form of 'pre-treatment' to ensure a near-identical starting point for each new measurement, i.e. that it's history is effectively erased. In both Jenike and Schulze shear cells, this is achieved through the process of *preshear*.

In order to achieve a reproducible preshear, the shear cell is first loaded with a known weight (F_n in figure 4.38) to produce a well-defined normal stress on the particle bed which we denote σ_{pre}. A shear stress is then applied to the cell by applying a force (F_s in figure 4.38) to the upper ring in a direction perpendicular to the normal force; the shear stress is applied at a constant normal force σ_{pre} (i.e. the weight on the system remains in place throughout this part of the test). The shear stress, τ, experienced by the system initially increases with time (as illustrated in figure 4.39), before the material eventually yields and begins undergoing plastic deformation, characterised by a plateau in the plot of shear stress versus time shown in figure 4.39. At this point, the system is undergoing *steady-state flow*—as the material is sheared further at constant $\sigma = \sigma_{\text{pre}}$, both the shear stress and bulk density remain constant. In this state, the particulate system can be referred to as '*critically consolidated with respect to normal stress,* σ_{pre}'. The shear stress value corresponding to the aforementioned plateau can be plotted on a graph of shear stress versus normal stress to give the

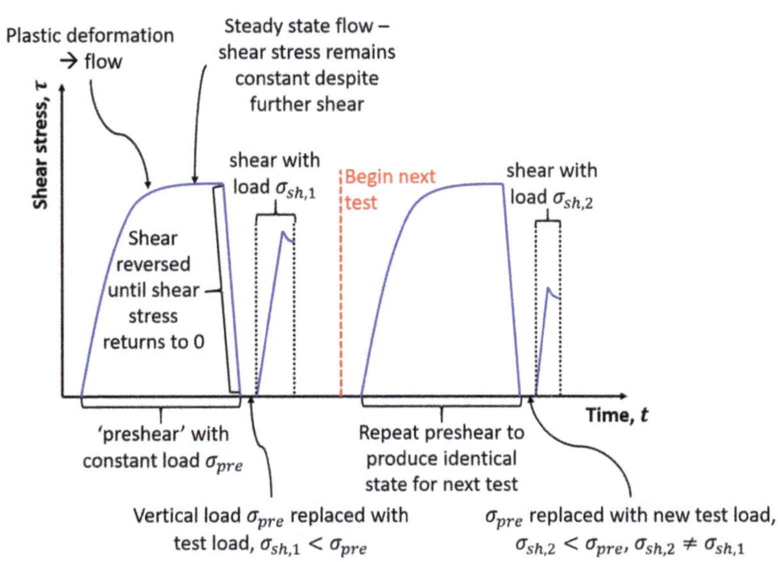

Figure 4.39. Plot of shear stress, τ, versus time, t, for the shear-testing process described in the main text.

'preshear point' for a given shear test set-up, as depicted in figure 4.40. Once the above-described steady state has been reached, the system is 'wound back' until the shear stress, τ, returns to a value of zero, as can again be seen in figure 4.39.

If the above procedure is performed correctly then, hypothetically, one should be able to create an identical initial state for any number of samples of a given particulate material prepared using the same value of σ_{pre}. It is important to note, however, that for this to hold true, the value of σ_{pre} used must be higher than any of the stresses, σ_{sh}, used in the shear tests themselves, which we describe next.

Once the system has been presheared, the consolidated sample is then sheared again, this time under a (decreased) normal load σ_{sh} (i.e. a smaller weight is placed on the top of the system than that used for σ_{pre}). Shear is applied until the sample begins to dilate and thus flow, this failure of the material being characterised as the point at which the gradient of the shear stress versus time plot shown in figure 4.39 switches from positive to negative. This '*shear point*' (or '*point of incipient flow*') can then be plotted against the relevant σ_{sh} value on a graph of τ versus σ. This result represents a single point on the yield locus of the material being tested. By repeating the full pre-shear and shear process (as illustrated in figures 4.39, 4.40) using the same normal loading σ_{pre} for pre-consolidation but different values of the test loading σ_{sh}, one may piece-wise build up a complete curve corresponding to the yield locus, as shown on the right-hand side of figure 4.40.

From a known yield locus, one may determine values of the unconfined yield strength, σ_c, the consolidation stress σ_1 and thus the flow factor ff_c, as well as the angle of internal friction, φ. For brevity, we will not describe here the full process through which this may be achieved; the interested reader may find full details of these processes in reference [161] and references therein. It is important to note here, however, that the values of σ_c and σ_1 achieved from a shear test may not be identical to those obtained from a uniaxial compression test [153, 161, 165, 166].

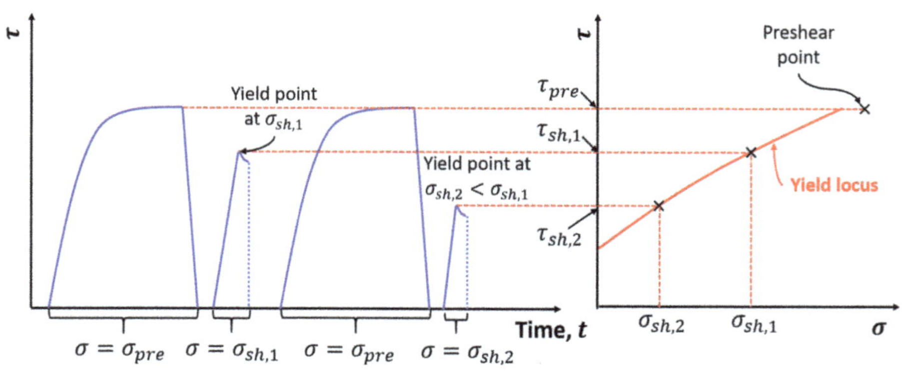

Figure 4.40. Left: Plot of shear stress, τ, versus time, t, for the shear-testing process described in the main text, highlighting the shear points, or points of incipient flow, for the two full shear tests conducted. Right: Plot of shear stress τ versus normal stress σ for the data shown in the left-hand graphic, with each point representing an individual yield point from the tests conducted (note that the preshear point should, by definition, remain the same for all preshear processes). The yield locus can be determined by drawing a line through all measured shear points.

A particular benefit from the use of a shear cell as opposed to a compression test is that one may gain information not just regarding inter-particle friction, but also *wall friction*. In order to investigate the latter, a slightly different set-up of the shear cell system—depicted schematically in figure 4.41—is used. Using this set-up, it is ensured that the applied shear stress cannot cause failure within the powder itself, but only between the lower surface of the powder and the wall material to be tested, which is placed underneath the cell itself.

The process used to acquire a wall yield locus is similar to that utilised to produce a powder yield locus, though with certain key differences (see figure 4.42). Much like the previously-described tests, a yield locus may be built up by performing a series of repeated tests under incrementally varying (typically decreasing [132, 155]) vertical loads F_n, corresponding in this case to normal wall stresses σ_w. In this case, however, the points on the shear versus stress plot (figure 4.42, right-hand panel) correspond not to points of incipient failure, but rather to points (τ_w, σ_w) where the system forms

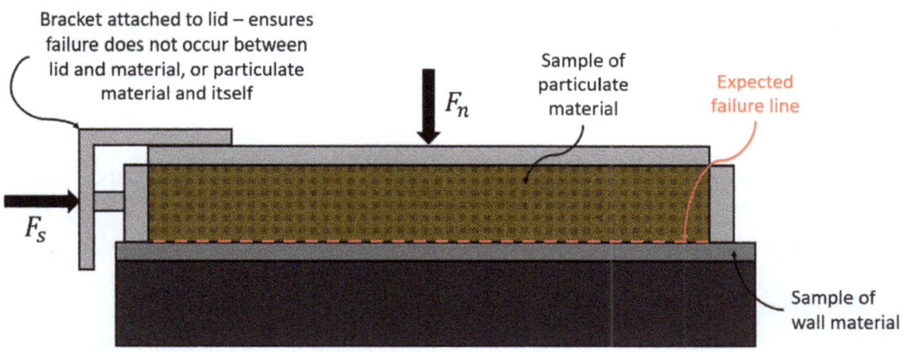

Figure 4.41. Simplified schematic showing how a Jenike shear tester [132] may be used to measure wall friction for a given combination of granular material and wall material.

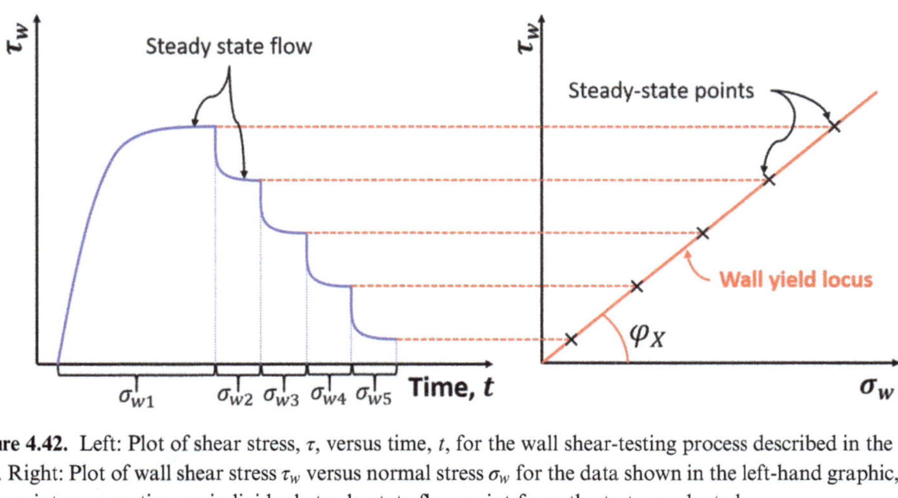

Figure 4.42. Left: Plot of shear stress, τ, versus time, t, for the wall shear-testing process described in the main text. Right: Plot of wall shear stress τ_w versus normal stress σ_w for the data shown in the left-hand graphic, with each point representing an individual steady-state flow point from the tests conducted.

a steady-state flow. For a given applied normal stress σ_w, τ_w, the *steady-state shear stress* corresponds to the necessary shear stress one must apply in order to move a bulk solid continuously across a wall under that applied normal stress.

The diagram shown in figure 4.42 offers a wealth of information regarding the materials tested. In addition to values of the steady-state shear stresses at various loadings and the wall yield locus itself, one may also determine the wall friction angle ϕ_X, given by the angle between the yield locus and the horizontal, and the wall friction coefficient, μ_w, given by $\mu_w = \frac{\tau_w}{\sigma_w}$, i.e. the gradient of the yield locus.

It should be noted that the diagram in figure 4.42 represents the relatively simple case of a non-cohesive material which obeys Coulomb friction, thus providing a straight line passing through the origin in a plot of τ_w versus σ_w. For many materials, however, one may obtain different wall friction coefficients under different loadings, σ_w, resulting in a curved yield locus (see figure 4.43). In this case, the specific wall friction angle and coefficient of wall friction for a given normal stress σ_{wi} may instead be determined from a straight line drawn between the origin and the relevant point (σ_{wi}, τ_{wi}), as depicted in figure 4.43. In addition, cohesive materials will show a positive offset, τ_{ad}, on the vertical axis, corresponding to the shear required to overcome cohesive forces and allow the particulate sample to begin sliding across the wall material.

Strengths:

- Provides relatively detailed and specific information compared to other powder characterisation techniques.
- Suitable for coarse and fine, cohesive and non-cohesive materials.

Limitations:

- Relatively involved operating procedure compared to other powder characterisation techniques.
- Provides only information relevant to quasi-static systems.

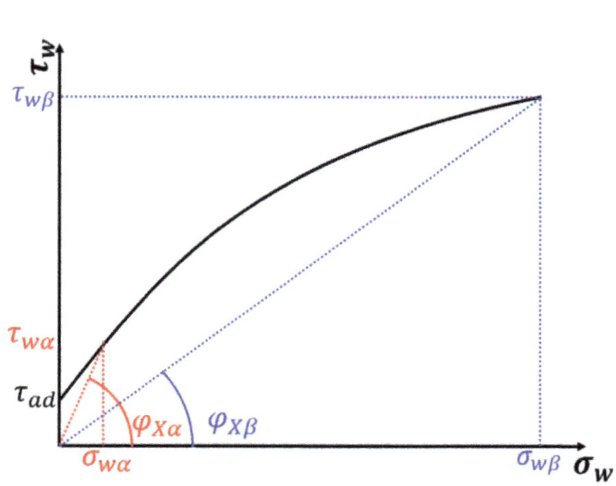

Figure 4.43. Wall yield locus for the case of a cohesive material which does not obey Coulomb friction.

Powder rheometry

As we have seen in the preceding section, shear testers can provide a significant amount of fundamental insight into the properties of particulate media. However, the tests described in the preceding section are all performed under *quasi-static conditions*, where frictional interactions between particles are dominant, and the behaviour of the materials being tested largely solid-like. As such, it is highly questionable whether the insight gained from these tests remains applicable for the case of *dynamic* systems, where particles exist in a more collisional regime [167], and behave en masse in a more fluid-like manner. That is to say, if we take a sample of a given powder and measure its strength and frictional properties using a conventional shear tester, will these measured properties still be meaningful if said powder is then placed in a highly-dynamic system such as a fluidised bed or a high-shear mixer?

In order to address the above issue, the late 1990s and early 2000s saw the development of the first *dynamic testers*, more commonly referred to as *powder rheometers*. Figure 4.44 shows an image of the Freeman Technology FT4, the most widely-used powder rheometer, alongside a schematic diagram illustrating its fundamental function.

It is important to note that there exist several different types of powder rheometer produced by different manufacturers, for example the Stable Microsystems Powder Flow Analyser and the Anton Paar Powder Flow Cell, all of which operate on the same general principle, but with certain non-trivial differences between their individual geometries and methodologies. In this text, we introduce the principle of powder rheometry using as our example the methodology of the Freeman FT4, as

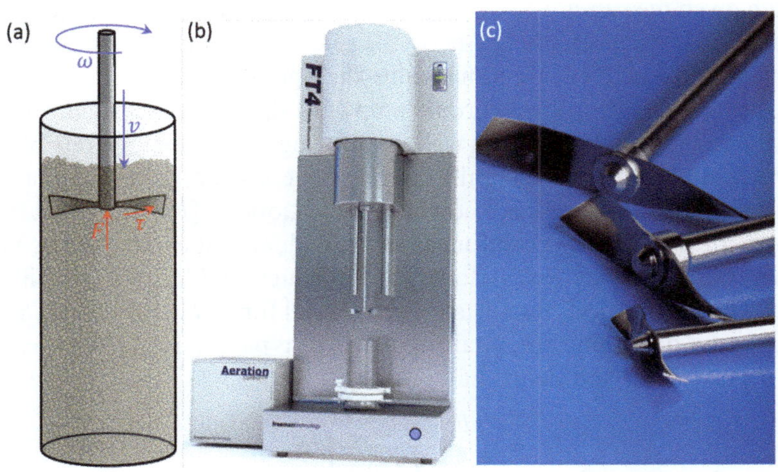

Figure 4.44. Panel (a) shows a simplified schematic of a canonical powder rheometer. When rotating at a constant rate ω and moving downward at a speed v, the powder will exert on the impeller a torque τ and a force F both acting in opposition to the impeller's motion. By measuring the torque and force exerted, one may determine the *'flow energy'* of the material being tested. Panels (b) and (c) show, respectively, an image of a Freeman Technology FT4 powder rheometer and the blades used thereby (images reproduced with the permission of Freeman Technology Ltd).

this is widely considered to be the industry standard for such measurements [168]. The FT4 can in fact perform a range of different tests, including consolidation tests providing information similar to that described in section 4.4, and shear cell and wall friction tests yielding data as described in section 4.4 [169, 170], as well as measurements of permeability [171] and '*aeration*' tests which directly measure a powder's cohesive strength [172]. In this brief overview, however, we focus on the dynamic flow tests unique to powder rheometry.

The basic principle of these dynamic flow tests is relatively simple: a sample of a particulate material to be investigated is placed within a cylindrical container, and a specially designed impeller or 'blade' (see figure 4.44(c)) is moved through the material at a specifically-controlled vertical and angular velocity (see figure 4.44(a)), thus creating a precise flow pattern within the system. As the blade moves through the system, the rheometer measures both the vertical force, F, transmitted to the bottom of the cell and the horizontal torque, τ, acting on the moving blade, providing a quantitative measurement of the resistance experienced in moving through the material, and thus an indication of its flowability (and indeed various other parameters), as we will describe in greater detail below.

The 'twisted' geometry of the blade (figure 4.44(c)) is deliberately designed so as to allow three distinct types of deformation to be imposed on a powder: (i) counter-clockwise rotation and downward motion, providing a relatively strong compaction of the material beneath; (ii) clockwise rotation and upward motion, causing slight compaction above the blade; (iii) clockwise rotation and downward motion, which acts to eliminate any prior compaction, and slightly aerates the bed [173].

Much like the case of shear cells discussed in the previous section, in order to ensure that tests are reproducible, some form of preparation must be performed before each given measurement so as to account for the hysteretic nature of granular media [169, 174, 175]. In the case of the shear cell this was achieved by pre-consolidation. For the FT4, the aim is quite the opposite: the blade is slowly rotated clockwise and moved first downward then upward through the system, with the goal of *removing* any precompaction, as well as ensuring any excess pockets of air within the system are eliminated, generally aiming to construct a loose, homogeneous particle packing. This process is known as 'conditioning' [169].

Once a sample has been conditioned, an individual 'test cycle' may be performed by moving the blade vertically downward through the system and imposing a counter-clockwise rotation. By measuring the axial force, F, and torque, τ discussed above, one may determine the total input work, typically referred to as the '*Flow Energy*', E [169, 174], and defined as:

$$E = \int_0^H \left(\frac{\tau}{R \tan \alpha}\right) dH, \tag{4.15}$$

where H is the penetration depth of the blade into the material, R is the radius of the blade and α is the '*helix angle*', whose definition can be seen in figure 4.45. As is clear from this figure, $\tan \alpha$ in equation (4.15) simply represents the ratio of vertical to rotational velocity. The flow energy in essence represents a given powder's resistance to being made to flow in a dynamic state.

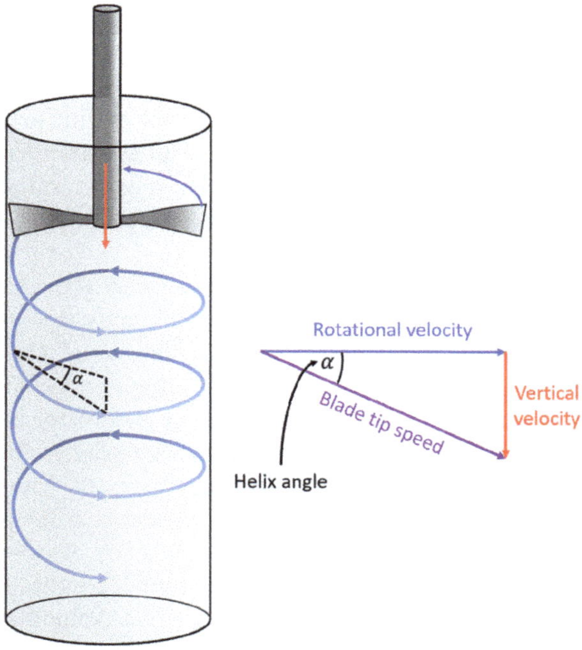

Figure 4.45. Left: Illustration of the counterclockwise helical motion exhibited by the FT4 during a typical test. Right: definition of the helix angle, α.

From this basic parameter, a series of more complex tests, comprising a number of such individual test cycles may be performed in order to extract a range of useful data. An example of such a test is the 'stability test', which involves the performance of seven complete test cycles, each preceded by a conditioning cycle, performed at a constant blade tip speed. From such a test, one may obtain two valuable characterisation parameters: the '*Stability Index*', SI, and the '*Basic Flow Energy*', BFE. The stability index is taken simply as the ratio of the flow energy measured during the last to that of the first test, i.e.

$$\mathrm{SI} = \frac{E_7}{E_1}. \tag{4.16}$$

This index provides insight into how exposure to flow and shear stresses (such as that experienced during the tests) may be expected to alter the (bulk) properties of the material under investigation. A powder that is perfectly 'stable' will possess a stability index of unity, while one whose properties are liable to be affected by flow will possess indices either greater or less than 1. A stability index $\mathrm{SI} > 1$ suggests that a powder may be subject to agglomeration, segregation and/or the accumulation of electrostatic charges, all of which may act to increase a material's resistance to flow. Conversely, a value $\mathrm{SI} < 1$ may indicate a tendency for attrition, fracture or deagglomeration.

The basic flow energy corresponds simply to the flow energy measured in the last of the seven repeated cycles, after which the system will generally have achieved a steady state (i.e. its properties change no further with additional exposure to flow and stress at the level applied in the test cycles). Like many of the indices discussed in this chapter, the BFE carries a dependence on a number of variables, including the size distribution, shape, density and cohesivity of the material in question, as well as external factors such as humidity, making it of limited use for a single-point measurement. However, it can be highly useful as part of a systematic study of a single parameter, for example studying the specific influence of humidity by performing repeated tests of identical samples under different moisture concentrations.

As may be expected from the above discourse, with the exception of certain special cases [169], a low value of BFE (i.e. the blade may easily pass through the material) corresponds to good flowability while a high value (the blade encounters significant resistance) suggests poor flowability (i.e. high cohesivity).

Another standard test performed with the powder rheometer is the '*variable flow rate test*', which can be used to determine the '*Flow Rate Index*', FRI, of a material, which is a measure of a powder's sensitivity to changes in flow rate. In order to determine the FRI, a stability test is first performed, as described above, at the highest tip speed to be used, to ensure that the system's flow properties have reached a steady state. Four subsequent tests, each preceded by a conditioning test, are then conducted at decreasing tip velocities. The FRI is then defined as the ratio of the flow energy measured in the final test (i.e. at minimum tip velocity) to that measured in the first test (i.e. at maximum velocity):

$$\mathrm{FRI} = \frac{E_{v_{\min}}}{E_{v_{\max}}}. \tag{4.17}$$

The FRI can be thought of in a very similar way to the Hausner ratio, with both indices yielding a value of unity for perfectly non-cohesive particulate media, with higher degrees of cohesivity leading to larger values of FRI.

4.5 Numerical simulation

The preceding sections describe a diverse array of (physical) experimental techniques which may, alongside PEPT, be used to extract information from particulate and multiphase systems. In the information age, however, it is important to discuss also the diverse range of *numerical* methods which are increasingly commonly used to study these systems. This is especially true in the present work, as these techniques—which can provide full details regarding every particle or volume of fluid throughout a system simultaneously, but without validation can produce inaccurate or even unphysical results—provide strong synergy with techniques such as PEPT—which cannot directly measure stresses or transients, but whose detailed, three-dimensional, outputs can be used to provide the rigorous validation required by these simulation models. In other words, by using PEPT in conjunction with widely-used simulation methodologies such as CFD and/or DEM (both of which we introduce below), we may achieve the 'best of both worlds'—full, detailed information with arbitrarily high spatial and temporal resolution, which we can fully trust to be quantitatively accurate and grounded in physical reality.

It is the authors' strong conviction that, in the future, this multi-modal, synergistic approach to the study of particulate and multiphase systems will be increasingly adopted by academia and industry alike—indeed, the beginnings of such a trend can already be seen [176–180]. As such, in the following sections we take the time to explore in detail some of the most promising numerical techniques currently available to us.

Though there exist many, diverse numerical modelling techniques which can be applied to particulate and multiphase systems, we focus here on the three most relevant to the present work: the discrete element method (DEM) and computational fluid dynamics (CFD), which are commonly used synergistically with PEPT [176–180] (as will be discussed extensively in chapter 8) and the Monte Carlo method, which plays a central role in the modelling of PEPT detector systems, as discussed in section 5.3.

4.5.1 The discrete element method (DEM)

While CFD (discussed in the next subsection) is indisputably the most commonly-used numerical modelling technique for fluids (and, to a lesser extent, multiphase flows), and indeed the one most likely to be known by a majority of scientists, engineers and industrialists, the discrete element method (DEM) (also known as the discrete *particle* method, DPM [181, 182]) is by far the most widely-adopted method for the modelling of *particulate* systems, with tens of thousands of publications utilising the technique since its inception in the 1970s [183, 184].

A major distinguishing factor between CFD and DEM is that the former is a *continuum method* whereas the latter, as its name implies, is a *discrete* simulation. Where CFD implements the relatively complex Navier–Stokes equations across a continuous domain, DEM instead utilises comparatively simple Newtonian physics,

Figure 4.46. DEM image of a simple rotating drum system.

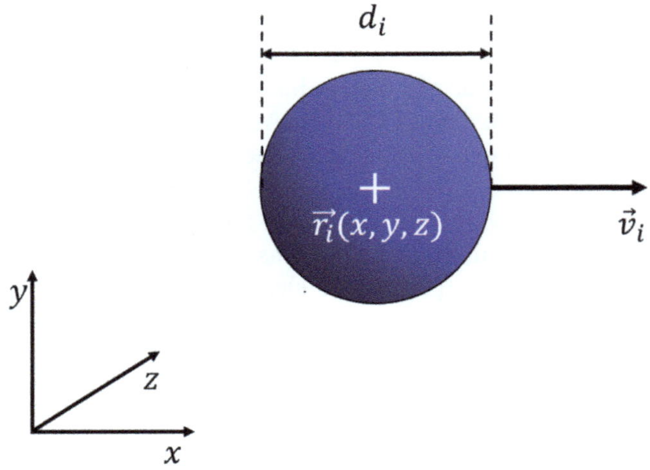

Figure 4.47. Schematic illustrating some of the fundamental properties possessed by a simulated DEM particle. As DEM simulations obey Newton's laws, a particle moving with velocity $\vec{v}_i$ with no external forces acting upon it will continue to travel at $\vec{v}_i$ ad infinitum.

but does so in a piece-wise manner, calculating the forces acting on each individual member of a system of discrete 'elements'—i.e. particles—and using these calculated forces to determine, through the integration of Newton's laws, the motions of these particles in subsequent timesteps.

A one-paragraph description of DEM such as that given above unavoidably makes the technique sound rather complex when, in reality, it is actually very intuitive and conceptually simple—albeit somewhat involved! A slightly slower, step-by-step explanation provides a much clearer picture of the DEM methodology.

Setting up a simulated system (calibration)

One of the great benefits of modelling particulate flows using DEM is that, unlike CFD, which must try to replicate the behaviour of discrete particles in a series of continuum equations, DEM works with a system of discrete objects which can, in essence, be considered as 'real' particles. That is to say, it considers a series of computational objects (see figure 4.47), each of which is assigned a position in three-dimensional space, $\vec{r}_i$, a velocity $\vec{v}_i$, and an acceleration $\vec{a}_i$, as well as a size d_i, a mass m_i, a frictional coefficient μ_i, a restitution coefficient ε_i, and other physical parameters representative of those possessed by real-world solid objects[19].

The modelling of particles as 'real' objects is helpful to us in many ways. Not only does it aid our ability to intuitively understand and visualise what we are modelling (see, as an example of a DEM simulation, figure 4.46), but it also means that these models can be calibrated directly using real experimental data corresponding to direct measurements of the real materials and geometries we wish to model. For example, if we want to model a

[19] DEM can also be used to model rotational motion, i.e. including also the particle's moment of inertia, I_i, angular velocity, ω_i and angular acceleration α. However, for simplicity, we neglect these terms in the elementary initial explanation given here.

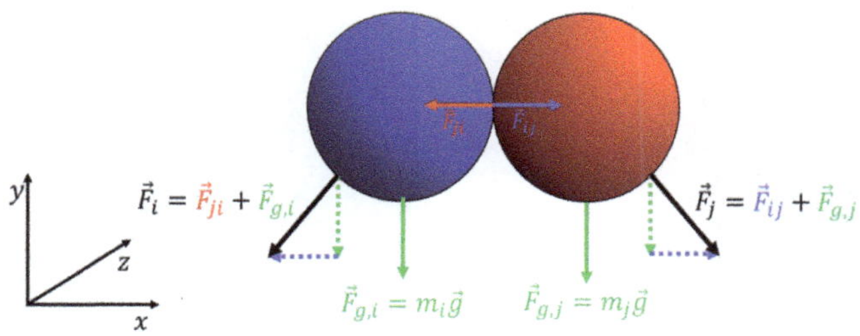

Figure 4.48. Schematic diagram of two colliding particles, illustrating how the net force, $\vec{F}_i$ acting on a given particle i at a given point in time can be determined from the vector sum of contact forces (here due to a collision with a second particle, j) and body forces (arising here due to gravity, g).

system of simple, spherical soda lime glass particles, we can simply look up their frictional coefficients, restitution coefficients, Young's moduli, etc online or in textbooks and implement the values we find directly in our simulations. For less commonly-used materials (or if we are skeptical of the published values!), we can measure the relevant parameters ourselves, and once again simply insert our measured values into our models[20].

Computing forces

For a known system of particles, we must next consider the forces acting on these particles. There are two main types of forces that we must consider: *contact forces*, F_{ji}, felt when a particle i collides with another particle or some other solid object j, and *body forces*, $F_{B,i}$, that act throughout the body of a system, arising for instance due to gravity or drag from air or another immersing fluid [189–191] (figure 4.48).

While most body forces are relatively simple to apply to DEM pseudo-particles—the gravitational force at a given moment in time, for example, can be obtained simply by multiplying the known mass m_i of a particle by the local gravitational acceleration vector $\vec{g}$—contact forces can be somewhat harder to realistically simulate. One of the simplest and best known frameworks for the modelling of inter-particle forces is the *linear spring-dashpot model* [184, 192, 193].

The first (and perhaps most problematic!) assumption of the linear spring-dashpot model[21] is that two colliding particles i and j do not deform, but rather 'overlap' (without changing shape) by a distance δ_{ij} (see figure 4.49) defined as:

$$\delta_{ij} = (a_i + a_j) - \left(\vec{r}_i - \vec{r}_j\right) \cdot \vec{n} \tag{4.18}$$

where $a_\chi = d_\chi/2$ is the radius of particle χ, and $\vec{n} = \vec{n}_{ij} = \frac{\vec{r}_i - \vec{r}_j}{|\vec{r}_i - \vec{r}_j|}$ is the unit normal vector pointing from the centre of particle j to the centre of particle i. As the two particles overlap (see figure 4.49), it is assumed that each experiences a repulsive normal force that is Hookean in nature, increasing linearly with δ_{ij}, much like a

[20] It should be noted that the calibration of DEM models is significantly more complicated than it may sound. Indeed, the development of an optimal method for calibrating DEM simulations is still a subject of significant, active research [185–188]!

[21] It should be noted, that this particular assumption is in fact shared by many, indeed the vast majority of, DEM models.

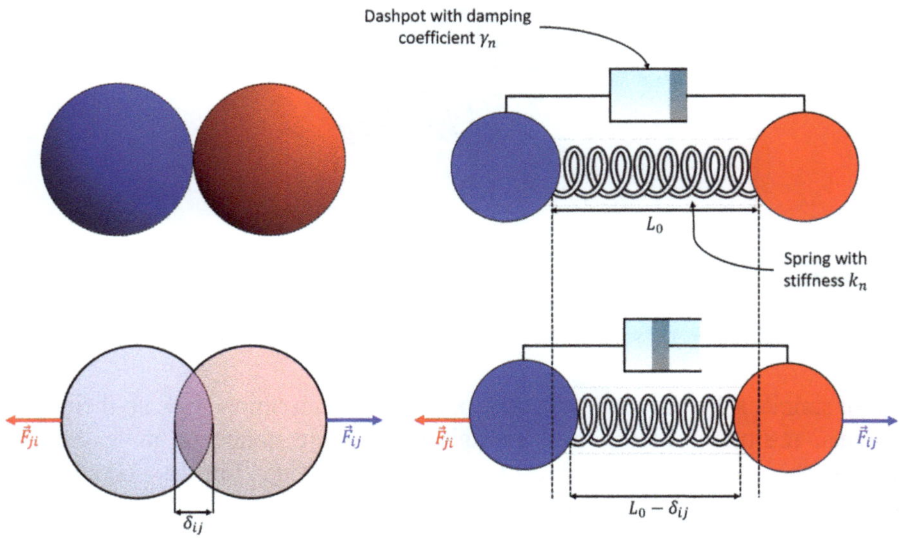

Figure 4.49. *Above:* A pair of particles whose surfaces are just touching (left) and the representation (right) of such a system as two masses attached by a spring and dashpot. *Below:* The same system (left) and its spring-dashpot representation (right) as the two particles are compressed together, creating an overlap δ_{ij} and thus a contact force between the pair.

compressed spring (hence the '*spring*' in the spring-dashpot model). Energy dissipation between the two colliding particles is modelled as linearly proportional to the relative velocity, v_n, between the two colliding particles—much like in the case of a mechanical dashpot (hence the spring-*dashpot* model!). At a given point in time, therefore, the force acting between two interacting particles can be written as:

$$F_{ij,n} = k_n\delta_{ij} + \gamma_n v_n \tag{4.19}$$

where k_n is an appropriately-chosen spring constant, representing the 'stiffness' of the particles (a proxy to their Young's modulus), and γ_n is a viscous damping coefficient. The scalar v_n represents the relative velocity between particles in the above-defined normal direction, and can be defined as [194]:

$$v_n = \frac{d\delta}{dt} = -\vec{v}_{ij} \cdot \vec{n} = -(\vec{v}_i - \vec{v}_i) \cdot \vec{n}. \tag{4.20}$$

The viscous damping coefficient, γ_n, can in fact be related to the (more easily experimentally-measurable) normal coefficient of restitution, ε_n, as:

$$\varepsilon_n = \exp\left(-\frac{\gamma_n t_c}{2m_{ij}}\right), \tag{4.21}$$

where t_c is the contact duration [195, 196] and m_{ij} is the *reduced mass* of the two interacting particles, defined as:

$$m_{ij} = \frac{m_i m_j}{m_i + m_j}. \tag{4.22}$$

The spring constant k_n can also be related to more easily experimentally-measurable parameters. A common method through which this may be achieved is to equate the maximum strain energy in a purely Hertzian [197] contact with that of the contact under consideration, yielding [198]:

$$k = \frac{16}{15} a_{ij}^{\frac{1}{2}} E_{ij} \left(\frac{15 m_{ij} V^2}{16 R_{ij}^{\frac{1}{2}} E_{ij}} \right)^{\frac{1}{5}}. \tag{4.23}$$

In the above, V is the typical impact velocity, a_{ij} is the equivalent radius, defined as

$$\frac{1}{a_{ij}} = \frac{1}{a_i} + \frac{1}{a_j}, \tag{4.24}$$

and E_{ij} is the equivalent Young's modulus, defined as

$$\frac{1}{E_{ij}} = \frac{1 - \nu_i^2}{E_i} + \frac{1 - \nu_j^2}{E_j}, \tag{4.25}$$

with ν_χ the Poisson ratio of species χ.

Though it may not seem it, the linear spring-dashpot model is in fact one of the least complex models commonly implemented in DEM. While this simplicity carries certain advantages in terms of ease of implementation and computational efficiency [199], it also carries certain drawbacks. As with most, if not all, DEM contact models, the underlying assumptions upon which the above-described model are built mean that its validity is limited to certain, specific conditions—in this case relatively large, non-deformable, non-cohesive particles interacting with comparatively large relative velocities. Of course, there exist many other models capable of operating outside this particular set of constraints. For example, Hertzian contact models [197] can more accurately reproduce the behaviours of finer particles, while the well-known JKR model [200]—and many others beside [201–204]—provide extra terms to simulate the effects of cohesion, and hysteretic models such as the Walton–Braun theory [193, 205, 206] and its derivatives [207, 208] can simulate the effects of plastic deformation. Though not crucial to the understanding of this book, the interested reader may find more detailed information regarding DEM force models in our references [192, 197, 209].

Integrating through time

Once we have calculated—using any of the above models—the net force, F, acting on a given particle at a given, discrete point in time, we can use this force to estimate its expected position and velocity at the next relevant point in time. Using Newton's second law, we can determine the net acceleration of our particle as:

$$\vec{a} = \frac{\vec{F}}{m}. \tag{4.26}$$

From this known acceleration, we can then simply integrate to find the relevant velocity, as we know that:

$$\vec{a} = \frac{d\vec{v}}{dt} \rightarrow \vec{v} = \int \vec{a}dt. \tag{4.27}$$

From this we may in turn determine the relevant position as:

$$\vec{v} = \frac{d\vec{r}}{dt} \rightarrow \vec{r} = \int \vec{v}dt. \tag{4.28}$$

In DEM, as with CFD (described below), we predict the dynamics of a system by moving through time in a series of discrete 'steps' of length Δt. If we know the net force, F_{t_n}, acting on a particle a given point in time t_n, we can use this net force, along with its known mass, m, and its known position, $\vec{r}_n$, and velocity, $\vec{r}_n$, at time t_n to predict its position at the next time step t_{n+1}.

This simplest manner in which this may be achieved is using the *Euler method* of numerical integration [210]. Applying the Euler method, equations (4.27) and (4.28) can be re-written as:

$$\vec{v}_{n+1} = \vec{v}_n + \vec{a}_n\Delta t \tag{4.29}$$

and

$$\vec{r}_{n+1} = \vec{r}_n + \vec{v}_n\Delta t \tag{4.30}$$

allowing us to calculate the particle's position and velocity, $\vec{r}_n$ and $\vec{r}_n$, at the next timestep, t_r. At this next timestep, the new forces acting on the particle are then determined, and the whole process repeated. By 'stepping through time' in this manner for all particles in our system we can, piece by piece, simulate the dynamics of a granular system.

The astute reader will notice a number of potential hazards in the above methodology. Chief among these is that the implementation of a finite timestep means that we cannot accurately account for any changes in the forces acting on particles between t_n and t_{n+1}. This can be expected to be particularly problematic in the case of colliding objects. Consider, as an example, a very simple simulation of a high-velocity particle in zero gravity impacting upon a flat wall (see figure 4.50(a)), for which a given timestep t_n falls immediately before the collision. During this timestep, the force acting on the particle is zero, meaning that there is no force retarding its motion.

Using a suitably small timestep, at t_{n+1} the particle will form a slight overlap with the wall, imparting a slight force in the direction opposite to the particle's motion, thus both reducing its velocity at the next timestep, and reducing also the distance that the particle overlaps with the wall. During the next timestep, the particle, now moving more slowly, will penetrate still less far into the wall, causing another slight increase in the opposing force, and reducing its velocity further. This process will continue until the motion of the particle is reversed, and it bounces away from the wall in a physical manner.

If an overly large timestep is used, however, the particle will maintain its initial velocity (and lack of retarding force) for a significant period of time without registering the presence of the wall, and thus maintain this velocity whilst travelling a significant distance. It will therefore appear to 'jump' from outside the wall to penetrating deeply within it (see

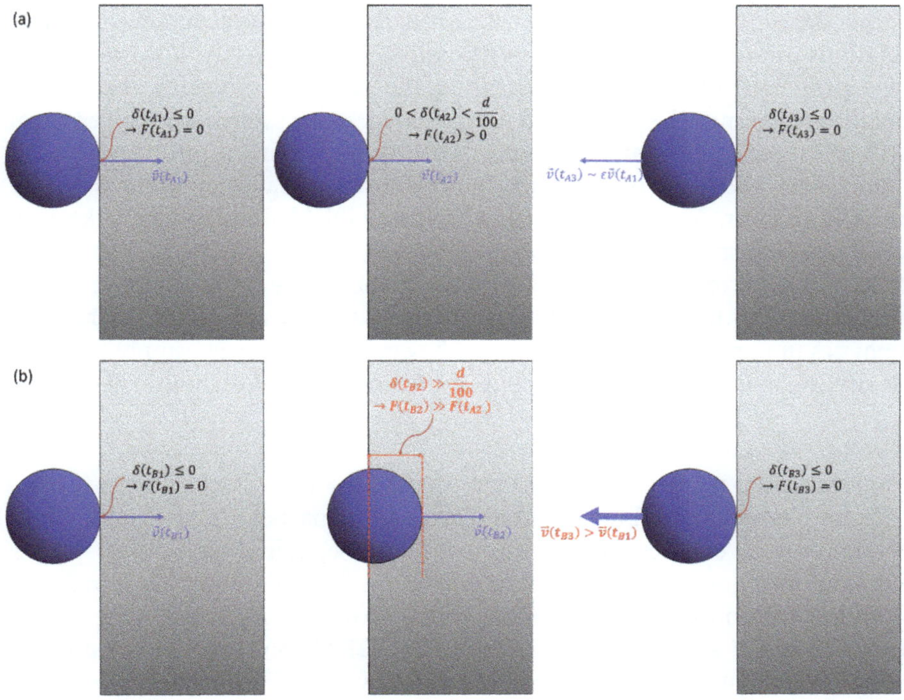

Figure 4.50. Simplified schematic illustrating a particle colliding with a static wall, simulated with (a) a suitably-chosen timestep and (b) an unsuitably long timestep, thus allowing an unreasonably large overlap between wall and particle, and ultimately an unphysical ($|v_f| > |v_i|$) post-collisional velocity. Note that in real DEM simulations collisions will—for a suitably-chosen dt(!)—typically last for several timesteps; in the illustrative images shown above, we can assume that we are 'skipping' some steps in between each sequential image.

figure 4.50(b)). This highly unrealistic behaviour will exert an enormous force on the particle due to the large overlap (see equation (4.19)), launching it back away from the wall at an unphysically high speed, potentially significantly larger than its initial impact velocity!

Though overly-large timesteps will not always necessarily produce behaviours as egregiously erroneous as that described above, this example provides a clear illustration of the problem at hand—and is in fact a commonly-observed problem encountered when developing and testing DEM models of new systems. Nonetheless, if our goal is the *quantitative prediction* of particle dynamics, then even much more minor issues will still cause unacceptable errors.

The use of a suitably small timestep also helps to resolve a related issue in DEM, which is the assumption that the forces exerted by particles, and similarly, the particles' velocities remain constant throughout each given timestep[22]. Clearly, this is not a true representation of the real world but, as $dt \to 0$, becomes an increasingly good approximation of reality.

[22] Note that while this holds exactly true for the Euler method, in other integration methods—such as the Velocity-Verlet scheme mentioned below—more than one value of velocity, acceleration, etc may be considered during a single timestep. Nonetheless, even in this case, the values used remain discrete and finite in number, and thus imperfectly representative of the true case in which position, velocity, force and acceleration values vary continuously in time.

From the above, it is clearly beneficial in terms of the accuracy of our DEM simulations to use the smallest timestep possible. Unfortunately, however, this is not always pragmatically possible: decreasing the timestep increases the number of calculations that must be performed to simulate a given system for a given length of time. For example, if a simulation runs in six minutes using a timestep $\Delta t = 10^{-6}$ s, the same simulation run on the same hardware with a timestep of $\Delta t = 10^{-7}$ s will likely take an hour. As such one must, when using DEM simulations, strike a careful balance between speed and accuracy.

A second issue relating to the integration methodology outlined above is that the Euler method does not conserve energy—an obvious problem if we are trying to model real physics in our systems! This issue can, however, be relatively simply overcome by implementing another integrator, such as the Velocity-Verlet algorithm [210][23].

Detecting collisions

As our simulated particles 'move about' during the above-described process, they will inevitably make and break contacts with other particles and objects. Since the determination of contact forces requires a non-negligible amount of processing power, and we know that when $\delta \leqslant 0$, $F = 0$, it makes sense to only implement the subroutines for contact force calculation if we know a given pair of particles are indeed in contact! As with so many aspects of DEM, however, this is much more complicated than it may seem when taken at face value. The most basic way to check for particle interactions would be to simply compare the positions of each particle in the system to each other particle in the system, checking for each pair if they overlap via equation (4.18). The number of such checks needed for a system, however, varies as the square of the total number, N_p, of particles in the system, meaning that the computation time required similarly scales as $\mathcal{O}(N_p^2)$. In other words, if we increase the number of particles in our system by a factor of 10, the computational time required for collision checking will increase by a factor of 100; if N_p is increased by a factor of 100, the time required will increase by a factor of 10,000 etc. etc.

A much more efficient method, known as the linked-cell method [211–213] reduces computational costs by only checking for particles in the same local 'neighbourhood'. For the simple case of uniformly-sized, spherical particles of diameter d which only interact collisionally (i.e. no long-range forces between particles), this is typically achieved by subdividing the simulation domain into a series of cells of size $L \geqslant d$ (see figure 4.51), and sorting the particles within the system into these cells. For each particle, instead of comparing its position to all other particles in the system, we instead compare only to particles in one of the N_n cells neighbouring the particle's location. For a 2D simulation (as shown in figure 4.51) $N_n = 9$, while $N = 27$ for a full 3D simulation. Even in the extremely

[23] Interestingly, a number of widely-used DEM codes (both open-source and commercial) opt to use the less-accurate Euler model over the more computationally-expensive Velocity-Verlet method so as to maximise simulation speed. As such, when using any such codes the reader is strongly advised to check that the timestep used in any given case is suitably small to ensure that said system does not unphysically gain energy during a simulation.

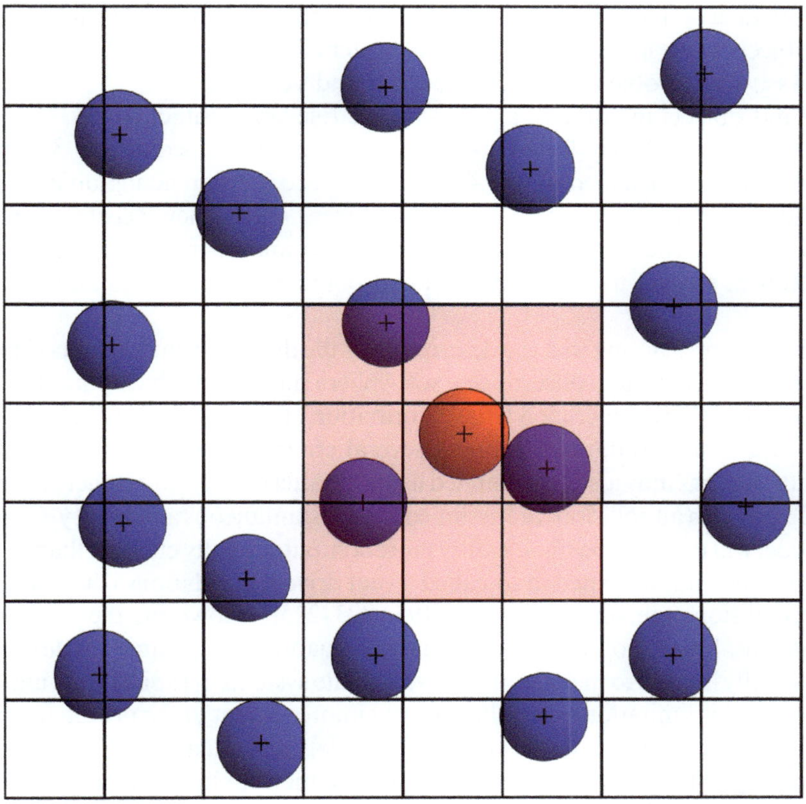

Figure 4.51. Simple schematic illustration of a linked-cell neighbourhood search. The red-shaded cells depict the neighbourhood of the red-coloured particle. In this example, therefore, only the three particles whose centres fall within this region would be considered when attempting to detect collisions with the red particle.

small-scale example shown in figure 4.51, the use of a neighbourhood search algorithm means that only 3—as opposed to 18—collision-detection checks need to be performed for the red particle shown, requiring only 1/6 of the computational time needed were it checked against all other particles. For the much larger systems typically simulated today, where N_p is often of the order of millions, the speed up achieved via the use of a linked-cell algorithm can be orders of magnitude greater still. In addition to the relatively simple linked-cell algorithm, there exist numerous other contact detection algorithms which are better optimised to handle, for example, systems in which particles are less homogeneously distributed [214], or which possess wide size distributions [215]. Though not strictly necessary to our understanding of DEM, the interested reader may find details of some such algorithms in our references [216–219].

Putting it all together

Now that we have a reasonable understanding of all the different aspects of DEM modelling, we can fully appreciate the manner in which they join together to allow us to simulate a dynamic particulate system:

1. Create a system of computational objects representing all particles, walls and objects corresponding to the 'real' system to be simulated.
2. Assign each object an initial position and velocity.
3. Find objects in contact, and determine relevant contact forces.
4. Apply sum of body forces and contact forces determined in step 3 to particles.
5. Use forces calculated in step 4 to compute acceleration acting on each particle.
6. Update particle velocities based on current and/or new acceleration.
7. Update particle positions based on current and/or new velocities.
8. Repeat steps 3 to 7 until desired end-point of simulation is reached.

It is finally worth noting that the description of the discrete element method presented in this section specifically concerns the 'soft-sphere model', in which particle collisions explicitly model (however crudely!) the deformation of colliding spheres as a continuous process with a finite duration spanning several computational time-steps [213, 220]. Particulate systems may also be simulated using the alternative 'hard-sphere model' [195, 221, 222], in which all collisions are assumed to be instantaneous and binary. Clearly, this latter model will typically be significantly more computationally efficient than the former (especially when implemented in so-called 'event driven' simulations where timesteps in between collisions are simply skipped [195, 223–225]). However, the assumption of instantaneous, binary collisions means that—if quantitative accuracy is desired—they are realistically limited to the case of relatively dilute systems of rapidly-moving particles [226]. Due to this limitation—and the ongoing improvement in the processing power of modern computers which allows soft-sphere models to simulate ever-larger systems—the popularity of hard-sphere models has waned somewhat in recent years, and as such they are not discussed in detail here. The interested reader may nonetheless find further details regarding these models and their applications in references [192, 213, 222].

Our discussion also focuses exclusively on the case of spherical particles (the simplest and most computationally-efficient scenario). However, various models also exist for simulating the dynamics of aspherical particles. Commonly used particle shape models include the 'glued sphere' or 'multisphere' method, where—as the name implies—several particles are, to put it crudely, stuck together to form a single, aspherical particle [227, 228], as well as shape representations based on superquadrics [229, 230] or polyhedra [231, 232].

Comparison with PEPT

Advantages:

- Full information regarding all particles within system at all times, including particle contacts and structural information.
- Full stress information, including force networks.
- Arbitrarily high spatial and temporal resolution, limited only by available processor power.

Disadvantages:

- Potential to produce unphysical results
- Outputs unreliable without rigorous calibration and experimental validation.

- Computationally expensive
    - Limitations on system size (i.e. particle number) that can be feasibly simulated.
    - Typically more time-intensive than practical experimentation[24].

4.5.1.1 Interactive example: a simple discrete element method simulation
The notebook files accompanying this section are available in the supplementary data online at https://iopscience.iop.org/book/978-0-7503-3071-8.

A simple discrete element method simulation

The discrete element method (DEM) is a powerful numerical technique that can be used to model the motion of (typically large numbers of) particles—or discrete elements. If correctly calibrated, DEM simulations can reproduce complex granular phenomena *with quantitative accuracy*—and for this reason it is widely used in e.g. the pharmaceutical, mining and food industries.

In a DEM simulation, each particle's state can be fixed using four variables: the spatial position $\vec{X}$ and linear momentum $\vec{P}$ (forming the translational, or Newton's, equations of motion), and the orientation $\vec{R}$ and angular momentum $\vec{L}$ (forming the rotational, or Euler's, equations of motion). Simulating a system of DEM particles is essentially solving a large system of differential equations, where each particle's variable is a single set of equations. Each 'set' of equations yields one differential per system dimension (e.g. the spatial position $\vec{X}$ is split into the r_x, r_y, and r_z coordinates)—so in a 3D system, simulating 100,000 particles involves solving $3 \times 4 \times 100,000 = 1,200,000$ differential equations in total!

In this notebook we will explore a very simple DEM simulation of a single 2D particle bouncing on a flat surface at $y = 0$.

[1]:
```python
# Importing necessary libraries for numerical array manipulation and plotting
import numpy as np
import matplotlib.pyplot as plt

# Set Matplotlib theme (try "darkgrid" too for a more modern look)
import seaborn
seaborn.set_style("whitegrid")
```

For illustrative purposes, only the translational equations of motion will be solved. Following Newton's second law, the rate of change in linear momentum $\vec{P}$ is equal to the sum of forces $\vec{F}$ acting on the object:

$$\frac{d\vec{P}}{dt} = \sum \vec{F} \frac{d\vec{X}}{dt} = \vec{V}.$$

[24] That is to say, specifically, the time required to *run* a given simulation/experiment—this is not taking into account issues regarding the time costs involved in the construction of experimental systems versus the development of a DEM model, or possible issues when performing experimental work relating to the availability of equipment, etc.

Where the linear momentum is equal to the object mass multiplied by its velocity $\vec{P} = m\vec{V}$. In our 2D system, the mass of our particle is constant, and only the gravitational force $F_{y,g} = mg$ and *contact forces* $F_{x,n}$ and $F_{y,n}$ are acting on the particle:

$$\frac{v_x}{dt} = \frac{1}{m}F_{x,n}\frac{v_y}{dt} = \frac{1}{m}F_{y,g} + F_{y,n}\frac{r_x}{dt} = v_x\frac{r_y}{dt} = v_y.$$

Again, for didactic purposes, the *contact force* is defined as a simple spring-dashpot model; for a simple flat surface, only the vertical contact force component is applicable:

$$F_{y,n} = k_n\delta_y + \gamma_n v_y.$$

Where k_n is the normal spring constant and γ_n is the viscous damping coefficient.

Let's define the simulated system's constants:

[2]:
```
# User-defined system constants
g = 9.81                                # Gravitational acceleration, m s^-2
rho = 2500                              # Particle (glass) density, kg m^-3
radius = 10e-3                          # Particle radius, m

spring_constant = 1000                  # Spring constant, N / m
damping_coefficient = 0.6               # Viscous damping coefficient, N / (m/s)

# Initial conditions
rx0 = 0                                 # Particle initial x position, m
ry0 = 1                                 # Particle initial y position, m

vx0 = 0.25                              # Particle initial x velocity, m / s
vy0 = 2                                 # Particle initial y velocity, m / s

# Simulation timespan
t0 = 0                                  # Starting simulation time, s
tf = 4                                  # Final simulation time, s

dt = 1e-5                               # Simulation advancing timestep, s


# Derived system constants
volume = np.pi * 4 / 3 * radius ** 3    # Particle volume, m^3
mass = volume * rho                     # Particle mass, kg

# All initial conditions together in a single system of differential equations
y0 = np.array([vx0, vy0, rx0, ry0])
```

Note that the 2D particle's state vector is simply (v_x, v_y, r_x, r_y).

Define system of differential equations

As it is common in ordinary differential equation (ODE) numerical solving, the ODE is encapsulated in a function of the form `def ode_rhs(t, y):`, where t is the current simulated time and y is a vector containing the current values of the state variables. The function evaluates the right-hand side of the ODE, returning the derivative of the state variables at the given point in space and time—which is then used to predict the system's state at future timesteps.

[3]:
```
# Define the system of differential equations describing the motion of a single
# 2D DEM particle bouncing on a flat surface at y = 0

def spring_dashpot(overlap, relative_velocity):
    '''Return the contact force experienced by a 2D DEM particle using the
    Spring-Dashpot model
    '''
    if overlap <= 0:
        return 0

    return spring_constant * overlap - damping_coefficient * relative_velocity


def ode_rhs(t, y):
    '''Evaluates the system of differential equation's RHS given a particular
    point in time `t` and current state variables' values `y`
    '''

    # Unpack the vector of state variables
    vx_current, vy_current, rx_current, ry_current = y

    # Gravitational force, only vertical component
    fgy = -mass * g

    # Contact force, only vertical component for simple wall at y = 0
    overlap = radius - ry_current
    relative_velocity = vy_current

    fny = spring_dashpot(overlap, relative_velocity)

    # Evaluate ODE RHS
    vx_diff = 0
    vy_diff = 1 / mass * (fgy + fny)

    rx_diff = vx_current
    ry_diff = vy_current

    rhs = np.array([vx_diff, vy_diff, rx_diff, ry_diff])
    return rhs
```

Numerically solving the ODE: Euler method

Now that we have a system of ordinary differential equations (ODEs) encapsulated in a Python function, we can solve it using a numerical integration scheme. In simple terms, a *numerical integrator*—in the context of solving ODEs—approximates a function given a set of initial conditions by only evaluating its derivative (i.e. the right-hand side of a differential). For a DEM system, predicting future states through a numerical integration scheme is equivalent to moving the simulation forward in time.

The most basic of integrators, the Euler method, simply approximates the differential's infinitesimal difference dt (i.e. derivative) as a finite difference Δt;

$$\frac{df}{dt} = y(t, f) \Longrightarrow \frac{\Delta f}{\Delta t} \approx y(t, f) \Longrightarrow f_{n+1} = f_n + \Delta t \ y_n.$$

Naturally, the smaller the finite difference, the better the approximation. However, the Euler method almost always 'overshoots' or 'undershoots' the function being approximated—and indeed, better techniques exist. However, through its simplicity, the Euler method is ideal as a first example:

[4]:
```
# Numerically solve (i.e. approximate) the evolution of a 2D DEM particle using
# the Euler method

# Keeping track of where our simulation is: current time t and state variables y
t_current = t0
y_current = y0

# Save the simulated times and state variables in growable lists
t = [t_current]
y = [y_current]

# Incrementally advance the simulation in time by predicting its future (n + 1)
# state based on the current one (n) and its derivative
while t_current < tf:
    y_next = y_current + dt * ode_rhs(t_current, y_current)     # Euler method
    t_next = t_current + dt

    # At the end of the loop, the "next" (n + 1) velocities / positions become
    # the current ones (n)
    y_current = y_next
    t_current = t_next

    # Save velocities / positions in our lists
    t.append(t_current)
    y.append(y_current)

print(f"Simulated the system using {len(t)} steps")

# Transform the lists into NumPy arrays
t = np.array(t)
y = np.array(y)

# Extract the particle XY positions from the state vector
rx = y[:, 2]
ry = y[:, 3]


# Plotting:
#   rx(t) vs. t       - normal plot, dependent var rx(t) vs. independent var t
#   ry(t) vs. t       - normal plot, dependent var ry(t) vs. independent var t
#   ry(t) vs. rx(t)   - phase plot, dependent var ry(t) vs. dependent var rx(t)
fig, axes = plt.subplots(1, 3)
fig.set_size_inches(18, 6)
```

```
axes[1].plot(t, ry)
axes[2].plot(rx, ry)

axes[0].set(xlabel="$t$ ($s$)", ylabel="$r_x(t)$ ($m$)")
axes[1].set(xlabel="$t$ ($s$)", ylabel="$r_y(t)$ ($m$)")
axes[2].set(xlabel="$r_x(t)$ ($m$)", ylabel="$r_y(t)$ ($m$)", aspect="equal");
```

```
Simulated the system using 400001 steps
```

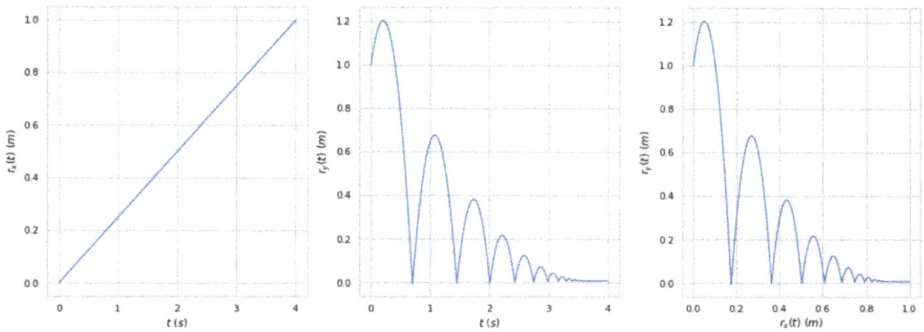

Numerically solving the ODE: adaptive integration schemes

The choice of integration scheme directly affects the accuracy of the simulation and the number of timesteps required (and hence the simulation running time)—the two most important things when simulating extremely large physical systems! However, being such important tools, numerical solvers are most often implemented in mature, sophisticated libraries; one of them, LSODA, automatically switches between multiple integration algorithms depending on how ‘easily predictable’ (or ‘non-stiff’ in mathematical parlance) the function is—which, in our case, depends on the complexity of the force models involved. The LSODA algorithm can be used through the user-friendly `solve_ivp` numerical system of ODEs solver from `SciPy`:

```
[5]: # Numerically solve (i.e. approximate) the evolution of a 2D DEM particle using
     # the LSODA integrator from SciPy's `solve_ivp` subroutine
     from scipy.integrate import solve_ivp

     # Time span of the simulation; save the system state at `t_eval` steps
     t_span = [t0, tf]
     t_eval = np.linspace(t_span[0], t_span[1], 1000)

     # Choose numerical integrator: LSODA, RK45, RK23, DOP853, Radau, BDF
     method = "LSODA"

     # Solve system of differential equations
```

```
print(f"Simulated the system using {sol.nfev} steps")

# Extract the saved timesteps and particle XY positions from the state vector
t = sol.t
rx = sol.y[2, :]
ry = sol.y[3, :]

# Plotting:
#    rx(t) vs. t        - normal plot, dependent var rx(t) vs. independent var t
#    ry(t) vs. t        - normal plot, dependent var ry(t) vs. independent var t
#    ry(t) vs. rx(t)    - phase plot, dependent var ry(t) vs. dependent var rx(t)
fig, axes = plt.subplots(1, 3)
fig.set_size_inches(18, 6)

axes[0].plot(t, rx)
axes[1].plot(t, ry)
axes[2].plot(rx, ry)

axes[0].set(xlabel="$t$ ($s$)", ylabel="$r_x(t)$ ($m$)")
axes[1].set(xlabel="$t$ ($s$)", ylabel="$r_y(t)$ ($m$)")
axes[2].set(xlabel="$r_x(t)$ ($m$)", ylabel="$r_y(t)$ ($m$)", aspect="equal");
```

```
Simulated the system using 2646 steps
```

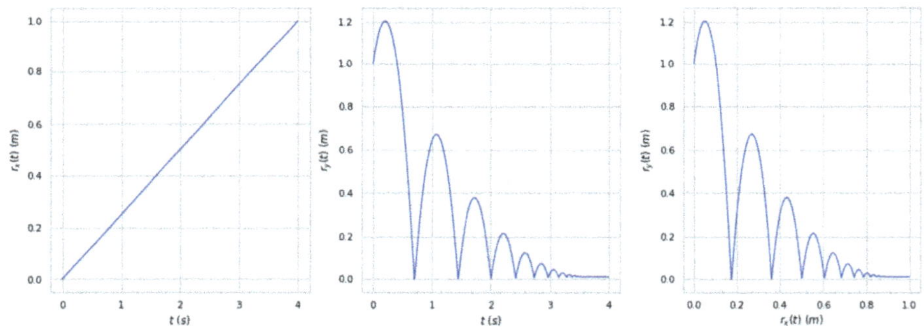

Notice that we could simulate 4 seconds of a 2D DEM particle bouncing on a flat surface in 2646 steps (or `ode_rhs` function evaluations); using the Euler method with equally-spaced timesteps, this would have been equivalent to using a timestep of 4/2646 = 0.0015 s; change the `dt` variable in the second code cell to this value and re-run the Euler method simulation to see the importance of a good numerical integration scheme!

Complementarity with PEPT

It is immediately noticeable from the preceding comparison that the advantages offered by DEM simulation go beyond any other technique (experimental or numerical) we have seen thus far—including CFD, which despite providing detailed, whole-field information, cannot resolve particle-level interactions (and therefore structures such as force chains), nor indeed many other crucial microscopic behaviours exhibited by particulate or particle-laden systems.

The disadvantages of DEM, however, are severe; indeed, used in isolation (i.e. with no experimental calibration or validation) the results of a DEM simulations simply cannot be trusted [199]. Nonetheless, the two most serious of the three disadvantages listed above can be eliminated through the application of rigorous and detailed experimental calibration and validation[25].

In light of the above, there exists the potential for an extraordinary synergy between DEM and PEPT: the former offers complete detail regarding the motion of every individual element of a particulate system at all times, yet may produce inaccurate or unphysical results; the latter is experimental and thus inherently provides physical results, and although it does not provide *complete* data, nonetheless yields a wealth of detailed, high-resolution, three-dimensional information which can be used for the in-depth calibration and validation of DEM. In other words, by combining both techniques, one may reliably use PEPT-validated simulations to 'fill in the gaps'—i.e. to gain information regarding, for example, particle collisions or transient behaviours which cannot be extracted from PEPT data alone, or to fill in regions of parameter space which cannot be directly explored in experiment. This approach, illustrated schematically in figure 4.52, has already proven highly successful (see chapter 8), and as both PEPT and DEM evolve further, the synergistic application of these techniques is likely to prove increasingly valuable in the future.

A wide variety of examples of cases where PEPT and DEM have been combined to great effect in the study of diverse industrial and scientific systems can be found in chapter 8.

4.5.2 Computational fluid dynamics (CFD)

As mentioned in 4.5.1 computational fluid dynamics (CFD) is the most commonly used technique when it comes to the numerical modelling of fluid or multiphase flows. CFD differs significantly from DEM in the sense that the frame of reference is *Eulerian*, meaning that we are interested in how a fluid evolves in a control volume (a fancy way of saying a volume of interest) rather than tracking the particles themselves (recall our discussion of PTV and PIV from section 4.1). The backbone of this approach is the application of the Navier–Stokes equation (NSE) and continuity equation [235]. As noted previously, DEM is more intuitive to set up, and the underlying equations are *significantly* less intimidating than those used in CFD, so what benefits does it have over DEM? As we will explain in this section, the main benefit is the reduced computing time for this method

[25] The third will naturally become less and less of a constraint as time progresses, with computer processing power continuing to improve and researchers developing ever-more efficient DEM algorithms and methods by which to apply them [230, 233, 234].

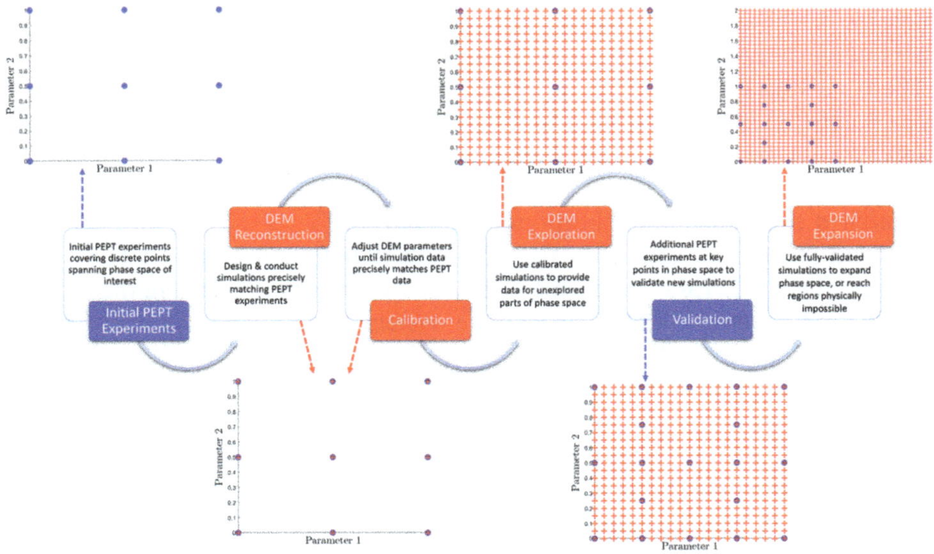

Figure 4.52. Flow diagram showing how PEPT and DEM can be used synergistically to produce comprehensive data sets of reliable, detailed information across a wide parameter space. The images represent the regions of a given, arbitrary parameter space explored experimentally (blue circles) and numerically (red crosses).

compared to DEM. In short, this is because CFD supports implicit and variable timestep schemes as well as the explicit schemes that are used within DEM, which allow for much larger timesteps than are permissible for DEM as collisions are not monitored in CFD [236]. The previous sentence will make more sense as we progress through this section. Additionally, the 'effort' required in CFD is generating the geometry and inputting physical constants that are more easily obtained than for DEM. One can use a fitted model from a rheometer-produced flow curve and simple density measurements to input all required physical data into a CFD simulation, which contrasts with the significant effort required to estimate DEM model parameters[26].

In this section, we will outline the formulation of the NSE, for compressible and incompressible Newtonian fluids. We will then cover solutions to CFD problems in laminar and turbulent flows. Setting up a CFD problem will then be outlined, considering how to numerically evaluate the problem and represent the simulation domain. Extensions to CFD such as the two-fluid model and free surface problems will also be (briefly) considered.

Before we begin, a forewarning to the reader: while in previous sections we have managed to keep the equations to a minimum, the theory and mathematics of CFD are inextricably linked, and as such this section is, unavoidably, significantly heavier on the maths.

[26] It is worth noting here that while this is certainly the case when modelling fluids (and some multiphase systems) with CFD, as predominantly discussed in this section, it is decidedly not the case when attempting to model the motion of particulate media.

Theory

There are many variants of the NSE, depending on the assumptions and problem scope during derivation, such as the body force terms to use, or whether the fluid being modelled is incompressible. The reason for the many variants is because of the complexity of the NSE to solve—as with DEM (and indeed many other problems!), the problem can be solved more easily by making broad assumptions, but said assumptions simultaneously make the result less realistic, or at least render it applicable to a less broad range of systems. In fact, it is still not known if in three dimensions the NSE always has a physically realistic solution[27]! Regardless of the form, the core assumptions that underpin the NSE are [235]:

1. The fluid is a *continuum*.
2. The fields of interest (e.g. pressure, velocity) are *differentiable*.

Let us unpick these assumptions to give some context. The continuum assumption implies that the fluid is not a collection of particles, but rather a *continuous medium*. Of course, this is technically not true, given that a fluid is made of atoms[28]; however, at the correct length-scale then this means that the fluid has field properties at infinitely small points as it is a continuous entity. This contrasts with DEM where the particles are discrete, as are properties such as velocity and pressure—if you were to examine the space between two DEM particles, it would be an empty void which is not the case for the NSE. The differentiable assumption is a mathematical necessity, in essence this assumption is saying that there are no singularities in the fields of interest (which in most cases, at least for simulations of everyday, earthbound systems, is typically not unreasonable). This means that they can be differentiated across the entire system domain. This will make more sense as we outline the NSE below. It is important to note that the derivation below is not rigorous as this is not a fluid mechanics textbook, but sufficient detail is included to understand the significance of the terms in the NSE and how these subsequently apply to CFD[29].

The NSE is, in essence, a momentum continuity equation for a continuum of fluid which in itself is a fancy way of using the fact that momentum is conserved to construct an equation to describe the forces acting upon the fluid. For a fluid, if working with control volumes then momentum ends up defined as momentum per volume. This is useful, as this allows the geometry to be problem specified instead of having numerous formulations for specific geometries! We start with the Cauchy momentum equation,

$$\frac{\partial(\rho\vec{u})}{\partial t} + \nabla \cdot (\rho\vec{u}\vec{u}) = \nabla \cdot \vec{\vec{\sigma}} + \rho\vec{F} \tag{4.31}$$

[27] For those who desire a challenge, a $1 million prize awaits those that can prove that the Navier–Stokes equation in three dimensions has smooth and physically reasonable solutions [237].

[28] This becomes even less true when applying, for example, the kinetic theory of granular flow, where we attempt to model particulate media as continua.

[29] For the reader who wishes to gain a more in depth understanding of the theory behind NSE, they are referred to a transport phenomena/fluid mechanics transport textbook such as [235].

where ρ is the fluid density, $\vec{u}$ is the fluid velocity, $\vec{\vec{\sigma}}$ is the stress tensor and $\vec{F}$ represents the total body forces acting upon the fluid (such as gravity). The symbol ∇ represents the divergence operator, which represents the field gradient based on its orientation. So, let's understand what each term in equation (4.31) means [238]:

- The momentum accumulation term ($\frac{\partial \vec{u}}{\partial t}$).
- The momentum advection term ($\nabla \cdot (\rho \vec{u}\vec{u})$).
- The stress tensor, $\vec{\vec{\sigma}}$, is a 3 ×3 matrix which considers normal and shear forces acting on a parcel of fluid's *surface.*
  - The diagonal elements represent normal forces (σ).
  - The non-diagonal elements represent shear forces (τ).
- The body force term, $\vec{F}$, considers body forces acting on a parcel of fluid's *volume.*

Before proceeding further, the momentum continuity equation is complemented by the mass continuity equation,

$$\frac{\partial \rho}{\partial t} + \nabla \cdot (\rho \vec{u}) = 0 \tag{4.32}$$

this equation enforces conservation of mass within a system. Using the product rule on the left hand side of equation (4.31),

$$\frac{\partial(\rho \vec{u})}{\partial t} + \nabla \cdot (\rho \vec{u}\vec{u}) = \rho \frac{\partial(\vec{u})}{\partial t} + \vec{u}\frac{\partial(\rho)}{\partial t} + \rho \vec{u} \cdot \nabla \vec{u} + \rho \vec{u} \nabla \cdot \vec{u} + \vec{u}\vec{u} \cdot \nabla \rho \tag{4.33}$$

$$= \vec{u}\left(\frac{\partial(\rho)}{\partial t} + \vec{u} \cdot \nabla \rho + \rho \nabla \cdot \vec{u}\right) + \rho\left(\frac{\partial(\vec{u})}{\partial t} + \vec{u} \cdot \nabla \vec{u}\right) \tag{4.34}$$

$$= \vec{u}\left(\frac{\partial(\rho)}{\partial t} + \nabla \cdot (\rho \vec{u})\right) + \rho\left(\frac{\partial(\vec{u})}{\partial t} + \vec{u} \cdot \nabla \vec{u}\right) \tag{4.35}$$

we see that equation (4.32) has appeared in the first bracketed term on the right hand side thus the value of this expression is zero, meaning that for either a compressible or incompressible fluid,

$$\frac{\partial(\rho \vec{u})}{\partial t} + \nabla \cdot (\rho \vec{u}\vec{u}) = \rho\left(\frac{\partial(\vec{u})}{\partial t} + \vec{u} \cdot \nabla \vec{u}\right) = \rho \frac{D\vec{u}}{Dt} \tag{4.36}$$

where the quantity $\rho \frac{D\vec{u}}{Dt}$ is known as the material derivative.

Obviously, the form of equation (4.31) isn't useful on its own—how do we define the stress tensor and body force terms?

If we write out the stress tensor, using Cartesian coordinates for simplicity,

$$\vec{\vec{\sigma}} = \begin{pmatrix} \sigma_{xx} & \tau_{xy} & \tau_{xz} \\ \tau_{yx} & \sigma_{yy} & \tau_{yz} \\ \tau_{zx} & \tau_{zy} & \sigma_{zz} \end{pmatrix} \tag{4.37}$$

where the first letter in the subscript denotes the axis perpendicular to the face of the control volume and the second denotes the direction that the force is acting. If we consider that pressure within continuum mechanics is defined as $P = -\frac{1}{3}(\sigma_{xx} + \sigma_{yy} + \sigma_{zz})$ then the stress tensor can be rewritten as shown below:

$$\vec{\vec{\sigma}} = -\begin{pmatrix} P & 0 & 0 \\ 0 & P & 0 \\ 0 & 0 & P \end{pmatrix} + \begin{pmatrix} \sigma_{xx} + P & \tau_{xy} & \tau_{xz} \\ \tau_{yx} & \sigma_{yy} + P & \tau_{yz} \\ \tau_{zx} & \tau_{zy} & \sigma_{zz} + P \end{pmatrix} = -p\vec{\vec{I}} + \vec{\vec{\tau}} \tag{4.38}$$

where $\vec{\vec{I}}$ is the identity matrix. This is more useful as pressure is of interest for a fluid, but we still have an unknown term, $\vec{\vec{\tau}}$ known as the deviatoric stress tensor [235]. What is left is to specify behaviour for the deviatoric stress tensor and the pressure. We shall focus on a Newtonian fluid that is compressible (this is more likely for gases), before considering the incompressible case (this is more likely for fluids). The Mach number can inform us as to how compressible a system is—for a large Mach number, a compressible formulation must be used[30] [238].

For a Newtonian fluid, it is assumed [238]:

- The shear stress tensor is linearly related to the strain rate tensor, $\vec{\vec{e}}$, thus $\vec{\vec{\tau}} \propto \vec{\vec{e}}$.
  - This is known as Newton's law of viscosity.
- The fluid is isotropic and at rest satisfies $\nabla \cdot \vec{\vec{\tau}} = 0$.
  - This is so that the only stress present at rest is hydrostatic pressure.

This leads to rewriting the deviatoric stress tensor (in Cartesian coordinates) as,

$$\vec{\vec{\tau}} = \mu(\nabla\vec{u} + (\nabla\vec{u})^{\mathsf{T}})) + \nabla \cdot \left[\lambda(\nabla \cdot \vec{u})\vec{\vec{I}}\right] \tag{4.39}$$

$$= 2\vec{\vec{e}} + \nabla \cdot \left[\lambda(\nabla \cdot \vec{u})\vec{\vec{I}}\right] \tag{4.40}$$

[30] If you are able to, it is imperative that you check how compressible the fluid that you are trying to simulate is. If the fluid shows noticeable density variations, then an incompressible approach will incorrectly predict the velocities and pressures for your fluid!

$$=\mu\begin{pmatrix} 2\frac{\partial u_x}{\partial x}+\lambda\frac{\partial u_x}{\partial x} & \frac{\partial u_x}{\partial y}+\frac{\partial u_y}{\partial x} & \frac{\partial u_x}{\partial z}+\frac{\partial u_z}{\partial x} \\ \frac{\partial u_y}{\partial x}+\frac{\partial u_x}{\partial y} & 2\frac{\partial u_y}{\partial y}+\lambda\frac{\partial u_y}{\partial y} & \frac{\partial u_y}{\partial z}+\frac{\partial u_z}{\partial y} \\ \frac{\partial u_z}{\partial x}+\frac{\partial u_x}{\partial z} & \frac{\partial u_z}{\partial y}+\frac{\partial u_z}{\partial y} & 2\frac{\partial u_z}{\partial z}+\lambda\frac{\partial u_z}{\partial z} \end{pmatrix} \tag{4.41}$$

where μ and λ are the dynamic and volume viscosities of the fluid respectively[31]. So, for a compressible, Newtonian, fluid we have,

$$\rho\frac{D\vec{u}}{Dt} = -\nabla P + \nabla\cdot[\mu(\nabla\vec{u} + (\nabla\vec{u})^{\mathrm{T}})] + \nabla\cdot\left[\lambda(\nabla\cdot\vec{u})\vec{I}\right] + \rho\vec{F} \tag{4.42}$$

as the fluid is compressible, then an appropriate equation of state (such as the ideal gas law) is required alongside an equation for energy conservation. This is because we must provide a closure for pressure and density variations, which in turn introduces a temperature dependence. However, for an incompressible fluid, these are not required and further simplifications can be made. The assumption of an incompressible fluid leads to three main assumptions [239]:

1. Dynamic viscosity is constant as it is no longer affected by pressure.
2. Volume viscosity is equal to 0 as there is no volume change to the fluid.
3. The mass continuity equation (equation (4.32)) is simplified to $\nabla\cdot\vec{u} = 0$.

This removes the non-diagonal terms of the $\nabla\cdot\mu(\nabla\vec{u} + (\nabla\vec{u})^{\mathrm{T}})$ term in equation (4.42) when expressed as the strain rate tensor, and as $\lambda = 0$,

$$\rho\frac{D\vec{u}}{Dt} = -\nabla P + \mu\nabla^2\vec{u} + \rho\vec{F} \tag{4.43}$$

or this can be thought of as:

$$\begin{gathered}\text{momentum accumulation} + \text{momentum advection} = \\ \text{pressure gradient} + \text{momentum diffusion} + \text{body forces}\end{gathered} \tag{4.44}$$

A simple analytical example

So, armed with the NSE we can now begin to understand how this equation can be used to model the motion of fluids. For simple geometries in the laminar regime, often there is the possibility of an analytical solution. Consider the case of fully

[31] Volume viscosity is mostly insignificant except in specific applications such as acoustics, modelling shockwaves or complex multiphase simulations with a liquid phase and numerous gas bubbles—if it *must* be used, then typically the value is taken as $-\frac{2}{3}\mu$ [235].

developed flow in a pipe of length, L, and radius, R, that is in the laminar regime and working in cylindrical coordinates and assuming the following:

1. The flow is steady state ($\frac{\partial \vec{u}_z}{\partial t} = 0$).
2. The flow only moves in the z-direction ($\vec{u}_\theta = \vec{u}_r = 0$, only z-component of NSE is relevant).
3. The flow is symmetrical about the z-axis (all $\frac{\partial}{\partial \theta}$ terms equal 0).
4. The flow is fully developed ($\frac{\partial \vec{u}_z}{\partial z} = 0$).

Thus, we observe,

$$\begin{aligned}&\rho\left(1.\ \frac{\partial \vec{u}_z}{\partial t} + 2.\ \vec{u}_r\frac{\partial \vec{u}_z}{\partial r} + 2.\ \frac{\vec{u}_\theta}{r}\frac{\partial \vec{u}_z}{\partial \theta} + 4.\ \vec{u}_z\frac{\partial \vec{u}_z}{\partial z}\right)\\ &= -\frac{\partial P}{\partial z} + 2.\ \rho g_z + \mu\left(\frac{1}{r}\frac{\partial}{\partial r}\left(\frac{r\partial \vec{u}_z}{\partial r}\right) + 3.\ \frac{1}{r^2}\frac{\partial^2 \vec{u}_z}{\partial \theta^2} + 4.\ \frac{\partial^2 \vec{u}_z}{\partial z^2}\right)\end{aligned} \tag{4.45}$$

leaving:

$$-\frac{\partial P}{\partial z} + \mu\frac{1}{r}\frac{\partial}{\partial r}\left(\frac{r\partial \vec{u}_z}{\partial r}\right) = 0 \tag{4.46}$$

and if we further assume that the pressure varies linearly along the pipe, then the pressure gradient term becomes $\frac{\Delta P}{L}$ which leaves equation (4.45) solely as a function of the pipe radius. The boundary condition for this system is no slip ($\vec{u}_z(R) = \vec{u}_{\text{wall}} = 0$) and a further constraint that the velocity must be finite at the centreline ($\vec{u}_z(0)$):

$$\mu\int\frac{\partial}{\partial r}\left(\frac{r\partial \vec{u}_z}{\partial r}\right)dr = \frac{\Delta P}{L}\int r dr \tag{4.47}$$

$$\mu\int\frac{\partial \vec{u}_z}{\partial r}dr = \frac{\Delta P}{2L}\int r dr + c_1\int\frac{1}{r}dr \tag{4.48}$$

$$\mu\vec{u}_z(r) = \frac{\Delta P}{4L}r^2 + c_1\ln r + c_2 \tag{4.49}$$

log terms are undefined at 0, thus the $c_1 \ln r$ term must be dropped

$$\mu\vec{u}_z(r) = \frac{\Delta P}{4L}r^2 + c_2 \tag{4.50}$$

applying the no slip condition

$$c_2 = \frac{\Delta P}{4\mu L}R^2 \tag{4.51}$$

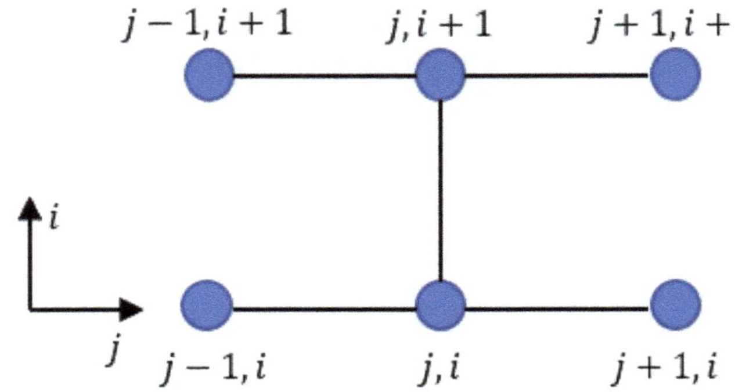

Figure 4.53. Stencil used for finite difference approximations, where *i* represents the timestep and *j* represents the node location.

thus

$$\vec{u}_z(r) = -\frac{\Delta P}{4\mu L}r^2 + \frac{\Delta P}{4\mu L}R^2 \tag{4.52}$$

some rearranging

$$\vec{u}_z(r) = \frac{1}{4\mu R^2}\frac{\Delta P}{L}\left(1 - \left(\frac{r^2}{R^2}\right)\right). \tag{4.53}$$

The reader more familiar with fluid mechanics may recognise this parabolic velocity profile as Hagen–Poiseuille flow, which will recover the Hagen–Poiseuille equation if integrating equation (4.53) with respect to the pipe cross section [240]. For a simple system such as a pipe in laminar flow, then CFD is likely not required as an analytical solution exists. The existence of such analytically-soluble cases is also very useful if we wish to test and validate CFD solvers and/or models—if the output of a (numerical) CFD simulation matches the expected analytical outcome, we can be confident that our set-up is correct.

Numerical schemes

Numerically approximating the NSE falls into two schemes, finite difference (FD) and finite volume (FV) which decompose the domain of interest into a collection of reference points known as a *mesh*. We first focus on how FD schemes work by decomposing the system domain into a grid with finite spacing, h, between grid nodes to spatially discretise the spatial terms of the NSE, and using a finite timestep to discretise the temporal terms The stencil shown in figure 4.53 shows the nodes necessary to construct forward, backward or central difference schemes relative to a node a location j for timesteps i and $i + 1$. Generally, the central difference scheme is the most accurate of the three schemes[32] as it is second order accurate (the other methods are first order accurate), and its use in CFD is outlined in detail by

[32] Advection terms can wreak havoc on a central difference scheme as the information comes from the opposite direction to flow, so where convection is significant, then it is preferable to use a forward or backward scheme depending on flow direction [238].

Patankar [239]. It is worth quickly addressing what we mean by order of accuracy. Simply put, finite difference schemes are created through the use of Taylor expansions to approximate derivatives, which approach the true value of a derivative as the order of terms used approaches infinity. Considering that we would like to spend a *finite* amount of time evaluating a CFD scheme, then the expansion's terms up to a certain order are utilised. This leaves a remainder, the difference between the true derivative and the approximated value which is a function of the step term and in numerical modelling, the lowest power of the step term determines the order of accuracy[33]. For example, a second order accurate scheme has a remainder term proportional to step raised to the power of 2. This type of error introduced into a numerical scheme is called *truncation error*[34].

So, let's define *what* the three finite difference schemes are for an arbitrary function that has been discretised as in figure 4.53:[35]

- Forward difference: $f'(j) = \frac{f(j+1) - f(j)}{h}$
- Backward difference: $f'(j) = \frac{f(j) - f(j-1)}{h}$
- Central difference[36]: $f'(j) = \frac{f(j+1) - f(j-1)}{2h}$.

A combination of these methods are used to discretise the NSE and mass continuity equation (equation (4.32)) which converts them from a partial differential equation (PDE) into a system of algebraic equations where pressure and velocity are still coupled. There is one caveat, however, in that convective terms introduce non-linearity into the system, and so to handle this, an FD algorithm 'guesses' values for velocity and pressure until the guess is within some allowable tolerance[37]. Such algorithms include the SIMPLE algorithm; for brevity we will not outline the finer points of such algorithms as they deserve a full explanation such as in Patankar [239]. When choosing a CFD package, it is advisable to check the algorithms implemented, and evaluate their benefits to choose a suitable scheme.

The advantages and disadvantages of FD schemes are as follows:

- Advantages
  - Simple formulation.
- Disadvantages
  - Susceptible to numerical diffusion[38] [239].
  - An improper set-up will violate conservation laws [241].
  - Struggles to cope with complex geometries due to the regular grid formation.

---

[33] As the step size invariably is less than 1, then increasing powers lead to increasingly small and insignificant values in terms of analysis. For example, if the step size is 0.01, then the first order term is 100 times larger than the second order term.

[34] This is *not* anything to do with numerical round-off for floating point numbers on a computer!

[35] Note that whilst these are defined based on a spatial basis, these are also applicable for a temporal basis.

[36] The astute reader may observe that if the grid is equally spaced, then the central difference is the mean of the forward and backward difference methods.

[37] In practice, this is done by a sophisticated iteration procedure.

[38] This is where due to truncation errors a diffusion greater than what is present in the "real" system being modelled is implied.

In contrast to the FD scheme, the finite volume (FV) scheme is inherently conservative by its very nature in which the domain is divided into cells with nodes in their centre [238]. The NSE are numerically evaluated by integrating across cell volumes. If we consider the incompressible, Newtonian case[39], and refer back to equation (4.43), and dividing through by density,

$$\frac{D\vec{u}}{Dt} = -\frac{1}{\rho}\nabla P + \nu\nabla^2\vec{u} + \vec{F} \tag{4.54}$$

integrating over cell volume, V, the unsteady and body force terms are assumed constant over the cell volume [239]

$$\frac{\partial(\vec{u})}{\partial t}\int_V dV + \int_V \vec{u}\cdot\nabla\vec{u}dV = -\frac{1}{\rho}\int_V \nabla P dV + \nu\int_V \nabla^2\vec{u}dV + \vec{F}\int_V dV \tag{4.55}$$

applying the divergence theorem to the advection, pressure and diffusion terms[40] [235]

$$\frac{\partial(\vec{u})}{\partial t}V + \int_S \vec{u}\vec{u}d\vec{S} = -\frac{1}{\rho}\int_S Pd\vec{S} + \nu\int_S \nabla\vec{u}d\vec{S} + \vec{F}V \tag{4.56}$$

for polyhedral cells with n sides, this is further divided into a sum across the surfaces, s, of each cell

$$\frac{\partial(\vec{u})}{\partial t}V + \sum^{n}\left(\int_{s_n} \vec{u}\vec{u}d\vec{s}\right) = -\frac{1}{\rho}\sum^{n}\left(\int_{s_n} Pd\vec{s}\right) + \nu\sum^{n}\left(\int_{s_n} \nabla\vec{u}d\vec{s}\right) + \vec{F}V \tag{4.57}$$

where ν is the kinematic viscosity of the fluid. Values obtained using this method are volume averages on a per cell basis, however, linear interpolation is used to relate volume averages and surface values for connected cells [241]. Unfortunately, this approach does not remove advection terms, so iterative procedures as with the FD approach are still required. The strength of this approach is shown in figure 4.54 where the circular geometry is easily captured by a small number of triangular elements. This can be further refined by using smaller triangles close to the circle surface to better capture the curvature—this is easily performed in the FV method.

The advantages and disadvantages of FV schemes are as follows:

- Advantages
  - Capable of handling complex geometries.
  - Inherently conservative scheme, so is robust to numerical diffusion.
  - Variable mesh resolution is computationally simple to generate[41].
  - Momentum transport across adjacent faces must be equal so for a structured mesh this significantly simplifies computation.

---

[39] This is chosen for simplicity, other forms of the NSE can also be used in FV schemes.

[40] This is a really clever theorem that allows us to convert from integrating all sources within a volume to calculating the rate at which a quantity travels through a surface boundary.

[41] This allows for coarse meshes where the finer details aren't as important, such as in the bulk flow and so finer meshes can be deployed only where required. This can *significantly* reduce computation time!

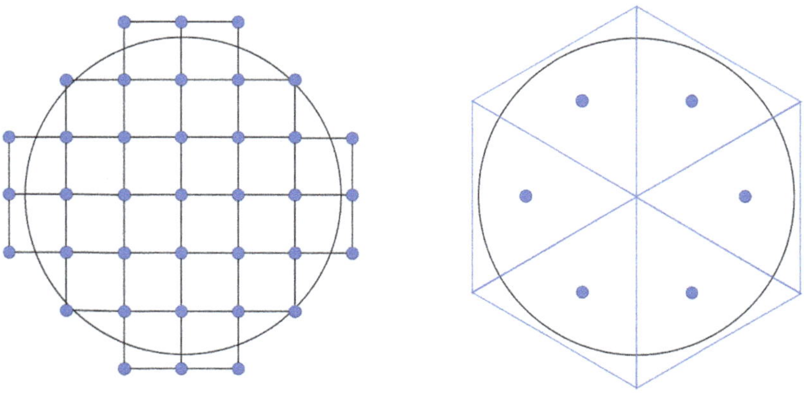

Figure 4.54. Comparison of FD (left) and FV (right) schemes for describing a circular geometry. Adapted from [238] 2017 © Springer International Publishing Switzerland 2017. With permission of Springer.

- Disadvantages
  - Accurately meshing the system domain is a problem within itself!
  - The implementation of an FV solver is noticeably more complex than an FD solver.
  - Unstructured meshes require significant effort to track and calculate fluxes properly.

At this point, it is worth spending some time addressing the meshing of a domain. Regardless of the numerical scheme used, a mesh *is required* to generate the spatial frame of reference to evaluate the NSE. The 'fineness' of the mesh which is essentially the spacing of the grid points (FD) or volume size[42] (FV) is akin to the timestep size used in the Euler method outlined in section 4.5.1. As was discussed in that section, a careful balance is drawn between simulation speed and simulation accuracy. In CFD, one must consider both spatial and temporal discretisation; as one may inherently expect, 'better' solutions are typically obtained using extremely fine grid spacing and small timesteps but, as with decreasing timesteps and increasing particle numbers in DEM, this comes at the cost of a rapidly increasing the number of points considered, and thus increasing computational expense. Figure 4.55 shows how for an FD scheme, halving the grid spacing doubles the number of nodes in each dimension, so for a 3D simulation there would be an eightfold increase in nodes. For an FV scheme, the actual increase is complex to determine beforehand but one must still pay careful attention to the sizes used.

Meshing for FV typically uses Delaunay mesh generation [242] which divides a complex geometry up into comparatively 'simple' shapes such as triangles, quadrilaterals or polyhedra. This is critical as we know how to perform operations on these simple shapes, but not necessarily for others. During the meshing process, certain criteria are necessary and some are user defined for the resultant mesh. As mentioned above about a variety of algorithms developed to evaluate the NSE

[42] This can also be defined in terms of the side lengths used if using polygonal elements.

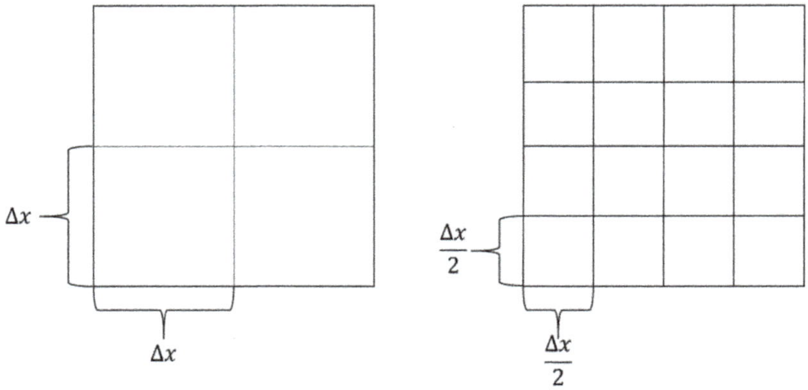

Figure 4.55. Illustration of increase in number of points as a function of grid spacing.

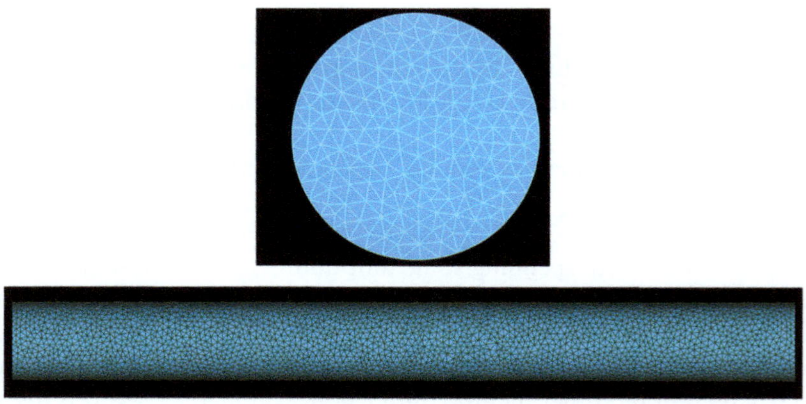

Figure 4.56. Example of a meshed pipe from (top) *x–z* perspective (bottom) *x–y* perspective.

and continuity equation that we will not go into detail with, this is the case also for meshing algorithms. In short, a meshing algorithm aims to provide mesh elements with reasonable internal angles so that numerical stability and accuracy is promoted. Extremely small internal angles lead to so-called 'sliver faces' which are *extremely* problematic in CFD and can even cause simulations to fail [243]. An example of representing an object as a mesh is shown below in figure 4.56.

Whilst many robust algorithms exist, it is ultimately down to the user to determine if the mesh quality is appropriate. This can be done at the meshing stage by checking quality metrics such as number of sliver faces and unreasonably small elements. When a mesh (or multiple meshes) has/have been deemed acceptable, a mesh independence study is performed. In a nutshell, this study determines the minimum fineness of a mesh beyond which the solution is essentially unchanged according to an acceptable level of error which means that an appropriate spatial discretisation has been achieved. This minimum fineness represents a compromise between accuracy and computational time as a mesh at this 'minimum fineness' represents the smallest possible fineness (i.e. the coarsest possible grid) to obtain the desired accuracy in the solution. The reader is directed to the following references that illustrate a mesh independence study [244–247].

The handling of time-evolution for a CFD system broadly falls into two approaches, regardless of whether an FD or FV scheme is used—explicit and implicit. It is easier to visualise both schemes by using the FD scheme (the node diagram is repeated for clarity) but the discussion below is also applicable to FV. We start with the explicit scheme as this is the more simple scheme. This was outlined previously in section 4.5.1, but for clarity and ease of comparison the key points are reiterated here.

An explicit time-stepping scheme uses the most recently calculated value to calculate the future value of a function. If we use a simple Euler integration scheme in time and a central difference scheme for the spatial second order derivative for the below system,

$$\frac{\partial \vec{u}_x}{\partial t} = \nu \frac{\partial^2 \vec{u}_x}{\partial x^2} \tag{4.58}$$

or, equation (4.43) if we assume all advection, pressure and body force terms are zero, and only the x-direction of flow is of interest. So, using the notation in figure 4.57, denoting the timestep as a superscript and the node position as a subscript then we can discretise equation (4.58) as,

$$\frac{\vec{u}_{x,j}^{i+1} - \vec{u}_{x,j}^{i}}{\Delta t} = \nu \frac{\vec{u}_{x,j+1}^{i} - 2\vec{u}_{x,j}^{i} + \vec{u}_{x,j-1}^{i}}{\Delta x^2} \tag{4.59}$$

which can be rearranged as

$$\vec{u}_{x,j}^{i+1} = (1 - 2k)\vec{u}_{x,j}^{i} + k(\vec{u}_{x,j-1}^{i} + \vec{u}_{x,j+1}^{i}) \tag{4.60}$$

where $k = \frac{\Delta t \nu}{\Delta x^2}$. This scheme is stable and will converge when $k \leqslant \frac{1}{2}$, i.e. when the first right hand side term is non-zero [248]. This is simple to set up and evaluate as the current values are used to determine the value at the $i + 1$ timestep. Within CFD,

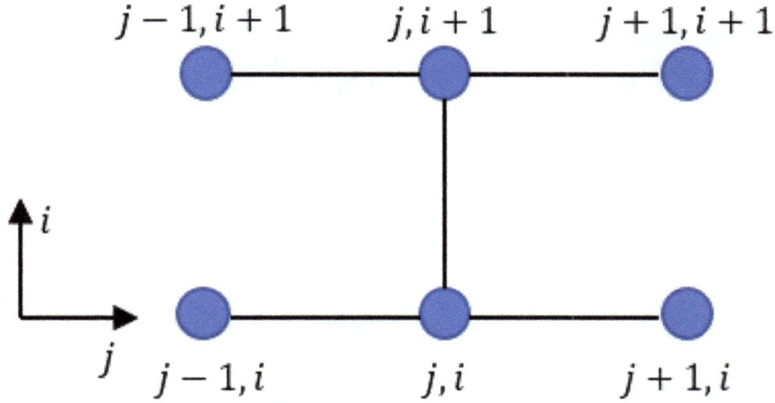

Figure 4.57. Repeat of figure 4.53, where i represents the timestep and j represents the node location.

there is an additional criterion alongside $k \leqslant \frac{1}{2}$, the Courant–Friedrichs–Lewy (CFL) condition which imposes a limit on the mesh size and timestep used. In short, this considers the ratio of the fluid speed to the ratio of the 'mesh speed', which is the ratio of mesh length to the timestep used. This is known as the Courant number, C. The mesh speed is a measure of how quickly the information about the fluid can travel from one mesh node to another, so this ratio compares the rate at which a hypothetical particle following the fluid exactly would travel to adjacent nodes and the rate at which information associated with it travels [249]. For an explicit time-stepping scheme, this ratio *must* be less than unity. In symbolic form with Cartesian coordinates,

$$C = \Delta t\left(\frac{\vec{u}_x}{\Delta x} + \frac{\vec{u}_y}{\Delta y} + \frac{\vec{u}_z}{\Delta z}\right) \leqslant C_{\max} \tag{4.61}$$

where Δx, Δy and Δz are not required to have the same value. If $C > 1$ with an explicit time-stepping scheme, then it is very likely that numerical instability will ensue.

An implicit time-stepping scheme essentially works in reverse, where the future values of the solution are used to calculate its future value. That sounds confusing, right—how can we use a value we currently don't have to work out the value we don't yet know? If we initially consider a simple function, $\frac{dy}{dt} = 2y$, then to use an implicit scheme such as the *Euler backward* method,

$$\frac{y(t + \Delta t) - y(t)}{\Delta t} = 2y(t + \Delta t) \tag{4.62}$$

we have an equation that is implicit in y[43]. An analytical solution is not guaranteed, so root-finding is required for the subsequent matrix of equations generated by this approach. This incurs more computational cost but provides more stability, as an implicit scheme can be stable for $C > 1$ due to the fact that an iterative procedure for all mesh nodes is required which couples them to each other [250]. However, this does *not* mean that a high Courant number makes for an accurate simulation! This is because using higher time-steps risks loss of information on transient behaviour[44]. Applying the implicit scheme to equation (4.58),

$$\frac{\vec{u}_{x,j}^{i+1} - \vec{u}_{x,j}^{i}}{\Delta t} = \nu\frac{\vec{u}_{x,j+1}^{i+1} - 2\vec{u}_{x,j}^{i+1} + \vec{u}_{x,j-1}^{i+1}}{\Delta x^2} \tag{4.63}$$

which can be rearranged as

$$\vec{u}_{x,j}^{i} = (1 + 2k)\vec{u}_{x,j}^{i+1} - k(\vec{u}_{x,j-1}^{i+1} + \vec{u}_{x,j+1}^{i+1}) \tag{4.64}$$

[43] This is a simple example, and obviously this *could* be solved by rearranging quite simply. But consider if the right hand side was a higher order term, or a trigonometric function instead. How could we efficiently apply various solution (if possible) methods case-wise? Thus, this form is implicit with respect to the future value.

[44] This is fine if only the steady state solution is of interest, however, there are more efficient algorithms to achieve this than waiting for a transient simulation to approach steady state.

which in this case gives a system of linear equations to solve. For both types of schemes, it is possible to use a Courant number condition to vary the timestep such that $C \leqslant 1$ to accelerate the simulation and maintain accuracy.

Flow regime

We now understand the fundamental equations and numerical schemes to model fluid flow, however, before we are truly ready to model fluid flow we first must consider the flow regimes possible and if any further requirements exist for successful modelling. Within fluid dynamics, there are two types of flow regime, laminar and turbulent (and an additional transitional state where the flow is neither purely laminar or turbulent, but this regime of flow is out of scope for this book). The laminar regime is characterised by the fluid streamlines following a regular, layered pattern. This is contrasted by turbulent flow where the streamlines appear to follow no such pattern and seem random. Schematic diagrams comparing these two main states are shown in figure 4.58. The type of regime present is inferred by the Reynolds number, Re,

$$Re = \frac{\text{inertial forces}}{\text{viscous forces}} = \frac{\rho L^3 \frac{\overline{\|\vec{u}\|}}{t}}{\mu \frac{u}{L} L^2} = \frac{\rho\, \overline{\|\vec{u}\|} L}{\mu} \tag{4.65}$$

where L is a characteristic length associated with the system, such as the diameter of a pipe and $\overline{\|\vec{u}\|}$ is the mean velocity magnitude. Low Reynolds numbers imply that a flow is dominated by viscous forces, giving rise to laminar flow and above a certain value (this is geometry dependent, such as 2300 for a pipe [235]) then inertial forces can be considered dominant, and the flow will become turbulent. If the Reynolds number of the system of interest suggests laminar flow, then no further considerations are required, and we can simulate a system with (relative) ease based on the information provided above.

If turbulent flow is expected, however, then some additional considerations are required for the model. Turbulence in fluids is a (significant) field of study in its own right, so only a brief explanation is provided here to contextualise the need for different turbulence modelling in CFD; for more detailed information, the reader is referred to [251–254]. A turbulent flow contains eddies (regions of fluid flow which go against the mean or 'net' flow of the system) of multiple scales. It is theorised that energy dissipation occurs at the smallest scales, i.e. where viscous forces dominate.

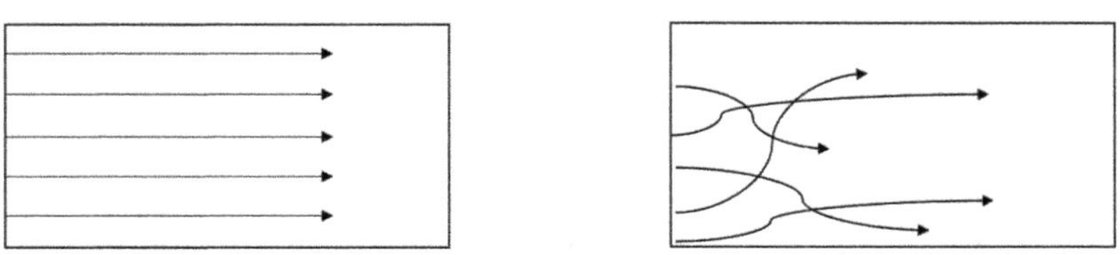

Figure 4.58. Simple schematic diagrams, with arrows indicating flow direction, comparing the expected flow fields for exemplary laminar and turbulent regimes.

This creates a cascade of energy from the largest eddy scales down to scales defined by the Kolmogarov length, η,

$$\eta = \left(\frac{\nu^3}{\epsilon}\right)^{\frac{1}{4}} \tag{4.66}$$

where ϵ is the rate of energy dissipation. One may see how this makes simulating turbulent flows complex as an extremely wide range of scales must be considered which leads to the need to generate very fine meshes, which in turn demands significant computing resources and constraints such as explicit time-stepping (due to the sheer memory requirements). In 2013, turbulent flow between two infinite plates[45] was directly resolved, which took *20 million* core hours to complete [255]. For context, on a 32 core machine then this simulation would take just over *71 years* to complete! Thus, alternative routes to simulating turbulence are required in the interest of runtime and resource availability. Before we outline some of the approaches to handle turbulence, we shall draw a distinction between what is meant by *resolving* and *modelling* turbulence. *Resolved* turbulence is where the discretisation is fine enough such that the turbulence is directly captured through the NSE—though this does not have to encompass *all* turbulence scales as we will outline below (i.e. some turbulence scales can be resolved and others not). In contrast, *modelled* turbulence employs a surrogate model that can take the form of an empirical equation or simplified analytical function that provides an estimate of the turbulence present. This is *not* to be confused with the fact that CFD models fluid flow and is rather a distinction between directly (resolving) or indirectly (modelling) treating turbulence!

CFD modelling of turbulent systems is typically performed using one of the three approaches listed below in order of increasing computational complexity[46] [239]:

1. Reynolds-Averaged Navier–Stokes (RANS) model turbulence using an 'eddy viscosity', μ_t—i.e. turbulence is entirely unresolved.
2. Large Eddy Simulations (LES) directly resolve turbulence down to a certain length-scale, below which it is modelled.
3. Direct Numerical Simulation (DNS) directly resolves *all* length-scales present.

These three approaches are in fact umbrella terms, each encompassing a family of methods to treat turbulence; in the interest of brevity, we shall only outline the principles behind each approach. The first approach splits the velocity into a time-averaged, $\bar{\vec{u}}(\vec{x})$ and fluctuating component, $\vec{u}'(\vec{x}, t)$, which has a mean of zero during the Reynolds-averaging process,

$$\vec{u}(\vec{x}, t) = \bar{\vec{u}}(\vec{x}) + \vec{u}'(\vec{x}, t) \tag{4.67}$$

[45] Periodic boundary conditions were used in the x and y direction to achieve this geometry.

[46] As one may expect from preceding discussions, this is also in order of increasing accuracy.

and thus the NSE takes on a different form to account for this[47] [256]:

$$\rho\nabla\bar{\vec{u}} - \nabla\bar{P} + \mu\nabla^2\bar{\vec{u}} - \rho\bar{\vec{F}} = -\nabla\vec{\vec{\tau}}_{\mathrm{RE}} \tag{4.68}$$

Where $\vec{\vec{\tau}}_{\mathrm{RE}}$ is known as the Reynolds stress, and is an additional degree of freedom introduced during the time averaging process. It may come as no surprise that since mean properties are used and the turbulence is *modelled* and not *resolved* then RANS demands the least in terms of computational resources [257], and so is amenable for use on an average machine—or to model very large systems on an above-average machine!

There are three methods for specifying the new stress involved in RANS, arranged in increasing order of complexity [257]:

1. Linear eddy viscosity.
2. Non-linear eddy viscosity.
3. Reynolds stress transport model.

The former two methods use the Boussinesq hypothesis, which relates the Reynolds stress and eddy viscosity to each other [258] whilst the latter method utilises a transport model for the Reynolds stress. A useful list of RANS models that utilise all three methods is maintained by NASA, and can be found in [259]. Linear eddy and non-linear eddy viscosity methods utilise the following relations, respectively [258]:

linear

$$\vec{\vec{\tau}}_{RE,L} = 2\mu_t\left(\bar{\vec{\vec{e}}} - \frac{1}{3}\nabla\vec{u}\vec{\vec{I}}\right) - \frac{2}{3}\rho k\vec{\vec{I}} \tag{4.69}$$

non-linear

$$\vec{\vec{\tau}}_{RE,NL} = \vec{\vec{\tau}}_{RE,L} + f(\bar{\vec{\vec{e}}}, \bar{\vec{\vec{\omega}}}) \tag{4.70}$$

and turbulent viscosity is given by

$$\mu_t = f\left(\frac{\rho k^2}{\epsilon}\right) \tag{4.71}$$

where k is the turbulent kinetic energy and $\bar{\vec{\vec{\Omega}}}$ is the mean rotation rate tensor (the rotational analogue to the strain rate tensor). The modern approach for determining μ_t is using transport equations for turbulent variables such as turbulent kinetic energy, dissipation, ϵ and specific dissipation into heat ω. Transport equations have replaced mixing length models (these treat eddy viscosity as a function of the local shear rate and mixing length was a constant that had to be estimated) due to their improved generality—models such as the $k - \epsilon$ model were even specifically derived

[47] Again, the incompressible NSE is used for simplicity.

to overcome the shortcomings of mixing length models [260]. The general form for the transport equations in RANS,

$$\text{accumulation} + \text{advection} = \text{diffusion} + \text{production rate} - \text{destruction rate} \tag{4.72}$$

is very similar to the text form of the NSE (equation (4.44)). Different methods focus on different turbulent variables which create different numbers of transport equations to evaluate alongside equation (4.68). To maintain the generality in this description, specific model formulations are not listed, but can be found at [259]. Within CFD, this is normally the first choice in the approach as they are well suited to a variety of systems as well as having an amenable computational load [239], with the most popular method being the $k - \epsilon$ model which uses two transport equations for k and ϵ respectively, that performs well in industrial problems [246, 261–263]. It is worth noting that despite the results obtained using methods derived from the Boussinesq hypothesis largely being seen as 'good enough', the underlying theory behind this hypothesis has been challenged as discussed by Schmitt [258]. In short, turbulence displays non-Gaussian statistics and so assumptions about the microscopic behaviour from a kinetic theory viewpoint break down. However, this is not to discourage the use of this type of RANS approach as the data generated by this method shows good experimental agreement, but it is important to understand the drawbacks of each turbulence approach to better understand which one(s) are appropriate for a given system.

Rather than assuming relationships between eddy viscosity and Reynolds stress, the Reynolds stress transport model is a transport equation that broadly solves for all components of Reynolds stress [239],

$$\rho \frac{D\vec{\vec{\tau}}_{\mathrm{RE}}}{Dt} = \rho(\vec{\vec{D}} + \vec{\vec{P}} + \vec{\vec{\Pi}} + \vec{\vec{M}} + \vec{\vec{\epsilon}}) \tag{4.73}$$

$$\begin{aligned} \text{accumulation} + \text{advection} = &\ \text{diffusion} + \text{production rate} - \text{destruction rate} \\ &+ \text{pressure–strain interaction transport} \\ &+ \text{rotational transport} \end{aligned} \tag{4.74}$$

which can handle complex flows and geometries. Only the production term is fully closed whilst the other terms require modelling. Some formulations for the unclosed terms can be found in [259]. This approach overcomes shortcomings of Boussinesq hypothesis inspired methods which assume isotropic turbulence, and struggle to accurately handle situations such as where flow separation and/or recirculation zones are present [239].

Boundary layers are of importance in turbulence[48], the model must be able to correctly describe boundary layer behaviour as this is where the resistance to mass, heat and momentum transfer is experienced and thus it is essential to accurately describe a boundary layer [264]. By correctly modelling the boundary layers, one can

[48] They are also important in laminar flow, but as the flow can be directly simulated in laminar flow, dealing with boundary layers is *considerably* easier than with turbulent flow.

elucidate things such as the effectiveness of a catalyst packing by understanding the mass transfer coefficient, the effectiveness of a heat exchanger geometry or the drag experienced by a racing car. Whilst this is important, if one is mainly interested in the bulk flow, such as for mixing, then an empirical approach can be used, where an empirical function known as a wall function is employed to model fluid behaviour in the boundary layer. This allows for a more coarse mesh, and leaves the turbulence model to handle bulk flow which suits models such as the $k - \epsilon$ model where k is undefined at the wall. But how do we know where the wall function is appropriate and where RANS should take over? To understand that, we define four important variables:

1. Wall shear stress, $\vec{\tau}_W = \mu \frac{\partial \vec{u}}{\partial y}|_{y=0}$
2. Friction velocity, $\vec{u}_\tau = \sqrt{\frac{\vec{\tau}_W}{\rho}}$
3. Dimensionless velocity, $\vec{u}^+ = \frac{\vec{u}}{\vec{u}_\tau}$
4. Dimensionless wall distance[49], $y^+ = \frac{y\rho\vec{u}_\tau}{\mu}$

where y is the distance from the wall. The definition of a dimensionless wall distance is important as this allows the boundary layer shown to be split into three regions where the dimensionless velocity changes with y^+ differently based on the region as the ratio between viscous and inertial effects changes [265]:

1. $y^+ < 5$: viscous sublayer, $\vec{u}^+ = y^+$.
2. $5 < y^+ < 30$: buffer layer, complex as viscous and inertial effects have a similar magnitude. Typically the viscous sublayer approach is more accurate before the curves intersect and the logarithmic approach after.
3. $y^+ > 30$: Logarithmic (or log-law) layer, $\vec{u}^+ = \frac{1}{\kappa}\log(y^+) + B$ (where $\frac{1}{\kappa}$ and B are 0.41 and 5.2, respectively).

Wall functions are most accurate when the centre of the mesh element closest to the wall is in the log-law layer ($y^+ > 30$) but with y^+ not being too large so that accuracy is maintained [266]. This *greatly* reduces the computational load as the mesh can be more coarse. However, if the forces acting upon the wall are of great interest, or boundary layer separation is expected, then a low-Reynolds approach is required. This avoids the use of wall functions and instead uses a fine mesh resolution near the wall to model the boundary layer [265]. It is important to note that this confusingly titled approach means that the local Reynolds number is low (i.e. within a cell) rather than the global Reynolds number (for the whole system)! This requires a mesh resolution such that the centre of the mesh element closest to the wall has a y^+ value of 1–5, i.e. the viscous sublayer is considered directly by the RANS model utilised.

---

[49] This can be thought of as a local Reynolds number as this equation shares similarities with the Reynolds number defined above.

If greater accuracy is required and/or the computational resources are available, then LES is the next candidate for the family of simulation methods to try. LES and DNS will be outlined in less detail than RANS, owing to the fact that they are not yet practical for a mainstream usage for industrial problems [267].

Like with the low-RE approach, a fine mesh is required, even in the bulk of the flow to resolve turbulence. However, the essence of the LES approach is that it models turbulence occurring at scales below the mesh resolution—essentially acting as a low-pass filter. This provides a greater level of accuracy than the RANS approach at the cost of increased computation resource demand[50]. Whereas RANS can typically run on most computers, LES simulations require at the minimum a high performance computer to handle the mesh resolution required, thus, either investment in a high-spec personal machine or access to a computing cluster.

For LES, the numerical treatment is divided into two parts during evaluation:

1. Directly resolving turbulence down to a length-scale, Δ and timescale, Δt_{LES}.
2. Modelling turbulence for length-scales Δ to η, i.e. down to the Kolmogarov lengthscale.

The direct resolution of turbulence is achieved by utilising a filtered form of the NSE, which for an incompressible and Newtonian fluid,

$$\rho\frac{D\bar{\vec{u}}}{Dt} - \nabla\bar{P} + \mu\nabla^2\bar{\vec{u}} - \rho\bar{\vec{F}} = -\nabla\vec{\vec{\tau}}_L \tag{4.75}$$

where $\vec{\vec{\tau}}_L$ is the Leonard tensor, which is the sum of pure large-scale (scale > Δ), sub-filter scale (scale < Δ) and cross-scale interactions—so essentially the LES analogue of the Reynolds stress tensor. This, as was the case with RANS, requires closure, which is a complex problem as multiscale interactions must be properly captured [268]. The overbars in equation (4.75) represent *filtered* quantities, and not Reynolds-averaged quantities like in equation (4.68). Typically, the Leonard tensor is closed using an eddy-viscosity assumption,

$$\vec{\vec{\tau}}_L = 2\mu_t\bar{\vec{\vec{e}}} + \frac{1}{3}\vec{\vec{I}}\vec{\vec{\tau}}_L. \tag{4.76}$$

The $\frac{1}{3}\vec{\vec{I}}\vec{\vec{\tau}}_L$ term is absorbed into the pressure term, meaning that the pressure is different to the static pressure normally calculated in the NSE. In this case, the eddy viscosity term is typically handled using the Smagorinsky model [269],

$$\mu_t = \rho(C_S\Delta)^2 S \tag{4.77}$$

[50] However, the computational demand is much less than for directly resolving turbulence.

where

$$S = \sqrt{2\vec{\vec{e}}\,\vec{\vec{e}}} \tag{4.78}$$

and

$$\Delta = (\Delta x \Delta y \Delta z)^{\frac{1}{3}} \tag{4.79}$$

where C_S is the Smagorinsky constant that depends on the system geometry. The subject of closures within LES is an active topic of research, with such improvements including adding local calculations for C_S [270, 271].

Boundary conditions for LES are complex, as the turbulence must be properly specified, unlike with RANS where the fluctuating components are averaged out. Synthesis techniques such as proper orthogonal decomposition [272] and digital filter methods [273] can be used to approximate the turbulence boundary conditions, but work is required to develop a method that robustly generates the desired turbulence properties for boundary conditions [267]. The complexity of forming LES boundary conditions means, that its usage is limited to experienced CFD users.

The final approach, direct numerical simulation (DNS), abandons any form of modelling of turbulence and instead directly resolves turbulence at *all* length and time-scales. The NSE are directly evaluated, which requires an extremely fine mesh to cope with the large gradients and fluctuations that occur near walls and to capture motion down to the Kolmogarov length-scale. DNS is normally performed for smaller geometries to provide validation data for the above approaches' models as it is impractical to perform for any industrially relevant problem. Unless limited to extremely small systems (i.e. of the order of 100 or fewer particles) this approach is realistically limited to high performance computing clusters. Explicit time-stepping is required for DNS to reduce the computational overhead [255]. Figure 4.59 briefly compares the three approaches described in this section.

Extensions to CFD

Here, we outline some of the more popular extensions to CFD, such as multiphase modelling, the two-fluid model and non-Newtonian fluids.

A multiphase system requires modification to the NSE to account for the presence of two different phases. This is typically achieved with the volume of fluid (VoF)

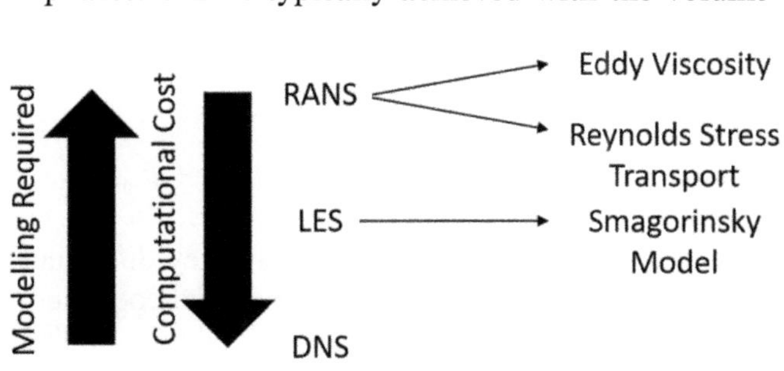

Figure 4.59. Illustration of typical CFD modelling approaches in order of increasing performance requirements.

approach, which introduces an additional parameter, α, which represents the volume fraction of a fluid within a cell [274] and is used to weight properties in a cell based on the fractions of fluid present. This enables a piecewise representation of the interface(s) between the different phases based on the fraction of fluid in each cell [275]. Whereas previously, we have not mentioned body force terms more than simply defining what they are, within multiphase systems, this term becomes crucial as it allows the phases to 'talk' to each other and exchange information to properly describe the state of the system. The current state of multiphase modelling cannot generally apply body force terms that work in all circumstances; as such, experience and knowledge of the flow regime is critical for accurate simulations [276]. Examples of possible body forces for a fluid include:

Gravity

$$\vec{F}_g = \alpha \rho g \tag{4.80}$$

surface tension

$$\vec{F}_{\mathrm{SV}} = \gamma \kappa \vec{n} \delta \tag{4.81}$$

where g is acceleration due to gravity, γ is the surface tension coefficient, κ is the mean surface curvature, $\vec{n}$ is the normal vector of the interface and δ is the Dirac delta function. This approach is used frequently for simulations of stirred tanks, where there is air in the tank headspace [277]. High ratios between the densities of the different phases is problematic for multiphase modelling, and only recently has been approached as discussed by Nangia *et al* [275].

Alongside considering multiple fluids, it is possible to consider a solid phase as a continuum and model consequences of collisions instead of directly resolving them like is done in DEM [261]. This approach, the so-called two-fluid model leverages kinetic theory of granular flows (KTGF) to create a solid 'fluid' that has solid and shear viscosities [196]. A brief description of KTGF is defining the particle distribution function, $f(\vec{v}, \vec{x}, t)$, in velocity, physical and time space. The second moment of this distribution is the KTGF analogue of the NSE, describing the particle momentum [278],

$$\rho \frac{\partial \vec{v}}{\partial t} + \nabla \cdot \vec{\vec{\sigma}} = 0 \tag{4.82}$$

where the stress tensor in KTGF is defined as

$$\vec{\vec{\sigma}} = -P + \mu_{\mathrm{shear}}(e_n, e_t)([\nabla \vec{v} + (\nabla \vec{v})^T] - \frac{2}{3}(\nabla \cdot \vec{v})\vec{\vec{I}}) + \mu_{\mathrm{solid}}(e_n, e_t)\nabla \vec{v}\vec{\vec{I}}. \tag{4.83}$$

This equation looks very similar to the NSE, with a key difference being that the viscosities here are functions of the normal and tangential coefficients of restitution. The two-fluid model can successfully handle dense [279] and fluidised [280] systems, but the regime must be clearly one of those two—a transitional regime is poorly handled by the two-fluid model [261]. This is in part due to the fact that the closures for the viscosities differ based on regime [281]. The two phases normally have a drag force closure to relate the presence of solids and fluids—other forces used depend on the nature of the system itself.

Putting it all together
Now that we have a reasonable understanding of the more popular methods utilised in CFD modelling, we can fully appreciate the manner in which they join together to allow us to simulate fluids:

1. Create mesh that represents the system domain.
2. Select the desired model to use.
3. Select the desired numerical schemes for spatial and temporal terms.
4. Generate boundary conditions for model.
5. Perform mesh independence study to verify that the resolution is acceptable.
6. Perform simulation and evaluate results.
7. Repeat steps 2–6 until satisfactory results obtained[51].

It is worth noting that despite outlining a number of models and numerical schemes in this section that there are numerous others that have been created to fulfil specific purposes. Thus, different CFD software codes contain a range of different models and so it is advisable that the reader checks the models implemented by particular CFD codes to see if this meets requirements for a simulation.

Comparison with PEPT
Advantages:

- Full (Eulerian) information regarding whole system, including transient dynamics.
- Full stress information.
- Arbitrarily high spatial and temporal resolution, limited only by available processor power.

Disadvantages:

- Potential to produce unphysical results
- Standard CFD provides only Eulerian information
- Outputs unreliable without rigorous calibration and experimental validation
- Computationally expensive in some cases[52].
    - Limitations on system size that can be feasibly simulated.
    - Typically more time-intensive than practical experimentation[53].

[51] During the course of a simulation you may find that the initial model chosen may not actually be suitable for the system—it is not always possible to know what model will be appropriate beforehand.
[52] The degree of computational expense varies significantly with the specific methodology chosen, as discussed in detail above. Using some methods, CFD is in fact capable of simulating *larger* systems than PEPT can feasibly measure.
[53] As with DEM, this refers specifically to the time required to *run* a given simulation/experiment, not taking into account issues regarding the time costs involved in the construction of experimental systems versus the development of a DEM model, or possible issues when performing experimental work relating to the availability of equipment, etc.

Complementarity with PEPT

As discussed in the preceding section regarding DEM, the combination of PEPT and CFD allows the significantly greater amount of information available from the latter technique to be rigorously validated against the data-rich output of the former, allowing the provision of deeper, more accurate and more detailed information than can be offered by the two techniques individually.

A significant difference between the use of PEPT to validate CFD (as compared to DEM), however, is that when measuring fluid flow, PEPT cannot be *truly* non-invasive, as it is when tracking particle motion. In other words, if we are to use PEPT as a (reliable) validation tool, we must be confident that the tracer is suitably small, and adequately well density-matched, that it faithfully follows the streamlines of the fluid it is intended to represent. Alternatively, validation may be achieved by adopting a coupled CFD-DEM approach and directly modelling the PEPT tracer in the simulation. In reality, for the range of tracer sizes currently feasible for use with PEPT (combined with its inherently limited temporal and spatial resolution), one can reasonably expect a well-designed tracer to follow the broad flow of a system, but it is unlikely to be able to map turbulent behaviour on the finest scales of many systems. Nonetheless, as a simple validation tool, this is typically sufficient, and the finer details can then be extrapolated from the CFD simulation (much as, when using DEM, we extrapolate forces, about which PEPT similarly offers no direct information).

As with DEM. a variety of studies in which PEPT and CFD have been successfully synergised can be found throughout chapter 8.

4.5.3 The Monte Carlo method

In this final subsection, we discuss briefly the Monte Carlo Method which though not, like CFD and DEM, used as a 'complementary method' to PEPT does have applications of direct relevance to this book: it can be used to simulate PEPT detector systems (a matter discussed in detail in section 5.3).

'Monte Carlo' is an umbrella term for a broad class of numerical methods which, to put it crudely, use large numbers of random 'guesses' constrained by a set of user-defined 'rules' to approximate the solution to a given problem. Though this rough description may make the method sound extremely haphazard, with suitably good statistics (i.e. adequately large numbers of 'guesses') Monte Carlo algorithms can converge to highly accurate solutions. An example showing the potential accuracy (but also, for unsuitably small numbers of guesses, potential *in*accuracy!) of the Monte Carlo method is provided in the interactive example below. For those unfamiliar with Monte Carlo simulations, we strongly recommend taking a look at this example *before* continuing to read this section, as in many ways the method is much easier to understand by seeing it in action than by having it described in prose.

The tutorial guides you through a classic example of the Monte Carlo method. The methodology employed is extremely simple: we consider a circle of unit radius, r, enclosed in a square of side $L = 2r$, centred at (0, 0) (see figure 4.60(a)). We then use a random number generator to randomly choose two numbers between -1 and 1, representing a pair of coordinates (our 'guesses' in the current,

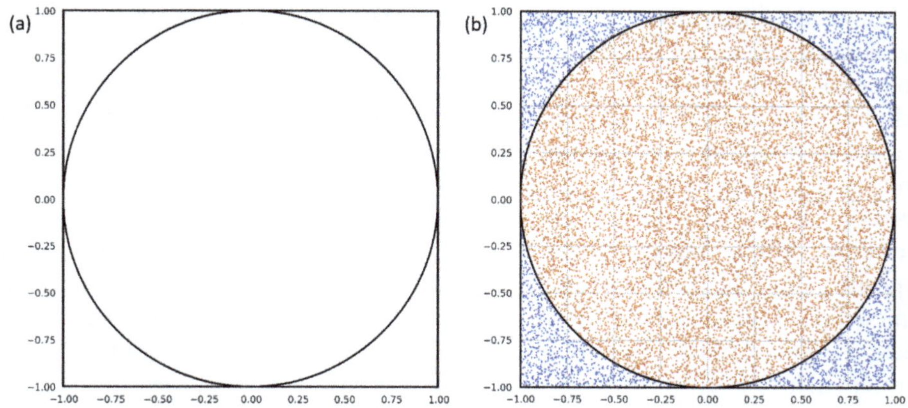

Figure 4.60. (a) A circle of radius r enclosed in a square of side length $L = 2r$. (b) Points corresponding to 15 000 pairs of random coordinates chosen from a uniform distribution between −1 and 1. Points falling inside the circle are coloured red, and those outside are coloured blue.

simple example). The random coordinates are then plotted within our square region of interest, and the process repeated a user-defined number of times (see figure 4.60(b)). From our known values of L and r, we know the area of our square must be equal to

$$A_{\text{square}} = L^2 = 4r^2, \tag{4.84}$$

and the area of our circle is equal to

$$A_{\text{circle}} = \pi r^2. \tag{4.85}$$

The ratio of the area of our circle to the area of our square is therefore

$$\frac{A_{\text{circle}}}{A_{\text{square}}} = \frac{\pi r^2}{4r^2} = \frac{\pi}{4} \quad \rightarrow \quad \pi = 4\frac{A_{\text{circle}}}{A_{\text{square}}}. \tag{4.86}$$

If the coordinates generated as described above are truly randomly distributed across our area of interest, then $\frac{\pi}{4}$ as many should land within the bounds of our circle as land within the bounds of our square. As such, the quantity $4\frac{A_{\text{circle}}}{A_{\text{square}}}$ should —for an infinitely large number of points—be equal to π. For $N_{\text{points}} < \infty$, however, we will clearly see some deviation due to the random nature of our numerical experiment, with the magnitude of this deviation increasing as the number of points we use decreases. Spend some time playing with the example at the end of this chapter to get a feel for what numbers of points yield 'good' and 'bad' statistics, and at what point the increased expense due to increasing the number of 'guesses' is no longer worth the ever-diminishing improvement in accuracy. As we will see in subsequent chapters, when dealing with PEPT data a feel for such things is highly useful when choosing, for example, the number of lines of response to use when determining a particle's position, or the size of the cells or slices to use when binning data to produce velocity or density distributions.

Though the illustrative example discussed above may be somewhat abstract, it nonetheless provides a clear exemplar of a methodology which may be applied to a number of problems of direct relevance to PEPT. A similar approach—drawing randomly from distributions of various forms—may be used, for example, to model the random decay of a beta-emitting isotope. Once a beta particle is emitted, it will then travel a given distance before annihilating with an electron. This distance travelled can similarly be considered as a quantity being drawn from a known distribution, whose precise form depends on various parameters, such as the energy of the positron and the density of electrons in the surrounding media, and thus can also be modelled using a Monte Carlo approach. Once annihilation has occurred, the direction of travel of the resultant photons can again be assumed random. If our beta-emitting particle is sat in the field of view of a positron camera, then this particular problem can be simplified down into something similar to the calculation of π above—either a photon will or won't hit our detector, with $\frac{P_{\text{hit}}}{P_{\text{miss}}}$ being determined by the solid angle subtended by the detectors. If a photon is indeed emitted in the direction of a detector, then yet another probabilistic calculation must be made as to whether or not a photon passing through the detector will register or not.

As is clear from the above, the process of detecting a line of response can in essence be considered as a chain of random events, each of whose probabilities can be modelled statistically, using probability distributions developed from our understanding of the processes' fundamental physics. As such, the Monte Carlo method, despite its simplicity, can be used to build up a surprisingly accurate picture of even relatively complex processes such as PEPT or PET imaging [282, 283], as will be discussed further in section 5.3.

Though there is much more to be said about Monte Carlo simulations—and considerably more mathematical detail to be included—our goal here is simply to give the reader a conceptual understanding of the technique, which will in turn aid understanding of section 5.3, where we describe the manner in which Monte Carlo simulations can be used to model a positron camera, and indeed a full PEPT experiment. The interested reader may find further and more detailed information regarding Monte Carlo methods in our references [284, 285].

4.5.3.1 Interactive example: a simple Monte Carlo simulation

The notebook files accompanying this section are available in the supplementary data online at https://iopscience.iop.org/book/978-0-7503-3071-8.

A simple Monte Carlo simulation

The Monte Carlo method is a powerful simulation technique which, in broad terms, is used to approximate complex phenomena using random guesses. It can be particularly useful in cases where the underlying physics are not known exactly—or are too computationally expensive to simulate—but random samples can be taken.

The canonical example of a simple Monte Carlo Simulation is the approximation of π by throwing N random darts at a 2×2 square $\{(x, y) \mid x, y \in (-1, 1)\}$; the probability of a dart falling into the unit circle $\{(x, y) \mid x^2 + y^2 < 1\}$ is equal to the area of the unit circle divided by the area of the square:

$$\frac{\text{Area of a unit circle}}{\text{Area of a 2x2 square}} = \frac{\pi}{4}$$

So if we throw a *large* number of darts at the 2×2 square and compute the ratio of darts falling into the unit circle we can approximate π! See the example below and vary the number of darts thrown (the `num_darts` variable)—how close is our approximation of π, and how does it change with the number of darts thrown?

[1]:
```python
# Importing necessary libraries for numerical array manipulation and plotting
import numpy as np
import matplotlib.pyplot as plt

# Set Matplotlib theme (try "darkgrid" too for a more modern look)
import seaborn
seaborn.set_style("whitegrid")
```

[2]:
```python
# Throw `num_darts` random darts at a 2x2 square {(x, y) | x, y in (-1, 1)}
num_darts = 5000
print(f"Throwing {num_darts} random darts at a 2x2 square...")

# Generate an array of numbers between (-1, 1), with `num_darts` rows
# corresponding to each dart and 2 columns for the (X, Y) coordinates.
#
# Note: numpy.random.random generates random numbers between [0, 1); we scale
# them between (lo, hi) using the `lo + (hi - lo) * random` formula
rng = np.random.default_rng(0)
darts = -1 + 2 * rng.random(size=(num_darts, 2))

# Calculate the distance from the origin to each dart
distances = np.sqrt(darts[:, 0] ** 2 + darts[:, 1] ** 2)

# Select the number of darts that fell inside the unit circle
darts_inside = darts[distances < 1]
darts_outside = darts[distances >= 1]

# Calculate the ratio of darts falling inside the unit circle
ratio = len(darts_inside) / len(darts)
print(f"Ratio of random darts falling inside the unit circle = {ratio}")
print(f"Ratio × 4 = {ratio * 4}")

# Plot the darts, colour-coding the ones falling inside the unit circle
fig, ax = plt.subplots()

ax.scatter(darts_outside[:, 0], darts_outside[:, 1], s=1)
ax.scatter(darts_inside[:, 0], darts_inside[:, 1], s=1)

ax.set_aspect("equal")
```

```
Throwing 5000 random darts at a 2x2 square...
Ratio of random darts falling inside the unit circle = 0.787
Ratio × 4 = 3.148
```

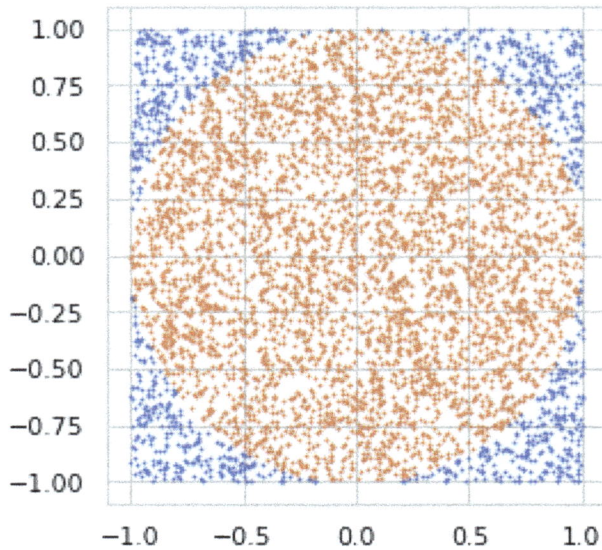

How many random samples?

We mentioned that we need a **large** number of darts to have an accurate approximation of π—but how large is large? The code cell below runs the Monte Carlo π approximation for varying numbers of darts—try changing the `numbers_of_darts` variable:

[3]:
```
# Create a random number generator with a fixed seed for deterministic values
rng = np.random.default_rng(0)

def throw_darts(num_darts):
    '''Throw `num_darts` random darts at a 2x2 square and return 4 times the
    ratio of darts falling inside the unit circle.'''

    darts = -1 + 2 * rng.random(size=(num_darts, 2))
    distances = np.sqrt(darts[:, 0] ** 2 + darts[:, 1] ** 2)
    darts_inside = darts[distances < 1]
    ratio = len(darts_inside) / len(darts)

    return 4 * ratio

# A vector of the numbers of darts to try the Monte Carlo Pi approximation for.
#
# Note: `numpy.arange(start, end, [step])` generates a vector with consecutive
# values betweeen `start` and `end` in steps of `step`
numbers_of_darts = np.arange(5, 2000)
ratios = [throw_darts(n) for n in numbers_of_darts]
```

```
# Compute the rolling average of the Pi approximations on a sliding "window" -
# change the window length below! Try 5, 10, 15, 20, 35, 50, 100
window = 20
rolling_average = np.convolve(ratios, np.ones(window) / window, mode="valid")


# Plot the ratios of darts falling inside the circle for the `numbers_of_darts`
# tried, plus the rolling average of the approximations and true Pi value
fig, ax = plt.subplots()

ax.scatter(numbers_of_darts, ratios, s=2, label = "Approximations")

# Plot only the valid region of the rolling average (i.e. with a window
# half-width trimmed from the edges)
ax.plot(numbers_of_darts[window // 2:-window // 2 + 1], rolling_average, 'k',
        label = "Rolling Average")
ax.plot([0, numbers_of_darts[-1]], [np.pi, np.pi], 'r',
        label = "True $\pi$ value")

ax.set(
    xlabel="Number of points",
    ylabel="$\pi$ approximation",
    ylim=[2.8, 3.5],
)

ax.legend();
```

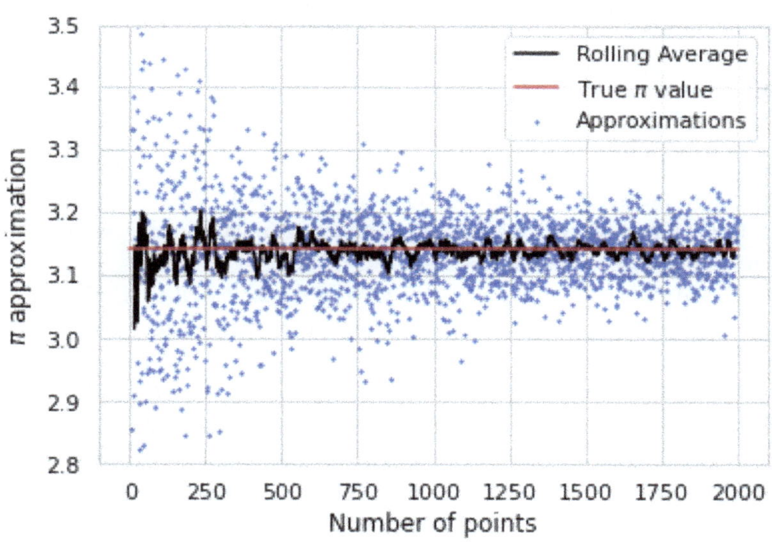

Notice how the approximations are distributed around the true π value: for small numbers of points (roughly <750) the approximations are quite spurious, but as the *numbers of samples* are increased, they converge around π—within a given margin. The rolling average (black line) depicts the spurious initial region with large jumps between values, before concentrating around π. Try changing the `window` variable to use different window lengths for the rolling average—how does the π approximation change?

By the end of this notebook you should have some insight into the power of random sampling—and how we can use randomness to predict constants or, by the same process, complex physics.

4.6 Other techniques

It is important to note here that this section is intended only to provide information concerning the techniques most widely used in the study of particulate and multiphase flows, and thus of the greatest relevance to the main focus of this book. That is to say, our primary interest lies in discussing techniques which you, as the reader, will easily be able to find a practitioner of, and thus potentially use in your own research. There exist many other techniques—both experimental and numerical—which are not mentioned in the brief overview provided here due to the fact that they are either too new to yet be widely available (and in many cases to be fully proven!), or too old and/or redundant to be of interest.

An overview of the 'bleeding-edge' methodologies currently being developed and which may, in the future, form a next generation of imaging techniques can be found in our reference [48]. More detailed information regarding these techniques, namely acoustic imaging [286], Compton imaging [287], divergent beam x-ray imaging [288], multi-axis x-ray imaging [289], radar imaging [290], Terahertz imaging [291] and x-ray rheography [292], may be found in the references provided.

References

[1] Jesuthasan N, Baliga B R and Savage S B 2006 Use of particle tracking velocimetry for measurements of granular flows: review and application—particle tracking velocimetry for granular flow measurements *KONA Powder Part. J.* **24** 15–26

[2] Maas H G, Gruen A and Papantoniou D 1993 Particle tracking velocimetry in three-dimensional flows *Exp. Fluids* **15** 133–46

[3] Dabiri D and Pecora C 2020 *Particle Tracking Velocimetry* (Bristol: IOP Publishing)

[4] Arroyo M P and Greated C A 1991 Stereoscopic particle image velocimetry *Meas. Sci. Technol.* **2** 1181

[5] Prasad A K 2000 Stereoscopic particle image velocimetry *Exp. Fluids* **29** 103–16

[6] Marion A 2013 *Introduction to Image Processing* (Berlin: Springer)

[7] van Ginkel M, Hendriks C L L and van Vliet L J 2004 A short introduction to the radon and hough transforms and how they relate to each other *Delft University of Technology*

[8] Atkins M D 2016 Velocity field measurement using particle image velocimetry (PIV) *Application of Thermo-Fluidic Measurement Techniques* (Amsterdam: Elsevier) pp 125–66

[9] Di Cristo C 2011 Particle imaging velocimetry and its applications in hydraulics: a state-of-the-art review *Experimental Methods in Hydraulic Research* (Berlin: Springer) pp 49–66

[10] Raffel M, Willert C E, Scarano F, Kähler C J, Wereley S T and Kompenhans J 2018 *Particle Image Velocimetry: A Practical Guide* (Berlin: Springer)

[11] Adrian R J 2005 Twenty years of particle image velocimetry *Exp. Fluids* **39** 159–69

[12] Bryanston-Cross P J, Towers C E, Judge T R, Towers D P, Harasgama S P and Hopwood S T 1992 The application of particle image velocimetry (PIV) in a short-duration transonic annular turbine cascade *J. Turbomach.* **114** 504–9

[13] Willert C, Raffel M, Kompenhans J, Stasicki B and Kähler C 1996 Recent applications of particle image velocimetry in aerodynamic research *Flow Meas. Instrum.* **7** 247–56

[14] van Dijkhuizen W, Bokkers G A, Deen N G, Van Sint Annaland M and Kuipers J A M 2007 Extension of PIV for measuring granular temperature field in dense fluidized beds *AIChE J.* **53** 108–18

[15] Steingart D A and Evans J W 2005 Measurements of granular flows in two-dimensional hoppers by particle image velocimetry. Part I: experimental method and results *Chem. Eng. Sci.* **60** 1043–51

[16] Ulissi F, Ippolito I and Calvo A 2009 PIV technique applied to granular flows in hoppers *J. Phys.: Conf. Ser.* **166** 012004

[17] Eckart W and Gray J M N T *et al* 2003 Particle image velocimetry (PIV) for granular avalanches on inclined planes *Dynamic Response of Granular and Porous Materials Under Large and Catastrophic Deformations* (Berlin: Springer) pp 195–218

[18] Puglisi A, Visco P, Barrat A, Trizac E and van Wijland F 2005 Fluctuations of internal energy flow in a vibrated granular gas *Phys. Rev. Lett.* **95** 110202

[19] Baxter G W and Olafsen J S 2007 Experimental evidence for molecular chaos in granular gases *Phys. Rev. Lett.* **99** 028001

[20] Windows-Yule C R K and Parker D J 2013 Boltzmann statistics in a three-dimensional vibrofluidized granular bed: idealizing the experimental system *Phys. Rev.* E **87** 022211

[21] Wakabayashi T 1950 Photo-elastic method for determination of stress in powdered mass *J. Phys. Soc. Japan* **5** 383–5

[22] Ajovalasit A, Barone S and Petrucci G 1998 A review of automated methods for the collection and analysis of photoelastic data *J. Strain Anal. Eng. Des* **33** 75–91

[23] Veje C T, Howell D W, Behringer R P, Schöllmann S, Luding S and Herrmann H J 1998 *Fluctuations and flow for granular shearing Physics of Dry Granular Media* (Berlin: Springer) pp 237–42

[24] Howell D, Behringer R P and Veje C 1999 Stress fluctuations in a 2D granular Couette experiment: a continuous transition *Phys. Rev. Lett.* **82** 5241

[25] Veje C T, Howell D W and Behringer R P 1999 Kinematics of a two-dimensional granular Couette experiment at the transition to shearing *Phys. Rev.* E **59** 739

[26] Leven M M 2013 *Photoelasticity: The Selected Scientific Papers of MM Frocht* (Amsterdam: Elsevier)

[27] Cloud G L and Pindera J T 1968 Techniques in infrared photoelasticity *Exp. Mech.* **8** 193–201

[28] Wood D M and Leśniewska D 2011 Stresses in granular materials *Granul. Matter* **13** 395–415

[29] Li F 2010 Study of Stress Measurement Using Polariscope *PhD Thesis* Georgia Institute of Technology

[30] Zadeh A A *et al* 2019 Enlightening force chains: a review of photoelasticimetry in granular matter *Granul. Matter* **21** 83

[31] Vriend N and Thomas A 2019 Stress and velocity fluctuations in photoelastic granular avalanches *Bulletin of the American Physical Society*

[32] Thomas A L, Tang Z, Daniels K E and Vriend N M 2019 Force fluctuations at the transition from quasi-static to inertial granular flow *Soft Matter* **15** 8532–42

[33] Majmudar T S 2006 *Experimental Studies of Two-dimensional Granular Systems using Grain-scale Contact Force Measurements* (Durham, NC: Duke University)

[34] Daniels K E, Kollmer J E and Puckett J G 2017 Photoelastic force measurements in granular materials *Rev. Sci. Instrum.* **88** 051808

[35] Farhadi S and Behringer R P 2014 Dynamics of sheared ellipses and circular disks: effects of particle shape *Phys. Rev. Lett.* **112** 148301

[36] Farhadi S, Zhu A Z and Behringer R P 2015 Stress relaxation for granular materials near jamming under cyclic compression *Phys. Rev. Lett.* **115** 188001

[37] Zhao Y, Zheng H, Wang D, Wang M and Behringer R P 2019 Particle scale force sensor based on intensity gradient method in granular photoelastic experiments *New J. Phys.* **21** 023009

[38] Majmudar T S and Behringer R P 2005 Contact force measurements and stress-induced anisotropy in granular materials *Nature* **435** 1079–82

[39] Franklin S V and Shattuck M D 2016 *Handbook of Granular Materials* (Boca Raton, FL: CRC Press)

[40] Puckett J G and Daniels K E 2013 Equilibrating temperature like variables in jammed granular subsystems *Phys. Rev. Lett.* **110** 058001

[41] Ren J, Dijksman J A and Behringer R P 2013 Reynolds pressure and relaxation in a sheared granular system *Phys. Rev. Lett.* **110** 018302

[42] Frocht M M 1941 *Photoelasticity* (Amsterdam: Elsevier)

[43] Timoshenko S P and Goodier J N 1970 *Theory of Elasticity* (New York: McGraw-Hill)

[44] Michell J H 1900 Some elementary distributions of stress in three dimensions *Proc. Lond. Math. Soc.* **1** 23–61

[45] Siewerdsen J H and Jaffray D A 2000 Optimization of x-ray imaging geometry (with specific application to flat-panel cone-beam computed tomography) *Med. Phys.* **27** 1903–14

[46] Windows-Yule C R K, Weinhart T, Parker D J and Thornton A R 2014 Effects of packing density on the segregative behaviors of granular systems *Phys. Rev. Lett.* **112** 098001

[47] Tai C H, Hsiau S S and Kruelle C A 2010 Density segregation in a vertically vibrated granular bed *Powder Technol.* **204** 255–62

[48] Rosato A and Windows-Yule K 2020 Investigative approaches I: experimental imaging techniques *Segregation in Vibrated Granular Systems* ed A Rosato and K Windows-Yule (New York: Academic) ch 3 pp 37–74

[49] Raparia D, Alessi J and Kponou A 1997 The algebraic reconstruction technique (ART) *Proc. of the 1997 Particle Accelerator Conf. (Cat. No. 97CH36167)* vol 2 (Piscataway, NJ: IEEE)

[50] Herman G T and Lent A 1976 Iterative reconstruction algorithms *Comput. Biol. Med.* **6** 273–94

[51] Beister M, Kolditz D and Kalender W A 2012 Iterative reconstruction methods in x-ray CT *Phys. Medica* **28** 94–108

[52] Stearns C W, Chesler D A and Brownell G L 1987 Three dimensional image reconstruction in the Fourier domain *IEEE Trans. Nucl. Sci.* **34** 374–8

[53] Wang G, Ye J C, Mueller K and Fessler J A 2018 Image reconstruction is a new frontier of machine learning *IEEE Trans. Med. Imaging* **37** 1289–96

[54] Willemink M J and Noël P B 2019 The evolution of image reconstruction for CT–from filtered back projection to artificial intelligence *Eur. Radiol.* **29** 2185–95

[55] Jain A K 1989 *Fundamentals of Digital Image Processing* (Englewood Cliffs, NJ: Prentice-Hall)

[56] Athanassiadis A G, La Rivière P J, Sidky E, Pelizzari C, Pan X and Jaeger H M 2014 X-ray tomography system to investigate granular materials during mechanical loading *Rev. Sci. Instrum.* **85** 083708

[57] Thompson W M, Lionheart W R B, Morton E J, Cunningham M and Luggar R D 2015 High speed imaging of dynamic processes with a switched source x-ray CT system *Meas. Sci. Technol.* **26** 055401

[58] Lionheart W R B and Coban S B 2017 Nonlinear problems in fast tomography *Developments in X-Ray Tomography XI* vol 10391 (Int. Society for Optics and Photonics) p 1039116

[59] Laruelle M *et al* 1993 SPECT imaging of dopamine and serotonin transporters with [123i] β-CIT: pharmacological characterization of brain uptake in nonhuman primates *Synapse* **13** 295–309

[60] Benamer H T S, Patterson J, Wyper D J, Hadley D M, Macphee G J A and Grosset D G 2000 Correlation of Parkinson's disease severity and duration with 123I-FP-CIT SPECT striatal uptake *Mov. Disord.* **15** 692–8

[61] Wilcox J 2012 *Carbon Capture* (Berlin: Springer)

[62] Iida H, Itoh H, Nakazawa M, Hatazawa J, Nishimura H, Onishi Y and Uemura K 1994 Quantitative mapping of regional cerebral blood flow using iodine-123-IMP and SPECT *J. Nucl. Med.* **35** 2019–30

[63] Israel O, Iosilevsky G, Bettman L, Frenkel A, Ish-Shalom S, Steiner M, Ben-Harush M and Kolodny G M *et al* 1990 SPECT quantitation of iodine-131 concentration in phantoms and human tumors *J. Nucl. Med.* **31** 1945–9

[64] Podreka I, Suess E, Goldenberg G, Steiner M, Brücke T, Müller C H, Lang W, Neirinckx R D and Deecke L 1987 Initial experience with technetium-99 m HM-PAO brain SPECT *J. Nucl. Med.* **28** 1657–66

[65] Madsen M T 2007 Recent advances in SPECT imaging *J. Nucl. Med.* **48** 661–73

[66] Slomka P J, Pan T, Berman D S and Germano G 2015 Advances in SPECT and PET hardware *Prog. Cardiovasc. Dis.* **57** 566–78

[67] Buck A K *et al* 2008 SPECT/CT *J. Nucl. Med.* **49** 1305–19

[68] Kinahan P E, Townsend D W, Beyer T and Sashin D 1998 Attenuation correction for a combined 3D PET/CT scanner *Med. Phys.* **25** 2046–53

[69] Patton J A and Turkington T G 2008 SPECT/CT physical principles and attenuation correction *J. Nucl. Sci. Technol.* **36** 1–10

[70] Cherry S R, Sorenson J A and Phelps M E 2012 Positron emission tomography *Physics in Nuclear Medicine* ed S R Cherry, J A Sorenson and M E Phelps 4th edn (Philadelphia, PA: Saunders) ch 18 pp 307–43

[71] Razifar P, Sandström M, Schnieder H, Laangström B, Maripuu E, Bengtsson E and Bergstrom M 2005 Noise correlation in PET, CT, SPECT and PET/CT data evaluated using autocorrelation function: a phantom study on data, reconstructed using FBP and OSEM *BMC Med. Imaging* **5** 5

[72] Soleimani M and Lionheart W R B 2005 Nonlinear image reconstruction for electrical capacitance tomography using experimental data *Meas. Sci. Technol.* **16** 1987

[73] Fang W 2004 A nonlinear image reconstruction algorithm for electrical capacitance tomography *Meas. Sci. Technol.* **15** 2124

[74] Marashdeh Q, Teixeira F L and Fan L-S 2015 *Electrical capacitance tomography Industrial Tomography* (Amsterdam: Elsevier) pp 3–21

[75] Marashdeh Q, Fan L-S, Du B and Warsito W 2008 Electrical capacitance tomography–a perspective *Ind. Eng. Chem. Res.* **47** 3708–19

[76] Darma P N, Baidillah M R, Sifuna M W and Takei M 2019 Improvement of image reconstruction in electrical capacitance tomography (ECT) by sectorial sensitivity matrix using a k-means clustering algorithm *Meas. Sci. Technol.* **30** 075402
[77] Yan H, Wang Y, Wang Y F and Zhou Y G 2020 Electrical capacitance tomography image reconstruction by improved orthogonal matching pursuit algorithm *IET Sci. Meas. Technol.* **14** 367–75
[78] Daily W, Ramirez A, Binley A and LeBrecque D 2004 Electrical resistance tomography *Lead. Edge* **23** 438–42
[79] Holder D S 2004 *Electrical Impedance Tomography: Methods, History and Applications* (Boca Raton, FL: CRC Press)
[80] Hahn E L 1950 Nuclear induction due to free Larmor precession *Phys. Rev.* **77** 297
[81] Hahn E L 1953 Free nuclear induction *Phys. Today* **6** 4–9
[82] Bernstein M A, King K F and Zhou X J 2004 *Handbook of MRI Pulse Sequences* (Amsterdam: Elsevier)
[83] Tubridy N and McKinstry C S 2000 Neuroradiological history: Sir Joseph Larmor and the basis of MRI physics *Neuroradiology* **42** 852–5
[84] Allisy-Roberts P J and Williams J 2007 *Farr's Physics for Medical Imaging* (Elsevier: Amsterdam)
[85] Cresser J D 2011 Quantum Physics Notes *Department of Physics, Macquarie University, Australia*
[86] Johnston W, Dybbs A and Edwards R 1975 Measurement of fluid velocity inside porous media with a laser anemometer *Phys. Fluids* **18** 913–4
[87] Huang A Y L, Huang M Y F, Capart H and Chen R-H 2008 Optical measurements of pore geometry and fluid velocity in a bed of irregularly packed spheres *Exp. Fluids* **45** 309–21
[88] van der Vaart K, Gajjar P, Epely-Chauvin G, Andreini N, Gray J M N T and Ancey C 2015 Underlying asymmetry within particle size segregation *Phys. Rev. Lett.* **114** 238001
[89] Dijksman J A, Rietz F, Lőrincz K A, van Hecke M and Losert W 2012 Invited article: refractive index matched scanning of dense granular materials *Rev. Sci. Instrum.* **83** 011301
[90] Dijksman J A, Brodu N and Behringer R P 2017 Refractive index matched scanning and detection of soft particles *Rev. Sci. Instrum.* **88** 051807
[91] Russ J C 1997 *The Image Processing Handbook* (Boca Raton, FL: CRC Press) pp 19–60
[92] Besseling R, Isa L, Weeks E R and Poon W C K 2009 Quantitative imaging of colloidal flows *Adv. Colloid Interface Sci.* **146** 1–17
[93] Crocker J C and Grier D G 1996 Methods of digital video microscopy for colloidal studies *J. Colloid Interface Sci.* **179** 298–310
[94] Olafsen J 2010 *Experimental and Computational Techniques in Soft Condensed Matter Physics* (Cambridge: Cambridge University Press)
[95] Wang Z, Lee K O and Gardner R P 2014 A dual system for monitoring the positions of multiple radioactive tracer pebbles in scaled pebble bed reactors *Nucl. Technol.* **185** 259–69
[96] Rasouli M, Bertrand F and Chaouki J 2015 A multiple radioactive particle tracking technique to investigate particulate flows *AIChE J.* **61** 384–94
[97] Larachi F, Chaouki J, Kennedy G and Dudukovic M P 1997 Radioactive particle tracking in multiphase reactors: principles and applications *Non-invasive Monitoring of Multiphase Flows* (Amsterdam: Elsevier) pp 335–406

[98] Roy S, Larachi F, Al-Dahhan M H and Dudukovic M P 2001 Resolution and sensitivity in computer-automated radioactive particle tracking (CARPT) *Process Imaging for Automatic Control* vol 4188 (Int. Society for Optics and Photonics) pp 122–33
[99] Lin J S, Chen M M and Chao B T 1985 A novel radioactive particle tracking facility for measurement of solids motion in gas fluidized beds *AIChE J.* **31** 465–73
[100] Gupta A and Roy S 2013 Euler-Euler simulation of bubbly flow in a rectangular bubble column: experimental validation with radioactive particle tracking *Chem. Eng. J.* **225** 818–36
[101] Kalaga D V, Pant H J, Dalvi S V, Joshi J B and Roy S 2017 Investigation of hydrodynamics in bubble column with internals using radioactive particle tracking (RPT) *AIChE J.* **63** 4881–94
[102] Pant H J 2021 Preparation of radioactive particles for particle tracking experiments *Appl. Radiat. Isot.* **168** 109380
[103] Karwi A A M 2012 Preparation radioactive particle technique used to track the motion of oil phase in high pressure tank *Int. J. Electron. Commun. Comput. Eng.* **3** 2278–4209
[104] Moslemian D, Devanathan N and Dudukovic M P 1992 Radioactive particle tracking technique for investigation of phase recirculation and turbulence in multiphase systems *Rev. Sci. Instrum.* **63** 4361–72
[105] Doucet J, Bertrand F and Chaouki J 2008 An extended radioactive particle tracking method for systems with irregular moving boundaries *Powder Technol.* **181** 195–204
[106] Roy S, Larachi F, Al-Dahhan M H and Duduković M P 2002 Optimal design of radioactive particle tracking experiments for flow mapping in opaque multiphase reactors *Appl. Radiat. Isot.* **56** 485–503
[107] Larachi F, Chaouki J and Kennedy G 1995 3-D mapping of solids flow fields in multiphase reactors with RPT *AIChE J.* **41** 439–43
[108] Nicuşan A L and Windows-Yule C R K 2020 Positron emission particle tracking using machine learning *Rev. Sci. Instrum.* **91** 013329
[109] Buist K A, van der Gaag A C, Deen N G and Kuipers J A M 2014 Improved magnetic particle tracking technique in dense gas fluidized beds *AIChE J.* **60** 3133–42
[110] Hu C, Li M, Song S, Zhang R and Meng M Q-H *et al* 2010 A cubic 3-axis magnetic sensor array for wirelessly tracking magnet position and orientation *IEEE Sens. J.* **10** 903–13
[111] Nocedal J and Wright S 2006 *Numerical Optimization* (Berlin: Springer)
[112] Degaleesan S, Dudukovic M P and Pan Y 2002 Application of wavelet filtering to the radioactive particle tracking technique *Flow Meas. Instrum.* **13** 31–43
[113] Buist K A, Van Erdewijk T W, Deen N G and Kuipers J A M 2015 Determination and comparison of rotational velocity in a pseudo 2-d fluidized bed using magnetic particle tracking and discrete particle modeling *AIChE J.* **61** 3198–207
[114] Rackl M, Grötsch F E, Rusch M and Fottner J 2017 Qualitative and quantitative assessment of 3D-scanned bulk solid heap data *Powder Technol.* **321** 105–18
[115] Cho G-C, Dodds J and Santamarina J C 2006 Particle shape effects on packing density, stiffness, and strength: natural and crushed sands *J. Geotech. Geoenviron. Eng.* **132** 591–602
[116] Sarraguça M C, Cruz A V, Soares S O, Amaral H R, Costa P C and Lopes J A 2010 Determination of flow properties of pharmaceutical powders by near infrared spectroscopy *J. Pharm. Biomed. Anal.* **52** 484–92
[117] Tan G, Morton D A V and Larson I 2015 On the methods to measure powder flow *Curr. Pharm. Des.* **21** 5751–65

[118] Huang Q, Zhang H and Zhu J 2009 Experimental study on fluidization of fine powders in rotating drums with various wall friction and baffled rotating drums *Chem. Eng. Sci.* **64** 2234–44

[119] Yang R Y, Yu A B, McElroy L and Bao J 2008 Numerical simulation of particle dynamics in different flow regimes in a rotating drum *Powder Technol.* **188** 170–7

[120] Walton O R and Braun R L 1993 Simulation of rotary-drum and repose tests for frictional spheres and rigid sphere clusters *Technical report* Lawrence Livermore National Lab., CA (United States)

[121] Maneval J E, Hill K M, Smith B E, Caprihan A and Fukushima E 2005 Effects of end wall friction in rotating cylinder granular flow experiments *Granul. Matter* **7** 199–202

[122] Arntz M M H D, den Otter W K, Briels W J, Bussmann P J T, Beeftink H H and Boom R M 2008 Granular mixing and segregation in a horizontal rotating drum: a simulation study on the impact of rotational speed and fill level *AIChE J.* **54** 3133–46

[123] Henein H, Brimacombe J K and Watkinson A P 1983 Experimental study of transverse bed motion in rotary kilns *Metall. Trans.* **B14** 191–205

[124] Ding Y L, Seville J P K, Forster R and Parker D J 2001 Solids motion in rolling mode rotating drums operated at low to medium rotational speeds *Chem. Eng. Sci.* **56** 1769–80

[125] Juarez G, Chen P and Lueptow R M 2011 Transition to centrifuging granular flow in rotating tumblers: a modified Froude number *New J. Phys.* **13** 053055

[126] Neveu A, Janssen P and Lumay G 2020 The flowability of lactose powders to optimise tableting processes *ONdrugDelivery*

[127] Miura K, Maeda K and Toki S 1997 Method of measurement for the angle of repose of sands *Soils Found.* **37** 89–96

[128] Yang R Y, Zou R P and Yu A B 2003 Microdynamic analysis of particle flow in a horizontal rotating drum *Powder Technol.* **130** 138–46

[129] Norouzi H R, Zarghami R and Mostoufi N 2015 Insights into the granular flow in rotating drums *Chem. Eng. Res. Des.* **102** 12–25

[130] Oko C O C, Diemuodeke E O and Akilande I S 2010 Design of hoppers using spreadsheet *Res. Agric. Eng.* **56** 53–8

[131] Fitzpatrick J J, Barringer S A and Iqbal T 2004 Flow property measurement of food powders and sensitivity of Jenike's hopper design methodology to the measured values *J. Food Eng.* **61** 399–405

[132] Jenike A W 1964 Storage and flow of solids *Bulletin No. 123, Utah State University*

[133] Nelson E 1955 Measurement of the repose angle of a tablet granulation *J. Am. Pharm. Assoc.* **44** 435–7

[134] Carr R L, Carr R L, Carr R I and Carr R 1965 Evaluating flow properties of solids *Chem. Eng.* **72** 163–8

[135] Riley R E and Hausner H H 1970 Effect of particle size distribution on the friction in a powder mass *Int. J. Powder Met* **6** 17–22

[136] American Society for Testing, Materials Committee D18 on Soil, and Rock 2008 *Standard Test Method for Bulk Solids Characterization by Carr Indices* (ASTM Int.)

[137] Al-Hashemi H M B and Al-Amoudi O S B 2018 A review on the angle of repose of granular materials *Powder Technol.* **330** 397–417

[138] Philippe P and Bideau D 2002 Compaction dynamics of a granular medium under vertical tapping *Europhys. Lett.* **60** 677

[139] Arsenović D, Vrhovac S B, Jakšić Z M, Budinski-Petković L and Belić A 2006 Simulation study of granular compaction dynamics under vertical tapping *Phys. Rev.* E **74** 061302

[140] Rosato A D, Zuo L, Blackmore D, Wu H, Horntrop D J, Parker D J and Windows-Yule C 2016 Tapped granular column dynamics: simulations, experiments and modeling *Computat. Part. Mech.* **3** 333–48

[141] Philippe P and Bideau D 2003 Granular medium under vertical tapping: change of compaction and convection dynamics around the Liftoff threshold *Phys. Rev. Lett.* **91** 104302

[142] Ludewig F, Dorbolo S, Gilet T and Vandewalle N 2008 Energetic approach for the characterization of taps in granular compaction *Europhys. Lett.* **84** 44001

[143] Zettler A, Hilden J, Koenig M, Breslin C, Aburub A, Allgeier M, Patel P and Mitra B 2016 Evaluation of small-scale powder flow characterization tests in the prediction of large-scale process failures *J. Pharm. Innov* **11** 189–99

[144] Chaib O, Achouri I E, Gosselin R and Abatzoglou N 2021 Gravity mass powder flow through conical hoppers–part I: a mathematical model predicting the radial velocity profiles of free-flowing granular systems as a function of cohesion and adhesion properties *Can. J. Chem. Eng.* **99** 1643–53

[145] Majmudar T S, Sperl M, Luding S and Behringer R P 2007 Jamming transition in granular systems *Phys. Rev. Lett.* **98** 058001

[146] Gioia A 1980 Intrinsic flowability: a new technology for powder-flowability classification *Pharm. Technol.*

[147] Larsson M and Brandner E 2013 Introduction to a new standardized test method for powder mixes, the Gustavsson flowmeter funnel *Proc. EURO PM* 2013 pp 15–8

[148] Schulze D 2008 Silo design for flow *Powders and Bulk Solids: Behavior, Characterization, Storage and Flow* (Berlin: Springer) pp 291–317

[149] Beverloo W A, Leniger H A and Van de Velde J 1961 The flow of granular solids through orifices *Chem. Eng. Sci.* **15** 260–9

[150] US Pharmacopoeia 2002 *Chapter 1174, Powder Flow* p 618 of PF 28(2) (Mar.–Apr. 2002)

[151] Makino K and Kuramitsu K 1986 Estimation method of powder yield locus *KONA Powder Part. J.* **4** 98–103

[152] Emery E, Oliver J, Pugsley T, Sharma J and Zhou J 2009 Flowability of moist pharmaceutical powders *Powder Technol.* **189** 409–15

[153] Schwedes J and Schulze D 1990 Measurement of flow properties of bulk solids *Powder Technol.* **61** 59–68

[154] Molerus O 1985 Kenntnisstand und entwicklungslinien der schüttgutmechanik *Chem. Ing. Tech.* **57** 1–7

[155] Schulze D 1990 *Standard Shear Testing Technique for Particulate Solids using the Jenike Shear Cell* (Rugby: Institution of Chemical Engineers)

[156] Schulze D 2008 Powders and bulk solids *Behaviour, Characterization, Storage and Flow* vol 22 (Springer: Berlin)

[157] Schulze D 1994 Development and application of a novel ring shear tester *Aufbereitungs Technik* **35** 524–35

[158] Schulze D 1994 A new ring shear tester for flowability and time consolidation measurements *Proc. 1st Int. Particle Technology Forum* pp 11–6

[159] Schulze D 1996 Flowability and time consolidation measurements using a ring shear tester *Powder Handl. Process.* **8** 221–6

[160] ASTM Standard 2015 *D6773: Standard Shear Test Method for Bulk Solids Using the Schulze Ring Shear Tester (ASTM Int.)*

[161] Schulze D 2006 *Flow Properties of Powders and Bulk Solids* (Braunschweig: University of Applied Sciences)

[162] Swize T, Osei-Yeboah F, Peterson M L and Boulas P 2019 Impact of shear history on powder flow characterization using a ring shear tester *J. Pharm. Sci.* **108** 750–4

[163] Teunou E and Fitzpatrick J J 2000 Effect of storage time and consolidation on food powder flowability *J. Food Eng.* **43** 97–101

[164] Salústio P J, Inácio C, Nunes T, Silva J P S e and Costa P C 2020 Flow characterization of a pharmaceutical excipient using the shear cell method *Pharm. Dev. Technol.* **25** 237–44

[165] Schulze D 1996 Flowability of bulk solids–definition and measuring techniques, Part I and II *Powder Bulk Eng.* **10** 45–61

[166] Schulze D 1998 Measurement of the flowability of bulk solids *Silos—Fundamentals of Theory, Behaviour and Design* (London and New York: E & FN Spon) pp 18–52

[167] Ancey C 2001 Dry granular flows down an inclined channel: experimental investigations on the frictional-collisional regime *Phys. Rev.* **E65** 011304

[168] ASTM Standard 2020 *ASTM d7891-15: Standard Test Method for Shear Testing of Powders Using the Freeman Technology FT4 Powder Rheometer Shear Cell (ASTM Int.)*

[169] Freeman R 2007 Measuring the flow properties of consolidated, conditioned and aerated powders–a comparative study using a powder rheometer and a rotational shear cell *Powder Technol.* **174** 25–33

[170] Freeman R E, Cooke J R and Schneider L C R 2009 Measuring shear properties and normal stresses generated within a rotational shear cell for consolidated and non-consolidated powders *Powder Technol.* **190** 65–9

[171] Cordts E and Steckel H 2012 Capabilities and limitations of using powder rheology and permeability to predict dry powder inhaler performance *Eur. J. Pharm. Biopharm.* **82** 417–23

[172] Gnagne E H, Petit J, Gaiani C, Scher J and Amani G N 2017 Characterisation of flow properties of foutou and foufou flours, staple foods in West Africa, using the FT4 powder rheometer *J. Food Meas. Charact* **11** 1128–36

[173] Li M 2017 Study of the FT4 powder rheometer: comparison of the test methods and optimization of the protocols *PhD Thesis* University of Compiègne

[174] Hare C, Zafar U, Ghadiri M, Freeman T, Clayton J and Murtagh M J 2015 Analysis of the dynamics of the FT4 powder rheometer *Powder Technol.* **285** 123–7

[175] Nan W, Ghadiri M and Wang Y 2017 Analysis of powder rheometry of FT4: effect of particle shape *Chem. Eng. Sci.* **173** 374–83

[176] Govender I, Cleary P W and Mainza A N 2013 Comparisons of PEPT derived charge features in wet milling environments with a friction-adjusted DEM model *Chem. Eng. Sci.* **97** 162–75

[177] Marigo M, Davies M, Leadbeater T, Cairns D L, Ingram A and Stitt E H 2013 Application of positron emission particle tracking (PEPT) to validate a discrete element method (DEM) model of granular flow and mixing in the turbula mixer *Int. J. Pharm.* **446** 46–58

[178] Windows-Yule C R K, Rosato A D, Rivas N and Parker D J 2014 Influence of initial conditions on granular dynamics near the jamming transition *New J. Phys.* **16** 063016

[179] Jayasundara C T, Yang R Y, Guo B Y, Yu A B, Govender I, Mainza A, van der Westhuizen A and Rubenstein J 2011 CFD-DEM modelling of particle flow in IsaMills–comparison between simulations and PEPT measurements *Miner. Eng.* **24** 181–7

[180] Eesa M and Barigou M 2008 Horizontal laminar flow of coarse nearly-neutrally buoyant particles in non-Newtonian conveying fluids: CFD and PEPT experiments compared *Int. J. Multiph. Flow* **34** 997–1007

[181] Xu B H and Yu A B 1997 Numerical simulation of the gas-solid flow in a fluidized bed by combining discrete particle method with computational fluid dynamics *Chem. Eng. Sci.* **52** 2785–809

[182] Thornton A R, Krijgsman D, te Voortwis A, Ogarko V, Luding S, Fransen R, Gonzalez S, Bokhove O, Imole O and Weinhart T 2013 A review of recent work on the Discrete Particle Method at the University of Twente: an introduction to the open-source package MercuryDPM *Discrete Element Methods* **6**

[183] Cundall P A 1971 A computer model for simulating progressive, large-scale movement in blocky rock system *Proc. of the Int. Symp. on Rock Mechanics*

[184] Cundall P A and Strack O D L 1979 A discrete numerical model for granular assemblies *Geotechnique* **29** 47–65

[185] Marigo M and Stitt E H 2015 Discrete element method (DEM) for industrial applications: comments on calibration and validation for the modelling of cylindrical pellets *KONA Powder Part. J.* **32** 236–52

[186] Coetzee C J 2016 Calibration of the discrete element method and the effect of particle shape *Powder Technol.* **297** 50–70

[187] Coetzee C J 2017 Calibration of the discrete element method *Powder Technol.* **310** 104–42

[188] Boikov A V, Savelev R V and Payor V A 2018 DEM calibration approach: design of experiment *J. Phys.: Conf. Ser.* **1015** 032017

[189] Marshall J S 2009 Discrete-element modeling of particulate aerosol flows *J. Comput. Phys.* **228** 1541–61

[190] Furtney J, Zhang F and Han Y 2013 Review of methods and applications for incorporating fluid flow in the discrete element method *Proc. of the 3rd Int. FLAC/DEM Symp., Hangzhou, China*

[191] Apostolou K and Hrymak A N 2008 Discrete element simulation of liquid-particle flows *Comput. Chem. Eng.* **32** 841–56

[192] Luding S 2008 Introduction to discrete element methods: basic of contact force models and how to perform the micro-macro transition to continuum theory *Eur. J. Environ. Civil Eng.* **12** 785–826

[193] Luding S 1998 Collisions & contacts between two particles *Physics of Dry Granular Media* (Berlin: Springer) pp 285–304

[194] Ting J M and Corkum B T 1992 Computational laboratory for discrete element geomechanics *J. Comput. Civil Eng.* **6** 129–46

[195] Luding S and McNamara S 1998 How to handle the inelastic collapse of a dissipative hard-sphere gas with the TC model *Granul. Matter* **1** 113–28

[196] Duan Y, Feng Z-G, Michaelides E E and Mao S 2017 Modified kinetic theory applied to the shear flows of granular materials *Phys. Fluids* **29** 043302

[197] Di Renzo A and Di Maio F P 2004 Comparison of contact-force models for the simulation of collisions in DEM-based granular flow codes *Chem. Eng. Sci.* **59** 525–41

[198] DEM Solutions 2014 *Edem 2.6 Theory Reference Guide* (Edinburgh: DEM Solutions)

[199] Windows-Yule C R K, Tunuguntla D R and Parker D J 2016 Numerical modelling of granular flows: a reality check *Computat. Part. Mech.* **3** 311–32

[200] Johnson K L, Kendall K and Roberts A A D 1971 Surface energy and the contact of elastic solids *Proc. R. Soc.* A **324** 301–13

[201] Luding S 2008 Cohesive, frictional powders: contact models for tension *Granul. Matter* **10** 235

[202] Walton O R and Johnson S M 2009 Simulating the effects of interparticle cohesion in micron-scale powders *AIP Conf. Proc.* vol 1145 pp 897–900

[203] Richefeu V, El Youssoufi M S, Peyroux R and Radjai F 2008 A model of capillary cohesion for numerical simulations of 3D polydisperse granular media *Int. J. Numer. Anal. Methods Geomech.* **32** 1365–83

[204] Carr M J, Chen W, Williams K W and Katterfeld A 2016 Comparative Investigation on Modelling Wet and Sticky Material Behaviours with a Simplified JKR Cohesion Model and Liquid Bridging Cohesion Model in DEM *ICBMH 2016 - 12th International Conference on Bulk Materials Storage, Handling and Transportation. ICBMH 2016 Conf. Proc. (Darwin, NT, 11-14 July, 2016)* pp 40–49

[205] Walton O R and Braun R L 1986 Viscosity, granular-temperature, and stress calculations for shearing assemblies of inelastic, frictional disks *J. Rheol.* **30** 949–80

[206] Tomas J 2000 Particle adhesion fundamentals and bulk powder consolidation *KONA Powder Part. J.* **18** 157–69

[207] Zhu C Y, Shukla A and Sadd M H 1991 Prediction of dynamic contact loads in granular assemblies *J. Appl. Mech.* **58** 341–6

[208] Sadd M H, Tai Q M and Shukla A 1993 Contact law effects on wave propagation in particulate materials using distinct element modeling *Int. J. Non-Linear Mech.* **28** 251–65

[209] Mishra B K 2003 A review of computer simulation of tumbling mills by the discrete element method: Part I–contact mechanics *Int. J. Miner. Process.* **71** 73–93

[210] Blanes S and Casas F 2017 *A Concise Introduction to Geometric Numerical Integration* (Boca Raton, FL: CRC Press)

[211] Welling U and Germano G 2011 Efficiency of linked cell algorithms *Comput. Phys. Commun.* **182** 611–5

[212] Hockney R W and Eastwood J W 1988 *Computer Simulation Using Particles* (Boca Raton, FL: CRC Press)

[213] Allen M P and Tildesley D J 2017 *Computer Simulation of Liquids* (Oxford: Oxford University Press)

[214] Ogarko V and Luding S 2012 A fast multilevel algorithm for contact detection of arbitrarily polydisperse objects *Comput. Phys. Commun.* **183** 931–6

[215] Iwai T, Hong C-W and Greil P 1999 Fast particle pair detection algorithms for particle simulations *Int. J. Mod. Phys.* C **10** 823–37

[216] Munjiza A A 2004 *The Combined Finite-Discrete Element Method* (New York: Wiley)

[217] Muth B, Müller M K, Eberhard P and Luding S 2002 Collision detection and administration for many colliding bodies *Proc. Discrete Elem. Methods* 7 1–18

[218] Ogarko V and Luding S 2010 Data structures and algorithms for contact detection in numerical simulation of discrete particle systems *Proc. of World Congress Particle Technology* vol 6 (Citeseer)

[219] Raschdorf S and Kolonko M 2011 A comparison of data structures for the simulation of polydisperse particle packings *Int. J. Numer. Methods Eng.* **85** 625–39

[220] Rapaport D C 2004 *The Art of Molecular Dynamics Simulation* (Cambridge: Cambridge University Press)

[221] Hoomans B P B, Kuipers J A M, Briels W J and van Swaaij W P M 1996 Discrete particle simulation of bubble and slug formation in a two-dimensional gas-fluidised bed: a hard-sphere approach *Chem. Eng. Sci.* **51** 99–118

[222] Alder B J and Wainwright T E 1957 Phase transition for a hard sphere system *J. Chem. Phys.* **27** 1208–9

[223] Lubachevsky B D 1991 How to simulate billiards and similar systems *J. Comput. Phys.* **94** 255–83

[224] Miller S 2004 Clusterbildung in Granularen Gasen *Dissertation* University of Stuttgart

[225] Miller S and Luding S 2004 Event-driven molecular dynamics in parallel *J. Comput. Phys.* **193** 306–16

[226] Luding S and Goldshtein A 2003 Collisional cooling with multi-particle interactions *Granul. Matter* **5** 159–63

[227] Kodam M, Bharadwaj R, Curtis J, Hancock B and Wassgren C 2009 Force model considerations for glued-sphere discrete element method simulations *Chem. Eng. Sci.* **64** 3466–75

[228] Markauskas D, Kačianauskas R, Džiugys A and Navakas R 2010 Investigation of adequacy of multi-sphere approximation of elliptical particles for DEM simulations *Granul. Matter* **12** 107–23

[229] Lu G, Third J R and Müller C R 2012 Critical assessment of two approaches for evaluating contacts between super-quadric shaped particles in DEM simulations *Chem. Eng. Sci.* **78** 226–35

[230] Podlozhnyuk A, Pirker S and Kloss C 2017 Efficient implementation of superquadric particles in discrete element method within an open-source framework *Computat. Part. Mech.* **4** 101–18

[231] Nassauer B, Liedke T and Kuna M 2013 Polyhedral particles for the discrete element method *Granul. Matter* **15** 85–93

[232] Boon C W, Houlsby G T and Utili S 2012 A new algorithm for contact detection between convex polygonal and polyhedral particles in the discrete element method *Comput. Geotech.* **44** 73–82

[233] Jiang M, Yu H-S and Leroueil S 2007 A simple and efficient approach to capturing bonding effect in naturally microstructured sands by discrete element method *Int. J. Numer. Methods Eng* **69** 1158–93

[234] Kruggel-Emden H, Sturm M, Wirtz S and Scherer V 2008 Selection of an appropriate time integration scheme for the discrete element method (DEM) *Comput. Chem. Eng.* **32** 2263–79

[235] Bird R B, Stewart W E and Lightfoot E N 2007 *Transport Phenomena* 2nd edn (New York: Wiley)

[236] O'Sullivan C and Bray J 2004 Selecting a suitable time step for discrete element simulations that use the central difference time integration scheme *Eng. Comput.* **21** 278–303

[237] Clay Mathematics Institute 2000 Navier-Stokes equation (https://www.claymath.org/millennium-problems/navier-stokes-equation)

[238] Krüger T, Kusumaatmaja H, Kuzmin A, Shardt O, Silva G and Viggen E M 2017 *The Lattice Boltzmann Method: Principles and Practice* (Berlin: Springer)

[239] Patankar S V 2018 *Numerical Heat Transfer and Fluid Flow* (Boca Raton, FL: CRC Press)

[240] White F M 2011 *Fluid MechanicsMcGraw-Hill Series in Mechanical Engineering* (New York: McGraw-Hill)

[241] Ferziger J H, Perić M and Street R L 2002 *Computational Methods for Fluid Dynamics* vol 3 (Berlin: Springer)

[242] Cheng S W, Dey T K and Shewchuk J 2012 *Delaunay Mesh Generation* (London: Taylor and Francis)

[243] Edelsbrunner H, Li X-Y, Miller G, Stathopoulos A, Talmor D, Teng S-H, Üngör A and Walkington N 2000 Smoothing and cleaning up slivers *Proc. of the Thirty-Second Annual ACM Symp. on Theory of Computing* p 273–77

[244] Sadrehaghighi I 2021 *Mesh Sensitivity & Mesh Independence Study*

[245] Kulkarni S S, Chapman C and Shah H 2016 *Computational Fluid Dynamics (CFD) Mesh Independency Study of a Straight Blade Horizontal Axis Tidal Turbine*

[246] Beinert S, Fragnière G, Schilde C and Kwade A 2015 Analysis and modelling of bead contacts in wet-operating stirred media and planetary ball mills with CFD-DEM simulations *Chem. Eng. Sci.* **134** 648–62

[247] Zhao Y and Su X 2019 IMM FSI model validations and applications for compressible flows *Computational Fluid–Structure Interaction* ed Y Zhao and X Su (New York: Academic) ch 12 pp 355–408

[248] Crank J 1979 *The Mathematics of Diffusion* (Oxford: Oxford University Press)

[249] Courant R, Friedrichs K and Lewy H 1928 Über die partiellen differenzengleichungen der mathematischen physik *Math. Ann.* **100** 32–74

[250] Bui T 2010 Explicit and implicit methods in solving differential equations *PhD Thesis* University of Connecticut

[251] McComb W D 1990 *The Physics of Fluid Turbulence* (Oxford: Oxford University Press)

[252] Wilcox D C *et al* 1998 *Turbulence Modeling for CFD* vol 2 (La Canada, CA: DCW Industries)

[253] Verma M K 2018 *Physics of Buoyant Flows: From Instabilities to Turbulence* (Singapore World Scientific)

[254] Schmidt O T, Towne A, Rigas G, Colonius T and Brès G A 2018 Spectral analysis of jet turbulence *J. Fluid Mech.* **855** 953–82

[255] Lee M, Malaya N and Moser R 2013 Petascale direct numerical simulation of turbulent channel flow on up to 786k cores *Proc. of the Int. Conf. on High Performance Computing Networking, Storage and Analysis* pp 1–11

[256] Tennekes H and Lumley J L 1992 *A First Course in Turbulence* (Cambridge, MA MIT Press)

[257] Jiang C, Mi J, Laima S and Li H 2020 A novel algebraic stress model with machine-learning-assisted parameterization *Energies* **13** 258

[258] Schmitt F G 2007 About Boussinesq's turbulent viscosity hypothesis: historical remarks and a direct evaluation of its validity *C. R. Méc.* **335** 617–27

[259] NASA 2021 Turbulence modeling resource (https://turbmodels.larc.nasa.gov/)

[260] Hanjalić K and Launder B E 1972 A Reynolds stress model of turbulence and its application to thin shear flows *J. Fluid Mech.* **52** 609–38

[261] Kieckhefen P, Pietsch S, Dosta M and Heinrich S 2020 Possibilities and limits of computational fluid dynamics–discrete element method simulations in process engineering: a review of recent advancements and future trends *Annu. Rev. Chem. Biomol. Eng.* **11** 397–422

[262] Versteeg H K and Malalasekera W 2007 *An Introduction to Computational Fluid Dynamics: The Finite Volume Method* (Harlow: Pearson Education)

[263] Delacroix B, Rastoueix J, Fradette L, Bertrand F and Blais B 2021 CFD-DEM simulations of solid-liquid flow in stirred tanks using a non-inertial frame of reference *Chem. Eng. Sci.* **230** 116137

[264] Bergman T L, Incropera F P, DeWitt D P and Lavine A S 2011 *Fundamentals of Heat and Mass Transfer* (New York: Wiley)

[265] Schlichting H and Gersten K 2017 *Boundary-Layer Theory* 9th edn (Berlin: Springer)

[266] Craft T J, Gerasimov A V, Iacovides H and Launder B E 2002 Progress in the generalization of wall-function treatments *Int. J. Heat Fluid Flow* **23** 148–60

[267] Zhiyin Y 2015 Large-eddy simulation: past, present and the future *Chin. J. Aeronaut.* 28 11–24

[268] Leonard A 1975 Energy cascade in large-eddy simulations of turbulent fluid flows *Advances in Geophysics* vol 18 (Amsterdam: Elsevier) pp 237–48

[269] Smagorinsky J 1963 General circulation experiments with the primitive equations: I. The basic experiment *Mon. Weather Rev.* **91** 99–164

[270] Germano M, Piomelli U, Moin P and Cabot W H 1991 A dynamic subgrid-scale eddy viscosity model *Phys. Fluids* **A3** 1760–5

[271] Singh S and You D 2013 A dynamic global-coefficient mixed subgrid-scale model for large-eddy simulation of turbulent flows *Int. J. Heat Fluid Flow* **42** 94–104

[272] Druault P, Lardeau S, Bonnet J-P, Coiffet F, Delville J, Lamballais E, Largeau J-F and Perret L 2004 Generation of three-dimensional turbulent inlet conditions for large-eddy simulation *AIAA J.* **42** 447–56

[273] Veloudis I, Yang Z, McGuirk J J, Page G J and Spencer A 2007 Novel implementation and assessment of a digital filter based approach for the generation of les inlet conditions *Flow Turbul. Combust.* **79** 1–24

[274] Kalland K M 2008 A Navier–Stokes solver for single- and two-phase flow *PhD Thesis* University of Oslo

[275] Nangia N, Griffith B E, Patankar N A and Bhalla A P S 2019 A robust incompressible Navier-Stokes solver for high density ratio multiphase flows *J. Comput. Phys.* **390** 548–94

[276] Brennen C E 2005 *Fundamentals of Multiphase Flows* (Cambridge: Cambridge University Press)

[277] Zhang S, Müller D, Arellano-Garcia H and Wozny G 2013 CFD simulation of the fluid hydrodynamics in a continuous stirred-tank reactor *Chem. Eng. Trans* **32** 1441–6

[278] Kumaran V 2015 Kinetic theory for sheared granular flows *C. R. Phys.* **16** 51–61 (Granular Physics/Physique des milieux granulaires)

[279] Chen P, Yuan Z, Chyang C-S and Zhuan F-X 2011 Sawdust discharge rate from aerated hoppers *Particuology* **9** 306–13

[280] Ozarkar S S, Yan X, Wang S, Milioli C C, Milioli F E and Sundaresan S 2015 Validation of filtered two-fluid models for gas-particle flows against experimental data from bubbling fluidized bed *Powder Technol.* **284** 159–69

[281] Musser J and Carney J 2020 Theoretical review of the MFIX fluid and two-fluid models *Technical Report* DOE/NETL-2020/2100TRN: US2103896 National Energy Technology Laboratory (NETL)

[282] S Jan et al 2004 Gate: a simulation toolkit for PET and SPECT *Phys. Med. Biol.* **49** 4543

[283] Strulab D, Santin G, Lazaro D, Breton V and Morel C 2003 GATE (geant4 application for tomographic emission): a PET/SPECT general-purpose simulation platform *Nucl. Phys.* **B125** 75–9

[284] Binder K, Heermann D, Roelofs L, Mallinckrodt A J and McKay S 1993 Monte Carlo simulation in statistical physics *Comput. Phys.* **7** 156–7

[285] Robert C and Casella G 2013 *Monte Carlo Statistical Methods* (Berlin: Springer)

[286] Forte G, Alberini F, Simmons M and Stitt H E 2021 Use of acoustic emission in combination with machine learning: monitoring of gas-liquid mixing in stirred tanks *J. Intell. Manuf.* **32** 633–47

[287] Yang Y F, Gono Y, Motomura S, Enomoto S and Yano Y 2001 A compton camera for multitracer imaging *IEEE Trans. Nucl. Sci.* **48** 656–61

[288] Andò E, Marks B and Roux S 2021 Single-projection reconstruction technique for positioning monodisperse spheres in 3D with a divergent x-ray beam *Meas. Sci. Technol.* **32** 095405

[289] Maranic Z, Guillard F, Baker J, Einav I and Marks B 2021 A granular thermometer *Granul. Matter* **23** 19

[290] Ott F, Herminghaus S and Huang K 2017 Radar for tracer particles *Rev. Sci. Instrum.* **88** 051801

[291] Born P and Holldack K 2017 Analysis of granular packing structure by scattering of THz radiation *Rev. Sci. Instrum.* **88** 051802

[292] Baker J, Guillard F, Marks B and Einav I 2018 x-ray rheography uncovers planar granular flows despite non-planar walls *Nat. Commun.* **9** 5119

Part II

The PEPT technique in detail

IOP Publishing

Positron Emission Particle Tracking
A comprehensive guide
Kit Windows-Yule, Leonard Nicuşan, Matthew T Herald, Samuel Manger and David Parker

Chapter 5

Tracers and detectors

To image the dynamics of a system using positron emission particle tracking, we require two main, physical components: a radioactively-labelled 'tracer particle', and a suitable detector system, or 'positron camera', to record the motion of said tracer. In this chapter, we discuss both of these items in detail. We begin by discussing what makes an ideal, non-invasive tracer for PEPT imaging, before detailing the two main manners in which a 'normal' material comprising stable atoms may be labelled with a suitable positron-emitting radioisotope. We then move on to discuss the detector systems used in PEPT, providing an overview of the main components from which they are built and the relevant physics thereof, before describing the various geometries into which these components may be formed, including an adaptable 'modular' geometry unique to PEPT imaging. The chapter ends with a discussion of how PEPT tracers, detectors, and even whole PEPT experiments may be numerically modelled using Monte Carlo simulations.

5.1 Creating tracers

5.1.1 Introduction—the ideal tracer

In an ideal world, a PEPT radiotracer should—besides emitting positrons—match the physical properties of the system being studied as closely as possible, while also avoiding leaching radioactivity into the surrounding system. For studies of particulate systems, this goal is in fact eminently achievable, as it is possible—using the methods outlined below—to induce β^+ activity in a wide variety of materials without meaningfully altering their physical properties, allowing for truly non-invasive measurements. For the study of fluid flow, however, a truly identical (i.e. gaseous or liquid) tracer would simply disperse into the system making it impossible to track using PEPT. Nonetheless, using suitably small[1], neutrally-buoyant tracers, PEPT has been used successfully on many occasions to track the motion of various Newtonian and non-Newtonian fluids (see chapter 8, in particular sections 8.4 and 8.6).

[1] The size of the tracer is important as it will determine its ability to precisely follow turbulent flow. The tracer used must adequately small that it can accurately map turbulent flows on the lengthscales of interest.

doi:10.1088/978-0-7503-3071-8ch5

In addition to being physically representative of the medium/media under investigation, PEPT tracers must also possess an activity which is adequately high for it to be tracked successfully, but suitably low that it does not ‘blind’ the detectors (details of which are discussed in section 5.2, below). For current detector systems, an activity between 2 and 20 MBq is typically suitable dependent on the precise materials and geometries being explored. The half-life of the radionuclide with which the tracer is labelled should also be adequately long that its activity does not change dramatically during the duration (typically of the order of hours) of an experiment, yet (ideally) short enough that it will decay to a safe level of radioactivity within a few days, thus not necessitating complicated (and potentially hazardous) storage and disposal methods.

A ‘bulk’ particle can also be directly activated by irradiation using a cyclotron or (less often) a nuclear reactor—a process known as *direct activation.* In other cases, the radioactive material is attached to a particle—typically by absorption from solution—which is referred to as *indirect activation.* The radionuclide needed for indirect activation can again be produced using a cyclotron or reactor; alternatively it may sometimes be available as the decay product of a longer-lived radioactive parent in a ‘generator’ system. In the following sections we first discuss the production of radionuclides (including direct activation), and subsequently the issues involved in indirectly activating particles.

5.1.2 Positron-emitting nuclides and direct activation

Positron-emitting radionuclides used for PEPT are characterised principally by their half-life, and also by the extent of interference from other gamma-ray emissions.

In beta decay, positrons are emitted as a continuous spectrum of kinetic energy up to the maximum (‘endpoint’) energy determined by their *Q*-value (or amount of energy released [1]) of the radioactive decay. The endpoint energy (which shows some degree of negative correlation with half-life) controls the maximum range of the emitted positrons (discussed further in section 5.3), which can in turn affect the ultimate spatial resolution achievable in PET imaging—essentially, the higher the endpoint energy, the further away from the particle being tracked the annihilation pair of photons used to track said particle are produced. However, this is not usually as significant concern in PEPT—a higher endpoint energy will in essence create a greater ‘spread’ of lines of response (LoR) origin points, but they will nonetheless (ignoring for the moment density inhomogeneities in the system of interest) remain approximately normally distributed. As such, their centroids (which is all that is required for a PEPT measurement) can nonetheless be determined with reasonable accuracy.

Assuming that radionuclides can be generated locally (i.e. they do not need to be transported over large distances, losing valuable activity), the ideal half-life for a nuclide to be used as a PEPT tracer is usually a few hours—long enough to perform the necessary studies but not so long that the materials used in a study remain radioactive for a long period afterwards. In some cases much longer-lived nuclides such as ^{22}Na (half-life 2.7 years) have been used; it is then necessary to retrieve the tracer particle after each study, and to provide safe storage for it between uses (as opposed to shorter-lived tracers such as ^{18}F, which become effectively inert after a day or so). Nonetheless, a few long-lived sources are often kept for calibrating detectors and locating reference points within the field of view, for example.

As in medical PET, the nuclide most commonly used for PEPT studies is ^{18}F, which has a half-life of approximately 110 minutes. ^{18}F is commonly produced using a cyclotron, either by bombarding material containing ^{18}O with protons, via the reaction

$$^{18}O(p,\ n)^{18}F, \tag{5.1}$$

or by bombarding natural ^{16}O with ^{3}He via the reaction

$$^{16}O(^{3}He,\ p)^{18}F. \tag{5.2}$$

In principle, a high energy ^{4}He beam may also be used, though this gives a lower yield of the beta-emitting nuclide desired. ^{18}O comprises only 0.2% of natural oxygen, so proton bombardment of natural oxygen results in a negligible yield of ^{18}F (instead producing copious quantities of the short-lived positron emitter ^{13}N, with half-life 10 min). Water enriched to >97% ^{18}O is widely available as a raw material for the synthesis of ^{18}F radiopharmaceuticals, and can be used to generate ^{18}F for the indirect activation of PEPT tracers.

Because radionuclides decay during production, the yield from a given production route is conventionally expressed as the 'saturated' activity which would be hypothetically obtained after an infinitely long bombardment (i.e. when the decay rate becomes equal to the production rate). In practice, bombardment for one half-life results in half the saturated activity. For the production of ^{18}F from water, as an example, bombardment of pure $H_2{}^{18}O$ with 15 MeV protons would give a saturated activity of around 8 GBq per μA of beam, while bombardment of natural water with 35 MeV ^{3}He gives a saturated activity approximately six times lower (1.3 GBq/μA). These yields are for the 'thick target' case where the beam is assumed to be completely stopped by the target; in practice this requires a water layer with a thickness of around 2.5 mm for protons or 1.2 mm for ^{3}He.

Because of the convenience of using natural water (as opposed to ^{18}O), the Birmingham group originally used mainly ^{3}He to produce ^{18}F from ^{16}O. However, as $H_2{}^{18}O$ has become more easily obtainable protons are now usually preferred. To offset the high price of the enriched water, it is generally diluted with natural water. Using water enriched to only 5% ^{18}O, bombardment with 10 μA of 10 MeV protons for 2 hours yields around 1 GBq ^{18}F, comparable to the saturated activity achieved via the ^{3}He bombardment of normal water.

The ^{3}He reaction on ^{16}O is still extensively used for the direct activation of materials other than water which contain oxygen. For example, various sands, or glass, alumina or zirconia beads (all of which are relevant to a range of natural and/or industrial processes) can be directly activated by bombardment with 35 MeV ^{3}He, generating ^{18}F which remains fixed within the bead's physical structure. Because of the intense energy deposited during bombardment this technique can only be used with high-melting point materials which are resistant to radiation damage. For a given bombarding flux, the activity which can be generated within a bead of diameter d is proportional to d^2 (if $d \gg T$) or to d^3 (if $d \ll T$), where T is the useful activation depth for the relevant beam in the relevant material. For 35 MeV ^{3}He on glass, $T \approx 0.6$ mm. Beads down to around 1 mm in diameter can readily be activated like this: a typical irradiation might use 3 μA of ^{3}He spread over a beam spot around 0.5 square centimetres. The current hitting a 1 mm diameter bead is then around 0.1 μA and irradiation for 2 hours will produce ^{18}F activity of up to 30 MBq. For beads much smaller than this, direct activation is impractical—even if a beam flux high enough to generate useful activity could be achieved, the resulting energy density would likely destroy the particle. Activation of larger particles is usually relatively straightforward, although damage can sometimes result from the differential heating as the beam deposits all of its energy in the front face. In order to ensure that tracers remain representative of the bulk medium they are being used to explore, it is important to avoid such damage. It should also be noted that direct activation of a particularly large particle can result in activity being concentrated on one side of the particle, so that in subsequent PEPT tracking it is this point which is tracked rather than the centre of the bead; however, such effects can be mitigated through the careful post-processing of the PEPT data acquired.

Other commonly-used cyclotron-produced PEPT nuclides include ^{55}Co and ^{64}Cu. ^{55}Co (half-life 17.5 hours) can be produced by proton bombardment of iron, via the reaction

$$^{56}\mathrm{Fe}(\mathrm{p}, 2\mathrm{n})^{55}\mathrm{Co}, \tag{5.3}$$

meaning that materials containing iron (for example steels, bauxite, and various other minerals of industrial interest) can be directly activated. During the process, several other shorter-lived positron emitters are produced at the same time, which must be allowed to decay away in order to make optimal use of the tracer produced. Materials containing copper can be directly activated by deuteron bombardment, generating ^{64}Cu (half-life 12.7 hours) via the reaction

$$^{63}\mathrm{Cu}(^{2}\mathrm{H}, \mathrm{p})^{64}\mathrm{Cu}. \tag{5.4}$$

The nuclide ^{64}Cu is unusual among positron emitters in that it can also be produced by neutron capture in the neutron flux from a nuclear reactor via the reaction

$$^{63}\mathrm{Cu}(\mathrm{n}, \gamma)^{64}\mathrm{Cu}. \tag{5.5}$$

The Delft PEPT group (having easier access to a reactor than a cyclotron) produced tracers comprising glass beads which incorporated up to 26% copper oxide by weight. After irradiation for 24 hours in the neutron flux of the Delft reactor (1.7×10^{13} cm^{-2}s^{-1}), this method was shown capable of providing a specific activity of up to 200 MBq $\cdot$ mm^{-3} in the tracers [2].

The most useful *generator* system for PEPT is the ^{68}Ge/^{68}Ga system. ^{68}Ge decays to ^{68}Ga with a half-life of 271 days. In turn, ^{68}Ga decays with a half-life of 68 minutes, emitting positrons. ^{68}Ga generators containing the parent 68Ge are commercially available, from which the daughter ^{68}Ga can be obtained when required by elution with hydrochloric acid. Techniques for labelling particles with this solution are described in the next section.

Most of the nuclides referred to above are essentially 'pure positron-emitters', in the sense that they do not emit any gamma-rays other than the 511 keV photons resulting from positron annihilation. The exceptions are ^{22}Na and ^{55}Co, both of which emit other high energy gammas. These associated gammas can be detected in coincidence with 511 keV photons, increasing the background of corrupt events and necessitating use of stricter methods for the 'cleaning' of data when processing for PEPT (see chapter 6).

5.1.3 Indirect activation

Direct activation, as described above, is a relatively simple, one-step process: a material containing an isotope that can be converted into a β^+ emitter via exposure to a suitable ion beam is exposed to such an ion beam, creating positron emitting isotopes directly within the material of interest. Indirect activation is a slightly more involved, two-stage process. Though there are various forms of indirect activation, the basic process can be broadly outlined as follows: a solution of a suitable positron-emitting radionuclide is first generated, as described in the previous section; a suitable particle is then immersed in this solution, allowing it to absorb radioactivity. The two nuclides which have mainly been used for indirect activation to date are ^{18}F and ^{68}Ga.

At Birmingham particles are predominantly labelled with ^{18}F, produced as described in the previous section. Up to 1 ml of ^{18}F solution (specific activity up to 500 MBq ml^{-1}), created as described above, is placed in a small pot and a few particles are added. The pot is shaken and simultaneously heated using an infra-red lamp such that the water gradually evaporates. For this particular methodology, the most effective labelling is achieved using anion exchange resins. Since the affinity of such resins for fluoride is lower than for many other anions, their presence must be minimised by using very pure water, and the resin must initially be in the hydroxide form. Fan *et al* [3] investigated the labelling of strong base anion exchange resins. Uptake of 40 MBq was achieved in a 600 μm particle, or 10 MBq in a 60–70 μm particle. Optimum labelling occurs after shaking for around 20–30 minutes—beyond this time, loss of activity due to radioactive decay exceeds any gain due to further uptake. The Bergen group used a similar technique for labelling ion exchange resin with ^{18}F. Beads of Amberlyst A26 resin (hydroxide form) approximately 600 μm diameter were labelled with up to 40 MBq ^{18}F [4].

At different times pots made of glass or of PTFE have been used for the above-described process; PTFE is typically considered superior due to its lack of affinity for ^{18}F. Before use, pots are cleaned using concentrated nitric acid and then rinsed in pure water, so as to avoid any surface contamination that may result in fluoride attaching to the pot rather than to the particles.

Of course, indirect activation is not limited solely to ion exchange resins—a wide variety of other materials can also be labelled using similar (or indeed identical) methodologies with varying degrees of success. Fan *et al* [5] reported attempts to label 210–50 μm particles of six different materials in this way. Hydroxyapatite was found to be the most promising candidate material, but even this took up only 8% of the ^{18}F absorbed by an equivalent-size ion exchange resin. Uptake could, however, be greatly enhanced by first modifying the surface of the particles by exposure to ferric (Fe^{3+}) ions—after treatment in this way, uptake of 19 MBq was achieved in a microcrystalline cellulose particle and 4 MBq in a polyethylene particle.

A significant issue affecting particles labelled in this way is that ^{18}F has a tendency to leach off when the particle is exposed to water, especially if this water is not extremely pure. To counter this, tracers intended for use in aqueous environments are often coated after labelling, as described in the next section.

Though, as mentioned above, most materials cannot compare to ion exchange resins in their ability to hold ^{18}F activity, in order to perform PEPT we do not require excellent uptake—rather, we only need to create a tracer with enough activity to facilitate tracking to a desired degree of accuracy! As such, the 40 MBq achieved even for 600 micron resin particles is in fact excessive, to the point of potentially being detrimental to the accuracy of readings. PEPT experiments have been performed successfully using a dizzying array of indirectly-activated materials, including pharmaceutical APIs and excipients [6–8], plastics [9, 10], and even coffee beans [11].

The Cape Town group has, to date, mainly performed indirect activation using ^{68}Ga eluted from a generator containing the parent ^{68}Ge. Elution is performed with dilute (typically 0.6 M) hydrochloric acid and results in a solution containing the Ga^{3+} ion. Cole *et al* [12] investigated labelling of two resins: Chelex 100, a chelating resin, and Purolite NRW100, a strong cation-exchange resin. Higher activities were obtained using the Purolite resin, and the resulting tracers were stable against leaching in neutral/weakly acidic conditions. Particles as small as 50 μm have been labelled with 0.4 MBq ^{68}Ga [12], while larger (300–400 μm) particles can be labelled with 20–25 MBq [13]. Whereas at Birmingham the pot containing the particles is generally heated until the solution has completely evaporated, at Cape Town the solution containing the particles is shaken for a suitable period (typically 60 min) and then the particles are recovered from the solution.

5.1.4 Handling and coating

For the smallest tracer particles, handling is a significant practical issue. Typically several small particles are simultaneously labelled, from which one (often the most active) must be selected and carefully transferred to the system under study. Since the particles may be too small to be seen with the naked eye, manipulation must be done under a microscope. Various techniques have been used for picking up small particles, including vacuum tweezers or relying on the electrostatic attraction to a metal tip such as a syringe needle. Certainly, there remains significant room for improvement in these relatively crude methodologies—as discussed further in chapter 9—though for applications thus far they have provided a suitable means to an end.

In some cases the labelled particle is coated before use, either to seal in a radionuclide which would otherwise leach out, or to make its surface characteristics match those of the system interest.

For the case in which we wish to seal in activity prior to use in aqueous environments, resin tracers are often coated with cellulose paint or epoxy resin.

A recent innovation is to use ultraviolet-curing epoxy resin, which has been shown to be very effective in preventing leaching of radioactivity, even in high impact and/or high-shear environments where other coatings may become damaged. Typically two or three coats are applied to ensure that activity is completely sealed in.

To make a tracer representative of PMMA spheres, the Cape Town group applied a surface coating of PMMA to their Purolite tracers [13]. An intermediate adhesive layer of epoxy resin (strengthened by the addition of nickel powder) was first applied, and then a layer of 1–20 μm PMMA powder.

For studying processes such as froth flotation where hydrophobic surface properties are crucial (see section 8.7.5), Cole *et al* [14] developed a technique for modifying the labelled Purolite tracer (which is normally hydrophilic) by coating with silanised silica, using cyanoacrylate as the adhesive.

5.2 Detector systems

5.2.1 Introduction

The technology behind the detector systems used for PEPT studies has largely remained the same since the early 2000s. At present PEPT is not a driving force behind detector development, but as the technique evolves, and expands into new fields, we expect more research to focus on improving the high-speed tracking of radioactive tracers (see chapter 9). On the other hand, the use of more mature technology means that the cost of acquiring a detector system capable of performing PEPT can be relatively low, especially when compared to the price-tags of state-of-the-art medical PET systems. In particular for PEPT, we can focus on enhancing data acquisition systems and system geometries to further the technique. One of the beauties of PEPT is that, due to the simplicity of its fundamental operating principles, it can achieve a lot even with decidedly mediocre equipment!

The characteristics of a good PEPT system are high detection efficiency, reasonable spatial resolution, and a fast response, allowing a high rate of coincidences to be detected before saturation (for those readers to whom these terms currently don't mean much, this sentence will become much clearer in the following pages). Typically, commercially available detector systems have been used for PEPT, although bespoke modifications have been performed in order to increase the detection efficiency and also to allow the capture of list-mode LoR data, which older PET scanners did not record due to the cost and availability of the required data storage.

The first generation of PEPT experiments were performed using multi-wire proportional counters (see also chapter 3), however, since the late 1990s scintillation detectors have been used, exclusively read out using photomultiplier tubes (PMTs). Here we will discuss the physics and principles of scintillation detectors, the characteristics of a large-scale detector system, and recent efforts to model PEPT experiments using Monte Carlo software, as well as offering an outlook into future detector development.

5.2.2 Scintillation detectors

The 511 keV photons yielded by positron–electron annihilation are an energetic and (thus) highly-penetrating form of radiation. In order to reliably and efficiently detect these photons, we use scintillation detectors. Scintillation detectors convert the incident radiation into visible light, that can then be captured to produce an electrical signal which is measured to determine the original energy of the incoming photon. The scintillation detector comprises the scintillator itself—generally an inorganic crystal that produces visible light in response to incoming radiation—and a photodetector. The scintillation detector is an indirect type of radiation detector—that is to say it measures a secondary signal generated in response to the primary photon, as opposed to directly measuring the primary photon. In an ideal detector, the amount of visible light produced by the scintillator is proportional to the energy of the incoming photon, and the size of the electrical signal is proportional to the light that is captured.

Scintillation detectors are ideal candidates for PEPT detectors due to the density and high Z-number of the scintillating crystal, resulting in a relatively high attenuation coefficient that makes them highly effective at capturing incident photons. On the other hand, the nature of the indirect method leads to some losses in the capture of visible light, meaning that other forms of detection can offer better energy resolution than scintillation detectors; however, the robustness of PEPT as a technique allows scintillation detectors to be used with very good results.

Principles of operation

The production of scintillation light in an inorganic scintillator occurs when incoming radiation excites electrons in the valence band of the scintillation crystal [15, 16]. The atomic electrons are only allowed to be in discrete energy states belonging to the crystal, and with sufficient energy they can jump the band gap to reach the conduction band. The electron will then de-excite and drop back into the valence band, emitting energy in the form of photons. The band structure of these crystals can be modified through doping; for example, sodium iodide (NaI) can be doped with thallium (Tl) to produce NaI(Tl), introducing new allowed energy levels in the band structures, called activator states. This reduces the effective band gap allowing the electrons to make smaller jumps, resulting in lower energy photons being produced. In the case of NaI(Tl) for example, the emitted photons now have energy in the visible light spectrum; this is illustrated in figure 5.1.

Besides producing scintillation light, electrons may de-excite through other methods such as heat dissipation. This is just one way in which we may lose some energy resolution from a scintillation detector. Energy loss that does not result in the production of scintillation light is known as *quenching*.

The scintillation light is then measured using a photodetector. Although the use of silicon photodiodes is becoming more widespread, the field of PEPT, at present,

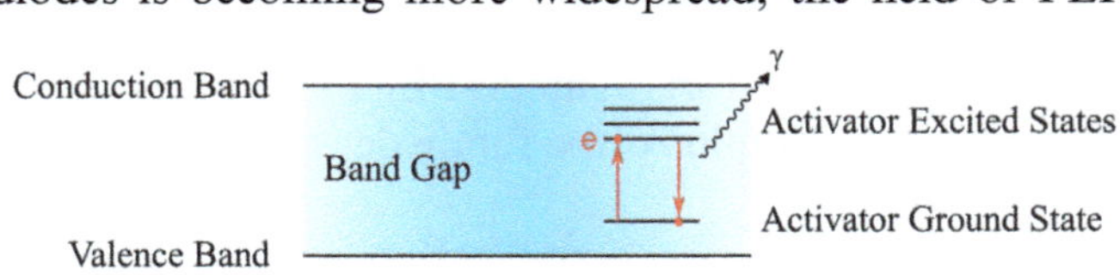

Figure 5.1. Energy levels of a doped scintillator. Adapted from [15] 1980, Copyright © 1980, Alan Martin and Samuel A Harbison.

almost exclusively relies on photomultiplier tubes (PMTs). A PMT is a vacuum tube containing a photocathode, a focussing electrode, a series of dynodes and an anode. Light entering the tube through an entrance window is incident to the photocathode, where the light is absorbed and electrons are released via the photoelectric effect. The focussing electrode channels the electrons towards the first dynode. The dynode has a positive potential applied, accelerating the electrons, giving it sufficient kinetic energy to cause the release of secondary electrons at the first dynode. These secondary electrons are then accelerated towards the next dynode, generating even more secondary electrons in a process that multiplies and amplifies the original signal, which is finally collected on an anode to be analysed and digitized by some readout electronics. This is illustrated schematically in figure 5.2 [16].

The exact readout electronics vary for each detector system, but normally one of the first stages will be the measurement and digitization of the signal from the PMT. Integrating the charge of a pulse from a single event allows the energy of the incoming gamma ray to be determined. The energy spectrum of a large number of pulses is shown in figure 5.3. In this figure, we can clearly see a peak around 511 keV, which we expect as this is the energy of the photon we are detecting. However, there is also a broader, shallower peak between 100 and 200 keV. These lower energy photons have generally lost energy between positron annihilation and absorption in the detector. The most common cause of this energy loss is Compton scattering, meaning that the photon will likely have deviated from its straight-line path and will produce an LoR that sits away from the point at which the annihilation event happened. For this

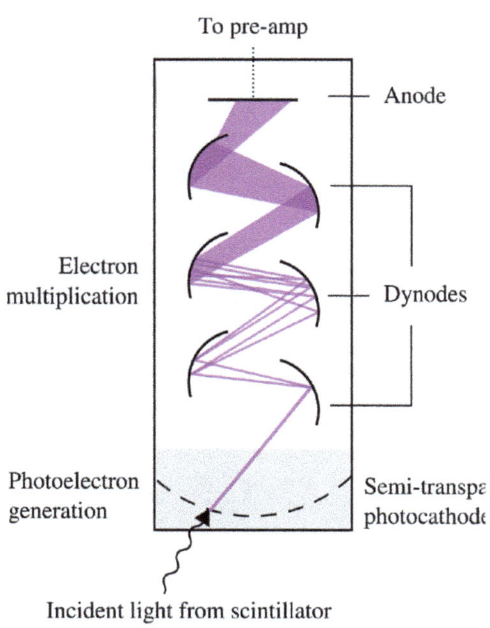

Figure 5.2. A schematic of a typical photomultiplier tube. Adapted from [16] 2005, © Springer-Verlag London Limited 2005, with permission of Springer. Photoelectrons are produced at the photocathode in response to incident light. Positively charged dynodes accelerate the photoelectron towards them generating sufficient kinetic energy to release further electrons, thus amplifying the signal. This is repeated over typically 10–12 dynodes where the signal is then read out.

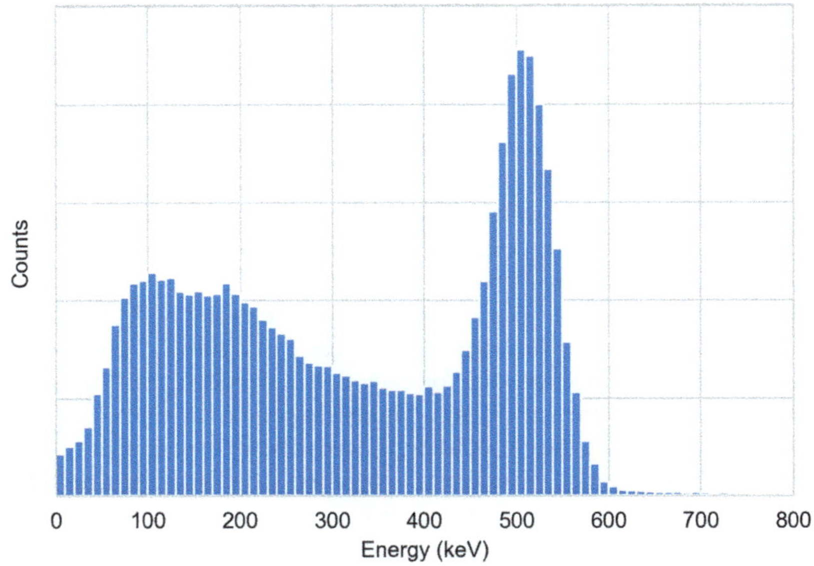

Figure 5.3. A typical singles energy spectrum from a scintillator detector in a PEPT experiment. A photopeak centred at 511 keV represents the detection of unperturbed gamma rays produced by annihilation, with some blurring in the detected energy due to the energy resolution limits of the detector. The broader peak at lower energies arises from detected gamma rays that have undergone scatter and lost a significant proportion of their initial energy.

reason, we usually choose to discard hits that have energy below a threshold using an energy discriminator. Discriminator levels are usually set at around 350 keV for the lower limit, and 650 keV for the upper limit, allowing for some energy blurring to account for the energy resolution of our detectors, however this is less crucial for PEPT than in PET as the PEPT technique is more robust to (false) LoRs caused by scattered events.

There are many opportunities throughout the detection process to lose energy from the incoming gamma ray. This might mean that the gamma ray detection might be lost entirely. The ratio of the number of detected photons, m versus the *expected* number of detected photons n can be represented by a value called the quantum efficiency, which we treat as a macroscopic property encompassing the efficiency of the scintillator, light transfer to the PMT and the PMT combined:

$$\varepsilon_q = \frac{m}{n}. \tag{5.6}$$

Because we are interested in detecting two photons and not just one for a coincidence event, this efficiency becomes even more significant. If we consider that the probability of detecting a single event is given as ε_q, then the probability of detecting a coincidence event becomes ε_q^2. Therefore, increasing the quantum efficiency of our detectors by a factor of 2 increases the probability of detecting a coincident event by a factor of 4.

Scintillator materials

Three materials have been primarily used in scintillator detectors for PEPT. Early scintillation detectors were constructed with thallium-doped sodium iodide (NaI (Tl)), which formed the main scintillator of the first *Anger camera*, named after its inventor Hal Anger [17]. The material offers a high light output and good energy resolution; however, it is fragile and also *hygroscopic*. Hygroscopic materials absorb moisture from the atmosphere, which can affect the uniformity of the detector crystal.

NaI(Tl) can be more sensitive, as detectors that use this material must be sealed in air-tight chambers; they often require much more care in manufacturing and servicing than those using alternative scintillator materials [15].

Bismuth germanate (BGO) scintillators for the detection of photons started to be used in the 1970s [18]. Generally more robust than NaI(Tl) and non-hygroscopic, BGO allowed the use of smaller scintillator elements which enabled the construction of the first ring positron cameras [19]. While the decay time of BGO is higher and its light output is worse than NaI(Tl), the high Z-number and thus density of BGO means that the stopping power of BGO is higher than with NaI(Tl) and thus thinner detector elements can be used. BGO was a prevalent detector material in PET scanners produced throughout the 1990s and 2000s [20, 21].

The third scintillator material used in PEPT detectors is lutetium oxyorthosilicate doped with cerium (LSO). LSO offers a significantly reduced decay time compared to both BGO and NaI(Tl), while still offering a good amount of light output. The density of LSO is similar to BGO, meaning that they have similar stopping powers. However, LSO contains a radioactive isotope decaying by β^- emission, resulting in the production of gamma rays with energies between 88 and 400 keV [16]; using energy discrimination on the detected pulse means that these photons are unlikely to significantly affect PEPT measurements. Considering the properties discussed above, LSO has an advantage over BGO in smaller, more densely packed detector systems, where the chances of the same detector being hit in quick succession are higher. The lower decay time reduces the chance of *pileup*, where an incoming gamma ray causes a scintillation light pulse to be produced before the previous scintillation light pulse has decayed. This can be a significant cause of dead time in detectors (discussed in section 5.2.3). The higher light output also allows LSO to be used in detectors with higher spatial resolution, as the light signal from smaller elements can be more easily shared amongst PMTs [21]; such high resolution PET systems using LSO began to appear in the late 1990s and early 2000s [23–25].

We summarise these scintillator properties in table 5.1.

Detector construction

The particular layout of a scintillation detector has significant consequence on the geometry of the detector system. Two layouts are used to make scintillation detectors for PEPT: the Anger camera is often employed to make large flat-panel detectors; the block detector, on the other hand, is a smaller detector typically used for ring geometries and modular systems. A simple schematic of these PMT arrangements is shown in figure 5.4.

Table 5.1. Some properties of common scintillator materials in PEPT, taken from [16].

Material	NaI(Tl)	BGO	LSO
Density (g cc^{-1})	3.67	7.13	7.4
Effective atomic number	50.6	74.2	65.5
Linear attenuation (cm^{-1}).	0.341 1	0.946	0.865 8
Decay constant (ns)	230	300	40
Light output (relative to NaTl(I))	1	0.15	0.75
Hygroscopic?	Yes	No	No

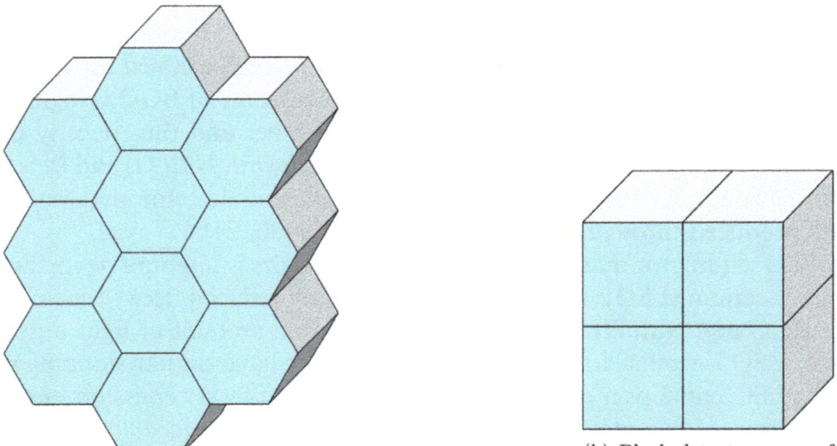

(a) Anger detector array of PMTs

(b) Block detector array of PMTs

Figure 5.4. A schematic of the PMT arrangements of the Anger detector and the block detector. In an Anger detector, a large array of PMTs is used to determine the location of each hit. In a block detector, a smaller array of PMTs are used to determine the location of each hit in a smaller area. Multiple block detectors are combined to cover the same area as an Anger array. (a) Anger detector array of PMTs. (b) Block detector array of PMTs.

The Anger camera is a detector that uses a single scintillating crystal paired to an array of PMTs. Originally designed in the 1950s, the PMT array allows the position of the incoming gamma ray to be determined by weighting the electrical signals from each of the individual PMTs. Through knowledge of the position of these PMTs, a centroid can be calculated [26, 27]. The high light output of NaI(Tl) makes it ideal for use in an Anger camera as the light signal is shared between many PMTs. However, for the same reason the system is susceptible to pileup leading to increased dead time compared to the block detector.

Invented by Casey and Nutt in the 1980s, the *block detector* comprises a segmented crystal coupled to a densely-packed array of PMTs [28]. The crystal is cut into discrete pixel-like elements, with a reduced version of Anger logic used to determine which segment has been hit by the incoming gamma ray, as shown in figure 5.5. Both BGO

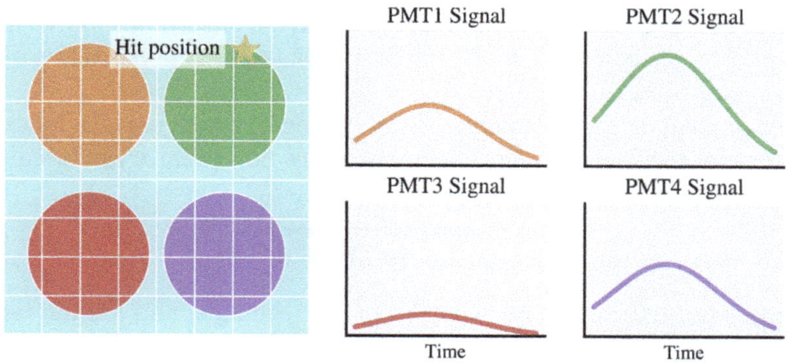

Figure 5.5. An illustration of the individual PMT signals in a block detector when a gamma ray is absorbed in a crystal element. Using Anger logic, the weights of the signals are used to determine the hit location.

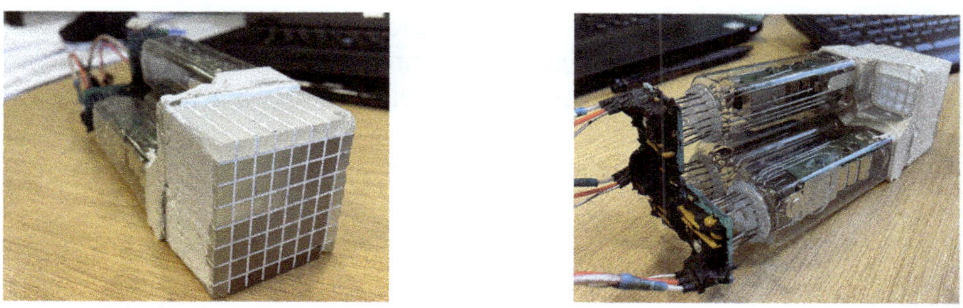

Figure 5.6. Photographs of a block detector with external casing and one photomultiplier tube removed. The cuts in the block crystal are clearly visible as are the dynodes in the photomultiplier tubes.

and LSO are ideal candidate materials for use in block detectors [29]. Compared to the Anger camera, the block detector is much more modular and flexible allowing the construction of ring detectors. Using separate block detectors reduces light-sharing and thus the global dead time of the system, if one block detector is hit and has a decaying light signal, the neighbouring detector is still active. A photograph of a block detector with its light-tight wrapping and one PMT removed is shown in figure 5.6.

5.2.3 PEPT system geometries

We have begun to consider the relative merits of the types of detectors used for PEPT, and we can now cover the topic of detector geometry. The PEPT technique relies on the detection of coincident, quasi-collinear gamma rays, and therefore the geometry of a detector system must encompass the object that we wish to study. To optimise a system for PEPT, we would typically seek to have a high geometric efficiency (described below) whilst retaining a usefully-sized field of view. How much of our PEPT detector system surrounds the object of interest can be described as a *geometric efficiency*, and is quite simply defined as the total solid angle that the detector system makes around a point in the centre, divided by the surface area of the unit sphere around the object (4π). The geometric efficiency is a useful parameter as it describes the percentage of gamma rays emitted from a point source that would be intercepted by the detectors, assuming that it was a perfect system (i.e. all incident gamma rays are successfully detected).

For a rectangular detector element, such as an Anger camera panel or block detector, the solid angle of the face of a detector focussed on a point source can be calculated as

$$\Omega_{\text{det}} = \frac{ab}{\sqrt{(a^2 + 4r^2)(b^2 + 4r^2)}} \tag{5.7}$$

where a and b are the lengths of two sides of the rectangle, and r is the radius from the source to the centre of the rectangular face of the detector [30].

The geometric efficiency is then defined

$$\varepsilon_{\text{geom}} = \frac{\sigma_{\text{det}}}{4\pi}. \tag{5.8}$$

At this point, recalling the quantum efficiency discussed previously, we should state that—ignoring dead time effects—the count rate m of a detector system with geometric efficiency $\varepsilon_{\text{geom}}$ and a quantum efficiency $\varepsilon_{\text{quantum}}$ in the presence of a positron source with activity A decays per second (Becquerels) is:

$$m = 2A\varepsilon_q\varepsilon_{\text{geom}} \tag{5.9}$$

the factor of 2 arises from the fact that each positron emitted from the source annihilates to produce two gamma rays.

The two geometries conventionally used in PEPT are the parallel-plane geometry and ring geometry, as illustrated in figure 5.7.

Parallel plane system

In a parallel-pane system, such as the ADAC Forté Gamma Camera that is used at the Birmingham Positron Imaging Centre, two opposing flat-panel detectors are

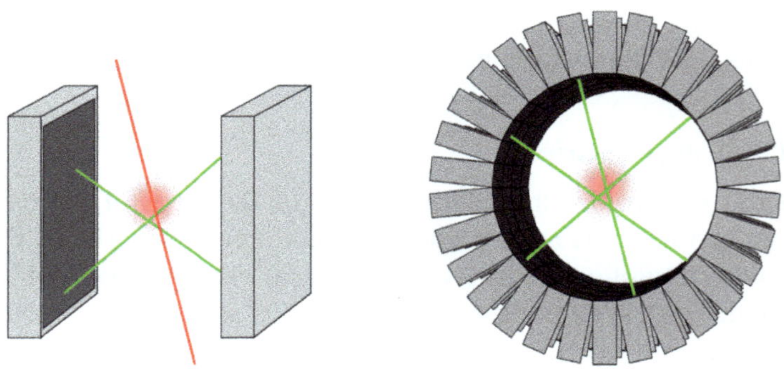

(a) Parallel Plane Detectors (b) Ring Detectors

Figure 5.7. A schematic of parallel plane and ring geometries. The ring detector generally offers more angular coverage of the emitted LORs compared to the parallel plane system, however, the large axial field-of-view of the parallel plane system means that its geometric efficiency remains surprisingly competitive. (a) Parallel Plane Detectors. (b) Ring Detectors.

used to capture coincidence events. Relatively large detectors are used to maximise the geometric efficiency of a such a system. An advantage of this setup is that the field-of-view is only constrained on two sides, and therefore the camera can be used to study quite large pieces of equipment.

Ring geometry system

In a ring geometry system, smaller block detectors are arranged in rings around the field-of-view. Typically, several rings are used along the axis. Examples include the Siemens ECAT HR++ system, used at iThemba LABS with a ring diameter of 83 cm, and the Concord MicroPET P4, with a ring diameter of 19 cm. These systems were originally built for studying humans and primates, respectively, so the sizes of their field of views differ quite significantly; however, their geometric efficiencies are quite similar due to the relatively large axial length of the HR++. The Siemens Inveon scanner, used at the University of Tennessee, has a an impressive geometric efficiency calculated to be 55%. Although the ADAC is capable of almost matching the geometric efficiency of the Inveon, the ring detector system offers a more uniform field of view. It is also worth noting that the quantum efficiency, spatial resolution and dead times of the Inveon scanner are significantly improved in comparison with the ADAC, so that overall the Inveon is significantly more sensitive.

The calculated geometric efficiencies for a number of PEPT systems currently in use are shown in figure 5.8.

Dead time

The scintillator detectors described in the previous section are used to identify single gamma events. We have briefly mentioned the concepts of dead-times and decay times,

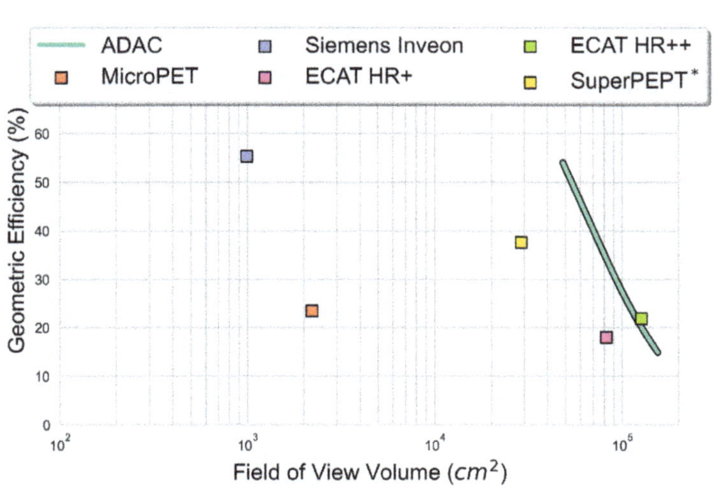

Figure 5.8. The geometric efficiency versus field of view of various systems that have been used for PEPT. The field of view of the ADAC can be adjusted by changing the detector head separation between 25 cm and 80 cm, affecting both the field of view volume and the geometric efficiency, hence it is plotted as a line. The SuperPEPT system is a novel geometry under development at the University of Birmingham, and only the central portion is represented here. Values are calculated from [23, 31–35].

but in this section will go into the concept in greater detail. These can be modelled as we begin to consider the effect of dead-times on the count rates that we can measure; in this sense, 'dead-time' is a generic term describing the interval during which a detector is inactive (i.e. cannot register any new events). With a scintillator detector, this dead time is generally dominated by the decay time of the scintillator; this type of dead-time is known as a *paralysable dead-time*. The dead-time begins when a gamma ray hits the detector and generates the light signal, which we measure and record as a count. If another gamma ray enters while this light is still decaying, we cannot identify and record this new count. However, the second gamma ray will in turn produce more light and extend the effective decay time, thus *paralysing* the detector. Therefore, at high tracer activity, when many gamma rays are incident on the detector, our measured count rate begins to decrease until our detector is paralysed and we see zero counts. In general, for a paralysable dead time τ_p, the measured count rate m can be described as a function of the true count rate n:

$$m = ne^{-n\tau_p}. \tag{5.10}$$

The other dead-time component that we must consider is the *non-paralysable dead time*; in our detectors, this is generally larger than the paralysable dead time but perhaps less dramatic. A non-paralysable dead time would be associated with the electronic components in the signal readout chain. For example, if the coincidence processor receives two hits in quick succession within the dead-time, the second hit may not be processed. This leads to the data rate saturating, hence the term *non-paralysing* dead-time. For a system with solely non-paralysable dead-time, the recorded count rate m can be described as a function of the input count rate n with non-paralysable deadtime τ_{np}:

$$m = \frac{n}{1 + n\tau_{\text{np}}}. \tag{5.11}$$

The effect of each dead-time component on the measured count rate is shown in figure 5.9. In this figure, we use the same time constant for the paralysable and non-paralysable component. We also show the performance of an ideal detector, which demonstrates how quickly the dead-time begins to have an effect on the count rate.

In practice, we rarely observe these dead-time components in isolation. We will often see a component that experiences paralysable dead-time feeding into a component that experiences non-paralysable dead-time. We can model this by substituting in equation (5.10) as an input to equation (5.11), resulting in:

$$m = \frac{ne^{-n\tau_p}}{1 + ne^{-n\tau_p}\tau_{\text{np}}}. \tag{5.12}$$

We show this in figure 5.10. We can use some example values here to get a feel for the effects described before: if we set a paralysable dead-time and a non-paralysable dead-time of 500 ns, we see that the peak of the non-paralysable dead-time curve occurs at a true count rate of 2 MHz, where we measure a count rate of 750 kHz immediately from the detector crystal. However, once we add the non-paralysable dead-time component—which can well be introduced by the analogue-digital-conversion of signal

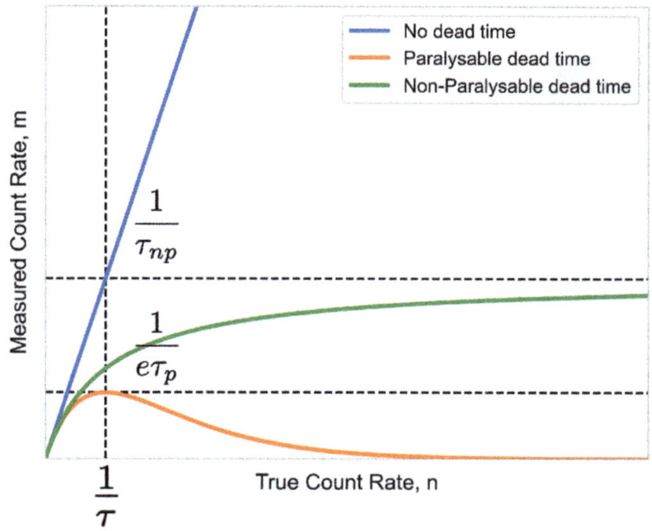

Figure 5.9. Count rate curves showing the measured count rate as a function of the true count rate in an ideal detector system, where the measured rate = true rate, and with paralysable and non-paralysable dead time applied according to equations (5.10) and (5.11). The dead time τ is indicated on the figure, with the paralysable dead time peaking at $1/e\tau$ and the non-paralysable curve reaching an asymptote at $1/\tau$.

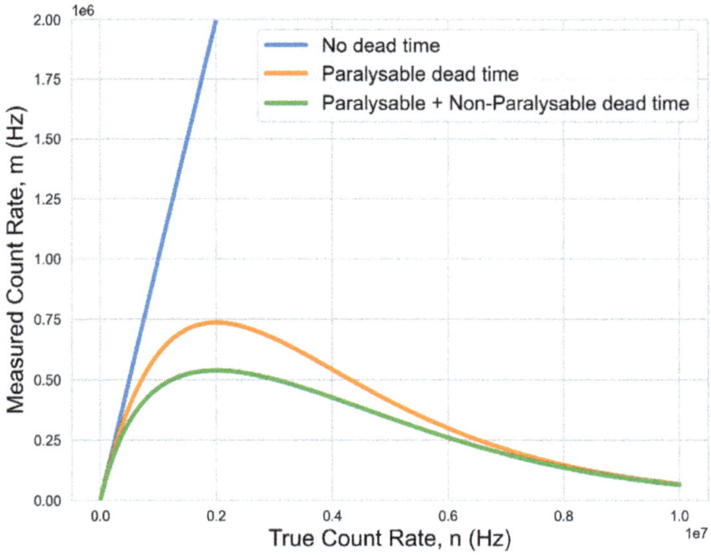

Figure 5.10. An example of a measured count rate as a function of true count rate for a detector with paralysable and non-paralysable dead time components of 500 ns applied in parallel. The non-paralysable dead time component, representative of the signal chain in the detector, further reduces the peak count rate seen at n = 2 MHz, with the measured count rate being 0.75 kHz.

to data—that rate is reduced to almost 500 kHz. In this plot, we can see that our measured count rate as a function of the true count rate is far from linear, and we demonstrate why the approach using modular components such as block detectors and on-board readout electronics is favourable to reduce the effective dead-time of a detector system and thus increase the count-rate efficiency of such a system.

It is rather straightforward to model the dead-times of single components. However, in a complete system dead-time is mitigated by, for example, duplicating the components susceptible to dead-time. In a coincidence processor, there will generally be several application-specific integrated circuits (ASICs) designed to identify single events, adding in redundancy and reducing the effective dead time of the system. Using block detectors means that the scintillator detector is much more modular, and therefore we can tolerate dead time in a block without paralysing a significant portion of the field of view. Extending the simple count rate model for all of these factors becomes quite involved—or downright impossible!—if trying to solve everything analytically. However, we can use Monte Carlo simulations to model such complex systems—which we describe in detail in section 5.3.

Coincidence processing

We now know how a typical PEPT system arrives at the detection of a single gamma ray event. To provide a full LoR, we need two coincident events occurring within a few nanoseconds. This coincidence window is sufficiently short that we can be confident that two detected gamma rays originated from the same positron annihilation event, yet must be long enough to allow the gamma rays to traverse the field of view, and to account for any timing uncertainty in the electronics. For a field of view with detectors of the order of 1 m apart, with the speed of light being 3×10^8 m s^{-1}, we can determine that the time taken for a gamma ray to traverse the field of view is around 3 ns. A typical coincidence window would be up to 15 ns, referred to in this text as 2τ.

Modern PET systems use hardware coincidence processors, using bespoke electronics comprising ASICs and FPGAs in order to deal with the high data rates seen in the detector systems. It is not unusual for a modern PET system to be able to cope with millions of coincidence events per second. However, at its heart the coincidence processor can be thought of as a simple logic device comprising an AND gate triggered by an incoming event, as illustrated in figure 5.11. A detected gamma ray in detector A at time t_A triggers an AND gate with a period of 2τ. If a gamma ray is detected in detector B at a time $t_A \pm \tau$, then a *prompt* coincidence is reported figure 5.12.

The coincidence processor is also capable of reporting *delayed* coincidences. The delayed coincidence AND gate behaves in the same way as the prompt AND gate except its start point is delayed by a period of around 4τ. This time delay means that any two gamma rays that are detected and cause a delayed coincidence are uncorrelated and won't be coming from the same positron annihilation. Therefore, these delayed coincidences can be interpreted as *random* coincidences. Knowing the random coincidence rate at any one time is extremely useful, because it

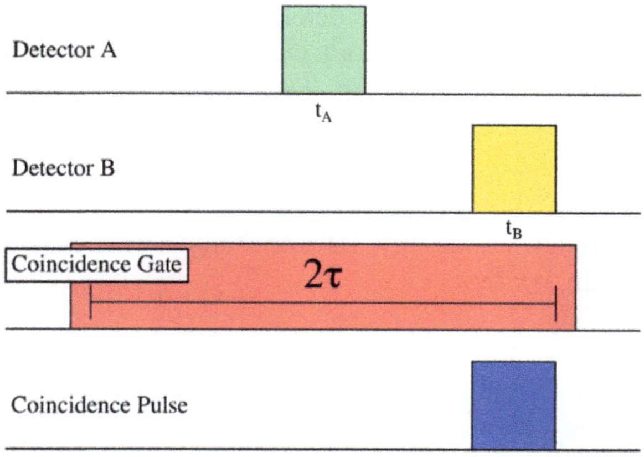

Figure 5.11. An illustration of the coincidence AND gate producing a pulse.

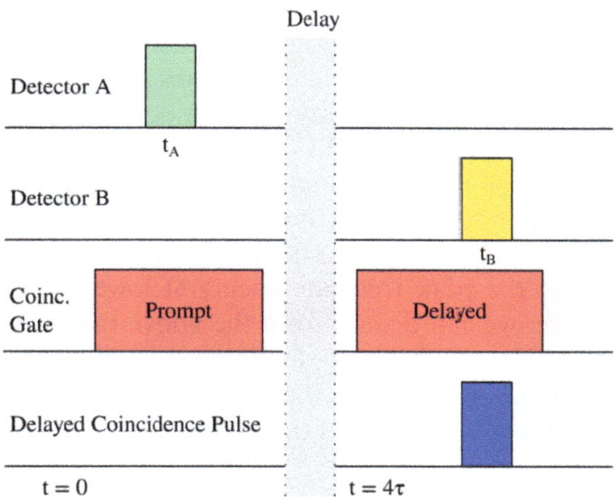

Figure 5.12. An illustration of the prompt and delayed coincidence gates.

will also give an estimate of the portion of prompt coincidences that are also random coincidences. If we want to know the *true*, correlated count rate then we can subtract the delayed coincidence rate from the prompt coincidence rate.

$$R_{\text{true}} = R_{\text{prompt}} - R_{\text{delayed}}. \tag{5.13}$$

Neglecting dead time, the instantaneous prompt coincidence count rate for a source of activity A encompassed by a detector system subtending the solid angle Ω can be calculated by multiplying the probabilities that a gamma ray will be detected in detector A and detector B, effectively the quantum efficiency of the detectors

$$R_{\text{true}} = \frac{A\Omega\varepsilon^2}{4\pi}. \tag{5.14}$$

We can also use the timing window to determine the rate of *random* coincidences. The probability of observing a random event increases with the square of the measured singles rate in each detector. The measured random coincidence rate R_{random} is given as

$$R_{\text{random}} = 2\tau m_A m_B. \tag{5.15}$$

If the measured singles count rate is the same in each detector such that $m_A = m_B = m_S$, we can say that

$$R_{\text{random}} = 2\tau m_s^2 = \frac{2\tau}{\varepsilon^2} R_{\text{true}}^2. \tag{5.16}$$

We can determine from these equations that while the true coincidence count rate increases linearly with source activity, the random coincidence count rate increases with the square of the source activity, meaning that as the activity increases, the fraction of prompt coincidences that are true decreases. At higher activities, the prompt counts become entirely dominated by random coincidences and we are unable to obtain anything meaningful from the measured LoRs.

Measuring the *prompt*, *random* and *true* count rates as functions of source activity allows us to determine the optimal activity that can be used with a detector system. As can be seen in figure 5.13, initially the prompt count rate almost wholly comprises true events. However, as the random event rate increases and we begin to see the effects of paralysable and non-paralysable deadtime saturating the detector, the prompt count rate begins to fall off, even as the random rate increases. In fact, we notice that the peak true rate occurs at lower activity than the peak prompt rate. Therefore, we can deduce from the above that at this source activity level, random coincidences will begin to dominate the LoRs and we can no longer reliably track our particle. Accordingly, there is a limit to the maximum amount of activity that we can expect to be able to track.

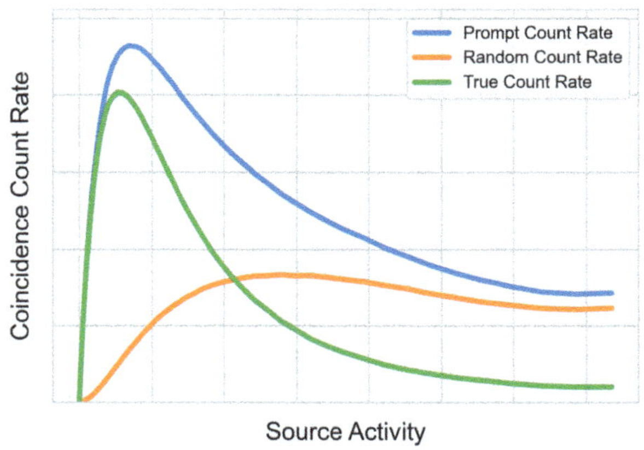

Figure 5.13. Reported coincidence count rates from a coincidence processor. The coincidence processor reports the prompt count rate and the random count rate. We calculate the true count rate by subtracting the random count rate from the prompt count rate.

When working at such high data rates, it is also possible that the system will see three or more single events within a coincidence window, thus defining a *multiple* LoR. Because it is impossible to determine which two events should be correlated to one another, these events are simply discarded.

We illustrate the prompt, random and true count rates as a function of source activity in figure 5.13.

The parallax effect

The parallax effect describes a spatial warping that affects the spatial resolution of PEPT. This effect arises when the emitted gamma ray from a positron annihilation does not interact in the first detector element that it encounters and instead travels into some neighbouring elements before being detected. This occurs when the incident angle into a detector crystal element is fairly shallow, meaning that the path-length the gamma ray takes in the first detector element is not sufficient to stop the gamma ray. This effect means that the detected LoR location is 'pulled' towards the centre of the field of view [16, 36].

Although the parallax effect is generally thought to be a radial distortion in ring scanners, in scanners with an extended axial field of view with elongated bores the parallax effect can also occur in the axial direction. This is illustrated in figure 5.14 for both the axial and radial dimension. Along the axial field of view, the effect can be compensated for by either discarding LoRs that have oblique angles either through the use of electronic logic or by reducing the angular acceptance window of the detector through the use of physical septa or collimators. While this approach is effective, it reduces the number of LoRs acquired by the system, reducing the overall efficiency.

A correction for the parallax effect in both the axial and radial dimension is suggested by Wiggins *et al* [37]. By using a well-located point source, a dewarping map can be produced by comparing the known point source location with a measured point source location, determined with a PEPT algorithm[2]. A third order polynomial fit is performed to the data in the radial and axial dimensions to produce a correction that can be applied to recorded PEPT data in post-processing.

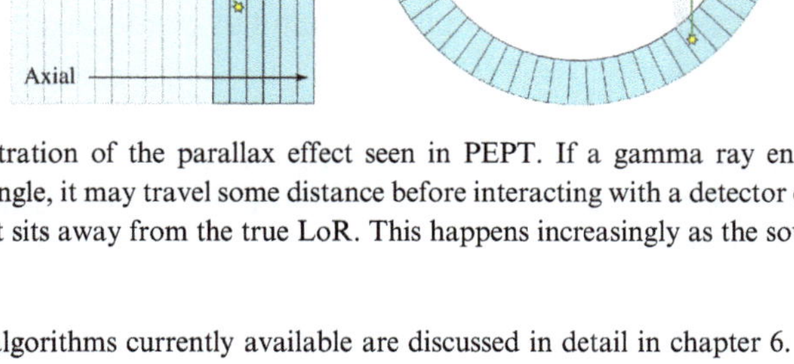

Figure 5.14. An illustration of the parallax effect seen in PEPT. If a gamma ray enters the detector at a particularly shallow angle, it may travel some distance before interacting with a detector element, meaning that its measurement point sits away from the true LoR. This happens increasingly as the source moves off-centre.

[2] The various PEPT algorithms currently available are discussed in detail in chapter 6.

An alternative solution to these methods that requires no post-processing—while still retaining efficiency—is through the use of depth-of-interaction detectors [38]. With extra information about the depth that the gamma ray interacted with the detector, the LoR measurement can be improved, reducing the so-called (radial) astigmatism of the system. While the technology has not yet made it to PEPT, there are a number of proposed schemes to determine the depth-of-interaction inside a scintillator crystal.

5.2.4 Examples of PEPT systems

Multi-wire proportional counter

The first type of gamma camera used for PEPT was a multi-wire proportional counter (MWPC), based on a design from Rutherford Appleton Labs. The first studies discussing the design of the camera date back to 1986, with PEPT first described in 1991 [39, 40]. The motivation for developing the MWPC rose from the high cost of scintillating crystals and photomultiplier tubes, which are now commonplace in modern detector systems.

The MWPC system comprises two detector planes with adjustable separation between them, with each plane having an active area measuring $60 \times 30\ \text{cm}^2$. The detector plane construction is made of alternating layers of lead cathodes and gold-plated tungsten wire anodes in a gas-filled chamber. The lead cathode layers have strips of lead measuring 2.2 mm wide separated by 0.8 mm designed to absorb the incoming gamma rays, and thus release free electrons. A high voltage (3.5 kV) is applied to the anode wires, accelerating the free electrons such that their kinetic energy exceeds the ionisation energy of the gas, liberating further electrons that are also accelerated towards the anode and thus amplifying the signal on the anode from a single gamma ray. The use of 20 detector layers means that the depth of interaction in the detector plane can also be measured, which in a detector plane with a thickness of 132 mm considerably reduces the parallax error in a measured line of response between the two planes.

Even though the location of each hit can be determined with a FWHM of 6 mm (comparable to the significantly more modern ADAC camera discussed previously), the detector is inefficient. A detection efficiency of 7% for single gamma rays leads to a coincidence detection rate of about 0.5%. The random coincidence detection rate begins to dominate the total coincidence rate when the source activity approaches 4 MBq, meaning that the true detection rate saturates at around 3–4 kHz. While these rates were acceptable at the time, and particles could be located and tracked within a few mm of accuracy at approximately 20 Hz, advances in detector technology have massively dwarfed these data rates and the system has been long retired.

ADAC Forté Gamma Camera

In 1999 the University of Birmingham acquired an ADAC Forté Gamma Camera for use as a PEPT instrument [33], shown in figure 5.15. The camera contains two sodium iodide (NaI(Tl)) crystal scintillator plates measuring approximately 47×59 cm and 16 mm thick on adjustable arms, such that the distance between them can be adjusted from 25 cm to 80 cm. This adjustment allows the geometric sensitivity of the

Figure 5.15. The ADAC Forté Gamma camera at the University of Birmingham.

camera to be altered, as well as allowing equipment of varying sizes into the field of view. A custom data acquisition system records coincidence events onto a PC allowing the LoRs to be reconstructed.

The scintillator detectors are read out by an array of PMTs in the Anger logic configuration, with 55 PMTs on each detector head. Each PMT is read out by its own analogue-to-digital converter (or ADC), with the signals fed into a computer on each detector head. A software routine on the computer determines the centroid of each hit by analysing the 55 channels, with the spatial resolution determined to have a FWHM of approximately 6 mm. A measurement of the singles count rate in one detector head was used to estimate the dead time of the detector at around 170 ns with the quantum efficiency determined to be 23%, leading to a coincidence detection efficiency of 5.3%, a factor of 10 higher than the MWPC. Combined with reduced dead time and faster read out electronics, the ADAC is capable of achieving coincidence rates of up to 300 kHz, although the data logging rate is limited at 100 kHz. Nonetheless, the system offers significant improvement over the MPWC for PEPT.

The ADAC system is still under routine use at The University of Birmingham, and a similar gamma camera, the ADAC Vertex, is in use at the University of Cape Town.

Modular camera

The University of Birmingham's Modular Camera is a flexible PEPT system with a geometry that can be adapted to suit the system being studied. The Modular Camera was built from components from Siemens ECAT 931 and 951 PET scanners, where the use of block detectors provides an ideal platform to adapt the components into bespoke geometries [41, 42]. More recent work at the University of Cape Town using similar components to the University of Birmingham is also underway to construct their own modular camera system using components from the ECAT 951 system [43]. The following text is relevant to both systems.

In the original configuration, the Siemens ECAT 931 and 951 scanners comprised two rings of 64 block detectors with a BGO scintillating crystal read out by four photo-multiplier tubes. Each block detector measures 5 cm wide and 5.4 cm tall, with a crystal depth of 3.0 cm. The 931 series block crystals are cut into a 4×8 array, while the 951 series are cut into an 8×8 array, providing elements of approximately $6.25 \times 6.75\ mm^2$. The rings had a diameter of 102 cm, with the total axial field of view being 12.4 cm.

The original scanner construction grouped four radial block detectors into a module known as a *bucket*. Each bucket comprises four block detectors sharing a controller board which performs calibration, addressing and read out of the individual block detectors.

Each bucket then serves as an input to a hardware coincidence processor, that outputs each LOR as a 32-bit data word comprising the bucket, block and crystal address of the first point and the bucket, block and crystal address of the second point. This act of data compression means that with sufficient knowledge of the position of the detector blocks and knowledge of the data format, we can take the data word from the coincidence processor and process it to generate an LoR in Euclidean space. This system design makes it ideal for a flexible geometry that can be optimised for any given system that the user wishes to study.

A detector bucket in the original system becomes a single ‘module’ in the modular camera. Several ECAT 931 and 951 scanners have been acquired by the University of Birmingham, with the components used to create two distinct modular camera systems. The first system retains the original detector bucket construction, with close-packed blocks in a radial arrangement. The second system uses reconstructed buckets that have been built into rectangular modules that are stackable and more easily transportable. The blocks are spaced further apart with a gap of 35 mm between them, however, they all lie on a parallel plane. Both of these buckets are shown in figure 5.16.

When operating the system using the rectangular modules, the gaps between the detectors mean that there is some variation in sensitivity over the field of view.

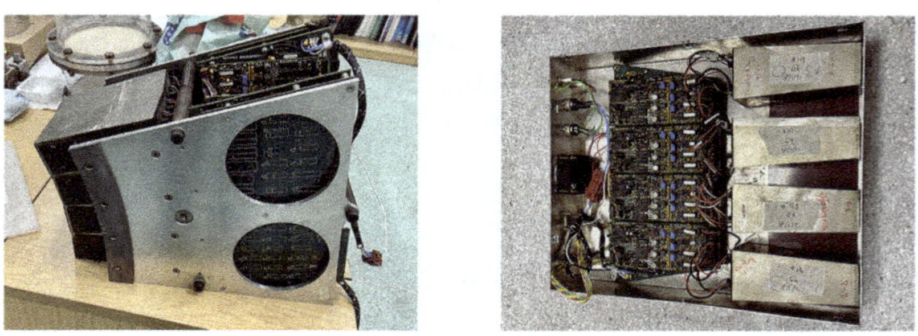

(a) Unmodified ECAT 951 Bucket (b) Square bucket

Figure 5.16. Photographs of the original Siemens ECAT 951 bucket, unmodified after removal from the PET scanner and the custom square bucket geometry. Both detector bucket types are used in the modular camera and they share common hardware, however, the square buckets are stackable and more versatile with a larger field of view, at the expense of there being reduced sensitivity between the detector gaps. (a) Unmodified ECAT 951 Bucket. (b) Square bucket.

The severity of this variation in sensitivity will increase as the detectors are brought closer together, which can result in a loss of either spatial or temporal resolution when tracking, depending on whether the data is tracked using constant time stamps, or fixed numbers of LoRs. This variation in the sensitivity is illustrated schematically in figure 5.17.

When arranging the modules into novel geometries, some thought needs to be given to the *allowed coincidences* that the hardware coincidence processor can produce. In the original configuration, the ECAT series PET scanners were designed to allow coincidences between opposing buckets and the neighbouring three buckets on each side. This is illustrated in figure 5.18, where bucket 0 is allowed to be in coincidence with its opposing bucket 8, as well as 5, 6, 7 and 9, 10, 11. Bucket 1 would then be allowed to be in coincidence with bucket 9, as well as 6, 7, 8 and 10, 11, 12, etc. Any coincidences not in an allowed combination are discarded, as they generally wouldn't have any use in PET reconstruction—that is to say, it is unlikely

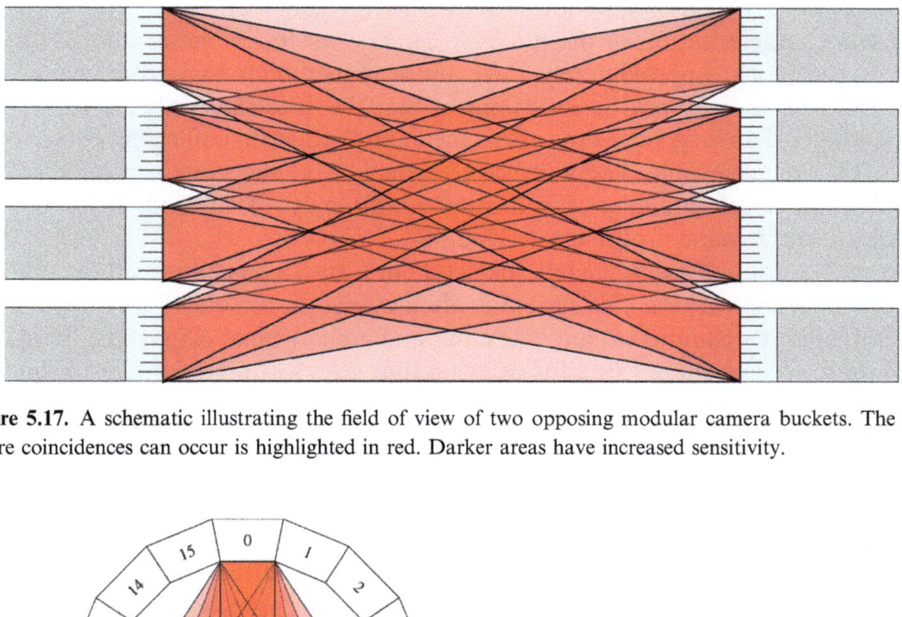

Figure 5.17. A schematic illustrating the field of view of two opposing modular camera buckets. The area where coincidences can occur is highlighted in red. Darker areas have increased sensitivity.

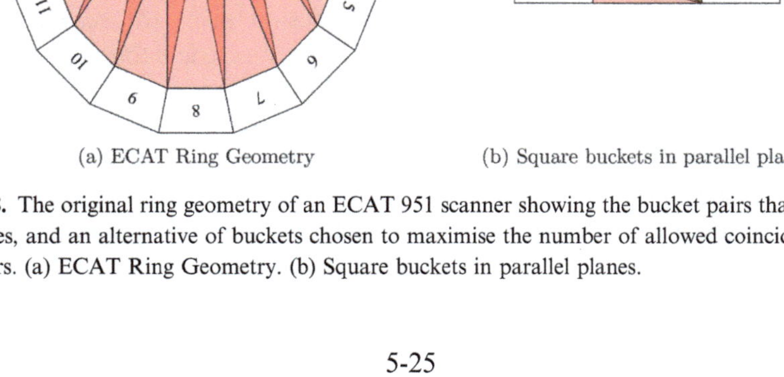

(a) ECAT Ring Geometry

(b) Square buckets in parallel planes

Figure 5.18. The original ring geometry of an ECAT 951 scanner showing the bucket pairs that have allowed coincidences, and an alternative of buckets chosen to maximise the number of allowed coincidences between bucket pairs. (a) ECAT Ring Geometry. (b) Square buckets in parallel planes.

that a pair of back-to-back photons would hit, for example, detector 0 and detector 1. This means that we have to be careful when defining our bucket numbers in novel geometries to ensure that we obtain allowed coincidences. An example is shown in figure 5.18(b) where we have two stacks of four detector buckets opposing each other. This is a common geometry, roughly mimicking the ADAC Camera. As we saw in the ring geometry, bucket 0 can only be in coincidence with buckets 5 through 9. Therefore, if we numbered our right-hand side 4, 5, 6, 7, we would lose one coincidence pair. Bucket 3, however, can only be in coincidence with buckets 8 through to 12. So by skipping bucket numbers 4, 5, 6 and 7 we ensure that we maximise the allowed coincidences between these two planes.

The flexibility of the camera geometry makes it a versatile tool. For pipe flow studies, the buckets can be mounted to an octagonal frame to make a small diameter ring system with a diameter of approximately 20.3 cm and an axial field of view of approximately 70 cm. The system can also be set up alongside the ADAC camera to extend the PEPT field of view for larger industrial equipment.

Initial characterisation of the modular camera system estimates a dead time of 600 ns per block, approximately three times higher than that from the ADAC. However, the use of block detectors means that the dead time only affects a single block, rather than larger detector areas, meaning that useful coincidence rates of up to 1 MHz are achievable. The intrinsic efficiency of the detectors is around 45% at the detector level implying a coincidence efficiency of around 20%, offering significant improvement over the ADAC [44].

As well as routine use in the lab, the Modular Camera has been used for studying industrial systems *in situ*. The system is often moved to the metallurgy department on the University campus to study the casting of molten metal [45], where the system is set up to give a square field of view using four stacks of modular camera buckets around the casting mould, as shown in figure 5.19. In another instance, the camera was used to study a large fluidised bed at a BP site in Hull [46].

Siemens Inveon

The Siemens Inveon small animal PET scanner has been used at the University of Tennessee, Knoxville for PEPT studies since the early 2010s. The detector system uses block detectors comprising an LSO scintillator measuring 32 × 32 mm^2, with the crystal segmented into 20 × 20 crystal elements. The scintillator crystal is 10 mm deep and uses a tapered lightguide to couple the crystal to a single position-sensitive photomultiplier tube [47]. The blocks are arranged into four rings giving the field of view an axial length of 127 mm, with each ring comprising 16 blocks giving a diameter of 161 mm. This relatively compact design gives the scanner excellent geometric efficiency and spatial resolution; however, it restricts the size of the system that can be studied with PEPT. In terms of performance, a maximum true coincidence count rate of 1.9 MHz is reported, using a coincidence window of 3.43 ns [48]. This high data rate capability means that a large number of LoRs can be recorded in the case that multiple radioactive tracers are in the field of view.

The scanner has a bore size of 120 mm diameter, which makes it ideal for studying dynamic pipe flow systems, as shown in figure 5.20. Recent work with the scanner

Figure 5.19. The Modular Camera set up for aluminium casting studies in the metallurgy department at the University of Birmingham.

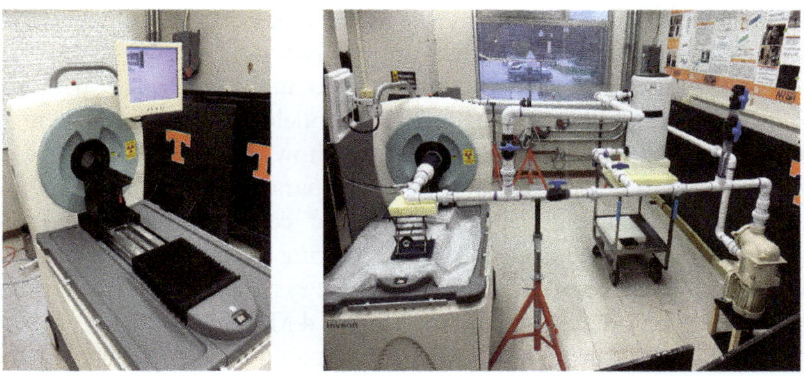

Figure 5.20. The Siemens Inveon scanner at the University of Tennessee, seen on its own and as part of a PEPT experiment studying pipe flow. Photos courtesy of Cody Wiggins, UoT.

characterised turbulent flow conditions using a multiparticle tracking algorithm with approximately 325 tracers in a flow loop, each with an activity of approximately 2 MBq. The PEPT spatial resolution of the experiment was of the order of 0.1 mm [37].

Siemens ECAT HR++

The Siemens ECAT HR++ (also called the ECAT3D in the literature) is a clinical PET scanner with an extended axial field of view compared to standard PET systems. One HR++ system is in use for PEPT, at iThemba LABS, South Africa operated by the University of Cape Town. The HR++ uses block detectors with crystals made

Figure 5.21. The Siemens ECAT HR++ scanner at the University of Cape Town with its covers removed. Photo reprinted with permission of the University of Cape Town [34].

from BGO, with each block measuring 38.8 × 36.0 mm × 30 mm^3 segmented into 8 × 8 crystal elements, with each element measuring 4.39 × 4.05 mm^2 [21].

The HR++ block detectors are arranged to make six rings, with each ring comprising 72 blocks. The axial length of the field of view is 23.4 cm with the detector rings having a diameter of 83 cm, as shown in figure 5.21. The extended axial field of view can be problematic due to the parallax effect; however, in use, the HR++ restricts the coincidence planes so that coincidences are only recorded between blocks in either the same or neighbouring rings, reducing the number of scattered events [49]. The spatial resolution of the system for PEPT is approximately 1 mm, with a maximum achievable data rate of 4 MHz [34].

The larger bore of the ECAT HR++ compared to the Siemens Inveon, for example, makes the scanner more useful for studying systems such as rotating drums, where the spatial resolution requirements are not as stringent as they may be for fluid flow.

5.2.5 Future developments

Many of the detector systems used for PEPT have been derived from clinical PET systems. Because of this, they generally have physical arrangements optimised for studying patients or small animals, and are designed to minimise the parallax effect. Generally, systems with a higher spatial resolution such as the Concord MicroPET P4 or Siemens Inveon are used around small bores measuring around 20 cm in diameter, and those with larger bores often have a relatively large radius-to-axial ratio.

For PEPT studies that may involve larger equipment, or studying microscopic processes, the geometry of current PET cameras is often not the optimal arrangement of the detector blocks. Expanding on the modular camera concept, custom geometries can be built using block detectors from a complete PET system to enhance the PEPT technique for certain geometries. Two such systems are currently under development at the University of Birmingham, microPEPT and superPEPT [50], which provide optimal geometries for studying equipment on the micro and macro-scale by rearranging components from a Concord MicroPET P4 and a combination of Siemens ECAT HR+, ECAT ART and ECAT EXACT 31 scanners.

An exciting development direction for PEPT is the use of time-of-flight (ToF) information. In a time-of-flight PET scanner, the location of the positron annihilation along an LoR is estimated by measuring the time difference between the two gamma ray hits, as shown in figure 5.22. Assuming that we have a distance R between the centre of the field of view and each detector, and that our positron annihilation event occurs at a point P lying a distance D from the centre of the field of view, we can infer that the distance i from detectors 1 and 2 to the annihilation are:

$$d_1 = R - D \tag{5.17}$$

$$d_2 = R + D. \tag{5.18}$$

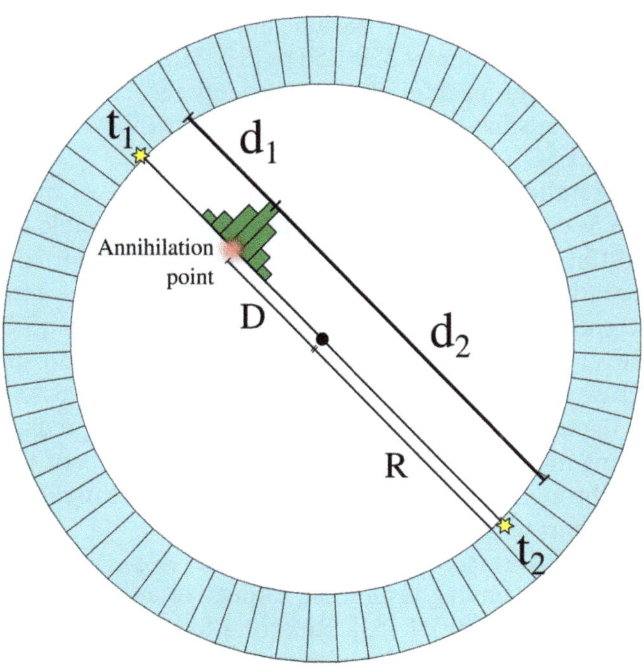

Figure 5.22. Schematic of ToF LOR detection.

Given that the speed of light i is well known we can see that

$$t_1 = \frac{d_1}{c} \tag{5.19}$$

$$t_2 = \frac{d_2}{c}. \tag{5.20}$$

Using this information, we can deduce that the time difference between the two detected hits, $\Delta t = t_2 - t_2$ is

$$\Delta t = \frac{2D}{c}. \tag{5.21}$$

Therefore, for an ideal system with perfect time resolution, we can fully resolve the location of the annihilation event. However, in practice we see that our systems have finite timing resolution, of the order of a few hundreds of ps. For a system with a timing resolution of 500 ps for example, we would be able to locate the point with an uncertainty of 7.5 cm.

ToF PET imaging was first investigated in the 1980s and is now used in commercial scanners. The obstacles towards developing instrumentation capable of performing ToF experiments were primarily in the choice of scintillating material for the detector. Referring to our previous section, we see that the decay times of NaI(Tl) and BGO are relatively poor compared to LSO. Initial experiments with ToF PET used scintillators such as barium fluoride (BaF_4) and caesium fluoride (CsF), achieving ToF resolution of 550–750 ps [51]. However, both materials have a relatively low light output meaning that their energy resolution is compromised. In addition, these materials have a lower density and thus they are less efficient than NaI(Tl), BGO and LSO.

The second generation of ToF scanners took advantage of the development of LSO and also LYSO and in the mid-2000s, achieving comparable timing resolution to the BaF_2 and CsF scanners, with values of 400–600 ps reported [52]. The latest scanners incorporating ToF still use lutetium-based scintillators; however, the most significant difference is in the use of silicon photomultipliers (SiPM) instead of PMTs. These novel photodetectors offer improved efficiency and spatial resolution over PMTs and can also incorporate read out electronics in a monolithic design. This means that the signal can be digitized at the detector level, reducing the need for complex processing and multiplexing electronics. The benefits of this new read out are that the data throughput in a detector system can be greatly increased, allowing larger detector systems to be built.

There is currently a research push towards developing the first scintillating detectors with a 10 ps timing resolution, orders of magnitude faster than the detectors currently used for PEPT [53]. Referring to our previous example, we see that we would be able to locate our annihilation point along a single LoR with a precision of 1.5 mm, offering PEPT-like precision with just a single detected LoR!

5.3 Modelling PEPT systems

The complicated nature of PEPT experiments makes them ideal candidates for Monte Carlo simulation (introduced briefly in section 4.5.3). While we discussed some simple models to calculate expected singles and coincidence count-rates and introduced dead time models, real PEPT systems often rely on many stages of processing, each with their own dead time components, making them too complex (or at least too involved) to model analytically. Digitization of the signals and the logic systems that allow certain coincidence combinations can affect expected count rates. A number of software packages have been used in nuclear physics for modelling imaging systems. One of the most popular packages for PET scanners and therefore PEPT is the Geant4 Application for Tomographic Emission, or GATE [54] [3].

GATE is a macro-based application built around the particle physics toolkit Geant4, built by CERN [55]. Geant4 contains a large number of physics libraries, validated for use over many applications. GATE provides a simplified user interface to Geant4 and allows the user to quickly model a PEPT detector system and obtain results similar to those in experiment [56]. As well as detailed modelling of the physics at play in PEPT, GATE also provides a digitizer module allowing the user to accurately replicate the readout chain from detector components, meaning that it is possible to accurately replicate the data produced by a PEPT system [57]. In the following sections we discuss how the GATE software works, available models of PEPT detectors, how to validate a new PEPT detector, modelling PEPT tracers, recreating a PEPT experiment, and finally how one may use simulated PEPT systems to improve PEPT research.

5.3.1 How it works

The Geant4 package and GATE simulate particles on an event-by-event basis. A primary particle is generated according to some initial parameters provided by the user. Given the type of particle, its initial position and momentum and placement in the world, the probability of it undergoing some physics processes are calculated. This moment is referred to as a *step*. A random number generator is used to decide which process it will undergo, and the position, momentum and state of the particle at the end of the step. This process is repeated until all of the energy from the primary particle has been distributed through the simulation, or if the particle has exited the region of interest and has been cut.

More specifically for PEPT, the primary particle generator can be an unstable isotope such as F-18. The primary particle is a positron emitted at a local time $t = 0$ with an energy sampled from the distribution of β^+ emission energies for F-18. During the next few steps, the positron will have a likelihood of interacting with surrounding material until it annihilates, generating two collinear photons. At this point, the software begins to calculate the *history* of the *secondary* particles, the photons. At each step, the chance of the particle undergoing Compton scatter or attenuation in any surrounding media is calculated and evaluated. If the photon reaches a volume designated as a detector, then the energy absorbed by the detector

[3] http://www.opengatecollaboration.org

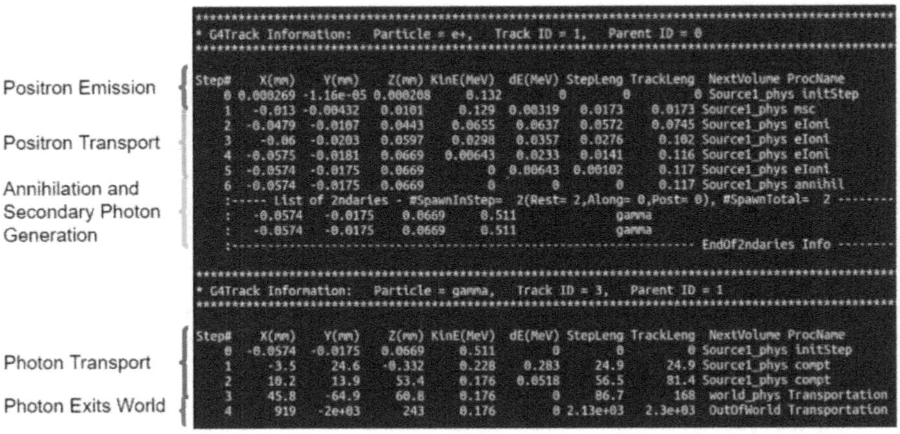

Figure 5.23. Track history of a positron being generated, travelling through the medium, annihilating with an electron, producing two secondary photons, and one photon being transported until it leaves the world.

as the photon interacts with it is recorded. Using this information, a simulated pulse can be generated and we define a simulated *hit* (figure 5.23).

Of course, this is a rather simplified explanation of the process, but we can see why this software can be incredibly useful if we know that the physics models used by GATE can be trusted to accurately model what we want to look at. If we are sure that the physics models are accurate, then through the use of *scoring* or *actors*, we can record many more properties about the primary and secondary particles that may be of use to us. For example, we might want to look at the calculated depth of interaction in a detector. We might want to see if any components in a simulated experiment cause excess scatter. We might even begin to use the simulation to train an AI to recognise a PEPT output from a specific input.

As is hopefully evident from the preceding discussion, Monte Carlo simulation of PEPT experiments opens up a whole new avenue of research, allowing us to more accurately quantify the many variables at play. It is expected that, in time, GATE will become a highly important tool in understanding and evaluating the sources of uncertainties in PEPT as well as in evaluating and enhancing the capabilities of novel PEPT algorithms [58].

5.3.2 Existing GATE models

The first step in modelling a PEPT experiment is to select a validated detector model or validate a new model. Monte Carlo modelling is of great usefulness to medical imaging and thus several detector models have been created in GATE and validated through experiments [59]. Since PEPT uses these same detectors, this work can be re-purposed to suit our means. Examples of detectors for which there currently exist validated models the Siemans ECAT EXACT HR++, the Siemans Inveon, and the Phillips Vereos [60–62]. The ECAT EXACT HR++ is a bismuth germanate based ring scanner used in clinical practices and represents the typical PET scanner one might encounter in hospital. The PEPT facility iThemba LABS in Cape Town, South Africa possess a model of this detector [63]. The GATE model for the ECAT EXACT HR++ can be found in 'GateContrib' folder of the OpenGATE GitHub at https://github.com/OpenGATE/GateContrib/tree/master/imaging/PET (figure 5.24).

The Siemans Inveon is a lutetium oxyorthosilicate based trimodal imaging system capable of PET, single photo emission computed tomography (SPECT), and x-ray computed tomography (CT). It is a pre-clinical ring scanner typically used for small-animal studies and was also used at the University of Tennessee for the PEPT studies of pipe flow discussed previously in this chapter [64]. The GATE model can downloaded from the authors' website at http://web.eecs.utk.edu/~jgregor/gate-model/siemens_inveon.php (figure 5.25).

Finally, the Phillips Vereos is a lutetium–yttrium oxyorthosilicate based ring scanner capable of ToF. Developed as recently as 2013, this detector is a new model with a ToF resolution of 220 ps. When both detection of the coincident gamma rays are detected, this leads to a coincidence time resolution of 311 ps. It also features silicon photomultipliers which convert the detection to digital signal without the need for an additional analogue-to-digital converter. The model was created in 2020 and is included in the 'GateBenchmarks' folder on the OpenGATE GitHub at https://github.com/OpenGATE/GateBenchmarks/tree/master/t4_necr (figure 5.26).

Any of these detectors can be used for real PEPT experiments, given that users have access to the detector and the experiment fits within the field of view. While originally designed for simulations of conventional medical imaging, these GATE

Figure 5.24. GATE model of the Siemens ECAT HR++ scanner.

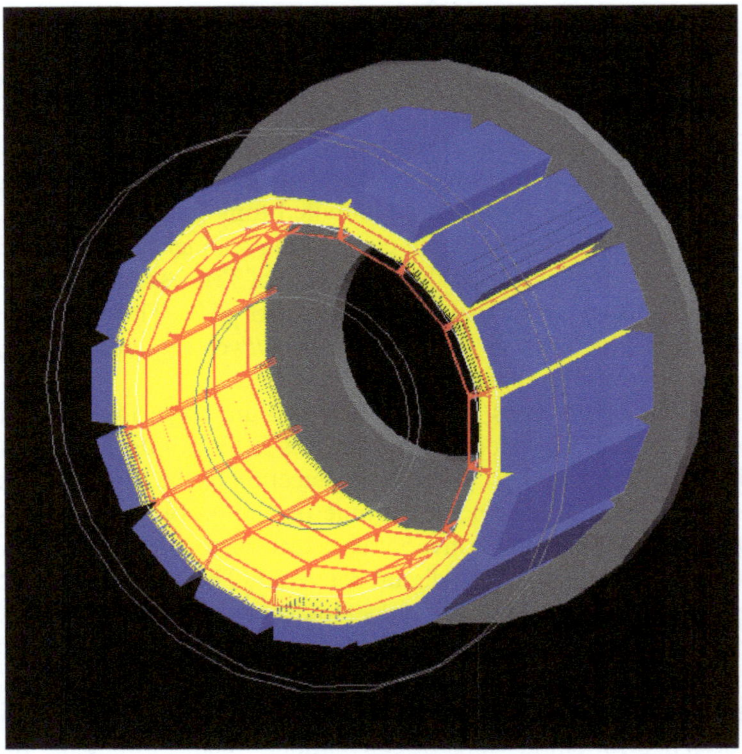

Figure 5.25. GATE model of the Siemens Inveon pre-clinical PET scanner.

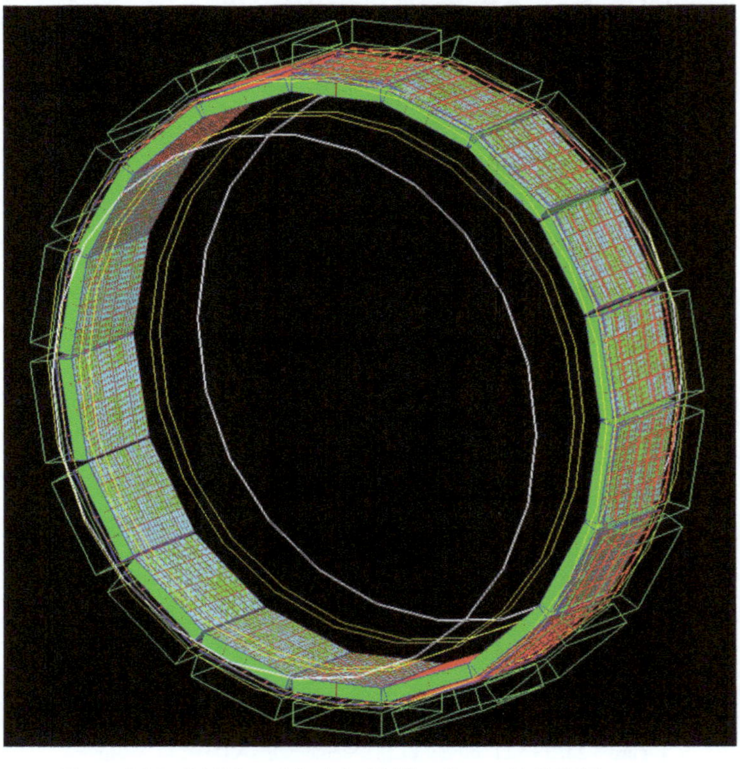

Figure 5.26. GATE model of the Phillips Vereos ToF PET scanner.

models are easily adaptable to PEPT research. The geometry of the scanner, physics modules, and data output never change from simulation to simulation. All a GATE user needs to do is define their own source and experimental geometry within the GATE environment. These steps will be discussed in sections 5.3.4 and 5.3.5.

5.3.3 Modelling a PEPT detector

Along with downloading a ready-to-use GATE model of a PEPT detector, creating a new model is a desirable option if the detector doesn't already have a GATE model or is a new design. The ADAC Forté Gamma Camera is an example of an older detector, acquired by the University of Birmingham in 2001, of which no GATE models existed prior to 2021 [56]. In this case, a structural model of the camera was built, and then validation experiments were conducted at the University of Birmingham's Positron Imaging Centre to calibrate a digitizer model which reproduced the experiments. To demonstrate how to build a model of a PEPT detector, the steps used to create and calibrate the ADAC Forté detector will be described below. These methods are applicable to a wide range of detectors systems used for PEPT measurements and can be repeated on new designs of detector as well.

Creating a geometric model of a detector involves taking physical measurements of the detector dimensions, identifying key components which will contribute to the scattering and detection of photons, and creating material definitions within GATE which reflect the materials used in the real detector. The most important geometric consideration for the detector is the dimensions of the scintillation crystal. This is the volume in which the detection of photons occurs. For the ADAC Forté this is a $470 \times 590 \times 16$ mm sodium iodide crystal [33].

After numerically reproducing the geometry of the scintillation crystal, to accurately capture the scattering of photons which aren't completely absorbed by the crystal a back-compartment and the rest of the major structural components of the detector must be added [65]. This includes the glass lightguide, photomultiplier tubes directly behind the crystal, as well as the plastic covering encasing the detector head aluminium gantry ring. While there are many other components in PEPT detectors, only the large components need to be modelled to achieve accurate results. The rest of the components contribute negligibly to the scattering behaviour (figure 5.27).

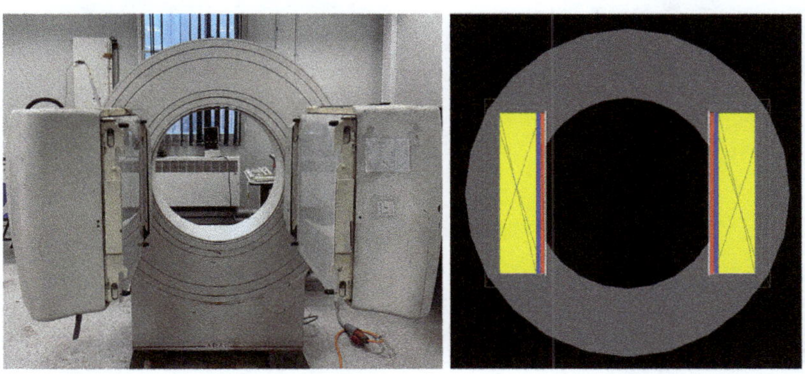

Figure 5.27. Picture of the ADAC Forté compared with the GATE model.

Once the geometric model is completed, experiments need to be run on the *real* detector to provide a basis for comparison. The four key characteristics of the detector which should be captured are the sensitivity, spatial resolution, scatter fraction, and count-rates over activity. These characteristics have been identified by the National Electronics Manufacturers Association in a standard for comparing the performance of PET scanners [66]. The standard also describes the acceptable methods by which these characteristics can be measured, including the acceptable equipment, source activities, and data processing. This standard and the specialised equipment was designed to reasonably compare the performance of different machines, however, in order to characterise the detector for calibrating a GATE model, only the source activities and data processing must be followed. Any type of source may be used for experiments, as long as this source is replicated in the GATE environment.

An example showing the agreement between a calibration experiment and the ADAC Forté GATE model is shown in figure 5.28. This figure shows the total, true, scattered, and random count-rates in the real experiment and GATE model. The GATE files for the ADAC Forté can be found at https://github.com/mxh1092/ForteCamera.

5.3.4 Defining a PEPT tracer

Once a detector system has been selected and the experimental system the user wants to image has been recreated within the GATE environment, the next task is to define a source which will act as the PEPT tracer. In real PEPT experiments tracers are designed to be representative of a particle in the system. This can be anything from a grain of sand in fluidized bed to a sphere of microcrystalline cellulose in a rotating drum to a coffee bean in a roaster [67]. Additionally, tracers may be activated with F-18 or other positron emitting isotopes may be coated in a layer of

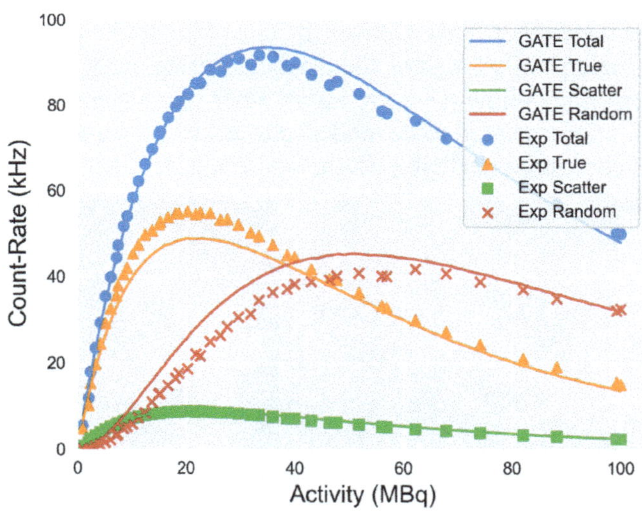

Figure 5.28. Measured count-rates of the ADAC Forté and corresponding GATE simulation of the same experiment.

positron-emitters or volumetrically activated, and can vary in activity as the source decays [68, 69]. As such, the user must recreate these sources using the tools available with GATE.

The workflow for creating a tracer in GATE begins with the geometric and material considerations of the tracer. The tracer may be modelled as simple spheres or other basic geometric objects using GATE's geometry modelling functions. These include: boxes, cones, cylinders, ellipsoids, and wedges. Additionally, more complex objects can be created by importing mesh files such as STL. These provide users with nearly unlimited freedom to design realistic geometries. For example, an STL file of a coffee bean can be imported into the GATE environment and visualised (figure 5.29).

Now that the coffee bean geometry is created inside GATE, for example, a user must define what material definition should be prescribed to the object. The GATE material definition library consists of many pre-compiled elements and materials which might be found within a medical context, such as plastics, different types of common scintillation crystals, and materials corresponding to different organs in the human body. However, coffee beans are (perhaps unsurprisingly) not included in this list so the user must create their own material definition. This process is very straightforward and general enough to create any material. The new material must be given a name, e.g. 'CoffeeBean', a density, here 0.7 g m^{-3}, how many elements or materials the coffee bean is comprised of, here four, and what physical state the material is in, here solid. The next step is to list the materials and define the fraction of each in the bean. Coffee beans composition can vary from bean to bean, species to species, and the density will change with varying amounts of water content during the roasting process, meaning this is only an estimate of the material definition, but is more accurate than using a material like water or plastic to model a coffee bean (figure 5.30).

Creating an accurate material definition is important for two main reasons. First, as the coincident gamma rays pass through the material of the tracer, they have a probability of scattering, which will affect the scatter fraction and overall count-rate. Secondly, when the positrons are emitted from within the tracer, they must interact

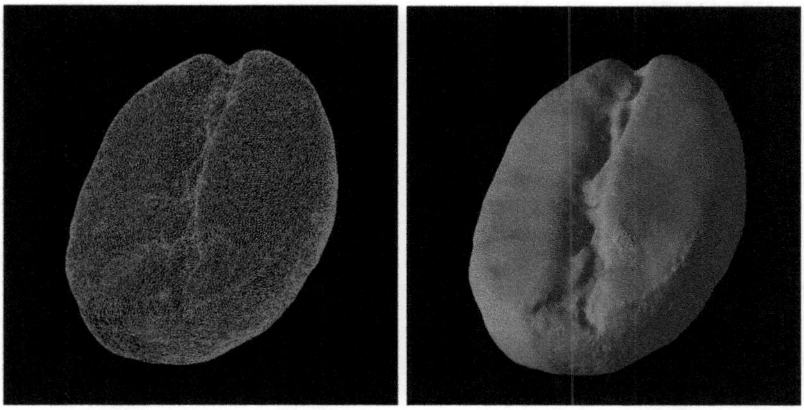

Figure 5.29. Left: The mesh of the coffee bean visualised in GATE. Right: Solid coffee bean geometry.

```
CoffeeBean: d=0.7 g/cm3; n=4; state=Solid
    +el: name=Carbon; f=0.308
    +el: name=Hydrogen; f=0.474
    +el: name=Oxygen; f=0.215
    +el: name=Nitrogen; f=0.003
```

Figure 5.30. Coffee bean material definition.

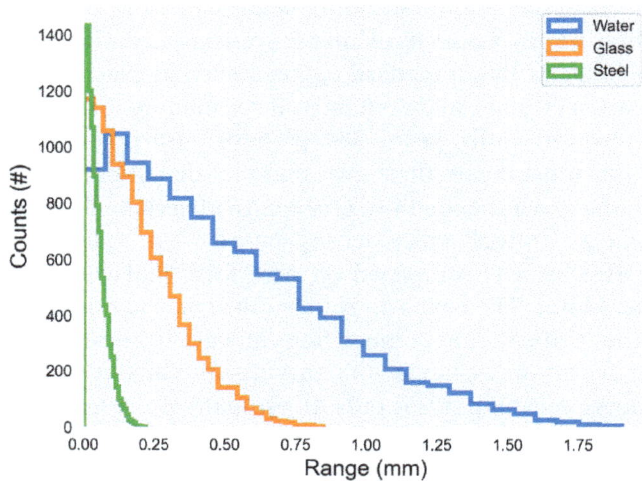

Figure 5.31. Positron range in three common materials: water, glass, and steel.

with an electron in order to create the coincident gamma rays. The density and the material composition can both drastically change the range in which positrons annihilate. To demonstrate this, a point source in a quasi-infinite sphere of material is simulated in GATE resulting in different positron ranges when the sphere material is changed. For each material, 10 000 positron emissions are simulated. As the density of material increases the positron range becomes shorter on average (figure 5.31).

Material	Density (g cm^{-3})	Mean range (mm)	Max range (mm)
Air	0.00129	454	1910
Water	1.000	0.499	1.903
Plastic	1.180	0.451	1.747
Glass	2.500	0.196	0.850
Aluminium	2.700	0.178	0.796
Steel	7.920	0.0440	0.220
Lead	11.40	0.0183	0.0967

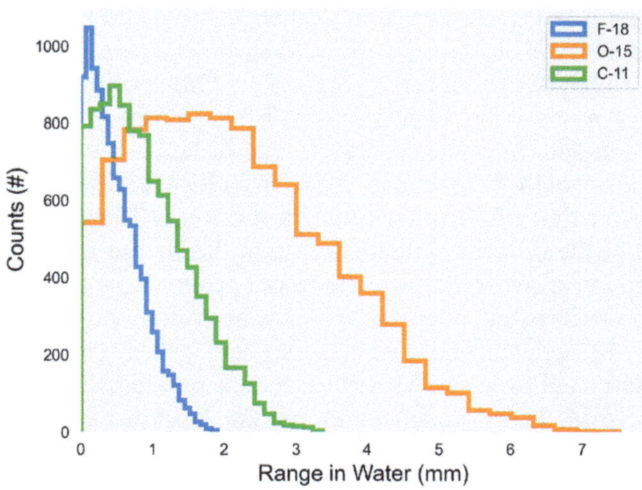

Figure 5.32. Positron range in water for three positron-emitters.

The range for individual positrons varies due to the probabilistic nature of interacting with an electron and also the energy at which a positron is emitted. Beta plus decay produces a recoiling nucleus, a positron, and a neutrino. The energy produced by the decay is shared between the three particles obeying energy and momentum conservation, however, there are many possible ways the energy can be shared, resulting in a spectrum of positron energies. Higher energy positrons can travel further through material before annihilating and thus have a higher range on average. However, for each positron emitted, the beta spectrum has a well-defined maximum energy when nearly all of the energy from the decay is carried away by the positron. Three common positron emitters used in PEPT are fluorine-18, oxygen-15, and carbon-11. The maximum energy of the positron for these three isotopes, respectively, are 0.634, 1.732, and 0.960 MeV. Thus, for a point source in a quasi-infinite medium of water one would expect the oxygen-15 source to have the furthest positron range, as we indeed see in figure 5.32. When a GATE simulation is run for 10 000 positron emissions for each of the three isotopes, the results in table 5.3.4 show the isotopes with the highest energy produce further positron ranges (figure 5.32). As we can see from the figure, of these common isotopes, fluorine-18 produces by far the least 'spread', one of many reasons for its previously-discussed popularity for use in both PET and PEPT.

Isotope	Maximum energy (MeV)	Mean range (mm)	Max range (mm)
Fluorine-18	0.634	0.4986	1.903
Oxygen-15	1.732	2.173	7.519
Carbon-11	0.960	0.921	3.370

Another consideration that will effect PEPT experiments is the size of the tracer and how it is activated. For example, the range of positrons emitted from a large volumetrically activated tracer and a very small tracer will differ if the surrounding medium is of different density than the tracer. Take, for example, a fluidized bed where a dense tracer may often be surrounded by voids of air. Since positron range in air is much larger than that of plastic, if the tracer size is small, many positrons may escape the tracer before annihilating and, in doing so, travel large distances away from the particle's centroid. An estimate of the number of positrons which leave the tracer can be seen in the simulation shown in figure 5.33 where a spherical tracer of 10 mm and 1 mm diameter are created. All of the positrons annihilate within the 10 mm tracer, while a handful escape the 1 mm. Positrons are shown in blue and gamma rays are shown in green.

This effect is exacerbated when the tracer is activated only on the surface since positrons have a higher chance of escaping. To demonstrate this, the same two sizes of tracers are activated only on the surface then simulated again. In this example, shown in figure 5.34, a few positrons escape the 10 mm diameter tracer and all escape the 1 mm tracer. Thus during PEPT experiments, if the tracers are small, the range of positrons must be considered, especially if the medium they are placed in contains large voids of air or the tracers are surface-activated.

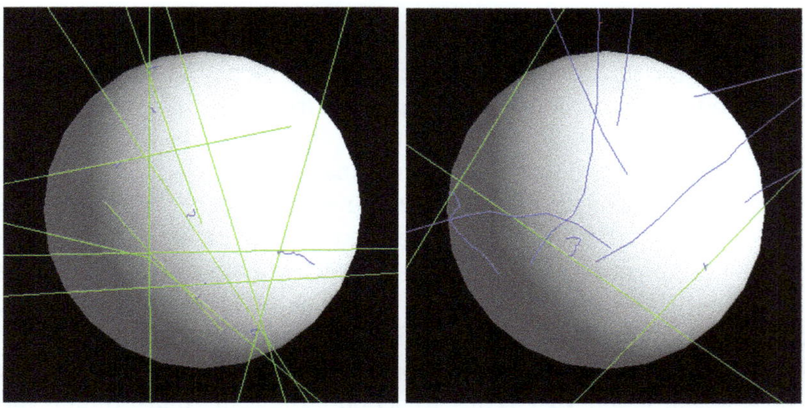

Figure 5.33. Positrons annihilating within volumetric plastic tracers of diameter 10 mm (left) and 1 mm (right).

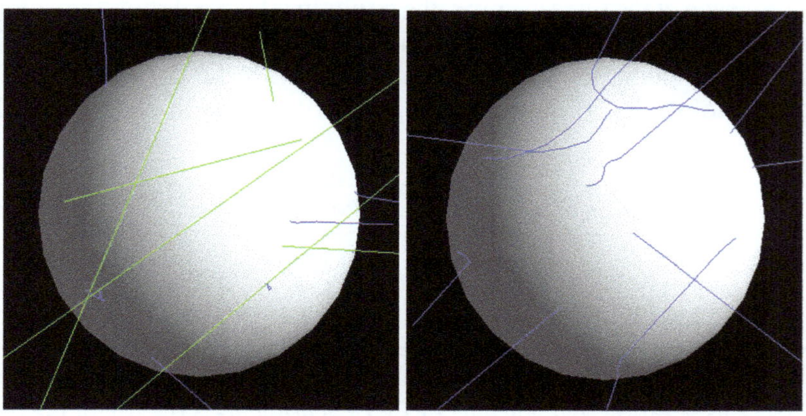

Figure 5.34. Positrons annihilating within surface-activated plastic tracers of diameter 10 mm (left) and 1 mm (right).

5.3.5 Recreating an experiment

To achieve realistic comparisons between PEPT experiments and GATE simulations, not only do the detectors and tracers need to be modelled, but also the systems in which the experiment as these too will induce scatter, and other phenomena that may influence results. PEPT experiments are conducted in a wide range of systems like rotating drums and fluidized bed (see chapter 8), as has been previously described. These can be recreated in GATE using the built-in geometry modelling tools, similar to how tracers were made. One of the simplest examples to demonstrate recreating a PEPT experimental system is a pipe-flow experiment. Pipes are relatively easy to create in GATE since these can be modelled using simple cylinders to represent the flow and pipe walls. A tracer can be placed inside the flow of the pipe and prescribed a velocity to mimic the tracer moving through the pipe. Using this system, a user can simulate pipes of different diameters and composed of different materials to see what fraction of the LoRs are scattered.

However, tracers moving in straight lines through a system is hardly representative of real tracer motion. In PEPT systems, tracer motion is often highly chaotic due to turbulence in a fluid system and collisions between particles in granular media. Thus the trajectory of tracer in GATE must be able to replicate this. Using computational fluid dynamics (CFD) or discrete element method simulations (DEM)[4], it is possible to generate a simulated tracer trajectory in a system of interest. Such software computes the forces acting on a particle in small timesteps to generate history of particle positions. These positions can be used to move one or more tracer in GATE using the 'GenericMove' function. During the GATE simulation, positions are read from a file and used to move the tracer. This approach has been used in previous work with a computational fluid dynamics simulation of water flowing through a pipe to estimate the uncertainty in a PEPT experiment studying pipe-flow [70] (figure 5.35).

To demonstrate how a DEM simulation can be used in conjunction with GATE, an example of a rotating drum will be used. In this system one trajectory for a particle moving in the drum is extracted from the simulation. The trajectory can then be

```
###### List of placement (translation and rotation)
###### Column 1 is Time in s (second)
###### Column 2 is rotationAngle in degree
###### Columns 3,4,5 are rotation axis
###### Columns 6,7,8 are translation in mm
Time s
Rotation deg
Translation mm
0.0      0    0 0 0    -0.417  12.156  -36.807
0.0001   0    0 0 0    -0.432  12.156  -36.803
0.0002   0    0 0 0    -0.447  12.156  -36.799
```

Figure 5.35. GATE placement file for moving a tracer.

[4] See chapter 4 for an introduction to these numerical techniques.

formatted so as to be read as a series of successive movements for a GATE modelled tracer. GATE simulations are broken into a series of discrete timesteps in which all geometries are stationary. If the timestep is sufficiently small, then small movements of the tracer will appear to be smooth movements in the same way that objects appear to move from discrete pictures on a TV screen. As the simulation progresses, the tracer is moved to the DEM position closest in time to the GATE timestep. When reconstructing the tracer trajectory from the GATE output, the tracer will appear to follow the trajectory of the DEM simulated particle in the rotating drum (figure 5.36).

A rotating drum is a multi-phase system usually consisting of solid and fluid phases [71]. The solid phase is usually much denser than the fluid phase (in this case air) and during a PEPT experiment one would expect this to have a strong effect on the positron range and overall scattering behaviour compared to a fluid like air. Thus, the ability to model a complex, moving geometry is needed since the particle motion alone tells an incomplete story. The structure of the rotating drum and housing in which it is encased can be modelled using simple boxes and cylinders, however, the material inside the drum is irregularly shaped and consists of a mixture of the solid and fluid phases. To replicate this geometry, a method called voxelization can be applied.

A voxel is the three-dimensional equivalent to a two-dimensional pixel. In GATE, arrays of voxels can be created with each one representing a different material, mixture, or density. Similar to how a two-dimensional image can be constructed using different coloured pixels, a three-dimensional object can be reconstructed using voxels. There are two ways to voxelize a system like a rotating drum, either you voxelize each timestep by computing the positions of all particles and calculating the instantaneous voxel representation of the system, or you can average the histories of all particles into a time-averaged representation of the drum. The latter approach is more computationally efficient, while the former is more accurate.

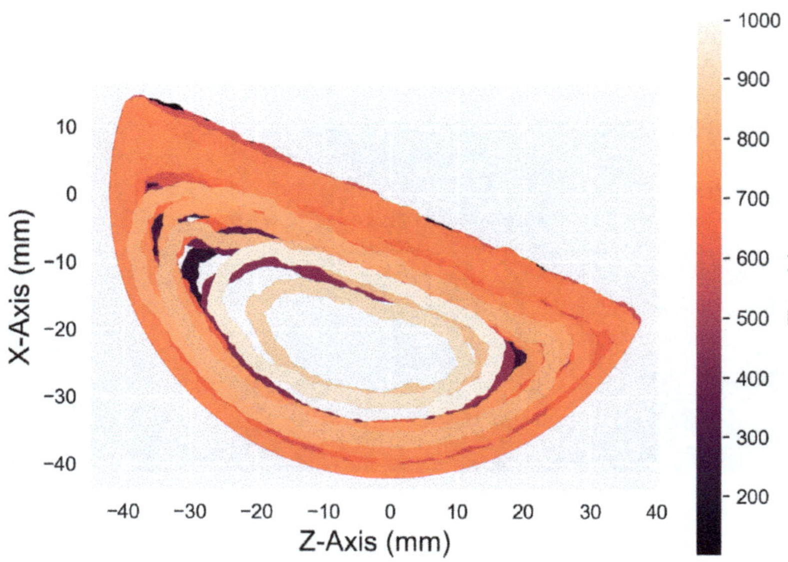

Figure 5.36. Trajectory of the DEM simulated particle in a rotating drum over time.

However, in this particular, steady-state system the voxel arrays do not change significantly between timesteps, thus the averaged approach can be safely taken.

To do the actual voxelization, the domain over which the DEM particles are simulated must be divided into cells, and then the number of particles in each cell is calculated. The volume of particles in the cell is subsequently divided by the cell's volume to calculate the fraction of the cell that contains particles (as opposed to the background fluid). The remaining fraction is thus that of air. Next, a series of materials in GATE should be created corresponding to different fractions of particles and air (figure 5.37).

A three-dimensional array should now be saved containing the fraction of particles in each cell. GATE is capable of reading a number of three-dimensional medical imaging formats such as DICOM (Digital Imaging and Communications in Medicine), Interfile, and MetaImage. This example saves the data in MetaImage format. The voxel array can be placed in the GATE world like any other geometry and moved around, but to link the voxel values with a GATE defined material, 'Range Translator' is needed. The 'Range Translator' is a file that maps ranges of voxel values to materials. With this, the voxel grid in GATE can now be initialised with the appropriate materials in each voxel, thus for a tracer inside the grid, the positron range will be similar to the range in the real rotating drum and a realistic scattering of photons will be generated (figure 5.38).

5.3.6 Studying a PEPT system using GATE

The previous sections have discussed how to recreate PEPT experiments by modelling detectors, tracers, and systems. However, once the virtual experiment is set-up, the real power in modelling is how you use the models to better understand PEPT algorithms, the limits of experiments, and the effect of varying experimental parameters on the simulated results. Calibrated GATE models provide data representative to that generated during a real experiment and, therefore, can be used to see how a real experiment would turn out. There is little time or resources to thoroughly test algorithms on real data when the positions of the tracer may have a

```
Drum0: d=0.00129 g/cm3 ; n=2 ; state=solid
    +mat: name=MCC_Bulk ; f=0.0
    +mat: name=Air ; f=1.0

Drum1: d=0.0212642 g/cm3 ; n=2 ; state=solid
    +mat: name=MCC_Bulk ; f=0.02
    +mat: name=Air ; f=0.98

Drum2: d=0.0412384 g/cm3 ; n=2 ; state=solid
    +mat: name=MCC_Bulk ; f=0.04
    +mat: name=Air ; f=0.96
```

Figure 5.37. GATE material definitions of different amounts of particle and air fractions in a rotating drum.

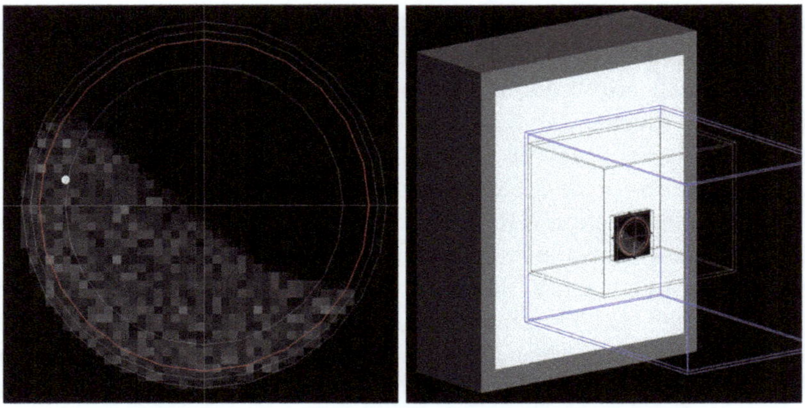

Figure 5.38. Left: A close-up view on the side of the voxel array representing the rotating drum. Right: voxel array placed in housing holding the rotating drum inside the ADAC Forté GATE model.

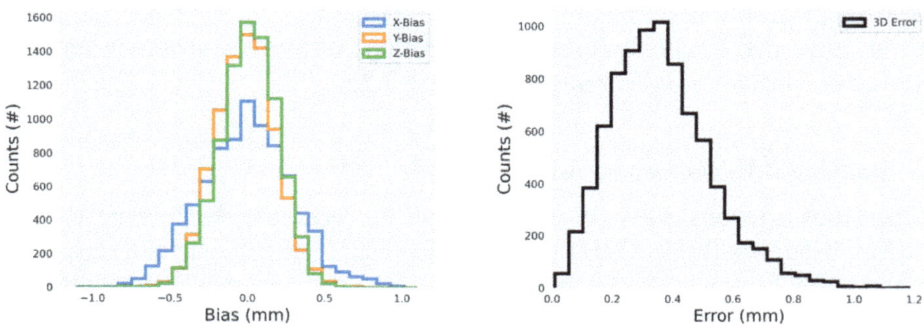

Figure 5.39. Left: bias is PEPT reconstructed positions. Right: 3D error of the PEPT tracer.

large uncertainty. Thus, there are few ways to rigorously validate that the tracer motion detected using PEPT is representative of real tracer motion. Further, finding the optimal experimental parameters would take multiple days of experiments for a single system, since a range of detector positions and tracer activities would need to be conducted. Using GATE as essentially an unlimited sandbox to generate synthetic data is the only way to answer within a feasible timescale, some of these questions and is generally a much more efficient use of time and resources.

Testing the capabilities of PEPT algorithms, for example, is a challenge uniquely suited to GATE simulations [58]. Since the position of a tracer can be prescribed with certainty in GATE, a direct comparison can be made between this and the position reconstructed using a PEPT algorithm. The distance between the detected and prescribed position can be analysed by looking at the biases in the x, y, and z components individually, or the three-dimensional distance between the two points as shown in figure 5.39. There are several PEPT algorithms in existence and by comparing their performance on a common dataset where the ground truth position of the tracer is known, the relative strengths and weaknesses of the algorithms can be—indeed recently has been—demonstrated [58].

Tests can designed to determine the maximum velocity which can be tracked using a particular PEPT algorithm or determine how the error decreases on a stationary tracer when more LoRs are used to calculate its position. In real PEPT experiments it has been shown that the uncertainty in position of a stationary PEPT tracer decreases proportional to the inverse square root of the number of LoRs used for clustering [72]. However, this is only the uncertainty, not the error between the ground truth tracer position and the PEPT detected positions. Using GATE this has been evaluated and shown to be true in figure 5.40.

Additionally, the second avenue of research with GATE models of experiments is for users to explore the limitations of their experiments, including whether a proposed experiment will achieve sufficient data quality. When looking at a trajectory extracted from a PEPT experiment, two qualities of the data are of paramount importance: the error and the time resolution. The error being the previously described three-dimensional distance between the detected position and the real positions of the tracer, but also how often the tracer can be detected. Ideally a PEPT experiment should return a trajectory that follows a tracer (a) accurately and (b) often enough to fully capture the behaviour in the system. This is very important for validating DEM or CFD simulations using PEPT experiments. A difference between the DEM simulation and the PEPT experiment may indicate a problem with the DEM model and the PEPT experiment, but ensuring that a reasonable estimate of the tracer trajectory can be tracked in a GATE simulation can rule out an error on the PEPT side. Additionally, for a proposed experiment, especially comprising dense materials, it is useful to run a GATE simulation of a static tracer inside a representation of an experimental system to ensure that there is no excessive scattering which would reduce the count-rate below a rate which would result in an adequate spatio-temporal resolution.

Interestingly, this approach can be taken a step further to explore a range of experimental parameters. Two of the most important parameters for a PEPT

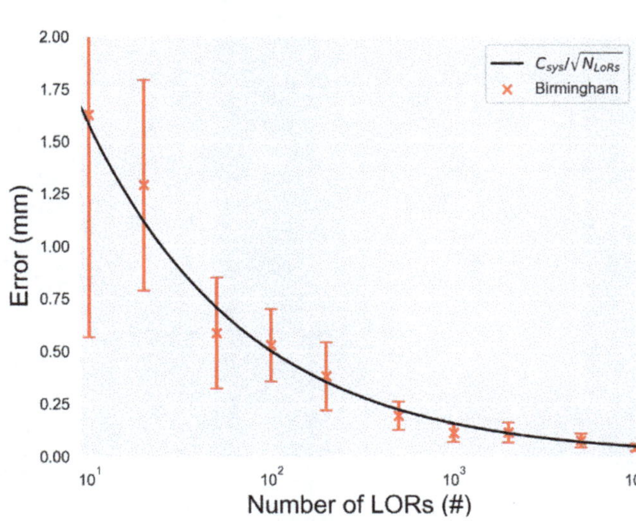

Figure 5.40. The error in the reconstructed position of a PEPT tracer decreases proportional to the inverse square root of the number of LoRs used to find the tracer.

experiment are the detector head-separation and the tracer activity. When real experiments are conducted, there is typically inadequate time or resource to run multiple experiments to find the optimum positioning of the detectors and experiment, or to select an optimum tracer activity which will yield the highest true LoR count-rate, and thus the best PEPT trajectories. Users are usually locked-in to a detector head separation the experimental system fits into and the activity of tracer that happened to be produced. There is no tailoring of these parameters to the optimum, simply because without GATE modelling capabilities the 'optimum parameters' for a given system are unknown.

By using GATE models of PEPT experiments, users are now able to run a series of simulations varying parameters then quantify the results. For example, an industrial powder mixer is recreated within GATE and a trajectory of a tracer spiralling through the mixer is prescribed to a source. The mixer geometry can be seen in figure 5.41.

As we saw in figure 5.28, the true LoR count-rate rises until it reaches a peak and then starts to decrease as the detector experiences high amounts of dead-time. The optimum count-rate occurs at a tracer activity that is unique to each head-separation. To find the optimum tracer activity and head-separation combination,

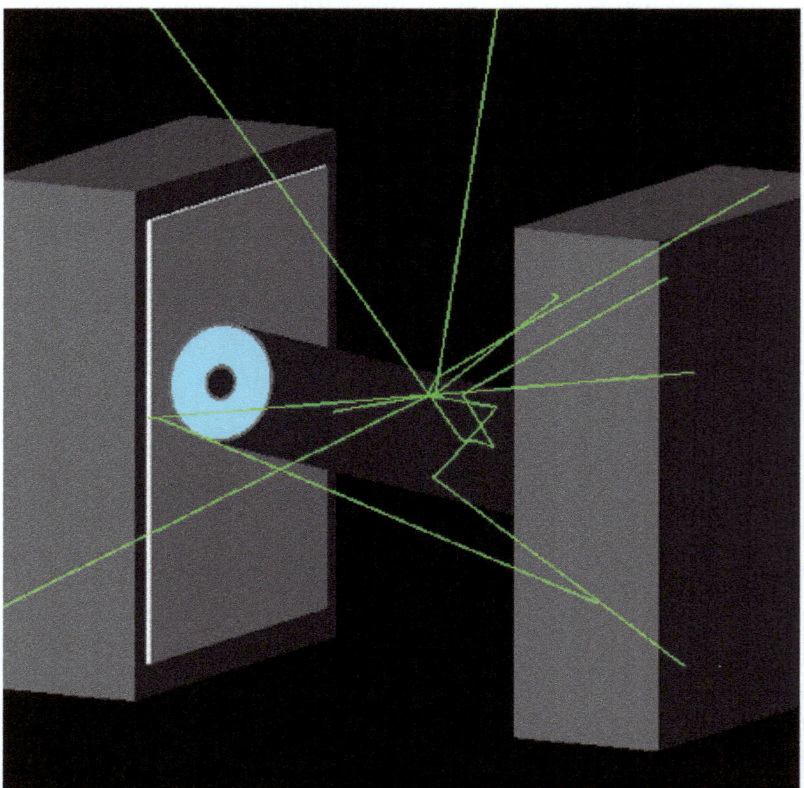

Figure 5.41. The simplified industrial powder mixer and LoRs being transported in the system.

simulations following a parameter sweep of different separations and activities can be conducted. The GATE simulation data then can be clustered using a PEPT algorithm and the spatio-temporal resolution of the trajectories compared. The spatio-temporal resolution results of a parameter sweep can be seen in figure 5.42.

The results show that there exists an optimum combination of tracer activities and detector head-separation at 7 MBq and 350 mm, respectively. The worst parameter combinations can be found, perhaps unsurprisingly from the preceding discussion in this chapter, at low activities and high separations and at high activities and low separations. To understand why this is the case, it's helpful to look at examples of PEPT trajectories extracted from these simulations in figure 5.43.

The trajectory returned by the PEPT algorithm at the optimal tracer activity and head-separation clearly show that the tracer is following the trajectory prescribed in GATE. However, the two other extremes from the parameter sweep show poorer

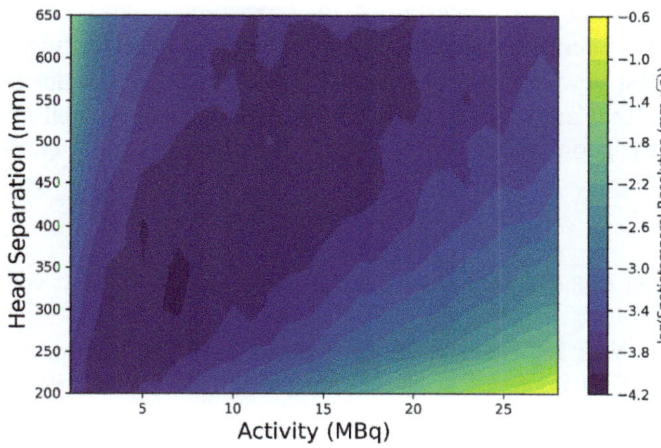

Figure 5.42. The spatio-temporal resolution as function of the two variables: activity and head-separation. There exists an optimum combination at 7 MBq and 350 mm.

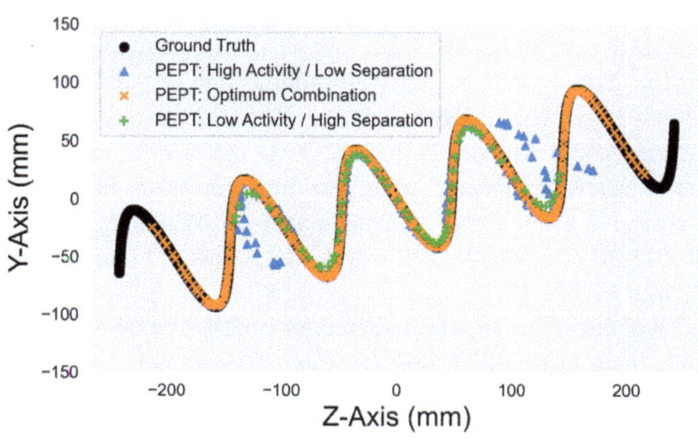

Figure 5.43. Examples of three trajectories extracted from GATE simulations of a simplified industrial mixer in the ADAC Forté GATE model.

performance, but in unique ways. For example, both the trajectories are sparsely populated with detection compared to the optimum trajectory. This is a result of the low count-rate. In the low activity case, the low count-rate is caused simply by few available LoRs to detect, but in the high activity case high amounts of dead-time cause the low count-rate. In addition to this, the simulation conducted at high tracer activity and low head-separation has severe distortion of detected positions, especially evident near the edges of the field of view. This is because at high activities a large number of the recorded LoRs are random LoRs, and thus are corrupted for use in PEPT. A distorted trajectory can be impossible to tell apart from real tracer motion in an experiment. In GATE, since we can compare the prescribed positions to the detected ones, it becomes clear that this combination of tracer activity and head-separation will lead to bad results.

These are only a few examples of how modelling PEPT systems can be useful in PEPT research. In the future, this work is expected to be expanded and used to further improve PEPT algorithms, estimate uncertainty in experiments, and select the optimum parameters in experiments. One exciting new application of modelling PEPT detectors is the ability to test new detector designs. This could lead to the development of new detector systems, that are optimised for PEPT experiments, as well as facilitating the optimisation of the geometrical arrangements of existing modular camera arrays.

References

[1] Krane K S and Halliday D *et al* 1988 *Introductory Nuclear Physics* vol 465 (New York: Wiley)

[2] Stellema C S, Vlek J, Mudde R F, De Goeij J J M and Van den Bleek C M 1998 Development of an improved positron emission particle tracking system *Nucl. Instrum. Methods Phys. Res.* A **404** 334–48

[3] Fan X, Parker D J and Smith M D 2006 Labelling a single particle for positron emission particle tracking using direct activation and ion-exchange techniques *Nucl. Instrum. Methods Phys. Res.* A **562** 345–50

[4] Chang Y-F, Ilea C G, Aasen Ø L and Hoffmann A C 2011 Particle flow in a hydrocyclone investigated by positron emission particle tracking *Chem. Eng. Sci.* **66** 4203–11

[5] Fan X, Parker D J and Smith M D 2006 Enhancing ^{18}F uptake in a single particle for positron emission particle tracking through modification of solid surface chemistry *Nucl. Instrum. Methods Phys. Res.* A **558** 542–6

[6] Lee K T, Ingram A and Rowson N A 2012 Twin screw wet granulation: the study of a continuous twin screw granulator using Positron Emission Particle Tracking (PEPT) technique *European J. Pharmaceutics Biopharmaceutics* **81** 666–73

[7] Liang L *et al* 2015 PEPT study of particle cycle and residence time distributions in a Wurster fluid bed *AIChE J* **61** 756–68

[8] Palmer S *et al* 2007 Investigation of the sources of variability in the Wurster coater: Analysis of particle cycle times using PEPT *12th Int. Conf. Fluidization*

[9] Windows-Yule C R K, Gibson S, Parker D J, Kokalova T Z and Seville J P K 2020 Effect of distributor design on particle distribution in a binary fluidised bed *Powder Technol.* **367** 1–9

[10] Windows-Yule C R K, Moore A, Wellard C, Werner D, Parker D J and Seville J P K 2020 Particle distributions in binary gas-fluidised beds: shape matters—but not much *Chem. Eng. Sci.* **216** 115440

[11] Al-Shemmeri M, Windows-Yule K, Lopez-Quiroga E and Fryer P J 2021 Coffee bean particle motion in a spouted bed measured using positron emission particle tracking (PEPT) *J. Food Eng.* **311** 110709

[12] Cole K E, Buffler A, Van der Meulen N P, Cilliers J J, Franzidis J P, Govender I, Liu C and van Heerden M R 2012 Positron emission particle tracking measurements with 50 micron tracers *Chem. Eng. Sci.* **75** 235–42

[13] Sommer A-E, Ortmann K, Van Heerden M, Richter T, Leadbeater T, Cole K, Heitkam S, Brito-Parada P R and Eckert K 2020 Application of positron emission particle tracking (PEPT) to measure the bubble-particle interaction in a turbulent and dense flow *Miner. Eng.* **156** 106410

[14] Cole K, Buffler A, Cilliers J J, Govender I, Heng J Y Y, Liu C, Parker D J, Shah U V, van Heerden M and Fan X 2014 A surface coating method to modify tracers for positron emission particle tracking (PEPT) measurements of froth flotation *Powder Technol.* **263** 26–30

[15] Knoll G 2010 *Scintillation detector principles Radiation Detection and Measurement* 4th edn (New York: Wiley) pp 223–74

[16] Bailey D L 2005 *Physics and Instrumentation in PET* (Berlin: Springer) pp 14–39

[17] Anger H O 1952 Use of a gamma-ray pinhole camera for *in vivo* studies *Nature* **170** 200–1

[18] Nestor O H and Huang C Y 1975 Bismuth germanate: a high-Z gamma-ray and charged particle detector *IEEE Trans. Nucl. Sci.* **22** 68–71

[19] Burnham C A, Bradshaw J, Kaufman D, Chesler D and Brownell G L 1984 A stationary positron emission ring tomograph using BGO detector and analog readout *IEEE Trans. Nucl. Sci.* **31** 632–6

[20] Spinks T J, Jones T, Gilardi M C and Heather J D 1988 Physical performance of the latest generation of commercial positron scanner *IEEE Trans. Nucl. Sci.* **35** 721–5

[21] Spinks T J *et al* 2000 Physical characteristics of the ECAT EXACT3D positron tomograph *Phys. Med. Biol.* **45** 2601

[22] Melcher C L 2000 Scintillation crystals for PET *J. Nucl. Med.* **41** 1051–5

[23] Tai Y-C, Chatziioannou A, Siegel S, Young J, Newport D, Goble R N, Nutt R E and Cherry S R 2001 Performance evaluation of the micropet p4: a PET system dedicated to animal imaging *Phys. Med. Biol.* **46** 1845

[24] Kemp B J, Hruska C B, McFarland A R, Lenox M W and Lowe V J 2009 NEMA NU 2-2007 performance measurements of the Siemens Inveon® preclinical small animal PET system *Phys. Med. Biol.* **54** 2359–76

[25] Amirrashedi M, Zaidi H and Ay M R 2020 Advances in preclinical PET instrumentation *PET Clin.* **15** 403–26

[26] Scintillation Camera 1958 *Rev. Sci. Instrum.* **29** 27–33

[27] Anger H O 1964 Scintillation camera with multiple collimators *J. Nucl. Med.* **5** 515–31

[28] Casey M E and Nutt R 1986 A multicrystal two dimensional BGO detector system for positron emission tomography *IEEE Trans. Nucl. Sci.* **33** 460–3

[29] Casey M E, Dautet H, Waechter D, Lecomte R, Eriksson L and Schmand M 1998 An LSO BLOCK detector for PET using an avalanche photodiode array *Nuclear Science Symp. Conf. Record. 1998 IEEE Nuclear Science Symp. and Medical Imaging Conf. (Cat. No. 98CH36255)* vol 2 (Piscataway, NJ: IEEE) pp 1105–8

[30] Gossman M S, Pahikkala A J, Rising M B and McGinley P H 2011 Letter to the editor *J. Appl. Clin. Med. Phys.* **12** 242–3

[31] Langford S, Wiggins C, Tenpenny D and Ruggles A 2016 Positron emission particle tracking (PEPT) for fluid flow measurements *Nucl. Eng. Des.* **302** 81–9

[32] Bao Q, Newport D, Chen M, Stout D B and Chatziioannou A F 2009 Performance evaluation of the Inveon dedicated PET preclinical tomograph based on the NEMA NU-4 standards *J. Nucl. Med.* **50** 401–8

[33] Parker D J, Forster R N, Fowles P and Takhar P S 2002 Positron emission particle tracking using the new Birmingham positron camera *Nucl. Instrum. Methods Phys. Res.* A **477** 540–5

[34] Buffler A, Cole K, Leadbeater T W and van Heerden M R 2018 Positron emission particle tracking: a powerful technique for flow studies *Int. J. Mod. Phys* **48** 1860113

[35] Brix G, Zaers J, Adam L E, Bellemann M E, Ostertag H, Trojan H, Haberkorn U, Doll J, Oberdorfer F and Lorenz W J 1997 Performance evaluation of a whole-body PET scanner using the NEMA protocol. National Electrical Manufacturers Association. *J. Nucl. Med.* **38** 1614–23

[36] Moses W W 2011 Fundamental limits of spatial resolution in PET *Nucl. Instrum. Methods Phys. Res.* A **648** S236–40

[37] Wiggins C, Patel N, Bingham Z and Ruggles A 2019 Qualification of multiple-particle positron emission particle tracking (M-PEPT) technique for measurements in turbulent wall-bounded flow *Chem. Eng. Sci.* **204** 246–56

[38] Mohammadi I, Castro I F C, Correia P M M, Silva A L M and Veloso J F C A 2019 Minimization of parallax error in positron emission tomography using depth of interaction capable detectors: methods and apparatus *Biomed. Phys. Eng. Express* **5** 062001

[39] Hawkesworth M R, O'Dwyer M A, Walker J, Fowles P, Heritage J, Stewart P A E, Witcomb R C, Bateman J E, Connolly J F and Stephenson R 1986 A positron camera for industrial application *Nucl. Instrum. Methods Phys. Res.* A **253** 145–57

[40] Hawkesworth M R, Parker D J, Fowles P, Crilly J F, Jefferies N L and Jonkers G 1991 Nonmedical applications of a positron camera *Nucl. Instrum. Methods Phys. Res.* **310** 423–34

[41] Sadrmomtaz A, Parker D J and Byars L G 2007 Modification of a medical PET scanner for PEPT studies *Nucl. Instrum. Methods Phys. Res.* **573** 91–4

[42] Parker D J, Leadbeater T W, Fan X, Hausard M N, Ingram A and Yang Z 2009 Positron emission particle tracking using a modular positron camera *Nucl. Instrum. Methods Phys. Res.* **604** 339–42

[43] Sovechles J M, Boucher D, Pax R, Leadbeater T, Sasmito A P and Waters K E 2017 Performance analysis of a new positron camera geometry for high speed, fine particle tracking *Meas. Sci. Technol.* **28** 095402

[44] Leadbeater T W, Parker D J and Gargiuli J 2011 Characterization of the latest Birmingham modular positron camera *Meas. Sci. Technol.* **22** 104017

[45] Griffiths W D, Beshay Y, Parker D J and Fan X 2008 The determination of inclusion movement in steel castings by positron emission particle tracking (PEPT) *J. Mater. Sci.* **43** 6853–6

[46] Ingram I, Hausard M, Fan X, Parker D J, Seville J P K, Finn N and Evans M 2007 Portable positron emission particle tracking (PEPT) for industrial use *12th Int. Conf. on Fluidization—New Horizons in Fluidization Engineering (Engineering Confs. Int. Symp. Series)*

[47] Mintzer R A and Siegel S B 2007 Design and performance of a new pixelated-LSO/PSPMT gamma-ray detector for high resolution PET imaging *IEEE Nuclear Science Symp. Conf. Record 5* pp 3418–22
[48] Visser E P, Disselhorst J A, Brom M, Laverman P, Gotthardt M, Oyen W J G and Boerman O C 2009 Spatial resolution and sensitivity of the Inveon small-animal PET scanner *J. Nucl. Med.* **50** 139–47
[49] Volkwyn T S, Buffler A, Govender I, Franzidis J-P, Morrison A J, Odo A, Van Der Meulen N P and Vermeulen C 2011 Studies of the effect of tracer activity on time-averaged positron emission particle tracking measurements on tumbling mills at PEPT Cape Town *Miner. Eng.* **24** 261–6
[50] Hampel D M *et al* 2021 SuperPEPT: A new tool for positron emission particle tracking; first results *Nucl. Instrum. Methods Phys. Res.* A **1028** 166254
[51] Lewellen T K 1998 Time-of-flight PET *Semin. Nucl. Med.* **28** 268–75
[52] Surti S and Karp J S 2020 Update on latest advances in time-of-flight PET *Phys. Medica* **80** 251–8
[53] Lecoq P *et al* 2020 Roadmap toward the 10 ps time-of-flight PET challenge *Phys. Med. Biol.* **65**
[54] Jan S *et al* 2004 GATE—Geant4 Application for Tomographic Emission: a simulation toolkit for PET and SPECT *Phys. Med. Biol.* **49** 4543–61
[55] Agostinelli S *et al* 2003 Geant4—a simulation toolkit *Nucl. Instrum. Methods Phys. Res.* A **506** 250–303
[56] Herald M, Wheldon T and Windows-Yule C 2021 Monte Carlo model validation of a detector system used for positron emission particle tracking *Nucl. Instrum. Methods Phys. Res.* A **993** 165073
[57] Kerhoas-Cavata S and Guez D 2006 Modeling electronic processing in GATE *Nucl. Instrum. Methods Phys. Res.* A **569** 330–4
[58] Windows-Yule C R, Herald M, Nicusan L, Wiggins C and Pratx G *et al* 2022 Recent advances in positron emission particle tracking: a comparative review *Rep. Prog. Phys.* **85** 016101
[59] Jan S *et al* 2011 Gate v6: a major enhancement of the gate simulation platform enabling modelling of CT and radiotherapy *Phys. Med. Biol.* **56** 881–901
[60] Jan S, Comtat C, Strul D, Santin G and Trebossen R 2005 Monte Carlo simulation for the ECAT exact HR+ system using gate *IEEE Trans. Nucl. Sci.* **52** 627–33
[61] Lee S, Gregor J and Osborne D 2013 Development and validation of a complete GATE model of the Siemens Inveon trimodal imaging platform *Mol. Imaging* **12** 434–45
[62] Salvadori J, Labour J, Odille F, Marie P-Y, Badel J-N, Imbert L and Sarrut D 2020 Monte Carlo simulation of digital photon counting PET *EJNMMI Physics* **7** 23
[63] Buffler A, Govinder I, Cilliers J J, Parker D J, Franzidis J-P, Mainza A, van der Westhuizen A, Newman R T and Powell M 2010 A new positron emission particle tracking facility at iThemba labs *Proc. of Int. Topical Meeting on Nuclear Research Applications and Utilization of Accelerators*
[64] Langford S T, Wiggins C S, Santos R, Hauser M, Becker J M and Ruggles A E 2017 Three-dimensional spatiotemporal tracking of fluorine-18 radiolabeled yeast cells via positron emission particle tracking *PLoS One* **12** e0180503

[65] Rault E, Staelens S, Van Holen R, De Beenhouwer J and Vandenberghe S 2011 Accurate Monte Carlo modelling of the back compartments of SPECT cameras *Phys. Med. Biol.* **56** 87–104

[66] National Electronics Manufacturer Association 2007 Performance measurements of positron emission tomographs *Technical Report* NU-2-2007 National Electronics Manufacturer Association

[67] Al-Shemmeri M, Windows-Yule K, Lopez-Quiroga E and Fryer P J 2021 Coffee bean particle motion in a spouted bed measured using positron emission particle tracking (PEPT) *J. Food Eng.* **311** 110709

[68] Boucher D, Jordens A, Sovechles J, Langlois R, Leadbeater T W, Rowson N A, Cilliers J J and Waters K E 2017 Direct mineral tracer activation in positron emission particle tracking of a flotation cell *Miner. Eng.* **100** 155–65

[69] Parker D J and Fan X 2008 Positron emission particle tracking—application and labelling techniques *Particuology* **6** 16–23 (Selected papers from 1st UK–China Particle Technology Forum)

[70] Herald M, Bingham Z, Santos R and Ruggles A 2018 Simulated time-dependent data to estimate uncertainty in fluid flow measurements *Nucl. Eng. Des.* **337** 221–7

[71] Morrison A J, Govender I, Mainza A N and Parker D J 2016 The shape and behaviour of a granular bed in a rotating drum using Eulerian flow fields obtained from PEPT *Chem. Eng. Sci.* **152** 186–98

[72] Parker D J, Broadbent C J, Fowles P, Hawkesworth M R and McNeil P 1993 Positron emission particle tracking—a technique for studying flow within engineering equipment *Nucl. Instrum. Methods Phys. Res.* A **326** 592–607

Chapter 6

Pre-processing: PEPT data and algorithms

There exist a number of 'PEPT Algorithms' whose job it is to convert the raw data acquired from a positron camera into an accurate, three-dimensional tracer trajectory. In this chapter, we provide an overview of all algorithms developed to date, from the original, iterative 'Birmingham method', whose development began in the late 1980s (but which is still widely used today), to the most recent techniques implementing cutting-edge machine learning methods. As with the different experimental techniques discussed in chapter 4, the different algorithms presented possess their own inherent strengths and weaknesses, which we highlight throughout the chapter, before summarising in the final section. We also include discussion of other key pre-processing methods, such as the delineation of distinct tracer trajectories from data sets tracking multiple particles, which are common to all of the algorithms discussed.

In the previous chapter, we discussed in detail the tracers and detectors necessary to perform PEPT. However, in addition to these physical components, a third ingredient is also required in order to perform PEPT: a suitable algorithm to convert the raw listmode data (i.e. a list of LoR interaction points, typically saved as binary 'data words') output by a PEPT detector into a particle location, and thus a trajectory. In this chapter, we discuss the various algorithms which one may use to perform this task, discussing in each case basic principles underlying the technique, the methods used and their major strengths and weaknesses. We also provide interactive examples which the user may use to gain a deeper understanding of the algorithms and their application.

6.1 Understanding PEPT data

The source data that allows us to perform PEPT is the coincident detection of a pair of gamma rays. Each detector system (see section 5.2) that allows us to perform PEPT reports these coincidence events in different ways. For example, the ADAC Gamma Camera writes several 32-bit data words to disc for each coincident event, using one each to describe the x-coordinate and y-coordinate of each event comprising the coincident event. On the other hand, some systems derived from medical PET hardware may use a single 32-bit data word to describe each coincident event by referring to a module pair number. This number allows us to use a look-up table to determine exactly

doi:10.1088/978-0-7503-3071-8ch6

which two detectors have been hit, as determined by a hardware coincidence processor. This method is highly efficient when considering disc space and also data transfer rates but means we have to allocate an x, y and z location for each detector, based on our knowledge of their position. As a result, we must use a number of scripts to decode the binary data that is collected from the detector systems and convert it to a text file.

The result of the above is a simple file containing the start and end points of all lines of response (LoRs) detected during a given experiment. As we know from chapter 2, however, this alone is not enough to tell us the trajectory of our tracer. In order to extract this information, we must apply a suitable algorithm to allow us to distinguish between 'true' and 'false' LoRs, and thus pinpoint the position of our tracer. There exist a number of algorithms through which this may be achieved, and it is these which form the main focus of this chapter.

6.1.1 Interactive example: PEPT data format

The notebook files accompanying this section are available in the supplementary data online at https://iopscience.iop.org/book/978-0-7503-3071-8.

PEPT Data Format

In this interactive example, we will understand the 'list-mode' data format that PET/PEPT scanners output, along with simple initialisation and visualisation steps using `NumPy` and the `pept` Python library. While not required *per se* for illustrating PEPT concepts, using `pept` significantly reduces the amount of repetitive code and visual noise, allowing the reader to focus on the high-level ideas.

In a 'production' environment, however, a researcher using PEPT would typically use the high-performance implementations from the `pept` library, rather than the more explicit, but much slower Python versions. In the spirit of completeness, the interactive examples depicting PEPT algorithms will contain both approaches where a high-performance algorithm is available.

An example dataset is used from an experiment run at the University of Birmingham Positron Imaging Centre using the ADAC Forté. It consists of a single 1 mm diameter MCC particle activated with fluorine-18 inside a bubbling fluidised bed. The fluidised bed was filled with 90% sand and 10% MCC; air was fed into the bottom of the bed at a rate of 37 litres per minute at 3.5 bar. This dataset was chosen for its high quality captured lines of response, with the tracer still depicting the random particle motion that is inherent to bubbling fluidised beds—and typical in Lagrangian particle tracking.

[1]:
```
# Importing necessary libraries for numerical array manipulation and plotting,
# plus the `pept` library for straightforward PEPT data processing
import numpy as np
import matplotlib.pyplot as plt
import pept

# Set Matplotlib theme (try "darkgrid" too for a more modern look)
import seaborn
seaborn.set_style("whitegrid")

# Set Plotly figure output (try "browser" too for interactive graphs)
import plotly
plotly.io.renderers.default = "png"
```

Initialise raw line of response data

In this initial step, we read in the 'raw' line of response (LoR) data from the ASCII list-mode output of an ADAC Forté parallel-screen gamma camera. This type of scanner has two detector *heads*, capturing each end of the line of response. Each row in the raw list-mode output contains a single LoR and is formatted into five columns: a timestamp (in milliseconds), the *X*, *Y* coordinates of the LoR on the first detector head, followed by the corresponding LoR's *X*, *Y* coordinates on the second head.

Importantly, the 3D coordinate convention used in PEPT is slightly different to the common mathematical convention: the *Y* and *Z* axes are normally switched, such that *Y* points 'upwards'; in the case of a parallel-screens PEPT detector, *X* and *Y* are parallel to the two detector heads, while the *Z* axis is perpendicular to them. While this may be confusing for people with a mathematical background, it has the advantage that 2D projections of tracer trajectories onto the *XY* plane yield the highest resolution (the *Z* axis having the largest 'spread' in tracer locations). The code cells below show this, along with a 3D plot illustrating the PEPT 3D axes convention.

[2]:
```
# Plot the first 20 rows from the experimental PEPT data sample
filepath = "Data/FluidisedBed1Tracer.csv"

# Print the first 20 lines in the file
with open(filepath) as f:
    for _ in range(20):
        print(f.readline(), end="")
```

```
University of Birmingham Positron Imaging Centre
```

This is a sample of an experiment run at the Univesity of Birmingham. It consists of a single 1 mm diameter MCC particle activated with fluorine-18 radioactive tracer material inside a bubbling fluidised bed. The fluidised bed was filled with 90% sand and 10% MCC; air was fed into the bottom of the bed at a rate of 37 litres per minute at 3.5 bar. Columns are formatted as [*t*, *x*1, *y*1, *x*2, *y*2], where the '*t*' represents the LoR timestamp (ms), *x*1 and *y*1 represent the LoR coordinates on the first detector screen (mm), and *x*2 and *y*2 represent the LoR coordinates on the second screen.

```
Separation= 600
f(opt) : 0.050 Displacement parameters : 300, 400, 1500
Fixed slices: 250 events/slice, 1 locations/slice
    0.0   190.0   168.7   346.3   142.8
    0.1   243.7   167.6   314.5   313.9
    0.1   194.1   410.0   417.1   240.1
    0.2   288.5   296.2   384.7   452.5
    0.2   164.6   115.1   382.9   353.4
```

[3]:
```
# Read in a sample of experimental PEPT data into a NumPy array. Skip the file
# header's first 15 lines
lors_raw = pept.utilities.read_csv("Data/FluidisedBed1Tracer.csv", skiprows=15)

print(f"First 10 rows from the raw LoR data:\n{lors_raw[:10]}\n")
```

```
First 10 rows from the raw LoR data:
 [[0.000e+00 1.900e+02 1.687e+02 3.463e+02 1.428e+02]
  [1.000e-01 2.437e+02 1.676e+02 3.145e+02 3.139e+02]
  [1.000e-01 1.941e+02 4.100e+02 4.171e+02 2.401e+02]
  [2.000e-01 2.885e+02 2.962e+02 3.847e+02 4.525e+02]
  [2.000e-01 1.646e+02 1.151e+02 3.829e+02 3.534e+02]
  [2.000e-01 1.286e+02 1.322e+02 4.649e+02 2.661e+02]
  [2.000e-01 3.339e+02 3.735e+02 3.316e+02 1.310e+02]
  [3.000e-01 1.640e+02 1.115e+02 3.794e+02 3.534e+02]
  [4.000e-01 2.620e+02 2.094e+02 2.932e+02 2.808e+02]
  [4.000e-01 1.304e+02 2.749e+02 2.543e+02 3.062e+02]]
```

Generally, each LoR is a 3D line, defined by two end-points on the two detector heads. For a parallel-screens gamma camera, the Z-coordinate of each line is implicit: the first head is at $Z_1 = 0$, with the second head at $Z_2 =$ *screen separation*. The screen separation is saved separately in the header of the ASCII file; the Z-coordinates of the LoRs must be inserted as extra columns in the `lors_raw` array:

```
[4]: # ADAC Forte's two detector heads' separation, in mm
     head_separation = 600

     lors = np.insert(lors_raw, 3, 0, axis = 1)                 # z1 = 0
     lors = np.insert(lors, 6, head_separation, axis = 1)       # z2 = head_separation

     # Print the first 10 rows from the line of response (LoR) data
     lors[:10]
```

```
[4]: array([[0.000e+00, 1.900e+02, 1.687e+02, 0.000e+00, 3.463e+02, 1.428e+02,
             6.000e+02],
            [1.000e-01, 2.437e+02, 1.676e+02, 0.000e+00, 3.145e+02, 3.139e+02,
             6.000e+02],
            [1.000e-01, 1.941e+02, 4.100e+02, 0.000e+00, 4.171e+02, 2.401e+02,
             6.000e+02],
            [2.000e-01, 2.885e+02, 2.962e+02, 0.000e+00, 3.847e+02, 4.525e+02,
             6.000e+02],
            [2.000e-01, 1.646e+02, 1.151e+02, 0.000e+00, 3.829e+02, 3.534e+02,
             6.000e+02],
            [2.000e-01, 1.286e+02, 1.322e+02, 0.000e+00, 4.649e+02, 2.661e+02,
             6.000e+02],
            [2.000e-01, 3.339e+02, 3.735e+02, 0.000e+00, 3.316e+02, 1.310e+02,
             6.000e+02],
            [3.000e-01, 1.640e+02, 1.115e+02, 0.000e+00, 3.794e+02, 3.534e+02,
             6.000e+02],
            [4.000e-01, 2.620e+02, 2.094e+02, 0.000e+00, 2.932e+02, 2.808e+02,
             6.000e+02],
            [4.000e-01, 1.304e+02, 2.749e+02, 0.000e+00, 2.543e+02, 3.062e+02,
             6.000e+02]])
```

Visualising a sample of LoRs

Now that the `lors` array contains the LoRs formatted as $[t, x_1, y_1, z_1, x_2, y_2, z_2]$ (units milliseconds and millimetres), we can visualise a *sample* of them to gain an intuition of what they look like. The `pept.plots` subpackage provides two PEPT-oriented interfaces to the Plotly visualisation library—`PlotlyGrapher` and `PlotlyGrapher2D`—containing straightforward plotting functions for PEPT data:

```
[5]: from pept.plots import PlotlyGrapher

     grapher = PlotlyGrapher()
     grapher.add_lines(lors[:200])
     grapher.show()
```

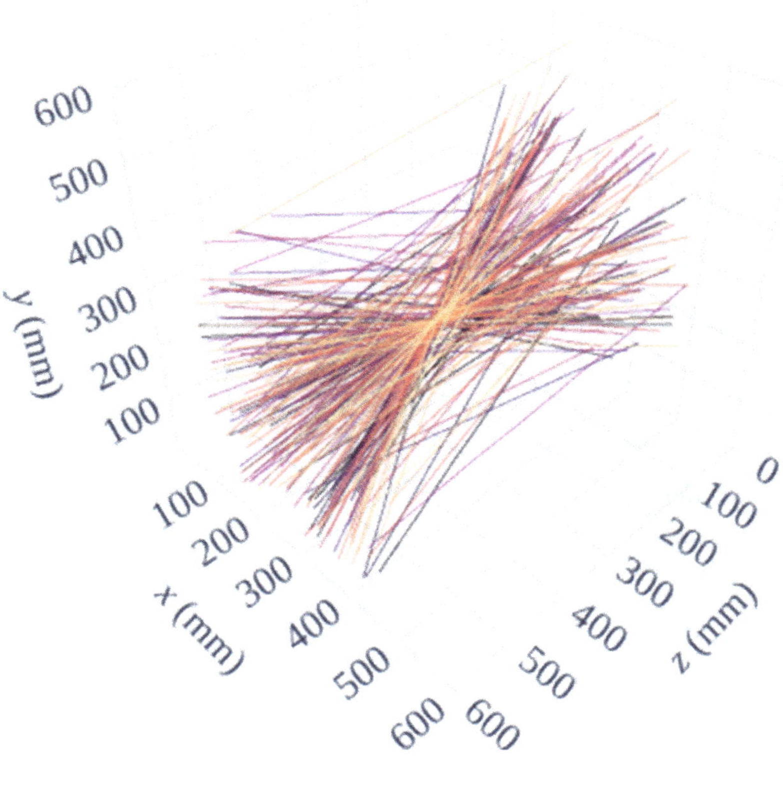

Note the PEPT 3D axes convention, differing from the common mathematical coordinates with *Y* and *Z* being switched, such that the *Y* axis points upwards, while the *Z* axis is perpendicular to the two detector heads.

Notice how the plotted lines concentrate around a central region in space—this is where the radioactive tracer is! But how to transform this intuitive region we see into an accurate tracer's location, with submillimetre precision, even in the presence of the scattered, erroneous LoRs that we see in the plot above? That is a PEPT *tracer locating algorithm's* job, which must find the tracer as accurately and quickly as possible, even inside dense scattering industrial equipment—and perhaps even more than one tracer—with as few LoRs as possible!

Even visually, we can gain an intuition of the effect of the number of LoRs used to find a tracer location:

[6]:
```
grapher = PlotlyGrapher(
    cols = 3,
    subplot_titles = ["10 LoRs", "100 LoRs", "1000 LoRs"],
)

grapher.add_lines(lors[:10])
grapher.add_lines(lors[:100], col = 2)
grapher.add_lines(lors[:1000], col = 3)

grapher.show()
```

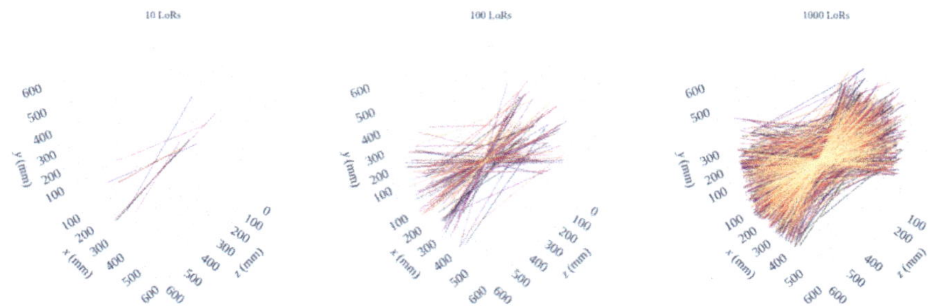

Which is easier to track? How would the spatial accuracy be affected by the number of LoRs used to find a single tracer location? What about the **temporal resolution**? How would the data look like for a tiny tracer that only has a small amount of radioactivity?

Temporal resolution?

Why is it important to use as few LoRs as possible to find our tracer? Let's look at how often we intercept LoRs—that is to say our data rate—in 100 ms intervals:

[7]:
```
# Extract the LoR timestamps from the first column
t = lors[:, 0]

# Plot histogram of the frequency of the timestamps in bins of 100 ms
bin_size = 100
bins = np.arange(t[0], t[-1], bin_size)

# Plotting a histogram of the LoRs' timestamps
fig, ax = plt.subplots()
counts, _, _ = ax.hist(t, bins=bins)
ax.set(xlabel='$t$ (ms)', ylabel='Number of LoRs intercepted')

# Compute average frequency of LoR acquisition
frequency = counts.mean() * (1000 / bin_size)     # number of counts / s = Hz

print(f"The average LoR acquisition rate is {frequency:0.1f} Hz")
```

```
The average LoR acquisition rate is 30201.2 Hz
```

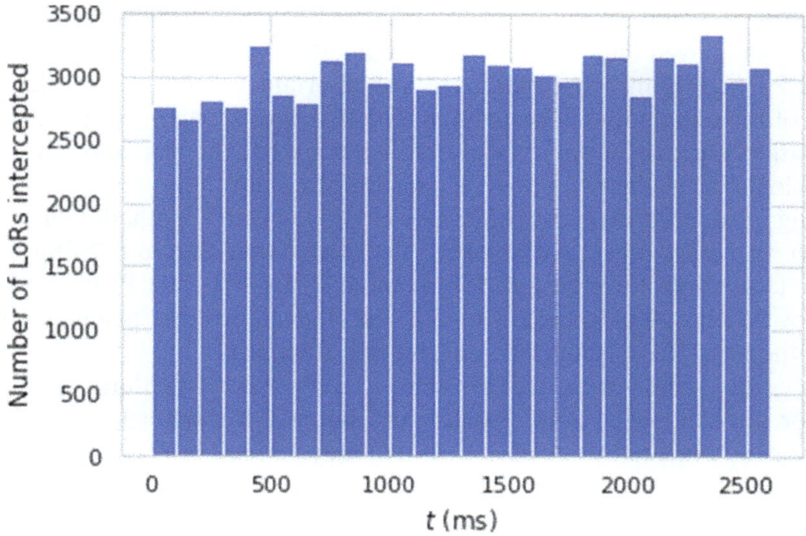

So how often can we track a tracer if we used 10, 100, or 1000 LoRs for a single location?

```
[8]: # Number of LoRs used to locate a tracer once - feel free to change it!
num_lors = 100

print((
    f"Using {num_lors} LoRs for a single location, a tracer can be located "
    f"{frequency / num_lors:0.1f} times per second"
))
```

```
Using 100 LoRs for a single location, a tracer can be located 302.0 times per
second
```

By the end of this tutorial, you have hopefully gained an intuition for the importance of a good PEPT tracer locating algorithm. In the next sections, we will take an in-depth look at different approaches to transforming raw lines of response into accurate particle trajectories, along with extending PEPT from single- to multiple-tracer tracking and solving non-trivial issues when separating particle locations into individual trajectories.

6.2 Available algorithms

In this section, we provide an overview of all PEPT algorithms currently available, summarising their method of operation, and their major strengths and limitations. In order to maximise the utility of this section for the reader, the available algorithms are not introduced in chronological order, but rather in order of their likely relevance. We thus begin by introducing the most commonly-used of all PEPT algorithms, the Birmingham algorithm, which has to date been applied in several hundred research papers. We then introduce the line-density method which, though less widely-employed than the Birmingham method, forms the basis of a number of other PEPT algorithms, thus making it an important foundation for our comprehension of the remainder of the chapter. The G-means clustering and feature point identification algorithms discussed thereafter represent the next most widely-used methodologies after the two previously-mentioned algorithms.

Having introduced the most commonly-employed PEPT methodologies, we next focus on the most recently-developed algorithms (the Stanford SBSR method, Voronoi tesselation, the triangulation method, PEPT-ML and PEPT-EM) which, though by definition not as heavily used as the older algorithms, potentially stand to become the next PEPT 'standard'—or, indeed, standards.

Finally, for completeness, we introduce a number of algorithms which for one reason or another never quite 'caught on', but which nonetheless contain some interesting and potentially valuable ideas that may perhaps be exploited in the future by PEPT developers.

6.2.1 The Birmingham algorithm

The Birmingham algorithm in brief

The original PEPT algorithm, widely known as the 'Birmingham algorithm' was first developed by Professor David Parker and Dr Michael Hawkesworth at the University of Birmingham [1, 2]. The basic operation of the Birmingham algorithm can be described as follows:

1. A '*sample*' of N_s individual, consecutive lines of response from a given PEPT data set (see chapter 2) is considered.
2. The '*minimum distance point*' (MDP)—i.e. the point in space for which the sum of distances between said point and all LoRs in the sample is at its minimum—is analytically determined for this sample.
3. The LoRs lying furthest from the MDP are discarded.
4. A new MDP is determined from the remaining LoRs.
5. Steps 3 and 4 are repeated until only a desired fraction, f, of the initial N_s LoRs remains.
6. The MDP of the final set of fN_s LoRs is taken as the particle's pseudo-instantaneous position for the current sample.

The steps outlined above are illustrated schematically in figure 6.1.

The Birmingham algorithm in detail

(1) Choosing a sample: As noted in point 1 above, our first task is to choose a suitably sized sample of LoRs. The optimal sample size depends on both the activity of the tracer and its typical velocity, which together determine the distance travelled by the tracer during the time required to collect a given number, N_s, of LoRs. If N_s is too large, the tracer is liable to move significantly during the time period considered, leading to distortion due to motion. Conversely, if N_s is too small, the lack of statistics will inherently lead to noisy data. A widely-utilised rule of thumb is that the

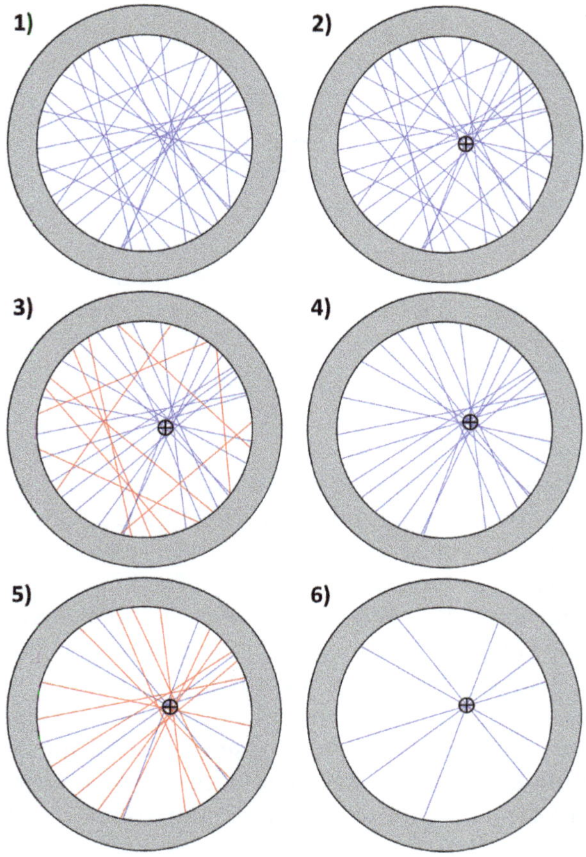

Figure 6.1. Schematic illustration showing the application of the Birmingham algorithm to a hypothetical data set with $N_s = 25$ and $f = 20$.

tracer should move no more than the intrinsic spatial resolution (see section 5.2) of the detector system being used[1].

(2) Calculating the MDP: As discussed in chapter 2, and in the interactive tutorial above, each LoR recorded in PEPT data corresponds to a line drawn between the interaction points, (x_1, y_1, z_1) and (x_2, y_2, z_2), of two photons detected by opposing detector blocks within a given resolving time. The perpendicular distance, δ, between this line and a given point in space, (x, y, z), can be analytically determined as:

$$r^2\delta^2 = [(x - x_2)z_{12} - (z - z_2)x_{12}]^2 + \left[(y - y_2)x_{12} - (x - x_2)y_{12}\right]^2 + \left[(z - z_2)y_{12} - (y - y_2)z_{12}\right]^2 \tag{6.1}$$

[1] Clearly, for a static tracer, $\bar{v} \rightarrow 0$, meaning that one can use an arbitrarily high value of N_s and thus, hypothetically, achieve arbitrarily high location accuracy!

where $\chi_{12} = \chi_1 - \chi_2$ ($\chi = x, y, z$), and $r = \sqrt{x_{12}^2 + y_{12}^2 + z_{12}^2}$. For a given sample of N_s LoRs, the root mean squared (RMS) distance, $D(x, y, z)$, between our arbitrary point (x, y, z) and all LoRs in said sample can be simply determined as:

$$D\left(x, y, z\right) = \sqrt{\frac{1}{N_s}\sum_{i=1}^{N_s} \delta_i(x, y, z)^2}. \tag{6.2}$$

The MDP, as its name implies, can be found by determining the values of x, y and z which minimise D, which may be achieved by simultaneously solving the set of differential equations:

$$\frac{\partial D}{\partial x} = 0, \frac{\partial D}{\partial y} = 0, \frac{\partial D}{\partial z} = 0, \tag{6.3}$$

(3) Discarding outliers: In any sample of raw LoR data, it is likely (indeed, under typical experimental conditions, practically certain!) that a number of the measured LoRs will be corrupt—i.e. correspond to false coincidences, as described in section 2.3. As such, once the MDP has been determined as outlined in point (2) above, all LoRs for which the individual MDP-LoR distance δ_i is greater than a predefined multiple of the root mean squared distance D are assumed to be corrupt, and thus discarded. In the majority of cases, the condition is set such that any LoR for which $\delta_i > 1.2D$ is discarded. The value of 1.2 has been empirically determined to be optimal for the vast majority of situations—at least for the case of single-particle tracking.

(4–5) Iterating toward a particle location: Once the N_o outlier points have been discarded from the original sample, a new minimum distance point is calculated for the remaining $N_s' = N_s - N_o$ LoRs, yielding a new, lower value of the RMS distance D—i.e. a more accurate estimate of the tracer's true position. The outlier-discarding process of point (3) is then repeated with these remaining points, and another MDP calculated. This process is repeated until $N_s' \leqslant fN_s$—i.e. only a fraction f of the original sample of LoRs remains. The value of f, like N_s, is typically user-defined. In order to maximise statistics whilst also ensuring that all corrupt events are discarded, the chosen value of f should correspond approximately to the expected fraction of 'true' events, thus depending on a variety of factors including the system material and geometry, the detector details and the source activity. As such, historically it has been very hard to determine *a priori* a suitable value of f. Thus, one of two approaches is typically adopted: firstly, a purely empirical, experimental approach may be applied in which one records data from a static source in the system of interest, processes said data for a range of f values, ultimately adopting that which gives the smallest spatial spread[2]. A second approach is to create a Monte Carlo (e.g. GEANT4—see section 5.3) model of the system of interest, and numerically determine the expected scatter fraction.

[2] Note that this approach is valid as the scatter fraction is not expected to depend in any meaningful way on the tracer velocity.

(6) Final location and error estimation: Once the iterative process described above has been completed, the particle position can be estimated as the MDP for the remaining fN data points. The final value of D, the RMS distance from the MDP to all remaining LoRs, can be taken as a measure of the error on the final position value obtained. Specifically, the uncertainty on location can be assumed to be of the order $\frac{D}{fN}$. The value of D can also be used as a validity test for the results obtained: if it is found that D is greater than the intrinsic spatial resolution of the cameras used, then it must typically be assumed that the reconstruction process has failed, either due to a poor choice of parameters, or because the tracer has left the camera's field of view.

Multiple particles

In the above, we have described how the Birmingham algorithm may be used to locate a single particle. However, with relatively minor adaptation, the algorithm may also be used to locate multiple, distinct tracers [3]. The basic method through which this is achieved for a system containing a total of N_t tracers is as follows:

1. The first (typically the most active) tracer is located using the algorithm described above.
2. The second tracer is then located by re-applying the algorithm to the data points discarded during (1).
3. The above process is repeated until all N_t tracers are located.

For each individual step above, the process remains largely identical to that described in the preceding subsection. The only notable difference is that a slightly higher threshold—typically $\delta_i > 1.5$—is set for the iterative discarding of corrupt events, meaning that supposedly false LoRs are discarded at a slower rate than for the single particle case.

Strengths and weaknesses

For the case of single-particle tracking, the Birmingham algorithm, despite its age, still holds up remarkably well against more modern algorithms in terms of both its accuracy and computational efficiency as demonstrated in a recent review of PEPT algorithms [54]; indeed, for the case of a single, static tracer it can hypothetically achieve arbitrarily high precision! Further, unlike several of the methods discussed below, it is 'mesh-free', meaning that it will not suffer from issues relating to pixelisation, for example.

The main limitations regarding this method arise when attempting to apply it to the case of multiple particle tracking. Firstly, the shift from single to multiple-particle tracking notably increases complexity, while the requisite of a higher threshold for LoR elimination inherently requires more iterations, meaning that as a whole the multiple-particle application of the Birmingham method leads to a considerable increase in computational expense. Nonetheless, the increase in computation time is not an insurmountable obstacle and, compared to some other algorithms mentioned below, is in fact quite reasonable. A more serious issue is that, as data must be discarded at every step, the statistics for each subsequent particle location inherently become increasingly poor, leading in turn to decreasing spatial resolution. Due largely to this fact, the method has thus far only been successfully used to track up to four particles [2]. Finally, the algorithm requires an *a priori* knowledge of the number of tracers within the system, such that it knows when to 'stop looking' for additional tracers. This restriction may prove problematic, for example, in the case of tracers leaving and/or entering the detectors' field of view, as is common when imaging a specific section of a large system.

The pros and cons of the Birmingham method are briefly summarised below:

Strengths
- Arbitrarily high accuracy for stationary tracer.
- Relatively computationally efficient.
- Mesh-free method.

Weaknesses
- Can only track limited number ($\lesssim$4) of individual tracers.
- Requires *a priori* knowledge of N_t.
- Accuracy decreases sharply with N_t.

6.2.2 Interactive example: the Birmingham algorithm

The notebook files accompanying this section are available in the supplementary data online at https://iopscience.iop.org/book/978-0-7503-3071-8.

Interactive PEPT analysis example using the Birmingham method [1]

This Jupyter Notebook

An example dataset is used from an experiment run at the University of Birmingham Positron Imaging Centre using the ADAC Forté. It consists of a single 1 mm diameter MCC particle activated with fluorine-18 inside a bubbling fluidised bed. The fluidised bed was filled with 90% sand and 10% MCC; air was fed into the bottom of the bed at a rate of 37 litres per minute at 3.5 bar. This dataset was chosen for its high quality captured lines of response, with the tracer still depicting the random particle motion that is inherent to bubbling fluidised beds—and typical in Lagrangian particle tracking.

The `pept` Python library is used for initialising and visualising PEPT data. While not required *per se* for illustrating PEPT algorithms' processing steps, it significantly reduces the amount of repetitive code and visual noise, allowing the reader to focus on the main concepts.

[1]:
```
# Importing necessary libraries for numerical array manipulation and plotting,
# plus the `pept` library for straightforward PEPT data processing
import numpy as np
import pept

# Set Plotly figure output (try "browser" too for interactive graphs)
import plotly
plotly.io.renderers.default = "png"
```

Initialise raw line of response data

In this initial step, we read in the 'raw' LoR data from which our tracer positions will be calculated. As illustrated in the 'Effect of sample size and overlap' notebook, varying the numbers of LoRs used to calculate a PEPT location can strongly affect the accuracy of the final position computed. Try altering the value `nrows` below to see how this influences your results.

```
[2]: # Read in a sample of experimental PEPT data into a NumPy array. Skip the file
     # header's first 15 lines, then read in the first `nrows` LoRs
     lors_raw = pept.utilities.read_csv(
         "Data/FluidisedBed1Tracer.csv",
         skiprows=15,
         nrows=50
     )

     # Convert the LoRs into the general 3D line format [t, x1, y1, z2, x2, y2, z2].
     # See the `PEPT Data Format` interactive example for detailed explanations

     # ADAC Forte's two detector heads' separation, in mm
     head_separation = 600

     lors = np.insert(lors_raw, 3, 0, axis = 1)                    # z1 = 0
     lors = np.insert(lors, 6, head_separation, axis = 1)          # z2 = head_separation

     # Print the first 10 rows from the line of response (LoR) data
     print(f"First 10 rows from the 3D-formatted LoR data:\n{lors[:10]}\n")
```

```
First 10 rows from the 3D-formatted LoR data:
[[0.000e+00 1.900e+02 1.687e+02 0.000e+00 3.463e+02 1.428e+02 6.000e+02]
 [1.000e-01 2.437e+02 1.676e+02 0.000e+00 3.145e+02 3.139e+02 6.000e+02]
 [1.000e-01 1.941e+02 4.100e+02 0.000e+00 4.171e+02 2.401e+02 6.000e+02]
 [2.000e-01 2.885e+02 2.962e+02 0.000e+00 3.847e+02 4.525e+02 6.000e+02]
 [2.000e-01 1.646e+02 1.151e+02 0.000e+00 3.829e+02 3.534e+02 6.000e+02]
 [2.000e-01 1.286e+02 1.322e+02 0.000e+00 4.649e+02 2.661e+02 6.000e+02]
 [2.000e-01 3.339e+02 3.735e+02 0.000e+00 3.316e+02 1.310e+02 6.000e+02]
 [3.000e-01 1.640e+02 1.115e+02 0.000e+00 3.794e+02 3.534e+02 6.000e+02]
 [4.000e-01 2.620e+02 2.094e+02 0.000e+00 2.932e+02 2.808e+02 6.000e+02]
 [4.000e-01 1.304e+02 2.749e+02 0.000e+00 2.543e+02 3.062e+02 6.000e+02]]
```

The `pept.plots` subpackage provides two PEPT-oriented interfaces to the Plotly visualisation library—`PlotlyGrapher` and `PlotlyGrapher2D`—containing straightforward plotting functions for PEPT data. Let's plot the lines of response from the `lors` array:

```
[3]: from pept.plots import PlotlyGrapher

     PlotlyGrapher().add_lines(lors).show()
```

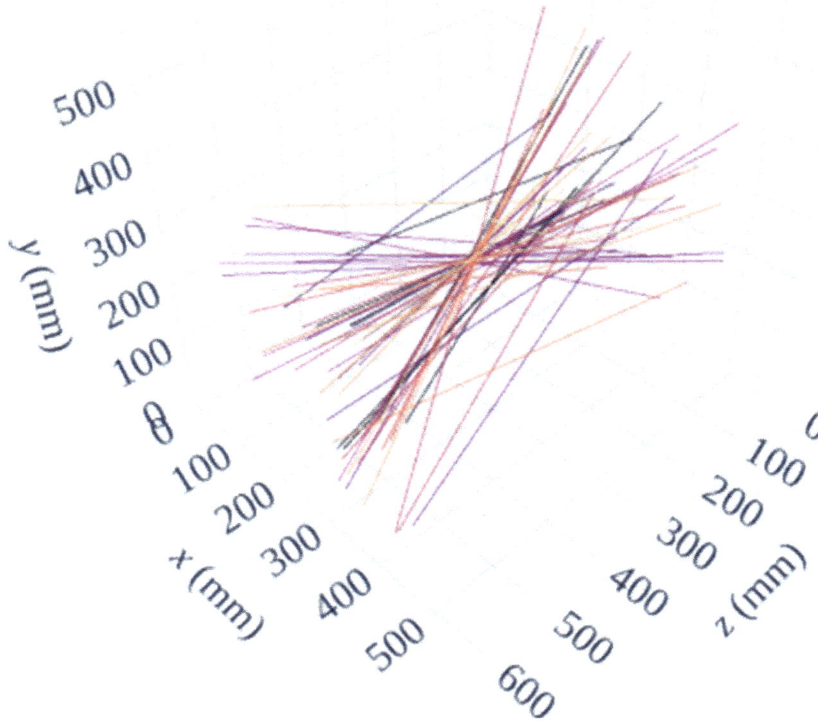

Note the PEPT 3D axes convention, differing from the common mathematical coordinates with Y and Z being switched, such that the Y axis points upwards, while the Z axis is perpendicular to the two detector heads.

Find minimum distance point

In this step, we find the three-dimensional point which minimises the distance to all LoRs in our sample, using the set of equations described in the main text.

```
[4]: # Find the minimum distance point (MDP) to all LoRs by minimising the sum of
     # distances to all lines.
     #
     # Note that the MDP can also be determined analytically (as is done in the C
     # implementation from the `pept` library), but the derivation is rather tedious
     # and takes away from the didactic purposes of this notebook
     from scipy.optimize import minimize

     def distances(point, lines):
         '''Return an array of distances from a `point` to an array of `lines`
         formatted as [t, x1, y1, z1, x2, y2, z2].'''
```

```
    p1 = lines[:, 1:4]
    p2 = lines[:, 4:7]

    d1 = np.linalg.norm(np.cross(point - p1, point - p2), axis=1)
    d2 = np.linalg.norm(p2 - p1, axis=1)

    return d1 / d2


def sum_distances(point, lines):
    '''Return the sum of distances from a `point` to an array of `lines`.'''
    return np.sum(distances(point, lines))


p0 = np.array([250, 250, 250])                    # initial guess
mdp = minimize(sum_distances, p0, lors).x

# Insert MDP timestamp at column 0
mdp = np.insert(mdp, 0, lors[:, 0].mean())
mdp
```

[4]: array([0.938   , 282.37288592, 242.9658988 , 317.53562967])

```
[5]: # Plot all LoRs and the minimum distance point (MDP)
     grapher = PlotlyGrapher()

     grapher.add_lines(lors)
     grapher.add_points([mdp], size=10)

     grapher.show()
```

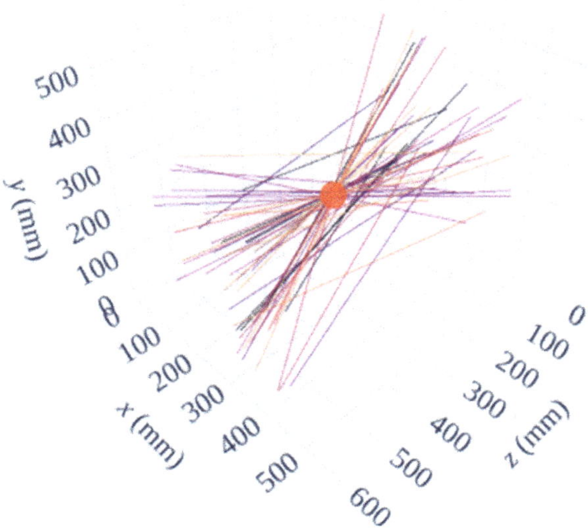

Remove the farthest lines of response

In this step, we remove the LoRs furthest removed from the calculated centre point—i.e. those most likely to be corrupt.

Try altering the value `nlors_remove`, which determines how many lines are removed in this step. If `nlors_remove` is too small, we can see it will take many iterations to reach only the true LoRs. Conversely, if we remove too many, you will notice the algorithm starts removing 'true' LoRs too.

[6]:
```
# Remove the nlors_remove farthest LoRs from the MDP to eliminate scattered
# LoRs, then recompute the MDP
nlors_remove = 15

# Compute all distances from the MDP to the LoRs, then find the *indices* of the
# sorted distances.
#
# The last 10 elements in `sorted_dists` are the indices of the farthest LoRs
dists = distances(mdp[1:], lors)        # Ignore the time column (index 0)
sorted_dists = np.argsort(dists)

scattered_lors = lors[sorted_dists[-nlors_remove:]]
good_lors = lors[sorted_dists[:-nlors_remove]]

# Recompute the MDP only from the retained ("good") LoRs
p0 = np.array([250, 250, 250])
new_mdp = minimize(sum_distances, p0, good_lors).x

# Insert MDP timestamp at column 0
new_mdp = np.insert(new_mdp, 0, good_lors[:, 0].mean())


# Plot the good and scattered LoRs with different colours, plus the initial and
# Birmingham Method-improved tracer location
grapher = PlotlyGrapher()

grapher.add_lines(good_lors, color="green")
grapher.add_lines(scattered_lors, color="black")

grapher.add_points([mdp], color="red", size=5)
grapher.add_points([new_mdp], color="blue", size=5)

grapher.show()
```

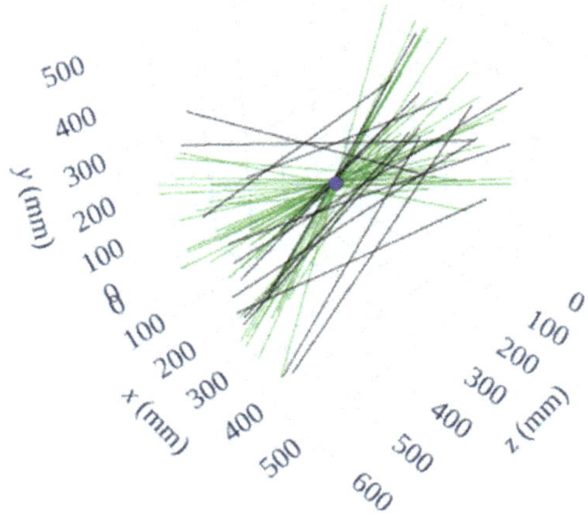

Iteratively remove the farthest LoRs and recompute MDP

In the code segment below, we can again adjust `nlors_remove` and directly observe the end effects of our choice. We can also adjust `nlors_target`—a higher value of this parameter will give us more LoRs, but a higher chance of false data; a lower value means we are more likely to have only true LoRs, but will have poorer statistics. Also try repeating the exercise with more or fewer initial LoRs to get a feel for how the correct value may vary with LoR activity.

[7]:
```
# The Birmingham method works by iteratively removing scattered LoRs:
#
#     1. Compute MDP
#     2. Remove `nlors_remove` LoRs that are farthest away from the MDP
#     3. Re-compute MDP
#
# Until `nlors_target` LoRs remain

nlors_remove = 5
nlors_target = 10

# Indices of "good" / unscattered LoRs
good_indices = np.arange(len(lors))

# Find the initial minimum distance point (MDP)
p0 = np.array([250, 250, 250])
cur_mdp = minimize(sum_distances, p0, lors[good_indices]).x

# Iteratively remove `nlors_remove` farthest LoRs at a time and recompute the
# MDP until `nlors_target` LoRs remain
while len(good_indices) > nlors_target:
    # Distances from MDP to all good LoRs
    dists = distances(cur_mdp, lors[good_indices])

    # Sorted indices of distances
    sorted_dists = np.argsort(dists)

    # Remove `nlors_remove` good LoR indices
    good_indices = good_indices[sorted_dists[:-nlors_remove]]

    # Current MDP
    cur_mdp = minimize(sum_distances, cur_mdp, lors[good_indices]).x

# Insert the tracer position's timestamp at column 0
position = np.insert(cur_mdp, 0, lors[good_indices, 0].mean())
```

[8]:
```
# Plot the remaining good LoRs and the final tracer location
grapher = PlotlyGrapher()

grapher.add_lines(lors[good_indices])
grapher.add_points([position], size=5)

grapher.show()
```

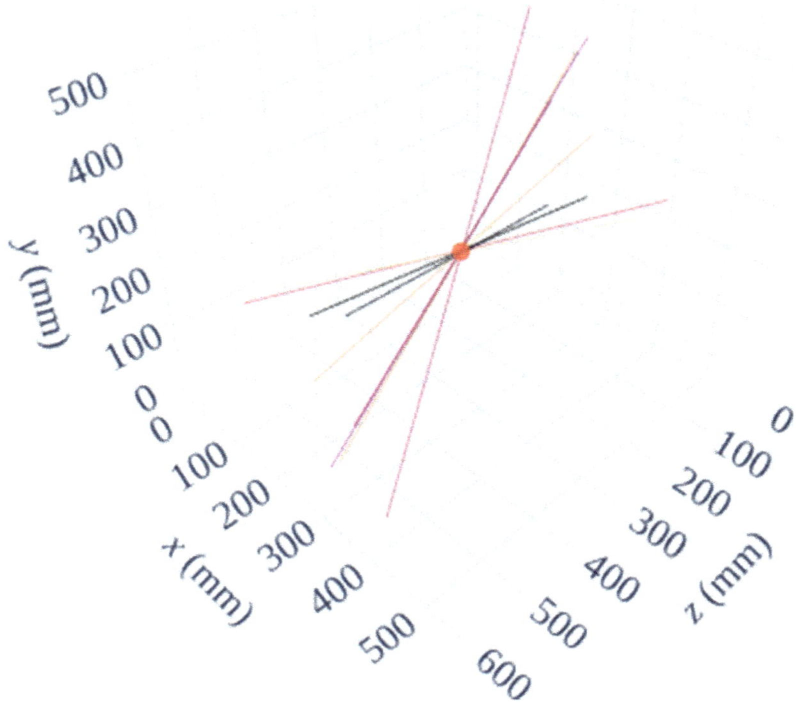

Complete Birmingham Method code

The code developed in the previous sections can be applied to consecutive 'samples' of LoRs to follow the tracer's movement through the fluidised bed, fully tracking its trajectory. A complete—albeit slow due to Python's dynamic nature—Birmingham Method implementation is given below.

First initialise a larger, multi-sample set of LoRs to analyse; change the `nrows` argument below to read in a different number of LoRs (the file included contains 80,000 LoRs in total), though keep in mind the computational expense of processing large numbers of LoRs!

```
[9]: # Read in a sample of experimental PEPT data into a NumPy array. Skip the file
     # header's first 15 lines
     lors_raw = pept.utilities.read_csv(
         "Data/FluidisedBed1Tracer.csv",
         skiprows=15,
         nrows=10_000,
     )
     # Convert the LoRs into the general 3D line format [t, x1, y1, z2, x2, y2, z2].
     # See the `PEPT Data Format` interactive example for detailed explanations
```

```
# ADAC Forte's two detector heads' separation, in mm
head_separation = 600

lors = np.insert(lors_raw, 3, 0, axis=1)                     # z1 = 0
lors = np.insert(lors, 6, head_separation, axis=1)           # z2 = head_separation

# Print the first 10 rows from the line of response (LoR) data
print(f"First 10 rows from the 3D-formatted LoR data:\n{lors[:10]}\n")
```

```
First 10 rows from the 3D-formatted LoR data:
[[0.000e+00 1.900e+02 1.687e+02 0.000e+00 3.463e+02 1.428e+02 6.000e+02]
 [1.000e-01 2.437e+02 1.676e+02 0.000e+00 3.145e+02 3.139e+02 6.000e+02]
 [1.000e-01 1.941e+02 4.100e+02 0.000e+00 4.171e+02 2.401e+02 6.000e+02]
 [2.000e-01 2.885e+02 2.962e+02 0.000e+00 3.847e+02 4.525e+02 6.000e+02]
 [2.000e-01 1.646e+02 1.151e+02 0.000e+00 3.829e+02 3.534e+02 6.000e+02]
 [2.000e-01 1.286e+02 1.322e+02 0.000e+00 4.649e+02 2.661e+02 6.000e+02]
 [2.000e-01 3.339e+02 3.735e+02 0.000e+00 3.316e+02 1.310e+02 6.000e+02]
 [3.000e-01 1.640e+02 1.115e+02 0.000e+00 3.794e+02 3.534e+02 6.000e+02]
 [4.000e-01 2.620e+02 2.094e+02 0.000e+00 2.932e+02 2.808e+02 6.000e+02]
 [4.000e-01 1.304e+02 2.749e+02 0.000e+00 2.543e+02 3.062e+02 6.000e+02]]
```

Iterating through each sample of LoRs, locate the tracer and append its location to a growing list of positions. As with the previous cells, you are invited to change the parameters highlighted below; details about each one are given in the comments.

```
[10]: # Use the Birmingham Method to track a moving tracer
from scipy.optimize import minimize


def distances(point, lines):
    '''Return an array of distances from a `point` to an array of `lines`
    formatted as [t, x1, y1, z1, x2, y2, z2].'''

    p1 = lines[:, 1:4]
    p2 = lines[:, 4:7]

    d1 = np.linalg.norm(np.cross(point - p1, point - p2), axis=1)
    d2 = np.linalg.norm(p2 - p1, axis=1)

    return d1 / d2
```

```
def sum_distances(point, lines):
    '''Return the sum of distances from a `point` to an array of `lines`.'''
    return np.sum(distances(point, lines))


# Birmingham method user-changeable parameters:
#     nlors_remove - the number of LoRs to remove at each Birmingham iteration
#     nlors_target - the target number of filtered LoRs to compute position
nlors_remove = 5
nlors_target = 10

# General PEPT tracking user-changeable parameters:
#     sample_start - the `lors` index at which to start tracking the tracer
#     sample_size - the number of LoRs used to compute a single location
#     overlap - the number of overlapping LoRs between consecutive samples
sample_start = 0
sample_size = 100
overlap = 10

# Tracked tracer locations will be appended to a growing list
positions = []

while sample_start + sample_size < len(lors):
    sample = lors[sample_start:sample_start + sample_size]

    # Indices of "good" / unscattered LoRs
    good_indices = np.arange(len(sample))

    # Find the initial minimum distance point (MDP)
    p0 = np.array([250, 250, 250])
    cur_mdp = minimize(sum_distances, p0, sample[good_indices]).x

    # Iteratively remove `nlors_remove` farthest LoRs at a time and recompute
    # the MDP until `nlors_target` LoRs remain
    while len(good_indices) > nlors_target:
        # Distances from MDP to all good LoRs
        dists = distances(cur_mdp, sample[good_indices])

        # Sorted indices of distances
        sorted_dists = np.argsort(dists)

        # Remove `nlors_remove` good LoR indices
        good_indices = good_indices[sorted_dists[:-nlors_remove]]

        # Current MDP
        cur_mdp = minimize(sum_distances, cur_mdp, sample[good_indices]).x
```

```
    # Insert the tracer position's timestamp at column 0
    position = np.insert(cur_mdp, 0, sample[good_indices, 0].mean())
    positions.append(position)

    sample_start += sample_size - overlap

positions = np.array(positions)

print(f"First 10 tracked positions:\n{positions[:10]}")
```

```
First 10 tracked positions:
[[  1.71        279.92078275 242.07864123 318.95713635]
 [  5.66        280.87726383 244.65962793 316.68206697]
 [  9.82        279.52281784 242.98812102 315.49789339]
 [ 12.43        281.76320403 242.66475282 317.01333052]
 [ 15.69        279.23841154 243.41776895 316.28551519]
 [ 18.95        280.43567995 245.75491284 318.52482284]
 [ 21.98        280.31275342 244.66141591 317.02544184]
 [ 25.53        281.43671858 243.54211038 316.02205684]
 [ 28.7         279.42095933 244.18914289 319.2710754 ]
 [ 31.75        280.86714385 245.05234477 317.98061415]]
```

[11]:
```
# Plot some of the LoRs used, plus the tracer locations we found
grapher = PlotlyGrapher(
    cols=2,
    subplot_titles=[
        "First 500 LoRs",
        f"Tracer locations tracked from {len(lors)} LoRs",
    ],
)

grapher.add_lines(lors[:500])
grapher.add_points(positions, col=2)

grapher.show()
```

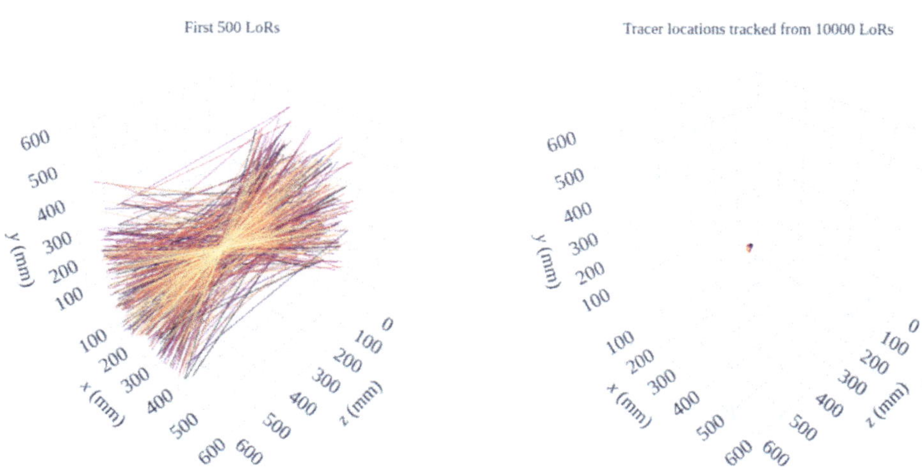

High-performance Birmingham Method implementation

The `pept` library contains a highly-optimised Birmingham Method implementation, which uses C code for the computationally-intensive operations and processes each sample of LoRs in parallel—therefore being able to fully utilise multi-CPU architectures in modern computers and clusters.

Most of the repetitive or tedious tasks above are automated, such as:

- Conversion from native PEPT scanner formats (e.g. binary, or parallel screens ASCII / CSV files) to the general 3D LoR format using the `pept.scanners` subpackage.
- Straightforward iteration of LoRs in terms of samples with optional overlap using the `pept.LineData` and `pept.PointData` classes.
- Terse usage of the PEPT tracking algorithms from the `pept.tracking` subpackage.
- PEPT-oriented plotting using the `pept.plots` subpackage.

[12]:

```
# Import the subroutines we will use to track our tracer
from pept.scanners import parallel_screens
from pept.tracking import BirminghamMethod, Stack
from pept.plots import PlotlyGrapher

# Initialise data from a Parallel Screens CSV format (i.e. [t, x1, y1, x2, y2])
# into the general "LineData" format ([t, x1, y1, z1, x2, y2, z2]) and set a
# sample size and overlap
lors = parallel_screens(
    "Data/FluidisedBed1Tracer.csv",
    screen_separation=600,
    skiprows=15,
    sample_size=200,
    overlap=100,
)

# Instantiate the `BirminghamMethod` tracker with the `fopt` value (the target
# fraction of LoRs used for a tracer's position) and "fit" each sample in `lors`
tracker = BirminghamMethod(fopt=0.15) + Stack()
positions = tracker.fit(lors)

# Plot all tracked positions
PlotlyGrapher().add_points(positions).show()
```

```
Initialised PEPT data in 0.036 s.

Batch 1 / 1 :: 100%|                              | 799/799
[00:01<00:00, 433.52it/s]

Processed samples in 1.8812768459320068 s
```

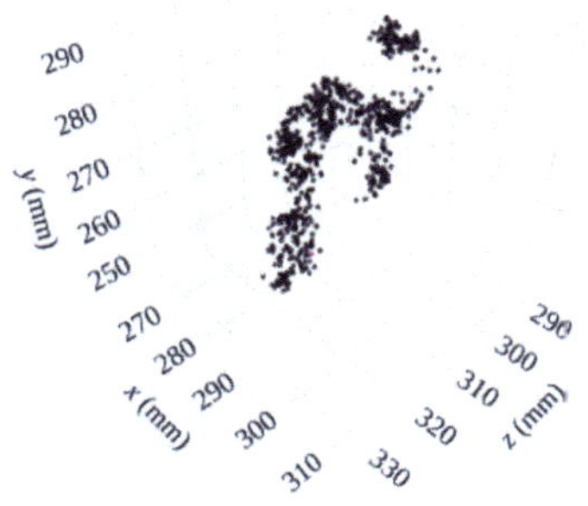


```
[13]: # Create a "timeseries" plot with the tracer positions vs. time
      from pept.plots import PlotlyGrapher2D

      PlotlyGrapher2D().add_timeseries(positions).show()
```

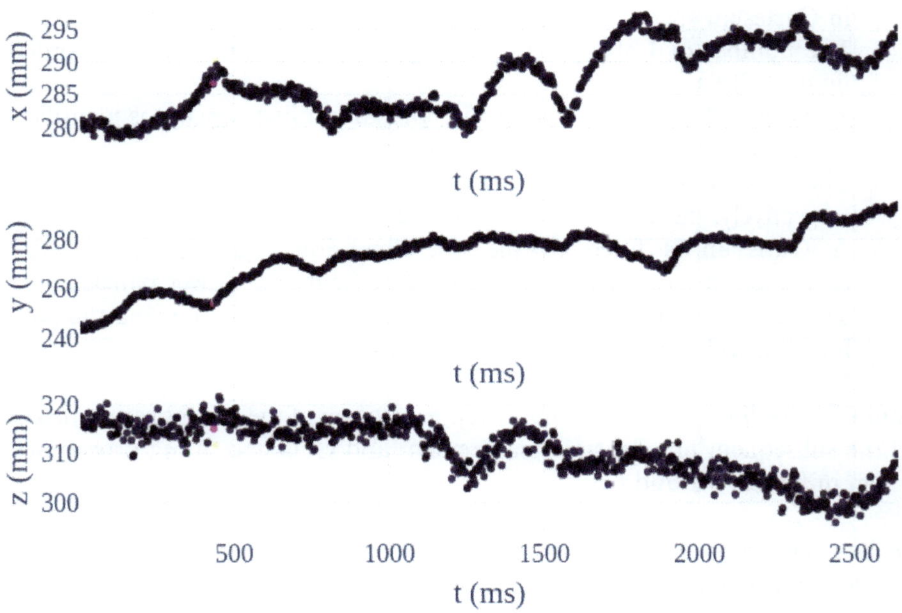

In what were essentially four Python statements, we processed 80,000 LoRs (in 0.032 s on the author's laptop) to track 2.6 s of a single tracer's trajectory through a fluidised bed, complete with loading bars and interactive plotting. Though `pept` automated most steps, there is no magic involved—just sensible defaults which can be changed for every subroutine called; plus, all code is public and free to use and modify from the `pept` repository at https://github.com/uob-positron-imaging-centre/pept.

6.2.3 The line-density method

The line-density method in brief

The line-density method [4], developed at the University of Cape Town, South Africa, represents a significant departure from the original Birmingham Method, which—as we will see in the following sections—has provided the foundation for a number of subsequently-developed algorithms. Instead of iteratively discarding LoRs as per the original method, the algorithm of Bickell, Buffler, Govender *et al* instead attempts to locate tracers by mapping the density of LoRs across space, under the assumption that any significant local maximum in LoR density must correspond to the location of a tracer.

The basic functionality of the line-density method can be briefly outlined, for the case of a single tracer particle, as follows:

1. A sample of N_s individual, consecutive lines of response from a given PEPT data set (see chapter 2) is considered.
2. The volume corresponding to the detector system's field of view is divided into a three-dimensional grid, typically comprising equally-sized, cubic cells in Cartesian space[3] (see figure 6.2(a)).
3. The number of LoRs passing through each individual cell is recorded (see figure 6.2(b)).
4. The cell with the highest count (i.e. the largest line density) is identified; it is assumed that the tracer particle falls within this cell.
5. The three one-dimensional rows of voxels passing through this central cell in, respectively, the x, y and z direction are identified (see figure 6.3, upper panel).
6. A Gaussian is fitted to the one-dimensional line-density distributions corresponding to each of these rows (see figure 6.3, lower panels).
7. The centroids of all three Gaussians are determined, allowing the particle's location to be estimated with sub-pixel resolution.

A simple, two-dimensional, schematic representation of the discretisation procedure, and the subsequent line density measurement and Gaussian fitting procedures, can be seen in figures 6.2 and 6.3.

The line-density method in detail

(1) Choosing a sample: As with the Birmingham method—and indeed most contemporary PEPT algorithms—our first task is again to choose a suitable sample of data from which to determine a pseudo-instantaneous particle location. In the work of Bickell *et al* [4], the authors suggest a fixed 'time slice' as opposed to a fixed number of LoRs, though this choice is not inherent to the method, just as choosing a fixed sample

[3] Note that there is no hard-written requisite that the grid must be Cartesian, nor that the cells must be equally sized, though this certainly both simplifies analysis and allows easier conceptualisation, thus for the purposes of this book we will adopt this framework.

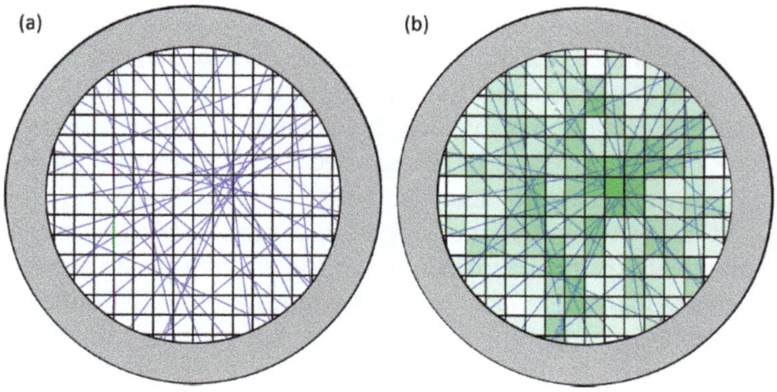

Figure 6.2. Schematic illustration showing, for the same hypothetical data set as shown in figure 6.1, (a) the subdivision of the experimental volume into individual cells and (b) the determination of line densities within these cells. In panel (b), the transparency of the cell illustrates the local line density, with a fully-transparent cell corresponding to a line density of zero, and a fully-opaque cell depicting the maximal line density achieved. Note that, for ease of presentation and conceptualisation, we depict here a 2D grid of pixels; in reality, however, a three-dimensional grid is used, though the fundamental principles remain the same.

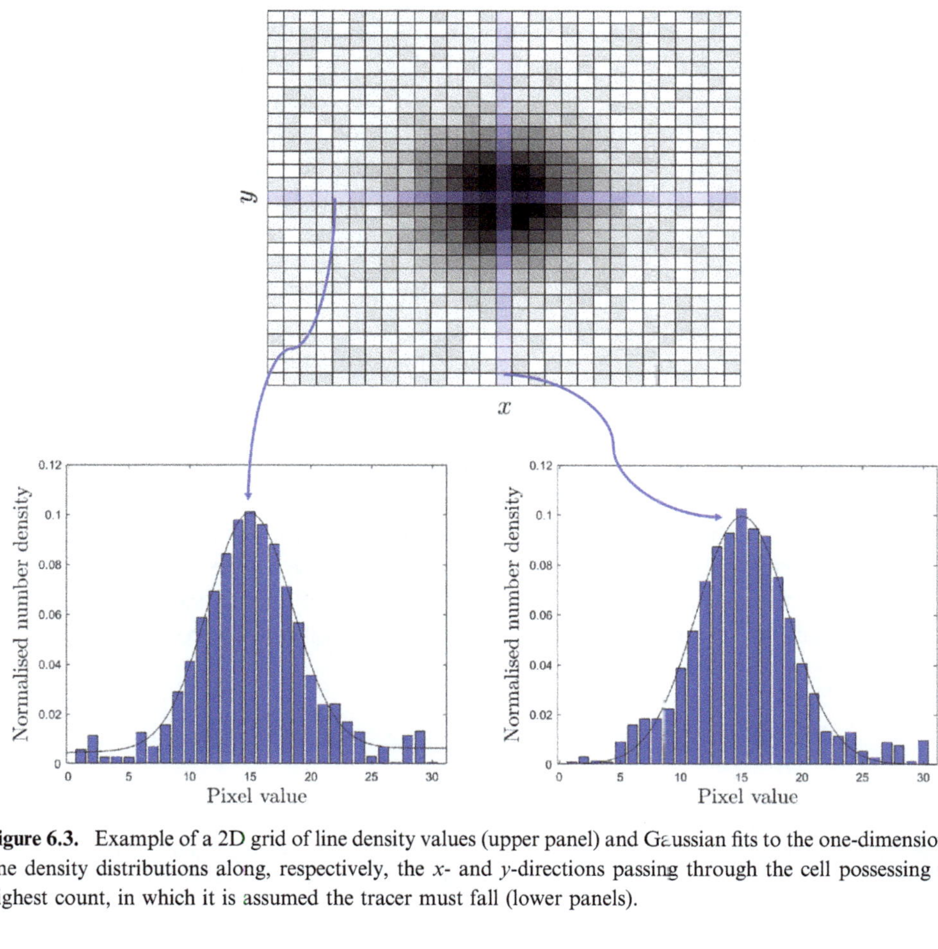

Figure 6.3. Example of a 2D grid of line density values (upper panel) and Gaussian fits to the one-dimensional line density distributions along, respectively, the x- and y-directions passing through the cell possessing the highest count, in which it is assumed the tracer must fall (lower panels).

size as opposed to a fixed sample duration is not a requisite for the Birmingham method. As with the choice of a suitable sample size discussed in the preceding section, the choice of time slice must be suitably short to ensure that distortion due to unsuitably large tracer motion is avoided, but suitably long that an adequately large number of LoRs can be expected to be recorded during any given time slice.

Aside: It is valuable here to take the opportunity to briefly discuss the major advantages and disadvantages of choosing a fixed sample *time* versus a fixed sample *size*. The main advantage of the former is that the use of a fixed timestep makes the calculation of many quantities which can be determined from PEPT data (see chapter 7) —in particular time-dependent quantities such as mean squared displacement or the velocity autocorrelation function—significantly easier. The major disadvantage is that as the activity of the source decreases, so too will the number of LoRs recorded per unit time, leading to a decrease in data quality with time. The two most commonly used isotopes—fluorine-18 and gallium-68—have half-lives of 110 minutes and 68 minutes, respectively. As a typical PEPT experiment for a moderately-sized system will require a duration of the order of hours to allow the tracer to fully explore the system [5] one can therefore expect a notable decrease in activity during such an experiment, and thus a non-negligible reduction in data quality between the first and last particle locations acquired. For highly heterogeneous systems—or scanners with sensitivity which varies strongly in space (see section 5.2)—a similar problem may be encountered even at fixed activity as the particle moves to different parts of the system, where the scanner may detect smaller or larger fractions of the emitted gamma rays, leading to continuously-fluctuating data quality. Of course, similar issues may potentially arise using a fixed sample size as the time taken to acquire the requisite number of LoRs varies with the tracer activity and/or local detector sensitivity, potentially leading to varying degrees of motion-distortion. However, with an adequately high baseline activity and a suitably chosen N_s, this issue can be relatively easily avoided.

(2) Discretisation: Once a suitable slice of data has been determined, the experimental volume containing this data must be discretised into a three-dimensional grid or 'mesh'. The resolution of the mesh—i.e. the size of the cells, L_c, into which the system is divided—can be defined by the user, and is typically chosen as a balance between accuracy and computational efficiency: smaller cells will typically allow a tracer to be located with greater precision (for reasons which will become clearer below!) but—as the number of required calculations scales as the number of cells and thus the inverse cube of the cell size—will also require a longer computation time[4]. In the tutorial exercise at the end of this section, you will have the opportunity to directly observe the effects of varying the chosen mesh resolution on both the accuracy of the tracer location and the computational time required.

[4] It should also be noted that even with an infinitely fast computer, smaller cells is not always better: for an overly small cell size, the distribution of line densities within adjacent cells will no longer be pseudo-continuous as pictured in figure 6.3, but will instead begin to show 'gaps', ultimately—as $L_C \to 0$—becoming a discrete distribution of ones and zeros!

(3–4) Rough tracer location: Once the experimental volume has been subdivided into a suitable grid, the spatial line-density distribution may be calculated. This is achieved simply through the process of 'backprojection'[5], which here simply involves, for each LoR in the sample, determining which voxels within the grid are passed through by said LoR, and accordingly incrementing a counter corresponding to each of these cells. Naturally, one may anticipate that cells closer to the tracer particle—i.e. the source of all our (true) LoRs—will be crossed by more LoRs (i.e. possess higher line densities), and similarly that the cell containing the tracer itself will have the most LoRs passing through it. As such, it is assumed that the tracer particle must be found in the cell containing the highest line density; the spatial coordinates of this cell therefore provide a rough, initial estimate of the tracer location.

(5–7) Fine tracer location: The above steps enable us to identify the voxel within which our tracer falls, thus allowing us to locate our particle's centroid with a resolution equal to the cell sized used in the above-described process. In order to achieve still finer—i.e. sub-voxel—resolution, we need to consider not just the voxel corresponding to the maximal line density, but also the surrounding voxels and the line densities thereof. Specifically, it is typical to consider three orthogonal, one-voxel-thick 'slices' through the maximum-density cell in each of the Cartesian x, y and z directions[6]. For each of these slices, we then consider the one-dimensional distribution of line densities, providing distributions such as those depicted in figure 6.3, for the current coordinate of interest. If the cell size, L_c, is suitably chosen—i.e. the size of the tracer is smaller than the L_c—one may reasonably apply a Gaussian fit to the discrete distribution obtained. Assuming both a suitable cell size and suitable statistics, the centroid of this Gaussian may be taken as the particle centroid position along the axis currently under investigation. By repeating this process for all three Cartesian coordinates, we may thus determine the tracer's three-dimensional location with sub-voxel resolution in all dimensions. The location error on this newly-defined particle centroid in each coordinate is typically taken as the full-width half-maximum (FWHM) of the Gaussian fit determined for that coordinate.

Multiple particles

In the case of multiple tracer particles, we must consider the existence of not one, global maximum in line density, but N_t individual, local maxima. The immediate problem faced in this case is how to distinguish between local maxima corresponding to a true tracer location, and local maxima (such as those evident in figure 6.2) which arise simply due to statistical fluctuations—i.e. cells where multiple false LoRs simply happen to cross paths. As different tracers will typically vary in activity and, as discussed in the previous section, the number of LoRs registered by the camera may vary significantly dependent on the tracer's position within the field of view, setting a 'threshold line density' is typically not a reliable option.

[5] Other forms of backprojection for non-PEPT applications are discussed in chapter 4.

[6] As previously noted, a similar analysis may be easily performed in other coordinate systems, though for the sake of simplicity and clarity we discuss here only the Cartesian.

The typical approach to multiple tracer location using the line density method is to begin at t_0 with a known number, N_t, of tracers occupying known approximate initial positions within the detector's field of view, and considering a separate, cubic 'region' of the field of view for each tracer. The algorithm isolates the cells falling within each of the regions determined, using in each case the relevant subset of LoRs to compute the location of the tracer lying within the relevant region in the manner described in the preceding subsection. For the next timestep t_1, the algorithm uses the tracer positions calculated at t_0 to determine a suitable set of regions for each tracer at t_1, and the process is repeated using these new regions. For subsequent timesteps t_n, if the tracer motion is suitably smooth, a predicted position for a given tracer at t_n can be extrapolated from its previous locations at t_{n-1} and t_{n-2}, this predicted position being used to establish the next set of cubic regions of consideration. This process is repeated until all locations have been determined.

In order for this scheme to work, the cubic regions used must possess a side length of at least $|a|(dt)^2$, where a is the particle acceleration and dt the timestep [4]. If the tracer motion is highly erratic—i.e. the above-mentioned extrapolation process cannot be meaningfully used—then one must typically use a larger region and/or (if feasible) a shorter timestep. The danger of using larger regions, however, is that it increases the likelihood that two (or more) tracers may fall in the same region, in which case only one location will be determined. In other words, the resolving power (i.e. the ability to distinguish two neighbouring particles) of the algorithm is directly related to the size of the cubic region used. For example, if a cube of volume $10 \times 10 \times 10$ mm^3 is utilised, one may achieve a resolving power of ~9 mm (i.e. particles lying closer than 9 mm cannot be successfully separated). The issue here though is that, as discussed above, for the algorithm to function correctly, the cubic region's dimension L_r must satisfy the inequality $L_r > |a|(dt)^2$. This means that for a typical timestep of 10 ms, if one wishes to achieve a resolving power of 9 mm, it must be safe to assume that the tracer particle will not experience any accelerations greater than 0.9 ms^{-2}—an assumption that for many experiments here on Earth may prove problematic! As such, when tracking multiple particles with the line-density method, one must strike a careful balance between resolving time and reliability.

The line density method has been demonstrated to track up to 16 tracer particles simultaneously, though this is not necessarily a hard limit.

Strengths and weaknesses

One of the major limitations of the line density method is that it requires not only *a priori* knowledge of the number of tracers (introducing the same issues as described for the case with the Birmingham method, above) but also of their approximate initial positions, inherently placing further restrictions on the use of the technique. A notable example of such a restriction is that in many industrial systems, access points are limited, meaning that particles (and thus tracers) must be poured into the system from a distance, as opposed to being manually placed, thus leading to an inherent lack of control regarding the initial tracer positions. Similar issues may be expected for the case in which it is necessary to introduce tracers when the system is already in a dynamic steady state.

In addition, as a mesh-based technique, the results obtained by the line-density are, as discussed above, grid-size dependent, and potentially subject to issues relating to pixelation [6, 7], problems not shared by mesh-free methods such as the Birmingham algorithm (and indeed the various other mesh-free methods described later in this section).

A notable strength of the technique compared to the Birmingham method (and, again, other methods yet to be discussed) is its computational efficiency in locating multiple tracers, as the algorithm assigns different regions of the experimental volume to different tracers, thus only scanning through each data set a single time.

Strengths

- Similar resolution in single-particle case to Birmingham method [4].
- Relatively computationally efficient.

Weaknesses

- Requires *a priori* knowledge of N_t.
- Requires *a priori* knowledge of initial tracer positions.
- Grid-size dependent results.
- Subject to pixelation for slower-moving tracers.

6.2.4 Interactive example: the line-density method

The notebook files accompanying this section are available in the supplementary data online at https://iopscience.iop.org/book/978-0-7503-3071-8.

Interactive PEPT analysis example using the line density algorithm [4]

This Jupyter Notebook

An example dataset is used from an experiment run at the University of Birmingham Positron Imaging Centre using the ADAC Forté. It consists of a single 1 mm diameter MCC particle activated with fluorine-18 inside a bubbling fluidised bed. The fluidised bed was filled with 90% sand and 10% MCC; air was fed into the bottom of the bed at a rate of 37 litres per minute at 3.5 bar. This dataset was chosen for its high quality captured lines of response, with the tracer still depicting the random particle motion that is inherent to bubbling fluidised beds—and typical in Lagrangian particle tracking.

The `pept` Python library is used for initialising and visualising PEPT data. While not required *per se* for illustrating PEPT algorithms' processing steps, it significantly reduces the amount of repetitive code and visual noise, allowing the reader to focus on the main conceptual procedures.

[1]:
```
# Importing necessary libraries for numerical array manipulation and plotting,
# plus the `pept` library for straightforward PEPT data processing
import numpy as np
import pept

# Set Plotly figure output (try "browser" too for interactive graphs)
import plotly
plotly.io.renderers.default = "png"
```

Initialise raw line of response data

In this initial step, we read in the 'raw' LoR data from which our tracer positions are calculated. As illustrated in the 'Effect of sample size and overlap' notebook, varying the numbers of LoRs used to calculate a PEPT location can strongly affect the accuracy of the final position computed. Try altering the value `nrows` below to see how this influences your results.

[2]:
```
# Read in a sample of experimental PEPT data into a NumPy array. Skip the file
# header's first 15 lines, then read in the first `nrows` LoRs
lors_raw = pept.utilities.read_csv(
    "Data/FluidisedBed1Tracer.csv",
    skiprows=15,
    nrows=100,
)

# Convert the LoRs into the general 3D line format [t, x1, y1, z2, x2, y2, z2].
# See the `PEPT Data Format` interactive example for detailed explanations

# ADAC Forte's two detector heads' separation, in mm
head_separation = 600

lors = np.insert(lors_raw, 3, 0, axis = 1)                  # z1 = 0
lors = np.insert(lors, 6, head_separation, axis = 1)        # z2 = head_separation

# Print the first 10 rows from the line of response (LoR) data
print(f"First 10 rows from the 3D-formatted LoR data:\n{lors[:10]}\n")
```

```
First 10 rows from the 3D-formatted LoR data:
[[0.000e+00 1.900e+02 1.687e+02 0.000e+00 3.463e+02 1.428e+02 6.000e+02]
 [1.000e-01 2.437e+02 1.676e+02 0.000e+00 3.145e+02 3.139e+02 6.000e+02]
 [1.000e-01 1.941e+02 4.100e+02 0.000e+00 4.171e+02 2.401e+02 6.000e+02]
 [2.000e-01 2.885e+02 2.962e+02 0.000e+00 3.847e+02 4.525e+02 6.000e+02]
 [2.000e-01 1.646e+02 1.151e+02 0.000e+00 3.829e+02 3.534e+02 6.000e+02]
 [2.000e-01 1.286e+02 1.322e+02 0.000e+00 4.649e+02 2.661e+02 6.000e+02]
 [2.000e-01 3.339e+02 3.735e+02 0.000e+00 3.316e+02 1.310e+02 6.000e+02]
 [3.000e-01 1.640e+02 1.115e+02 0.000e+00 3.794e+02 3.534e+02 6.000e+02]
 [4.000e-01 2.620e+02 2.094e+02 0.000e+00 2.932e+02 2.808e+02 6.000e+02]
 [4.000e-01 1.304e+02 2.749e+02 0.000e+00 2.543e+02 3.062e+02 6.000e+02]]
```

The `pept.plots` subpackage provides two PEPT-oriented interfaces to the Plotly visualisation library—`PlotlyGrapher` and `PlotlyGrapher2D`—containing straightforward plotting functions for PEPT data. Let's plot the lines of response from the `lors` array:

[3]:
```
# Plot the 3D LoRs that were initialised
from pept.plots import PlotlyGrapher

PlotlyGrapher().add_lines(lors).show()
```

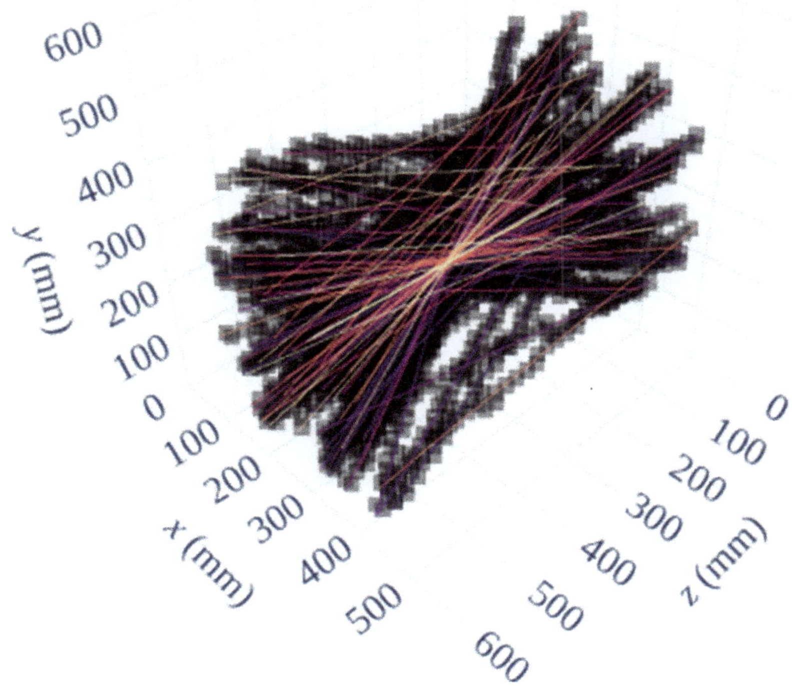

Note the PEPT 3D axes convention, differing from the common mathematical coordinates with *Y* and *Z* being switched, such that the *Y* axis points upwards, while the *Z* axis is perpendicular to the two detector heads.

Voxelise the lines of response

The next step is to 'voxelise' the LoRs—that is, to create a grid of square or rectangular cells, and count the number of lines passing through each. In the code below you can set the resolution—i.e. number of voxels in each dimension—by varying the pair of values corresponding to the `resolution` parameter. Typing, for example, `(20, 20, 20)` will give a 20 by 20 by 20 grid of voxels.

Try some different values for the `resolution` parameter. As discussed in the main text, larger values—i.e. a higher resolution, corresponding to smaller voxels—will allow the centroid of our particle to be determined more precisely. However, you may notice that as you increase these values your processor will take longer to execute the calculation and plotting steps below.

Now consider that this is just one of many thousands of locations we typically need to process, and you may get an idea of the issue of using an overly-fine grid. When using algorithms relying on voxelisation, there is a fine balance to be made between precision and compuational efficiency.

[4]:
```
resolution = (30, 30, 30)
voxels = pept.Voxels.from_lines(lors, resolution)

voxels
```

```
Initialised Voxels class in 0.0004482269287109375 s.
```

[4]:
```
Voxels
------
[[[0. 0. 0. … 0. 0. 0.]
  [0. 0. 0. … 0. 0. 0.]
  [0. 0. 0. … 0. 0. 0.]
  …
  [0. 0. 0. … 0. 0. 0.]
  [0. 0. 0. … 0. 0. 0.]
  [0. 0. 0. … 0. 0. 0.]]

 [[0. 0. 0. … 0. 0. 0.]
  [0. 0. 0. … 0. 0. 0.]
  [0. 0. 0. … 0. 0. 0.]
  …
  [0. 0. 0. … 0. 0. 0.]
  [0. 0. 0. … 0. 0. 0.]
  [0. 0. 0. … 0. 0. 0.]]

 [[0. 0. 0. … 0. 0. 0.]
  [0. 0. 0. … 0. 0. 0.]
  [0. 0. 0. … 0. 0. 0.]
  …
  [0. 0. 0. … 0. 0. 0.]
  [0. 0. 0. … 0. 0. 0.]
  [0. 0. 0. … 0. 0. 0.]]

 …

 [[0. 0. 0. … 0. 0. 0.]
  [0. 0. 0. … 0. 0. 0.]
  [0. 0. 0. … 0. 0. 0.]
  …
  [0. 0. 0. … 0. 0. 0.]
  [0. 0. 0. … 0. 0. 0.]
  [0. 0. 0. … 0. 0. 0.]]

 [[0. 0. 0. … 0. 0. 0.]
  [0. 0. 0. … 0. 0. 0.]
  [0. 0. 0. … 0. 0. 0.]
  …
  [0. 0. 0. … 0. 0. 0.]
  [0. 0. 0. … 0. 0. 0.]
```

```
[[0. 0. 0. ... 0. 0. 0.]
 [0. 0. 0. ... 0. 0. 0.]
 [0. 0. 0. ... 0. 0. 0.]
 ...
 [0. 0. 0. ... 0. 0. 0.]
 [0. 0. 0. ... 0. 0. 0.]
 [0. 0. 0. ... 0. 0. 0.]]]

number_of_voxels =      (30, 30, 30)
voxel_size =            [12.35333333 16.99333333 20.        ]

xlim =                  [110.3 480.9]
ylim =                  [ 47.2 557. ]
zlim =                  [  0. 600.]

voxel_grids:
([110.3 ... 480.9],
 [47.2 ... 557.0],
 [0.0 ... 600.0])
```

[5]:
```
# Plot all LoRs and all voxels with non-zero values (i.e. more than one LoR
# passing through)
PlotlyGrapher().add_voxels(voxels).add_lines(lors).show()
```

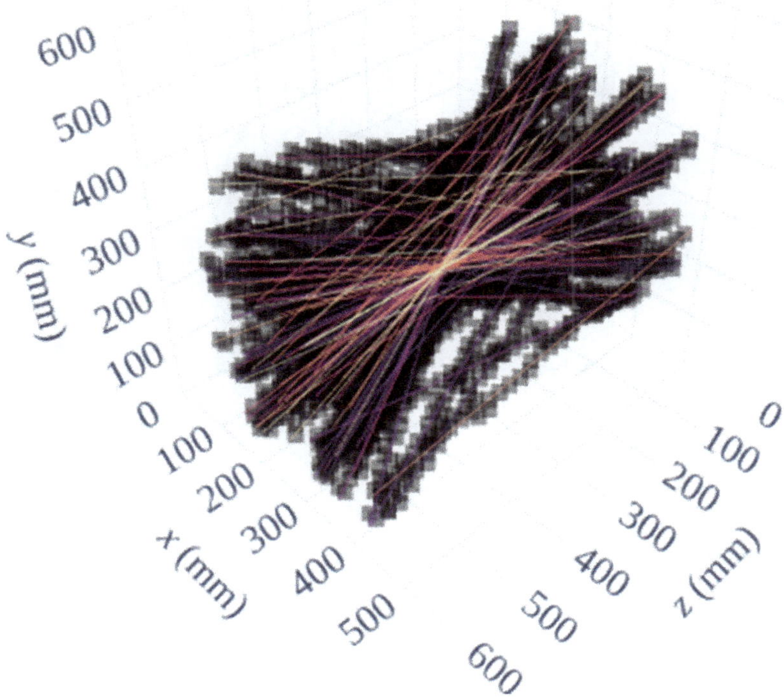

Fit 1D Gaussians around the peak

Having created the line density grid as shown above, the next step is simply to fit a Gaussian curve to the data, as described in the main text—a process which requires no additional user-defined variables!

The means of the fitted Gaussians, corresponding to the estimated particle positions, are then simply returned by the algorithm.

[6]:
```
# Find indices of maximum `voxels` value and plot a 2D slice of the voxel grid
from pept.plots import PlotlyGrapher2D

# Find the 3D voxel indices through which the maximum number of LoRs pass
max_indices = np.unravel_index(voxels.argmax(), voxels.shape)

grapher = PlotlyGrapher2D(
    subplot_titles = [f"Voxel slice through max Y index = {max_indices[1]}"],
)

grapher.add_trace(voxels.heatmap_trace(iy=max_indices[1]))
grapher.show()
```

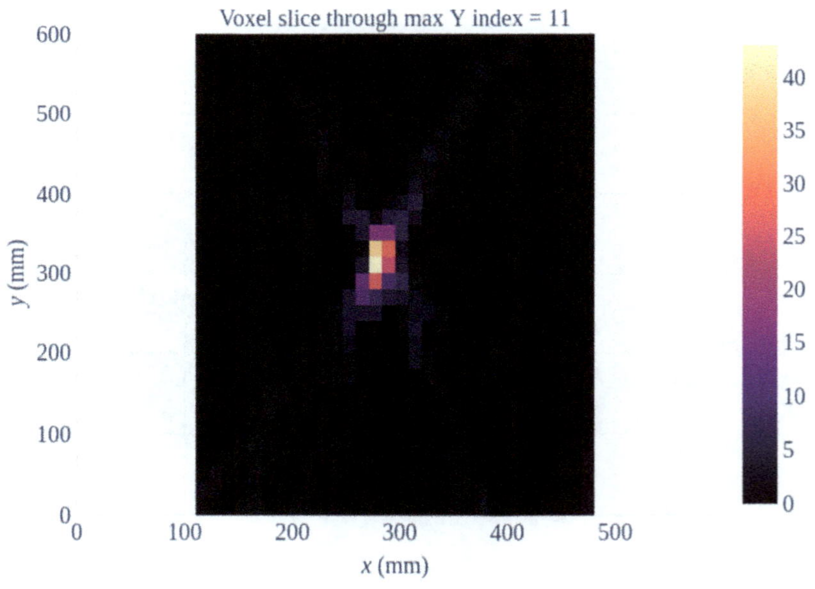

[7]:
```
# Fit 1D Gaussians through the voxel lines passing through the maximum voxel
# value (the indices of which were found in the previous code cell)
from scipy.optimize import curve_fit

# Define scaled Gauss model function to fit to data
def gauss(x, mu, sigma, scale):
    a = scale / sigma / np.sqrt(2 * np.pi)
    return a * np.exp(-0.5 * (x - mu)**2 / sigma**2)
```

```
# Coordinates of the voxel centres for each voxel line
voxels_x = np.linspace(voxels.xlim[0] + voxels.voxel_size[0] / 2,
                       voxels.xlim[1] - voxels.voxel_size[0] / 2,
                       voxels.shape[0])

voxels_y = np.linspace(voxels.ylim[0] + voxels.voxel_size[1] / 2,
                       voxels.ylim[1] - voxels.voxel_size[1] / 2,
                       voxels.shape[1])

voxels_z = np.linspace(voxels.zlim[0] + voxels.voxel_size[2] / 2,
                       voxels.zlim[1] - voxels.voxel_size[2] / 2,
                       voxels.shape[2])

# Initial guesses for fitting the Gaussian curves are simply the maximum voxel's
# X, Y, and Z coordinates, respectively
p0_x = np.array([voxels_x[max_indices[0]], voxels.voxel_size[0], 1.])
p0_y = np.array([voxels_y[max_indices[1]], voxels.voxel_size[1], 1.])
p0_z = np.array([voxels_z[max_indices[2]], voxels.voxel_size[2], 1.])

# Fit Gaussian curves to each "line of voxels" passing through the maximum
voxel_line_x = voxels[:, max_indices[1], max_indices[2]]
voxel_line_y = voxels[max_indices[0], :, max_indices[2]]
voxel_line_z = voxels[max_indices[0], max_indices[1], :]

opt_x, _ = curve_fit(gauss, voxels_x, voxel_line_x, p0_x)
opt_y, _ = curve_fit(gauss, voxels_y, voxel_line_y, p0_y)
opt_z, _ = curve_fit(gauss, voxels_z, voxel_line_z, p0_z)

# The tracer's location is given by the Gaussians' centres
location = np.array([opt_x[0], opt_y[0], opt_z[0]])

# Insert the tracer positions' timestamp at column 0
location = np.insert(location, 0, lors[:, 0].mean())

location
```

```
/home/andreinicusan/anaconda3/envs/peptbook/lib/python3.7/site-
packages/scipy/optimize/minpack.py:834: OptimizeWarning:

Covariance of the parameters could not be estimated
```

```
[7]: array([  1.931     , 276.97857002, 242.59937205, 309.77838094])
```

```
[8]: # Plot the voxel grid and the tracer location found via the Line Density method
grapher = PlotlyGrapher(cols = 2)

grapher.add_lines(lors)
grapher.add_voxels(voxels)

grapher.add_lines(lors, col=2)
grapher.add_points([location], size=10, col=2)

grapher.show()
```

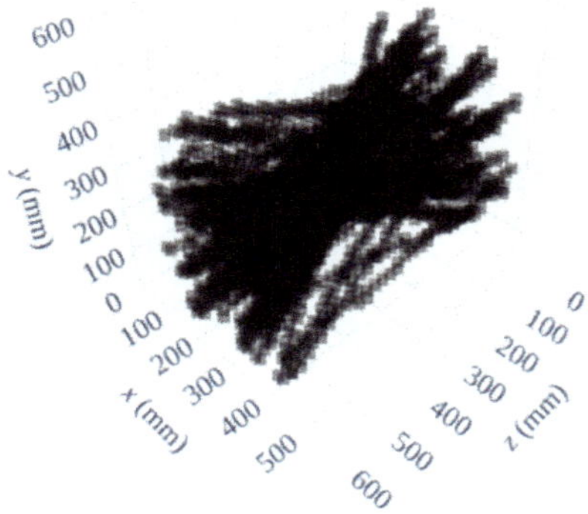

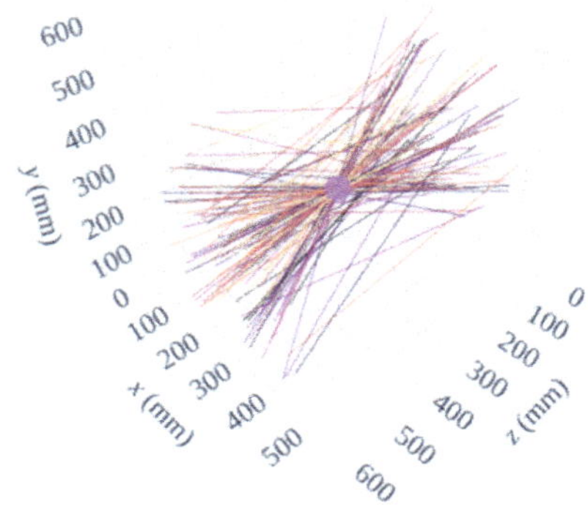

It is interesting to note that, compared to other methods found in these examples, the line-density method actually possesses relatively few parameters; however, it relies on non-linear curve fitting, which may fail depending on the starting guess.

Complete line density method code

The code developed in the previous sections can be applied to consecutive 'samples' of LoRs to follow the tracer's movement through the fluidised bed, fully tracking its trajectory. A complete—albeit slow due to Python's dynamic nature—line density method implementation is given below.

First initialise a larger, multi-sample set of LoRs to analyse; change the `nrows` argument below to read in a different number of LoRs (the file included contains 80,000 LoRs in total), though keep in mind the computational expense of processing large numbers of LoRs!

[9]:
```
# Read in a sample of experimental PEPT data into a NumPy array. Skip the file
# header's first 15 lines
lors_raw = pept.utilities.read_csv(
    "Data/FluidisedBed1Tracer.csv",
    skiprows=15,
    nrows=10_000,
)

# Convert the LoRs into the general 3D line format [t, x1, y1, z2, x2, y2, z2].
# See the `PEPT Data Format` interactive example for detailed explanations

# ADAC Forte's two detector heads' separation, in mm
head_separation = 600

lors = np.insert(lors_raw, 3, 0, axis=1)                    # z1 = 0
lors = np.insert(lors, 6, head_separation, axis=1)          # z2 = head_separation

# Print the first 10 rows from the line of response (LoR) data
print(f"First 10 rows from the 3D-formatted LoR data:\n{lors[:10]}\n")
```

```
First 10 rows from the 3D-formatted LoR data:
[[0.000e+00 1.900e+02 1.687e+02 0.000e+00 3.463e+02 1.428e+02 6.000e+02]
 [1.000e-01 2.437e+02 1.676e+02 0.000e+00 3.145e+02 3.139e+02 6.000e+02]
 [1.000e-01 1.941e+02 4.100e+02 0.000e+00 4.171e+02 2.401e+02 6.000e+02]
 [2.000e-01 2.885e+02 2.962e+02 0.000e+00 3.847e+02 4.525e+02 6.000e+02]
 [2.000e-01 1.646e+02 1.151e+02 0.000e+00 3.829e+02 3.534e+02 6.000e+02]
 [2.000e-01 1.286e+02 1.322e+02 0.000e+00 4.649e+02 2.661e+02 6.000e+02]
 [2.000e-01 3.339e+02 3.735e+02 0.000e+00 3.316e+02 1.310e+02 6.000e+02]
 [3.000e-01 1.640e+02 1.115e+02 0.000e+00 3.794e+02 3.534e+02 6.000e+02]
 [4.000e-01 2.620e+02 2.094e+02 0.000e+00 2.932e+02 2.808e+02 6.000e+02]
 [4.000e-01 1.304e+02 2.749e+02 0.000e+00 2.543e+02 3.062e+02 6.000e+02]]
```

Iterating through each sample of LoRs, the code locates the tracer and appends its location to a growing list of positions. As with the previous cells, you are invited to change the parameters highlighted below; details about each one are given in the comments in the code itself.

```
[10]: # Use the Line Density algorithm to track a moving tracer
from scipy.optimize import curve_fit

# Line Density Method user-changeable parameters:
#     resolution - the number of voxels in each dimension for the LoR voxel grid
resolution = (100, 100, 100)

# General PEPT tracking user-changeable parameters:
#     sample_start - the `lors` index at which to start tracking the tracer
#     sample_size - the number of LoRs used to compute a single location
#     overlap - the number of overlapping LoRs between consecutive samples
sample_start = 0
sample_size = 200
overlap = 10

# Tracked tracer locations will be appended to a growing list
positions = []

# Define scaled Gauss model function to fit to data
def gauss(x, mu, sigma, scale):
    a = scale / sigma / np.sqrt(2 * np.pi)
    return a * np.exp(-0.5 * (x - mu)**2 / sigma**2)


while sample_start + sample_size < len(lors):
    sample = lors[sample_start:sample_start + sample_size]
    voxels = pept.Voxels.from_lines(sample, resolution, verbose = False)

    # Coordinates of the voxel centres for each voxel line
    voxels_x = np.linspace(voxels.xlim[0] + voxels.voxel_size[0] / 2,
                           voxels.xlim[1] - voxels.voxel_size[0] / 2,
                           voxels.shape[0])

    voxels_y = np.linspace(voxels.ylim[0] + voxels.voxel_size[1] / 2,
                           voxels.ylim[1] - voxels.voxel_size[1] / 2,
                           voxels.shape[1])
```

```
    voxels_z = np.linspace(voxels.zlim[0] + voxels.voxel_size[2] / 2,
                           voxels.zlim[1] - voxels.voxel_size[2] / 2,
                           voxels.shape[2])

    # Initial guesses for fitting the Gaussian curves are simply the maximum
    # voxel's X, Y, and Z coordinates, respectively
    p0_x = np.array([voxels_x[max_indices[0]], voxels.voxel_size[0], 1.])
    p0_y = np.array([voxels_y[max_indices[1]], voxels.voxel_size[1], 1.])
    p0_z = np.array([voxels_z[max_indices[2]], voxels.voxel_size[2], 1.])

    # Fit Gaussian curves to each "line of voxels" passing through the maximum
    voxel_line_x = voxels[:, max_indices[1], max_indices[2]]
    voxel_line_y = voxels[max_indices[0], :, max_indices[2]]
    voxel_line_z = voxels[max_indices[0], max_indices[1], :]

    opt_x, _ = curve_fit(gauss, voxels_x, voxel_line_x, p0_x, maxfev=10000)
    opt_y, _ = curve_fit(gauss, voxels_y, voxel_line_y, p0_y, maxfev=10000)
    opt_z, _ = curve_fit(gauss, voxels_z, voxel_line_z, p0_z, maxfev=10000)

    # The tracer's location is given by the Gaussians' centres
    location = np.array([opt_x[0], opt_y[0], opt_z[0]])

    # Insert the tracer positions' timestamp at column 0
    location = np.insert(location, 0, lors[:, 0].mean())

    positions.append(location)

    sample_start += sample_size - overlap

positions = np.array(positions)
print(f"First 10 tracked positions:\n{positions[:10]}")
```

```
First 10 tracked positions:
[[  3.9045      161.843      103.70299998  93.          ]
 [ 11.4085      161.6        103.89849998  93.          ]
 [ 18.2215      161.1605     106.082        93.          ]
 [ 25.357       161.6465     109.56300002  93.          ]
 [ 31.6025      162.16550002 102.8175      93.          ]
 [ 37.323       161.51899998 103.5345      93.          ]
 [ 43.6595      163.67600374 104.6655      93.          ]
 [ 50.3885      170.95507464 104.9605      93.          ]
 [ 57.193       161.324      103.70299998  93.          ]
 [ 64.          173.07745433 104.29149998  93.          ]]
```

[11]:
```
# Plot some of the LoRs used, plus the tracer locations we found
from pept.plots import PlotlyGrapher, PlotlyGrapher2D

grapher = PlotlyGrapher(
    cols=2,
    subplot_titles=[
        "First 500 LoRs",
        f"Tracer locations tracked from {len(lors)} LoRs",
    ],
)

grapher.add_lines(lors[:500])
grapher.add_points(positions, col=2)

grapher.show()
```

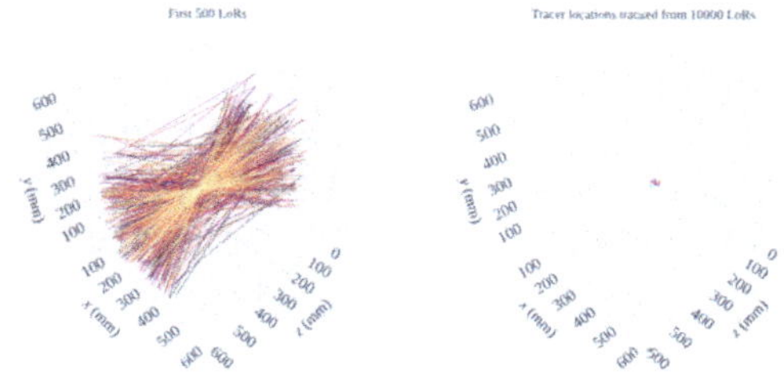

[12]:
```
# Create a "timeseries" plot with the tracer positions vs. time
PlotlyGrapher2D().add_timeseries(positions).show()
```

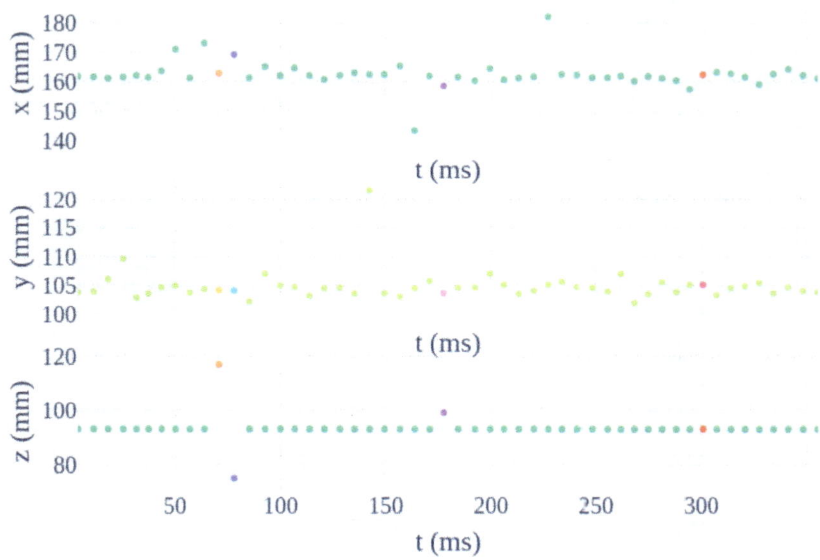

6.2.5 The G-means clustering algorithm

The G-means method in brief

The G-means algorithm of Wiggins, Santos and Ruggles [6] builds on the prior work of Bickell *et al* [4] described in section 6.2.3, integrating the use of line-density matrices and clustering algorithms (discussed further below) so as to allow the detection of multiple PEPT tracers with no *a priori* knowledge of the number thereof, nor their initial positions. As previously, a brief outline of the method is presented below, with a more detailed explanation provided in the following section.

1. A sample of N_s individual, consecutive lines of response from a given PEPT data set (see chapter 2) is considered.
2. The volume corresponding to the detector system's field of view is divided into a three-dimensional grid, typically comprising equally-sized, cubic cells in Cartesian space (see figure 6.2(a)).
3. The number of LoRs passing through each individual cell is recorded (see figure 6.2(b)).

4. The data is checked to see if any cell possesses a line density greater than a user-defined threshold value, thus establishing whether there exist any tracers in the scanner's field of view. If no tracers are found, we return to step (1) for a new sample. If not, we continue to point (5).
5. The maximal value of line density, $\rho_l^{\max}$, for any cell within the field of view is identified. A preset fraction f_m of this value is then subtracted from all cells in the three-dimensional grid.
6. Any cell for which $\rho_l - f_m\rho_l^{\max} < 0$ is set to zero, thus removing a degree of background noise (see figure 6.4(b)).
7. The remaining cells are treated as discrete points in three-dimensional space, with each point possessing a multiplicity equal to the count within said cell—i.e. if N_l LoRs pass through a given cell, that cell is considered as N_l distinct points, all placed at the cell's spatial location (see figure 6.4(c),(d)).

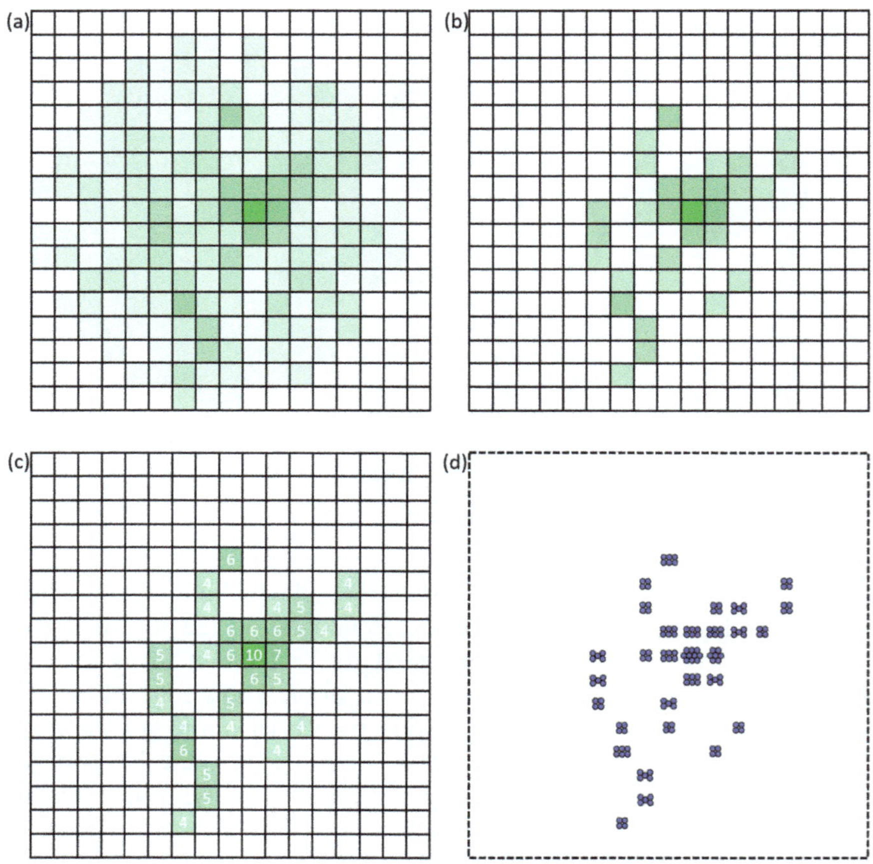

Figure 6.4. (a) Hypothetical example of a line-density grid, as described in the preceding section. (b) The same grid after it has been 'filtered' to remove all cells whose line density is less than 0.4 of the maximum cell density for the sample. (c) The same grid explicitly specifying the numbers of LoRs in each grid cell. (d) The grid from (c) converted into a series of points, as described in step (7) above. Note that, for the purposes of illustration, the N_l points in each cell are shown side-by-side, whereas in reality they would all be assigned to a single point in space.

8. G-means clustering [6, 8] is used to group the resultant points into one (or more) distinct clusters, the centroid(s) of which are taken as the tracer's location.

The G-means method in detail

(1–3) Determining a line density matrix: The G-means method uses as its starting point a three-dimensional grid of line-densities (or 'count matrix') identical to those used in the method of Bickell *et al* [4] described in section 6.2.3. The steps (1) to (3) outlined above are therefore in essence identical to those discussed in section 6.2.3, and as such we will not repeat the full details thereof here. After this point, however, the two methods diverge significantly.

(4) Checking for particles: Before the process proceeds any further, a simple check is performed to ensure that there is evidence of at least one particle within the field of view, and thus that computational time is not wasted trying to locate particles where there are none! This check is performed using a user-defined minimum line count, $N_l^{\text{thresh.}}$: if no cells can be found within the system possessing counts $N_l > N_l^{\text{thresh}}$, then it is assumed that there are no tracers present in the current sample, and the algorithm simply discards this data and moves onto the next sample of LoRs. If there exists at least one voxel for which $N_l > N_l^{\text{thresh}}$, then the algorithm moves on to the next step with the current sample.

It is important to note here that even for a single system—indeed a single sample of LoRs taken from a single system—there does not exist a universally-valid value of N_l^{thresh}. Rather, the optimal value (or range of values) for N_l^{thresh} is mesh-size dependent as the average number of lines intersected by a given voxel will inherently increase as the size of the voxel increases.

(5–6) Filtering: Once a matrix of line-density data has been created and it has been ensured that at least one candidate particle is present in the field of view, the next step is to filter this data—i.e. to attempt to minimise the 'background noise' present due to corrupt LoRs. In order to do so, the cell containing the maximal line density is located, and the local value thereof (ρ_l^{max}) recorded. A user-defined fraction, f_m, of this maximal value is then subtracted from the line density values of all other cells within the field of view, and all values subsequently possessing values lower than zero are set to zero; in other words, all cells with line density values lower than $f_m \rho_l^{\text{max}}$ are effectively removed from future calculations. The intention of this process is to eliminate all cells containing only noise due to corrupt data. As such, the value of f_m should be chosen to be representative of the particle's activity relative to the background. For a typical system, $f_m \in [0.25, 0.5]$, though higher or lower values may potentially be needed for, respectively, systems with particularly high or low degrees of scattering and/or background radiation (e.g. due to secondary emission).

(7–8) Clustering: Once the data has been filtered, the matrix of values must next be converted into a series of discrete points, so that they can be clustered—as illustrated schematically in figure 6.4(d). Specifically, for each grid cell, a number of points corresponding to the line count within said cell are created and placed at a location in three-dimensional space corresponding to the position of the voxel from which they originated. With the data in this form, it is now possible to apply a suitable clustering algorithm, in this case G-means clustering, of which we give a brief overview below.

G-means clustering

G-means (Gaussian-means) clustering is a derivative of the well-known and widely-used k-means clustering algorithm [9–11]. The k-means algorithm—to give a somewhat reductive overview—iteratively subdivides a given data set into a series of distinct clusters, assigning individual points to the cluster with the nearest centroid, typically (though not necessarily) based on their simple Euclidean distance. The G-means algorithm utilises k-means clustering combined with 'goodness-of-fit' testing to autonomously determine both the number and (crucially) positions of clusters within a sample, the centroids of which, in the current application, can be taken as our tracer positions.

G-means begins by using k-means clustering to search for an initial number $k = 1$ of clusters within a given data set[7], thus attempting to identify a single, 'universal' cluster. An attempt is then made to fit a Gaussian to this cluster, based on the knowledge that if the cluster found represents a genuine, single particle location, the distribution of line densities should adhere to a Gaussian form, as illustrated in the preceding section. If (based on a suitable statistical test, in the case of [6] a modified one-dimensional Anderson–Darling goodness of fit test [12, 13]) the Gaussian fit is judged to be suitably 'good'—i.e. the data appear to be normally distributed—the cluster is 'accepted'. The centroid of this cluster is then taken as the spatial location of the single particle found.

If the data is not found to adhere to a Gaussian fit, then we try again with $k = 2$, attempting now to find two distinct clusters in the data. Once two candidate clusters have been determined, we attempt to fit a Gaussian distribution to each; if both fits are judged adequate, then it is assumed that our system contains two tracers (one corresponding to each cluster) and, just as before, the centroids of these clusters are taken as the two tracer positions. Conversely, if one or both of the clusters is found to be non-Gaussian, then another attempt is made with $k = 3$. This entire process is repeated until a suitable value of k is obtained such that all clusters into which the data are subdivided show a suitable degree of normality, implying that all tracers within the field of view have been successfully located.

The interested reader may find more information regarding G-means clustering, the Anderson–Darling test, and the application of these techniques to PEPT, respectively, in our references [8, 12, 13] and [6].

Strengths and weaknesses

Arguably the most significant strength of Wiggins' G-means method compared to the previously-described Birmingham and line-density methods is that, as is clear from the description provided above, the G-means algorithm is able to autonomously determine the number of particles within the system at any point in time, thus removing the need for any *a priori* information regarding N_t.

As, like the line-density method, the G-means method relies on the use of a mesh to calculate line-density distributions, it is similarly subject to issues such as the grid-size

[7] If the number, N_t, of tracers in the sample is known *a priori*, then one can instead save computational effort by beginning with $k = N_t$. Similarly, if a lower bound $N_t \geqslant l$ on the number of tracers is known, then the algorithm can be initiated with $k = l$.

dependence of results, and the possibility of pixelation; however, with a well-chosen voxel size and careful sensitivity analysis, these issues can be largely mitigated.

An issue more unique to the present method is that the filtering process described in point (5) above can occasionally act to cause occlusion (i.e. the algorithm can 'miss' particles). Specifically, if a tracer moves to a region of a given system in which gamma attenuation is high and/or camera sensitivity is low[8] the line-density registered by the voxel containing said tracer may be low enough that it falls below the threshold $f_m \rho_l^{\max}$ and therefore is not considered in subsequent computation for a given timestep (or indeed multiple timesteps!). The effects of such over-filtering can be minimised through the careful choice of f_m, though it may not be possible to entirely eliminate the issue in all cases.

In the case of relatively large tracer numbers ($N_t \gtrsim 10$), the algorithm has also been observed to both over- and under-estimate the number of particles in the system, an issue which can only be dealt with in post-processing. The algorithm has nonetheless been demonstrated to successfully track up to 17 particles in test situations [6, 15], and 15 particles 'in anger' [16].

Strengths
- Can autonomously determine the number of tracers in a system, requiring no *a priori* knowledge.
- Particles entering and leaving the field of view can be successfully tracked.

Weaknesses
- Lower spatial resolution than the Birmingham method for the single particle case [54].
- Grid-size dependent results.
- Subject to pixelation for slower-moving tracers.
- Tracers may be 'lost' due to filtering in systems where attenuation and/or camera sensitivity are strongly heterogeneous.
- algorithm may misinterpret the number of tracers present for $N_t \gtrsim 10$.

6.2.6 Interactive example: the G-means algorithm

The notebook files accompanying this section are available in the supplementary data online at https://iopscience.iop.org/book/978-0-7503-3071-8.

Interactive PEPT analysis example using the clustering (G-means) algorithm [6]

This Jupyter Notebook

An example dataset is used from an experiment run at the University of Birmingham Positron Imaging Centre using the ADAC Forté. It consists of a single 1 mm diameter MCC particle activated with fluorine-18 inside a bubbling fluidised bed.

[8] Positron cameras, including the specific system used by the authors of [6], typically possess non-negligible gradients in sensitivity across their field of view [14].

The fluidised bed was filled with 90% sand and 10% MCC; air was fed into the bottom of the bed at a rate of 37 litres per minute at 3.5 bar. This dataset was chosen for its high quality captured lines of response, with the tracer still depicting the random particle motion that is inherent to bubbling fluidised beds—and typical in Lagrangian particle tracking.

The pept Python library is used for initialising and visualising PEPT data. While not required *per se* for illustrating PEPT algorithms' processing steps, it significantly reduces the amount of repetitive code and visual noise, allowing the reader to focus on the main concepts.

[1]:
```
# Importing necessary libraries for numerical array manipulation and plotting,
# plus the `pept` library for straightforward PEPT data processing
import numpy as np
import pept

# Use the G-Means implementation from the `pyclustering` library
import pyclustering

# Set Plotly figure output (try "browser" too for interactive graphs)
import plotly
plotly.io.renderers.default = "png"
```

Initialise raw line of response data

In this initial step, we read in the 'raw' LoR data from which our tracer positions are calculated. As discussed in the 'Effect of sample size and overlap' notebook, varying the numbers of LoRs used to calculate a PEPT location can strongly affect the accuracy of the final position computed. Try altering the value nrows below to see how this influences your results.

[2]:
```
# Read in a sample of experimental PEPT data into a NumPy array. Skip the file
# header's first 15 lines, then read in the first `nrows` LoRs
lors_raw = pept.utilities.read_csv(
    "Data/FluidisedBed1Tracer.csv",
    skiprows=15,
    nrows=100,
)

# Convert the LoRs into the general 3D line format [t, x1, y1, z2, x2, y2, z2].
# See the `PEPT Data Format` interactive example for detailed explanations

# ADAC Forte's two detector heads' separation, in mm
head_separation = 600

lors = np.insert(lors_raw, 3, 0, axis = 1)                   # z1 = 0
lors = np.insert(lors, 6, head_separation, axis = 1)      # z2 = head_separation

# Print the first 10 rows from the line of response (LoR) data
print(f"First 10 rows from the 3D-formatted LoR data:\n{lors[:10]}\n")
```

```
First 10 rows from the 3D-formatted LoR data:
[[0.000e+00 1.900e+02 1.687e+02 0.000e+00 3.463e+02 1.428e+02 6.000e+02]
 [1.000e-01 2.437e+02 1.676e+02 0.000e+00 3.145e+02 3.139e+02 6.000e+02]
 [1.000e-01 1.941e+02 4.100e+02 0.000e+00 4.171e+02 2.401e+02 6.000e+02]
 [2.000e-01 2.885e+02 2.962e+02 0.000e+00 3.847e+02 4.525e+02 6.000e+02]
 [2.000e-01 1.646e+02 1.151e+02 0.000e+00 3.829e+02 3.534e+02 6.000e+02]
 [2.000e-01 1.286e+02 1.322e+02 0.000e+00 4.649e+02 2.661e+02 6.000e+02]
 [2.000e-01 3.339e+02 3.735e+02 0.000e+00 3.316e+02 1.310e+02 6.000e+02]
 [3.000e-01 1.640e+02 1.115e+02 0.000e+00 3.794e+02 3.534e+02 6.000e+02]
 [4.000e-01 2.620e+02 2.094e+02 0.000e+00 2.932e+02 2.808e+02 6.000e+02]
 [4.000e-01 1.304e+02 2.749e+02 0.000e+00 2.543e+02 3.062e+02 6.000e+02]]
```

The `pept.plots` subpackage provides two PEPT-oriented interfaces to the Plotly visualisation library—`PlotlyGrapher` and `PlotlyGrapher2D`—containing straightforward plotting functions for PEPT data. Let's plot the lines of response from the `lors` array:

```
[3]: # Plot the 3D LoRs that were initialised
     from pept.plots import PlotlyGrapher

     PlotlyGrapher().add_lines(lors).show()
```

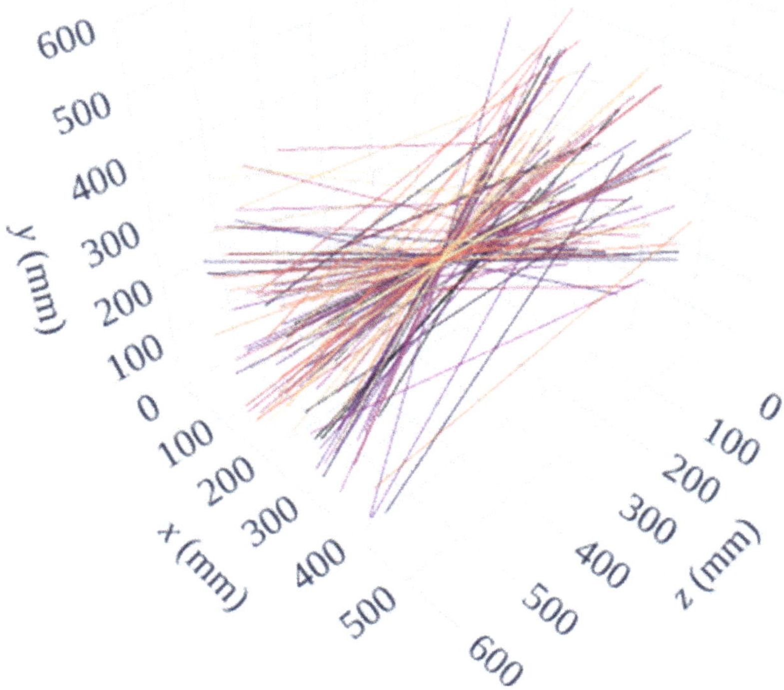

Note the PEPT 3D axes convention, differing from the common mathematical coordinates with *Y* and *Z* being switched, such that the *Y* axis points upwards, while the *Z* axis is perpendicular to the two detector heads.

Voxelise the lines of response

The next step is to 'voxelise' the LoRs—that is, to create a grid of square or rectangular cells, and count the number of lines passing through each. In the code below you can set the resolution—i.e. number of voxels in each dimension—by varying the pair of values corresponding to the `resolution` parameter. Typing, for example, `(20, 20, 20)` will give a 20 by 20 by 20 grid of voxels.

Try some different values for the `resolution` parameter. As discussed in the main text, larger values—i.e. a higher resolution, smaller voxels—will allow the centroid of our particle to be determined more precisely. However, you may notice that as you increase these values your processor will take longer to execute the calculation and plotting steps below.

Now consider that this is just one of many thousands of locations we typically need to process, and you may get an idea of the issue of using an overly-fine grid. When using algorithms relying on voxelisation, there is a fine balance to be made between precision and compuational efficiency.

[4]:
```
resolution = (30, 30, 30)
voxels = pept.Voxels.from_lines(lors, resolution)

voxels
```

```
Initialised Voxels class in 0.0007414817810058594 s.
```

[4]:
```
Voxels
------
[[[0. 0. 0. ... 0. 0. 0.]
  [0. 0. 0. ... 0. 0. 0.]
  [0. 0. 0. ... 0. 0. 0.]
  ...
  [0. 0. 0. ... 0. 0. 0.]
  [0. 0. 0. ... 0. 0. 0.]
  [0. 0. 0. ... 0. 0. 0.]]

 [[0. 0. 0. ... 0. 0. 0.]
  [0. 0. 0. ... 0. 0. 0.]
  [0. 0. 0. ... 0. 0. 0.]
  ...
  [0. 0. 0. ... 0. 0. 0.]
  [0. 0. 0. ... 0. 0. 0.]
  [0. 0. 0. ... 0. 0. 0.]]

 [[0. 0. 0. ... 0. 0. 0.]
  [0. 0. 0. ... 0. 0. 0.]
  [0. 0. 0. ... 0. 0. 0.]
  ...
  [0. 0. 0. ... 0. 0. 0.]
  [0. 0. 0. ... 0. 0. 0.]
  [0. 0. 0. ... 0. 0. 0.]]

 ...
```

```
[[0. 0. 0. … 0. 0. 0.]
 [0. 0. 0. … 0. 0. 0.]
 [0. 0. 0. … 0. 0. 0.]
 …
 [0. 0. 0. … 0. 0. 0.]
 [0. 0. 0. … 0. 0. 0.]
 [0. 0. 0. … 0. 0. 0.]]

[[0. 0. 0. … 0. 0. 0.]
 [0. 0. 0. … 0. 0. 0.]
 [0. 0. 0. … 0. 0. 0.]
 …
 [0. 0. 0. … 0. 0. 0.]
 [0. 0. 0. … 0. 0. 0.]
 [0. 0. 0. … 0. 0. 0.]]

[[0. 0. 0. … 0. 0. 0.]
 [0. 0. 0. … 0. 0. 0.]
 [0. 0. 0. … 0. 0. 0.]
 …
 [0. 0. 0. … 0. 0. 0.]
 [0. 0. 0. … 0. 0. 0.]
 [0. 0. 0. … 0. 0. 0.]]]

number_of_voxels =    (30, 30, 30)
voxel_size =          [12.35333333 16.99333333 20.        ]

xlim =                [110.3 480.9]
ylim =                [ 47.2 557. ]
zlim =                [  0. 600.]

voxel_grids:
([110.3 … 480.9],
 [47.2 … 557.0],
 [0.0 … 600.0])
```

[5]:

```
# Plot all LoRs and all voxels with non-zero values (i.e. more than one LoR
# passing through)
PlotlyGrapher().add_voxels(voxels).add_lines(lors).show()
```

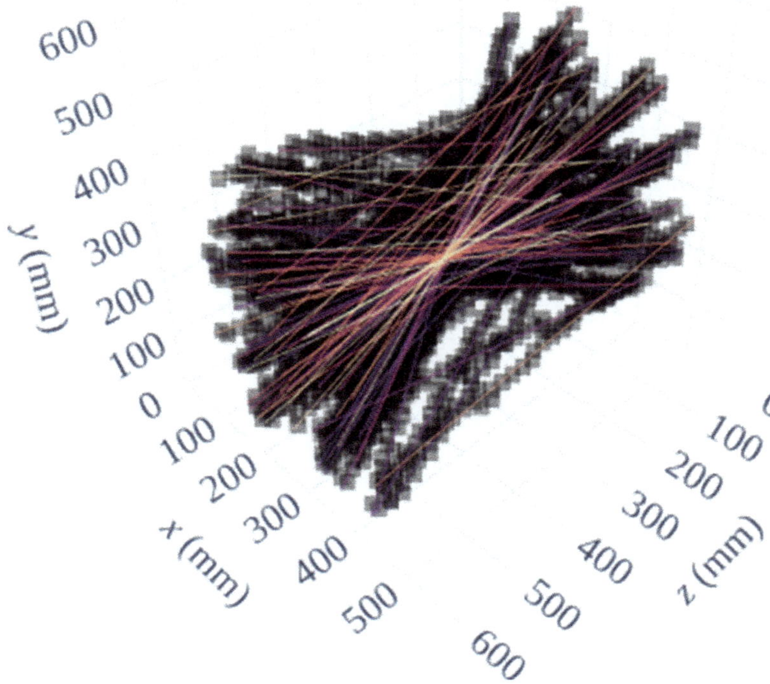

High pass filter

As shown in the image above, our data possesses a lot of background noise away from where we would expect the tracer's centre to be. The G-means PEPT algorithm takes a relatively straightforward approach to removing this noise, simply setting a threshold voxel value (i.e. line density) below which all data is removed. Specifically, a `fraction` of the maximum voxel value for a given sample of LoRs is chosen.

Try varying the `fraction` parameter below to see how this choice affects your results.

```
[6]: # Set all pixel values smaller than a `fraction` of the maximum `voxels` value
     # to zero
     fraction = 0.1
     threshold = fraction * float(voxels.max())

     voxels[voxels < threshold] = 0.
```

```
[7]: # Plot thresholded voxels, colour-coded by the number of LoRs passing through
     # each of them
     PlotlyGrapher().add_voxels(voxels).show()
```

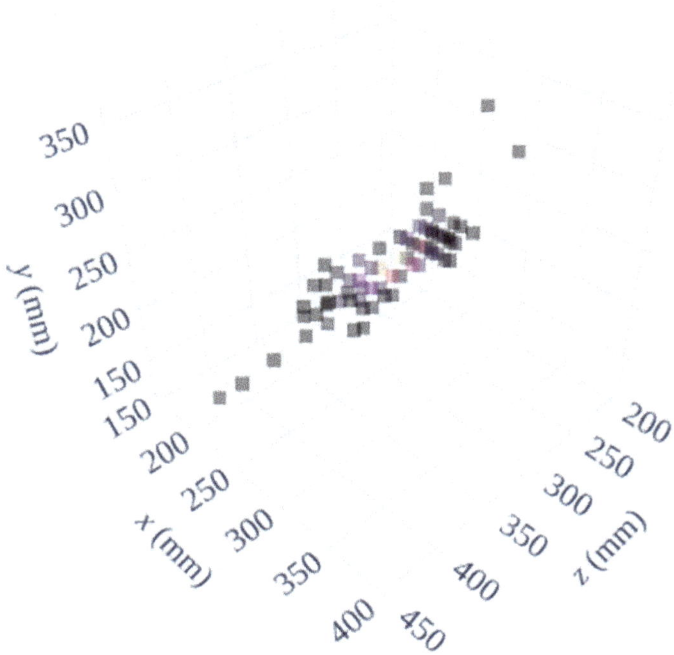

Cluster voxels with G-means

In this step, the 'G-Means' implementation from the `pyclustering` Python library is used.

To cluster the voxels, they must first be written as a data table, such that the columns are the data features (here a voxel centre's Euclidean position) and the rows are the data samples (here the non-zero voxels). To add more weight to voxels with larger values, each data sample is repeated by the number of LoRs passing through.

```
[8]: # Write non-zero voxels as a data table, where the feature columns are the XYZ
     # voxel coordinates.
     origin = np.array([voxels.xlim[0], voxels.ylim[0], voxels.zlim[0]])

     # Indices of each non-zero voxel
     nonzero_indices = np.argwhere(voxels != 0.)

     # Physical coordinates of each non-zero voxel's centre
     nonzero_coordinates = origin + voxels.voxel_size * nonzero_indices
     nonzero_coordinates += voxels.voxel_size * 0.5

     # Repeat each voxel's coordinates by the number of LoRs passing through it
     num_repeats = [voxels[tuple(i)] for i in nonzero_indices]
     samples_table = np.repeat(nonzero_coordinates, num_repeats, axis=0)

     print(f"The first 10 rows in the `samples_table`:\n{samples_table}")
```

```
The first 10 rows in the `samples_table`:
[[240.01        174.65        430.        ]
 [240.01        174.65        430.        ]
 [240.01        174.65        430.        ]
 ...
 [314.13        310.59666667 230.        ]
 [314.13        310.59666667 230.        ]
 [314.13        310.59666667 230.        ]]
```

[9]:
```
# Run G-Means to cluster data into Gaussian partitions
from pyclustering.cluster.gmeans import gmeans

gmeans_instance = gmeans(samples_table, repeat = 20).process()

# Extract clustering results: clusters and their centers
clusters = gmeans_instance.get_clusters()
centres = gmeans_instance.get_centers()
centres = np.array([np.array(c) for c in centres])
```

By default, with data presented as repeated, discrete voxel centres, the G-means algorithm will return a cluster *in every voxel*! As such, an extra step must be added to merge the centres of clusters. Specifically, any centres less than two voxels apart are merged into a single cluster.

[10]:
```
# Merge centres that are less than 2 voxel sizes apart
from scipy.cluster.hierarchy import linkage, fcluster

max_distance = 2 * np.linalg.norm(voxels.voxel_size)
centres_merged_labels = fcluster(linkage(centres), t = max_distance,
                                 criterion='distance')

centres_merged = []
for label in range(1, centres_merged_labels.max() + 1):
    centre = centres[centres_merged_labels == label].mean(axis = 0)
    centres_merged.append(centre)

centres_merged = np.array(centres_merged)

# Insert the tracer positions' timestamp at column 0
centres_merged = np.insert(centres_merged, 0, lors[:, 0].mean(), axis = 1)
samples_time = np.insert(samples_table, 0, lors[:, 0].mean(), axis = 1)

print(f"Final merged cluster centres:\n{centres_merged}")
```

```
Final merged cluster centres:
[[  1.931       278.02025641 238.96323077 324.76923077]
 [  1.931       314.13       310.59666667 230.        ]
 [  1.931       252.36333333 310.59666667 190.        ]]
```

[11]:
```
# Plot all thresholded voxels and merged cluster centres
grapher = PlotlyGrapher()

grapher.add_voxels(voxels)
grapher.add_points(centres_merged, size=10)

grapher.show()
```

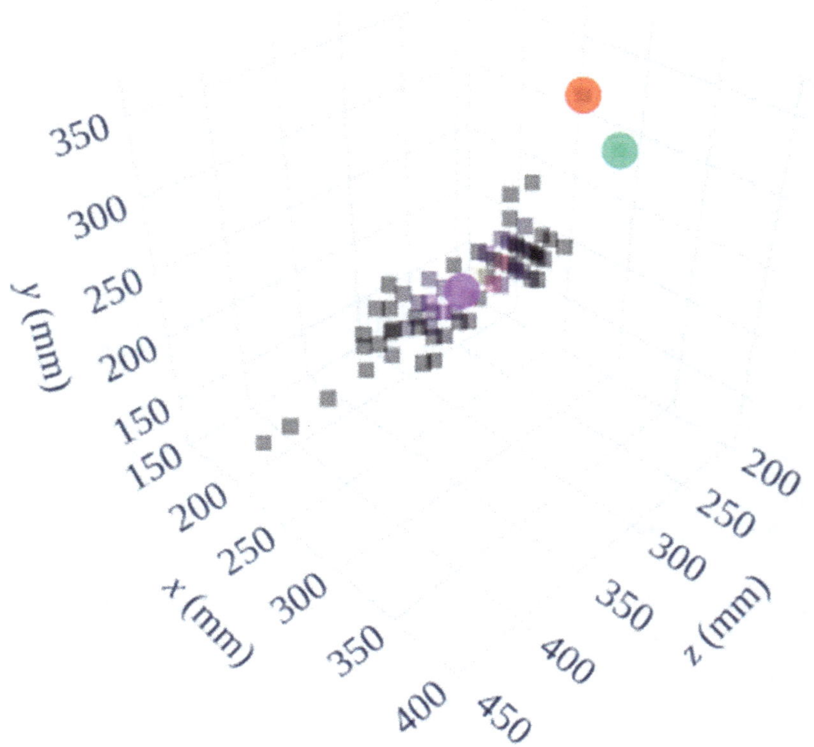

Notice that the G-Means algorithm also considered the two outlier voxels remaining after the thersholding step as individual clusters, meaning it is quite sensitive to noise.

Complete clustering (G-means) algorithm code

The code developed in the previous sections can be applied to consecutive 'samples' of LoRs to follow the tracer's movement through the fluidised bed, fully tracking its trajectory. A complete—albeit slow due to Python's dynamic nature—Clustering Algorithm implementation is given below.

First initialise a larger, multi-sample set of LoRs to analyse; change the `nrows` argument below to read in a different number of LoRs (the file included contains 80,000 LoRs in total), though keep in mind the computational expense of processing large numbers of LoRs!

[12]:
```
# Read in a sample of experimental PEPT data into a NumPy array. Skip the file
# header's first 15 lines
lors_raw = pept.utilities.read_csv(
    "Data/FluidisedBed1Tracer.csv",
    skiprows=15,
    nrows=10_000,
)
```

```
# Convert the LoRs into the general 3D line format [t, x1, y1, z2, x2, y2, z2].
# See the `PEPT Data Format` interactive example for detailed explanations

# ADAC Forte's two detector heads' separation, in mm
head_separation = 600

lors = np.insert(lors_raw, 3, 0, axis=1)                    # z1 = 0
lors = np.insert(lors, 6, head_separation, axis=1)          # z2 = head_separation

# Print the first 10 rows from the line of response (LoR) data
print(f"First 10 rows from the 3D-formatted LoR data:\n{lors[:10]}\n")
```

```
First 10 rows from the 3D-formatted LoR data:
[[0.000e+00 1.900e+02 1.687e+02 0.000e+00 3.463e+02 1.428e+02 6.000e+02]
 [1.000e-01 2.437e+02 1.676e+02 0.000e+00 3.145e+02 3.139e+02 6.000e+02]
 [1.000e-01 1.941e+02 4.100e+02 0.000e+00 4.171e+02 2.401e+02 6.000e+02]
 [2.000e-01 2.885e+02 2.962e+02 0.000e+00 3.847e+02 4.525e+02 6.000e+02]
 [2.000e-01 1.646e+02 1.151e+02 0.000e+00 3.829e+02 3.534e+02 6.000e+02]
 [2.000e-01 1.286e+02 1.322e+02 0.000e+00 4.649e+02 2.661e+02 6.000e+02]
 [2.000e-01 3.339e+02 3.735e+02 0.000e+00 3.316e+02 1.310e+02 6.000e+02]
 [3.000e-01 1.640e+02 1.115e+02 0.000e+00 3.794e+02 3.534e+02 6.000e+02]
 [4.000e-01 2.620e+02 2.094e+02 0.000e+00 2.932e+02 2.808e+02 6.000e+02]
 [4.000e-01 1.304e+02 2.749e+02 0.000e+00 2.543e+02 3.062e+02 6.000e+02]]
```

Iterating through each sample of LoRs, locate the tracer and append its location to a growing list of positions. As with the previous cells, you are invited to change the parameters highlighted below; details about each one are given in the comments.

```
[13]: # Use the G-Means algorithm to track moving tracer
from pyclustering.cluster.gmeans import gmeans
from scipy.cluster.hierarchy import linkage, fcluster


# Line Density Method user-changeable parameters:
#     fraction - fraction of the maximum voxel value for thresholding voxels
#     resolution - the number of voxels in each dimension for the LoR voxel grid
fraction = 0.1
resolution = (100, 100, 100)

# General PEPT tracking user-changeable parameters:
#     sample_start - the `lors` index at which to start tracking the tracer
#     sample_size - the number of LoRs used to compute a single location
#     overlap - the number of overlapping LoRs between consecutive samples
sample_start = 0
sample_size = 200
overlap = 10

# Tracked tracer locations will be appended to a growing list
positions = []

while sample_start + sample_size < len(lors):
    sample = lors[sample_start:sample_start + sample_size]
    voxels = pept.Voxels.from_lines(sample, resolution, verbose=False)
    # Set all pixel values smaller than a `fraction` of the maximum `voxels`
```

```
    # value to zero
    fraction = 0.1
    threshold = fraction * float(voxels.max())

    voxels[voxels < threshold] = 0.

    # Write non-zero voxels as a data table, where the feature columns are the
    # XYZ voxels coordinates
    origin = np.array([voxels.xlim[0], voxels.ylim[0], voxels.zlim[0]])

    # Indices of each non-zero voxel
    nonzero_indices = np.argwhere(voxels != 0.)

    # Physical coordinates of each non-zero voxel's centre
    nonzero_coordinates = origin + voxels.voxel_size * nonzero_indices
    nonzero_coordinates += voxels.voxel_size * 0.5

    # Repeat each voxel's coordinates by the number of LoRs passing through it
    num_repeats = [voxels[tuple(i)] for i in nonzero_indices]
    samples_table = np.repeat(nonzero_coordinates, num_repeats, axis=0)

    # Run G-Means to cluster data into Gaussian partitions
    gmeans_instance = gmeans(samples_table, repeat=20).process()

    # Extract clustering results: clusters and their centers
    clusters = gmeans_instance.get_clusters()
    centres = gmeans_instance.get_centers()
    centres = np.array([np.array(c) for c in centres])
    # Merge centres that are less than 2 voxel sizes apart
    max_distance = 2 * np.linalg.norm(voxels.voxel_size)
    centres_merged_labels = fcluster(linkage(centres), t=max_distance,
                                     criterion='distance')

    centres_merged = []
    for label in range(1, centres_merged_labels.max() + 1):
        centre = centres[centres_merged_labels == label].mean(axis=0)
        centres_merged.append(centre)

    centres_merged = np.array(centres_merged)

    # Insert the tracer positions' timestamp at column 0
    centres_merged = np.insert(centres_merged, 0, sample[:, 0].mean(), axis=1)
    samples_time = np.insert(samples_table, 0, sample[:, 0].mean(), axis=1)

    positions.extend(centres_merged)

    sample_start += sample_size - overlap

positions = np.array(positions)

print(f"First 10 tracked positions:\n{positions[:10]}")
```

```
First 10 tracked positions:
[[  3.9045     345.107      282.973     486.        ]
 [  3.9045     279.9026267  244.77890307 317.10879512]
 [  3.9045     261.111      221.509     267.        ]
 [  3.9045     280.201      257.363     381.        ]
 [  3.9045     303.109      227.9115    378.        ]
 [  3.9045     314.563      267.607     411.        ]
 [  3.9045     276.383      221.509     237.        ]
 [  3.9045     314.563      231.753     267.        ]
 [  3.9045     238.203      226.631     213.        ]
 [  3.9045     356.561      221.509     189.        ]]
```

[14]:
```
# Plot some of the LoRs used, plus the tracer locations we found
grapher = PlotlyGrapher(
    cols=2,
    subplot_titles=[
        "First 500 LoRs",
        f"Tracer locations tracked from {len(lors)} LoRs",
    ],
)

grapher.add_lines(lors[:500])
grapher.add_points(positions, col=2)

grapher.show()
```

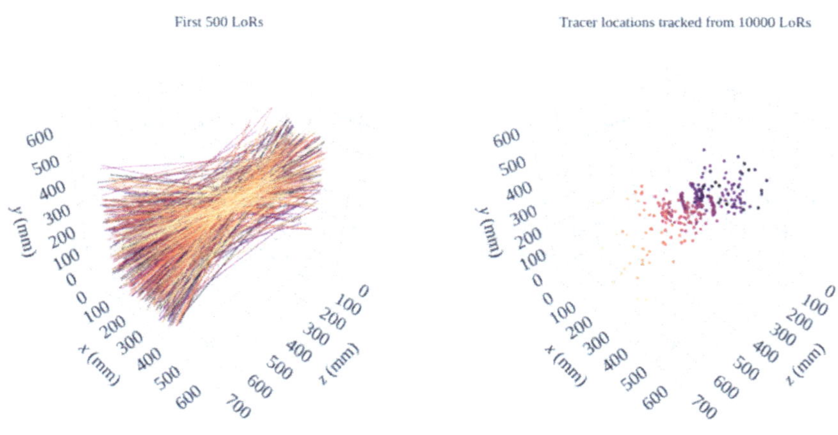

```
[15]: # Create a "timeseries" plot with the tracer positions vs. time
      from pept.plots import PlotlyGrapher2D

      PlotlyGrapher2D().add_timeseries(positions).show()
```

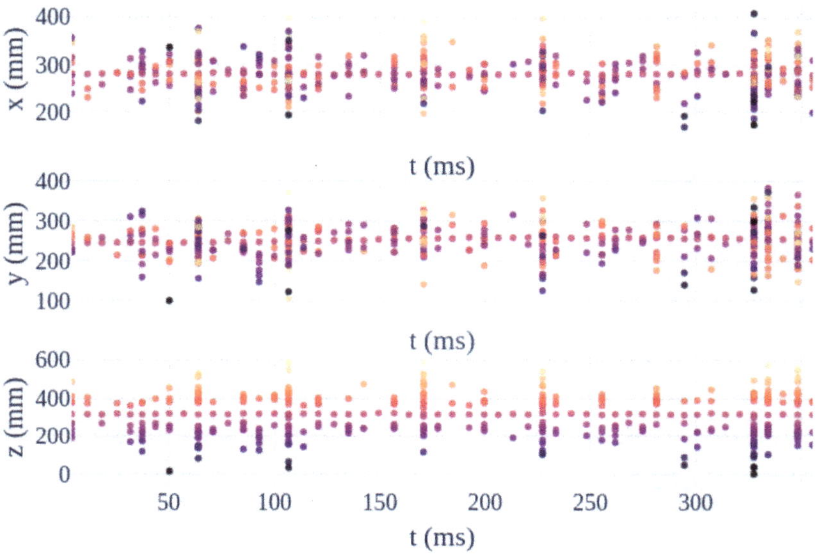

Again, notice the noisy points due to the lone voxel outliers being clustered by G-Means.

Multiple particle tracking

The complete code developed above can also be applied to a simple multiple tracer experiment. Here we use a sample of an experiment run by Dr Kit Windows-Yule at the University of Birmingham, consisting of two 1 mm diameter alumina particles with fluorine-18 radioactive tracer material. The particles are fixed at the two ends of an arm rotating at 42 rpm, their trajectories describing full circles.

[16]:
```
# Read in a sample of experimental PEPT data into a NumPy array. Skip the file
# header's first 15 lines
lors_raw = pept.utilities.read_csv(
    "Data/RotatingArm2Tracers.csv",
    skiprows=15,
    nrows=10_000,
)

# Convert the LoRs into the general 3D line format [t, x1, y1, z2, x2, y2, z2].
# See the `PEPT Data Format` interactive example for detailed explanations

# ADAC Forte's two detector heads' separation, in mm
head_separation = 712

lors = np.insert(lors_raw, 3, 0, axis=1)                    # z1 = 0
lors = np.insert(lors, 6, head_separation, axis=1)          # z2 = head_separation

# Print the first 10 rows from the line of response (LoR) data
print(f"First 10 rows from the 3D-formatted LoR data:\n{lors[:10]}\n")
```

```
First 10 rows from the 3D-formatted LoR data:
[[0.000e+00 2.094e+02 2.797e+02 0.000e+00 2.738e+02 9.500e+01 7.120e+02]
 [1.000e-01 4.325e+02 4.395e+02 0.000e+00 2.006e+02 1.929e+02 7.120e+02]
 [1.000e-01 2.513e+02 1.109e+02 0.000e+00 2.177e+02 3.469e+02 7.120e+02]
 [1.000e-01 3.227e+02 4.372e+02 0.000e+00 3.853e+02 1.097e+02 7.120e+02]
 [1.000e-01 1.894e+02 9.910e+01 0.000e+00 4.213e+02 1.044e+02 7.120e+02]
 [2.000e-01 3.658e+02 2.230e+02 0.000e+00 3.198e+02 5.003e+02 7.120e+02]
 [2.000e-01 2.543e+02 4.690e+02 0.000e+00 4.873e+02 1.257e+02 7.120e+02]
 [2.000e-01 3.339e+02 4.962e+02 0.000e+00 3.646e+02 8.970e+01 7.120e+02]
 [2.000e-01 2.844e+02 2.130e+02 0.000e+00 4.466e+02 5.322e+02 7.120e+02]
 [2.000e-01 2.053e+02 1.906e+02 0.000e+00 2.785e+02 2.224e+02 7.120e+02]]
```

Iterating through each sample of LoRs, we locate the tracers and append their locations to a growing list of positions. As with the previous cells, you are invited to change the parameters highlighted below; details about each one are given in the comments.

Importantly, the `sample_size` must, in the multiple particle case, be larger so as to include enough LoRs for each tracer—in cases using large numbers of LoRs, the `overlap` becomes increasingly important to yield *dense* points on each trajectory; see the 'Effect of sample size and overlap' section of chapter 6 for a detailed explanation.

[17]:
```
# Use the G-Means algorithm to track moving tracer
from pyclustering.cluster.gmeans import gmeans
from scipy.cluster.hierarchy import linkage, fcluster

# Line Density Method user-changeable parameters:
#     fraction - fraction of the maximum voxel value for thresholding voxels
#     resolution - the number of voxels in each dimension for the LoR voxel grid
fraction = 0.2
resolution = (100, 100, 100)
```

```
# General PEPT tracking user-changeable parameters:
#     sample_start - the `lors` index at which to start tracking the tracer
#     sample_size - the number of LoRs used to compute a single location
#     overlap - the number of overlapping LoRs between consecutive samples
sample_start = 0
sample_size = 200 * 2
overlap = 10

# Tracked tracer locations will be appended to a growing list
positions = []

while sample_start + sample_size < len(lors):
    sample = lors[sample_start:sample_start + sample_size]
    voxels = pept.Voxels.from_lines(sample, resolution, verbose = False)

    # Set all pixel values smaller than a `fraction` of the maximum `voxels`
    # value to zero
    fraction = 0.1
    threshold = fraction * float(voxels.max())

    voxels[voxels < threshold] = 0.

    # Write non-zero voxels as a data table, where the feature columns are the
    # XYZ voxels coordinates
    origin = np.array([voxels.xlim[0], voxels.ylim[0], voxels.zlim[0]])

    # Indices of each non-zero voxel
    nonzero_indices = np.argwhere(voxels != 0.)

    # Physical coordinates of each non-zero voxel's centre
    nonzero_coordinates = origin + voxels.voxel_size * nonzero_indices
    nonzero_coordinates += voxels.voxel_size * 0.5

    # Repeat each voxel's coordinates by the number of LoRs passing through it
    num_repeats = [voxels[tuple(i)] for i in nonzero_indices]
    samples_table = np.repeat(nonzero_coordinates, num_repeats, axis=0)

    # Run G-Means to cluster data into Gaussian partitions
    gmeans_instance = gmeans(samples_table, repeat=20).process()

    # Extract clustering results: clusters and their centers
    clusters = gmeans_instance.get_clusters()
    centres = gmeans_instance.get_centers()
    centres = np.array([np.array(c) for c in centres])

    # Merge centres that are less than 2 voxel sizes apart
    max_distance = 2 * np.linalg.norm(voxels.voxel_size)
    centres_merged_labels = fcluster(linkage(centres), t=max_distance,
                                     criterion='distance')
    centres_merged = []
    for label in range(1, centres_merged_labels.max() + 1):
        centre = centres[centres_merged_labels == label].mean(axis=0)
        centres_merged.append(centre)
```

```
    centres_merged = np.array(centres_merged)

    # Insert the tracer positions' timestamp at column 0
    centres_merged = np.insert(centres_merged, 0, sample[:, 0].mean(), axis=1)
    samples_time = np.insert(samples_table, 0, sample[:, 0].mean(), axis=1)

    positions.extend(centres_merged)

    sample_start += sample_size - overlap

positions = np.array(positions)

print(f"First 10 tracked positions:\n{positions[:10]}")
```

```
First 10 tracked positions:
[[  4.60925    291.34       166.298      555.36      ]
 [  4.60925    287.516      171.366      530.44      ]
 [  4.60925    276.044      196.706      537.56      ]
 [  4.60925    184.268      181.502      135.28      ]
 [  4.60925    231.65518182 202.50468466 273.99863636]
 [  4.60925    211.036      216.978      167.32      ]
 [  4.60925    230.156      191.638      195.8       ]
 [  4.60925    245.452      163.764      181.56      ]
 [  4.60925    172.796      222.046      117.48      ]
 [  4.60925    298.988      242.318      452.12      ]]
```

[18]:

```
# Plot some of the LoRs used, plus the tracer locations we found
grapher = PlotlyGrapher(
    cols=2,
    subplot_titles=[
        "First 500 LoRs",
        f"Tracer locations tracked from {len(lors)} LoRs",
    ],
)

grapher.add_lines(lors[:500])
grapher.add_points(positions, col=2)

grapher.show()
```

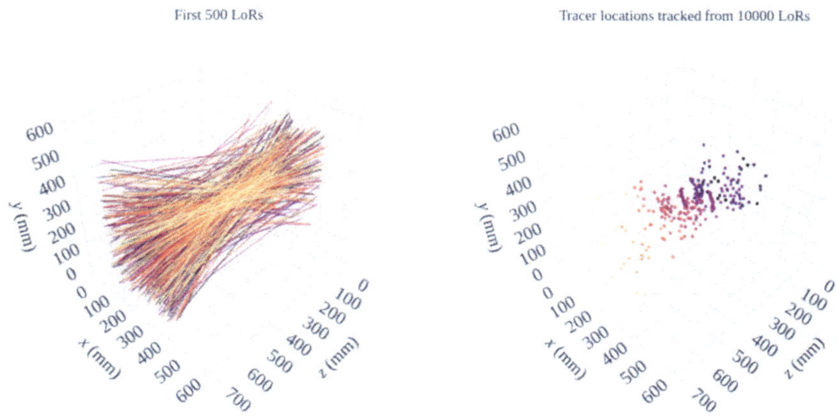


```
[19]: # Create a "timeseries" plot with the tracer positions vs. time
      from pept.plots import PlotlyGrapher2D

      PlotlyGrapher2D().add_timeseries(positions).show()
```

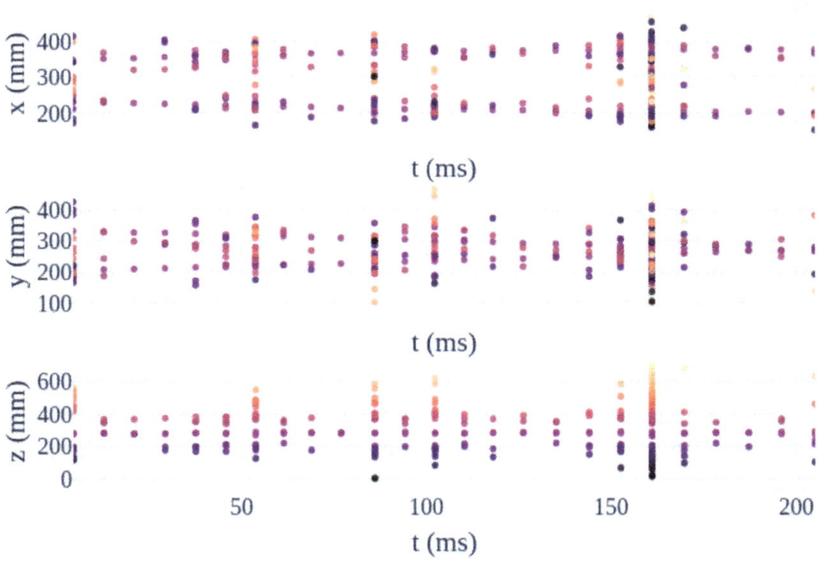

Besides the noisy lone voxels towards the edges of the detectors and a concentrated region in the middle, two circular sections being described by the two tracers can be visually distinguished. However, separating the two trajectories might prove difficult in the presence of noisy points 'connecting' them. Try adjusting the algorithm's parameters to see how much you can 'clean up' the data such that only the circle segments belonging to the two tracers remain.

6.2.7 Feature-point identification (FPI)

The FPI method in brief

The FPI method [17] is another algorithm derived from the line-density method of Bickell *et al* [4], developed by the same authors as the G-means algorithm [6] described previously. Where the previous method uses G-means clustering to extract particle positions from line density matrices, the FPI method uses tools adapted from the field of image processing—namely the eponymous feature point identification technique [18, 19].

The basic operation of the FPI method is outlined below, with a more detailed description provided in the next part of this subsection. As the FPI method was developed with the explicit intention of tracking multiple particles, in this instance rather than initially discussing the single particle case, we begin immediately with a discussion of the multiple-particle case:

1. A sample of N_s individual, consecutive lines of response from a given PEPT data set (see chapter 2) is considered.
2. The volume corresponding to the detector system's field of view is divided into a three-dimensional grid, typically comprising equally-sized, cubic cells in Cartesian space (see figure 6.2(a)).
3. The number of LoRs passing through each individual cell is recorded (see figure 6.2(b)).
4. The data are smoothed using a user-defined smoothing width f_s.
5. Local maxima in line density are recorded; these local maxima are taken as tracer positions if:
   (a) Their line-density values fall within the rth percentile of the sample (time-step) in which they occur, where r is a user-defined value.
   (b) Their values are larger than those of any other voxels within a cube of width $2w + 1$, where w is another user-defined parameter.
6. A refined tracer location is determined using a centroid calculation involving all voxels falling within the cube defined in step (5).

The FPI method in detail

(1–3) Determining a line density matrix: The first three steps of the process, used to create a three-dimensional grid of line-density values to be used in subsequent calculations, remain identical to those of the original line-density method, which are described in detail in section 6.2.3, and therefore will not be further discussed here.

(4) Smoothing: Where the original line-density algorithm of Bickell *et al* [4] simply considered the numerical matrix developed in steps 1–3, and the G-means method [6] converted the numerical values within the matrix to a series of discrete data points, the FPI method [17] adopts yet another approach. In the case of [17], the numerical data stored within the individual voxels of the matrix are considered in essence as grey values [20, 21] for a hypothetical, three-dimensional electronic image. For the sake of brevity, we will not discuss in detail the fundamentals of digital image processing here, as these constitute a monograph in their own right! We refer the interested reader to references [20, 21].

The first step of Wiggins' pseudo-image-processing procedure is to smooth the data, specifically using a box-car average [22] over a region of size $2f_s + 1$, where f_s is the smoothing width. If $\rho_l(x, y, z)$ is the local line density at a spatial position (x, y, z), then the smoothed value, $\rho_l'(x, y, z)$, thereof can be determined as [17]:

$$\rho_l'\left(x, y, z\right) = \frac{1}{(2f + 1)^3} \sum_{i=x-f}^{x+f} \sum_{i=y-f}^{y+f} \sum_{i=z-f}^{z+f} \rho_l\left(i, j, k\right). \tag{6.4}$$

In the present application, the amount of smoothing performed is typically highly limited ($f_s \lesssim 1$).

(5) Finding approximate tracer positions: Having smoothed the pseudo-image corresponding to our line density matrix, the next step is to determine the positions of local maxima in line density. To do so, we must firstly set a limit on what we consider to be a 'true' maximum (i.e. one corresponding to a genuine tracer location) and what we consider to be a false maximum (corresponding simply to a random statistical fluctuation). This is achieved in a manner similar to that utilised in the G-means algorithm, namely by considering in subsequent calculations only those peaks falling above the rth percentile for the current set of LoRs under consideration, rejecting all others. As with the earlier method, the value of r is typically chosen based on the expected level of noise (i.e. background and/or scattered radiation) to which the experiment is subject, and the activities of the tracers used.

Next, we must define what we consider to be 'local'—that is to say, choose the size of the separate spatial regions considered. Again as with the G-means approach described previously, for each candidate peak detected, we consider a cubic region, here of dimension $2w + 1$ (where w is a user-defined parameter) surrounding the voxel containing said peak. If the value of ρ_l within the central voxel is higher than that of any other voxel within the cubic region defined, then the peak is accepted as a true tracer location; if a 'brighter' voxel exists, then it is rejected. Clearly, then, one must be careful when choosing the size of w, as an overly-large value in a system with a high tracer density may lead to the less-active tracers being regularly 'discarded' when their 'regions' overlap with those of more active particles.

(6) Refining: The process described in step (5) yields an approximate, candidate position $\vec{r} = (x, y, z)^T$ for each successfully-detected tracer. In the final step of the process, this position estimate is refined using a centroid calculation across the cubic region defined above. The final, refined location is determined as:

$$\begin{pmatrix} x' \\ y' \\ z' \end{pmatrix} = \frac{1}{K} \sum_{i=x-w}^{x+w} \sum_{i=y-w}^{y+w} \sum_{i=z-w}^{z+w} \begin{pmatrix} i \\ j \\ k \end{pmatrix} \rho_l'\left(i, j, k\right) \tag{6.5}$$

where the normalisation constant K is given by:

$$K = \sum_{i=x-w}^{x+w} \sum_{i=y-w}^{y+w} \sum_{i=z-w}^{z+w} \rho_l'\left(i, j, k\right). \tag{6.6}$$

As with other methods, the process described above is repeated for each subsequent timestep or sample of LoRs. In the published work of Wiggins *et al* [17], particles are 'followed' through consecutive frames using the same 'nearest-neighbours' method described for the G-means algorithm above. However, more complex algorithms—such as the multi-frame tracking method of [23]—can also be implemented in order to achieve an improved resolving power (i.e. to more effectively distinguish particles lying in close proximity to one another), though for low-tracer-density applications the more efficient algorithm of [6] is typically found to suffice [15, 24, 25].

Strengths and weaknesses

As, like the G-means and line-density algorithms, the FPI method utilises as its foundation a grid of finite-size voxels, it is—like these other methods—subject to previously-mentioned issues relating to pixelation and grid-size-dependent results.

Compared to its predecessor the G-means algorithm [6], the FPI method not only allows more particles to be simultaneously detected, but also offers greater computational efficiency, requires fewer LoRs for successful location, and typically provides higher spatial resolution—with the exception for the single-tracer case, where the G-means algorithm can still offer higher precision.

Perhaps the most immediately impressive feature of the FPI method, however, is that it can successfully locate significantly larger numbers of tracers than previous methodologies—the algorithm is demonstrated to locate $N_t = 100$ tracers with ~0.5 mm precision in the original paper [17], though there is no reason to expect that this is a hard limit. Nonetheless, the algorithm is somewhat more limited in terms of maximal tracer density, showing a steep increase in location error for tracers falling within 7 mm of one another, despite the relatively sharp (1.75 mm [14]) intrinsic resolution of the positron camera used.

Strengths
- Significantly improved spatial resolution compared to G-means algorithm.
- More computationally efficient than G-means algorithm, allowing near-real-time analysis.
- Can successfully locate large numbers ($N_t \sim 100$) of tracers with sub-millimetre accuracy.
- Can autonomously determine the number of tracers in a system, requiring no *a priori* knowledge.
- Particles entering and leaving the field of view can be successfully tracked.

Weaknesses
- Grid-size dependent results.
- Subject to pixelation for slower-moving tracers.
- Sharp increase in error for small tracer separations.

6.2.8 Interactive example: FPI

The notebook files accompanying this section are available in the supplementary data online at https://iopscience.iop.org/book/978-0-7503-3071-8.

Interactive PEPT analysis example using the feature point identification algorithm [17]

This Jupyter Notebook

An example dataset is used from an experiment run at the University of Birmingham Positron Imaging Centre using the ADAC Forté. It consists of a single 1 mm diameter MCC particle activated with fluorine-18 inside a bubbling fluidised bed. The fluidised bed was filled with 90% sand and 10% MCC; air was fed into the bottom of the bed at a rate of 37 litres per minute at 3.5 bar. This dataset was chosen for its high quality captured lines of response, with the tracer still depicting the random particle motion that is inherent to bubbling fluidised beds—and typical in Lagrangian particle tracking.

The `pept` Python library is used for initialising and visualising PEPT data. While not required *per se* for illustrating PEPT algorithms' processing steps, it significantly reduces the amount of repetitive code and visual noise, allowing the reader to focus on the main conceptual procedures.

[1]:
```
# Importing necessary libraries for numerical array manipulation and plotting,
# plus the `pept` library for straightforward PEPT data processing
import numpy as np
import pept

# Set Plotly figure output (try "browser" too for interactive graphs)
import plotly
plotly.io.renderers.default = "png"
```

Initialise raw line of response data

In this initial step, we read in the 'raw' LoR data from which our tracer positions will be calculated. As illustrated in the 'Effect of sample size and overlap' notebook, varying the numbers of LoRs used to calculate a PEPT location can strongly affect the accuracy of the final position computed. Try altering the value `nrows` below to see how this influences your results.

```
[2]: # Read in a sample of experimental PEPT data into a NumPy array. Skip the file
     # header's first 15 lines, then read in the first `nrows` LoRs
     lors_raw = pept.utilities.read_csv(
         "Data/FluidisedBed1Tracer.csv",
         skiprows=15,
         nrows=100,
     )

     # Convert the LoRs into the general 3D line format [t, x1, y1, z2, x2, y2, z2].
     # See the `PEPT Data Format` interactive example for detailed explanations

     # ADAC Forte's two detector heads' separation, in mm
     head_separation = 600

     lors = np.insert(lors_raw, 3, 0, axis = 1)                  # z1 = 0
     lors = np.insert(lors, 6, head_separation, axis = 1)      # z2 = head_separation

     # Print the first 10 rows from the line of response (LoR) data
     print(f"First 10 rows from the 3D-formatted LoR data:\n{lors[:10]}\n")
```

```
First 10 rows from the 3D-formatted LoR data:
[[0.000e+00 1.900e+02 1.687e+02 0.000e+00 3.463e+02 1.428e+02 6.000e+02]
 [1.000e-01 2.437e+02 1.676e+02 0.000e+00 3.145e+02 3.139e+02 6.000e+02]
 [1.000e-01 1.941e+02 4.100e+02 0.000e+00 4.171e+02 2.401e+02 6.000e+02]
 [2.000e-01 2.885e+02 2.962e+02 0.000e+00 3.847e+02 4.525e+02 6.000e+02]
 [2.000e-01 1.646e+02 1.151e+02 0.000e+00 3.829e+02 3.534e+02 6.000e+02]
 [2.000e-01 1.286e+02 1.322e+02 0.000e+00 4.649e+02 2.661e+02 6.000e+02]
 [2.000e-01 3.339e+02 3.735e+02 0.000e+00 3.316e+02 1.310e+02 6.000e+02]
 [3.000e-01 1.640e+02 1.115e+02 0.000e+00 3.794e+02 3.534e+02 6.000e+02]
 [4.000e-01 2.620e+02 2.094e+02 0.000e+00 2.932e+02 2.808e+02 6.000e+02]
 [4.000e-01 1.304e+02 2.749e+02 0.000e+00 2.543e+02 3.062e+02 6.000e+02]]
```

The `pept.plots` subpackage provides two PEPT-oriented interfaces to the Plotly visualisation library—`PlotlyGrapher` and `PlotlyGrapher2D`—containing straightforward plotting functions for PEPT data. Let's plot the lines of response from the `lors` array:

```
[3]: # Plot the 3D LoRs that were initialised
     from pept.plots import PlotlyGrapher

     PlotlyGrapher().add_lines(lors).show()
```

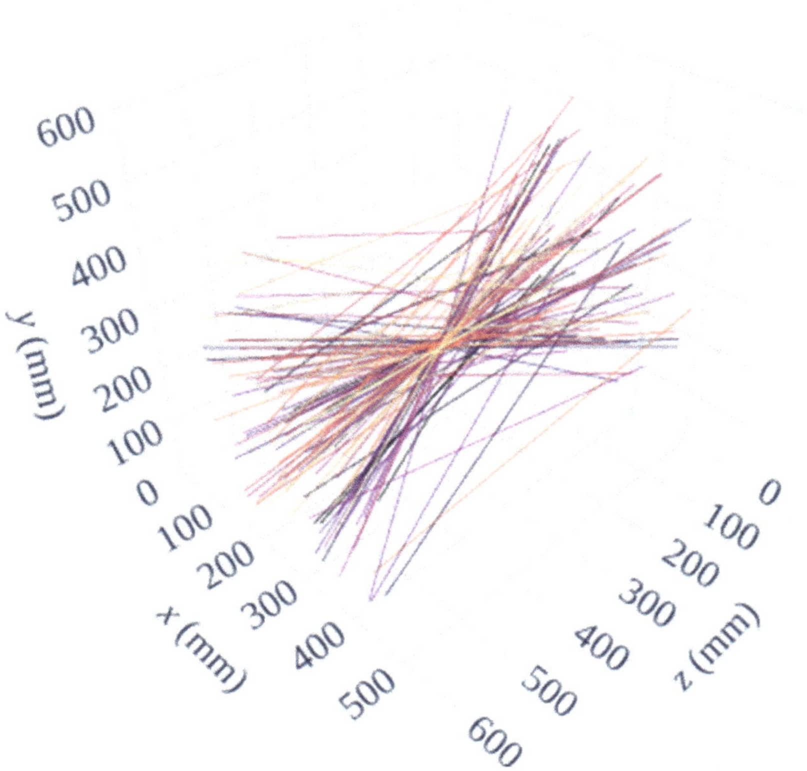

Note the PEPT 3D axes convention, differing from the common mathematical coordinates with *Y* and *Z* being switched, such that the *Y* axis points upwards, while the *Z* axis is perpendicular to the two detector heads.

Voxelise the lines of response

The next step is to 'voxelise' the LoRs—that is, to create a grid of rectangular cells, and count the number of lines passing through each. In the code below you can set the resolution—i.e. number of voxels in each dimension—by varying the pair of values corresponding to the `resolution` parameter. Typing, for example, `(20, 20, 20)` will result in a 20 by 20 by 20 grid of voxels.

Try some different values for the `resolution` parameter. As discussed in the main text, larger values—i.e. a higher resolution, smaller voxels—will allow the centroid of our particle to be determined more precisely. However, you may notice that as you increase these values your processor may take longer to execute the calculation and plotting steps below.

Now consider that this is just one of many thousands of locations we typically need to process, and you may get an idea of the issue of using an overly-fine grid. When using algorithms relying on voxelisation, there is a fine balance to be made between precision and compuational efficiency.

[4]:
```
# Voxellise sample of LoRs
resolution = (30, 30, 30)
voxels = pept.Voxels.from_lines(lors, resolution)

voxels
```

```
Initialised Voxels class in 0.0006213188171386719 s.
```

[4]:
```
Voxels
------
[[[0. 0. 0. … 0. 0. 0.]
  [0. 0. 0. … 0. 0. 0.]
  [0. 0. 0. … 0. 0. 0.]
  …
  [0. 0. 0. … 0. 0. 0.]
  [0. 0. 0. … 0. 0. 0.]
  [0. 0. 0. … 0. 0. 0.]]

 [[0. 0. 0. … 0. 0. 0.]
  [0. 0. 0. … 0. 0. 0.]
  [0. 0. 0. … 0. 0. 0.]
  …
  [0. 0. 0. … 0. 0. 0.]
  [0. 0. 0. … 0. 0. 0.]
  [0. 0. 0. … 0. 0. 0.]]

 [[0. 0. 0. … 0. 0. 0.]
  [0. 0. 0. … 0. 0. 0.]
  [0. 0. 0. … 0. 0. 0.]
  …
  [0. 0. 0. … 0. 0. 0.]
  [0. 0. 0. … 0. 0. 0.]
  [0. 0. 0. … 0. 0. 0.]]

 …

 [[0. 0. 0. … 0. 0. 0.]
  [0. 0. 0. … 0. 0. 0.]
  [0. 0. 0. … 0. 0. 0.]
  …
  [0. 0. 0. … 0. 0. 0.]
  [0. 0. 0. … 0. 0. 0.]
  [0. 0. 0. … 0. 0. 0.]]

 [[0. 0. 0. … 0. 0. 0.]
  [0. 0. 0. … 0. 0. 0.]
  [0. 0. 0. … 0. 0. 0.]
  …
  [0. 0. 0. … 0. 0. 0.]
```

```
 [0. 0. 0. ... 0. 0. 0.]
 [0. 0. 0. ... 0. 0. 0.]]

[[0. 0. 0. ... 0. 0. 0.]
 [0. 0. 0. ... 0. 0. 0.]
 [0. 0. 0. ... 0. 0. 0.]
 ...
 [0. 0. 0. ... 0. 0. 0.]
 [0. 0. 0. ... 0. 0. 0.]
 [0. 0. 0. ... 0. 0. 0.]]]

number_of_voxels =    (30, 30, 30)
voxel_size =          [12.35333333 16.99333333 20.        ]

xlim =                [110.3 480.9]
ylim =                [ 47.2 557. ]
zlim =                [  0. 600.]

voxel_grids:
([110.3 ... 480.9],
 [47.2 ... 557.0],
 [0.0 ... 600.0])
```

[5]:
```
# Plot all LoRs and all voxels with non-zero values (i.e. more than one LoR
# passing through)
PlotlyGrapher().add_voxels(voxels).add_lines(lors).show()
```

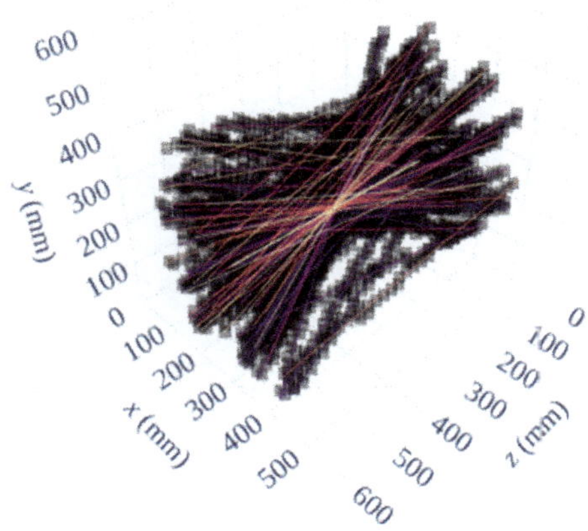

Subtract convolved matrix and blur

Having created a grid of voxels, we can now, in essence, treat these just like a digital image, as discussed in the main text.

In the first step below, we are applying a simple smoothing filter (namely a mean filter). Try changing the degree of smoothing exhibited by altering the the values in the line `kernel = np.ones((3, 3, 3))`—higher values will apply the mean filter over a wider area, giving a more significant smoothing effect, and vice versa.

The smoothed image is then subtracted from the original, 'un-smoothed' image to remove background noise, as depicted in the images at the end of this section.

In step 2, a Gaussian blur is then applied to the resultant image to allow the subsequent clustering routine to work more effectively. Changing the value of `sigma` will alter the width of the Gaussian, i.e. increasing or decreasing the extent of blurring. Try some different values and see what influence they have on the images output.

[6]:
```
# Apply a simple mean filter with the `convolve` subroutine using a simple
# boxcar kernel and subtract it from the original voxel grid
from scipy.signal import convolve

kernel = np.ones((3, 3, 3))
kernel /= np.sum(kernel)

convolved = convolve(voxels, kernel, mode = "same")
voxels_sub = voxels - convolved
voxels_sub[voxels_sub < 0] = 0.
```

[7]:
```
# Construct a 3D Gaussian kernel and convolve the voxel grid with it
#
# Create a centred (M, M, M) cube for the kernel, filled with values along a 3D
# Gaussian distribution centred at (0, 0, 0)
shape = (3, 3, 3)
x, y, z = np.meshgrid(
    np.linspace(-1, 1, shape[0]),
    np.linspace(-1, 1, shape[1]),
    np.linspace(-1, 1, shape[2]),
)

# Distance from origin to each voxel in the kernel cube
d = np.sqrt(x * x + y * y + z * z)

# Calculate kernel values from axis-wise Gaussian distributions
sigma = 0.7
mu = 0.0

alpha = 1 / sigma / np.sqrt(2 * np.pi)
gaussian_kernel = alpha * np.exp(-0.5 * (d - mu)**2 / sigma**2)

# Convolve the voxel grid using the Gaussian distribution
voxels_blurred = convolve(voxels_sub, gaussian_kernel, mode = "same")
voxels_blurred = pept.Voxels(
    voxels_blurred,
    voxels.xlim,
    voxels.ylim,
    voxels.zlim,
)

voxels_blurred
```

[7]: **Voxels**

```
------
[[[ 1.66803210e-16  1.47673562e-16  1.48829894e-16 ...  1.02057617e-16
    1.45359764e-17  3.26497073e-18]
  [ 1.86195591e-16  1.96331505e-16  1.58537300e-16 ...  1.77378509e-16
    2.32482455e-16  1.15747838e-16]
  [ 1.84070610e-16  1.84791014e-16  1.60432835e-16 ...  4.61583822e-18
    1.26484530e-16  1.32107896e-16]
  ...
  [-2.74337703e-17  2.49726649e-16  3.23411873e-16 ...  6.75773516e-17
    7.60465465e-17 -2.65885600e-17]
  [ 4.71566685e-18  1.79480985e-16  1.90026551e-16 ...  1.25257059e-16
    2.17664165e-16  4.60917552e-17]
  [-5.87711452e-17  4.06428741e-18  2.91141148e-17 ...  1.76380516e-16
    2.66136453e-16  1.18897871e-16]]

 [[ 2.50096027e-16  3.22223638e-16  2.98378193e-16 ...  3.89047097e-17
    5.63599031e-17  4.89743843e-17]
  [ 2.70246084e-16  3.27477136e-16  2.29608583e-16 ...  6.46108107e-17
    2.16349786e-16  1.10677357e-16]
  [ 5.25767477e-17  6.25771146e-17  3.77395745e-17 ... -1.12074882e-16
    6.71956428e-17  4.08964691e-17]
  ...
  [-5.64845793e-17  1.08318454e-16  3.16627609e-16 ... -1.09159152e-17
   -6.13112353e-17 -1.33662205e-16]
  [ 1.06903111e-16  1.74385763e-16  2.55332467e-16 ...  8.03152367e-17
    4.54841671e-17 -3.68412633e-18]
  [ 3.90898473e-17  9.24386513e-17  1.14347141e-16 ...  1.08703666e-16
    5.98550004e-17  2.72142631e-17]]

 [[ 3.02557879e-17  9.19228218e-17  1.49337820e-16 ...  2.51621968e-17
    7.94491375e-17  3.51332386e-17]
  [ 9.46763452e-17  1.34671147e-16  9.38148073e-17 ...  6.02645212e-17
    1.47016155e-16  6.93683874e-17]
  [-6.02054360e-17 -6.77877106e-17 -1.45098917e-17 ... -5.23312159e-17
   -3.13005617e-17 -4.68548005e-17]
  ...
  [ 1.00312608e-16  1.23005112e-16  8.17154647e-17 ...  1.74080958e-16
    1.54952756e-16  6.89423877e-17]
  [ 1.26799049e-16  1.28010423e-16  7.82059284e-17 ... -3.60832064e-17
    2.50880557e-17  4.40582703e-17]
  [ 1.11615820e-16  7.49261702e-17  7.50738509e-17 ...  1.02835755e-16
    5.75151430e-17  1.38201346e-17]]

 ...

 [[ 3.03576608e-17  4.88141167e-17  3.48917567e-17 ...  1.77929461e-02
    9.58228084e-02  1.46686397e-01]
  [ 1.42247325e-16  8.54510991e-17  5.64071130e-17 ...  4.93634492e-02
    2.65843796e-01  4.06956018e-01]
  [ 2.05564732e-16  1.63943472e-16  9.25194233e-17 ...  3.85513832e-02
    1.76148930e-01  2.30520352e-01]
  ...
  [ 1.05646680e-16  1.81681933e-16  2.34254641e-16 ...  7.33728430e-16
    6.37085824e-16  3.33450973e-16]
  [ 1.75386294e-16  1.19546371e-16  1.25888424e-16 ...  4.70174989e-16
```

```
  3.50748446e-16  1.97423533e-16]
 [ 1.04515752e-16  5.70385642e-17  1.17251240e-16 … 2.83411160e-16
   2.82397126e-16  2.44985571e-16]]

[[ 1.32269223e-16  1.51528675e-16  3.24792547e-17 … 2.15370143e-16
   1.61599943e-16  7.42065226e-17]
 [ 5.86413093e-17  1.44397811e-16  8.91681540e-17 … 1.61391058e-16
   1.93114013e-16  1.39339661e-16]
 [-8.84708973e-17  5.67935803e-17  6.59279685e-17 … 5.75906907e-02
   2.22850921e-01  2.32582799e-01]
 …
 [-3.56924448e-16 -2.72719763e-16 -2.11201491e-16 … 3.85788991e-16
   3.08546220e-16 -3.04991635e-17]
 [-6.05904016e-17 -9.82016331e-17 -6.84461115e-17 … 2.27582516e-16
   1.65348066e-16 -1.60701325e-17]
 [ 8.15299216e-17  4.18076730e-17  8.85091387e-17 … 2.11575076e-16
   2.10654469e-16  1.10781886e-16]]

[[-1.03493562e-16 -3.60747019e-17  4.37698484e-17 … 1.83117980e-16
   2.39679771e-16  1.98674080e-16]
 [-1.42766344e-16 -1.31894216e-16 -1.58881293e-16 … 1.71018183e-16
   1.81473455e-16  1.44753145e-16]
 [-3.92914867e-16 -2.97225816e-16 -2.67359477e-16 … 2.07584371e-02
   8.03261218e-02  8.38339555e-02]
 …
 [-1.54562999e-16 -2.25316581e-17 -2.88036074e-17 … 1.09592247e-16
   8.24448809e-17 -9.24106883e-17]
 [ 4.76565597e-17  1.79367647e-16  2.27714434e-16 … 1.25351925e-16
   2.05464580e-16  1.22787329e-16]
 [ 2.48474163e-16  3.49903420e-16  3.81882009e-16 … 3.59627372e-16
   4.05346592e-16  3.11053975e-16]]]

number_of_voxels =    (30, 30, 30)
voxel_size =          [12.35333333 16.99333333 20.        ]

xlim =                [110.3 480.9]
ylim =                [ 47.2 557. ]
zlim =                [  0. 600.]

voxel_grids:
([110.3 … 480.9],
 [47.2 … 557.0],
 [0.0 … 600.0])
```

[8]:
```
# Plot each step from the voxel grid processing pipeline. Change the `opacity`
# to view further "inside" the 3D grid
grapher = PlotlyGrapher(
    cols=3,
    subplot_titles=[
        "Original Voxellised LoRs",
        "Subtracted Convolution",
        "Gaussian Blur",
    ],
)

opacity = 0.05

grapher.add_voxels(voxels, opacity=opacity)
grapher.add_voxels(voxels_sub, opacity=opacity, col=2)
grapher.add_voxels(voxels_blurred, opacity=opacity, col=3)

# Plot tracer positions, colour-coding the last column - i.e. the spatial error
grapher.show()
```

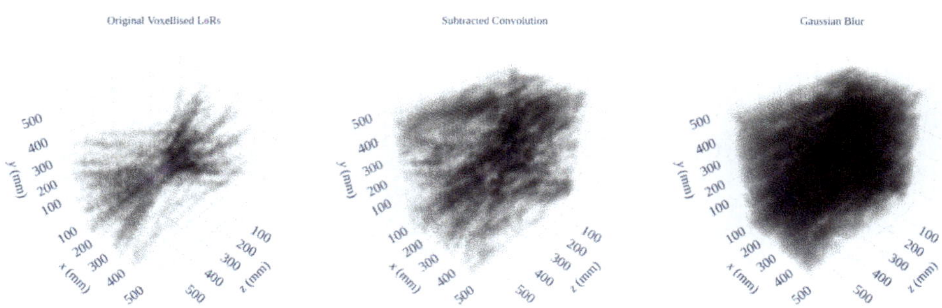

Extract voxel peaks

Finally, from the blurred and subtracted image we extract the peaks which (if our parameters are correct) will represent particle positions. Specifically, we search for any peaks lying in the upper `r_percent`-th percentile of voxel values (i.e. line densities)—for example, if `r_percent = 40`, we consider only voxels whose values lie in the top 40%.

In the case of multiple particles, we are not only looking for one global maximum, but multiple local maxima. In this case, we look for peaks within individual 'regions' of size `(2 * w_voxels + 1)`.

Try varying the size of `r_percent` and `w_voxels` and see the effect it has on your results. You will notice for the default settings that the code gives three locations —one true, and two false. You will find that the erroneous peak can be removed by adjusting both `w_voxels` and `r_percent`. Give some thought as to which of these options would be most effective in different real-world situations.

[9]:
```
# Extract peaks in the upper `r_percent`-th percentile, which also have the
# maximum value in a voxel cube of length (2 * `w_voxels` + 1)
r_percent = 40
w_voxels = 3

r_upper_quantile = np.quantile(voxels_blurred, 1 - r_percent / 100)
likely_indices = np.argwhere(voxels_blurred > r_upper_quantile)

peaks = []
for ix, iy, iz in likely_indices:
    # Vicinity voxel cube indices, truncated at the grid's edges
    min_ix = max(ix - 2 * w_voxels, 0)
    max_ix = min(ix + 2 * w_voxels + 1, voxels_blurred.shape[0])

    min_iy = max(iy - 2 * w_voxels, 0)
    max_iy = min(iy + 2 * w_voxels + 1, voxels_blurred.shape[1])

    min_iz = max(iz - 2 * w_voxels, 0)
    max_iz = min(iz + 2 * w_voxels + 1, voxels_blurred.shape[2])

    # If the current voxel at (ix, iy) is the maximum in the vicinity cube,
    # it is a peak
    cube = voxels_blurred[min_ix:max_ix, min_iy:max_iy, min_iz:max_iz]
    if voxels_blurred[ix, iy, iz] == cube.max():
        peaks.append((ix, iy, iz))

peaks = np.array(peaks)

# Transform peaks from voxel indices to physical coordinates
origin = np.array([voxels.xlim[0], voxels.ylim[0], voxels.zlim[0]])
positions = origin + peaks * voxels.voxel_size + 0.5 * voxels.voxel_size

# Insert the tracer position's timestamp at column 0
positions = np.insert(positions, 0, lors[:, 0].mean(), axis = 1)
positions
```

[9]:
```
array([[  1.931     , 227.65666667, 174.65      ,  30.        ],
       [  1.931     , 277.07      , 242.62333333, 310.        ],
       [  1.931     , 425.31      ,  72.69      , 570.        ]])
```

[10]:
```
# Plot the processed voxels, LoRs and tracer locations found
grapher = PlotlyGrapher(cols=2)

grapher.add_voxels(voxels_blurred, opacity=0.05)
grapher.add_points(positions, size=10, color="red")

grapher.add_lines(lors, col=2)
grapher.add_points(positions, size=10, color="red", col=2)

# Plot tracer positions, colour-coding the last column - i.e. the spatial error
grapher.show()
```

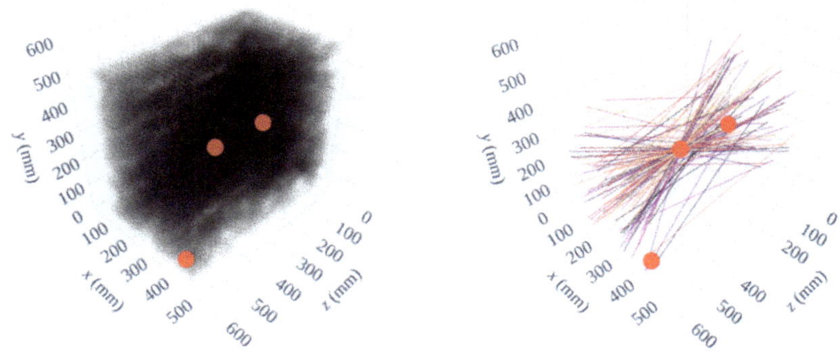

Complete FPI algorithm code

The code developed in the previous sections can be applied to consecutive 'samples' of LoRs to follow the tracer's movement through the fluidised bed, fully tracking its trajectory. A complete—albeit slow due to Python's dynamic nature—feature point identification algorithm implementation is given below.

First initialise a larger, multi-sample set of LoRs to analyse; change the `nrows` argument below to read in a different number of LoRs (the file included contains 80,000 LoRs in total), though keep in mind the computational expense of processing large numbers of LoRs!

[11]:
```
# Read in a sample of experimental PEPT data into a NumPy array. Skip the file
# header's first 15 lines and read in the first `nrows` LoRs
lors_raw = pept.utilities.read_csv(
    "Data/FluidisedBed1Tracer.csv",
    skiprows=15,
    nrows=5000,
)

# Convert the LoRs into the general 3D line format [t, x1, y1, z2, x2, y2, z2].
# See the `PEPT Data Format` interactive example for detailed explanations

# ADAC Forte's two detector heads' separation, in mm
head_separation = 600

lors = np.insert(lors_raw, 3, 0, axis=1)                    # z1 = 0
lors = np.insert(lors, 6, head_separation, axis=1)          # z2 = head_separation

# Print the first 10 rows from the line of response (LoR) data
print(f"First 10 rows from the 3D-formatted LoR data:\n{lors[:10]}\n")
```

```
First 10 rows from the 3D-formatted LoR data:
[[0.000e+00 1.900e+02 1.687e+02 0.000e+00 3.463e+02 1.428e+02 6.000e+02]
 [1.000e-01 2.437e+02 1.676e+02 0.000e+00 3.145e+02 3.139e+02 6.000e+02]
 [1.000e-01 1.941e+02 4.100e+02 0.000e+00 4.171e+02 2.401e+02 6.000e+02]
 [2.000e-01 2.885e+02 2.962e+02 0.000e+00 3.847e+02 4.525e+02 6.000e+02]
 [2.000e-01 1.646e+02 1.151e+02 0.000e+00 3.829e+02 3.534e+02 6.000e+02]
 [2.000e-01 1.286e+02 1.322e+02 0.000e+00 4.649e+02 2.661e+02 6.000e+02]
 [2.000e-01 3.339e+02 3.735e+02 0.000e+00 3.316e+02 1.310e+02 6.000e+02]
 [3.000e-01 1.640e+02 1.115e+02 0.000e+00 3.794e+02 3.534e+02 6.000e+02]
 [4.000e-01 2.620e+02 2.094e+02 0.000e+00 2.932e+02 2.808e+02 6.000e+02]
 [4.000e-01 1.304e+02 2.749e+02 0.000e+00 2.543e+02 3.062e+02 6.000e+02]]
```

Iterating through each sample of LoRs, locate the tracer and append its location to a growing list of positions. As with the previous cells, you are invited to change the parameters highlighted below; details about each one are given in the comments.

[12]:
```
# Use the Feature Point Identification (FPI) algorithm to track a moving tracer
from scipy.signal import convolve

# FPI Method user-changeable parameters:
#     resolution - the number of voxels in each dimension for the LoR voxel grid
#     shape - size of the convolution / blur kernels
#     sigma - standard deviation in the Gaussian blur kernel
#     r_percent - select the upper `r_percent` voxel values
#     w_voxels - cube of length (2 * w + 1) for selecting voxel peak
resolution = (30, 30, 30)
shape = (3, 3, 3)
sigma = 0.7
r_percent = 40
w_voxels = 3

# General PEPT tracking user-changeable parameters:
#     sample_start - the `lors` index at which to start tracking the tracer
#     sample_size - the number of LoRs used to compute a single location
#     overlap - the number of overlapping LoRs between consecutive samples
sample_start = 0
sample_size = 200
overlap = 10
# Tracked tracer locations will be appended to a growing list
positions = []

while sample_start + sample_size < len(lors):
    sample = lors[sample_start:sample_start + sample_size]

    # Voxellise sample of LoRs
    voxels = pept.Voxels.from_lines(sample, resolution, verbose = False)

    # Apply a simple mean filter with the `convolve` subroutine using a simple
    # boxcar kernel and subtract it from the original voxel grid
    kernel = np.ones((3, 3, 3))
    kernel /= np.sum(kernel)

    convolved = convolve(voxels, kernel, mode = "same")
    voxels_sub = voxels - convolved
    voxels_sub[voxels_sub < 0] = 0.
    # Construct a 3D Gaussian kernel and convolve the voxel grid with it
    #
    # Create a centred (M, M, M) cube for the kernel, filled with values along a
    # 3D Gaussian distribution centred at (0, 0, 0)
    x, y, z = np.meshgrid(
        np.linspace(-1, 1, shape[0]),
        np.linspace(-1, 1, shape[1]),
        np.linspace(-1, 1, shape[2]),
    )

    # Distance from origin to each voxel in the kernel cube
    d = np.sqrt(x * x + y * y + z * z)
```

```python
    # Calculate kernel values from axis-wise Gaussian distributions
    mu = 0.0

    alpha = 1 / sigma / np.sqrt(2 * np.pi)
    gaussian_kernel = alpha * np.exp(-0.5 * (d - mu)**2 / sigma**2)

    # Convolve the voxel grid using the Gaussian distribution
    voxels_blurred = convolve(voxels_sub, gaussian_kernel, mode = "same")
    voxels_blurred = pept.Voxels(
        voxels_blurred,
        voxels.xlim,
        voxels.ylim,
        voxels.zlim,
    )
    # Extract peaks in the upper `r_percent`-th percentile, which also have the
    # maximum value in a voxel cube of length (2 * `w_voxels` + 1)
    r_upper_quantile = np.quantile(voxels_blurred, 1 - r_percent / 100)
    likely_indices = np.argwhere(voxels_blurred > r_upper_quantile)

    peaks = []
    for ix, iy, iz in likely_indices:
        # Vicinity voxel cube indices, truncated at the grid's edges
        min_ix = max(ix - 2 * w_voxels, 0)
        max_ix = min(ix + 2 * w_voxels + 1, voxels_blurred.shape[0])

        min_iy = max(iy - 2 * w_voxels, 0)
        max_iy = min(iy + 2 * w_voxels + 1, voxels_blurred.shape[1])

        min_iz = max(iz - 2 * w_voxels, 0)
        max_iz = min(iz + 2 * w_voxels + 1, voxels_blurred.shape[2])

        # If the current voxel at (ix, iy) is the maximum in the vicinity cube,
        # it is a peak
        cube = voxels_blurred[min_ix:max_ix, min_iy:max_iy, min_iz:max_iz]
        if voxels_blurred[ix, iy, iz] == cube.max():
            peaks.append((ix, iy, iz))

    peaks = np.array(peaks)
    # Transform peaks from voxel indices to physical coordinates
    origin = np.array([voxels.xlim[0], voxels.ylim[0], voxels.zlim[0]])
    centres = origin + peaks * voxels.voxel_size + 0.5 * voxels.voxel_size

    # Insert the tracer position's timestamp at column 0
    centres = np.insert(centres, 0, sample[:, 0].mean(), axis = 1)
    positions.extend(centres)

    sample_start += sample_size - overlap

positions = np.array(positions)

print(f"First 10 tracked positions:\n{positions[:10]}")
```

```
First 10 tracked positions:
[[  3.9045     282.11        241.14333333 310.         ]
 [ 11.4085     141.96666667 498.745       570.         ]
 [ 11.4085     167.3        276.055       530.         ]
 [ 11.4085     281.3        241.795       310.         ]
 [ 11.4085     293.96666667  70.495        30.         ]
```

```
 [ 18.2215      129.515      292.75333333  30.        ]
 [ 18.2215      278.435      242.07333333 310.        ]
 [ 18.2215      315.665      478.58       530.        ]
 [ 25.357       280.055      242.71       310.        ]
 [ 31.6025      280.385      238.19166667 310.        ]]
```

[13]:
```
# Plot some of the LoRs used, plus the tracer locations we found
grapher = PlotlyGrapher(
    cols=2,
    subplot_titles=[
        "First 500 LoRs",
        f"Tracer locations tracked from {len(lors)} LoRs",
    ],
)

grapher.add_lines(lors[:500])
grapher.add_points(positions, col=2)

# Plot tracer positions, colour-coding the last column - i.e. the spatial error
grapher.show()
```

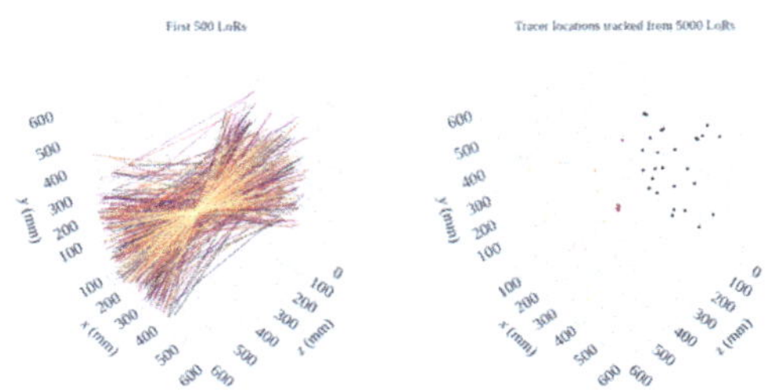

[14]:
```
# Create a "timeseries" plot with the tracer positions vs. time
from pept.plots import PlotlyGrapher2D

PlotlyGrapher2D().add_timeseries(positions).show()
```

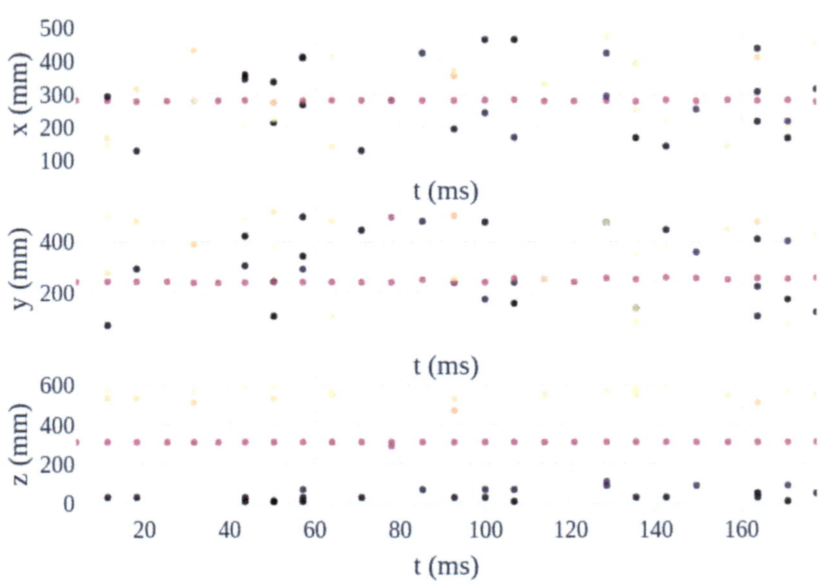

Again, the default parameters yield some erroneous points towards the edges of the experimental volume—based on the experience you've gained in the previous notebook sections, try varying the FPI parameters to remove the noise.

Multiple particle tracking

The complete code developed above can also be applied to a simple multiple tracer experiment. Here we use a sample of an experiment, consisting of two 1 mm diameter alumina particles with fluorine-18 radioactive tracer material. The particles are fixed at the two ends of an arm rotating at 42 rpm, their trajectories describing full circles.

[15]:
```
# Read in a sample of experimental PEPT data into a NumPy array. Skip the file
# header's first 15 lines
lors_raw = pept.utilities.read_csv(
    "Data/RotatingArm2Tracers.csv",
    skiprows=15,
    nrows=10_000,
)

# Convert the LoRs into the general 3D line format [t, x1, y1, z2, x2, y2, z2].
# See the `PEPT Data Format` interactive example for detailed explanations

# ADAC Forte's two detector heads' separation, in mm
head_separation = 712

lors = np.insert(lors_raw, 3, 0, axis=1)                    # z1 = 0
lors = np.insert(lors, 6, head_separation, axis=1)          # z2 = head_separation

# Print the first 10 rows from the line of response (LoR) data
print(f"First 10 rows from the 3D-formatted LoR data:\n{lors[:10]}\n")
```

```
First 10 rows from the 3D-formatted LoR data:
[[0.000e+00 2.094e+02 2.797e+02 0.000e+00 2.738e+02 9.500e+01 7.120e+02]
 [1.000e-01 4.325e+02 4.395e+02 0.000e+00 2.006e+02 1.929e+02 7.120e+02]

 [1.000e-01 2.513e+02 1.109e+02 0.000e+00 2.177e+02 3.469e+02 7.120e+02]
 [1.000e-01 3.227e+02 4.372e+02 0.000e+00 3.853e+02 1.097e+02 7.120e+02]
 [1.000e-01 1.894e+02 9.910e+01 0.000e+00 4.213e+02 1.044e+02 7.120e+02]
 [2.000e-01 3.658e+02 2.230e+02 0.000e+00 3.198e+02 5.003e+02 7.120e+02]
 [2.000e-01 2.543e+02 4.690e+02 0.000e+00 4.873e+02 1.257e+02 7.120e+02]
 [2.000e-01 3.339e+02 4.962e+02 0.000e+00 3.646e+02 8.970e+01 7.120e+02]
 [2.000e-01 2.844e+02 2.130e+02 0.000e+00 4.466e+02 5.322e+02 7.120e+02]
 [2.000e-01 2.053e+02 1.906e+02 0.000e+00 2.785e+02 2.224e+02 7.120e+02]]
```

Iterating through each sample of LoRs, we locate the tracers and append their locations to a growing list of positions. As with the previous cells, you are invited to change the parameters highlighted below; details about each one are given in the comments.

Importantly, the `sample_size` must be larger to include enough LoRs for each tracer—in cases using large numbers of LoRs, the `overlap` becomes increasingly important to yield *dense* points on each trajectory; see the 'Effect of sample size and overlap' section for a detailed explanation.

[16]:
```
# Use the Feature Point Identification (FPI) algorithm to track a moving tracer
from scipy.signal import convolve

# FPI Method user-changeable parameters:
#     resolution - the number of voxels in each dimension for the LoR voxel grid
#     shape - size of the convolution / blur kernels
#     sigma - standard deviation in the Gaussian blur kernel
#     r_percent - select the upper `r_percent` voxel values
#     w_voxels - cube of length (2 * w + 1) for selecting voxel peak
resolution = (30, 30, 30)
shape = (3, 3, 3)
sigma = 0.7
r_percent = 40
w_voxels = 5

# General PEPT tracking user-changeable parameters:
#     sample_start - the `lors` index at which to start tracking the tracer
#     sample_size - the number of LoRs used to compute a single location
#     overlap - the number of overlapping LoRs between consecutive samples
sample_start = 0
sample_size = 200
overlap = 10

# Tracked tracer locations will be appended to a growing list
positions = []

while sample_start + sample_size < len(lors):
    sample = lors[sample_start:sample_start + sample_size]

    # Voxellise sample of LoRs
    voxels = pept.Voxels.from_lines(sample, resolution, verbose = False)

    # Apply a simple mean filter with the `convolve` subroutine using a simple
    # boxcar kernel and subtract it from the original voxel grid
    kernel = np.ones((3, 3, 3))
    kernel /= np.sum(kernel)
```

```python
convolved = convolve(voxels, kernel, mode = "same")
voxels_sub = voxels - convolved
voxels_sub[voxels_sub < 0] = 0.

# Construct a 3D Gaussian kernel and convolve the voxel grid with it
#
# Create a centred (M, M, M) cube for the kernel, filled with values along a
# 3D Gaussian distribution centred at (0, 0, 0)
x, y, z = np.meshgrid(
    np.linspace(-1, 1, shape[0]),
    np.linspace(-1, 1, shape[1]),
    np.linspace(-1, 1, shape[2]),
)

# Distance from origin to each voxel in the kernel cube
d = np.sqrt(x * x + y * y + z * z)

# Calculate kernel values from axis-wise Gaussian distributions
mu = 0.0

alpha = 1 / sigma / np.sqrt(2 * np.pi)
gaussian_kernel = alpha * np.exp(-0.5 * (d - mu)**2 / sigma**2)

# Convolve the voxel grid using the Gaussian distribution
voxels_blurred = convolve(voxels_sub, gaussian_kernel, mode = "same")
voxels_blurred = pept.Voxels(
    voxels_blurred,
    voxels.xlim,
    voxels.ylim,
    voxels.zlim,
)

# Extract peaks in the upper `r_percent`-th percentile, which also have the
# maximum value in a voxel cube of length (2 * `w_voxels` + 1)
r_upper_quantile = np.quantile(voxels_blurred, 1 - r_percent / 100)
likely_indices = np.argwhere(voxels_blurred > r_upper_quantile)

peaks = []
for ix, iy, iz in likely_indices:
    # Vicinity voxel cube indices, truncated at the grid's edges
    min_ix = max(ix - 2 * w_voxels, 0)
    max_ix = min(ix + 2 * w_voxels + 1, voxels_blurred.shape[0])

    min_iy = max(iy - 2 * w_voxels, 0)
    max_iy = min(iy + 2 * w_voxels + 1, voxels_blurred.shape[1])

    min_iz = max(iz - 2 * w_voxels, 0)
    max_iz = min(iz + 2 * w_voxels + 1, voxels_blurred.shape[2])

    # If the current voxel at (ix, iy) is the maximum in the vicinity cube,
    # it is a peak
    cube = voxels_blurred[min_ix:max_ix, min_iy:max_iy, min_iz:max_iz]
    if voxels_blurred[ix, iy, iz] == cube.max():
        peaks.append((ix, iy, iz))

peaks = np.array(peaks)
```

```
    # Transform peaks from voxel indices to physical coordinates
    origin = np.array([voxels.xlim[0], voxels.ylim[0], voxels.zlim[0]])
    centres = origin + peaks * voxels.voxel_size + 0.5 * voxels.voxel_size

    # Insert the tracer position's timestamp at column 0
    centres = np.insert(centres, 0, sample[:, 0].mean(), axis = 1)
    positions.extend(centres)

    sample_start += sample_size - overlap

positions = np.array(positions)

print(f"First 10 tracked positions:\n{positions[:10]}")
```

```
First 10 tracked positions:
[[  2.365      229.84333333 207.68666667 272.93333333]
 [  6.6425     230.79333333 205.5        272.93333333]
 [ 10.615      345.82166667 327.715      272.93333333]
 [ 10.615      384.11166667  71.065      676.4       ]
 [ 14.4775     346.89333333 328.315      272.93333333]
 [ 18.3525     231.14166667 207.725      272.93333333]
 [ 22.4385     231.86333333 209.16666667 272.93333333]
 [ 26.6675     360.15       327.655      272.93333333]
 [ 30.6965     220.67833333 206.57833333 272.93333333]
 [ 30.6965     360.415      324.015      272.93333333]]
```

[17]:

```
# Plot some of the LoRs used, plus the tracer locations we found
grapher = PlotlyGrapher(
    cols=2,
    subplot_titles=[
        "First 500 LoRs",
        f"Tracer locations tracked from {len(lors)} LoRs",
    ],
)

grapher.add_lines(lors[:500])
grapher.add_points(positions, col=2)

grapher.show()
```

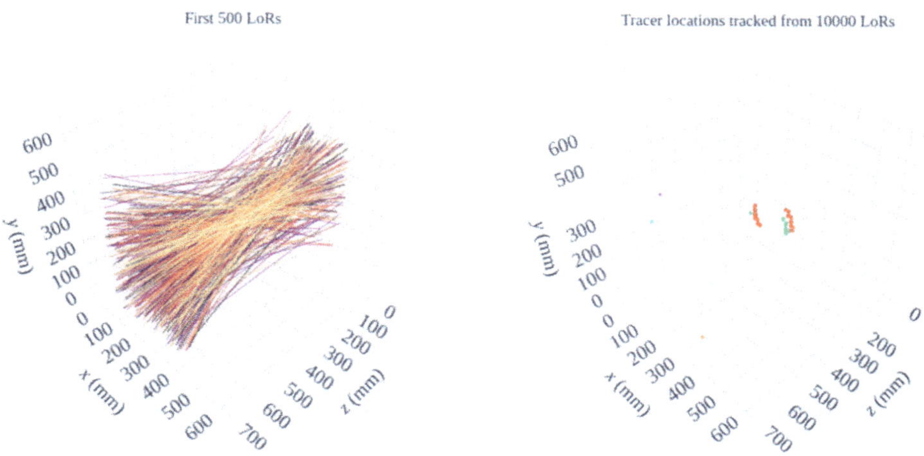

[18]:
```python
# Create a "timeseries" plot with the tracer positions vs. time
from pept.plots import PlotlyGrapher2D

PlotlyGrapher2D().add_timeseries(positions).show()
```

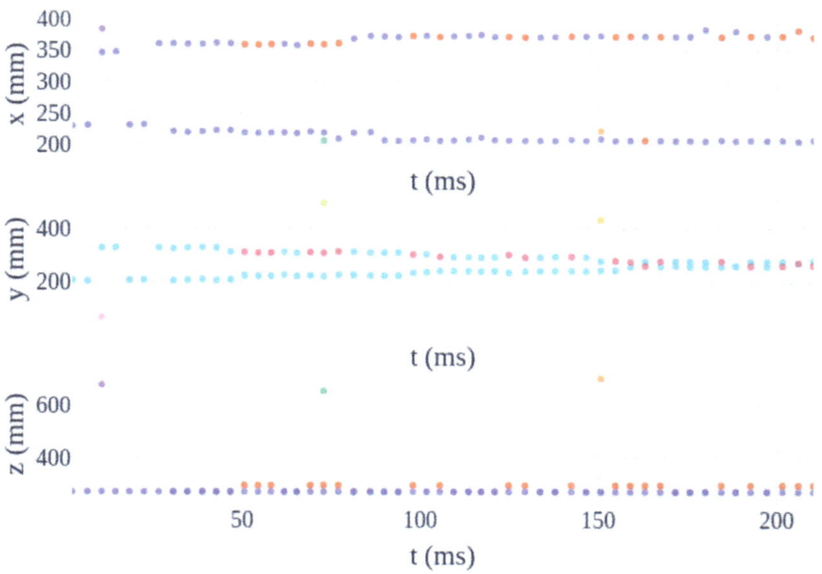

High-performance FPI algorithm implementation

The `pept` library also contains a highly-optimised FPI Algorithm implementation, which uses C++ code for the computationally-intensive operations—kindly contributed to the `pept` library by the original author of the algorithm, Dr Cody Wiggins—and processes each sample of LoRs in parallel—therefore being able to fully utilise multi-CPU architectures in modern computers and clusters.

Most of the repetitive or tedious tasks above are automated, such as:

- Conversion from native PEPT scanner formats (e.g. binary, or parallel screens ASCII / CSV files) to the general 3D LoR format using the `pept.scanners` subpackage.
- Straightforward iteration of LoRs in terms of samples with optional overlap using the `pept.LineData` and `pept.PointData` classes.
- Terse usage of the PEPT tracking algorithms from the `pept.tracking` subpackage.
- PEPT-oriented plotting using the `pept.plots` subpackage.

[19]:
```
# Import the subroutines we will use to track our tracer
from pept.tracking import Voxelize, FPI, Stack
from pept.plots import PlotlyGrapher, PlotlyGrapher2D

# Initialise data from a Parallel Screens CSV format (i.e. [t, x1, y1, x2, y2])
# into the general "LineData" format ([t, x1, y1, z1, x2, y2, z2]) and set a
# sample size and overlap
lors = pept.scanners.parallel_screens(
    "Data/FluidisedBed1Tracer.csv",
    screen_separation=600,
    skiprows=15,
    sample_size=200,
    overlap=100,
)

# We can chain a sequence of "filters" in a pept.Pipeline that will process all
# samples in parallel, while minimising memory consumption
#
# Use a finer voxel grid because the C/C++ implementation is much faster
#
# Instantiate the `FPI` tracker with the voxel cube half-length `w` value and
# voxel selection upper `r` ratio - then "fit" each sample in `lors`
pipeline = pept.Pipeline([
    Voxelize(number_of_voxels=(50, 50, 50)),
    FPI(w=5, r=0.4),
    # pept.tracking.BirminghamMethod(),
    Stack(),
])

positions = pipeline.fit(lors)

# Plot all tracked positions
PlotlyGrapher().add_points(positions).show()
```

```
Initialised PEPT data in 0.030 s.

Batch 1 / 1 :: 100%|                               | 799/799
[00:02<00:00, 383.69it/s]

Processed samples in 2.203626871109009 s
```

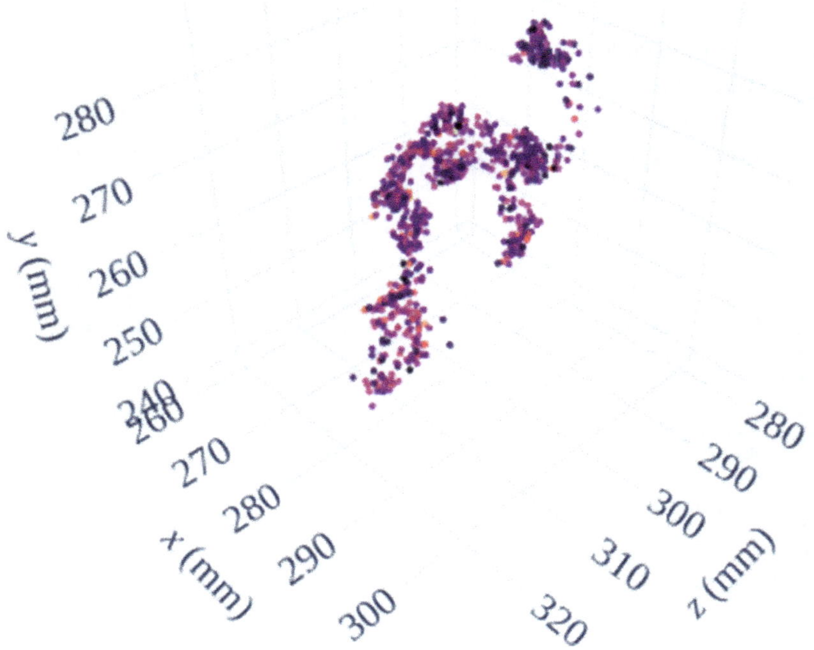

[20]:
```
# Create a "timeseries" plot with the tracer positions vs. time
PlotlyGrapher2D().add_timeseries(positions).show()
```

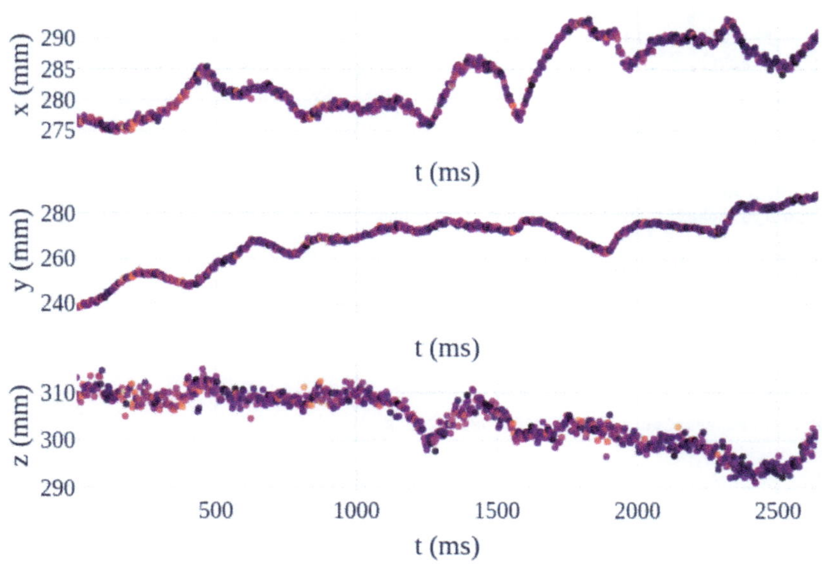

In what were essentially five Python statements, we processed 80,000 LoRs (in 0.71 s on the author's laptop) to track 2.6 s of a single tracer's trajectory through a fluidised bed, complete with loading bars and interactive plotting. Though `pept` automated most steps, there is no magic involved—just sensible defaults which can be changed for every subroutine called; plus, all code is public and free to use and modify from the `pept` repository.

The same algorithm can applied to the two-tracer experiment, with the only change being the larger `sample_size` to allow the same number of LoRs *per tracer*.

[21]:
```
# Import the subroutines we will use to track our tracer
from pept.tracking import Voxelize, FPI, Stack
from pept.plots import PlotlyGrapher, PlotlyGrapher2D

# Initialise data from a Parallel Screens CSV format (i.e. [t, x1, y1, x2, y2])
# into the general "LineData" format ([t, x1, y1, z1, x2, y2, z2]) and set a
# sample size and overlap
lors = pept.scanners.parallel_screens(
    "Data/RotatingArm2Tracers.csv",
    screen_separation=712,
    skiprows=15,
    sample_size=400,
    overlap=100,
)

# We can chain a sequence of "filters" in a pept.Pipeline that will process all
# samples in parallel, while minimising memory consumption
#
# Use a finer voxel grid because the C/C++ implementation is much faster
#
# Instantiate the `FPI` tracker with the voxel cube half-length `w` value and
# voxel selection upper `r` ratio - then "fit" each sample in `lors`
pipeline = pept.Pipeline([
    Voxelize(number_of_voxels=(100, 100, 100)),
    FPI(w=3, r=0.4),
    Stack(),
])

positions = pipeline.fit(lors)

# Plot all tracked positions
PlotlyGrapher().add_points(positions).show()
```

```
Initialised PEPT data in 0.030 s.

Batch 1 / 1 :: 100%|                              | 266/266
[00:00<00:00, 1575.18it/s]


Processed samples in 0.2708463668823242 s
```

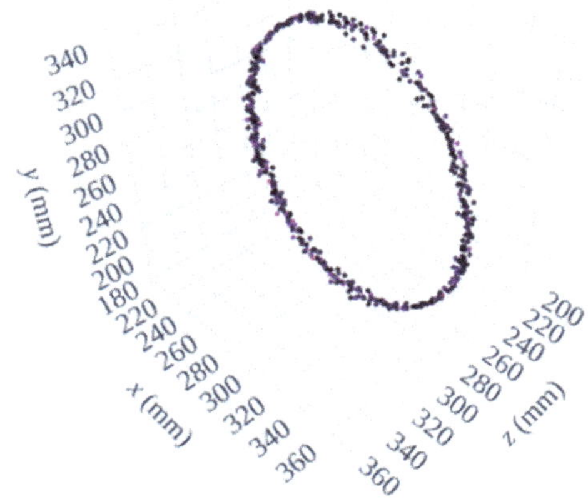


```
[22]: # Create a "timeseries" plot with the tracer positions vs. time
      PlotlyGrapher2D().add_timeseries(positions).show()
```

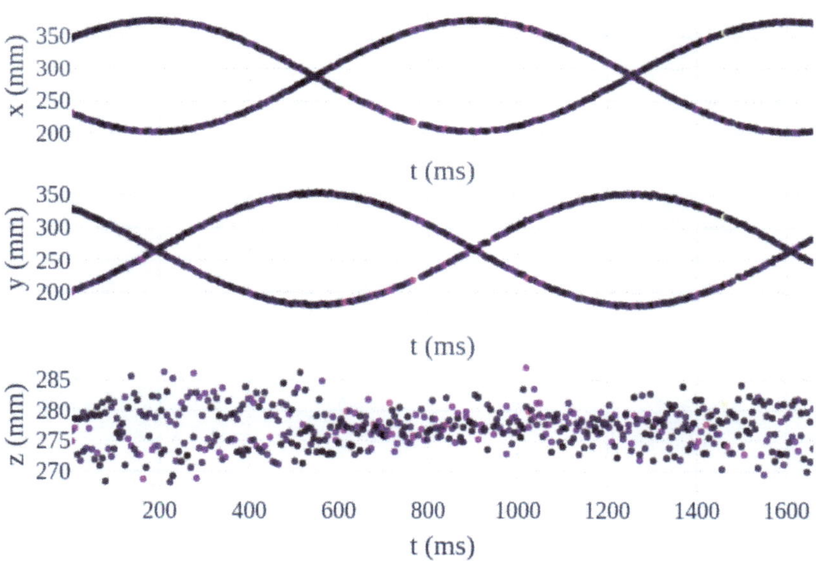

The tracers' movement in the XY plane is spread out enough that the timeseries graphs look effectively continuous. As the axis of rotation is perpendicular to the Z dimension, the tracer trajectories should be almost flat - therefore, the spread in the bottom subplot is mainly due to data acquisition errors, both due to the parallel screens scanner geometry and the tracer location algorithm.

6.2.9 Spatiotemporal B-spline reconstruction (SBSR)

The SBSR method in brief

The SBSR method [26], developed at Stanford University, again represents a marked departure from previous PEPT algorithms, most notably in that particles are tracked not through the determination of individual, distinct and independent locations at discrete points in time, but rather through the fitting of splines, which carry a dependence on both antecedent and descendent points in time. The SBSR algorithm is also distinct from those preceding it in that it was designed with primarily medical, as opposed to industrial, applications in mind.

The operation of the SBSR method can be described, in simple terms, as follows:

1. A series of individual LoRs is considered.
2. The tracer trajectory is modelled as a three-dimensional B-spline [27] of the continuous variable t (time), for which each spatial coordinate comprises a user-defined number N of individual basis functions.
3. Optimisation algorithms are used to determine the weighted sum of the N basis functions which most closely matches the raw PEPT data—i.e. it determines the single path that passes closest to all LoRs in the system.
4. The B-spline yielding the lowest overall error is taken as the tracer's trajectory (figure 6.5).

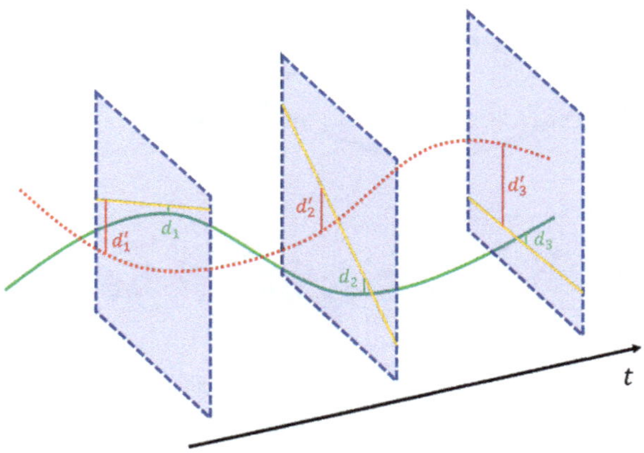

Figure 6.5. Simplified schematic illustrating the methodology SBSR algorithm. The algorithm determines the optimal trajectory by finding the 3D spline which minimised the mean of the squared distances, d, between the spline and the LoRs (orange lines) passed at each point in time. Examples of both a well-optimised (solid green line) and poorly-optimised (dotted red line) trajectory are shown. For ease of visualisation, we consider only two-dimensional space, though in reality the algorithm operates in three dimensions.

The SBSR method in detail

(1) Individual LoRs: The SBSR method is somewhat unique amongst PEPT algorithms in that it doesn't explicitly divide LoRs into samples in the temporal domain, as described for both previous methods. Instead of attempting to divide the temporal domain into a series of slices and, for each slice, attempting to determine the pseudo-instantaneous location of a tracer under the assumption that *dt* is small enough for it to be assumed static, the SBSR method instead, in essence, attempts to follow a continuous stream of LoRs through time and space.

(2) B-spline representation of a trajectory: The trajectory of a tracer moving through space is modelled as a three-dimensional spline [27]—that is, a continuous function formed piece-wise by a series of polynomials. Specifically, the authors represent their tracer trajectories as a B-Spline [27] of the continuous temporal variable t, which models each of the three spatial coordinates (x, y, z) as a weighted sum of N individual spline basis functions, $B_i(t)$, where the value of N can be chosen by the user. A given trajectory, $\vec{r}(t)$, can thus be represented by three sets of basis coefficients, $\vec{a}_i = (a_i^x, a_i^y, a_i^z)$. To put it mathematically,

$$\vec{r}(t) = \sum_{i=1}^{N} \vec{a}_i B_i(t), \; \vec{a}_i = (a_i^x, a_i^y, a_i^z). \tag{6.7}$$

In the above, B_i represents the ith one-dimensional spline basis function and $\vec{a}_i$ the relevant spline coefficients.

(3,4) Optimisation: Assuming that all LoRs in a data set are 'true' (see section 2.3) and that the tracer can reasonably be considered point-like, it follows that the tracer's true trajectory must pass through each LoR in the data set. Based on this fundamental principle, the SBSR algorithm attempts to recreate this trajectory by determining the set of spline coefficients for which the distance between the resultant B-spline and each LoR is minimised. Specifically, the algorithm determines a series of points, $\vec{M}_j$, corresponding to the spatial coordinates of a given B-spline 'candidate trajectory' falling at times t_j corresponding to individual LoR detection events. For each event, the minimum distance, d_j, between the candidate point M_j and the relevant LoR is determined. A cost function quantifying the 'goodness of fit' of the current candidate trajectory is then taken as the mean square distance,

$$I(\vec{a}) = \frac{1}{K} \sum_{j}^{K} d_j^2, \tag{6.8}$$

where K is the total number of LoRs in the data set. The most suitable set of basis coefficients—i.e. the best representation of the particle's true trajectory—can thus be determined by minimising this cost function using standard convex optimisation methods.

Of course, the above represents—as ever—somewhat of a simplification of the real situation. In reality, we must consider also the presence of corrupt data which—as the cost function utilises the square of the distance between LoR and B-spline—can potentially have a significant impact on the algorithm's results. To mitigate the influence of such points, the distance penalty is capped beyond a maximum threshold [28], which is set according to the spatial resolution of the PET scanner. One must also consider the possibility of non-unique solutions to the above-described optimisation problem, which can be especially problematic for experiments with low data rates (as are often experienced in the biomedical applications for which this algorithm was designed [26]). In the case that multiple solutions exist, the algorithm will favour shorter trajectories.

As the SBSR algorithm is IP protected, we do not include an interactive example for this method.

Strengths and weaknesses

Certainly the greatest strength of the SBSR method is that, as alluded to above, it is capable of accurately tracking particles possessing extremely low levels of activity. As such, it is able to track the motion of small particles such as cells—indeed, the technique has recently been applied successfully in the *in vivo* tracking of cancer cells [29]. Like the Birmingham method, the SBSR algorithm is also mesh-free, meaning that it does not suffer from grid-size dependence or possible pixelation effects like the line density method, or indeed other grid-based methods described in the sections below.

The only notable disadvantage of the SBSR method in its current incarnation is that it can only track a single particle.

Strengths
- Capable of tracking low-activity particles.
- Mesh-free method.

Weaknesses
- Currently unable to track multiple tracers.

6.2.10 Voronoi tesselation method

The Voronoi tesselation method in brief

Where previously discussed methods are either entirely mesh-free [1], or utilise a simple, fixed, cubic mesh [4, 6, 17], the Voronoi tesselation method adopts an alternative approach, using the patterns of LoRs within the system to create a mesh whose cell size varies inversely with the local LoR density. This approach, though somewhat less computationally efficient than previously discussed methodologies, carries the potential to provide a high degree of location accuracy.

As previously, since the Voronoi method is predominantly intended as a multiple particle tracking algorithm, we shall introduce the multiple-particle case from the outset.

1. A sample—referred to by the authors of [30] as a 'frame'—of N_s individual, consecutive lines of response from a given PEPT data set is considered (see figure 6.6(a)).
2. Each LoR in the current sample is subdivided into a series of discrete, equidistant points separated by a distance ϵ_{sep} (see figure 6.6(b)).
3. The resultant points in three-dimensional space are used as seed points to create a Voronoi tesselation [31, 32] (see figure 6.6(c,d)).

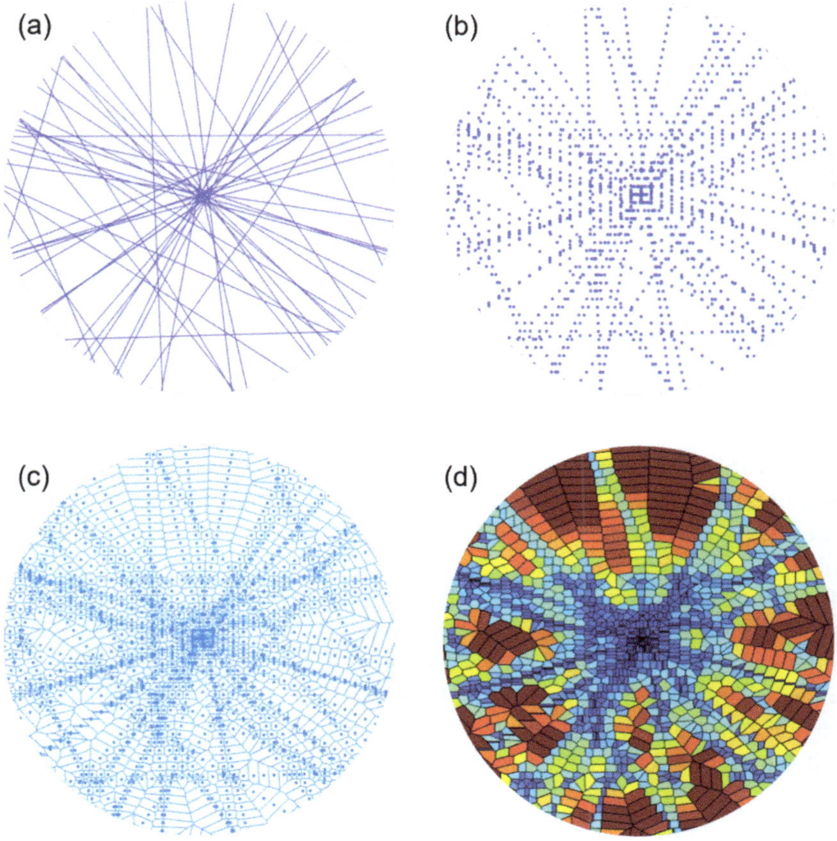

Figure 6.6. (a) Hypothetical sample of LoRs. (b) The same set of LoRs with each separated into a series of discrete, collinear, equidistant points. (c) A Voronoi tesselation using the points from (b) as seeds. (d) The tesselation from (c), with each cell coloured according to its size, as measured by the mean vertex distance, δ_{MVD}. Colours toward the blue (red) end of the spectrum correspond to smaller (larger) cells.

4. For each discretised LoR, the point corresponding to the smallest Voronoi cell is considered as a candidate tracer location and recorded as a 'point under consideration' (PUC).
5. Local filtering removes any outliers in the neighbourhood of each PUC.
6. Global filtering removes any remaining points falling within the upper *n*th percentile of Voronoi cell sizes. This is achieved by:
   (a) Discarding any cells whose size lies more than two standard deviations above the mean.
   (b) Of the data remaining after step 6(a), discarding any cells with sizes more than 0.2 standard deviations above the (new) mean.
   (c) Repeating the above in a trial-and-error process until only the desired fraction of data is eliminated.

7. The DBSCAN clustering algorithm [33, 34] is then applied to the filtered data, dividing the points into distinct clusters, each (if the algorithm has run correctly) corresponding to a distinct tracer.
8. The geometric mean of each cluster is computed, and taken as the centroid location of the correspondent tracer.

The Voronoi tesselation method in detail

(1) Choosing a sample: As with the majority of PEPT algorithms, the first requisite step is to choose a suitably-sized sample of LoRs. Where the authors' previous (line density) method utilised a fixed time slice, in the present algorithm they use instead a fixed sample size, as in the original Birmingham algorithm. As with previous PEPT algorithms, the error on location is observed to decrease as $\frac{1}{\sqrt{N_S}}$, i.e. a larger sample (more LoRs) allows for a more precise tracer location, so long as the tracer does not move significantly during the sample. Thus, the choice of sample size requires—as discussed in greater detail in section 6.2.1—a careful balance between ensuring Ns is large enough to gather suitable statistics, but small enough to avoid motion distortion.

(2) Creating seed points: Following the by now very familiar step (1), the protocol at this point notably diverges from prior methodologies. Specifically, each LoR within the sample is not considered as a single line, but rather 'broken up' into a series of n_p infinitesimal points lying along the relevant line through three-dimensional space, with each pair of consecutive points separated by a distance ϵ_{sep}. That is to say, if a given LoR is defined by the equation

$$\vec{r} = \vec{a} + \lambda(\vec{b} - \vec{a}), \tag{6.9}$$

where $\vec{a}$ and $\vec{b}$ are the vector spatial coordinates representing its start and end-points (i.e. its two interaction points with the PEPT detector system), the desired series of equidistant points can be obtained by setting

$$\lambda \equiv \lambda_i = \frac{i\epsilon_{\text{sep}}}{|\vec{b} - \vec{a}|} \tag{6.10}$$

where $i \in \mathbb{Z}$, $0 \leqslant i \leqslant n_p$ and $\lambda \in \mathbb{R}$, $0 \leqslant \lambda \leqslant 1$. The number of points lying along a given LoR can be determined simply as:

$$n_p = \left\lceil \frac{|\vec{b} - \vec{a}|}{\epsilon_{\text{sep}}} \right\rceil \tag{6.11}$$

where the ceiling function $\lceil \chi \rceil$ is used to denote the rounding-up of a value χ to the nearest integer. In the original work of Blakemore *et al* [30] it was determined heuristically that near-optimal results could be achieved by setting the separation distance to the tracer particle radius, i.e. $\epsilon_{\text{sep}} = \frac{d}{2}$. It is important to note, however, that while this is the case for the specific tracers and systems used by the authors, it may not necessarily hold true for all systems. It was also noted by the authors that—

in the practically inevitable case that LoRs differ in length—the use of a constant ϵ_{sep} was found to produce better results than a constant n_p.

(3) Voronoi tesselation: The points in three-dimensional space created in step (2) are used as seed points to produce a three-dimensional Voronoi tesselation. Clearly (as we have discussed in previous sections) one may expect the line density—and thus, in this case, the point density—to increase in the vicinity of a tracer, where large numbers of LoRs intersect. Thus, one may expect to find the smallest Voronoi cells in the region of a tracer and, conversely, the regions furthest from the tracer to yield the largest cells. In the following steps, this cell size information is exploited in order to filter the data, and ultimately determine the tracer position(s).

(4) Finding candidate locations: As noted in step (3), one may expect the size of Voronoi cells to decrease as we approach a tracer. Taking this logic to its natural conclusion, the centroid position of a tracer should thus correspond to a (local) minimum in cell size. The algorithm therefore scans along each LoR in the current sample, finding the smallest Voronoi cell lying in its path. The centroid position of this cell is then recorded as a 'point under consideration' (PUC), i.e. a candidate for a tracer location. This process is repeated for each LoR in the system, thus yielding up to N_s PUCs.

In order to perform the above process, we need to determine a rigorous definition of the size of a Voronoi cell which—due to their inherently complex polyhedral geometries—is not as trivial a task as it may sound. The authors of [30] utilise a metric which they term the 'mean vertex distance', δ_{MVD}, defined as the mean distance from the cell's geometric centroid, $\vec{c}$, to each of its n_V vertices, $\vec{V}_i$, i.e.

$$\delta_{\text{MVD}} = \frac{1}{n_V} \sum_{i=1}^{n_V} \left| \vec{V}_i - \vec{c} \right|, \tag{6.12}$$

where the centroid position $\vec{c}$ can be determined from

$$\vec{c} \frac{1}{n_V} \sum_{i=1}^{n_V} \vec{V}_i. \tag{6.13}$$

(5) Local filtering: The extensive filtering process adopted by the Voronoi method begins by attempting to remove any of the PUCs recorded in step (4) which represent local outliers. This is performed using the 'Local Outlier Factor' (LOF) framework of Breunig *et al* [35] to assign to each value (here, each Voronoi cell) in a data set a numerical value quantifying its 'outlier-ness' relative to other data in its local neighbourhood[9]. PUCs whose LOF (outlier-ness) values lie in the top X% of all recorded values are assumed to be outliers and thus discarded. In the work of Blakemore *et al* [30], the threshold value is set to $X = 50\%$, though as with previously-discussed parameters, the optimal value may vary with differing systems, tracers, detectors, and combinations thereof.

[9] The size of the neighbourhood can be chosen by the user, though the authors of [30] use a fixed value of $k_{\text{LOF}} = 4$ as the nearest-neighbour search parameter.

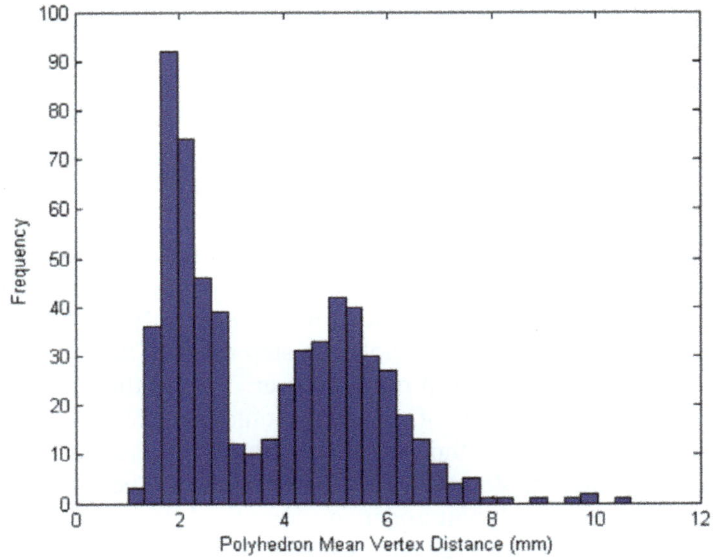

Figure 6.7. Typical example of the bimodal distribution of Voronoi cell sizes (mean vertex distances) following the LOF filtering process described in step (5). Image reproduced from [30], Copyright 2019, with permission from Elsevier.

(6) Global filtering: Once local filtering has been applied during step (5), additional data-cleaning is performed on the global scale. The PUC data remaining after step (5) are observed to exhibit a bimodal distribution of Voronoi cell sizes (or, more specifically, mean vertex distances), as illustrated in figure 6.7. The global filtering step corresponds simply to removing all Voronoi cells whose volumes lie in a range corresponding to the greatest 40%[10] of those remaining after the initial LOF filtering step. This is achieved through a trial and error process, first removing any cells with a size δ_{MVD} more than two standard deviations greater than the current mean volume, and subsequently repeating the process for the remaining cells, this time removing any whose values lie more than 0.2 standard deviations above the new mean.

(7) Clustering: Once noise within the data has been reduced using the above-described filtering processes, the penultimate step of the process is to divide the remaining data into a series of distinct clusters, each (if all has worked correctly!) corresponding to a tracer particle. The clustering process used by Blakemore *et al* [30] differs from the approach of Wiggins *et al* [6] in two main manners: firstly, the former clusters data points corresponding to the above-discussed Voronoi seeds as opposed to the weighted voxel centres used in the latter. Secondly, Blakemore *et al* [30] utilise the DBSCAN (Density-based spatial clustering of applications with noise) algorithm [33, 34, 36] as opposed to G-means clustering in order to partition

[10] As with other user-defined parameters, the authors choose a specific value here of 40%, though the optimal value is likely to vary dependent on a variety of factors relating to the physical system, detector system, and tracer properties.

the data. The main, generic advantage of the DBSCAN algorithm over G-means is that it is somewhat less sensitive to noise, and better at handling outlier points [33, 34, 36] [11]. A more specific advantage of using DBSCAN in this particular case is that the two input parameters required—k_{min}, the minimum number of neighbours required for a given point to be considered a 'core data point' and ϵ, the maximum separation at which two points can be considered to belong to the same cluster—correspond closely to the two parameters k_{LOF} and ϵ_{sep} discussed above. Indeed, in their implementation the authors of [30] simply use $k_{\text{min}} = k_{\text{LOF}}$ and $\epsilon = \epsilon_{\text{sep}}$.

(8) Tracer location: Once the clusters corresponding to each tracer within the positron camera's field of view have been determined, the centroid positions of said clusters are determined via a simple geometric mean over the relevant data points, and the values computed taken as the spatial locations of said tracers.

Strengths and weaknesses

Like the G-means algorithm discussed in section 6.2.5, the DBSCAN algorithm is designed to autonomously determine the number of clusters in an arbitrary system, and is thus able to autonomously determine the number of tracers in our specific system. Thus, the Voronoi method can be easily applied to the case in which tracers regularly enter and leave the field of view of our detector system.

The algorithm has also been shown to successfully track $N_t = 20$ tracers in experiment. Though other algorithms have exceeded this value in simulation, this is —to date—the largest number of experimentally tracked particles recorded in the literature. It is also worth noting that the quoted upper limit of $N_t = 20$ corresponds to a limitation in the number of tracers available to the authors, not a limitation in the abilities of the algorithm; as such, it is possible (indeed likely) that the algorithm is capable of successfully tracking a greater number of tracers than this figure suggests.

One of the main potential limitations of the code is the large number of free parameters involved—including the sample size, the choice of the separation distance ε_{sep}, the neighbourhood size for local filtering, k_{lof}, the threshold LOF value for local filtering, the threshold cell size for global filtering, and the k and ϵ parameters imposed for the DBSCAN algorithm—all of which may significantly alter the effectiveness and accuracy of the algorithm. Though the authors offer suggestions for optimal choices of these parameters, the vast majority of these are heuristic, and thus cannot necessarily be expected to hold for different tracers, experimental systems, detector systems and combinations thereof. As such, one may expect that the use of the Voronoi methodology on any new system will require significant input in terms of calibration and optimisation.

The algorithm is also found to be highly computationally expensive and, further, the computational cost scales with the number of tracers present as $\mathcal{O}(N_t \log N_t)$. As

[11] The DBSCAN algorithm is also more efficient when working in higher dimensions, though as in the present application we are very much limited to three-dimensional space, this is not a relevant issue!

an example, the authors note that the algorithm required approximately 1 second per frame to locate 2 tracers, and 14 seconds per frame for the case of 20 tracers[12]. In other words, by this measure, an hour-long PEPT experiment with an average frame rate of 10 ms involving $N_t = 20$ tracers would require approximately two months(!) to process. This extreme computational time may potentially be mitigated, however, for example by creating an implementation of the program in a lower-level language[13] and/or parallelising the code.

Strengths

- Can autonomously determine the number of tracers in a system, requiring no *a priori* knowledge.
- Particles entering and leaving the field of view can be successfully tracked.
- Can robustly track at least 20 particles[14].

Weaknesses

- Lower spatial resolution than the Birmingham method for the single particle case.
- Extremely computationally expensive.
- Quality of results reliant on several free parameters

6.2.11 Interactive example: Voronoi tesselation

The notebook files accompanying this section are available in the supplementary data online at https://iopscience.iop.org/book/978-0-7503-3071-8.

Interactive PEPT analysis example using the Voronoi tesselation method [30]

This Jupyter Notebook

An example dataset is used from an experiment run at the University of Birmingham Positron Imaging Centre using the ADAC Forté by Matthew Herald. It consists of a single 1 mm diameter MCC particle activated with fluorine-18 radioactive tracer material inside a bubbling fluidised bed. The fluidised bed was filled with 90% sand and 10% MCC; air was fed into the bottom of the bed at a rate of 37 litres per minute at 3.5 bar. This dataset was chosen for its high quality captured lines of response, with the tracer still depicting the random particle motion that is inherent to bubbling fluidised beds—and typical in Lagrangian particle tracking.

The `pept` Python library is used for initialising and visualising PEPT data. While not required *per se* for illustrating PEPT algorithms' processing steps, it significantly reduces the amount of repetitive code and visual noise, allowing the reader to focus on the main conceptual procedures.

[12] Note that the hardware used for these computations is not specified, nor is it clear whether the authors are referring to wall clock time or CPU time.

[13] The implementation discussed in [30] is written entirely in Python.

[14] Note that the algorithm can most likely track significantly larger numbers of tracers; the quoted figure of 20 corresponds simply to the maximum number of tracers available to the authors.

```
[1]: # Importing necessary libraries for numerical array manipulation and plotting,
     # plus the `pept` library for straightforward PEPT data processing
     import numpy as np
     import matplotlib.pyplot as plt
     import pept

     # This algorithm implementation uses SciPy for the Voronoi computation and
     # SciKit-Learn for the DBSCAN clustering algorithm
     import scipy
     import sklearn

     # Set Matplotlib theme (try "darkgrid" too for a more modern look)
     import seaborn
     seaborn.set_style("whitegrid")

     # Set Plotly figure output (try "browser" too for interactive graphs)
     import plotly
     plotly.io.renderers.default = "png"
```

Initialise raw line of response data

In this initial step, we read in the 'raw' LoR data from which our tracer positions will be calculated. As illustrated in the 'Effect of sample size and overlap' notebook, varying the numbers of LoRs used to calculate a PEPT location can strongly affect the accuracy of the final position computed. Try altering the value `nrows` below to see how this influences your results.

```
[2]: # Read in a sample of experimental PEPT data into a NumPy array. Skip the file
     # header's first 15 lines, then read in the first `nrows` LoRs
     lors_raw = pept.utilities.read_csv(
         "Data/FluidisedBed1Tracer.csv",
         skiprows=15,
         nrows=100,
     )

     # Convert the LoRs into the general 3D line format [t, x1, y1, z2, x2, y2, z2].
     # See the `PEPT Data Format` interactive example for detailed explanations

     # ADAC Forte's two detector heads' separation, in mm
     head_separation = 600

     lors = np.insert(lors_raw, 3, 0, axis = 1)                # z1 = 0
     lors = np.insert(lors, 6, head_separation, axis = 1)      # z2 = head_separation

     # Print the first 10 rows from the line of response (LoR) data
     print(f"First 10 rows from the 3D-formatted LoR data:\n{lors[:10]}\n")
```

```
First 10 rows from the 3D-formatted LoR data:
[[0.000e+00 1.900e+02 1.687e+02 0.000e+00 3.463e+02 1.428e+02 6.000e+02]
 [1.000e-01 2.437e+02 1.676e+02 0.000e+00 3.145e+02 3.139e+02 6.000e+02]
 [1.000e-01 1.941e+02 4.100e+02 0.000e+00 4.171e+02 2.401e+02 6.000e+02]
 [2.000e-01 2.885e+02 2.962e+02 0.000e+00 3.847e+02 4.525e+02 6.000e+02]
 [2.000e-01 1.646e+02 1.151e+02 0.000e+00 3.829e+02 3.534e+02 6.000e+02]
 [2.000e-01 1.286e+02 1.322e+02 0.000e+00 4.649e+02 2.661e+02 6.000e+02]
 [2.000e-01 3.339e+02 3.735e+02 0.000e+00 3.316e+02 1.310e+02 6.000e+02]
 [3.000e-01 1.640e+02 1.115e+02 0.000e+00 3.794e+02 3.534e+02 6.000e+02]
 [4.000e-01 2.620e+02 2.094e+02 0.000e+00 2.932e+02 2.808e+02 6.000e+02]
 [4.000e-01 1.304e+02 2.749e+02 0.000e+00 2.543e+02 3.062e+02 6.000e+02]]
```

The `pept.plots` subpackage provides two PEPT-oriented interfaces to the Plotly visualisation library—`PlotlyGrapher` and `PlotlyGrapher2D`—containing straightforward plotting functions for PEPT data. Let's plot the lines of response from the `lors` array:

```
[3]: # Plot the 3D LoRs that were initialised
     from pept.plots import PlotlyGrapher

     grapher = PlotlyGrapher()
     grapher.add_lines(lors)
     grapher.show()
```

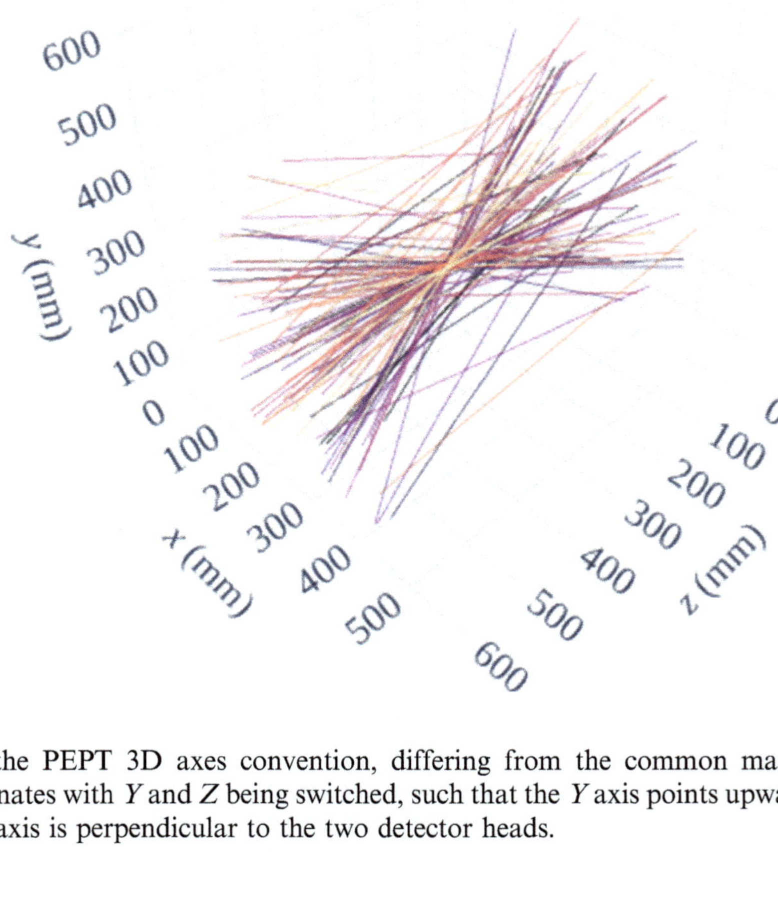

Note the PEPT 3D axes convention, differing from the common mathematical coordinates with Y and Z being switched, such that the Y axis points upwards, while the Z axis is perpendicular to the two detector heads.

Discretise the lines of response

The next step is to discretise the lines of response into equidistant points along each line, which will serve as the *seeds* for the Voronoi tesselation. In the original paper all points are separated by a distance equal to the tracer radius, $\epsilon_{sep} = r_{tracer}$ [1].

[4]:
```
# Discretise each LoR into equidistant points along each line, separated by
# `eps` mm (corresponding to the tracer radius)
eps = 0.5

# Write LoRs as line vectors: L(t) = S + t U
origin = lors[:, [1, 2, 3]]
direction = lors[:, [4, 5, 6]] - lors[:, [1, 2, 3]]

# For each LoR, compute the discretised points and append each array of discrete
# points to a list (whose length will therefore be equal to the number of LoRs)
discrete_lors = []
```

```
for i in range(len(lors)):
    # Compute the discrete_t values which yield equidistant points on
    # L(t) = S + t U
    discrete_ts = np.arange(0, 1, eps / np.linalg.norm(direction[i]))
    points = origin[i] + discrete_ts[:, np.newaxis] * direction[i]

    # Insert the LoR's timestamp into the points and add them to the list
    points = np.insert(points, 0, lors[i, 0], axis=1)
    discrete_lors.append(points)

all_discrete_lors = np.vstack(discrete_lors)
```

[5]:
```
# Plot all LoRs and all voxels with non-zero values (i.e. more than one LoR
# passing through)
PlotlyGrapher().add_points(all_discrete_lors).show()
```

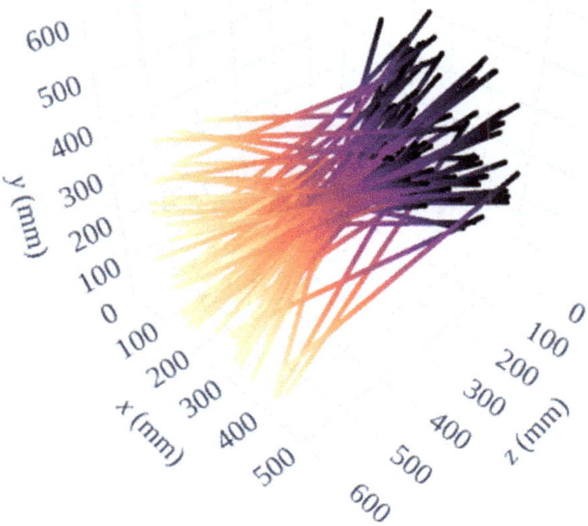

Voronoi tesselation

The namesake of the algorithm, and the heart of it, is represented by the Voronoi tesselation of the discretised LoRs' points. A Voronoi tesselation is a partitioning of a plane (in 2D), space (in 3D), or hyperspace into regions—called Voronoi cells—that are closest to a set of points—called seeds. In other words, each seed point yields a cell whose interior points are closer to the seed than to any other.

For ease of visualisation, let's first compute the Voronoi tesselation for the 2D *YZ* projection of the discretised LoRs' points:

```
[6]: # Plot the Voronoi tesselation of the 2D projection of the discretised LoRs
     from scipy.spatial import Voronoi, voronoi_plot_2d

     tesselation2d = Voronoi(all_discrete_lors[:, [3, 2]])

     fig = voronoi_plot_2d(tesselation2d, show_vertices=False, show_points=False,
                           line_width=0.05)
```

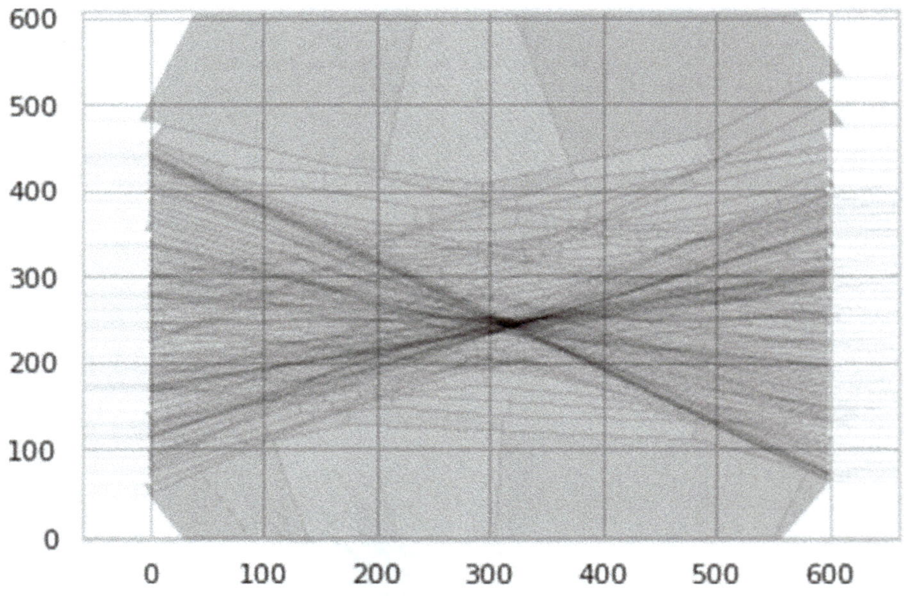

Notice that the cells become increasingly small in the regions where the LoRs become more concentrated. Now compute the equivalent Voronoi tesselation for the full 3D discretised LoRs:

```
[7]: tesselation = Voronoi(all_discrete_lors[:, 1:4])
```

Gather points under consideration (PUCs)

Now for each LoR, find the geometric centre of the cell with the smallest mean distance from each cell's vertices to the initial seed point on the LoR.

[8]:
```
# Find Points under Consideration - this process is rather involved, so follow
# the explicit steps done in the code below

# List of the number of points for each discretised LoR
lor_num_points = [len(discrete_lor) for discrete_lor in discrete_lors]

# Pre-allocate array of PUCs and array of mean vertex distances for each PUC
pucs = np.full((len(discrete_lors), 4), np.nan)
puc_distance = np.full(len(discrete_lors), np.nan)

# Find the PUC for each discretised LoR
for i in range(len(discrete_lors)):
    # The seeds used in the tesselation (`all_discrete_lors`) are just all the
    # discrete points, for all LoRs, stacked. Find the starting point index of
    # the current LoR in that stacked array
    start_index = sum(lor_num_points[:i])
    discrete_lor = discrete_lors[i]

    # For each point on the discretised LoR, find the mean distance to each
    # Voronoi region's vertex. Keep track of each Voronoi cell's geometric
    # centroid
    #
    # Pre-allocate array for the mean distances and cell centroids
    mean_distances = np.full(len(discrete_lor), np.nan)
    cell_centroids = np.full((len(discrete_lor), 3), np.nan)

    for j in range(len(discrete_lor)):
        # The index of the Voronoi region (cell) around the current point
        region_index = tesselation.point_region[start_index + j]

        # The indices of the vertices forming the Voronoi region
        vertices_indices = tesselation.regions[region_index]

        # If a vertex index -1 is present, then the region might be unbounded -
        # this happens around the edges; skip this point
        if -1 in vertices_indices:
            continue

        # Select the vertices around this Voronoi region and compute the
        # distance from each vertex to the seed point on the discrete LoR
        vertices = tesselation.vertices[vertices_indices]
        distances = np.linalg.norm(vertices - discrete_lor[j, 1:4], axis=1)

        # Compute the mean vertex distance and the Voronoi cell's centroid
        mean_distances[j] = distances.mean()
        cell_centroids[j, :] = vertices.mean(axis=0)

    # If all vertices were NaNs (due to the LoR being very far), then this
    # LoR does not have a PUC, and the corresponding distance is infinity
    if np.all(np.isnan(mean_distances)):
        pucs[i, :] = np.nan
```

```
            puc_distance[i] = np.inf
            continue

        # Find this LoR's seed point index with the smallest mean vertex distance
        seed_index = np.nanargmin(mean_distances)

        # The Point Under Consideration is then the corresponding Voronoi cell's
        # centroid. The PUC's timestamp corresponds to the LoR's time
        pucs[i, 1:] = cell_centroids[seed_index]
        pucs[i, 0] = lors[i, 0]

        # Record the mean vertex distance for this PUC as it will be used in the
        # next local filtering step
        puc_distance[i] = mean_distances[seed_index]

# Remove NaNs from the PUCs
nanfilter = np.isfinite(puc_distance)
pucs = pucs[nanfilter]
puc_distance = puc_distance[nanfilter]

print(f"First 10 PUCs found:\n{pucs[:10]}")
```

```
First 10 PUCs found:
[[0.00000000e+00 3.10493648e+02 1.47930528e+02 4.68299663e+02]
 [1.00000000e-01 2.80702401e+02 2.44883498e+02 3.16655536e+02]
 [1.00000000e-01 3.59438430e+02 2.85447483e+02 4.41370457e+02]
 [2.00000000e-01 3.14247922e+02 3.42318510e+02 1.74183996e+02]
 [2.00000000e-01 2.80559488e+02 2.41325863e+02 3.17723753e+02]
 [2.00000000e-01 3.03398531e+02 2.02783048e+02 3.07882137e+02]
 [2.00000000e-01 3.32080585e+02 1.89669678e+02 4.57997536e+02]
 [3.00000000e-01 2.79491855e+02 2.40991468e+02 3.21712155e+02]
 [4.00000000e-01 2.78461813e+02 2.46668653e+02 3.12348716e+02]
 [4.00000000e-01 1.55359231e+02 2.81177473e+02 1.04939343e+02]]
```

```
[9]: # Plot 3D LoRs and the corresponding PUCs found, colour-coded by their
     # corresponding mean vertex distance
     grapher = PlotlyGrapher(cols=2)

     grapher.add_lines(lors)
     grapher.add_points(pucs, color=puc_distance, col=2)

     grapher.show()
```

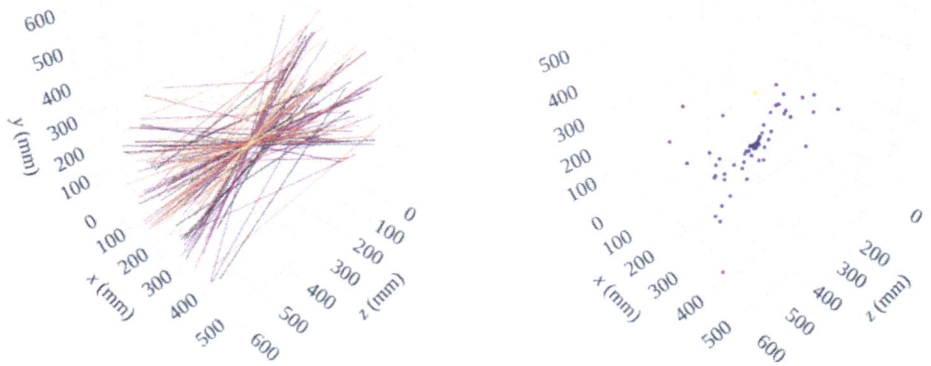

Local filtering based on the local outlier factor

Now filter the PUC based on the LOF of Breuning *et al*—then select all PUCs whose mean vertex distance's LOF is below 50%. Here, the popular `sklearn` machine-learning library's LOF implementation is used:

```
[10]: # Select only PUC's whose corresponding mean vertex distance's LOF is lower
      # than `threshold` - which is normally 0.5
      from sklearn.neighbors import LocalOutlierFactor

      threshold = 0.5
      lof_clusterer = LocalOutlierFactor(n_neighbors = 4, contamination=threshold)

      # The `LocalOutlierFactor` class automatically labels each PUC with 1 (inlier)
      # and 0 (outlier)
      labels = lof_clusterer.fit_predict(puc_distance[:, np.newaxis])

      # Select only inlier PUCs
      pucs_inlier = pucs[labels == 1]
      puc_inlier_distance = puc_distance[labels == 1]
```

```
[11]: # Plot all initial PUCs, then the locally-filtered PUCs
      grapher = PlotlyGrapher(cols=2)

      grapher.add_points(pucs, color=puc_distance)
      grapher.add_points(pucs_inlier, color=puc_inlier_distance, col=2)

      grapher.show()
```

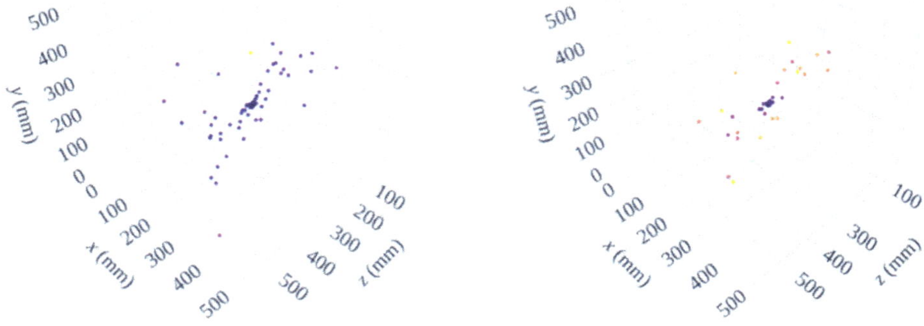

Global filtering

The final filtering step is applied to the remaining PUCs which have been seen to form a bimodal distribution: first, all PUCs whose mean vertex distance is more than two standard deviations larger than the mean are removed, then all remaining PUCs whose mean vertex distances are more than 0.2 times the standard deviation larger than the mean are removed.

[12]:
```
# Plot a histogram of the inlier (locally-filtered) PUC distances
fig, ax = plt.subplots()

ax.hist(puc_inlier_distance, density=True)
ax.set(xlabel="Mean Vertex Distance", ylabel="Frequency");
```

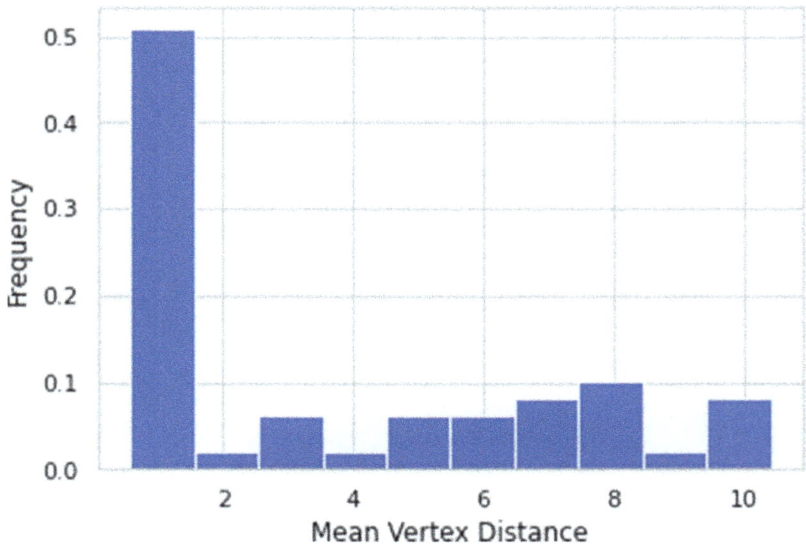

For this particular dataset, the second global filtering substep—the removal of all remaining PUCs whose mean vertex distances is more than 0.2 times the standard deviation larger than the mean—removes all remaining PUCs. For completeness, this substep is included in the code cell below, but it is commented out.

```
[13]: # Remove PUC's with mean vertex distances larger than 2 std-devs
std_dev = puc_inlier_distance.std()
puc_mean = puc_inlier_distance.mean()
selection_indices = (
    (puc_mean - 2 * std_dev < puc_inlier_distance) &
    (puc_mean + 2 * std_dev > puc_inlier_distance)
)

pucs_filtered1 = pucs_inlier[selection_indices]
pucs_distances_filtered1 = puc_inlier_distance[selection_indices]

# Remove PUC's with mean vertex distances larger than 0.2 std-devs
# std_dev = pucs_distances_filtered1.std()
# puc_mean = pucs_distances_filtered1.mean()
# selection_indices = (
#     (puc_mean - 0.2 * std_dev < pucs_distances_filtered1) &
#     (puc_mean + 0.2 * std_dev > pucs_distances_filtered1)
# )
#
# pucs_filtered2 = pucs_filtered1[selection_indices]
# pucs_distances_filtered2 = pucs_distances_filtered1[selection_indices]
```

```
[14]: # Plot the locally-filtered PUCs, then the globally-filtered ones
grapher = PlotlyGrapher(cols=2)

grapher.add_points(pucs_inlier, color=puc_inlier_distance)
grapher.add_points(pucs_filtered1, color=pucs_distances_filtered1, col=2)

grapher.show()
```

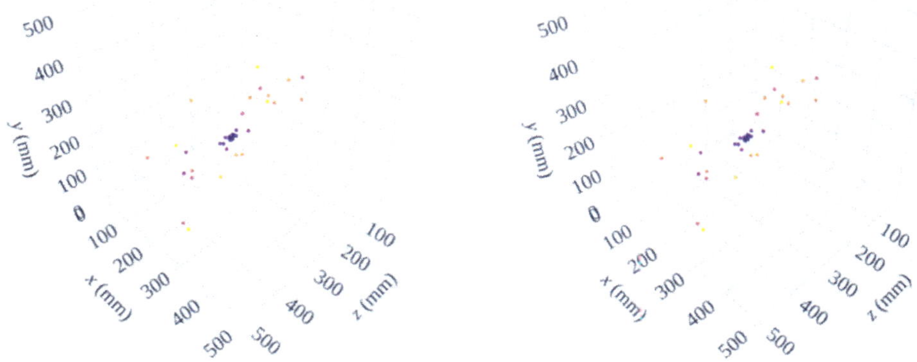

Clustering the remaining PUCs

Finally, the PUCs are *clustered* to reveal the points concentrated around the tracer location using the density-based spatial clustering of applications with noise (DBSCAN). The two most important clustering parameters are by default set based on two previously-used constants: ϵ, the maximum distance between two points to be considered as part of a cluster is set to $\epsilon_{\text{tracer}} = r$, the tracer diameter; the minimum number of neighbours required for a point to be considered a 'core data point', k_{min} is set to the value 4 that was used before.

Again, the feature-complete `sklearn` Python library is used for the specific DBSCAN implementation. Note that the default `eps` value of 0.5 yields no clusters, but a larger value finds the tracer in the code below—hence the importance of varying the parameters used!

[15]:
```
# Cluster the remaining PUCs using the DBSCAN library
from sklearn.cluster import DBSCAN

eps = 5
neighbors = 4

clusterer = DBSCAN(eps=eps, min_samples=neighbors)
labels = clusterer.fit_predict(pucs_filtered1[:, 1:4])

# Append the labels as the last column in the `pucs_filtered1`
pucs_filtered1 = np.c_[pucs_filtered1, labels]

print(f"Clustered PUCs' labels:\n{labels}")
```

```
Clustered PUCs' labels:
[ 0 -1  0 -1  1 -1  1  1  1 -1  0 -1  0  0 -1 -1 -1 -1  1 -1 -1 -1 -1  1
 -1 -1 -1  0  0 -1  1 -1 -1  0 -1 -1 -1  0 -1 -1 -1 -1 -1 -1  0  0 -1 -1
 -1 -1]
```

The labels are cluster IDs, indexed starting from zero (i.e. 0 for the first cluster, 1 for the second cluster, etc), with the value –1 representing noise. Let's plot the filtered PUCs colour-coded by their label:

[16]:
```
# Plot the clustered PUCs
grapher = PlotlyGrapher()

# Plot noise in black
grapher.add_points(pucs_filtered1[labels == -1], color="black")

# Plot each cluster separately
for i in range(labels.max() + 1):
    grapher.add_points(pucs_filtered1[labels == i])

grapher.show()
```

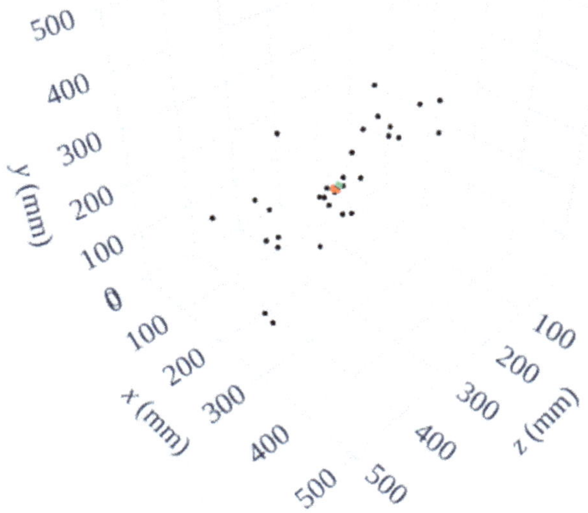

Extract final tracer locations

The final step is to take each cluster of PUCs and compute their centroid to yield the final tracer location; the centroid is simply the column- (or dimension-) wise mean:

[17]:
```
# For each cluster, extract the geometric centroid of PUCs as tracer locations
centres = []

for i in range(labels.max() + 1):
    centres.append(pucs_filtered1[labels == i].mean(axis=0))

centres = np.array(centres)
print(f"Final tracer locations found:\n{centres}")
```

```
Final tracer locations found:
[[  1.76363636 282.0155202  243.63801098 321.39241      0.        ]
 [  1.05714286 281.17928543 246.80192179 312.04081042   1.        ]]
```

[18]:
```
# Plot the LoRs used, the filtered PUCs and the final tracer location
grapher = PlotlyGrapher(
    cols=2,
    subplot_titles=["Sample of LoRs used", "Clustered PUCs and Tracer Location"]
)

grapher.add_lines(lors)
grapher.add_points(pucs_filtered1, col=2)
grapher.add_points(centres, size=5, col=2)

grapher.show()
```

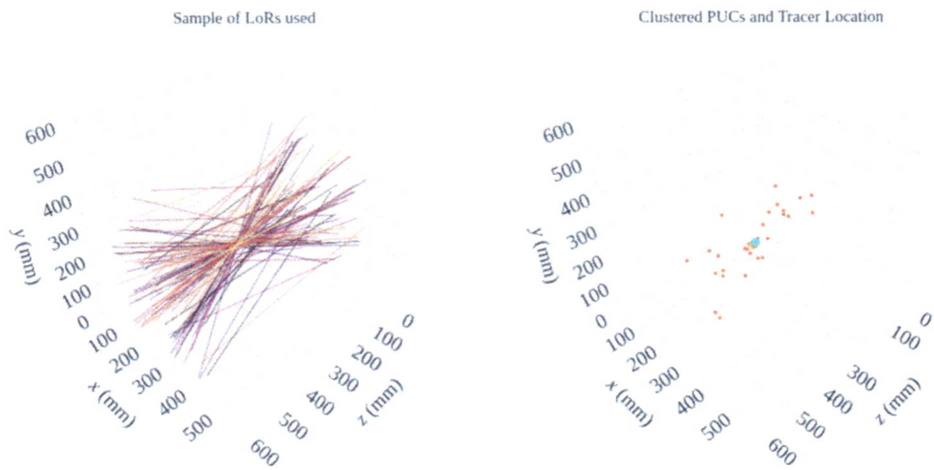

Complete Voronoi tesselation algorithm code

The code developed in the previous sections can be applied to consecutive 'samples' of LoRs to follow the tracer's movement through the fluidised bed, fully tracking its trajectory. A complete Voronoi tesselation algorithm implementation is given below; please note that due to the extreme computational expense of the method (in particular the Voronoi tesselation and PUC extraction steps) only a small number of LoRs can be tracked in a reasonable time.

In the code below we first initialise a larger, multi-sample set of LoRs to analyse; change the `nrows` argument below to read in a different number of LoRs (the file included contains 80,000 LoRs in total), though—as noted above—keep in mind the computational expense of processing large numbers of LoRs!

[19]:
```
# Read in a sample of experimental PEPT data into a NumPy array. Skip the file
# header's first 15 lines and read in the first `nrows` LoRs
lors_raw = pept.utilities.read_csv(
    "Data/FluidisedBed1Tracer.csv",
    skiprows=15,
    nrows=200,
)

# Convert the LoRs into the general 3D line format [t, x1, y1, z2, x2, y2, z2].
# See the `PEPT Data Format` interactive example for detailed explanations

# ADAC Forte's two detector heads' separation, in mm
head_separation = 600

lors = np.insert(lors_raw, 3, 0, axis=1)                    # z1 = 0
lors = np.insert(lors, 6, head_separation, axis=1)          # z2 = head_separation

# Print the first 10 rows from the line of response (LoR) data
print(f"First 10 rows from the 3D-formatted LoR data:\n{lors[:10]}\n")
```

```
First 10 rows from the 3D-formatted LoR data:
[[0.000e+00 1.900e+02 1.687e+02 0.000e+00 3.463e+02 1.428e+02 6.000e+02]
 [1.000e-01 2.437e+02 1.676e+02 0.000e+00 3.145e+02 3.139e+02 6.000e+02]
 [1.000e-01 1.941e+02 4.100e+02 0.000e+00 4.171e+02 2.401e+02 6.000e+02]
 [2.000e-01 2.885e+02 2.962e+02 0.000e+00 3.847e+02 4.525e+02 6.000e+02]
 [2.000e-01 1.646e+02 1.151e+02 0.000e+00 3.829e+02 3.534e+02 6.000e+02]
 [2.000e-01 1.286e+02 1.322e+02 0.000e+00 4.649e+02 2.661e+02 6.000e+02]
 [2.000e-01 3.339e+02 3.735e+02 0.000e+00 3.316e+02 1.310e+02 6.000e+02]
 [3.000e-01 1.640e+02 1.115e+02 0.000e+00 3.794e+02 3.534e+02 6.000e+02]
 [4.000e-01 2.620e+02 2.094e+02 0.000e+00 2.932e+02 2.808e+02 6.000e+02]
 [4.000e-01 1.304e+02 2.749e+02 0.000e+00 2.543e+02 3.062e+02 6.000e+02]]
```

Iterating through each sample of LoRs, locate the tracer and append its location to a growing list of positions. As with the previous cells, you are invited to change the parameters highlighted below; details about each one are given in the comments.

[20]:
```
# Use the Voronoi Tesselation Algorithm to track a moving tracer
from scipy.spatial import Voronoi
from sklearn.neighbors import LocalOutlierFactor
from sklearn.cluster import DBSCAN

# Voronoi Tesselation user-changeable parameters:
#     eps - distance between the discretised LoR points, equal to tracer radius
#     threshold - the local outlier factor/LOF threshold percentage for inliers
#     neighbors - the number of neighbors to search for in the LOF filtering
#     dbscan_eps - the maximum distance between two PUCs to form a cluster
eps = 0.5
threshold = 0.5
neighbors = 4
dbscan_eps = 5.

# General PEPT tracking user-changeable parameters:
#     sample_start - the `lors` index at which to start tracking the tracer
#     sample_size - the number of LoRs used to compute a single location
#     overlap - the number of overlapping LoRs between consecutive samples
sample_start = 0
sample_size = 100
overlap = 0

# Tracked tracer locations will be appended to a growing list
positions = []

while sample_start + sample_size <= len(lors):
    sample = lors[sample_start:sample_start + sample_size]

    # Discretise each LoR into equidistant points along each line, separated by
    # `eps` mm (corresponding to the tracer radius)
    # Write LoRs as line vectors: L(t) = S + t U
    origin = sample[:, [1, 2, 3]]
    direction = sample[:, [4, 5, 6]] - sample[:, [1, 2, 3]]
```

```python
# For each LoR, compute the discretised points and append each array of
# discrete points to a list (whose length will therefore be equal to the
# number of LoRs)
discrete_lors = []

for i in range(len(sample)):
    # Compute the discrete_t values which yield equidistant points on
    # L(t) = S + t U
    discrete_ts = np.arange(0, 1, eps / np.linalg.norm(direction[i]))
    points = origin[i] + discrete_ts[:, np.newaxis] * direction[i]

    # Insert the LoR's timestamp into the points and add them to the list
    points = np.insert(points, 0, sample[i, 0], axis=1)
    discrete_lors.append(points)

# Stack all discretised LoRs' points and compute the Voronoi tesselation
# using their XYZ-coordinates
all_discrete_lors = np.vstack(discrete_lors)
tesselation = Voronoi(all_discrete_lors[:, 1:4])

# Find Points under Consideration - this process is rather involved, so
# follow the explicit steps done in the code below

# List of the number of points for each discretised LoR
lor_num_points = [len(discrete_lor) for discrete_lor in discrete_lors]

# Pre-allocate array of PUCs and array of mean vertex distances for each PUC
pucs = np.full((len(discrete_lors), 4), np.nan)
puc_distance = np.full(len(discrete_lors), np.nan)

# Find the PUC for each discretised LoR
for i in range(len(discrete_lors)):
    # The seeds used in the tesselation (`all_discrete_lors`) are just all
    # the discrete points, for all LoRs, stacked. Find the starting point
    # index of the current LoR in that stacked array
    start_index = sum(lor_num_points[:i])
    discrete_lor = discrete_lors[i]

    # For each point on the discretised LoR, find the mean distance to each
    # Voronoi region's vertex. Keep track of each Voronoi cell's geometric
    # centroid
    #
    # Pre-allocate array for the mean distances and cell centroids
    mean_distances = np.full(len(discrete_lor), np.nan)
    cell_centroids = np.full((len(discrete_lor), 3), np.nan)

    for j in range(len(discrete_lor)):
        # The index of the Voronoi region (cell) around the current point
        region_index = tesselation.point_region[start_index + j]

        # The indices of the vertices forming the Voronoi region
        vertices_indices = tesselation.regions[region_index]

        # If a vertex index -1 is present, then the region might be
        # unbounded - this happens around the edges; skip this point
        if -1 in vertices_indices:
            continue
```

```
            # Select the vertices around this Voronoi region and compute the
            # distance from each vertex to the seed point on the discrete LoR
            vertices = tesselation.vertices[vertices_indices]
            distances = np.linalg.norm(vertices - discrete_lor[j, 1:4], axis=1)

            # Compute the mean vertex distance and the Voronoi cell's centroid
            mean_distances[j] = distances.mean()
            cell_centroids[j, :] = vertices.mean(axis=0)

        # If all vertices were NaNs (due to the LoR being very far), then this
        # LoR does not have a PUC, and the corresponding distance is infinity
        if np.all(np.isnan(mean_distances)):
            pucs[i, :] = np.nan
            puc_distance[i] = np.inf
            continue

        # Find this LoR's seed point index with the smallest mean vertex
        # distance
        seed_index = np.nanargmin(mean_distances)

        # The Point Under Consideration is then the corresponding Voronoi cell's
        # centroid. The PUC's timestamp corresponds to the LoR's time
        pucs[i, 1:] = cell_centroids[seed_index]
        pucs[i, 0] = sample[i, 0]

        # Record the mean vertex distance for this PUC as it will be used in the
        # next local filtering step
        puc_distance[i] = mean_distances[seed_index]

    # Remove NaNs from the PUCs
    nanfilter = np.isfinite(puc_distance)
    pucs = pucs[nanfilter]
    puc_distance = puc_distance[nanfilter]

    # Select only PUCs whose corresponding mean vertex distance's LOF is lower
    # than `threshold` - which is normally 0.5
    lof_clusterer = LocalOutlierFactor(n_neighbors=neighbors,
                                       contamination=threshold)

    # The `LocalOutlierFactor` class automatically labels each PUC with 1
    # (inlier) and 0 (outlier)
    labels = lof_clusterer.fit_predict(puc_distance[:, np.newaxis])

    # Select only inlier PUCs
    pucs_inlier = pucs[labels == 1]
    puc_inlier_distance = puc_distance[labels == 1]

    # Remove PUC's with mean vertex distances larger than 2 std-devs
    std_dev = puc_inlier_distance.std()
    puc_mean = puc_inlier_distance.mean()
    selection_indices = (
        (puc_mean - 2 * std_dev < puc_inlier_distance) &
        (puc_mean + 2 * std_dev > puc_inlier_distance)
    )
```

```
    pucs_filtered1 = pucs_inlier[selection_indices]
    pucs_distances_filtered1 = puc_inlier_distance[selection_indices]

    # Remove PUC's with mean vertex distances larger than 0.2 std-devs
    # std_dev = pucs_distances_filtered1.std()
    # puc_mean = pucs_distances_filtered1.mean()
    # selection_indices = (
    #     (puc_mean - 0.2 * std_dev < pucs_distances_filtered1) &
    #     (puc_mean + 0.2 * std_dev > pucs_distances_filtered1)
    # )
    #
    # pucs_filtered2 = pucs_filtered1[selection_indices]
    # pucs_distances_filtered2 = pucs_distances_filtered1[selection_indices]

    # Cluster the remaining PUCs using the DBSCAN library
    clusterer = DBSCAN(eps=dbscan_eps, min_samples=neighbors)
    labels = clusterer.fit_predict(pucs_filtered1[:, 1:4])

    # For each cluster, extract the geometric centroid of PUCs as tracer
    # locations
    for i in range(labels.max() + 1):
        positions.append(pucs_filtered1[labels == i].mean(axis=0))

    sample_start += sample_size - overlap

positions = np.array(positions)

print(f"First 10 tracked positions:\n{positions[:10]}")
```

```
First 10 tracked positions:
[[  1.76363636 282.0155202  243.63801098 321.39241    ]
 [  1.05714286 281.17928543 246.80192179 312.04081042]
 [  6.07333333 279.50292241 245.65723981 319.38949709]
 [  5.6        280.49227904 242.95837209 305.09506217]]
```

```
[21]: # Plot some of the LoRs used, plus the tracer locations we found
grapher = PlotlyGrapher(
    cols=2,
    subplot_titles=[
        "First 500 LoRs",
        f"Tracer locations tracked from {len(lors)} LoRs",
    ],
)

grapher.add_lines(lors[:500])
grapher.add_points(positions, col=2)

grapher.show()
```

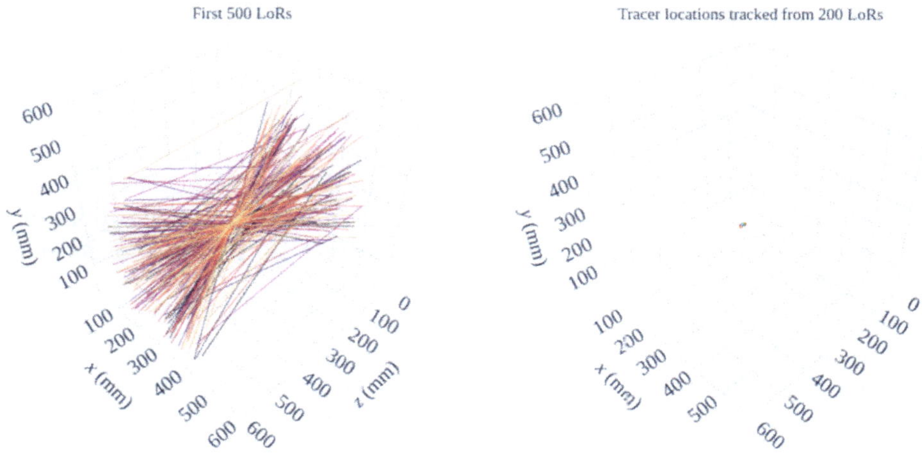

6.2.12 The triangulation method

The triangulation method in brief

The triangulation method of Odo *et al* [37], despite originating from the same group as the line density [4] and Voronoi [30] methods is, at its heart, much more similar to the original Birmingham algorithm than either of these later methods. Like the Birmingham method, it operates by iteratively discarding corrupt LoRs until only supposedly true trajectories remain, using these remaining data to triangulate the tracer position. The main distinguishing factor between the two methods is that where the Birmingham algorithm discards a user-defined fraction of LoRs, Odo's triangulation method utilises *a priori* information regarding the tracer to delineate corrupted and uncorrupted LoRs, thus hypothetically guaranteeing the complete, autonomous elimination of false coincidence events. Though the algorithm has thus far been tested only for the case of simple, spherically-symmetric tracers (discussed further in the next subsection), in the below we provide an overview of the algorithm's general methodology, which may hypothetically be applied to a far wider range of tracer geometries:

1. A set of heuristic conditions are established through which uncorrupted LoRs may be distinguished from corrupted LoRs[15].
2. As with the previously-described PEPT algorithms, the data are subdivided into a series of samples. For each LoR in a given sample, the points of closest approach with all other LoRs in the sample are determined. Any two LoRs passing within a predetermined distance *R*, related to the tracer's size, are considered as 'nearest neighbours'.

[15] The specific conditions for the case of a spherical tracer are discussed in detail in the next subsection.

3. For each LoR, the locations along said LoR of all nearest neighbours' points of closest approach are recorded, and the centroid of these locations determined. Any neighbours whose interaction points lie further than a distance R away from this centroid are assumed corrupt.
4. After the above-described refinement process, the LoR with the largest number of remaining neighbours is taken as the final 'reference LoR'.
5. The tracer centroid is computed as the mean of the midpoint positions of the minimum distance lines connecting the reference LoR to each of its nearest neighbours.

The triangulation method in detail

(1) Establishing criteria for uncorrupted LoRs: As the entire triangulation method is predicated on the ability to autonomously and reliably distinguish 'true' LoRs from 'false' LoRs, our first, central task is to establish a suitably rigorous set of criteria through which this may be achieved. Here, as in the case of the original work of Odo *et al* [37], we simplify matters by considering only the case of spherically-symmetric tracers. In particular, we focus on two distinct cases, both of which are illustrated in figure 6.8: a spherical tracer containing a central 'core' of positron emitting material (figure 6.8(a)) and a similar tracer upon the surface of which a thin layer of β^+-emitting material has been applied. Nonetheless, there is no reason why the general methodology presented here could not in the future be expanded to more complex cases.

For the types of tracer detailed above, a suitable set of heuristic criteria for the identification of true (uncorrupted) LoRs can be outlined as follows:

i. If the mean free path, λ, of positrons within the tracer medium is smaller than the tracer radius (i.e. $\lambda < R$), then one may reasonably assume a higher

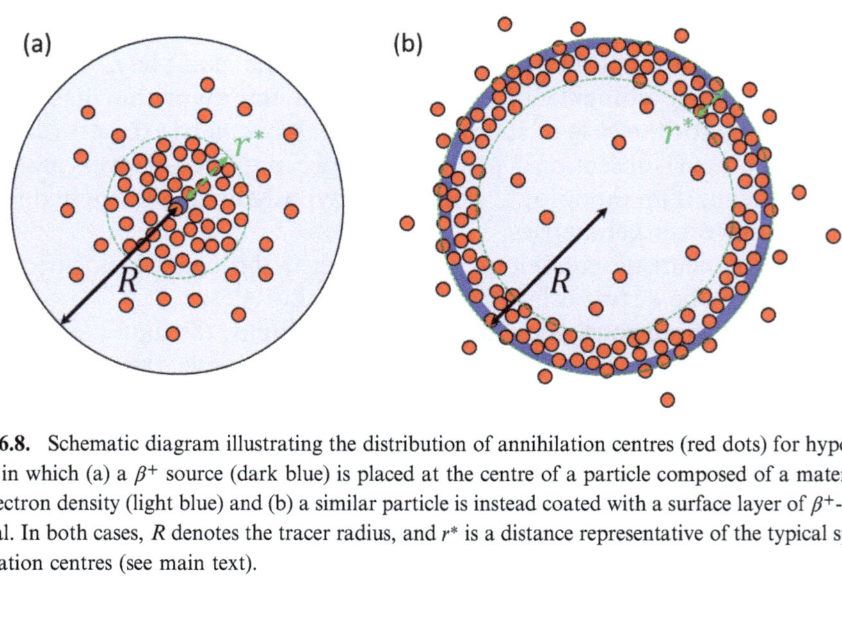

Figure 6.8. Schematic diagram illustrating the distribution of annihilation centres (red dots) for hypothetical tracers in which (a) a β^+ source (dark blue) is placed at the centre of a particle composed of a material with high electron density (light blue) and (b) a similar particle is instead coated with a surface layer of β^+-emitting material. In both cases, R denotes the tracer radius, and r^* is a distance representative of the typical spread of annihilation centres (see main text).

average density of positron-electron annihilation events or 'annihilation centres' within the tracer volume.

ii. For the case of a central core of radioactive material (as illustrated schematically in figure 6.8(a)), it can be assumed that the annihilation centres will be approximately uniformly distributed within a spherical shell of radius r^* centred at the core itself.

iii. For the case of a homogeneous surface layer of active material, if (a) positron emission is equally likely in all directions and (b) the tracer possesses a higher electron density than the surrounding medium, then one may assume that annihilation centres will be predominantly concentrated just inside the surface of the tracer (see figure 6.8(a)).

(2) Finding nearest neighbours: Having established a suitable set of rules which must be obeyed by uncorrupted LoRs, we must now set about finding a manner in which to test whether a given LoR adheres to said rules. Let us consider here the simplest case of a single tracer containing a small core of β^+-emitting material. Following criterion (ii) above, we know that there must exist a region in space for which a maximal concentration of true LoRs all fall within a distance r^* of one another and, further, that this region corresponds to the centre of the tracer volume. In order to determine the location of this region, we first require a metric through which to quantify the distances between LoRs. The minimal distance, $d_{i,j}$ between two LoRs L_i and L_j can be calculated as:

$$d_{i,j} = \frac{\left|\vec{P_i}\vec{P_j} \cdot \vec{n}\right|}{\|\vec{n}\|} \tag{6.14}$$

where $\vec{P_i}$, $\vec{P_j}$ are the points of closest approach of the two lines and $\vec{n}$ is a unit vector perpendicular to both L_i and L_j. Using this metric and the known dimension, R, of our tracer, we can determine—for a given 'reference' LoR, L_{ref}—which other LoRs within a sample satisfy the constraint $d_{\mathrm{ref},i} \leqslant R$. Clearly, if our reference LoR was known to represent the line of closest approach to the tracer's true centre, then any 'nearest neighbour' LoRs satisfying the aforementioned condition could, via criterion (ii), be reasonably assumed to represent uncorrupted LoRs. In reality, however, we cannot identify *a priori* the single LoR in our sample which represents the line of closest approach; instead, we determine the nearest neighbours for all LoRs, and use the a priori knowledge we do possess regarding the tracer size and annihilation centre distribution to refine our search. Figure 6.9 shows, for a hypothetical set of LoRs in two-dimensional space, the 'subgroup' of nearest neighbours belonging to each individual LoR.

(3) Refining the subgroups: From the data presented in figure 6.9(b), it would be tempting to assume that one may eliminate corrupt LoRs based solely on the calculated number of nearest neighbours, as one may intuitively expect that any (corrupt) LoR not passing through the true centre of activity would inherently

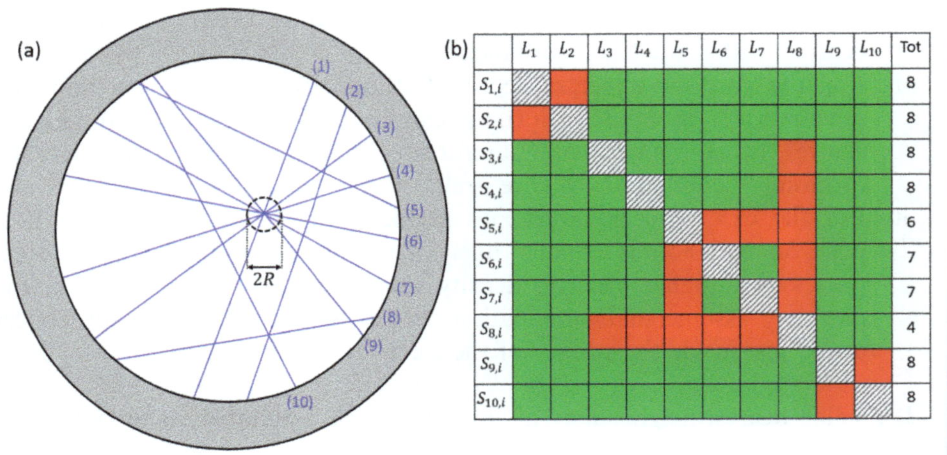

Figure 6.9. (a) Schematic depiction of a sample of $N_s = 10$ LoRs in a 2D plane. (b) table showing, for all LoRs within the sample, which lines L_i can be considered to fall within each subgroup $S_{j,i}$, alongside the total number of nearest neighbours for each line.

encounter relatively few other LoRs. If we look closely at figure 6.9, however, we see that this is not necessarily the case: the subgroup $S_{10,i}$ corresponding to the corrupt line (10), for example, includes as many nearest neighbours as any of the uncorrupted LoRs in the data set—indeed more than lines (6) and (7), both of which pass through the particle centre!

In order to create a more refined estimate of the number of neighbours belonging to all subgroups we begin by determining, for each LoR in our sample, the centroid—i.e. the mean position of all points of closest approach, $\vec{P}_i$, lying along said LoR—is determined as:

$$\vec{C}_i\left(x, y, z\right) = \sum_{j=1}^{n_n} \frac{P_i^j(x, y, z)}{n_n} \tag{6.15}$$

where n_n is the number of nearest neighbours initially calculated. Returning to our heuristic criteria, we can now leverage the *a priori* knowledge of our particle radius, R, and the assumption that our annihilation centres are distributed approximately uniformly within a region of dimension $r^* < R$ to discount any of a given LoR's nearest neighbours lying more than a distance R from said LoR's calculated centroid (see figure 6.10(a)). Figure 6.10(b) shows the remaining nearest neighbours—which we term 'true neighbours'—for each LoR. The revised values clearly show a clearer delineation between 'true' (uncorrupted) and 'false' (corrupted) LoRs.

(4) Identifying the optimal subgroup: From the refined data presented in figure 6.10, it is now reasonable to assume that—so long as there exists a tracer within the detector's field of view—the subgroup $S_{j,i}$ containing the largest number of true neighbours will yield a reasonable estimate of the tracer's true

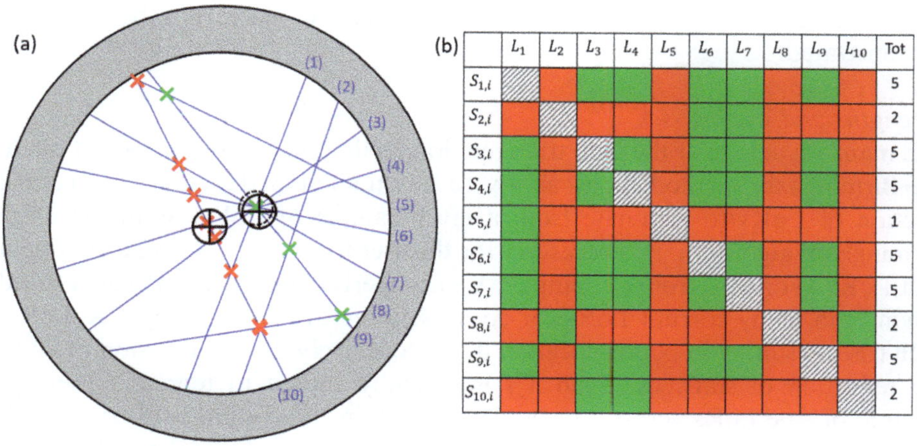

Figure 6.10. (a) Schematic diagram from 6.9(a) updated to show, for the corrupted LoR 10 and the uncorrupted LoR 9, all points of intersection with other LoRs (coloured crosses) as well as the centroid positions of each set of intersection points (black crosses) and the circle of radius R about these points within which intersection points are recorded as 'true neighbours'. Note that both lines possess eight points of intersection, but that in some cases these points overlap. (b) table from 6.9(b) updated to include only 'true neighbours'.

position. If, as in the current case, there exist multiple subgroups possessing an equal, maximal number of true neighbours, then the first such subgroup determined is chosen (as can be seen from figure 6.10(a), this choice will have no real impact on our results, as the centroids of all relevant samples are approximately coincident).

(5) Computing the tracer centroid: With a suitable subgroup of intersecting LoRs determined, the final task is to determine a value for the tracer location from these data. In the original Birmingham method, the tracer location would be simply taken as the minimum distance point of all LoRs in the chosen subgroup[16]. In the currently-described method, however, a slightly more refined approach is taken. Firstly, the minimum distance line (MDL) between the reference LoR of our chosen subgroup and every other line in said subgroup is determined. Secondly, the midpoint of each of the calculated MDLs (commonly referred to as a 'cutpoint') is determined. Finally, the particle centroid location is simply taken as the mean of all cutpoints determined.

It should be noted that the description provided above concerns the case of a single tracer. Though the triangulation method may potentially be extended to the

[16] Note that, in reality, the subgroup will contain a much larger number of LoRs than our highly simplified example!

case of multiple tracers, such an extension has, to date, not been explicitly demonstrated.

Strengths and weaknesses

The main advantage of the Odo triangulation methodology over its closest relation, the Birmingham method, is its ability to autonomously determine the necessary fraction of LoRs to eliminate from a given sample of LoRs. With the original Birmingham algorithm, it is necessary for the user to choose a single, fixed fraction, f, of LoRs to discard from all samples. If improperly chosen, this may lead to either leaving too many corrupt LoRs, or conversely removing too many true LoRs, in both cases resulting in poorer location due, respectively, to the inclusion of noise or the reduction of statistics. Even if f is carefully chosen to represent the typical fraction of true LoRs across a given data set, the relative number of true and false coincidences for each given sample within said data set will still inherently fluctuate leading to sub-optimal results. With the triangulation method, the optimal fraction of LoRs to eliminate (or, equivalently, retain) is determined for each sample individually, ensuring—if it operates as intended—optimal statistics for all individual particle locations computed.

Of course, this approach also inevitably carries certain downsides. Firstly, in order to perform such autonomous optimisation, one requires a considerable amount of *a priori* information regarding the particles used, thus limiting any given implementation of the algorithm to a specific particle geometry and labelling method. If one wishes to apply the technique to a new type of particle, one must formulate a new set of criteria by which to delineate true and false LoRs (see step (1), above). At present, only the relatively simple case of spherical tracers has been explicitly considered. Further, the necessity of determining a new optimal fraction for each timestep (sample) renders this method significantly more computationally expensive than the Birmingham algorithm.

It is finally of importance to note that, unlike other algorithms previously discussed, the accuracy of the triangulation method is directly sensitive to the physical size of the tracer compared to the particle, as one may perhaps anticipate from our description of the methodology above.

Strengths

- With adequate *a priori* information, can autonomously eliminate all corrupted LoRs, thus ensuring maximal accuracy for the final computed tracer centroid.

Weaknesses

- Requires *a priori* knowledge regarding tracer details.
- Currently only applicable to simple, spherical tracers labelled in specific manners.
- Comparatively computationally expensive.

6.2.13 Interactive example: triangulation method

The notebook files accompanying this section are available in the supplementary data online at https://iopscience.iop.org/book/978-0-7503-3071-8.

Interactive PEPT analysis example using the triangulation method [37]

This Jupyter Notebook

An example dataset is used from an experiment run at the University of Birmingham Positron Imaging Centre using the ADAC Forté. It consists of a single 1 mm diameter MCC particle activated with fluorine-18 inside a bubbling fluidised bed. The fluidised bed was filled with 90% sand and 10% MCC; air was fed into the bottom of the bed at a rate of 37 litres per minute at 3.5 bar. This dataset was chosen for its high quality captured lines of response, with the tracer still depicting the random particle motion that is inherent to bubbling fluidised beds—and typical in Lagrangian particle tracking.

The `pept` Python library is used for initialising and visualising PEPT data. While not required *per se* for illustrating PEPT algorithms' processing steps, it significantly reduces the amount of repetitive code and visual noise, allowing the reader to focus on the main conceptual procedures.

```
[1]: # Importing necessary libraries for numerical array manipulation and plotting,
     # plus the `pept` library for straightforward PEPT data processing
     import numpy as np
     import pept

     # Set Plotly figure output (try "browser" too for interactive graphs)
     import plotly
     plotly.io.renderers.default = "png"
```

Initialise raw line of response data

In this initial step, we read in the 'raw' LoR data from which our tracer positions are calculated. As illustrated in the 'Effect of sample size and overlap' notebook, varying the numbers of LoRs used to calculate a PEPT location can affect the accuracy of the final position computed. Try altering the value `nrows` below to see how this influences your results.

```
[2]: # Read in a sample of experimental PEPT data into a NumPy array. Skip the file
     # header's first 15 lines, then read in the first `nrows` LoRs
     lors_raw = pept.utilities.read_csv(
         "Data/FluidisedBed1Tracer.csv",
         skiprows=15,
         nrows=100,
     )

     # Convert the LoRs into the general 3D line format [t, x1, y1, z2, x2, y2, z2].
     # See the `PEPT Data Format` interactive example for detailed explanations

     # ADAC Forte's two detector heads' separation, in mm
     head_separation = 600
```

```
lors = np.insert(lors_raw, 3, 0, axis = 1)                  # z1 = 0
lors = np.insert(lors, 6, head_separation, axis = 1)        # z2 = head_separation

# Print the first 10 rows from the line of response (LoR) data
print(f"First 10 rows from the 3D-formatted LoR data:\n{lors[:10]}\n")
```

```
First 10 rows from the 3D-formatted LoR data:
[[0.000e+00 1.900e+02 1.687e+02 0.000e+00 3.463e+02 1.428e+02 6.000e+02]
 [1.000e-01 2.437e+02 1.676e+02 0.000e+00 3.145e+02 3.139e+02 6.000e+02]
 [1.000e-01 1.941e+02 4.100e+02 0.000e+00 4.171e+02 2.401e+02 6.000e+02]
 [2.000e-01 2.885e+02 2.962e+02 0.000e+00 3.847e+02 4.525e+02 6.000e+02]
 [2.000e-01 1.646e+02 1.151e+02 0.000e+00 3.829e+02 3.534e+02 6.000e+02]
 [2.000e-01 1.286e+02 1.322e+02 0.000e+00 4.649e+02 2.661e+02 6.000e+02]
 [2.000e-01 3.339e+02 3.735e+02 0.000e+00 3.316e+02 1.310e+02 6.000e+02]
 [3.000e-01 1.640e+02 1.115e+02 0.000e+00 3.794e+02 3.534e+02 6.000e+02]
 [4.000e-01 2.620e+02 2.094e+02 0.000e+00 2.932e+02 2.808e+02 6.000e+02]
 [4.000e-01 1.304e+02 2.749e+02 0.000e+00 2.543e+02 3.062e+02 6.000e+02]]
```

The `pept.plots` subpackage provides two PEPT-oriented interfaces to the Plotly visualisation library—`PlotlyGrapher` and `PlotlyGrapher2D`—containing straightforward plotting functions for PEPT data. Let's plot the lines of response from the `lors` array:

```
[3]: # Plot the 3D LoRs that were initialised
     from pept.plots import PlotlyGrapher

     grapher = PlotlyGrapher()
     grapher.add_lines(lors)
     grapher.show()
```

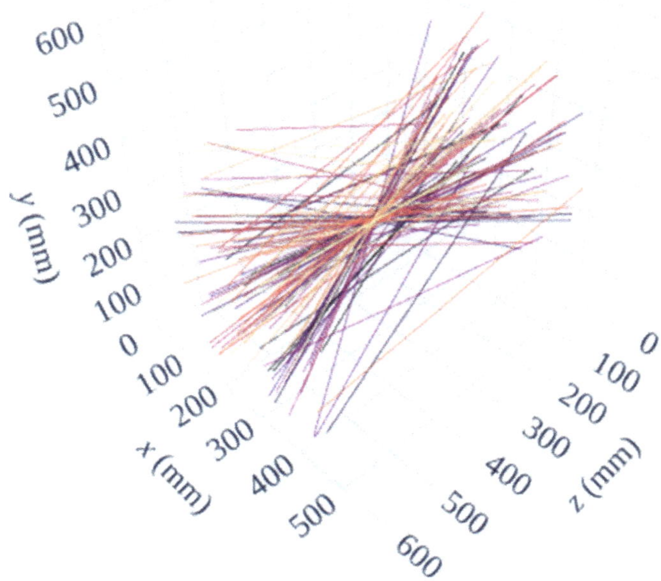

Note the PEPT 3D axes convention, differing from the common mathematical coordinates with Y and Z being switched, such that the Y axis points upwards, while the Z axis is perpendicular to the two detector heads.

Calculate the LoR distance matrix

Having read in our LoR data, the next step is to locate the minimum distances between all LoRs present, and save these distance values as a matrix.

```
[4]: def distance_lines(line1, line2):
         '''Calculate distance between two 3D lines `line1` and `line2`, each defined
         by a timestamp and two 3D points, so the lines are vectors formatted as
         [time, x1, y1, z1, x2, y2, z2].
         '''
         # Direction vectors:
         e1 = line1[4:7] - line1[1:4]
         e2 = line2[4:7] - line2[1:4]

         # Vector perpendicular to both lines
         n = np.cross(e1, e2)
         nd = np.linalg.norm(n)

         # If the lines are parallel, nd is zero
         if nd < 1e-8:
             e1 /= np.linalg.norm(e1)
             dist_vec = np.cross(e1, line2[1:4] - line1[1:4])
             return np.linalg.norm(dist_vec)

         dist = np.dot(n, line1[1:4] - line2[1:4]) / np.linalg.norm(n)
         return np.abs(dist)


     distance_matrix = np.zeros((len(lors), len(lors)))
     for i in range(len(lors)):
         for j in range(i + 1, len(lors)):
             distance_matrix[i, j] = distance_lines(lors[i], lors[j])
             distance_matrix[j, i] = distance_matrix[i, j]

     distance_matrix
```

```
[4]: array([[  0.        ,  46.55313541, 102.72026331, ...,  38.63438816,
             105.24919401,  39.0903317 ],
            [ 46.55313541,   0.        ,  59.12484141, ...,   0.82083753,
               2.65171848,   5.70048779],
            [102.72026331,  59.12484141,   0.        , ...,  49.94452411,
              56.091038  ,  62.61364673],
            ...,
            [ 38.63438816,   0.82083753,  49.94452411, ...,   0.        ,
               6.24979944,   6.08236823],
            [105.24919401,   2.65171848,  56.091038  , ...,   6.24979944,
               0.        ,  34.24092176],
            [ 39.0903317 ,   5.70048779,  62.61364673, ...,   6.08236823,
              34.24092176,   0.        ]])
```

Cluster LoRs closer than the tracer radius

Next, we search through the matrix created in the previous step in order to find sets of LoRs which are all separated by distances smaller than the `tracer_radius`. Following this, any of these sets containing fewer than `min_lors_cluster` lines are removed.

Try exploring different values of `tracer_radius` and `min_lors_cluster` and see the influence they have on the plot below, where discarded and true LoRs are represented, respectively, by dark and bright lines. You will find that a smaller `tracer_radius` and/or a larger `min_lors_cluster` will both give a harsher clustering, as we might expect from our discussion in the main text.

[5]:
```
# "Cluster" all LoRs that are closer than `tracer_radius` mm and form clusters
# with more than `min_lors_cluster` LoRs
from hdbscan import RobustSingleLinkage

tracer_radius = 0.5
min_lors_cluster = 7

single_linkage = RobustSingleLinkage(tracer_radius, min_lors_cluster,
                                     metric = "precomputed")
labels = single_linkage.fit_predict(distance_matrix)

print(f"Labels found for each clustered LoR:\n{labels}")
```

```
Labels found for each clustered LoR:
[-1 -1 -1 -1 -1 -1 -1 -1  0 -1 -1 -1  0 -1 -1 -1 -1 -1 -1 -1 -1 -1 -1 -1
 -1 -1 -1 -1  0 -1 -1 -1 -1 -1 -1 -1  0 -1 -1 -1 -1 -1 -1 -1 -1 -1 -1 -1
 -1 -1 -1 -1  0 -1 -1 -1 -1 -1 -1 -1 -1 -1 -1 -1 -1 -1 -1 -1 -1 -1 -1 -1
 -1 -1 -1 -1 -1 -1 -1 -1 -1 -1 -1 -1 -1 -1 -1 -1 -1 -1 -1 -1 -1 -1 -1 -1
 -1 -1 -1 -1]
```

Labels are integer IDs assigned starting from 0 (label 0 is the first cluster, label 1 is the second, etc) with –1 representing noise. Each cluster of LoRs is plotted separately below:

[6]:
```
# Plot LoRs separated by the label
from pept.plots import PlotlyGrapher

grapher = PlotlyGrapher()

# Add LoRs classified as noise (i.e. label = -1)
lors_noise = lors[labels == -1]
grapher.add_lines(lors_noise, color="black")

# Add the LoRs that are part of each label in turn
for i in range(labels.max() + 1):
    lors_cluster = lors[labels == i]
    grapher.add_lines(lors_cluster)

grapher.show()
```

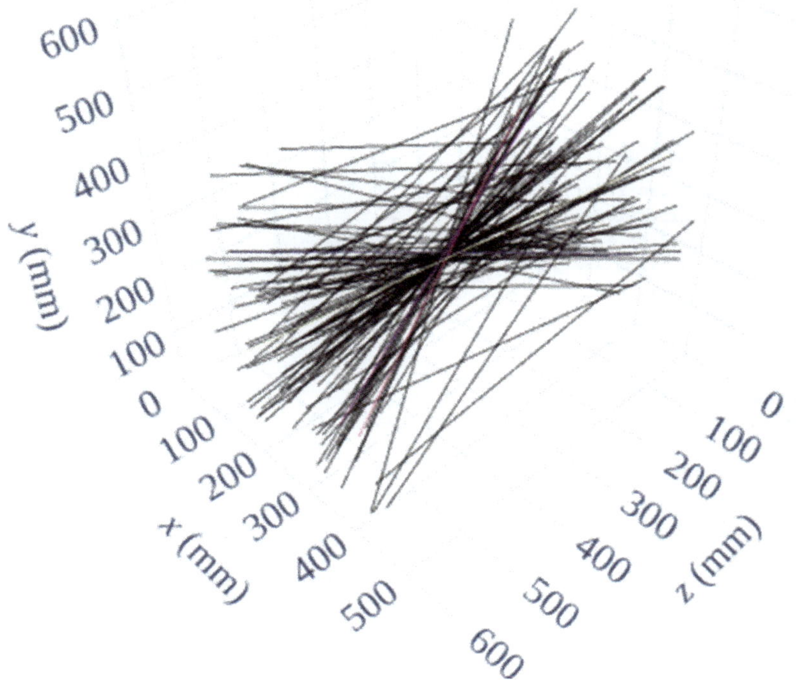

Find centroids of clustered LoRs' cutpoints

Now that we have in essence clustered our LoRs, our final step is to locate a particle centroid from this information.

To do so, we find the point of closest approach for each pair of LoRs and take the midpoint between the two lines at this point as a 'cutpoint'. The centroid (i.e. geometric mean position) of this constellation of cutpoints is then taken as our tracer position.

```
[7]: # For each label, collect the associated LoRs and compute the cutpoints
     from pept.tracking.peptml import find_cutpoints

     positions = []
     for label in range(labels.max() + 1):
         cluster = lors[labels == label]
         cutpoints = find_cutpoints(cluster, max_distance=np.inf).points
         centre = cutpoints.mean(axis=0)

         positions.append(centre)

     positions = np.array(positions)
     print(f"Tracer locations found as the cutpoints' centroids:\n{positions}")
```

```
Tracer locations found as the cutpoints' centroids:
[[  1.06       278.04016708 244.96323205 308.03406612]]
```

```
[8]: # Plot the clustered LoRs and the cutpoints for each cluster, plus the centres
grapher = PlotlyGrapher(
    cols = 2,
    subplot_titles = [
        "Clustered LoRs",
        "Clustered LoRs' Cutpoints and Centroid",
    ],
)

# Add LoRs classified as noise (i.e. label = -1)
lors_noise = lors[labels == -1]
grapher.add_lines(lors_noise, color="black")

# Add the LoRs that are part of each label in turn
for i in range(labels.max() + 1):
    lors_cluster = lors[labels == i]
    grapher.add_lines(lors_cluster)

grapher.add_points(cutpoints, col = 2)
grapher.add_points(positions, size = 5, col = 2)

grapher.show()
```

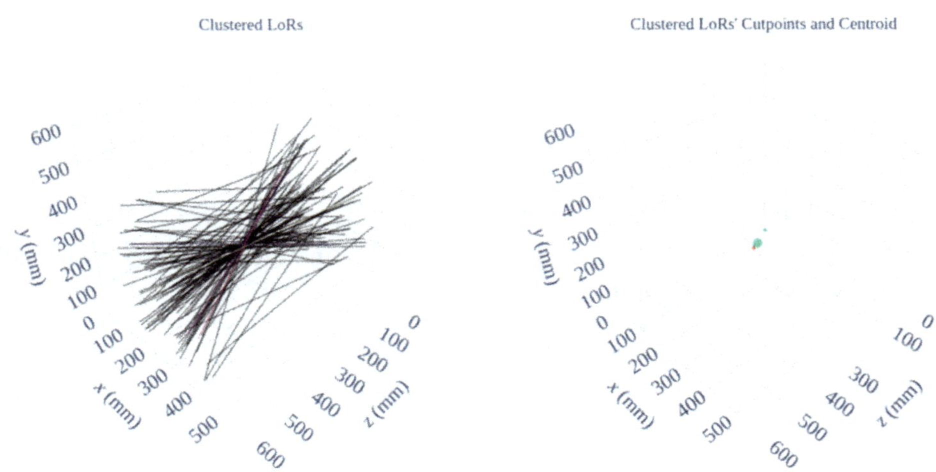

Complete triangulation method code

The code developed in the previous sections can be applied to consecutive 'samples' of LoRs to follow the tracer's movement through the fluidised bed, fully tracking its trajectory. A complete—albeit slow due to Python's dynamic nature—triangulation method implementation is given below.

First initialise a larger, multi-sample set of LoRs to analyse; change the `nrows` argument below to read in a different number of LoRs (the file included contains 80,000 LoRs in total), though keep in mind the computational expense of processing large numbers of LoRs!

```
[9]: # Read in a sample of experimental PEPT data into a NumPy array. Skip the file
     # header's first 15 lines and read in the first `nrows` LoRs
     lors_raw = pept.utilities.read_csv(
         "Data/FluidisedBed1Tracer.csv",
         skiprows=15,
         nrows=5000,
     )

     # Convert the LoRs into the general 3D line format [t, x1, y1, z2, x2, y2, z2].
     # See the `PEPT Data Format` interactive example for detailed explanations

     # ADAC Forte's two detector heads' separation, in mm
     head_separation = 600

     lors = np.insert(lors_raw, 3, 0, axis=1)                    # z1 = 0
     lors = np.insert(lors, 6, head_separation, axis=1)          # z2 = head_separation

     # Print the first 10 rows from the line of response (LoR) data
     print(f"First 10 rows from the 3D-formatted LoR data:\n{lors[:10]}\n")
```

```
First 10 rows from the 3D-formatted LoR data:
[[0.000e+00 1.900e+02 1.687e+02 0.000e+00 3.463e+02 1.428e+02 6.000e+02]
 [1.000e-01 2.437e+02 1.676e+02 0.000e+00 3.145e+02 3.139e+02 6.000e+02]
 [1.000e-01 1.941e+02 4.100e+02 0.000e+00 4.171e+02 2.401e+02 6.000e+02]
 [2.000e-01 2.885e+02 2.962e+02 0.000e+00 3.847e+02 4.525e+02 6.000e+02]
 [2.000e-01 1.646e+02 1.151e+02 0.000e+00 3.829e+02 3.534e+02 6.000e+02]
 [2.000e-01 1.286e+02 1.322e+02 0.000e+00 4.649e+02 2.561e+02 6.000e+02]
 [2.000e-01 3.339e+02 3.735e+02 0.000e+00 3.316e+02 1.310e+02 6.000e+02]
 [3.000e-01 1.640e+02 1.115e+02 0.000e+00 3.794e+02 3.534e+02 6.000e+02]
 [4.000e-01 2.620e+02 2.094e+02 0.000e+00 2.932e+02 2.308e+02 6.000e+02]
 [4.000e-01 1.304e+02 2.749e+02 0.000e+00 2.543e+02 3.062e+02 6.000e+02]]
```

Iterating through each sample of LoRs, locate the tracer and append its location to a growing list of positions. As with the previous cells, you are invited to change the parameters highlighted below; details about each one are given in the comments.

```
[10]: # Use the Triangulation Method to track moving tracer
      from hdbscan import RobustSingleLinkage
      from pept.tracking.peptml import find_cutpoints

      # Triangulation Method user-changeable parameters:
      #     tracer_radius -"cluster" all LoRs that are closer than `tracer_radius` mm
      tracer_radius = 0.5
      min_lors_cluster = 5

      # General PEPT tracking user-changeable parameters:
      #     sample_start - the `lors` index at which to start tracking the tracer
      #     sample_size - the number of LoRs used to compute a single location
      #     overlap - the number of overlapping LoRs between consecutive samples
      sample_start = 0
      sample_size = 200
      overlap = 10

      # Tracked tracer locations will be appended to a growing list
      positions = []
```

```
def distance_lines(line1, line2):
    '''Calculate distance between two 3D lines `line1` and `line2`, each defined
    by a timestamp and two 3D points, so the lines are vectors formatted as
    [time, x1, y1, z1, x2, y2, z2].
    '''
    # Direction vectors:
    e1 = line1[4:7] - line1[1:4]
    e2 = line2[4:7] - line2[1:4]

    # Vector perpendicular to both lines
    n = np.cross(e1, e2)
    nd = np.linalg.norm(n)

    # If the lines are parallel, nd is zero
    if nd < 1e-8:
        e1 /= np.linalg.norm(e1)
        dist_vec = np.cross(e1, line2[1:4] - line1[1:4])
        return np.linalg.norm(dist_vec)

    dist = np.dot(n, line1[1:4] - line2[1:4]) / np.linalg.norm(n)
    return np.abs(dist)


while sample_start + sample_size < len(lors):
    sample = lors[sample_start:sample_start + sample_size]

    # Compute matrix of distances between each pair of LoRs

    distance_matrix = np.zeros((len(sample), len(sample)))
    for i in range(len(sample)):
        for j in range(i + 1, len(sample)):
            distance_matrix[i, j] = distance_lines(sample[i], sample[j])
            distance_matrix[j, i] = distance_matrix[i, j]

    # "Cluster" all LoRs that are closer than `tracer_radius` mm and form
    # clusters with more than `min_lors_cluster` LoRs
    single_linkage = RobustSingleLinkage(tracer_radius, min_lors_cluster,
                                         metric="precomputed")
    labels = single_linkage.fit_predict(distance_matrix)

    # For each cluster of LoRs, take their cutpoints' centroid as the tracer
    # locations
    for label in range(labels.max() + 1):
        cluster = sample[labels == label]
        cutpoints = find_cutpoints(cluster, max_distance=np.inf).points
        centre = cutpoints.mean(axis=0)

        positions.append(centre)

    sample_start += sample_size - overlap

positions = np.array(positions)

print(f"First 10 tracked positions:\n{positions[:10]}")
```

```
First 10 tracked positions:
[[  3.94735919 279.75777575 243.88478195 315.08413242]
 [ 11.40070804 279.75756993 244.09093234 315.85728847]
 [ 18.19420928 280.29497034 243.78997384 316.65700007]
 [ 25.40332255 280.95506073 244.24750982 318.92534418]
 [ 31.58392189 280.67910817 245.34682077 317.00352711]
 [ 37.50646962 280.69455462 244.21221744 317.03740438]
 [ 43.44901017 279.65960249 244.24502638 316.24058565]
 [ 50.39202105 280.24158517 244.28966847 317.54008422]
 [ 57.21102171 280.77976076 245.51940003 316.56287519]
 [ 64.07956989 280.56439328 245.24521992 314.55198353]]
```

[11]:
```
# Plot some of the LoRs used, plus the tracer locations we found
from pept.plots import PlotlyGrapher, PlotlyGrapher2D

grapher = PlotlyGrapher(
    cols=3,
    subplot_titles=[
        "Last Sample's Clustered LoRs",
        "Last Sample's Clustered LoRs' Cutpoints",
        "Tracer Positions",
    ],
)

labels_colour = np.repeat(labels, 3)

grapher.add_lines(lors[sample_start - sample_size:sample_start],
                  color=labels_colour)
grapher.add_points(cutpoints, col=2)
grapher.add_points(positions, col=3)

grapher.show()
```

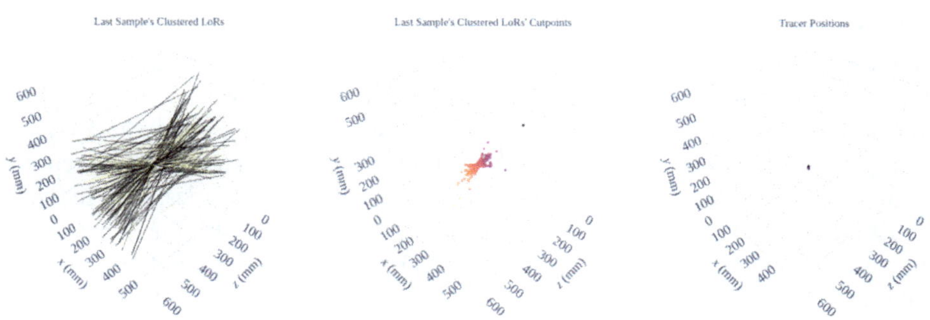

[12]:
```
# Create a "timeseries" plot with the tracer positions vs. time
PlotlyGrapher2D().add_timeseries(positions).show()
```

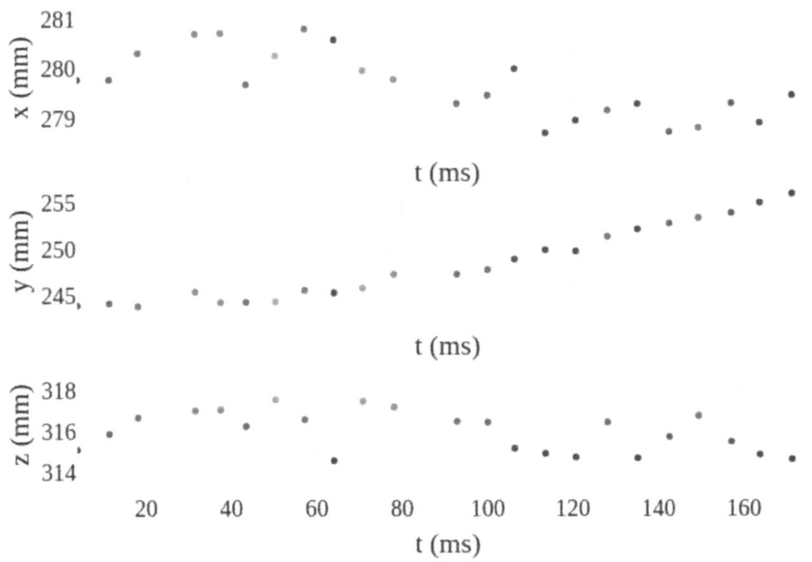

6.2.14 PEPT using machine learning (PEPT-ML)

PEPT-ML in brief

PEPT using machine learning [38]—or PEPT-ML—is a multiple particle tracking algorithm that was specifically developed to locate tracers even in the presence of significant amounts of noise with minimum numbers of LoRs to allow the imaging of dense systems or miniature tracers. It uses multiple techniques to maximise temporal resolution while yielding tight, dense locations on particle trajectories, even in the case of multiple freely-moving tracers that might cross or collide. PEPT-ML, even when used with older PEPT scanners—such as the ADAC Forté at Birmingham—is able to distinguish tracers at separations smaller than the detector's intrinsic resolution, and in a recent PEPT algorithm benchmark it was found to provide the best overall spatio-temporal resolution of currently-available algorithms [54].

PEPT-ML makes little distinction between the single- and multiple-particle tracking cases, and as such the steps outlined below are applicable to both:

1. A sample of individual, consecutive lines of response from a given PEPT data set is considered. For each pair of LoRs in the sample that are closer than a given maximum distance, their cutpoint—or the point of closest approach between the two LoRs—is determined; any cutpoints falling outside the detector's field of view are also discarded.
2. The Hierarchical Density-Based Clustering of Applications with Noise (HDBSCAN) algorithm [39] is applied to the cutpoints to separate out individual clusters of cutpoints from noise.
3. For each cluster, the cluster exemplars, or most significant intra-cluster cutpoints, are used to compute a geometric centroid corresponding to a single tracer location.
4. Optionally, a second pass of clustering can be applied to the already-found tracer locations to tighten the particle trajectories. An extremely small

sample size—with maximum overlap between consecutive samples—is used to minimise smoothing effects.

5. The number of cutpoints in each cluster is closely related to the tracer activity and shielding around it, and can be used as a tracer signature when reconstructing trajectories.

For an example trajectory separation algorithm that is often used in conjunction with PEPT-ML, see section 6.3.

PEPT-ML in detail

(1) Transforming LoRs into cutpoints: First, a sample of lines of response is selected—a typical number would be 200 LoRs per tracer, but it can easily be reduced to 100 or fewer for increased temporal resolution, or up to 600 for slow-moving tracers. For each pair of LoRs, their cutpoint, or point of closest approach is computed; for two non-parallel 3D lines, this simply corresponds to the middle of the unique segment that is perpendicular to both lines. In order to minimise 'cross-talk' between tracers (as seen in figure 6.11 panel (*d*)), all cutpoints coming from pairs of LoRs that are further than a fixed maximum distance are removed. Finally, all cutpoints falling outside the detectors—i.e. the box spanning all LoRs—are filtered out. See figure 6.11 for an illustration of the effect of the maximum distance; the optimum value depends strongly on the experiment conditions, and commonly a visual trade-off must be achieved between too few accurate cutpoints (at small maximum distances) and many noisy cutpoints (at large maximum distances). In practice the quality of the tracer trajectories found will not be very sensitive to this parameter as long as a sensibly large value is used. Some noise will always be present, but it can robustly be eliminated during the clustering step.

(2) Clustering cutpoints: Once we have a sample of cutpoints, we can separate out dense regions (corresponding to cutpoints around tracers) from noise. The clustering algorithm employed must be able to robustly filter out noisy points, be reasonably fast, and importantly, be able to find clusters with different densities of points; for example, if two tracers are used, the one with higher activity will emit more LoRs, and hence will have more cutpoints around it—which can be problematic for more classical partitioning methods. A powerful algorithm, HDBSCAN [39] builds upon the robustness of the well-established DBSCAN by specifically allowing clusters of

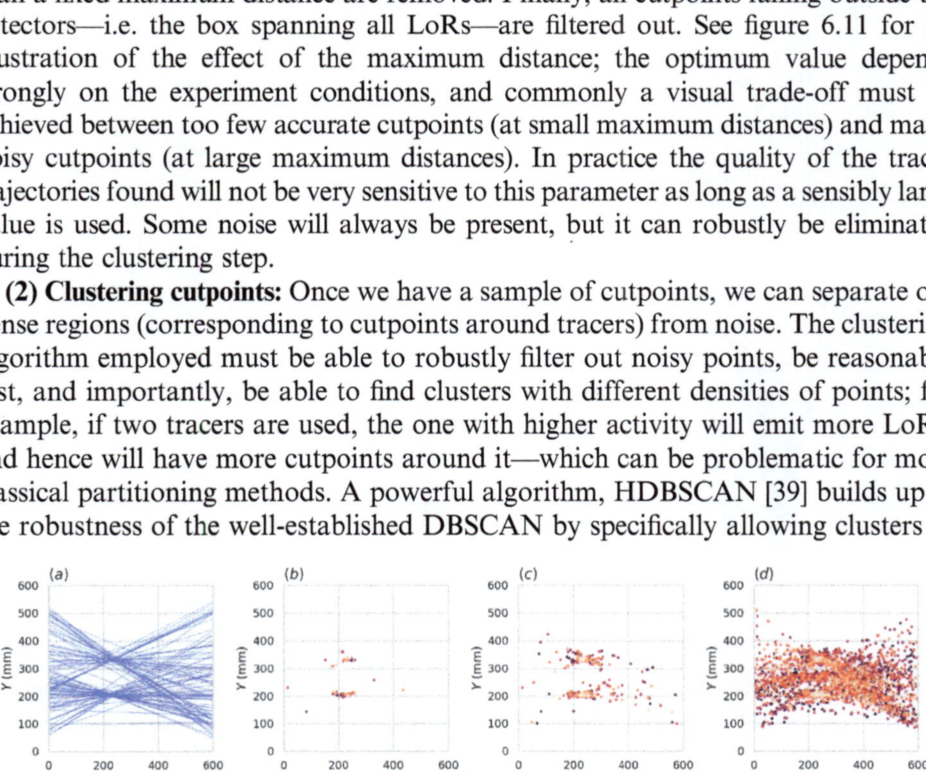

Figure 6.11. Panel (*a*) depicts the *YZ*-projection of a sample of 100 LoRs. Panel (*b*) shows the corresponding cutpoints computed for a maximum distance between pairs of LoRs of 0.5 mm; panel (*c*) depicts the cutpoints for a maximum distance of 5 mm. Panel (*d*) shows all possible cutpoints, with no maximum distance filtering. For ease of visualisation, the cutpoints are colour-coded by the mean timestamp of their constituent LoRs.

varying density, while also having more intuitive free parameters. PEPT-ML benefits from a high-performance HDBSCAN implementation of McInnes *et al* [40, 41] with a Python interface.

HDBSCAN starts by creating a minimum spanning tree (MST, or minimum distance path) connecting all cutpoints. This MST is used to 'cut' out individual clusters depending on the distance from one another; however, this method (called single linkage clustering in machine learning literature) is susceptible to erroneous cluster merging if there are a few, sparse, noisy points connecting them. To solve this, the older DBSCAN algorithm introduced a special metric, which computes the 'core distance' between two points as the maximum between their actual (Euclidean) distance and the distance to a kth nearest neighbour; this metric effectively 'pushes' away points from low-density regions, such as the sparse noisy points between real clusters. Panel (*b*) from figure 6.12 depicts the minimum spanning tree connecting all cutpoints, with all paths being colour-coded by their core distance; notice that paths in

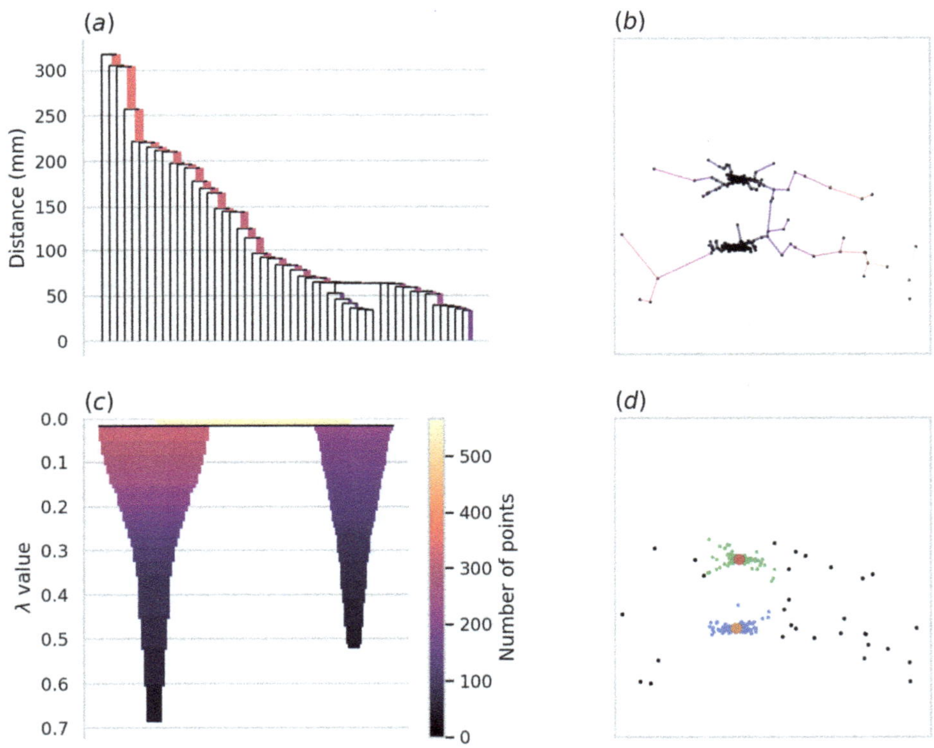

Figure 6.12. Illustration of the main steps applied by PEPT-ML to find tracer locations from a sample of cutpoints. Panel (*a*) shows the dendrogram representation of the minimum spanning tree, with only the first 50 levels depicted for ease of visualisation. Panel (*b*) depicts the minimum spanning tree—or minimum distance path—connecting all cutpoints; the paths are colour-coded by their distance, computed using the k-nearest neighbour-based metric of Campello *et al* [39]. Panel (c) shows the condensed dendrogram representation introduced in the same paper, providing a measure $\lambda = \text{distance}^{-1}$ of how far the points falling out of each cluster are. Panel (*d*) depicts the cutpoints colour-coded by their cluster labelling, with black signifying noise; each cluster's centroid is shown as a large circle.

sparse regions have larger core distances (signified by brighter colours) than points that may be spatially closer, but are within denser regions (i.e. have more neighbours).

Once the minimum spanning tree has been constructed, it can be converted to a dendrogram (panel (*a*) from figure 6.12), which represents all paths between pairs of points in the MST as 'forks' whose join height is the path length; the points are simply laid out at the bottom of the dendrogram. DBSCAN would simply cut the dendrogram at a given height and return all remaining clusters with more than a given number of points; however, in cases where clusters with significant differences in density are present (i.e. the points within a cluster are much closer than those in another), this fixed cut height can merge or split them. The addition of the 'H' in HDBSCAN stands for Hierarchical, relating to the condensed dendrogram that is constructed (shown in panel (*c*) from figure 6.12). Intuitively, a real cluster would have many paths with comparable lengths; this would mean that, as one traverses the dendrogram, there would be points 'falling out of the cluster', one by one (see the stairs-like shape in panel (*a*) from figure 6.12). This is quantified by HDBSCAN by traversing all forks in the dendrogram from the bottom-up; if a fork splits fewer points than a given minimum cluster size, they are considered to be part of a cluster; otherwise, the cluster is split. Panel (*c*) in figure 6.12 depicts this condensed dendrogram, showing the two clusters naturally formed; the *y*-axis depicts $\lambda = \text{distance}^{-1}$, a measure of how far from the maximum path length the points are falling out of the cluster. The width and colour represents the number of such points; finally, the most stable clusters are extracted from this data representation.

(3) Extracting tracer locations: The cutpoints within each extracted cluster are used to find each tracer's accurate location; however, due to the asymmetric distribution of LoRs (called parallax in the literature) which concentrates them towards the centre of the detectors, the clusters of cutpoints will be denser towards the scanner's centre too. Therefore, the simple geometric centroid of these cutpoints would be skewed. To counter this, the dendrogram is traversed again, selecting the cluster exemplars, or the most significant points from each cluster; in practice, these exemplars are uniformly distributed along the cluster, no matter its shape, being agnostic to parallax. The tracer locations are then computed as the centroid of each cluster's exemplars.

The previous steps are repeated for all samples of LoRs, consecutively. A technique that—even though it is algorithm-agnostic—is used extensively by PEPT-ML in particular, is the use of overlap between consecutive samples of LoRs, effectively increasing the number of points on each particle's trajectory (though without increasing temporal resolution per se). For a detailed discussion on the choice of sample size and overlap and their effects, see section 6.3. This overlap becomes essential when tightening trajectories using a second pass of clustering (discussed below).

(4) Second pass of clustering: The tracer positions found after clustering cutpoints will almost always exhibit some degree of scattering around the true tracers' trajectories. These trajectories can be tightened through the use of a second pass of clustering, where the locations previously found are iterated over again in samples which are re-clustered to yield more accurate centroids. Through the careful use of extremely small samples (as there will be very few noisy points now) and maximum overlap, there can be minimal loss in temporal resolution, but the spatial accuracy can be significantly improved. For fast moving tracers, some overlap can be used during the first pass of clustering too, to yield virtually no loss in temporal resolution. The effect of the second pass of clustering is explored in detail in the interactive notebook at the end of this section.

Notice that PEPT-ML contains many processing steps to yield accurate, dense tracer trajectories even in the case of significant amounts of noise, or tracers with relatively large differences in activity. Through the use of a second pass of clustering and large overlaps, the computational expense of the method becomes seemingly extreme. However, through the careful implementation of the computationally-intensive steps in highly optimised low-level C code (which is compiled directly to vectorised CPU instructions) and the aggressive parallelisation of each sample's processing to fully utilise modern multithreaded CPU architectures, PEPT-ML is able to process millions of LoRs in a matter of seconds—clearly showing the value of high quality algorithm implementations!

(5) Trajectory separation: In the case of multiple tracer tracking, the resulting locations for all tracers are simply sorted by time, and therefore intertwined. The trajectory separation technique normally employed after using PEPT-ML—though by no means tied to any specific tracer locating algorithm—is a two-step process: a simple spatio-temporal trajectory segregation step (which is completely tracking algorithm-agnostic), followed by a tracer signature-based trajectory linking step. The first is effectively a single linkage clustering algorithm—as mentioned in the paragraphs above—which uses a custom spatio-temporal metric to differentiate between tracer locations that are spatially close, but temporally far; the minimum spanning tree constructed using this metric is then simply 'cut' at a given maximum distance, and all remaining clusters of trajectories with more than a given minimum trajectory size are returned. However, it can happen that a tracer is lost for a longer period or distance than allowed by the aforementioned maximum distance—in which case a tracer signature can be used to link back some trajectories. This linking step simply checks the mean value of some tracer signature of each trajectory segment's two ends; if any pair of trajectories have a mean tracer signature difference smaller than a given minimum signature difference, they are linked together. PEPT-ML keeps track of each tracer location's number of constituent cutpoints; this cutpoints cluster size is proportional to the number of LoRs emitted by each tracer, which in turn depends on the tracer activity and spatial location within the experimental volume (see chapter 5 for a detailed discussion on this); this cluster size can therefore be used as a spatially-varying tracer signature. As trajectory separation is agnostic to the exact tracer locating algorithm used, it is more fully explained in detail in section 6.3, along with example implementations.

Strengths and weaknesses

PEPT-ML uses a vast array of machine learning-based techniques to yield the current best overall spatio-temporal resolution, as found in a recent blinded PEPT algorithm benchmark [54]. It produces dense, tight, accurate tracer trajectories for single- and multiple-particle tracking alike. An experienced PEPT-ML user can optimise the spatial and temporal resolution achieved for a given experiment; an advantage of the method is that it can naturally be applied to data from arbitrary radioactive sources, tracer dimensions or gamma detectors. On the other hand, its free parameters—namely the LoRs' sample size, overlap, cutpoints' maximum distance and HDBSCAN's minimum cluster size—can require some degree of experimentation to achieve optimum results; in this sense, PEPT-ML is harder to use than the Birmingham Method, which is often used as a 'plug and play' subroutine—still, these free parameters are fewer and more intuitive than that of other contemporary algorithms, such as the Voronoi tesselation (section 6–2.10). In exchange for this added complexity however, PEPT-ML can handle much smaller sample sizes and larger amounts of noise than older techniques, being suitable for use with miniature, fast tracers or in conjunction with densely-shielded systems—which is where PEPT as a technique truly shines in comparison with other imaging methods!

As PEPT-ML is a point-cloud-based method, it does not suffer from the grid-size-dependent results that voxel-based techniques normally have. On the other hand, the clouds of points are not limited in terms of size, potentially yielding unbounded memory usage for extreme numbers of LoRs; however, below 20,000 LoRs, no problems are typically encountered, even on consumer laptops. The algorithm has been successfully used to locate up to 128 tracers in its original paper, though this limit was found due to PEPT-ML's computational expense, not algorithm limitations.

Despite PEPT-ML being one of the more complex algorithms in terms of processing steps, it is competitive even with much simpler methods such as the Birmingham method in terms of speed due to its highly-optimised implementation. Finally, the existence of a PEPT-ML Python interface means that it is highly user-friendly, allowing the exploratory type of analysis currently being popularised by data science to be applied to PEPT as well; the existence of a wide scientific Python ecosystem (such as NumPy [42] and SciPy [43]) which PEPT-ML builds upon and extends, means that it can provide a one-stop solution for PEPT data extraction, processing, analysis and visualisation.

Strengths

- Yields tight, dense tracer trajectories.
- Can handle single- and multiple-tracer tracking without *a priori* knowledge of the number of tracers present in the system.
- Can use very small numbers of LoRs to locate tracers.
- Robustly handles noise and tracers with different activities.
- Has high-performance, parallelised implementation with a Python interface.

Weaknesses

- Free parameters can be less intuitive and require user experimentation.
- High memory usage for extreme numbers of LoRs.

6.2.15 Interactive example: PEPT-ML

The notebook files accompanying this section are available in the supplementary data online at https://iopscience.iop.org/book/978-0-7503-3071-8.

Interactive PEPT analysis example using the PEPT-ML algorithm [38]

This Jupyter Notebook

An example dataset is used from an experiment run at the University of Birmingham Positron Imaging Centre using the ADAC Forté. It consists of a single 1 mm diameter MCC particle activated with fluorine-18 inside a bubbling fluidised bed. The fluidised bed was filled with 90% sand and 10% MCC; air was fed into the bottom of the bed at a rate of 37 litres per minute at 3.5 bar. This dataset was chosen for its high quality captured lines of response, with the tracer still depicting the random particle motion that is inherent to bubbling fluidised beds—and typical in Lagrangian particle tracking.

The `pept` Python library is used for initialising and visualising PEPT data. While not required *per se* for illustrating PEPT algorithms' processing steps, it significantly reduces the amount of repetitive code and visual noise, allowing the reader to focus on the main conceptual procedures.

[1]:
```
# Importing necessary libraries for numerical array manipulation and plotting,
# plus the `pept` library for straightforward PEPT data processing
import numpy as np
import pept

# Set Plotly figure output (try "browser" too for interactive graphs)
import plotly
plotly.io.renderers.default = "png"
```

Initialise raw line of response data

In this initial step, we read in the 'raw' LoR data from which our tracer positions will be calculated. As illustrated in the 'Effect of sample size and overlap' notebook, varying the numbers of LoRs used to calculate a PEPT location can strongly affect the accuracy of the final position computed. Try altering the value `nrows` below to see how this influences your results.

[2]:
```
# Read in a sample of experimental PEPT data into a NumPy array. Skip the file
# header's first 15 lines, then read in the first `nrows` LoRs
lors_raw = pept.utilities.read_csv(
    "Data/FluidisedBed1Tracer.csv",
    skiprows=15,
    nrows=100,
)

# Convert the LoRs into the general 3D line format [t, x1, y1, z2, x2, y2, z2].
# See the `PEPT Data Format` interactive example for detailed explanations

# ADAC Forte's two detector heads' separation, in mm
head_separation = 600

lors = np.insert(lors_raw, 3, 0, axis = 1)                    # z1 = 0
lors = np.insert(lors, 6, head_separation, axis = 1)          # z2 = head_separation

# Print the first 10 rows from the line of response (LoR) data
print(f"First 10 rows from the 3D-formatted LoR data:\n{lors[:10]}\n")
```

```
First 10 rows from the 3D-formatted LoR data:
[[0.000e+00 1.900e+02 1.687e+02 0.000e+00 3.463e+02 1.428e+02 6.000e+02]
 [1.000e-01 2.437e+02 1.676e+02 0.000e+00 3.145e+02 3.139e+02 6.000e+02]
 [1.000e-01 1.941e+02 4.100e+02 0.000e+00 4.171e+02 2.401e+02 6.000e+02]
 [2.000e-01 2.885e+02 2.962e+02 0.000e+00 3.847e+02 4.525e+02 6.000e+02]
 [2.000e-01 1.646e+02 1.151e+02 0.000e+00 3.829e+02 3.534e+02 6.000e+02]
 [2.000e-01 1.286e+02 1.322e+02 0.000e+00 4.649e+02 2.661e+02 6.000e+02]
 [2.000e-01 3.339e+02 3.735e+02 0.000e+00 3.316e+02 1.310e+02 6.000e+02]
 [3.000e-01 1.640e+02 1.115e+02 0.000e+00 3.794e+02 3.534e+02 6.000e+02]
 [4.000e-01 2.620e+02 2.094e+02 0.000e+00 2.932e+02 2.808e+02 6.000e+02]
 [4.000e-01 1.304e+02 2.749e+02 0.000e+00 2.543e+02 3.062e+02 6.000e+02]]
```

The `pept.plots` subpackage provides two PEPT-oriented interfaces to the Plotly visualisation library— `PlotlyGrapher` and `PlotlyGrapher2D`—containing straightforward plotting functions for PEPT data. Let's plot the lines of response from the `lors` array:

[3]:
```
# Plot the 3D LoRs that were initialised
from pept.plots import PlotlyGrapher

PlotlyGrapher().add_lines(lors).show()
```

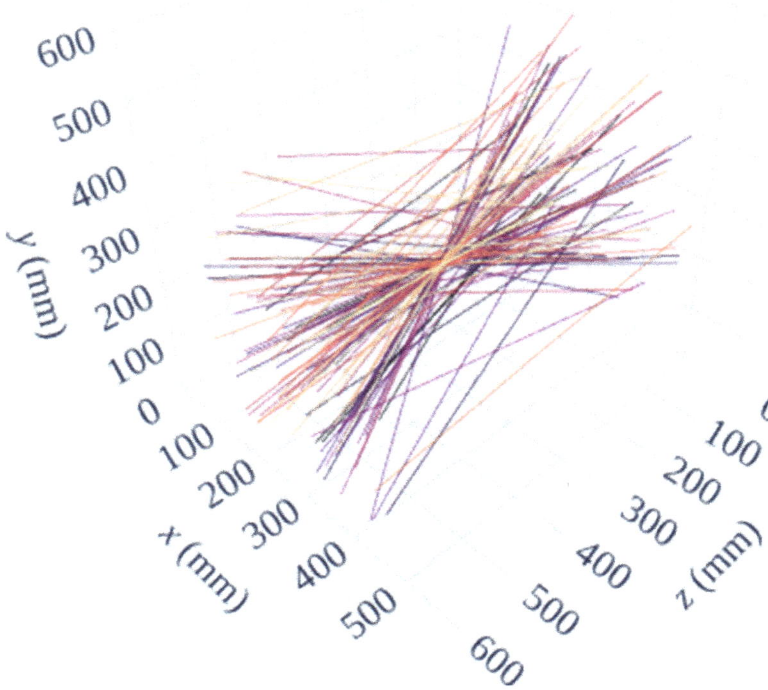

Note the PEPT 3D axes convention, differing from the common mathematical coordinates with *Y* and *Z* being switched, such that the *Y* axis points upwards, while the *Z* axis is perpendicular to the two detector heads.

Find cutpoints

In this step, we find the point of closest approach for each pair of LoRs and take the midpoint between the two lines at this point as a 'cutpoint'. Of course, for the purposes of PEPT, we only want to do this for LoRs that are being emitted from our tracer, and ignore false LoRs as far as possible. We do this by defining a `max_distance`, beyond which we can assume two LoRs do not pass through our particle, and simply not calculating a cutpoint for any two LoRs separated by more that this distance.

Try adjusting the value of `max_distance` and see what effect it has on the distribution of cutpoints produced.

As with the `nlors_remove` parameter in the Birmingham method example, being more restrictive with your choice of `max_distance` will remove more erroneous points, but yield less data, while being less restrictive will give you more data, but a higher probability of including erroneous values.

[4]:
```
# For all pairs of lines closer than `max_distance`, find their minimum distance
# points (cutpoints)
from pept.tracking import peptml

max_distance = 0.25
cutpoints = peptml.find_cutpoints(lors, max_distance).points
```

[5]:
```
# Plot this sample's LoRs and their corresponding cutpoints
grapher = PlotlyGrapher(cols = 2)

grapher.add_lines(lors)
grapher.add_points(cutpoints, col = 2)

grapher.show()
```

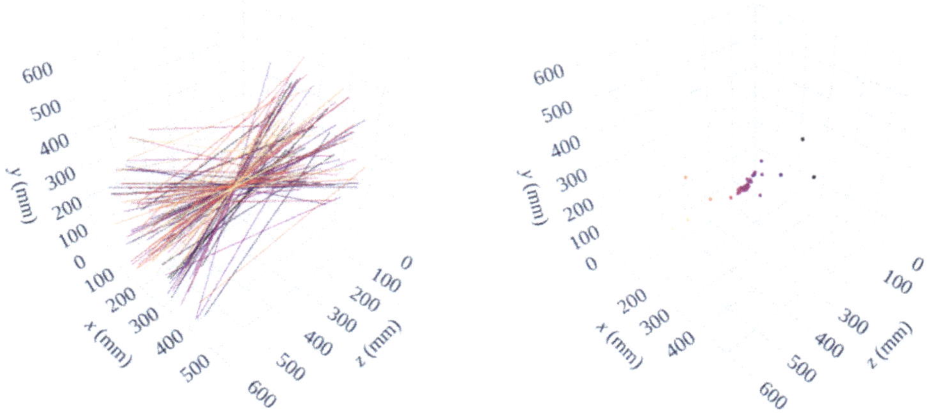

Notice the main characteristics of the cutpoints—importantly, they are more concentrated around our tracer, but they also include a significant amount of noise as yielded by larger `max_distance` values. We can extract the denser cutpoint region around the tracer using a density-based clustering algorithm that must be as robust as possible with respect to noise.

Cluster cutpoints with HDBSCAN

Having converted our LoRs into cutpoints, the next step is to create one or more clusters from these cutpoints using the Hierarchical density-based spatial clustering for applications with noise algorithm—or *HDBSCAN*—details of which you can find in the main text.

In order to optimise the clustering procedure, you can adjust the `min_cluster_size` parameter, which sets the minimum number of cutpoints that have to form a high-density region in space in order for the algorithm to view them as a true cluster. A larger value of `min_cluster_size` will give harsher clustering, with more points being classified as noise.

Try some different values and see the influence it has on your results.

[6]:
```
# Use the HDBSCAN algorithm to find clusters in the cutpoints
from hdbscan import HDBSCAN

min_cluster_size = 20
min_samples = 20

# Cluster the *spatial* coordinates of the cutpoints - i.e. columns [1, 2, 3]
hdbscan = HDBSCAN(min_cluster_size, min_samples, allow_single_cluster = True)
labels = hdbscan.fit_predict(cutpoints[:, 1:4])

print(f"Labels found for each clustered cutpoint:\n{labels}")
```

```
Labels found for each clustered cutpoint:
[ 0  0 -1  0 -1 -1  0 -1  0  0 -1  0  0  0 -1  0 -1 -1  0 -1 -1 -1 -1 -1
 -1  0 -1 -1 -1 -1 -1 -1  0  0 -1 -1 -1  0  0 -1 -1 -1 -1 -1  0  0 -1 -1
 -1  0 -1 -1 -1 -1 -1 -1 -1 -1  0 -1 -1 -1]
```

[7]:
```
# Plot cutpoints, coloured by their label index (numbered from 0, with -1
# representing noise)
PlotlyGrapher().add_points(cutpoints, color = labels).show()
```

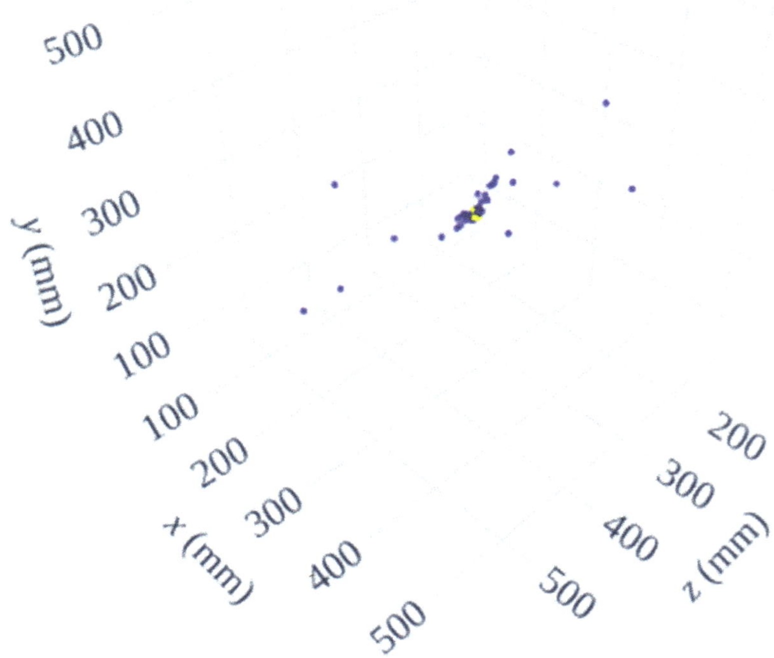

Compute cluster centres

The final step in this basic illustration of the PEPT-ML algorithm is to find the centres of these clusters, which should—if we have chosen our parameters correctly—correspond to the centre(s) of our tracer(s).

[8]:
```
# Compute the cutpoints' centroid from each cluster - should only be one!
centres = []

for i in range(labels.max() + 1):
    cluster = cutpoints[labels == 0]
    centres.append(cluster.mean(axis = 0))

centres = np.array(centres)
```

[9]:
```
# Plot the clustered cutpoints and the cluster centroids (corresponding to the
# tracer location)
grapher = PlotlyGrapher()

grapher.add_points(cutpoints, color = labels)
grapher.add_points(centres, size = 5)

grapher.show()
```

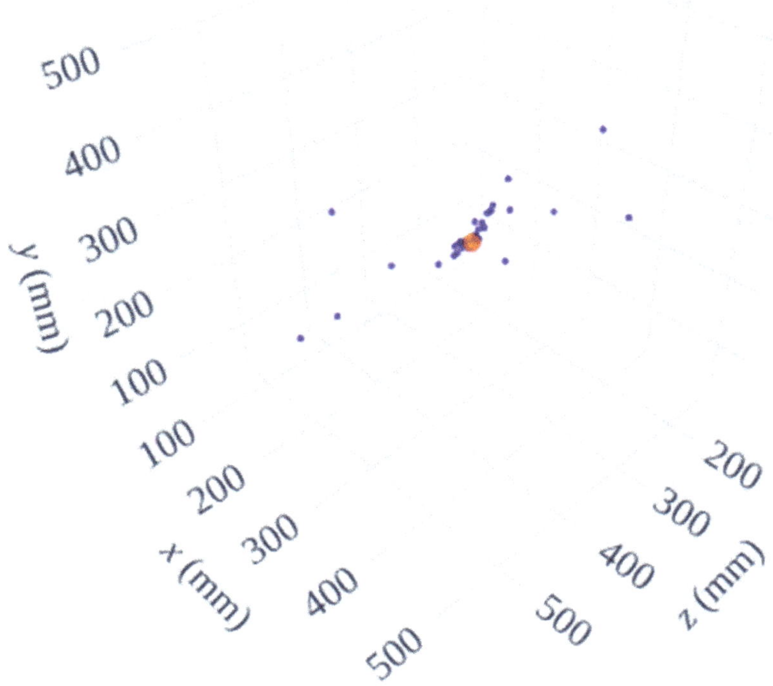

Complete PEPT-ML algorithm code

The code developed in the previous sections can be applied to consecutive 'samples' of LoRs to follow the tracer's movement through the fluidised bed, fully tracking its trajectory. A complete PEPT-ML algorithm implementation is given below. Because the most computationally-intensive steps are delegated to specialised subroutines (i.e. cutpoints' computation and HDBSCAN clustering), even this explicit version is relatively fast.

First initialise a larger, multi-sample set of LoRs to analyse; change the `nrows` argument below to read in a different number of LoRs (the file included contains 80,000 LoRs in total), though keep in mind the computational expense of processing large numbers of LoRs!

[10]:
```
# Read in a sample of experimental PEPT data into a NumPy array. Skip the file
# header's first 15 lines and read in the first `nrows` LoRs
lors_raw = pept.utilities.read_csv(
    "Data/FluidisedBed1Tracer.csv",
    skiprows=15,
    nrows=80_000,
)

# Convert the LoRs into the general 3D line format [t, x1, y1, z2, x2, y2, z2].
# See the `PEPT Data Format` interactive example for detailed explanations

# ADAC Forte's two detector heads' separation, in mm
head_separation = 600

lors = np.insert(lors_raw, 3, 0, axis=1)                # z1 = 0
lors = np.insert(lors, 6, head_separation, axis=1)      # z2 = head_separation

# Print the first 10 rows from the line of response (LoR) data
print(f"First 10 rows from the 3D-formatted LoR data:\n{lors[:10]}\n")
```

```
First 10 rows from the 3D-formatted LoR data:
[[0.000e+00 1.900e+02 1.687e+02 0.000e+00 3.463e+02 1.428e+02 6.000e+02]
 [1.000e-01 2.437e+02 1.676e+02 0.000e+00 3.145e+02 3.139e+02 6.000e+02]
 [1.000e-01 1.941e+02 4.100e+02 0.000e+00 4.171e+02 2.401e+02 6.000e+02]
 [2.000e-01 2.885e+02 2.962e+02 0.000e+00 3.847e+02 4.525e+02 6.000e+02]
 [2.000e-01 1.646e+02 1.151e+02 0.000e+00 3.829e+02 3.534e+02 6.000e+02]
 [2.000e-01 1.286e+02 1.322e+02 0.000e+00 4.649e+02 2.661e+02 6.000e+02]
 [2.000e-01 3.339e+02 3.735e+02 0.000e+00 3.316e+02 1.310e+02 6.000e+02]
 [3.000e-01 1.640e+02 1.115e+02 0.000e+00 3.794e+02 3.534e+02 6.000e+02]
 [4.000e-01 2.620e+02 2.094e+02 0.000e+00 2.932e+02 2.808e+02 6.000e+02]
 [4.000e-01 1.304e+02 2.749e+02 0.000e+00 2.543e+02 3.062e+02 6.000e+02]]
```

Iterating through each sample of LoRs, locate the tracer and append its location to a growing list of positions. As with the previous cells, you are invited to change the parameters highlighted below; details about each one are given in the comments.

[11]:
```
# Use PEPT-ML to track a moving tracer
from pept.tracking.peptml import find_cutpoints
from hdbscan import HDBSCAN
```

```
# PEPT-ML user-changeable parameters:
#     max_distance - only consider cutpoints from LoR pairs closer than this
#     min_cluster_size - minimum number of dense cutpoints to consider a cluster
#     min_samples - include more cutpoints in a cluster, larger = harsher
max_distance = 0.2
min_cluster_size = 30
min_samples = 30

# General PEPT tracking user-changeable parameters:
#     sample_start - the `lors` index at which to start tracking the tracer
#     sample_size - the number of LoRs used to compute a single location
#     overlap - the number of overlapping LoRs between consecutive samples
sample_start = 0
sample_size = 200
overlap = 100

# Tracked tracer locations will be appended to a growing list
positions = []

# Instantiate an HDBSCAN clusterer
hdbscan = HDBSCAN(min_cluster_size, min_samples, allow_single_cluster=True)

while sample_start + sample_size <= len(lors):
    sample = lors[sample_start:sample_start + sample_size]

    # Compute cutpoints for the selected sample of LoRs
    cutpoints = find_cutpoints(sample, max_distance).points

    # Cluster cutpoints and get each cutpoint's cluster label
    labels = hdbscan.fit_predict(cutpoints[:, 1:4])

    # For every cluster (i.e. label value >= 0), compute the centroid and append
    # it to the list of positions
    for label in range(labels.max() + 1):
        cluster = cutpoints[labels == label]
        centre = cluster.mean(axis = 0)
        positions.append(centre)

    sample_start += sample_size - overlap

positions = np.array(positions)

print(f"First 10 tracked positions:\n{positions[:10]}")
```

```
First 10 tracked positions:
[[  4.66833333 279.79580404 245.39652813 317.9314431 ]
 [  8.05333333 280.3674078  243.98765352 318.27365033]
 [ 12.32833333 280.95651503 243.22002371 316.52477974]
 [ 15.415      280.62009742 243.6831551  316.18364396]
 [ 19.24       279.81907652 243.59423609 316.48240956]
 [ 22.63166667 280.46093424 243.46769624 316.15026488]
 [ 26.48166667 281.27730732 243.82537652 315.56665014]
 [ 30.615      280.49704696 244.96138714 317.63686647]
 [ 32.31       280.52424758 245.01889696 318.17174687]
 [ 35.86666667 280.19679011 244.77658704 315.86035334]]
```

Second pass of clustering

The tracer locations found after the clustering of each sample of cutpoints inherently have some spread; the tracer trajectory can be 'tightened' by applying a 'second pass of clustering' to the already-found locations, acting as a robust trajectory smoothing step that does not need *a priori* information about the number of tracers.

The tracer locations are acting like the cutpoints from the first pass of clustering, so we iterate them one sample at a time and re-apply the clustering and centroid-computing steps. However, it is important to minimise the temporal resolution loss (or smoothing), so we use very small samples and extremely large overlaps. Indeed, this is a computationally-intensive step, but it yields much denser, tighter tracer trajectories—and using the algorithm is feasible time-wise due to the high-quality subroutine implementations.

[12]:
```
# Re-cluster the `positions` already found to "tighten" the tracer trajectory.
# Use a much smaller sample size and maximum overlap to minimise temporal
# resolution loss
sample_start = 0
sample_size = 30
overlap = 29

# The HDBSCAN parameters can be larger relative to the sample size as the data
# has already been "cleaned up" - the locations after the first pass of
# clustering should have almost no noise!
min_cluster_size = 15
min_samples = 15

hdbscan = HDBSCAN(min_cluster_size, min_samples, allow_single_cluster = True)
positions_2pass = []

while sample_start + sample_size <= len(positions):
    # Directly cluster the `positions` yielded by the first pass of clustering
    sample = positions[sample_start:sample_start + sample_size]
    labels = hdbscan.fit_predict(sample[:, 1:4])

    for label in range(labels.max() + 1):
        cluster = sample[labels == label]
        centre = cluster.mean(axis = 0)
        positions_2pass.append(centre)

    sample_start += sample_size - overlap

positions_2pass = np.array(positions_2pass)

print("First 10 tracked positions after the second pass of clustering:",
      f"\n{positions_2pass[:10]}")
```

```
First 10 tracked positions after the second pass of clustering:
[[ 44.50167025 280.3841331  245.10183439 317.06807257]
 [ 45.69922581 280.42847512 244.9732456  316.94932736]
 [ 48.68267025 280.50290993 245.06139348 316.93949213]
 [ 48.68267025 280.50290993 245.06139348 316.93949213]
 [ 53.40667025 280.54679975 245.21843075 317.05246258]
 [ 58.90900358 280.55385589 245.4923892  316.93949769]
 [ 58.90900358 280.55385589 245.4923892  316.93949769]
 [ 63.38855914 280.38596809 245.73911157 317.09762263]
 [ 67.62055914 280.24708084 245.9211691  316.77055953]
 [ 78.3871147  279.99835943 246.64981423 316.37109298]]
```

```
[13]: # Plot the tracer locations after the first and second passes of clustering
grapher = PlotlyGrapher(
    cols = 2,
    subplot_titles = [
        "Tracer locations - 1 pass clustering",
        "Tracer locations - 2 pass clustering",
    ]
)

grapher.add_points(positions, col = 1)
grapher.add_points(positions_2pass, col = 2)

grapher.show()
```

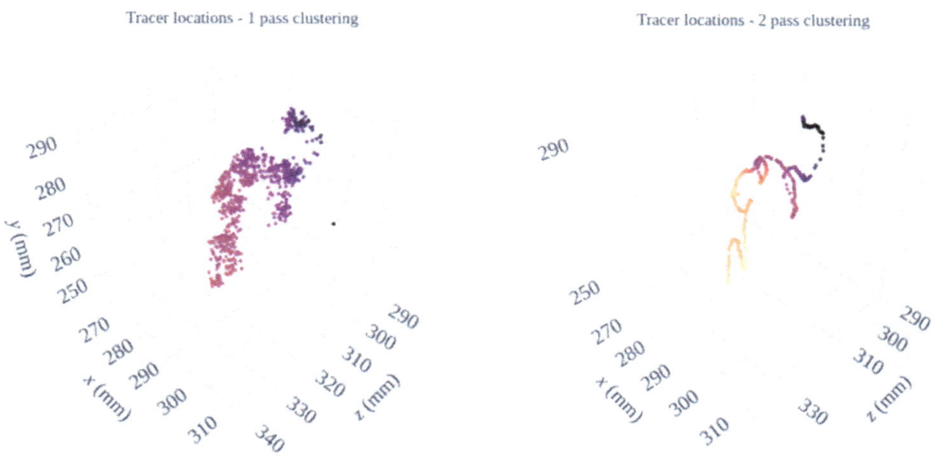

Note the tight tracer trajectory after the second pass of clustering (the right panel), revealing even the fast turns in the tracer's random motion inside the bubbling fluidised bed. Importantly, this precision was attained solely by computational means, without any hardware / data acquisition improvements—proving the importance of high quality algorithm implementations.

Multiple particle tracking

The complete code developed above can also be applied to a simple multiple tracer experiment. Here we use a sample of an experiment consisting of two 1 mm diameter alumina particles labelled with fluorine-18. The particles are fixed at the two ends of an arm rotating at 42 rpm, their trajectories describing full circles.

[14]:
```
# Read in a sample of experimental PEPT data into a NumPy array. Skip the file
# header's first 15 lines
lors_raw = pept.utilities.read_csv(
    "Data/RotatingArm2Tracers.csv",
    skiprows=15,
    nrows=30_000,
)

# Convert the LoRs into the general 3D line format [t, x1, y1, z2, x2, y2, z2].
# See the `PEPT Data Format` interactive example for detailed explanations

# ADAC Forte's two detector heads' separation, in mm
head_separation = 712

lors = np.insert(lors_raw, 3, 0, axis=1)                      # z1 = 0
lors = np.insert(lors, 6, head_separation, axis=1)            # z2 = head_separation

# Print the first 10 rows from the line of response (LoR) data
print(f"First 10 rows from the 3D-formatted LoR data:\n{lors[:10]}\n")
```

```
First 10 rows from the 3D-formatted LoR data:
[[0.000e+00 2.094e+02 2.797e+02 0.000e+00 2.738e+02 9.500e+01 7.120e+02]
 [1.000e-01 4.325e+02 4.395e+02 0.000e+00 2.006e+02 1.929e+02 7.120e+02]
 [1.000e-01 2.513e+02 1.109e+02 0.000e+00 2.177e+02 3.469e+02 7.120e+02]
 [1.000e-01 3.227e+02 4.372e+02 0.000e+00 3.853e+02 1.097e+02 7.120e+02]
 [1.000e-01 1.894e+02 9.910e+01 0.000e+00 4.213e+02 1.044e+02 7.120e+02]
 [2.000e-01 3.658e+02 2.230e+02 0.000e+00 3.198e+02 5.003e+02 7.120e+02]
 [2.000e-01 2.543e+02 4.690e+02 0.000e+00 4.873e+02 1.257e+02 7.120e+02]
 [2.000e-01 3.339e+02 4.962e+02 0.000e+00 3.646e+02 8.970e+01 7.120e+02]
 [2.000e-01 2.844e+02 2.130e+02 0.000e+00 4.466e+02 5.322e+02 7.120e+02]
 [2.000e-01 2.053e+02 1.906e+02 0.000e+00 2.785e+02 2.224e+02 7.120e+02]]
```

Iterating through each sample of LoRs, locate the tracers and append their locations to a growing list of positions. As with the previous cells, you are invited to change the parameters highlighted below; details about each one are given in the comments.

Importantly, the `sample_size` must be larger to include enough LoRs for each tracer—in cases using large numbers of LoRs, the `overlap` becomes increasingly important to yield *dense* points on each trajectory; see the 'Effect of sample size and overlap' book section for a detailed explanation.

[15]:
```
# Use PEPT-ML to track moving tracer
from pept.tracking.peptml import find_cutpoints
from hdbscan import HDBSCAN

# Triangulation Method user-changeable parameters:
#     max_distance - only consider cutpoints from LoR pairs closer than this
#     min_cluster_size - minimum number of dense cutpoints to consider a cluster
#     min_samples - include more cutpoints in a cluster, larger = harsher
max_distance = 0.2
min_cluster_size = 30
min_samples = 30

# General PEPT tracking user-changeable parameters:
#     sample_start - the `lors` index at which to start tracking the tracer
#     sample_size - the number of LoRs used to compute a single location
#     overlap - the number of overlapping LoRs between consecutive samples
sample_start = 0
sample_size = 200 * 2
overlap = 100 * 2


#
# First pass of clustering
```

```
#


# Tracked tracer locations will be appended to a growing list
positions = []

# Instantiate an HDBSCAN clusterer
hdbscan = HDBSCAN(min_cluster_size, min_samples, allow_single_cluster=True)

while sample_start + sample_size < len(lors):
    sample = lors[sample_start:sample_start + sample_size]

    # Compute cutpoints for the selected sample of LoRs
    cutpoints = find_cutpoints(sample, max_distance).points

    # Cluster cutpoints and get each cutpoint's cluster label
    labels = hdbscan.fit_predict(cutpoints[:, 1:4])

    # For every cluster (i.e. label value >= 0), compute the centroid and append
    # it to the list of positions
    for label in range(labels.max() + 1):
        cluster = cutpoints[labels == label]
        centre = cluster.mean(axis=0)
        positions.append(centre)

    sample_start += sample_size - overlap

positions = np.array(positions)
print(f"First 10 tracked positions:\n{positions[:10]}\n")


#
# Second pass of clustering
#


# Re-cluster the `positions` already found to "tighten" the tracer trajectory.
# Use a much smaller sample size and maximum overlap to minimise temporal
# resolution loss
sample_start = 0
sample_size = 30 * 2
overlap = 29 * 2

# The HDBSCAN parameters can be larger relative to the sample size as the data
# has already been "cleaned up" - the locations after the first pass of
# clustering should have almost no noise!
min_cluster_size = 15
```

```
min_samples = 15

hdbscan = HDBSCAN(min_cluster_size, min_samples, allow_single_cluster=True)
positions_2pass = []

while sample_start + sample_size < len(positions):
    # Directly cluster the `positions` yielded by the first pass of clustering
    sample = positions[sample_start:sample_start + sample_size]
    labels = hdbscan.fit_predict(sample[:, 1:4])

    for label in range(labels.max() + 1):
        cluster = sample[labels == label]
        centre = cluster.mean(axis=0)
        positions_2pass.append(centre)

    sample_start += sample_size - overlap

positions_2pass = np.array(positions_2pass)

print("First 10 tracked positions after the second pass of clustering:",
      f"\n{positions_2pass[:10]}")
```

```
First 10 tracked positions:
[[  4.37390438 348.37319728 331.60809859 283.01330909]
 [  4.37380137 231.75652958 202.8892693  275.72268136]
 [  9.16443662 349.69046344 329.26507905 280.71767941]
 [  8.50552486 230.54467512 205.26520912 280.8893907 ]
 [ 13.2669697  231.36451913 207.03165053 281.38385925]
 [ 13.26116838 351.90946537 328.38554752 282.41162758]
 [ 17.36978723 229.50741873 207.13145864 277.39143636]
 [ 16.93096085 351.79046384 328.24770468 285.52251595]
 [ 21.15979592 228.5719728  208.24368987 277.07763703]
 [ 21.65992647 352.96093026 327.64203089 284.37243882]]

First 10 tracked positions after the second pass of clustering:
[[ 65.13959907 218.4349655  222.41722211 277.75731498]
 [ 65.18954224 362.30332017 313.10107804 284.38491204]
 [ 69.35923888 363.10679702 311.75544009 284.59930902]
 [ 69.33324149 217.61773104 223.81125384 277.97149663]
 [ 73.50150865 216.83081229 225.16371065 277.89919899]
 [ 73.53037132 363.8511403  310.44497581 284.7712775 ]
 [ 77.68718772 215.99158052 226.52206172 277.75487159]
 [ 77.71598848 364.55238426 309.10252855 284.84176331]
 [ 81.92850698 365.2965647  307.73395108 284.85319451]
 [ 81.90504497 215.22233495 227.89376519 277.75810967]]
```

```
[16]: # Plot some of the LoRs used, then the tracer locations after the first and
      # second passes of clustering
      grapher = PlotlyGrapher(
          cols = 3,
          subplot_titles = [
              "First 500 LoRs",
              "Tracer locations - 1 pass clustering",
              "Tracer locations - 2 pass clustering",
          ]
      )

      grapher.add_lines(lors[:500])
      grapher.add_points(positions, col = 2)
      grapher.add_points(positions_2pass, col = 3)

      grapher.show()
```

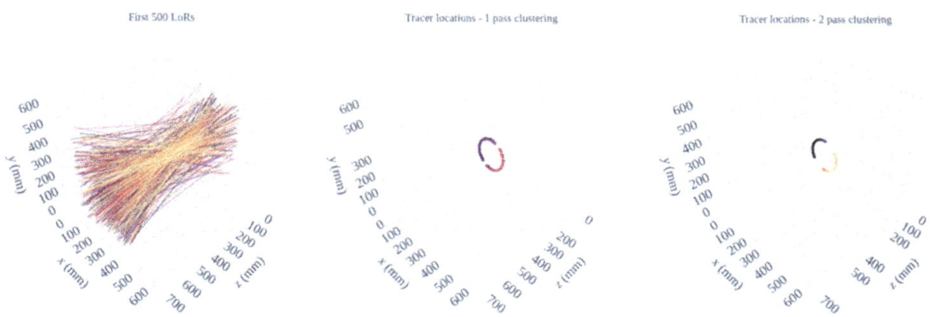

In this case, however, the second pass of clustering depicts an expected processing artifact due to the use of re-clustering with overlaps: the two ends of each tracer's trajectory seem 'trimmed off'. Depending on the experiment length, this may not be problematic, but it is an important issue to consider in terms of smoothing effects nonetheless.

High-performance PEPT-ML algorithm implementation

The `pept` library contains a highly-optimised PEPT-ML algorithm implementation, which uses C code for computationally-intensive operations and processes each sample of LoRs in parallel—therefore being able to fully utilise multi-CPU architectures in modern computers and clusters.

Most of the repetitive or tedious tasks above are automated, such as:

- Conversion from native PEPT scanner formats (e.g. binary, or parallel screens ASCII / CSV files) to the general 3D LoR format using the `pept.scanners` subpackage.
- Straightforward iteration of LoRs in terms of samples with optional overlap using the `pept.LineData` and `pept.PointData` classes.
- Terse usage of the PEPT tracking algorithms from the `pept.tracking` subpackage.

- PEPT-oriented plotting using the `pept.plots` subpackage.

```
[17]: # Import the subroutines we will use to track our tracer
import pept
from pept.tracking import *
from pept.plots import PlotlyGrapher, PlotlyGrapher2D

# Initialise data from a Parallel Screens CSV format (i.e. [t, x1, y1, x2, y2])
# into the general "LineData" format ([t, x1, y1, z1, x2, y2, z2]) and set a
# sample size and overlap
lors = pept.scanners.parallel_screens(
    "Data/FluidisedBed1Tracer.csv",
    screen_separation=600,
    skiprows=15,
    sample_size=200,
    overlap=120,
)

# Create PEPT-ML processing pipeline
pipeline = pept.Pipeline([
    # First pass of clustering
    Cutpoints(max_distance = 0.2),
    HDBSCAN(true_fraction = 0.2),
    SplitLabels() + Centroids(),

    # Second pass of clustering
    Stack(sample_size = 30, overlap = 29),
    HDBSCAN(true_fraction = 0.6),
    SplitLabels() + Centroids(),

    # Collect results
    Stack(),
])

# Use PEPT-ML pipeline to transform LoRs into accurate trajectories
positions = pipeline.fit(lors)

# Plot all tracked positions
PlotlyGrapher().add_points(positions).show()
```

```
Initialised PEPT data in 0.031 s.

Batch 1 / 2 :: 100%|                    | 998/998
[00:02<00:00, 490.18it/s]
Batch 2 / 2 :: 100%|                    | 943/943
[00:00<00:00, 2823.74it/s]
Processed samples in 3.1155920028686523 s
```

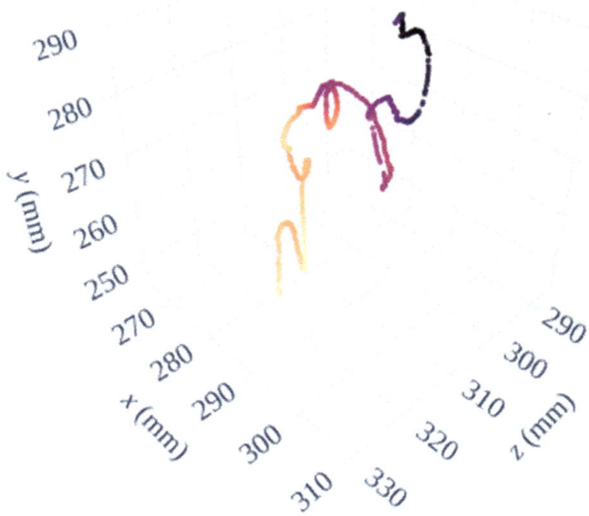


```
[18]: # Create a "timeseries" plot with the tracer positions vs. time
      PlotlyGrapher2D().add_timeseries(positions).show()
```

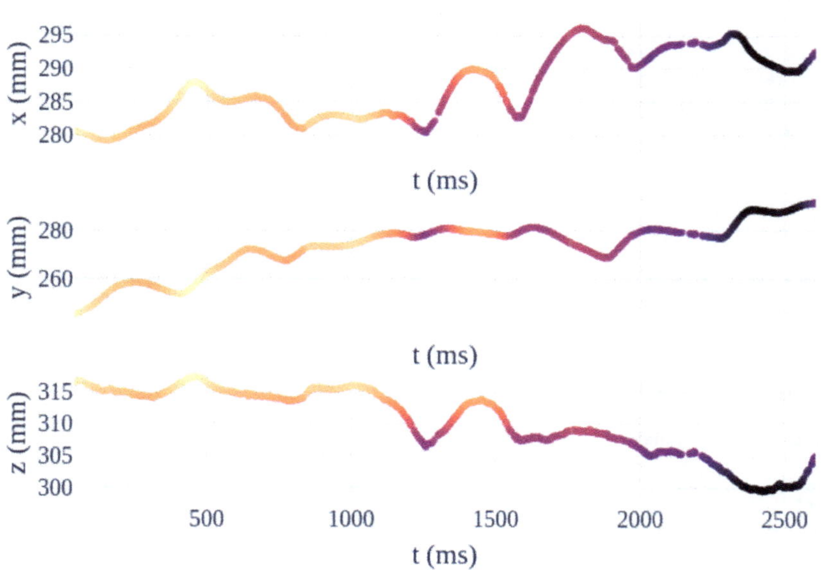

In what were essentially four Python statements, we processed 80,000 LoRs (in 1.3 s on the author's laptop) to track 2.6 s of a single tracer's trajectory through a fluidised bed, complete with loading bars and interactive plotting. Though `pept` automated most steps, there is no magic involved—just sensible defaults which can be changed for every subroutine called; plus, all code is public and free to use and modify from the `pept` repository.

The same algorithm can be applied to the two-tracer experiment, with the only change being the larger `sample_size` to allow the same number of LoRs *per tracer*.

[19]:
```
# Import the subroutines we will use to track our tracer
import pept
from pept.tracking import *
from pept.plots import PlotlyGrapher, PlotlyGrapher2D

# Initialise data from a Parallel Screens CSV format (i.e. [t, x1, y1, x2, y2])
# into the general "LineData" format ([t, x1, y1, z1, x2, y2, z2]) and set a
# sample size and overlap
lors = pept.scanners.parallel_screens(
    "Data/RotatingArm2Tracers.csv",
    screen_separation=600,
    skiprows=15,
    sample_size=2 * 200,
    overlap=2 * 120,
)
```

```
# Create PEPT-ML processing pipeline
pipeline = pept.Pipeline([
    # First pass of clustering
    Cutpoints(max_distance = 0.2),
    HDBSCAN(true_fraction = 0.1),
    SplitLabels() + Centroids(),

    # Second pass of clustering
    Stack(sample_size = 30, overlap = 29),
    HDBSCAN(true_fraction = 0.3),
    SplitLabels() + Centroids(),

    # Collect results
    Stack(),
])

# Use PEPT-ML pipeline to transform LoRs into accurate trajectories
positions = pipeline.fit(lors)

# Plot all tracked positions
PlotlyGrapher().add_points(positions).show()
```

```
Initialised PEPT data in 0.033 s.

Batch 1 / 2 :: 100%|                          | 498/498
[00:00<00:00, 1124.18it/s]
Batch 2 / 2 :: 100%|                          | 959/959
[00:00<00:00, 2840.90it/s]

Processed samples in 1.3432307243347168 s
```

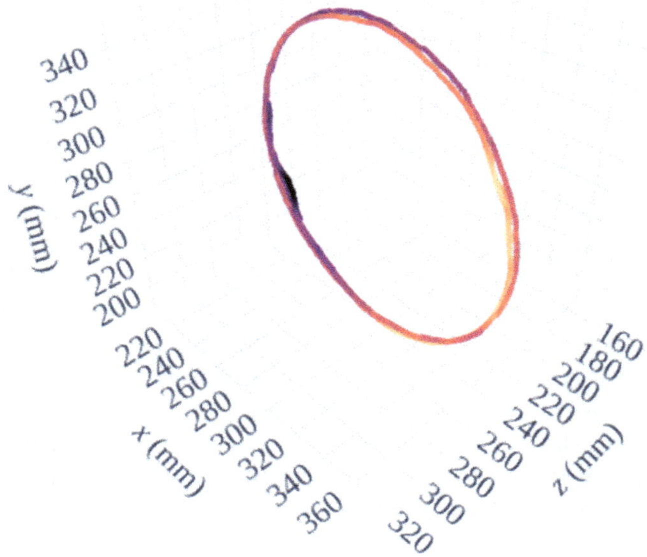

Notice that the tracer trajectories are accurate enough to depict the slight tilt in the rotating arm's axle, yielding slightly offset circular motions!

```
[20]: # Create a "timeseries" plot with the tracer positions vs. time
      PlotlyGrapher2D().add_timeseries(positions).show()
```

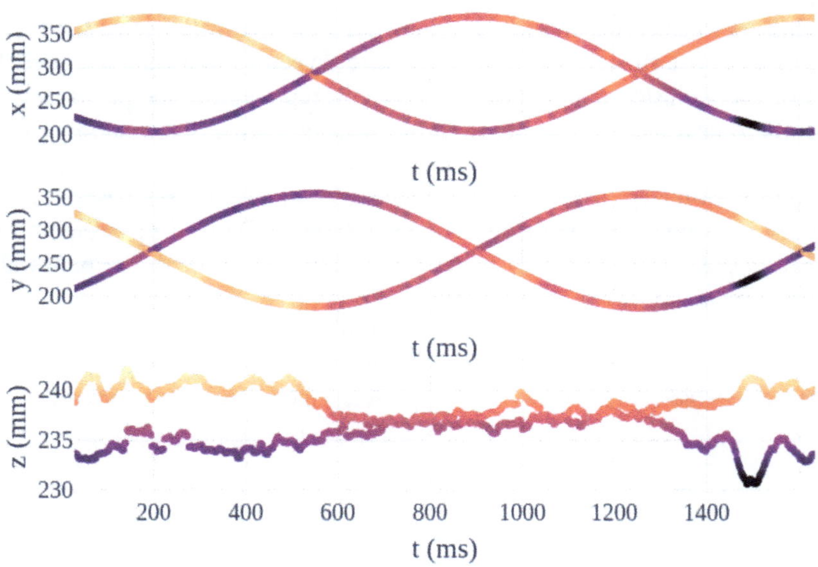

6.2.16 PEPT using expectation-maximisation (PEPT-EM)

PEPT-EM in brief

The PEPT using expectation-maximisation (EM) [44]—or PEPT-EM for short—is an algorithm that was developed primarily at the University of Edinburgh in collaboration with the University of Birmingham. It is a hybrid between the Birmingham method—in the sense that it aims to separate out scattered events from true LoRs—and an iterative clustering approach which assigns weights to partitioned LoRs until converging on the centroids of 3D Gaussian-approximated tracers. It uses an EM algorithm to estimate the parameters of a statistical model of the positron-electron annihilation events around the tracer. PEPT-EM was initially developed for accurately tracking the velocity of particles flowing through pipes, and as such, it is able to continuously follow the velocity and acceleration of tracers as they are being tracked, without post-processing steps required; moreover, it can use this information as an aid during trajectory reconstruction if the tracers are moving slowly enough. Like the other algorithms using the recorded LoRs directly to locate particles such as the Birmingham method (section 6.2.1) and the triangulation method (section 6.2.12), it requires larger samples of LoRs, reducing its temporal resolution. Moreover, the EM steps may converge on local minima, or take many iterations to find the real tracers. However, for applications where the tracers are not moving extremely fast, PEPT-EM was often found to have the best—or close to the best—spatial accuracy in a recent PEPT algorithms benchmark [54]. As in the previous sections, we first outline the high-level algorithmic steps, before describing them in detail and offering an interactive implementation in a notebook at the end of this section.

1. A sample of individual, consecutive lines of response from a given PEPT data set is considered, along with initial guesses for the locations of some number of clusters to find (which can be larger than the real number of tracers), plus a larger 'ghost cluster' spanning the entire detector collecting scattered LoRs.
2. For every cluster, assign all LoRs a weight representing the probability of each LoR to be a true event coming from the given cluster. Compute the centroid of each cluster as the minimum distance point (MDP) to the weighted LoRs.
3. Expectation step: for each cluster, recompute LoR weights based on the intra-cluster variance of the distances between the MDP and the weighted LoRs.
4. Maximisation step: recompute each cluster's MDP from the new weights of the LoRs.
5. The EM steps are repeated for a fixed number of iterations; for the first sample, a few thousands are used to ensure convergence. In the next samples, the previous tracer positions are used as initial guesses, which accelerates convergence.

PEPT-EM in detail

(1–2) Setting up the tracer-finding problem: PEPT-EM models the task of finding tracer locations as a Gaussian mixture problem, where the points of closest approach between the real tracer locations and their constituent LoRs are distributed according to a Gaussian distribution. Given a set of LoRs $\mathcal{L}$ and K clusters to find, in this formulation, the end-goal is to maximise the log-likelihood L of finding each of the K given clusters:

$$L\left(\left\{\vec{x}_k, \sigma_k^2, \rho_k\right\}_{1\leqslant k\leqslant K}\middle| \mathcal{L}\right) = \sum_{\ell\in\mathcal{L}} \log\left(\rho_0\alpha^{-1} + \sum_{k=1}^{K} \rho_k \sigma_k^{-2} \mathrm{e}^{-D^2(\vec{x}_k,\, \ell)/2\sigma_k^2}\right). \tag{6.16}$$

Where $\vec{x}_k$ is a cluster's centroid, σ_k^2 is the Gaussian distribution's variance, ρ_k is the cluster's weight, α is the variance of a large 'ghost cluster' spanning the field of view that will contain all scattered LoRs, and $D(\vec{x}_k, \ell)$ is the distance from the cluster's centroid to a given LoR. This Gaussian mixture problem can be solved iteratively using consecutive EM steps.

(3) Expectation step: The essence of PEPT-EM is, for each cluster, to find its constituent lines of response. To this end, each cluster assigns weights to all the LoRs in the sample, which will be used to compute the weighted minimum distance point representing each cluster's centroid; the weights and centroids will be consecutively recomputed until convergence. The LoR weights for each cluster k—or latent weights—are calculated based on the formula given below:

$$w_{\ell,k}\left(\left\{\vec{x}_{k'}, \sigma_{k'}^2, \rho_{k'}\right\}_{1\leqslant k'\leqslant K}\right) = \frac{\rho_k \sigma_k^{-2} \mathrm{e}^{-D^2(\vec{x}_k,\, \ell)/2\sigma_k^2}}{\rho_0\alpha + \sum_{k'=1}^{K} \rho_{k'} \sigma_{k'}^{-2} \mathrm{e}^{-D^2\left(\vec{x}_{k'},\, \ell\right)/2\sigma_{k'}^2}} \tag{6.17}$$

For the ghost cluster $k = 0$, the latent weights are computed as $w_{\ell,0} = 1 - \sum_{k=1}^{K} w_{\ell,k}$. Note that the calculation of the latent weights requires some guess of the tracer location $\vec{x}_k$—weights which in turn change the tracer location!—which is why a two-step refinement scheme is required.

(4–5) Maximisation step and iterative refinement: Given the newly-calculated LoR weights for each cluster, the centroids $\vec{x}_k$ can be recomputed, along with each cluster's variance σ_k and weight ρ_k:

$$\vec{x}_k = \left(\sum_{\ell\in\mathcal{L}} w_{\ell,k}(\mathbb{I}_3 - \vec{n}_\ell \otimes \vec{n}_\ell)\right)^{-1} \sum_{\ell\in\mathcal{L}} w_{\ell,k}(\mathbb{I}_3 - \vec{n}_\ell \otimes \vec{n}_\ell)\, \vec{y}_\ell \tag{6.18}$$

$$\sigma_k^2 = \frac{\sum_{\ell\in\mathcal{L}} w_{\ell,k} D^2 \vec{x}_k}{2\sum_{\ell\in\mathcal{L}} w_{\ell,k}}. \tag{6.19}$$

$$\rho_k = \frac{1}{N}\sum_{\ell\in L} w_{\ell,k}, \tag{6.20}$$

Where $\mathbb{I}_3$ is the identity matrix and each of the N LoRs ℓ is defined by a position vector $\vec{y}_\ell$ and a unit direction vector $\vec{n}_\ell$. The weight of the ghost cluster is $\rho_0 = 1 - \sum_{k=1}^{K} \rho_k$.

While mathematically involved, PEPT-EM can be best understood by keeping track of the conceptual steps we are following: for each tracer, we are finding the LoRs that are most likely to have originated from a 3D Gaussian distribution of LoRs around it (closest points on the LoRs to each tracer to be precise). We do this by assigning weights to each LoR, which we use to recompute the weighted cluster centroid—which we then use to refine the weights, and the two-step process is repeated for a fixed number of times. Finally, a Taylor expansion of the tracer centroids within a single sample of LoRs can optionally be computed with respect to time, allowing the calculation of derivatives of the tracer locations over time—i.e. velocity, acceleration. This information can be used to predict the tracers' locations in the next sample of LoRs, which can serve two purposes: a natural trajectory identification technique—done directly as the tracers are being located—and being a very good starting guess when repeating the EM steps for the next sample, which can greatly accelerate the iterative refinement of the tracer locations.

Strengths and weaknesses

PEPT-EM develops a mathematically-consistent model for the distribution of positron–electron annihilations around the positron-emitting tracers which can be iteratively fitted to a sample of LoRs using EM steps. It uses a number of algorithmic proxies to facilitate the location of multiple tracers in the presence of noise, such as a large 'ghost cluster' spanning the detectors' field of view to collect scattered LoRs; this way, the Gaussian mixture model can partition all LoRs into a fixed number of clusters, including the erroneous ones. However, some of PEPT-EM's techniques can sometimes make it more difficult to apply in practice: it requires a fixed numbers of clusters to partition the LoRs into; any cluster that did not converge onto a region in space then must be filtered out based on its larger variance; however, choosing the appropriate threshold for divergent clusters is not intuitive. Furthermore, choosing an appropriate variance of the ghost cluster—though known to be large enough to span the field of view—does require a fair amount of experimentation. Therefore, in order to apply PEPT-EM to experiments with differing conditions—e.g. more shielding, larger tracers—often requires an intimate knowledge of the mathematical formulation of the algorithm. However, in the hands of an experimented PEPT practitioner, PEPT-EM is able to find tracer locations with outstanding accuracy, as was found in a recent algorithm benchmark, where PEPT-EM often offered the best—or close to the best—spatial accuracy across multiple tests. On the other hand, the need for large samples of LoRs, like other similar algorithms, results in a decreased temporal accuracy. Finally, PEPT-EM provides a powerful mathematical framework for PEPT that can be further developed to overcome its current limitations.

Strengths
- Can yield remarkable spatial accuracy.
- Can track large numbers of tracers.
- Can yield velocity and acceleration without post-processing.
- Offers simple sample-by-sample trajectory separation.

Weaknesses
- The free parameters are non-intuitive and might need intimate knowledge of the mathematical formulation of the algorithm.
- Temporal resolution can suffer due to the need for large sample sizes.
- The iterative refinement step can be relatively computationally inefficient.

6.2.17 Interactive example: PEPT-EM

The notebook files accompanying this section are available in the supplementary data online at https://iopscience.iop.org/book/978-0-7503-3071-8.

Interactive PEPT analysis examples using PEPT-EM [44]

This Jupyter Notebook

An example dataset is used from an experiment run at the University of Birmingham Positron Imaging Centre using the ADAC Forté. It consists of a single 1 mm diameter MCC particle activated with fluorine-18 inside a bubbling fluidised bed. The fluidised bed was filled with 90% sand and 10% MCC; air was fed into the bottom of the bed at a rate of 37 litres per minute at 3.5 bar. This dataset was chosen for its high quality captured lines of response, with the tracer still depicting the random particle motion that is inherent to bubbling fluidised beds—and typical in Lagrangian particle tracking.

The `pept` Python library is used for initialising and visualising PEPT data. While not required *per se* for illustrating PEPT algorithms' processing steps, it significantly reduces the amount of repetitive code and visual noise, allowing the reader to focus on the main conceptual procedures.

```
[1]: # Importing necessary libraries for numerical array manipulation and plotting,
     # plus the `pept` library for straightforward PEPT data processing
     import numpy as np
     import pept

     # Set Plotly figure output (try "browser" too for interactive graphs)
     import plotly
     plotly.io.renderers.default = "png"
```

Initialise Raw Line of Response Data

In this initial step, we read in the 'raw' LoR data from which our tracer positions will be calculated. As illustrated in the 'Effect of sample size and overlap' notebook, varying the numbers of LoRs used to calculate a PEPT location can strongly affect the accuracy of the final position computed. Try altering the value `nrows` below to see how this influences your results.

[2]:
```
# Read in a sample of experimental PEPT data into a NumPy array. Skip the file
# header's first 15 lines, then read in the first `nrows` LoRs
lors_raw = pept.utilities.read_csv(
    "Data/FluidisedBed1Tracer.csv",
    skiprows=15,
    nrows=100,
)

# Convert the LoRs into the general 3D line format [t, x1, y1, z2, x2, y2, z2].
# See the `PEPT Data Format` interactive example for detailed explanations

# ADAC Forte's two detector heads' separation, in mm
head_separation = 600

lors = np.insert(lors_raw, 3, 0, axis = 1)                    # z1 = 0
lors = np.insert(lors, 6, head_separation, axis = 1)     # z2 = head_separation

# Print the first 10 rows from the line of response (LoR) data
print(f"First 10 rows from the 3D-formatted LoR data:\n{lors[:10]}\n")
```

```
First 10 rows from the 3D-formatted LoR data:
[[0.000e+00 1.900e+02 1.687e+02 0.000e+00 3.463e+02 1.428e+02 6.000e+02]
 [1.000e-01 2.437e+02 1.676e+02 0.000e+00 3.145e+02 3.139e+02 6.000e+02]
 [1.000e-01 1.941e+02 4.100e+02 0.000e+00 4.171e+02 2.401e+02 6.000e+02]
 [2.000e-01 2.885e+02 2.962e+02 0.000e+00 3.847e+02 4.525e+02 6.000e+02]
 [2.000e-01 1.646e+02 1.151e+02 0.000e+00 3.829e+02 3.534e+02 6.000e+02]
 [2.000e-01 1.286e+02 1.322e+02 0.000e+00 4.649e+02 2.661e+02 6.000e+02]
 [2.000e-01 3.339e+02 3.735e+02 0.000e+00 3.316e+02 1.310e+02 6.000e+02]
 [3.000e-01 1.640e+02 1.115e+02 0.000e+00 3.794e+02 3.534e+02 6.000e+02]
 [4.000e-01 2.620e+02 2.094e+02 0.000e+00 2.932e+02 2.808e+02 6.000e+02]
 [4.000e-01 1.304e+02 2.749e+02 0.000e+00 2.543e+02 3.062e+02 6.000e+02]]
```

In order to keep this notebook didactic and minimise the amount of mathematical derivations necessary, a 2D PEPT-EM implementation is provided, applied to the ZY projection of the LoRs:

[3]:
```
# Project the 3D lines onto the YZ plane for ease of analysis - i.e. select only
# columns [time, y1, z1, y2, z2] and flip columns to get [time, x1, y1, x2, y2]
lors = np.array(lors[:, [0, 3, 2, 6, 5]])
print(f"First 5 rows from the 2D YZ-projected LoR data:\n{lors[:5]}\n")
```

```
First 5 rows from the 2D YZ-projected LoR data:
[[0.000e+00 0.000e+00 1.687e+02 6.000e+02 1.428e+02]
 [1.000e-01 0.000e+00 1.676e+02 6.000e+02 3.139e+02]
 [1.000e-01 0.000e+00 4.100e+02 6.000e+02 2.401e+02]
 [2.000e-01 0.000e+00 2.962e+02 6.000e+02 4.525e+02]
 [2.000e-01 0.000e+00 1.151e+02 6.000e+02 3.534e+02]]
```

The `pept.plots` subpackage provides two PEPT-oriented interfaces to the Plotly visualisation library—`PlotlyGrapher` and `PlotlyGrapher2D`—containing straightforward plotting functions for PEPT data. Let's plot the lines of response from the `lors` array:

```
[4]: # Plot the 3D LoRs that were initialised
from pept.plots import PlotlyGrapher2D

grapher = PlotlyGrapher2D()
grapher.add_lines(lors)
grapher.show()
```

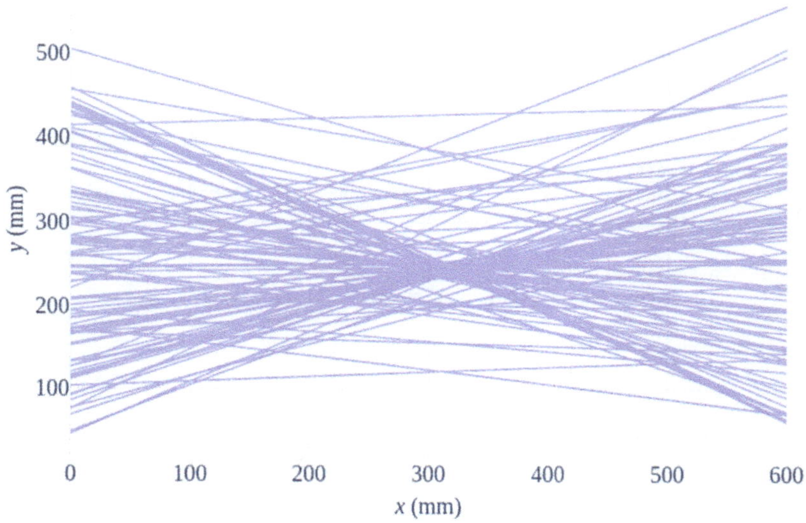

Calculate MDP and assign a weight to each LoR

The next step of the process is to calculate the minimum distance point (MDP) for the set of LoRs currently under consideration, akin to the Birmingham algorithm (see the main text). Then, using this information, each LoR is assigned a weight according to its distance from the MDP, with lines lying closer to this point being assigned a greater weight, and those further away a lesser weight.

In the plot below, the weight of a line is represented by its opacity, with more heavily-weighted LoRs being more opaque, and less-heavily-weighted LoRs being more transparent. The size of the red circle representing the minimum distance point corresponds to its variance, a measure of how strongly different LoRs' distances from the MDP vary—if all LoRs lie close to the MDP (as one would expect if only true LoRs were present) one would expect a small variance.

[5]:
```
# Rewrite LoRs in the vectorial form y(x) = position + x * direction
lors_original = np.copy(lors)

lors[:, 3] = lors[:, 3] - lors[:, 1]
lors[:, 4] = lors[:, 4] - lors[:, 2]

lors[:, 3:] /= np.linalg.norm(lors[:, 3:], axis=-1)[:, np.newaxis]
```

[6]:
```
# Calculate an initial guess for the location (essentially a minimum distance
# point) for these LoRs
from numpy import newaxis as nx


def centroid(lors, weights):
    m = (np.identity(2)[nx, :, :] - lors[:, nx, 3:] * lors[:, 3:, nx])
    m *= weights[:, nx, nx]

    M = np.sum(m, axis=0)
    V = np.sum(np.sum(m * lors[:, np.newaxis, 1:3], axis=-1), axis=0)

    return np.matmul(np.linalg.inv(M), V)


def dist_matrix(x, lors):
    X = x[np.newaxis, :3] - lors[:, 1:3]
    return np.sum(X**2, axis=-1) - np.sum(X * lors[:, 3:], axis=-1)**2


def latent_weights(d2, s, eps=0, r=None):
    if r is None:
        w = np.exp(-d2 / 2 / s**2) / s**2 + 10**(-20)
        w /= np.sum(w) + eps
    else:
        w = np.exp(-d2 / 2 / s**2) * r / s**2 + 10**(-20)
        w /= np.sum(w) + eps * (1 - np.sum(r))
    return w

# Begin with equal weights for all LoRs
weights = np.ones(len(lors))

# Calculate MDP and distance from it to all LoRs
x = centroid(lors, weights)
d2 = dist_matrix(x, lors)
variance = np.sqrt(np.mean(d2))

# Then recalculate the weight of each LoR for the given centroid
weights = latent_weights(d2, variance, eps=0.0001)

# Set the centroid's timestamp as the mean timestamp of the LoRs used
centre = np.insert(x, 0, lors[:, 0].mean())

print(f"Before iterating: Centroid = {x} mm, Variance = {variance} mm")
```

```
Before iterating: Centroid = [297.30957115 253.2870968 ] mm, Variance =
49.221029645817175 mm
```

```
[7]: # Plot LoRs, setting each line's opacity as its relative weight. The centroid's
     # size is given by its variance
     grapher = PlotlyGrapher2D()

     for i in range(len(lors_original)):
         grapher.add_lines(
             lors_original[i:i+1, :],
             opacity=0.6 * weights[i] / weights.max(),
             color='blue',
         )

     grapher.add_points([centre], size=variance)

     grapher.show()
```

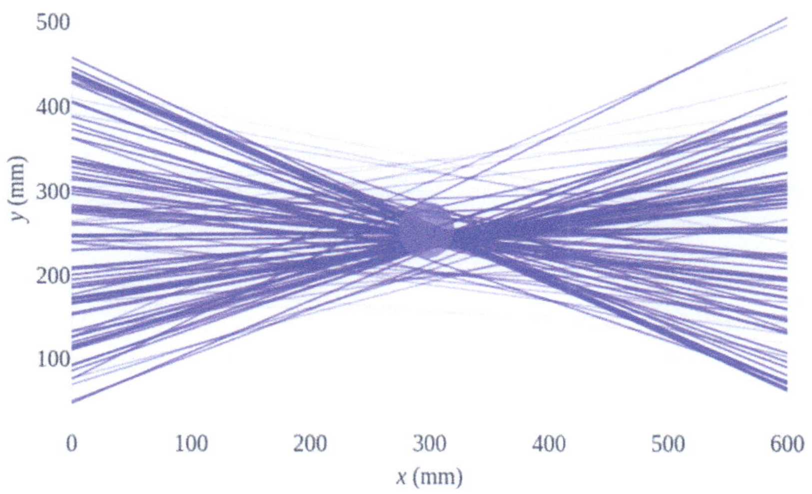

Recalculate MDP with previous weights

The above step is then repeated, iteratively recalculating the centroid and variance from the weighted LoRs, which themselves are also recalculated at each step. As can be seen both from the variance printed at the end of the section below and the relative size of the red dot compared to the first iteration, this process markedly reduces the impact of 'outlier' LoRs.

Once finished iterating, the particle position can be taken simply as the finally-computed MDP.

```
[8]: x = centroid(lors, weights)
     d2 = dist_matrix(x, lors)
     variance = np.sqrt(np.mean(d2 * weights))
     weights = latent_weights(d2, variance, eps=0.0001)

     print(f"After iterating: Centroid = {x} mm, Variance = {variance} mm")

     # Set the centroid's timestamp as the mean timestamp of the LoRs used
     centre = np.insert(x, 0, lors[:, 0].mean())
```

```
After iterating: Centroid = [307.1742582  246.89900838] mm, Variance =
2.496738080556994 mm
```

```
[9]: # Plot LoRs, setting each line's opacity as its relative weight. The centroid's
     # size is given by its variance
     grapher = PlotlyGrapher2D()

     for i in range(len(lors_original)):
         grapher.add_lines(
             lors_original[i:i + 1, :],
             opacity=0.6 * weights[i] / weights.max(),
             color='blue',
         )

     grapher.add_points([centre], size=variance)

     grapher.show()
```

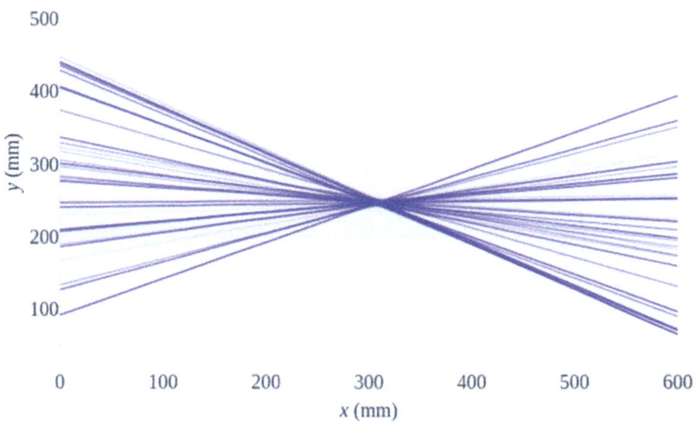


```
[10]: weights
```

```
[10]: array([2.93839347e-21, 1.08869661e-02, 2.93839347e-21, 2.93839347e-21,
             6.08003739e-05, 2.93839347e-21, 3.11476521e-02, 4.69393079e-06,
             4.39210507e-02, 2.93839347e-21, 2.93839347e-21, 3.78099188e-02,
             2.80015791e-02, 2.93839347e-21, 4.49988001e-02, 4.21540040e-02,
             7.86064944e-06, 9.01549032e-03, 7.83724877e-05, 2.06077428e-04,
             2.93839347e-21, 8.86225826e-06, 6.67392310e-07, 2.56150770e-04,
             3.58119025e-02, 2.93841837e-21, 7.11420532e-07, 2.20109239e-02,
             2.49297032e-02, 2.93839347e-21, 1.90961204e-02, 2.93839347e-21,
             2.93839347e-21, 2.23000205e-05, 7.93508335e-03, 4.46589345e-02,
             2.37014522e-02, 2.93848439e-21, 4.47137760e-02, 2.93839347e-21,
             4.70514178e-02, 4.37910097e-02, 2.93839349e-21, 2.93839347e-21,
             3.40340977e-02, 2.93839347e-21, 2.93839347e-21, 1.00855024e-03,
             2.60680229e-03, 2.93839347e-21, 2.93839347e-21, 5.22118427e-05,
             4.22693075e-02, 2.83081148e-02, 2.93839347e-21, 2.45916385e-02,
             2.93839347e-21, 3.40142507e-02, 4.50154537e-02, 2.93839347e-21,
             8.95917719e-03, 4.13017229e-02, 7.41202405e-05, 2.93839347e-21,
             2.93839347e-21, 3.00102488e-04, 3.02926512e-04, 2.93839347e-21,
             2.93839347e-21, 2.93839347e-21, 6.06707538e-03, 8.00626032e-04,
             2.93839347e-21, 3.30141437e-02, 1.34938554e-05, 9.48651717e-17,
             3.31681876e-06, 4.65762889e-02, 2.93839347e-21, 3.64921878e-20,
             1.88522460e-02, 1.51205757e-07, 1.03270314e-03, 2.93839347e-21,
             2.93839347e-21, 2.93839347e-21, 2.93839347e-21, 2.94409116e-03,
             2.93839347e-21, 1.07261036e-02, 1.73036210e-02, 1.69665539e-04,
             1.81430735e-02, 2.93839347e-21, 3.02824135e-08, 4.84379002e-03,
             2.93839347e-21, 6.18260205e-03, 8.14229361e-03, 3.45421193e-05])
```

In this instance, we see that the point converges very quickly and the PEPT-EM algorithm reduces the weight of outlier LORs. In the case of multiple particles, we may need more iterations in order to converge on the correct points.

Complete PEPT-EM algorithm code

The code developed in the previous sections can be applied to consecutive 'samples' of LoRs to follow the tracer's movement through the fluidised bed, fully tracking its trajectory. A complete PEPT-EM algorithm implementation is given below.

First initialise a larger, multi-sample set of LoRs to analyse; change the `nrows` argument below to read in a different number of LoRs (the file included contains 80,000 LoRs in total), though keep in mind the computational expense of processing large numbers of LoRs!

[11]:
```
# Read in a sample of experimental PEPT data into a NumPy array. Skip the file
# header's first 15 lines and read in the first `nrows` LoRs
lors_raw = pept.utilities.read_csv(
    "Data/FluidisedBed1Tracer.csv",
    skiprows=15,
    nrows=1000,
)

# Convert the LoRs into the general 3D line format [t, x1, y1, z2, x2, y2, z2].
# See the `PEPT Data Format` interactive example for detailed explanations

# ADAC Forte's two detector heads' separation, in mm
head_separation = 600

lors = np.insert(lors_raw, 3, 0, axis=1)                    # z1 = 0
lors = np.insert(lors, 6, head_separation, axis=1)          # z2 = head_separation

# Project the 3D lines onto the YZ plane for ease of analysis - i.e. select only
# columns [time, y1, z1, y2, z2] and flip columns to get [time, x1, y1, x2, y2]
lors = np.array(lors[:, [0, 3, 2, 6, 5]])

# Print the first 10 rows from the line of response (LoR) data
print(f"First 10 rows from the 2D YZ-projected LoR data:\n{lors[:10]}\n")
```

```
First 10 rows from the 2D YZ-projected LoR data:
[[0.000e+00 0.000e+00 1.687e+02 6.000e+02 1.428e+02]
 [1.000e-01 0.000e+00 1.676e+02 6.000e+02 3.139e+02]
 [1.000e-01 0.000e+00 4.100e+02 6.000e+02 2.401e+02]
 [2.000e-01 0.000e+00 2.962e+02 6.000e+02 4.525e+02]
 [2.000e-01 0.000e+00 1.151e+02 6.000e+02 3.534e+02]
 [2.000e-01 0.000e+00 1.322e+02 6.000e+02 2.661e+02]
 [2.000e-01 0.000e+00 3.735e+02 6.000e+02 1.310e+02]
 [3.000e-01 0.000e+00 1.115e+02 6.000e+02 3.534e+02]
 [4.000e-01 0.000e+00 2.094e+02 6.000e+02 2.808e+02]
 [4.000e-01 0.000e+00 2.749e+02 6.000e+02 3.062e+02]]
```

Iterating through each sample of LoRs, locate the tracer and append its location to a growing list of positions. As with the previous cells, you are invited to change the parameters highlighted below; details about each one are given in the comments.

[12]:
```
# Use PEPT-EM to track a moving tracer
from numpy import newaxis as nx

# PEPT-EM user-changeable parameters:
#     target_variance - target variance of tracer location found, in mm
target_variance = 0.1

# General PEPT tracking user-changeable parameters:
#     sample_start - the `lors` index at which to start tracking the tracer
#     sample_size - the number of LoRs used to compute a single location
#     overlap - the number of overlapping LoRs between consecutive samples
sample_start = 0
sample_size = 200
overlap = 100

# Tracked tracer locations will be appended to a growing list
positions = []


def centroid(lors, weights):
    m = (np.identity(2)[nx, :, :] - lors[:, nx, 3:] * lors[:, 3:, nx])
    m *= weights[:, nx, nx]

    M = np.sum(m, axis=0)
    V = np.sum(np.sum(m * lors[:, np.newaxis, 1:3], axis=-1), axis=0)

    return np.matmul(np.linalg.inv(M), V)


def dist_matrix(x, lors):
    X = x[np.newaxis, :3] - lors[:, 1:3]
    return np.sum(X**2, axis=-1) - np.sum(X * lors[:, 3:], axis=-1)**2

def latent_weights(d2, s, eps=0, r=None):
    if r is None:
        w = np.exp(-d2 / 2 / s**2) / s**2 + 10**(-20)
        w /= np.sum(w) + eps
    else:
        w = np.exp(-d2 / 2 / s**2) * r / s**2 + 10**(-20)
        w /= np.sum(w) + eps * (1 - np.sum(r))
    return w
```

```
# Rewrite LoRs in the vectorial form y(x) = position + x * direction
lors_original = np.copy(lors)

lors[:, 3] = lors[:, 3] - lors[:, 1]
lors[:, 4] = lors[:, 4] - lors[:, 2]

lors[:, 3:] /= np.linalg.norm(lors[:,3:], axis=-1)[:,np.newaxis]

while sample_start + sample_size < len(lors):
    sample = lors[sample_start:sample_start + sample_size]

    # Begin with equal weights for all LoRs in sample
    weights = np.ones(len(sample))

    # Calculate MDP and distance from it to all LoRs in sample
    x = centroid(sample, weights)
    d2 = dist_matrix(x, sample)
    variance = np.sqrt(np.mean(d2))

    # Then recalculate the weight of each LoR for the given centroid
    weights = latent_weights(d2, variance, eps=0.0001)

    # Iteratively recalculate MDP, LoR weights and variance of tracer location
    it = 0
    while variance < target_variance:
        x = centroid(sample, weights)
        d2 = dist_matrix(x, sample)
        variance = np.sqrt(np.mean(d2*weights))
        weights = latent_weights(d2, variance, eps=0.0001)

        it += 1

    # Set the centroid's timestamp as the mean timestamp of the LoRs used
    centre = np.insert(x, 0, sample[:, 0].mean())
    positions.append(centre)

    sample_start += sample_size - overlap

positions = np.array(positions)
print(f"First 10 tracked positions:\n{positions[:10]}")
```

```
First 10 tracked positions:
[[  3.9045     318.63524612 253.5024463 ]
 [  7.922      322.13210474 261.15287682]
 [ 11.777      304.93327131 261.90323644]
 [ 15.336      301.03956333 253.5327003 ]
 [ 18.9635     287.38610671 256.16911127]
 [ 22.7215     278.54317502 258.16693051]
 [ 26.422      295.67594028 258.96956364]
 [ 29.8        289.89532378 261.34140247]]
```

```
[13]: # Plot some of the LoRs used, plus the tracer locations we found
grapher = PlotlyGrapher2D(
    cols=2,
    subplot_titles=[
        "First 500 LoRs",
        f"Tracer locations tracked from {len(lors)} LoRs",
    ],
)

grapher.add_lines(lors_original[:500])
grapher.add_points(positions, col=2)

# Set figure size to fit subplots
grapher.fig.update_layout(height=600, width=1700)
grapher.show()
```

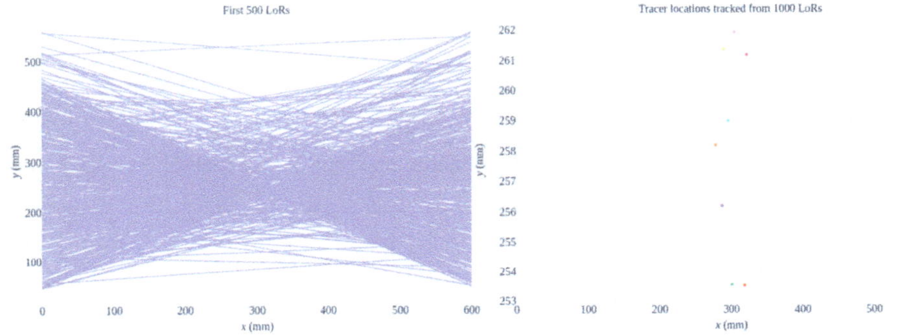

6.2.18 The K-medoids method

The K-medoids method in brief

The 2004 paper of Gundogdu *et al* [45], which preceded both PEPT-ML [38] and the Wiggins G-means clustering algorithm [6], was in fact the first algorithm to consider the use of 'cutpoints' in lieu of the iterative removal of LoRs or the implementation of line density maps, as well as the first to exploit the use of clustering algorithms in order to determine the positions of particle centroids. However, the methodology was never widely adopted in the field, and has since been superseded by the algorithms which followed it, both in terms of location accuracy, and the number of tracers which may be successfully tracked. Nonetheless, many of the concepts introduced by the paper are still being drawn upon to develop new algorithms [37], and as such it is certainly valuable to include in our discussion within this book. As with the other methods described in this chapter, we begin with a simple overview of the algorithm, before explaining each individual step thereof in greater detail.

1. A sample of N_s individual, consecutive lines of response from a given PEPT data set (see chapter 2) is considered.
2. For each pair of LoRs in the sample, the cutpoint—the midpoint of the minimum-distance line between the two LoRs—is determined.

3. Any cutpoints falling outside the detector's field of view are discarded, as well as any cutpoints yielded by pairs of LoRs whose minimum separation is more than 10 mm.
4. Cutpoints identified as outliers—determined based on their distances from their nearest neighbours [46–48]—are removed.
5. The remaining cutpoints are separated into a number N_C clusters using the Partitioning Around Medoids (PAM) algorithm [47, 48].
6. The *silhouette coefficient*, $s(i)$, is determined for each object in each cluster, and used to determine the average silhouette value, $\bar{SC}$ for each cluster.
7. Steps (6) and (7) are repeated for different values of N_C.
8. The average silhouette coefficient values for each N_C tested are compared. The value of N_C yielding the highest average silhouette coefficient is taken as the number of 'natural clusters'—i.e. the true number of tracers in the system.
9. The centroids of the clusters finally obtained are taken as the positions of the tracer particles within the system.

The K-medoids method in detail

(1–3) Computing cutpoints: For a given sample of lines of response (LoRs), the cutpoints—or points of closest approach—are determined for each pair of LoRs. For two non-parallel 3D lines, this corresponds to the middle of the unique segment that is perpendicular to both lines. In order to remove some of the 'cross-talk' between LoRs that come from regions that are far apart, only the cutpoints from pairs of LoRs that are closer than 10 mm are considered. Finally, all cutpoints falling outside the scanner's field of view—computed as simply the smallest box containing all LoRs—are removed. For an example of cutpoints computed from a one-tracer dataset, see panel (*a*) from figure 6.13.

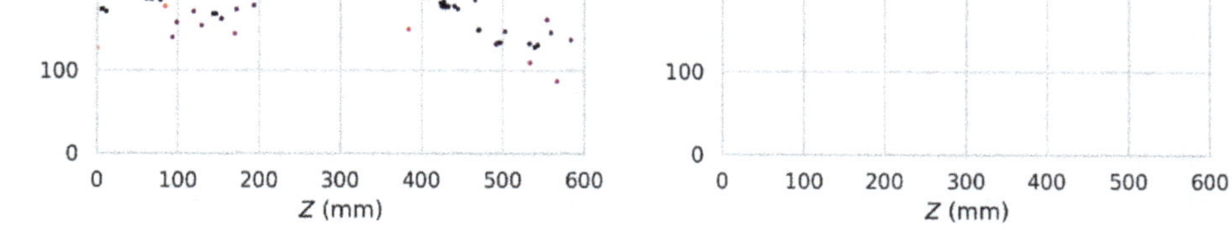

Figure 6.13. Panel (*a*) depicts all cutpoints found for a maximum distance between pairs of LoRs of 10 mm; the cutpoints are colour-coded by the distance to their nearest neighbour. Panel (*b*) shows the remaining cutpoints after removing all points with a nearest neighbour distance larger than 2.479 mm (as described in the original paper [45]); the cutpoints are again colour-coded by the distance to their remaining nearest neighbour.

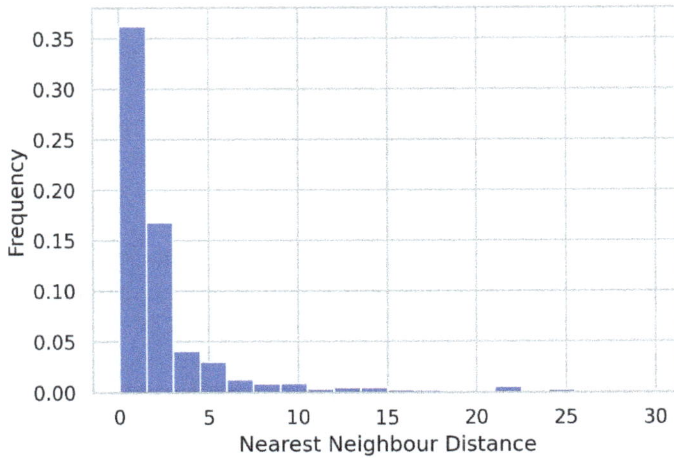

Figure 6.14. Normalised histogram depicting the distribution of the cutpoints' nearest neighbour distance.

(4) Filtering cutpoints: As the clustering algorithm employed can be quite sensitive to sparse points 'connecting' different clusters, a filtering step is applied to remove some of the sparse cutpoints. First, the distance from each cutpoint to their nearest neighbour is computed; as seen in panel (*b*) from figure 6.14, the distribution of nearest neighbour distances tails off at large values. Some of the outliers can be removed by simply filtering out all cutpoints whose nearest neighbour is further than a fixed threshold; in the original paper, a value of 2.479 was determined to be the ideal (in the interactive notebook at the end of this section, the reader can change this threshold to visualise the effect of harsher—or more lenient—outlier removal). The remaining cutpoints, as depicted in panel (*b*) from figure 6.13, show the successful filtering of many outliers; however, some noisy points may happen to randomly concentrate in some regions—in this case, the *k*th nearest neighbour distance may be used as a more robust method.

(5–8) Clustering filtered cutpoints: Perhaps the most famous clustering method is the k-means algorithm, which iteratively partitions a dataset into clusters around their geometric centroids. A natural, more robust—but much more computationally involved—extension to k-means is the k-medoids (or partitioning around medoids—PAM [47, 48]) algorithm, which chooses actual points from the dataset as the cluster centres. Given n cutpoints and k clusters, the goal is to select k cutpoints as the medoids that locally minimise the sum of distances from each cutpoint to their closest medoid. After an initial random assignment of k medoids and association of all cutpoints with their closest medoid, all swaps possible between the medoids and their partition's points are considered; the swap which minimises the sum of distances is performed, and the previous steps are repeated. If no other swap minimises the sum of distances, the algorithm terminates. Notice, however, that the number of partitions k must be known *a priori*, and in the case of PEPT data, it may not necessarily be that the optimum k is equal to the number of tracers in the system!

The *silhouette coefficient* (*SC*) provides a measure of how 'natural' a particular clustering is. For a given partitioning of n points into k clusters, for each point, the mean distance to all points within the cluster a and mean distance to all points within

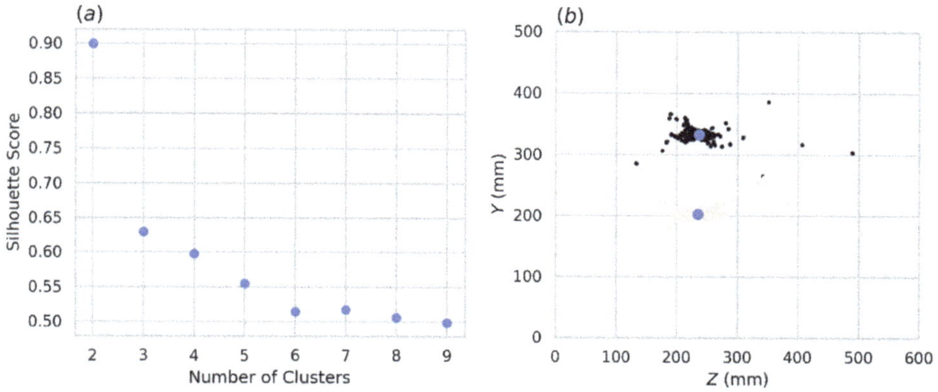

Figure 6.15. Panel (*a*) shows the silhouette score for the k-medoids clustering results at varying numbers of clusters. Panel (*b*) depicts the clustered cutpoints for the best silhouette score—in this case found to be 2—with points colour-coded by their cluster label; the final tracer locations, corresponding to the geometric centroid of each cluster, are plotted as large, blue circles.

the nearest distinct cluster b are computed; the silhouette value is then simply $(b - a)/\max(a, b)$. As can be seen from the definition, it varies between –1 (for 'wrong' clusterings), 0 (for overlapping clusters) and 1 (ideal, or natural clustering). The silhouette coefficient SC is then simply the average silhouette value for each cluster. In the case of PEPT, the SC is computed for different numbers of clusters k, with the best k being selected as simply the value that maximises the silhouette coefficient. For an example of SC values computed for a two-tracer PEPT dataset, see panel (*a*) in figure 6.15. Panel (*b*) from the same figure shows the clustered cutpoints—colour-coded by their cluster label—as given by the 'most natural' k-medoids partitioning. However, a downside of the silhouette coefficient approach is that the latter is only defined for two or more clusters—which means that in the case of single-particle tracking, the sole 'natural' cluster around the actual tracer will always be artificially split.

(9) Tracking tracer locations: Once the ideal k-medoids partitioning for a given sample of LoRs has been found, the tracer locations are simply computed as the geometric centroid of each cluster. The same steps—cutpoints, filtering, finding most natural partitioning—are then applied to consecutive samples of LoRs to dynamically track the tracers across multiple time frames.

Strengths and weaknesses

While the k-medoids PEPT algorithm may not use the most sophisticated techniques in comparison with methods developed in the last decade, it nonetheless paved the way for machine learning approaches in PEPT. The original paper explored multiple early data science techniques for determining the optimum number of clusters without *a priori* knowledge of the number of tracers in the system; in this sense, the k-medoids algorithm proves to be pioneering work for more modern approaches. However, due to its relative computational expense—especially considering its time—and high probability of missing tracers, it has not seen extensive use.

Strengths
- Autonomously finds the 'most natural' k-medoids-based clustering of the tracer data.

Weaknesses
- Splits single tracers into multiple artificial clusters.
- The k-medoids algorithm only partitions data—i.e. it does not remove noise—and the 'most natural' clustering may not find the actual tracers.

6.2.19 Interactive example: K-medoids method

The notebook files accompanying this section are available in the supplementary data online at https://iopscience.iop.org/book/978-0-7503-3071-8.

Interactive PEPT analysis example using the K-medoids method [45]

This Jupyter Notebook

An example dataset is used from an experiment run at the University of Birmingham Positron Imaging Centre using the ADAC Forté. It consists of a single 1 mm diameter MCC particle activated with fluorine-18 inside a bubbling fluidised bed. The fluidised bed was filled with 90% sand and 10% MCC; air was fed into the bottom of the bed at a rate of 37 litres per minute at 3.5 bar. This dataset was chosen for its high quality captured lines of response, with the tracer still depicting the random particle motion that is inherent to bubbling fluidised beds—and typical in Lagrangian particle tracking.

The `pept` Python library is used for initialising and visualising PEPT data. While not required *per se* for illustrating PEPT algorithms' processing steps, it significantly reduces the amount of repetitive code and visual noise, allowing the reader to focus on the main conceptual procedures.

[1]:
```python
# Importing necessary libraries for numerical array manipulation and plotting,
# plus the `pept` library for straightforward PEPT data processing
import numpy as np
import matplotlib.pyplot as plt
import pept

# This algorithm implementation uses K-Medoids from the `sklearn_extra` library
# and the Silhouette Coefficient calculation from `sklearn`
import sklearn
import sklearn_extra

# Set Matplotlib theme (try "darkgrid" too for a more modern look)
import seaborn
seaborn.set_style("whitegrid")

# Set Plotly figure output (try "browser" too for interactive graphs)
import plotly
plotly.io.renderers.default = "png"
```

Initialise raw line of response data

In this initial step, we read in the 'raw' LoR data from which our tracer positions are calculated. As illustrated in the 'Effect of sample size and overlap' notebook, varying the numbers of LoRs used to calculate a PEPT location can strongly affect the accuracy of the final position computed. Try altering the value `nrows` below to see how this influences your results.

[2]:
```
# Read in a sample of experimental PEPT data into a NumPy array. Skip the file
# header's first 15 lines, then read in the first `nrows` LoRs
lors_raw = pept.utilities.read_csv(
    "Data/FluidisedBed1Tracer.csv",
    skiprows=15,
    nrows=200,
)

# Convert the LoRs into the general 3D line format [t, x1, y1, z2, x2, y2, z2].
# See the `PEPT Data Format` interactive example for detailed explanations

# ADAC Forte's two detector heads' separation, in mm
head_separation = 600

lors = np.insert(lors_raw, 3, 0, axis = 1)                  # z1 = 0
lors = np.insert(lors, 6, head_separation, axis = 1)        # z2 = head_separation

# Print the first 10 rows from the line of response (LoR) data
print(f"First 10 rows from the 3D-formatted LoR data:\n{lors[:10]}\n")
```

```
First 10 rows from the 3D-formatted LoR data:
[[0.000e+00 1.900e+02 1.687e+02 0.000e+00 3.463e+02 1.428e+02 6.000e+02]
 [1.000e-01 2.437e+02 1.676e+02 0.000e+00 3.145e+02 3.139e+02 6.000e+02]
 [1.000e-01 1.941e+02 4.100e+02 0.000e+00 4.171e+02 2.401e+02 6.000e+02]
 [2.000e-01 2.885e+02 2.962e+02 0.000e+00 3.847e+02 4.525e+02 6.000e+02]
 [2.000e-01 1.646e+02 1.151e+02 0.000e+00 3.829e+02 3.534e+02 6.000e+02]
 [2.000e-01 1.286e+02 1.322e+02 0.000e+00 4.649e+02 2.661e+02 6.000e+02]
 [2.000e-01 3.339e+02 3.735e+02 0.000e+00 3.316e+02 1.310e+02 6.000e+02]
 [3.000e-01 1.640e+02 1.115e+02 0.000e+00 3.794e+02 3.534e+02 6.000e+02]
 [4.000e-01 2.620e+02 2.094e+02 0.000e+00 2.932e+02 2.808e+02 6.000e+02]
 [4.000e-01 1.304e+02 2.749e+02 0.000e+00 2.543e+02 3.062e+02 6.000e+02]]
```

The `pept.plots` subpackage provides two PEPT-oriented interfaces to the Plotly visualisation library–`PlotlyGrapher` and `PlotlyGrapher2D`–containing straightforward plotting functions for PEPT data. Let's plot the lines of response from the `lors` array:

[3]:
```
# Plot the 3D LoRs that were initialised
from pept.plots import PlotlyGrapher

grapher = PlotlyGrapher()
grapher.add_lines(lors)
grapher.show()
```

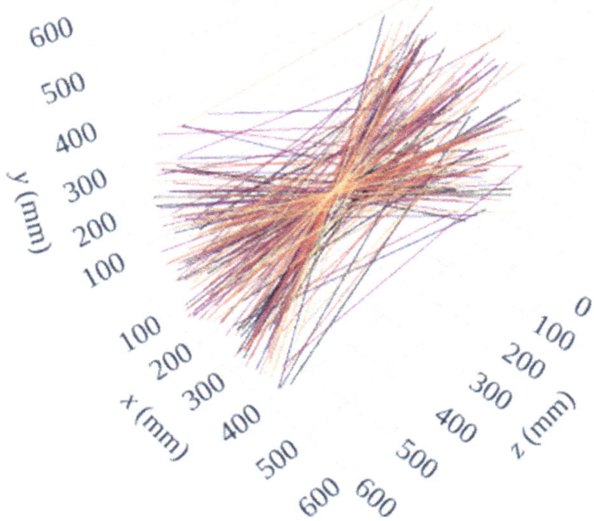

Note the PEPT 3D axes convention, differing from the common mathematical coordinates with Y and Z being switched, such that the Y axis points upwards, while the Z axis is perpendicular to the two detector heads.

Find cutpoints

In this step, we find the point of closest approach for each pair of LoRs and take the midpoint between the two lines as a 'cutpoint'. Of course, for the purposes of PEPT, we only want to do this for LoRs that are being emitted from our tracer, and ignore false LoRs as far as possible. We do this by defining a `max_distance`, beyond which we can assume two LoRs do not pass through our particle, and simply not calculating a cutpoint for any two LoRs separated by more that this distance.

Try adjusting the value of `max_distance` and see what effect it has on the distribution of cutpoints produced.

As with the `nlors_remove` parameter in the Birmingham method example, being more restrictive with your choice of `max_distance` will remove more erroneous points, but yield less data, while being less conservative will give you more data, but a higher probability of including erroneous values.

```
[4]: # For all pairs of lines closer than `max_distance`, find their minimum distance
     # points (cutpoints). As described in the original paper, the cutpoints falling
     # outside the experimental volume containing the LoRs are discarded (this is
     # done by default by the `find_cutpoints` subroutine) and a maximum distance
     # allowed between pairs of LoRs of 10 mm - but it is recommended to change it
     # and see its effect!
     from pept.tracking.peptml import find_cutpoints

     max_distance = 10.
     cutpoints = find_cutpoints(lors, max_distance).points
```

```
[5]: # Plot this sample's LoRs and their corresponding cutpoints
grapher = PlotlyGrapher(cols = 2)

grapher.add_lines(lors)
grapher.add_points(cutpoints, col = 2)

grapher.show()
```

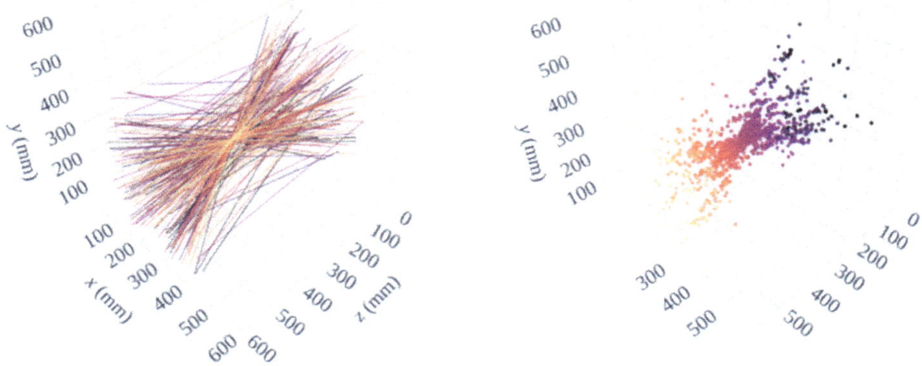

Notice the main characteristics of the cutpoints—importantly, they are more concentrated around our tracer, but they also include a significant amount of noise (as yielded by larger `max_distance`).

Filter cutpoints with far nearest neighbors

A simple filtering step is used to remove some of the outlier cutpoints, as an aid to the clustering algorithm employed in the later sections. For each cutpoint, the distance to its nearest neighbour is computed; then, all cutpoints whose nearest neighbor is further than 2.479 mm are discarded—this value was given in the original paper, but `neighbour_distance` is left as a free parameter in the implementation below as it is expected to vary with different experimental setups.

```
[6]: # Filter all cutpoints whose nearest neighbours are further than
# `neighbour_distance` mm away
#
# Use SciPy's high quality kD-tree implementation to speed up nearest neighbor
# finding; build the tree our of the cutpoints, then query the *two* nearest
# neighbours for each cutpoint (the first "neighbour" will be the cutpoint
# itself)
from scipy.spatial import KDTree

tree = KDTree(cutpoints)
distances, indices = tree.query(cutpoints, 2)

# The first "neighbour" is just the given cutpoint (hence their distance is 0)
distances = distances[:, 1]
indices = indices[:, 1]
```

```
[7]: # Remove all cutpoints whose nearest neighbor distance is larger than
     # `neigbour_distance`
     neighbour_distance = 2.479
     cutpoints_filtered = cutpoints[distances < neighbour_distance]
```

```
[8]: # Plot the cutpoints before and after filtering, colour-coded by the nearest
     # neighbour distance
     grapher = PlotlyGrapher(cols = 2)

     grapher.add_points(cutpoints, color=distances)
     grapher.add_points(cutpoints_filtered, col = 2)

     grapher.show()
```

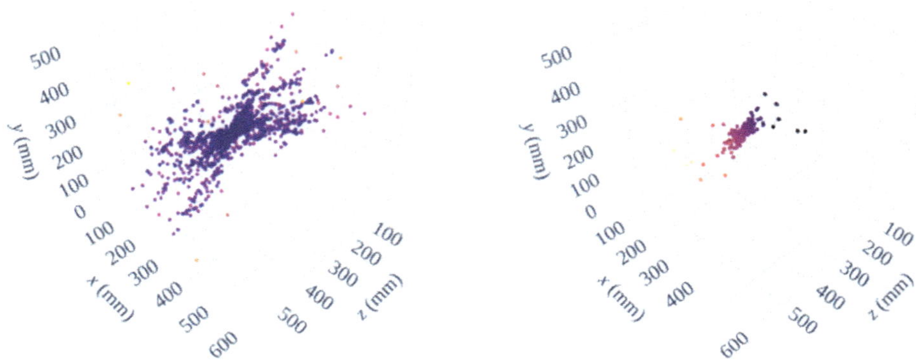

Cluster filtered cutpoints using K-medoids

Having converted our LoRs into cutpoints and filtered out the most extreme outliers, the next step is to partition the remaining cutpoints into *clusters*, with the tracer location —hopefully—being in the centre of the densest cluster. The clustering algorithm employed by the original authors is K-medoids (sometimes called partitioning around medoids, or PAM), which requires an *a priori* number of clusters `num_clusters` to partition the points into. However, the `num_clusters` is not trivial to choose, as ideally we would like the noise to be separated out from the denser cutpoints region around the tracer. In this case, the silhouette coefficient (or score) is computed for varying numbers of `num_clusters`; the number of clusters corresponding to the largest silhouette coefficient is then taken as the *natural number of clusters*.

Though K-medoids is very close to the well-known K-means clustering method, algorithmically it is much more complex; as such, there is no 'standard' implementation in `SciPy` or the standard `SciKit-Learn`. In this notebook, the `SciKit-Learn`-like `sklearn_extra` library is used for the particular K-medoids implementation. The silhouette score is computed using `sklearn`; do note that the former is only defined for more than two clusters.

[9]:
```
# Try KMedoids (from the `sklearn_extra` library) for different numbers of
# clusters to maximise the silhouette score (from the `sklearn` library)
from sklearn_extra.cluster import KMedoids
from sklearn.metrics import silhouette_score

num_clusters = np.arange(2, 10)
silhouette_scores = []

for nc in num_clusters:
    clusterer = KMedoids(nc)
    labels = clusterer.fit_predict(cutpoints_filtered[:, 1:4])
    sc = silhouette_score(cutpoints_filtered[:, 1:4], labels)
    silhouette_scores.append(sc)

silhouette_scores = np.array(silhouette_scores)
best_nc = num_clusters[silhouette_scores.argmax()]

print(f"Silhouette scores found:\n{silhouette_scores}\n")
print(f"Best silhouette score was found for {best_nc} clusters")
```

```
Silhouette scores found:
[0.35399861 0.27465399 0.20317895 0.23527612 0.18977794 0.17676967
 0.1697545  0.16072266]

Best silhouette score was found for 2 clusters
```

[10]:
```
# Cluster the filtered cutpoints with the "best" number of clusters found and
# plot them
clusterer = KMedoids(best_nc)
labels = clusterer.fit_predict(cutpoints_filtered[:, 1:4])

# Append the label to the filtered cutpoints
clustered_cutpoints = np.c_[cutpoints_filtered, labels]

PlotlyGrapher().add_points(clustered_cutpoints).show()
```

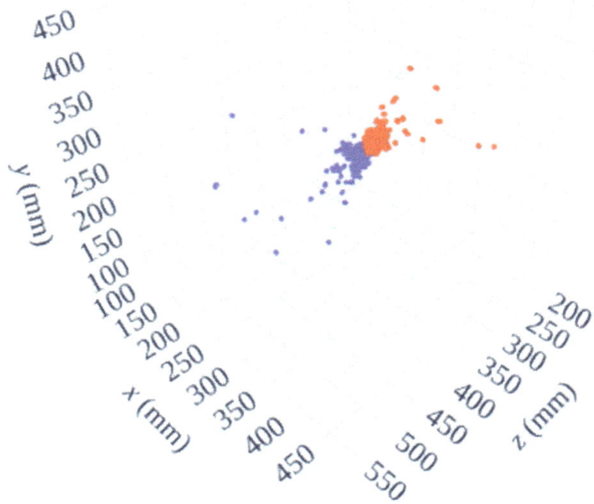

In this particular case, the 'natural number of clusters' as found using the silhouette coefficient yielded an 'artificial split' inside the cluster we visually distinguish. This is a shortcoming of the silhouette coefficient approach, which is only defined for more than two clusters.

For completeness, in the next code cells we compute the predicted cluster centroids, then provide the complete K-medoids implementation and finally apply it to a two-tracer dataset (with better results!).

Compute clusters' centroids

The final step, as discussed above, is to compute the locations of the tracers are computed as the geometric centroids of each cluster.

```
[11]: # Compute the centroid of each cluster
centres = []

for label in range(labels.max() + 1):
    centres.append(cutpoints_filtered[labels == label].mean(axis=0))

centres = np.array(centres)

print(f"Cluster centres found:\n{centres}")
```

```
Cluster centres found:
[[  3.81271038 281.3510202  243.84332778 326.66662573]
 [  3.91186702 279.7116615  244.65398317 305.38857113]]
```

[12]:
```
# Plot the clustered cutpoints and the cluster centroids found
grapher = PlotlyGrapher()
grapher.add_points(clustered_cutpoints)
grapher.add_points(centres, size=5)
grapher.show()
```

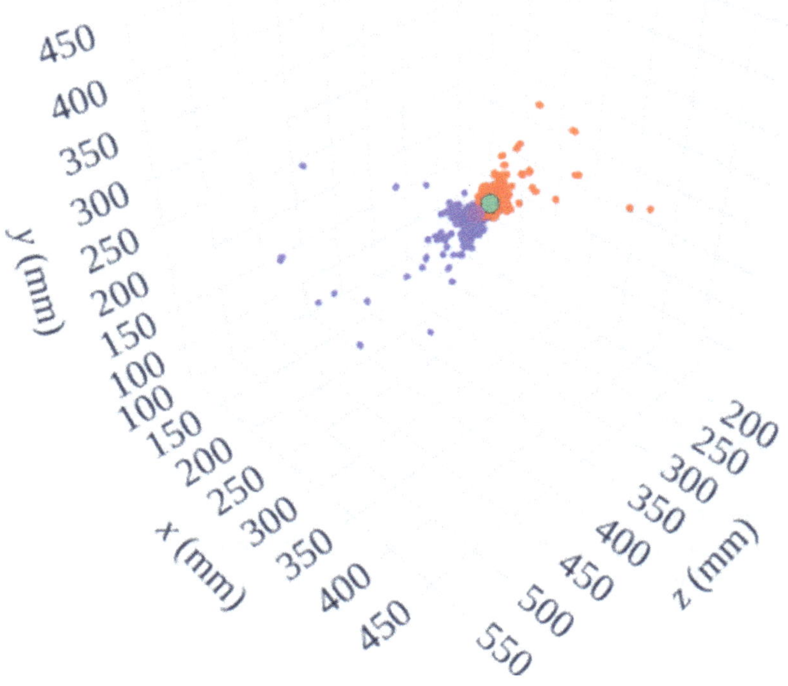

Complete K-medoids method mode

The code developed in the previous sections can be applied to consecutive ‘samples’ of LoRs to follow the tracer’s movement through the fluidised bed, fully tracking its trajectory. A complete—albeit slow due to Python’s dynamic nature—K-medoids method implementation is given below.

First initialise a larger, multi-sample set of LoRs to analyse; change the `nrows` argument below to read in a different number of LoRs (the file included contains 80,000 LoRs in total), though keep in mind the computational expense of processing large numbers of LoRs!

```
[13]: # Read in a sample of experimental PEPT data into a NumPy array. Skip the file
      # header's first 15 lines and read in the first `nrows` LoRs
      lors_raw = pept.utilities.read_csv(
          "Data/FluidisedBed1Tracer.csv",
          skiprows=15,
          nrows=1000,
      )

      # Convert the LoRs into the general 3D line format [t, x1, y1, z2, x2, y2, z2].
      # See the `PEPT Data Format` interactive example for detailed explanations

      # ADAC Forte's two detector heads' separation, in mm
      head_separation = 600

      lors = np.insert(lors_raw, 3, 0, axis=1)                    # z1 = 0
      lors = np.insert(lors, 6, head_separation, axis=1)          # z2 = head_separation

      # Print the first 10 rows from the line of response (LoR) data
      print(f"First 10 rows from the 3D-formatted LoR data:\n{lors[:10]}\n")
```

```
First 10 rows from the 3D-formatted LoR data:
[[0.000e+00 1.900e+02 1.687e+02 0.000e+00 3.463e+02 1.428e+02 6.000e+02]
 [1.000e-01 2.437e+02 1.676e+02 0.000e+00 3.145e+02 3.139e+02 6.000e+02]
 [1.000e-01 1.941e+02 4.100e+02 0.000e+00 4.171e+02 2.401e+02 6.000e+02]
 [2.000e-01 2.885e+02 2.962e+02 0.000e+00 3.847e+02 4.525e+02 6.000e+02]
 [2.000e-01 1.646e+02 1.151e+02 0.000e+00 3.829e+02 3.534e+02 6.000e+02]
 [2.000e-01 1.286e+02 1.322e+02 0.000e+00 4.649e+02 2.661e+02 6.000e+02]
 [2.000e-01 3.339e+02 3.735e+02 0.000e+00 3.316e+02 1.310e+02 6.000e+02]
 [3.000e-01 1.640e+02 1.115e+02 0.000e+00 3.794e+02 3.534e+02 6.000e+02]
 [4.000e-01 2.620e+02 2.094e+02 0.000e+00 2.932e+02 2.808e+02 6.000e+02]
 [4.000e-01 1.304e+02 2.749e+02 0.000e+00 2.543e+02 3.062e+02 6.000e+02]]
```

Iterating through each sample of LoRs, locate the tracer and append its location to a growing list of positions. As with the previous cells, you are invited to change the parameters highlighted below; details about each one are given in the comments.

```
[14]: # Use the K-Medoids Method to track a moving tracer
      from pept.tracking.peptml import find_cutpoints
      from scipy.spatial import KDTree
      from sklearn_extra.cluster import KMedoids
      from sklearn.metrics import silhouette_score

      # K-Medoids Method user-changeable parameters:
      #     max_distance - only consider cutpoints from LoR pairs closer than this
```

```
#     neighbor_distance - filter out cutpoints whose nearest neighbor is further
#     max_num_clusters - the maximum number of clusters to try K-Medoids for
max_distance = 10
neighbour_distance = 2.479
max_num_clusters = 10

# General PEPT tracking user-changeable parameters:
#     sample_start - the `lors` index at which to start tracking the tracer
#     sample_size - the number of LoRs used to compute a single location
#     overlap - the number of overlapping LoRs between consecutive samples
sample_start = 0
sample_size = 100
overlap = 0

# Tracked tracer locations will be appended to a growing list
positions = []

while sample_start + sample_size <= len(lors):
    sample = lors[sample_start:sample_start + sample_size]

    # For all pairs of lines closer than `max_distance`, find their minimum
    # distance points (cutpoints).
    cutpoints = find_cutpoints(sample, max_distance).points

    # Filter all cutpoints whose nearest neighbours are further than
    # `neighbour_distance` mm away
    tree = KDTree(cutpoints)
    distances, indices = tree.query(cutpoints, 2)

    # The first "neighbour" is just the given cutpoint (hence their distance=0)
    distances = distances[:, 1]
    indices = indices[:, 1]

    # Remove all cutpoints whose nearest neighbor distance is larger than
    # `neigbour_distance`
    cutpoints_filtered = cutpoints[distances < neighbour_distance]

    # Try KMedoids (from the `sklearn_extra` library) for different numbers of
    # clusters to maximise the silhouette score (from the `sklearn` library)
    num_clusters = np.arange(2, max_num_clusters)
    silhouette_scores = []

    for nc in num_clusters:
        clusterer = KMedoids(nc)
        labels = clusterer.fit_predict(cutpoints_filtered[:, 1:4])
        sc = silhouette_score(cutpoints_filtered[:, 1:4], labels)
        silhouette_scores.append(sc)
```

```
    silhouette_scores = np.array(silhouette_scores)
    best_nc = num_clusters[silhouette_scores.argmax()]

    # Cluster the filtered cutpoints using the "best" number of clusters found
    clusterer = KMedoids(best_nc)
    labels = clusterer.fit_predict(cutpoints_filtered[:, 1:4])

    # For every cluster (i.e. label value >= 0), compute the centroid and append
    # it to the list of positions
    for label in range(labels.max() + 1):
        cluster = cutpoints_filtered[labels == label]
        centre = cluster.mean(axis=0)
        positions.append(centre)

    sample_start += sample_size - overlap

positions = np.array(positions)

print(f"First 10 tracked positions:\n{positions[:10]}")
```

```
First 10 tracked positions:
[[  1.85839416 282.28218804 244.14623576 327.48097551]
 [  1.75659449 278.78497053 243.83515082 306.8848631 ]
 [  5.93922287 280.54265678 244.3515917  324.68771995]
 [  5.8006654  279.85961939 245.21408279 304.40650316]
 [ 10.18547794 281.31374474 243.96270185 309.37619642]
 [  9.85931122 280.0382506  242.56872999 326.01332913]
 [ 13.61687063 279.31490932 244.31292291 309.03303927]
 [ 13.40525727 278.91247101 243.83936332 326.18002369]
 [ 17.26538462 280.69820831 242.15683884 327.32946359]
 [ 17.15006143 280.35083694 245.85394099 307.72094808]]
```

[15]:
```
# Plot some of the LoRs used, plus the tracer locations we found
grapher = PlotlyGrapher(
    cols=2,
    subplot_titles=[
        "First 500 LoRs",
        f"Tracer locations tracked from {len(lors)} LoRs",
    ],
)

grapher.add_lines(lors[:500])
grapher.add_points(positions, col=2)

grapher.show()
```

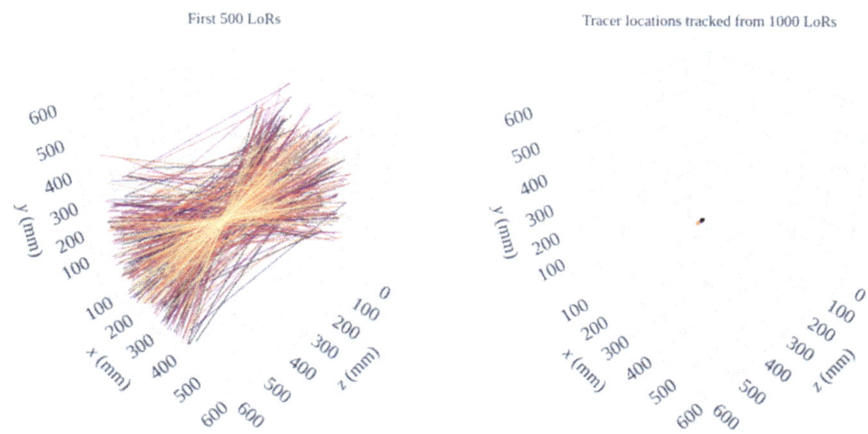

Multiple particle pracking

The complete code developed above can also be applied to a simple multiple tracer experiment. Here we use a sample of an experiment consisting of two 1 mm diameter alumina particles with fluorine-18 radioactive tracer material. The particles are fixed at the two ends of an arm rotating at 42 rpm, their trajectories describing full circles.

[16]:
```
# Read in a sample of experimental PEPT data into a NumPy array. Skip the file
# header's first 15 lines
lors_raw = pept.utilities.read_csv(
    "Data/RotatingArm2Tracers.csv",
    skiprows=15,
    nrows=10_000,
)

# Convert the LoRs into the general 3D line format [t, x1, y1, z2, x2, y2, z2].
# See the `PEPT Data Format` interactive example for detailed explanations

# ADAC Forte's two detector heads' separation, in mm
head_separation = 712

lors = np.insert(lors_raw, 3, 0, axis=1)                # z1 = 0
lors = np.insert(lors, 6, head_separation, axis=1)      # z2 = head_separation

# Print the first 10 rows from the line of response (LoR) data
print(f"First 10 rows from the 3D-formatted LoR data:\n{lors[:10]}\n")
```

```
First 10 rows from the 3D-formatted LoR data:
[[0.000e+00 2.094e+02 2.797e+02 0.000e+00 2.738e+02 9.500e+01 7.120e+02]
 [1.000e-01 4.325e+02 4.395e+02 0.000e+00 2.006e+02 1.929e+02 7.120e+02]
 [1.000e-01 2.513e+02 1.109e+02 0.000e+00 2.177e+02 3.469e+02 7.120e+02]
 [1.000e-01 3.227e+02 4.372e+02 0.000e+00 3.853e+02 1.097e+02 7.120e+02]
 [1.000e-01 1.894e+02 9.910e+01 0.000e+00 4.213e+02 1.044e+02 7.120e+02]
 [2.000e-01 3.658e+02 2.230e+02 0.000e+00 3.198e+02 5.003e+02 7.120e+02]
 [2.000e-01 2.543e+02 4.690e+02 0.000e+00 4.873e+02 1.257e+02 7.120e+02]
 [2.000e-01 3.339e+02 4.962e+02 0.000e+00 3.646e+02 8.970e+01 7.120e+02]
 [2.000e-01 2.844e+02 2.130e+02 0.000e+00 4.466e+02 5.322e+02 7.120e+02]
 [2.000e-01 2.053e+02 1.906e+02 0.000e+00 2.785e+02 2.224e+02 7.120e+02]]
```

Iterating through each sample of LoRs, locate the tracers and append their locations to a growing list of positions. As with the previous cells, you are invited to change the parameters highlighted below; details about each one are given in the comments.

Importantly, the `sample_size` must be larger to include enough LoRs for each tracer—in cases using large numbers of LoRs, the `overlap` becomes increasingly important to yield *dense* points on each trajectory; see the 'Effect of sample size and overlap' book section for a detailed explanation.

```
[17]: # Use the K-Medoids Method to track a moving tracer
from pept.tracking import find_cutpoints
from scipy.spatial import KDTree
from sklearn_extra.cluster import KMedoids
from sklearn.metrics import silhouette_score

# K-Medoids Method user-changeable parameters:
#     max_distance - only consider cutpoints from LoR pairs closer than this
#     neighbor_distance - filter out cutpoints whose nearest neighbor is further
#     max_num_clusters - the maximum number of clusters to try K-Medoids for
max_distance = 10
neighbour_distance = 4    # Needs to be larger than the default value
max_num_clusters = 10

# General PEPT tracking user-changeable parameters:
#     sample_start - the `lors` index at which to start tracking the tracer
#     sample_size - the number of LoRs used to compute a single location
#     overlap - the number of overlapping LoRs between consecutive samples
sample_start = 0
sample_size = 100
overlap = 0

# Tracked tracer locations will be appended to a growing list
positions = []

while sample_start + sample_size <= len(lors):
    sample = lors[sample_start:sample_start + sample_size]

    # For all pairs of lines closer than `max_distance`, find their minimum
    # distance points (cutpoints).
    cutpoints = find_cutpoints(sample, max_distance)

    # Filter all cutpoints whose nearest neighbours are further than
    # `neighbour_distance` mm away
    tree = KDTree(cutpoints)
    distances, indices = tree.query(cutpoints, 2)

    # The first "neighbour" is just the given cutpoint (hence their distance=0)
    distances = distances[:, 1]
    indices = indices[:, 1]
```

```
    # Remove all cutpoints whose nearest neighbor distance is larger than
    # `neigbour_distance`
    cutpoints_filtered = cutpoints[distances < neighbour_distance]

    # Try KMedoids (from the `sklearn_extra` library) for different numbers of
    # clusters to maximise the silhouette score (from the `sklearn` library)
    num_clusters = np.arange(2, max_num_clusters)
    silhouette_scores = []

    for nc in num_clusters:
        clusterer = KMedoids(nc)
        labels = clusterer.fit_predict(cutpoints_filtered[:, 1:4])
        sc = silhouette_score(cutpoints_filtered[:, 1:4], labels)
        silhouette_scores.append(sc)

    silhouette_scores = np.array(silhouette_scores)
    best_nc = num_clusters[silhouette_scores.argmax()]

    # Cluster the filtered cutpoints using the "best" number of clusters found
    clusterer = KMedoids(best_nc)
    labels = clusterer.fit_predict(cutpoints_filtered[:, 1:4])

    # For every cluster (i.e. label value >= 0), compute the centroid and append
    # it to the list of positions
    for label in range(labels.max() + 1):
        cluster = cutpoints_filtered[labels == label]
        centre = cluster.mean(axis = 0)
        positions.append(centre)

    sample_start += sample_size - overlap

positions = np.array(positions)

print(f"First 10 tracked positions:\n{positions[:10]}")
```

```
First 10 tracked positions:
[[  1.19680639 346.29335459 332.59356478 278.92626079]
 [  1.2312253  234.54012993 202.96021133 277.38812995]
 [  3.41871166 348.59324015 332.46838301 280.43849823]
 [  3.50190058 232.94286634 202.40163163 280.00090276]
 [  5.96768448 232.49445959 202.67389573 280.80050744]
 [  5.81373762 349.48608022 330.44543573 283.51908648]
 [  7.85340206 348.62158173 330.54404212 285.59309677]
 [  7.80695538 232.32881879 205.78973378 280.62983094]
 [  9.48321678 349.18408973 331.31526202 288.76893841]
 [  9.62462121 231.01150853 206.27971752 278.14605436]]
```

```
[18]: # Plot some of the LoRs used, plus the tracer locations we found
grapher = PlotlyGrapher(
    cols=2,
    subplot_titles=[
        "First 500 LoRs",
        f"Tracer locations tracked from {len(lors)} LoRs",
    ],
)

grapher.add_lines(lors[:500])
grapher.add_points(positions, col=2)

grapher.show()
```

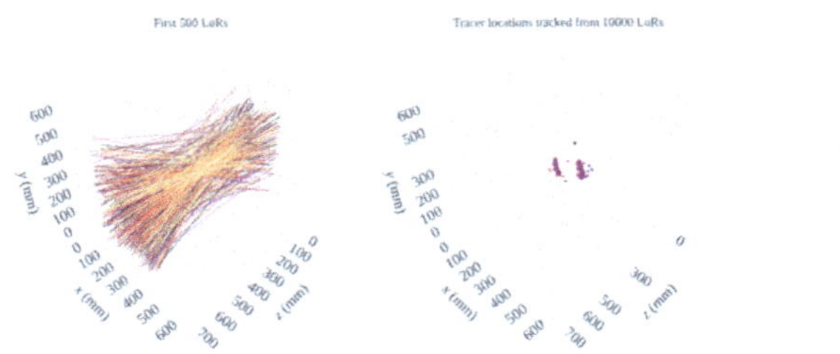

In the two-tracer case, the K-medoids method works much better, clearly depicting the two segments of the circular motion done by the two tracers, along with a few erroneous points which may potentially be removed by tweaking the relevant parameters—we invite the reader to have a go at doing so!

```
[19]: # Create a "timeseries" plot with the tracer positions vs. time
from pept.plots import PlotlyGrapher2D

PlotlyGrapher2D().add_timeseries(positions).show()
```

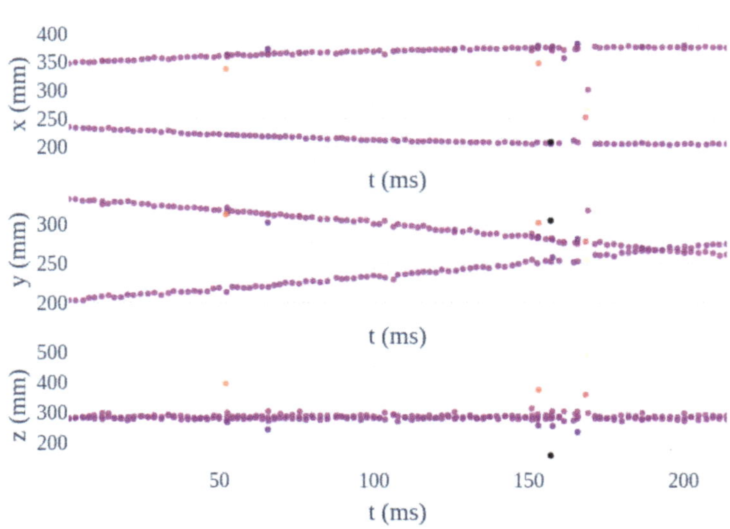

The tracers' movement in the XY plane is spread out enough that the timeseries graphs look effectively continuous. As the axis of rotation is perpendicular to the Z dimension, the tracer trajectories should be almost flat—therefore, the spread in the bottom subplot is mainly due to data acquisition errors, both due to the parallel screens scanner geometry and the tracer location algorithm.

6.2.20 The multiple location-allocation algorithm (MLAA)

The MLAA method in brief

The multiple location-allocation algorithm (MLAA) of Gundogdu *et al* [49] is another piece of early pioneering work in the application of algorithms developed in other fields—in this case, resource management—to PEPT. It is the first published paper using a voxelised representation of LoRs to isolate tracer locations using a location-allocation problem solver. The location-allocation problem is concerned with finding the optimum placement of n providers to minimise the cost of transporting goods to m consumers; the rather creative application of MLAA to PEPT finds the n tracer locations that minimise the weighted sum of distances to their closest voxels, in essence finding the spatial locations which are closest to the regions with high line density. The original paper explored many image processing techniques, such as adaptive thresholding, whose more sophisticated versions were used in modern voxel-based PEPT algorithms, such as the feature point identification method [17]. However, due to the extreme computational expense of the algorithm—the location-allocation problem is impossible to solve efficiently without strong heuristics—the methodology was never widely adopted in the field. Nonetheless, due to the pioneering work conducted by Gundogdu, it is indeed very valuable to include in this book; as with the other PEPT location algorithms, we begin with a summary of the steps employed by the algorithm, before providing a more detailed description and an interactive implementation at the end of the section.

1. A sample of N_s individual, consecutive lines of response (LoRs) from a given PEPT experiment is considered.
2. The LoRs are voxelised onto a 3D grid of equally-sized voxels, with each voxel recording the number of LoRs passing through.
3. The voxels are globally thresholded, with all voxels containing fewer LoRs than a given fraction of the highest-density voxel being set to zero.
4. The multiple location-allocation problem is solved to find the spatial positions of k tracers that minimise the weighted sum of distances from each voxel to its closest tracer; the weights are simply the voxel values.

The MLAA method in detail

(1–2) Voxelising a sample of LoRs: The experimental volume is divided into a three-dimensional grid of cells—called voxels in computer vision literature—with each cell recording the number of LoRs passing through; an example 2D projection of such a line density map is depicted in panel (*b*) of figure 6.16. The method is already described in section 6.2.3 and, in the interest of brevity, will not be repeated here; an interactive voxelisation step is included in the notebook at the end of this section, complete with 3D visualisations.

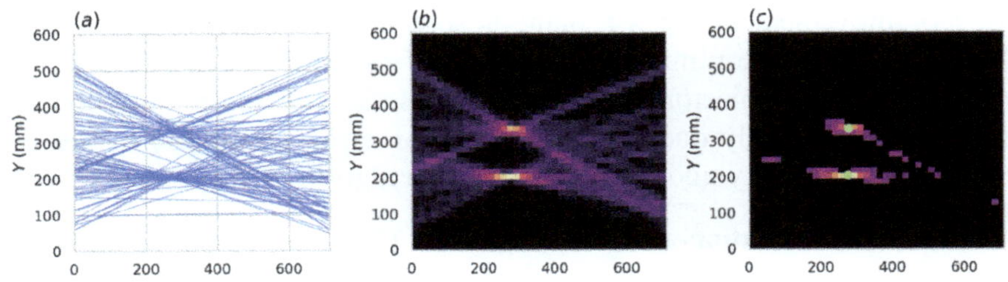

Figure 6.16. Panel (*a*) shows the *YZ* projection of a sample of 100 LoRs. Panel (*b*) depicts the voxelised—or pixellised in the case of 2D projections—representation of the sample of LoRs, with each cell being colour-coded by their line density. Panel (*c*) shows the thresholded cells for a fraction value $f = 0.3$ and the final tracer locations found using the MLAA algorithm.

(3) Thresholding voxels: Many voxels will only contain a single LoR passing through; as solving the location-allocation problem is extremely computationally intensive, it is important to reduce the number of voxels we handle. In the original paper, multiple approaches are explored, such as histogramming all voxel values and smoothing them, or variance thresholding [49]. In the end, the authors found that simple, global thresholding worked best for their purposes. A fixed fraction f of the largest voxel value—i.e. line density—is taken; all voxels with line densities below this fraction are then simply set to zero. As seen in panel (*c*) in figure 6.16, this thresholding successfully removes some of the outlier cells.

(4) Solving the location-allocation problem: Finally, the namesake of the algorithm will be applied. The goal is to find the k tracer locations that minimise the weighted sum of distances from each non-zero voxel to its closest tracer; the weight is simply the value stored in each voxel—i.e. the number of LoRs passing through—which intuitively gives more 'weight' to regions of high line density. This is a very non-trivial problem which, save for strong heuristics—which are commonly applied to geographical data—cannot be solved exactly without brute forcing all solutions [50, 51]. An alternative formulation views location-allocation as an optimisation problem, where the spatial coordinates of all tracers are the free parameters that must be varied in order to minimise the cost function mentioned above. The MLAA author uses a k-medoids partitioning (which is discussed in more detail in section 6.2.18) to find an initial guess of the tracer locations, which is then given to a gradient-based optimiser to refine into the locally-optimum locations.

It is useful to note that every single cost function evaluation done by the optimisation subroutine must find each non-zero voxel's nearest tracer location—that is currently being tried—and compute the required sum of weighted distances. Nearest-neighbour finding is a relatively computationally expensive step and, while previous algorithms only run a single nearest-neighbour finding subroutine per sample of LoRs, the MLAA method must do it multiple times (of the order of tens or hundreds) to find the optimum location of the tracers. Afterwards, the same steps outlined above are applied to each sample of LoRs to dynamically track the moving tracers on a frame-by-frame basis.

Strengths and weaknesses
The MLAA method, while relatively successful in finding the tracer locations—if the number thereof is known *a priori*—imposes an extreme computational cost to solve the location-allocation problem formulation of tracer finding. Moreover, knowing the exact number of tracers in a given sample of LoRs is oftentimes a strong limitation, as many experimental systems imaged with PEPT are larger than the scanners used, and therefore tracers are expected to leave and enter the detectors' field of view at different times. However, the techniques explored and applied by the MLAA authors in their paper were robust enough that they have found use in many of the modern voxel-based algorithms, for which MLAA is often viewed as an early pioneering work of applying techniques developed in other fields to PEPT.

Strengths

- With adequate *a priori* information, can find relatively accurate tracer locations even in the presence of noise.

Weaknesses

- Requires *a priori* knowledge of the exact number of tracers within the field of view in a given sample of LoRs.
- Extremely computationally expensive.

6.2.21 Interactive example: the multiple location-allocation algorithm (MLAA)

The notebook files accompanying this section are available in the supplementary data online at https://iopscience.iop.org/book/978-0-7503-3071-8.

Interactive PEPT analysis example using the multiple location-allocation algorithm (MLAA) [49]

This Jupyter Notebook

An example dataset is used from an experiment run at the University of Birmingham Positron Imaging Centre using the ADAC Forté. It consists of a single 1 mm diameter MCC particle activated with fluorine-18 inside a bubbling fluidised bed. The fluidised bed was filled with 90% sand and 10% MCC; air was fed into the bottom of the bed at a rate of 37 litres per minute at 3.5 bar. This dataset was chosen for its high quality captured lines of response, with the tracer still depicting the random particle motion that is inherent to bubbling fluidised beds—and typical in Lagrangian particle tracking.

The `pept` Python library is used for initialising and visualising PEPT data. While not required *per se* for illustrating PEPT algorithms' processing steps, it significantly reduces the amount of repetitive code and visual noise, allowing the reader to focus on the main conceptual procedures.

[1]:
```
# Importing necessary libraries for numerical array manipulation and plotting,
# plus the `pept` library for straightforward PEPT data processing
import numpy as np
import matplotlib.pyplot as plt
import pept

# Set Matplotlib theme (try "darkgrid" too for a more modern look)
import seaborn
seaborn.set_style("whitegrid")

# This notebook's implementation uses the KMeans implementation from the
# `sklearn` library, plus optimisation and spatial indexing from `SciPy`
import sklearn
import scipy

# Set Plotly figure output (try "browser" too for interactive graphs)
import plotly
plotly.io.renderers.default = "png"
```

Initialise raw line of response data

In this initial step, we read in the 'raw' LoR data from which our tracer positions will be calculated. As illustrated in the 'Effect of sample size and overlap' notebook, varying the numbers of LoRs used to calculate a PEPT location can strongly affect the accuracy of the final position computed. Try altering the value `nrows` below to see how this influences your results.

[2]:
```
# Read in a sample of experimental PEPT data into a NumPy array. Skip the file
# header's first 15 lines, then read in the first `nrows` LoRs
lors_raw = pept.utilities.read_csv(
    "Data/FluidisedBed1Tracer.csv",
    skiprows=15,
    nrows=200,
)

# Convert the LoRs into the general 3D line format [t, x1, y1, z2, x2, y2, z2].
# See the `PEPT Data Format` interactive example for detailed explanations

# ADAC Forte's two detector heads' separation, in mm
head_separation = 600

lors = np.insert(lors_raw, 3, 0, axis = 1)                # z1 = 0
lors = np.insert(lors, 6, head_separation, axis = 1)      # z2 = head_separation

# Print the first 10 rows from the line of response (LoR) data
print(f"First 10 rows from the 3D-formatted LoR data:\n{lors[:10]}\n")
```

```
First 10 rows from the 3D-formatted LoR data:
[[0.000e+00 1.900e+02 1.687e+02 0.000e+00 3.463e+02 1.428e+02 6.000e+02]
 [1.000e-01 2.437e+02 1.676e+02 0.000e+00 3.145e+02 3.139e+02 6.000e+02]
 [1.000e-01 1.941e+02 4.100e+02 0.000e+00 4.171e+02 2.401e+02 6.000e+02]
 [2.000e-01 2.885e+02 2.962e+02 0.000e+00 3.847e+02 4.525e+02 6.000e+02]
 [2.000e-01 1.646e+02 1.151e+02 0.000e+00 3.829e+02 3.534e+02 6.000e+02]
 [2.000e-01 1.286e+02 1.322e+02 0.000e+00 4.649e+02 2.661e+02 6.000e+02]
 [2.000e-01 3.339e+02 3.735e+02 0.000e+00 3.316e+02 1.310e+02 6.000e+02]
 [3.000e-01 1.640e+02 1.115e+02 0.000e+00 3.794e+02 3.534e+02 6.000e+02]
 [4.000e-01 2.620e+02 2.094e+02 0.000e+00 2.932e+02 2.808e+02 6.000e+02]
 [4.000e-01 1.304e+02 2.749e+02 0.000e+00 2.543e+02 3.062e+02 6.000e+02]]
```

The `pept.plots` subpackage provides two PEPT-oriented interfaces to the Plotly visualisation library—`PlotlyGrapher` and `PlotlyGrapher2D`—containing straightforward plotting functions for PEPT data. Let's plot the lines of response from the `lors` array:

```
[3]: # Plot the 3D LoRs that were initialised
     from pept.plots import PlotlyGrapher

     PlotlyGrapher().add_lines(lors).show()
```

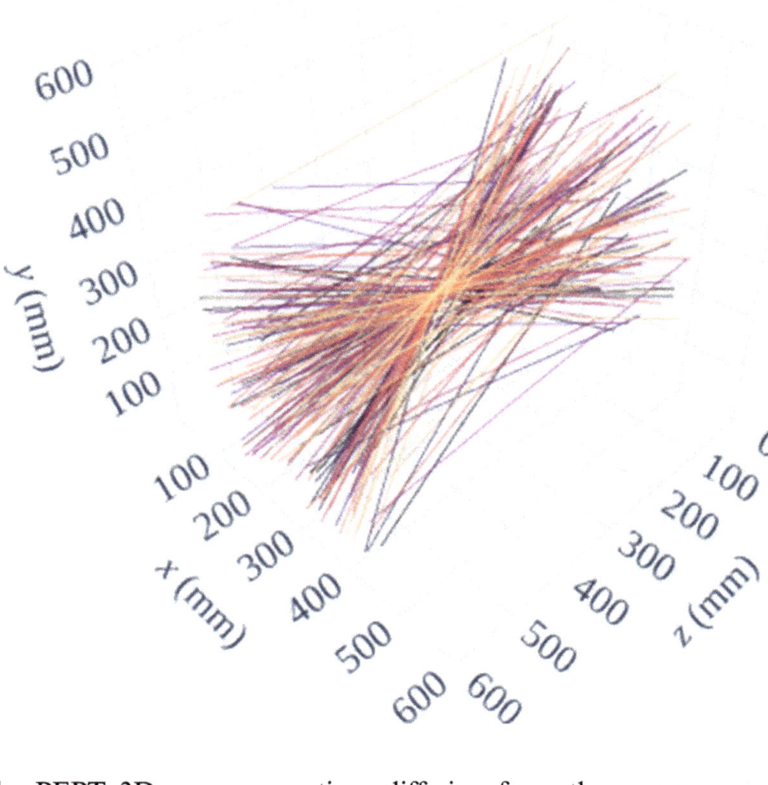

Note the PEPT 3D axes convention, differing from the common mathematical coordinates with *Y* and *Z* being switched, such that the *Y* axis points upwards, while the *Z* axis is perpendicular to the two detector heads.

Voxelise the lines of response

The next step is to 'voxelise' the LoRs—that is, to create a grid of square or rectangular cells, and count the number of lines passing through each cell. In the code below you can set the resolution—i.e. number of voxels in each dimension—by varying the pair of values corresponding to the `resolution` parameter. Typing, for example, `(20, 20, 20)` will result in a 20 by 20 by 20 grid of voxels.

Try some different values for the `resolution` parameter. As discussed in the main text, larger values—i.e. a higher resolution, smaller voxels—will allow the centroid of our particle to be determined more precisely. However, you may notice that as you increase these values your processor will take longer to execute the calculation and plotting steps below.

Now consider that this is just one of many thousands of locations we typically need to process, and you may get an idea of the issue of using an overly-fine grid. When using algorithms relying on voxelisation, there is a fine balance to be made between precision and compuational efficiency.

```
[4]: # Voxellise sample of LoRs
     resolution = (50, 50, 50)
     voxels = pept.Voxels.from_lines(lors, resolution)

     voxels
```

```
Initialised Voxels class in 0.0008099079132080078 s.
```

```
[4]: Voxels
     ------
     [[[0. 0. 0. … 0. 0. 0.]
       [0. 0. 0. … 0. 0. 0.]
       [0. 0. 0. … 0. 0. 0.]
       …
       [0. 0. 0. … 0. 0. 0.]
       [0. 0. 0. … 0. 0. 0.]
       [0. 0. 0. … 0. 0. 0.]]

      [[0. 0. 0. … 0. 0. 0.]
       [0. 0. 0. … 0. 0. 0.]
       [0. 0. 0. … 0. 0. 0.]
       …
       [0. 0. 0. … 0. 0. 0.]
       [0. 0. 0. … 0. 0. 0.]
       [0. 0. 0. … 0. 0. 0.]]

      [[0. 0. 0. … 0. 0. 0.]
       [0. 0. 0. … 0. 0. 0.]
       [0. 0. 0. … 0. 0. 0.]
       …
       [0. 0. 0. … 0. 0. 0.]
       [0. 0. 0. … 0. 0. 0.]
       [0. 0. 0. … 0. 0. 0.]]

      …
```

```
[[0. 0. 0. … 0. 0. 0.]
 [0. 0. 0. … 0. 0. 0.]
 [0. 0. 0. … 0. 0. 0.]
 …
 [0. 0. 0. … 0. 0. 0.]
 [0. 0. 0. … 0. 0. 0.]
 [0. 0. 0. … 0. 0. 0.]]

[[0. 0. 0. … 0. 0. 0.]
 [0. 0. 0. … 0. 0. 0.]
 [0. 0. 0. … 0. 0. 0.]
 …
 [0. 0. 0. … 0. 0. 0.]
[[0. 0. 0. ... 0. 0. 0.]
 [0. 0. 0. ... 0. 0. 0.]
 [0. 0. 0. ... 0. 0. 0.]
 ...
 [0. 0. 0. ... 0. 0. 0.]
 [0. 0. 0. ... 0. 0. 0.]
 [0. 0. 0. ... 0. 0. 0.]]]

number_of_voxels =    (50, 50, 50)
voxel_size =          [ 7.636 10.244 12.    ]

xlim =                [110.3 492.1]
ylim =                [ 44.8 557. ]
zlim =                [  0. 600.]

voxel_grids:
([110.3 ... 492.1],
 [44.8 ... 557.0],
 [0.0 ... 600.0])
```

[5]:
```
# Plot all LoRs and all voxels with non-zero values (i.e. more than one LoR
# passing through)
PlotlyGrapher().add_voxels(voxels).add_lines(lors).show()
```

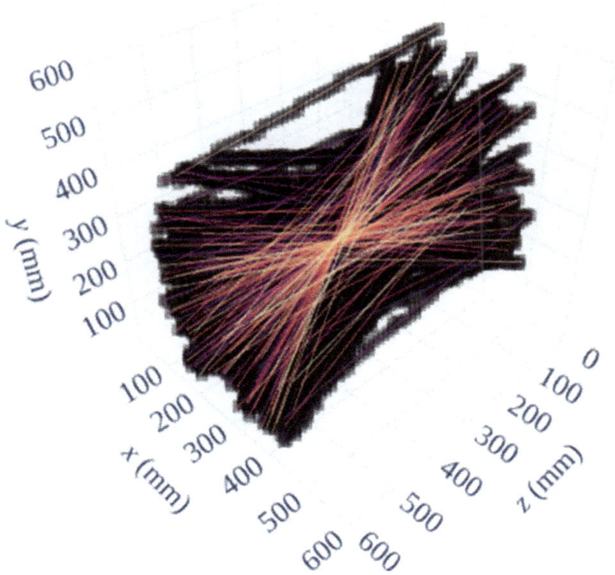

Voxel global thresholding

As is clear from the above image, our data possesses a lot of background noise away from where we would expect the tracer's centre to be. The MLAA PEPT algorithm takes a relatively straightforward approach to removing this noise, simply setting a threshold voxel value below which all data are removed. Specifically, a `fraction` of the maximum voxel value for a given sample of LoRs is chosen.

Try varying the `fraction` parameter below to see how this choice affects your data.

[6]:
```
# Set all pixel values smaller than a `fraction` of the maximum `voxels` value
# to zero
fraction = 0.15
threshold = fraction * float(voxels.max())

voxels[voxels < threshold] = 0.
```

[7]:
```
# Plot thresholded voxels, colour-coded by the number of LoRs passing through
# each of them. If only a small number of voxels are plotted, we can use
# draw each voxel as a volumetric cube (note that this only works for roughly
# <100 voxels due to the computational burden of plotting volumes)
PlotlyGrapher().add_traces(voxels.cubes_traces()).show()

# For large numbers of voxels
# PlotlyGrapher().add_voxels(voxels).show()
```

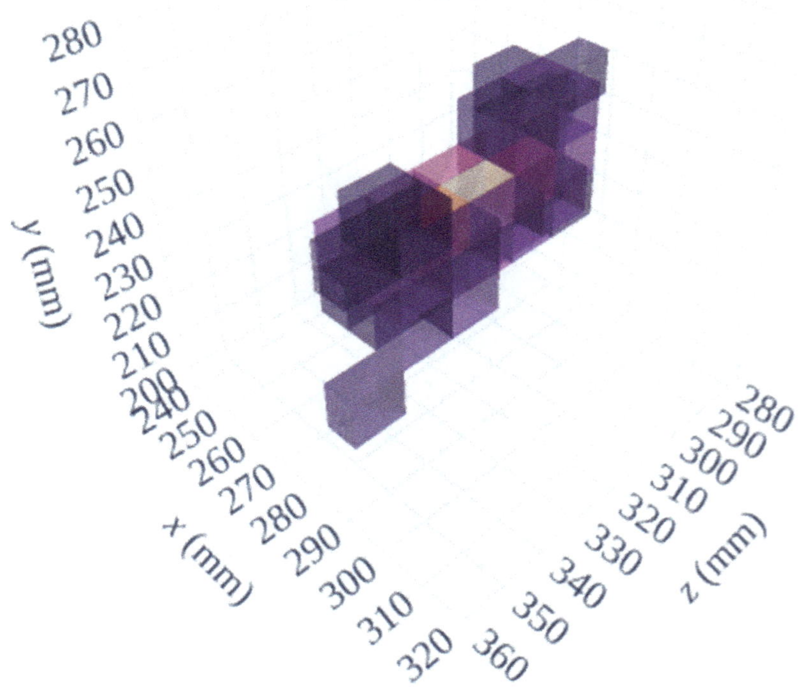

The location-allocation algorithm

The location-allocation problem is well-known in the fields of operational research, being mainly concerned with the optimal placement of N sources (e.g. warehouses) that must minimise the cost of transporting goods to M sinks (e.g. consumers). However, there is no analytical solution to this problem, and the true *best* placement of the *sources* most often involves checking all possible solutions—becoming computationally impossible even for relatively small N. As such, multiple other *heuristics* and different problem formulations have been developed.

Applied to the problem of finding tracer locations, MLAA can be reformulated such that the *sources* are the tracers, and the sum of weighted distances from each voxel to their closest tracer must be minimised; the weight of each distance is simply the number of LoRs passing through the corresponding voxel—therefore being more important for a source to be close to voxels with high line densities.

In this notebook, we will solve the MLAA problem using a global optimiser; the optimisation problem can be formulated as follows: find the spatial coordinates of N sources that minimise the sum of weighted distances to each source's closest voxel. Therefore, we have $3 \times N$ variables to optimise (three coordinates for each source); this optimisation problem is not convex, so we require a global optimisation subroutine from the `SciPy` library. In global optimisation problems, the starting guesses play essential roles, so—following the author's paper—the output from an initial clustering run will be used.

[8]:
```
# Solve the Location-Allocation problem for a PEPT system, where the sources
# are the tracers and the sinks are the voxel coordinates. This is framed as
# an optimisation problem: find the coordinates of `num_tracers` sources that
# minimise the weighted sum of distances from each voxel to their closest
# source; the weight is simply the number of LoRs passing through that voxel
from sklearn.cluster import KMeans
from scipy.spatial import KDTree
from scipy.optimize import basinhopping

# The number of tracers in the system, equal to the number N of sources
num_tracers = 1

# Get all nonzero voxels' physical coordinates
indices = np.argwhere(voxels != 0)
origin = np.array([voxels.xlim[0], voxels.ylim[0], voxels.zlim[0]])

# Compute the physical coordinates of the voxel centres
sinks = origin + voxels.voxel_size * indices + voxels.voxel_size * 0.5

# Get the number of LoRs passing through each sink - will be used as weights
weights = np.array([voxels[tuple(i)] for i in indices])
```

```
# First partition voxels using K-Means to get a starting guess
labels = KMeans(num_tracers).fit_predict(sinks)

# The optimisation starting guesses will then be the centroid of all points in
# each partition
guesses = []
for label in range(labels.max() + 1):
    guesses.append(sinks[labels == label].mean(axis=0))

# Now flatten all coordinates into a (3 * N) vector
guesses = np.ravel(guesses)


def mlaa_cost(sources):
    '''The MLAA cost function, representing the sum of weighted distances from
    each source in `sources` to their closest voxels (including all voxels)'''

    # `sources` is a flattened array of coordinates, reshape it into (N, 3)
    sources = sources.reshape(num_tracers, 3)

    # Speed up nearest neighbour lookup by constructing a KD-Tree of the sources
    tree = KDTree(sources)

    # Compute distances from each sink to their closest source
    distances, indices = tree.query(sinks)

    # The cost is simply the sum of distances multiplied by the corresponding
    # voxel's line density
    cost = (distances * weights).sum()
    return cost


# Run the Basin Hopping global optimiser to find the sources - i.e. the spatial
# coordinates of the tracers!
opt_result = basinhopping(mlaa_cost, guesses)
positions = opt_result.x.reshape(num_tracers, 3)

# Insert a timestamp for the tracer locations found - that is, the mean
# timestamp of the LoRs used in this sample
positions = np.insert(positions, 0, lors[:, 0].mean(), axis=1)

print(f"Sink locations found:\n{positions}")
```

```
Sink locations found:
[[  3.9045     281.58886558 244.12650251 317.90591948]]
```

[9]:

```
# Plot the LoRs used, the thresholded voxels and tracer positions found
grapher = PlotlyGrapher(cols=2)

grapher.add_lines(lors)
grapher.add_voxels(voxels, col=2)
grapher.add_points(positions, size=5, col=2)

grapher.show()
```

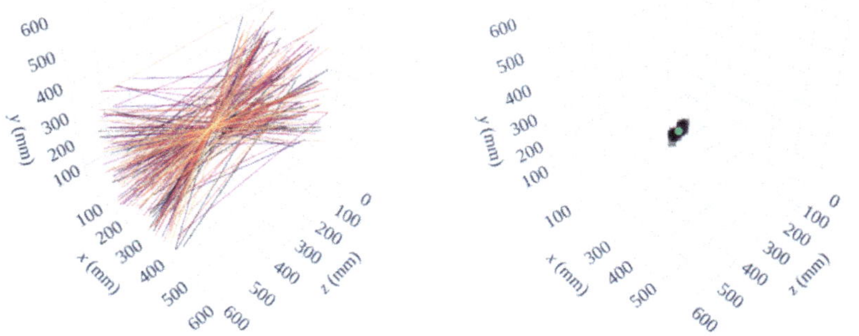

Complete multiple location-allocation algorithm code

The code developed in the previous sections can be applied to consecutive 'samples' of LoRs to follow the tracer's movement through the fluidised bed, fully tracking its trajectory. A complete MLAA method implementation is given below. Please note that due to the inherent computational expense of this method, only smaller numbers of LoRs can be processed in a reasonable amount of time.

First initialise a larger, multi-sample set of LoRs to analyse; change the `nrows` argument below to read in a different number of LoRs (the file included contains 80,000 LoRs in total), though, as noted above, keep in mind the computational expense of processing large numbers of LoRs!

[10]:
```
# Read in a sample of experimental PEPT data into a NumPy array. Skip the file
# header's first 15 lines and read in the first `nrows` LoRs
lors_raw = pept.utilities.read_csv(
    "Data/FluidisedBed1Tracer.csv",
    skiprows=15,
    nrows=2000,
)
# Convert the LoRs into the general 3D line format [t, x1, y1, z2, x2, y2, z2].
# See the `PEPT Data Format` interactive example for detailed explanations

# ADAC Forte's two detector heads' separation, in mm
head_separation = 600

lors = np.insert(lors_raw, 3, 0, axis=1)                    # z1 = 0
lors = np.insert(lors, 6, head_separation, axis=1)          # z2 = head_separation

# Print the first 10 rows from the line of response (LoR) data
print(f"First 10 rows from the 3D-formatted LoR data:\n{lors[:10]}\n")
```

```
First 10 rows from the 3D-formatted LoR data:
[[0.000e+00 1.900e+02 1.687e+02 0.000e+00 3.463e+02 1.428e+02 6.000e+02]
 [1.000e-01 2.437e+02 1.676e+02 0.000e+00 3.145e+02 3.139e+02 6.000e+02]
 [1.000e-01 1.941e+02 4.100e+02 0.000e+00 4.171e+02 2.401e+02 6.000e+02]
 [2.000e-01 2.885e+02 2.962e+02 0.000e+00 3.847e+02 4.525e+02 6.000e+02]
 [2.000e-01 1.646e+02 1.151e+02 0.000e+00 3.829e+02 3.534e+02 6.000e+02]
 [2.000e-01 1.286e+02 1.322e+02 0.000e+00 4.649e+02 2.661e+02 6.000e+02]
 [2.000e-01 3.339e+02 3.735e+02 0.000e+00 3.316e+02 1.310e+02 6.000e+02]
 [3.000e-01 1.640e+02 1.115e+02 0.000e+00 3.794e+02 3.534e+02 6.000e+02]
 [4.000e-01 2.620e+02 2.094e+02 0.000e+00 2.932e+02 2.808e+02 6.000e+02]
 [4.000e-01 1.304e+02 2.749e+02 0.000e+00 2.543e+02 3.062e+02 6.000e+02]]
```

Iterating through each sample of LoRs, locate the tracer and append its location to a growing list of positions. As with the previous cells, you are invited to change the parameters highlighted below; details about each one are given in the comments.

[11]:
```
# Use the MLAA Method to track a moving tracer
from sklearn.cluster import KMeans
from scipy.spatial import KDTree
from scipy.optimize import basinhopping

# MLAA Method user-changeable parameters:
#     resolution - the number of voxels to use in each dimension
#     fraction - threshold all voxels with values smaller than fraction * max
#     num_tracers - the number of tracers in the system, equal to the N sources
resolution = (50, 50, 50)
fraction = 0.15
num_tracers = 1

# General PEPT tracking user-changeable parameters:
#     sample_start - the `lors` index at which to start tracking the tracer
#     sample_size - the number of LoRs used to compute a single location
#     overlap - the number of overlapping LoRs between consecutive samples
sample_start = 0
sample_size = 200
overlap = 0

# Tracked tracer locations will be appended to a growing list
positions = []


def mlaa_cost(sources, weights):
    '''The MLAA cost function, representing the sum of weighted distances from
    each voxel to their closest source in `sources`'''

    # `sources` is a flattened array of coordinates, reshape it into (N, 3)
    sources = sources.reshape(num_tracers, 3)

    # Speed up nearest neighbour lookup by constructing a KD-Tree of the sources
    tree = KDTree(sources)

    # Compute distances from each sink to their closest source
    distances, indices = tree.query(sinks)

    # The cost is simply the sum of distances multiplied by the corresponding
    # voxel's line density
    cost = distances * weights
    return cost.sum()
```

```
while sample_start + sample_size <= len(lors):
    sample = lors[sample_start:sample_start + sample_size]

    # Voxellise the sample of LoRs
    voxels = pept.Voxels.from_lines(sample, resolution, verbose=False)

    # Set all pixel values smaller than a `fraction` of the maximum `voxels`
    # value to zero
    threshold = fraction * float(voxels.max())
    voxels[voxels < threshold] = 0.

    # Solve the Location-Allocation problem for a PEPT system, where the sources
    # are the tracers and the sinks are the voxel coordinates. This is framed as
    # an optimisation problem: find the coordinates of `num_tracers` sources to
    # minimise the weighted sum of distances from each voxel to their closest
    # source; the weight is simply the number of LoRs passing through that voxel

    # Get all nonzero voxels' physical coordinates
    indices = np.argwhere(voxels != 0)
    origin = np.array([voxels.xlim[0], voxels.ylim[0], voxels.zlim[0]])

    # Compute the physical coordinates of the voxel centres
    sinks = origin + voxels.voxel_size * indices + voxels.voxel_size * 0.5

    # Get the number of LoRs passing through each sink - will be used as weights
    weights = np.array([voxels[tuple(i)] for i in indices])

    # First partition voxels using K-Means to get a starting guess
    labels = KMeans(num_tracers).fit_predict(sinks)

    # The optimisation starting guesses will then be the centroid of all points
    # in each partition
    guesses = []
    for label in range(labels.max() + 1):
        guesses.append(sinks[labels == label].mean(axis=0))

    # Now flatten all coordinates into a (3 * N) vector
    guesses = np.ravel(guesses)

    # Run the Basin Hopping global optimiser to find the sources - i.e. the
    # spatial coordinates of the tracers!
    # Note: pass the weights as an extra argument to the cost function
    cost_function_args = dict(args=(weights,))
    opt_result = basinhopping(mlaa_cost, guesses,
                              minimizer_kwargs=cost_function_args)
    centres = opt_result.x.reshape(num_tracers, 3)
```

```
    # Insert a timestamp for the tracer locations found - that is, the mean
    # timestamp of the LoRs used in this sample
    centres = np.insert(centres, 0, sample[:, 0].mean(), axis=1)

    positions.extend(centres)

    sample_start += sample_size - overlap

positions = np.array(positions)

print(f"First 10 tracked positions:\n{positions[:10]}")
```

```
First 10 tracked positions:
[[  3.9045     281.58886562 244.12650258 317.9059194 ]
 [ 11.777      280.97544399 244.65891513 318.14455561]
 [ 18.9635     279.29649423 244.64811687 317.07583543]
 [ 26.422      280.715      245.15399999 318.        ]
 [ 32.805      280.44919406 244.90405886 315.41484026]
 [ 38.9795     281.44003172 244.39737773 317.73879803]
 [ 45.756      280.22479629 243.9107559  317.5757091 ]
 [ 52.8735     281.90997861 245.95314578 317.98391882]
 [ 60.0475     281.32584671 245.56246644 317.5493184 ]
 [ 67.1895     280.8680605  246.06779444 316.70722653]]
```

```
[12]: # Plot some of the LoRs used, plus the tracer locations we found
from pept.plots import PlotlyGrapher, PlotlyGrapher2D

grapher = PlotlyGrapher(
    cols=2,
    subplot_titles=[
        "First 500 LoRs",
        f"Tracer locations tracked from {len(lors)} LoRs",
    ],
)

grapher.add_lines(lors[:500])
grapher.add_points(positions, col=2)

grapher.show()
```

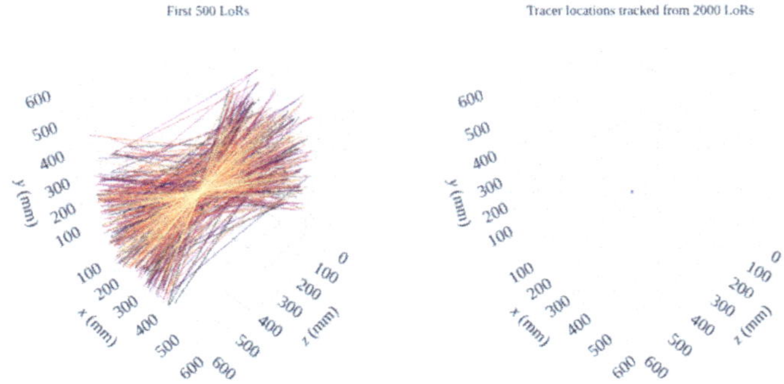

```
[13]: # Create a "timeseries" plot with the tracer positions vs. time
      PlotlyGrapher2D().add_timeseries(positions).show()
```

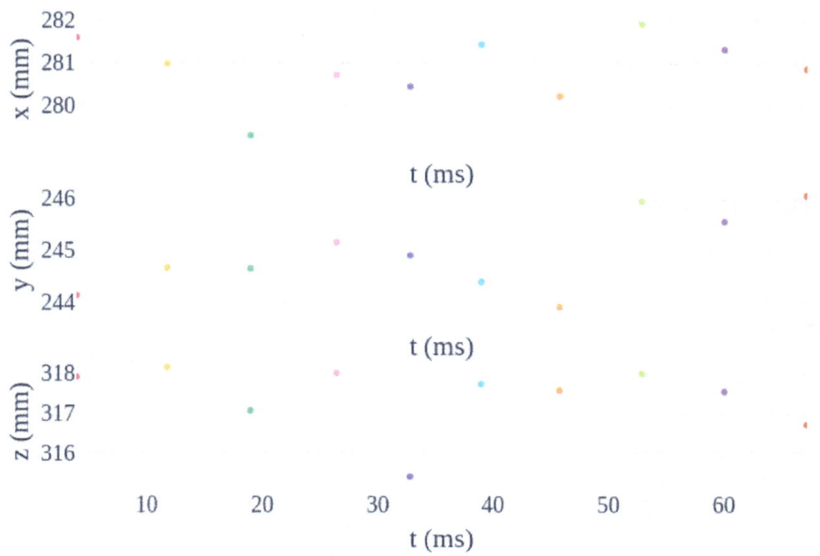

Multiple particle tracking

The complete code developed above can also be applied to a simple multiple tracer experiment. Here we use a sample of an experiment consisting of two 1 mm diameter alumina particles labelled with fluorine-18 radioactive tracer material. The particles are fixed at the two ends of an arm rotating at 42 rpm, their trajectories describing full circles.

```
[14]: # Read in a sample of experimental PEPT data into a NumPy array. Skip the file
      # header's first 15 lines
      lors_raw = pept.utilities.read_csv(
          "Data/RotatingArm2Tracers.csv",
          skiprows=15,
          nrows=1000,
      )

      # Convert the LoRs into the general 3D line format [t, x1, y1, z2, x2, y2, z2].
      # See the `PEPT Data Format` interactive example for detailed explanations

      # ADAC Forte's two detector heads' separation, in mm
      head_separation = 712

      lors = np.insert(lors_raw, 3, 0, axis=1)                # z1 = 0
      lors = np.insert(lors, 6, head_separation, axis=1)      # z2 = head_separation

      # Print the first 10 rows from the line of response (LoR) data
      print(f"First 10 rows from the 3D-formatted LoR data:\n{lors[:10]}\n")
```

```
First 10 rows from the 3D-formatted LoR data:
[[0.000e+00 2.094e+02 2.797e+02 0.000e+00 2.738e+02 9.500e+01 7.120e+02]
 [1.000e-01 4.325e+02 4.395e+02 0.000e+00 2.006e+02 1.929e+02 7.120e+02]
 [1.000e-01 2.513e+02 1.109e+02 0.000e+00 2.177e+02 3.469e+02 7.120e+02]
 [1.000e-01 3.227e+02 4.372e+02 0.000e+00 3.853e+02 1.097e+02 7.120e+02]
 [1.000e-01 1.894e+02 9.910e+01 0.000e+00 4.213e+02 1.044e+02 7.120e+02]
 [2.000e-01 3.658e+02 2.230e+02 0.000e+00 3.198e+02 5.003e+02 7.120e+02]
 [2.000e-01 2.543e+02 4.690e+02 0.000e+00 4.873e+02 1.257e+02 7.120e+02]
 [2.000e-01 3.339e+02 4.962e+02 0.000e+00 3.646e+02 8.970e+01 7.120e+02]
 [2.000e-01 2.844e+02 2.130e+02 0.000e+00 4.466e+02 5.322e+02 7.120e+02]
 [2.000e-01 2.053e+02 1.906e+02 0.000e+00 2.785e+02 2.224e+02 7.120e+02]]
```

Iterating through each sample of LoRs, locate the tracers and append their locations to a growing list of positions. As with the previous cells, you are invited to change the parameters highlighted below; details about each one are given in the comments.

Importantly, the `sample_size` must be larger to include enough LoRs for each tracer—in cases using large numbers of LoRs, the `overlap` becomes increasingly important to yield *dense* points on each trajectory; see the ‘Effect of sample size and overlap’ book section for a detailed explanation.

```
[15]: # Use the MLAA Method to track a moving tracer
      from sklearn.cluster import KMeans
      from scipy.spatial import KDTree
      from scipy.optimize import basinhopping

      # MLAA Method user-changeable parameters:
      #     resolution - the number of voxels to use in each dimension
      #     fraction - threshold all voxels with values smaller than fraction * max
      #     num_tracers - the number of tracers in the system, equal to the N sources
      resolution = (50, 50, 50)
      fraction = 0.15
      num_tracers = 2

      # General PEPT tracking user-changeable parameters:
      #     sample_start - the `lors` index at which to start tracking the tracer
      #     sample_size - the number of LoRs used to compute a single location
      #     overlap - the number of overlapping LoRs between consecutive samples
      sample_start = 0
      sample_size = 200
```

```python
overlap = 0

# Tracked tracer locations will be appended to a growing list
positions = []


def mlaa_cost(sources, weights):
    '''The MLAA cost function, representing the sum of weighted distances from
    each voxel to their closest source in `sources`'''

    # `sources` is a flattened array of coordinates, reshape it into (N, 3)
    sources = sources.reshape(num_tracers, 3)

    # Speed up nearest neighbour lookup by constructing a KD-Tree of the sources
    tree = KDTree(sources)

    # Compute distances from each sink to their closest source
    distances, indices = tree.query(sinks)

    # The cost is simply the sum of distances multiplied by the corresponding
    # voxel's line density
    cost = distances * weights
    return cost.sum()


while sample_start + sample_size <= len(lors):
    sample = lors[sample_start:sample_start + sample_size]

    # Voxellise the sample of LoRs
    voxels = pept.Voxels.from_lines(sample, resolution, verbose=False)

    # Set all pixel values smaller than a `fraction` of the maximum `voxels`
    # value to zero
    threshold = fraction * float(voxels.max())
    voxels[voxels < threshold] = 0.

    # Solve the Location-Allocation problem for a PEPT system, where the sources
    # are the tracers and the sinks are the voxel coordinates. This is framed as
    # an optimisation problem: find the coordinates of `num_tracers` sources to
    # minimise the weighted sum of distances from each voxel to their closest
    # source; the weight is simply the number of LoRs passing through that voxel

    # Get all nonzero voxels' physical coordinates
    indices = np.argwhere(voxels != 0)
    origin = np.array([voxels.xlim[0], voxels.ylim[0], voxels.zlim[0]])

    # Compute the physical coordinates of the voxel centres
```

```
    sinks = origin + voxels.voxel_size * indices + voxels.voxel_size * 0.5

    # Get the number of LoRs passing through each sink - will be used as weights
    weights = np.array([voxels[tuple(i)] for i in indices])

    # First partition voxels using K-Means to get a starting guess
    labels = KMeans(num_tracers).fit_predict(sinks)

    # The optimisation starting guesses will then be the centroid of all points
    # in each partition
    guesses = []
    for label in range(labels.max() + 1):
        guesses.append(sinks[labels == label].mean(axis=0))

    # Now flatten all coordinates into a (3 * N) vector
    guesses = np.ravel(guesses)

    # Run the Basin Hopping global optimiser to find the sources - i.e. the
    # spatial coordinates of the tracers!
    # Note: pass the weights as an extra argument to the cost function
    cost_function_args = dict(args=(weights,))
    opt_result = basinhopping(mlaa_cost, guesses,
                              minimizer_kwargs=cost_function_args)
    centres = opt_result.x.reshape(num_tracers, 3)

    # Insert a timestamp for the tracer locations found - that is, the mean
    # timestamp of the LoRs used in this sample
    centres = np.insert(centres, 0, sample[:, 0].mean(), axis=1)

    positions.extend(centres)

    sample_start += sample_size - overlap

positions = np.array(positions)

print(f"First 10 tracked positions:\n{positions[:10]}")
```

```
First 10 tracked positions:
[[  2.365      347.58228158 333.89826676 279.34404418]
 [  2.365      233.54349126 203.31638647 277.47397495]
 [  6.8535     349.78840856 331.62825834 281.53625315]
 [  6.8535     230.57763615 203.23236615 276.62805177]
 [ 11.0465     229.94642473 206.14056615 278.47490459]
 [ 11.0465     351.02543031 328.15200611 278.62336706]
 [ 15.0505     229.1678532  207.81683443 281.61833184]
 [ 15.0505     352.637      328.315      277.68      ]
 [ 19.242      228.89191212 206.51936843 275.3790468 ]
 [ 19.242      352.11311225 328.23444964 279.37719397]]
```

[16]:
```
# Plot some of the LoRs used, plus the tracer locations we found
grapher = PlotlyGrapher(
    cols=2,
    subplot_titles=[
        "First 500 LoRs",
        f"Tracer locations tracked from {len(lors)} LoRs",
    ],
)

grapher.add_lines(lors[:500])
grapher.add_points(positions, col=2)

grapher.show()
```

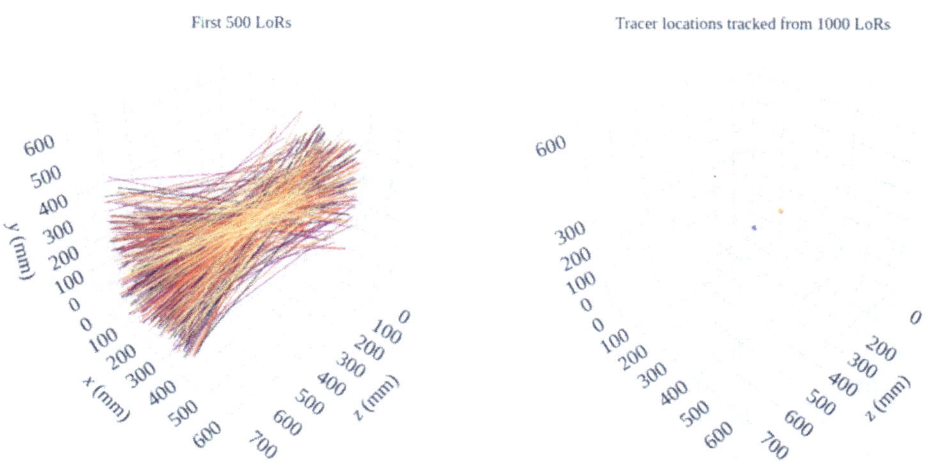

[17]:
```
# Create a "timeseries" plot with the tracer positions vs. time
PlotlyGrapher2D().add_timeseries(positions).show()
```

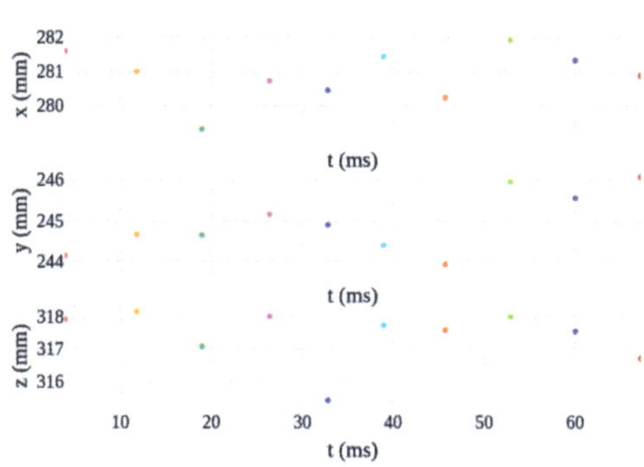

6.3 From finding tracers to tracking trajectories

In the previous section, we have taken an in-depth look at some of the most important PEPT algorithms, from the original Birmingham method to state-of-the-art point-cloud- and voxel-based techniques such as PEPT-ML, the triangulation method and FPI. The

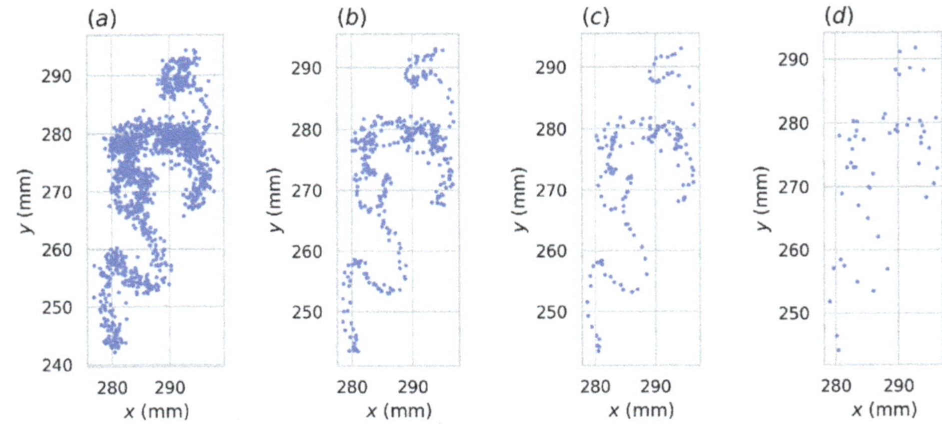

Figure 6.17. Tracer locations found using the Birmingham method for varying sample sizes: 50 LoRs (panel (*a*)), 200 LoRs (panel (*b*)), 500 LoRs (panel (*b*)), 1500 LoRs (panel (*d*)). No overlap has been used.

main—decidedly non-trivial—task accomplished by these algorithms is finding tracer locations within a single sample of LoRs. However, before arriving at continuous tracer trajectories tracked through a long experiment, there are three more important steps still:

- Achieving accurate, dense points on tracer trajectories—in practice, this means finding a balance between the temporal resolution (how often we find the tracer) and spatial resolution (how accurately we find the tracer).
- Filtering out noisy or erroneous locations—while algorithms are becoming more sophisticated and robust with respect to noise, some location artefacts are still expected to appear in long experiments, even with well optimised parameters used in the algorithms' initial application.
- In the case of multiple tracer tracking, a final trajectory separation step must be applied to isolate each individual tracer's trajectory from the unsorted locations found in each sample.

The first step is most often achieved through the careful calibration of the sample size, or number of LoRs used to find the tracers' locations in a single frame. As depicted in figure 6.17, small sample sizes yield very dense points (i.e. finding the tracer very often) but with low precision; at the other extreme, too large sample sizes may yield locations with significant smoothing effects (i.e. loss of fine details of the particle's motion)—as shown in panel (*d*) of figure 6.17 where quick turns in the tracer trajectories are no longer visible for 1500 LoRs per sample. In the middle, there is a trade-off between how often we find the tracer and how accurately we do it. This choice is non-trivial and indeed depends on the particular application—for fast-moving tracers, we would ideally maximise our temporal resolution by decreasing the sample sizes; however, if there are significant numbers of corrupted LoRs due to e.g. imaging dense, shielded industrial equipment, the sample size must be increased to still allow accurate tracking. Finally, the algorithm used may also affect the 'ideal' sample size—some algorithms were specifically developed to handle small numbers of LoRs but become computationally inefficient at large sample sizes (e.g. cutpoints-based methods such as PEPT-ML), while

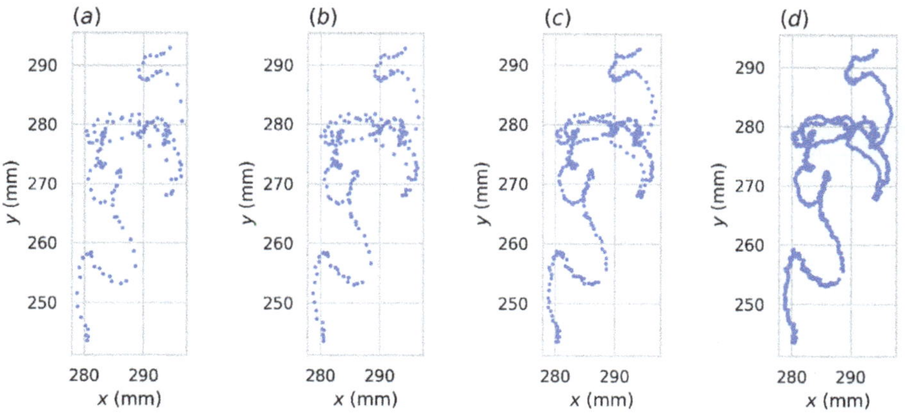

Figure 6.18. Tracer locations found using the Birmingham method for varying overlap: 50 LoRs (panel (*a*)), 150 LoRs (panel (*b*)), 300 LoRs (panel (*b*)), 450 LoRs (panel (*d*)). The sample size was fixed to 500 LoRs per sample.

others remain relatively consistent in terms of computational resources but require more LoRs as a minimum (e.g. FPI).

In the PEPT literature, there are two main ways of selecting samples of LoRs to locate tracers: constant numbers of LoRs or constant time windows. The first method, used primarily by Birmingham-based algorithms, e.g. Birmingham method [1, 2] and PEPT-ML [38], yields more constant LoR densities around tracers—which can help tracking algorithms with more predictable, or constant LoR data—with the downside of having varying temporal resolution when tracers travel through denser/less-dense, or more/less shielded areas. The number of LoRs acquired from a tracer of constant activity also changes with its position in the experimental volume (e.g. detectors have lower geometric efficiencies towards their edges—for a more detailed discussion on detector systems, see section 5.2), meaning a single sample of LoRs may span a shorter (longer) time window when the tracer is closer to (further from) the centre of the scanner. The second sampling method, used primarily by the voxel-based algorithms, e.g. line density [4] and FPI [17], selects LoRs acquired within a given time window that is moved forward—which can help post-processing algorithms with constant time intervals between frames—with the downside of having varying spatial resolution, as tracer locations in denser areas will have fewer constituent LoRs. This can be especially problematic for tracking algorithms which require finer parameter tuning, as the LoR density around tracers will vary significantly with this sampling method. To summarise, samples of constant numbers of LoRs can be expected to yield approximately constant spatial resolution, while samples of constant time intervals yield constant temporal resolution. At the time of writing, no systematic study has been published quantitatively comparing the two sampling methods. However, as their effects on the quality of the tracer trajectories found can be dramatic (as can be seen in figure 6.17), it is expected that more sophisticated sampling methods will be developed in the future, such as automatic selection of the sample size, or robust hybrids between the two methods outlined above.

A complementary technique to the LoR sampling is the use of overlap between consecutive samples; effectively being a sliding window approach to locating tracer locations across individual frames. While not changing the temporal resolution per se, the use of overlap yields denser points on tracer trajectories—as seen in figure 6.18—which can significantly improve trajectory separation and post-processing, as this minimises the 'jumps' between consecutive tracer locations. However, it can also dramatically increase the computational expense of tracking the tracers: for example, processing 100,000 LoRs in samples of 200 LoRs at a time would require tracking 500 samples with no overlap, 999 samples with an overlap of 100 LoRs, or 1997 samples with an overlap of 150 LoRs! The interactive notebook in the next subsection explores the effects of sample size and overlap on the tracer locations recorded using the Birmingham method, as well as its computational expense.

6.3.1 Interactive example: the effects of sample size and overlap

The notebook files accompanying this section are available in the supplementary data online at https://iopscience.iop.org/book/978-0-7503-3071-8.

Effect of sample size and overlap

An important aspect of processing consecutive samples of lines of response (LoRs) to track the frame-by-frame tracer locations is the **sample size** and **overlap**. Both were available as free parameters in the previous sections, but their effects—and importance—were not explored in detail. In this notebook we will make an important step from *locating* tracers—which was the focus of the PEPT algorithms—to *tracking* them.

As mentioned in the 'PEPT Data Format' notebook, the **temporal resolution** of PEPT is inversely proportional to the number of LoRs used to locate tracers in a single sample, while the **spatial resolution** is directly proportional to the sample size. In this tutorial we will directly see how the *precision* and *density* of tracked points vary with sample size and overlap—and how the careful choice of these two parameters can yield *dense, precise* tracer trajectories.

[1]:
```
# Importing necessary libraries for numerical array manipulation and plotting,
# plus the `pept` library for straightforward PEPT data processing
import numpy as np
import matplotlib.pyplot as plt
import pept

# Set Matplotlib theme (try "darkgrid" too for a more modern look)
import seaborn
seaborn.set_style("whitegrid")

# Set Plotly figure output (try "browser" too for interactive graphs)
import plotly
plotly.io.renderers.default = "png"
```

Prelude

In order to minimise visual noise and focus on the concepts relevant to this tutorial (as well as be able to track *many* samples), the PEPT algorithm implementations from the `pept` library will be used. In the code cells below we will have a short introduction to the handling of `sample_size` and `overlap` in the `pept` library:

[2]:
```
# Create an array of dummy values following the LoR data format, i.e.
# [timestamp, x1, y1, z1, x2, y2, z2]
lors_raw = np.arange(70).reshape(10, 7)

print(f"Array of dummy values following the 7-column LoR format:\n{lors_raw}")
```

```
Array of dummy values following the 7-column LoR format:
[[ 0  1  2  3  4  5  6]
 [ 7  8  9 10 11 12 13]
 [14 15 16 17 18 19 20]
 [21 22 23 24 25 26 27]
 [28 29 30 31 32 33 34]
 [35 36 37 38 39 40 41]
 [42 43 44 45 46 47 48]
 [49 50 51 52 53 54 55]
 [56 57 58 59 60 61 62]
 [63 64 65 66 67 68 69]]
```

Accessing samples of a few LoRs at a time via *slicing* is tedious and error-prone:

```
# Selecting samples of LoRs via slicing
sample_size = 5
sample1 = lors_raw[0:sample_size]
sample2 = lors_raw[sample_size:2 * sample_size]
sample3 = lors_raw[2 * sample_size:3 * sample_size]
# Processing consecutive samples
sample_start = 0
while sample_start + sample_size <= len(lors_raw):
    sample = lors_raw[sample_start:sample_start + sample_size]
    # ... processing the sample of LoRs ...
    sample_start += sample_size
```

It would be much easier if we could access samples of LoRs via simple *indexing*:

```
# Ideal selection of samples of LoRs
sample1 = lors[0]
sample2 = lors[1]
sample3 = lors[2]
# Ideal iteration of samples of LoRs
for sample in lors:
    # ... processing the sample of LoRs ...
```

This functionality is offered by the `pept` library in its 'base classes'. The most important data structure in `pept` is the `pept.LineData` class, which encapsulates a NumPy array like the `lors_raw` above and allows accessing *samples* of a few LoRs at a time with simple indexing (e.g. `lors[0]` yields the first sample of LoRs):

[3]:
```
# Encapsulate the "raw" NumPy array of LoRs in a `pept.LineData` class for easy
# accessing of samples
lors = pept.LineData(lors_raw, sample_size=4)

print(f"The first sample in `lors` - i.e. `lors[0]`:\n{lors[0]}\n")
print(f"The second sample in `lors` - i.e. `lors[1]`:\n{lors[1]}\n")

for sample in lors:
    print(f"Iterating through samples:\n{sample}")
```

```
The first sample in `lors` - i.e. `lors[0]`:
pept.LineData (samples: 1)
--------------------------
sample_size = 4
overlap = 0
lines =
  (rows: 4, columns: 7)
  [[ 0.  1. … 5.  6.]
   [ 7.  8. … 12. 13.]
   [14. 15. … 19. 20.]
   [21. 22. … 26. 27.]]
columns = ['t', 'x1', 'y1', 'z1', 'x2', 'y2', 'z2']
attrs = {}


The second sample in `lors` - i.e. `lors[1]`:
pept.LineData (samples: 1)
--------------------------
sample_size = 4
overlap = 0
lines =
  (rows: 4, columns: 7)
  [[28. 29. … 33. 34.]
   [35. 36. … 40. 41.]
   [42. 43. … 47. 48.]
   [49. 50. … 54. 55.]]
columns = ['t', 'x1', 'y1', 'z1', 'x2', 'y2', 'z2']
attrs = {}


Iterating through samples:
pept.LineData (samples: 1)
--------------------------
sample_size = 4
overlap = 0
lines =
  (rows: 4, columns: 7)
  [[ 0.  1. … 5.  6.]
   [ 7.  8. … 12. 13.]
   [14. 15. … 19. 20.]
   [21. 22. … 26. 27.]]
columns = ['t', 'x1', 'y1', 'z1', 'x2', 'y2', 'z2']
attrs = {}

Iterating through samples:
pept.LineData (samples: 1)
--------------------------
sample_size = 4
overlap = 0
lines =
  (rows: 4, columns: 7)
  [[28. 29. … 33. 34.]

  [35. 36. … 40. 41.]
  [42. 43. … 47. 48.]
  [49. 50. … 54. 55.]]
columns = ['t', 'x1', 'y1', 'z1', 'x2', 'y2', 'z2']
attrs = {}
```

```
[4]: # We can easily print more information about the `lors`
     lors
```

```
[4]: pept.LineData (samples: 2)
     --------------------------
     sample_size = 4
     overlap = 0
     lines =
       (rows: 10, columns: 7)
       [[ 0.  1. … 5.  6.]
        [ 7.  8. … 12. 13.]
        …
        [56. 57. … 61. 62.]
        [63. 64. … 68. 69.]]
     columns = ['t', 'x1', 'y1', 'z1', 'x2', 'y2', 'z2']
     attrs = {}
```

And that's all that `pept.LineData` is—just an interface to NumPy arrays allowing simple selection of samples. However, this abstraction proved quite powerful, as virtually *all* PEPT algorithms require this type of iteration—and is indeed the 'glue' between all PEPT-oriented algorithms in the `pept` library.

Finally, changing the `sample_size` and `overlap` on the fly is quite simple:

```
[5]: print(f"    sample_size = {lors.sample_size}\n    overlap = {lors.overlap}")
     print(f"-> number of samples = {len(lors)}\n")

     print("Changing the `sample_size` and `overlap`...\n")
     lors.sample_size = 3
     lors.overlap = 2

     print(f"    sample_size = {lors.sample_size}\n    overlap = {lors.overlap}")
     print(f"-> number of samples = {len(lors)}\n")

     print(f"First two samples - note the overlap!\n{lors[0]}\n\n{lors[1]}")
```

```
    sample_size = 4
    overlap = 0
-> number of samples = 2

Changing the `sample_size` and `overlap`…

    sample_size = 3
    overlap = 2
-> number of samples = 8

First two samples - note the overlap!
pept.LineData (samples: 1)
--------------------------
sample_size = 3
overlap = 0
lines =
  (rows: 3, columns: 7)
  [[ 0.  1. … 5.  6.]
   [ 7.  8. … 12. 13.]
   [14. 15. … 19. 20.]]
columns = ['t', 'x1', 'y1', 'z1', 'x2', 'y2', 'z2']
attrs = {}
```

```
pept.LineData (samples: 1)
--------------------------
sample_size = 3
overlap = 0
lines =
  (rows: 3, columns: 7)
  [[ 7.  8. … 12. 13.]
   [14. 15. … 19. 20.]
   [21. 22. … 26. 27.]]
columns = ['t', 'x1', 'y1', 'z1', 'x2', 'y2', 'z2']
attrs = {}
```

Initialise raw line of response data

An example dataset is used from an experiment run at the University of Birmingham Positron Imaging Centre using the ADAC Forté. It consists of a single 1 mm diameter MCC particle activated with fluorine-18 inside a bubbling fluidised bed. The fluidised bed was filled with 90% sand and 10% MCC; air was fed into the bottom of the bed at a rate of 37 litres per minute at 3.5 bar. This dataset was chosen for its high quality captured lines of response, with the tracer still depicting the random particle motion that is inherent to bubbling fluidised beds—and typical in Lagrangian particle tracking.

In this initial step, we read in the 'raw' LoR data from which our tracer positions will be calculated. The `pept.scanners` subpackage contains converters between scanner-specific data formats (e.g. binary files, parallel screens detectors) and the general 3D LoR format `[t, x1, y1, z1, x2, y2, z2]`, encapsulated in `pept.LineData` classes:

```
[6]: # Initialise the LoRs from the Parallel Screens format into a LineData
     lors = pept.scanners.parallel_screens(
         "Data/FluidisedBed1Tracer.csv",
         screen_separation=600,
         skiprows=15,
         sample_size=200,
     )

     print(f"The LoRs initialised into a `pept.LineData`:\n{lors}\n")
```

```
Initialised PEPT data in 0.032 s.

The LoRs initialised into a `pept.LineData`:
pept.LineData (samples: 400)
----------------------------
sample_size = 200
overlap = 0
lines =
  (rows: 80000, columns: 7)
  [[0.0000e+00 1.9000e+02 … 1.4280e+02 6.0000e+02]
   [1.0000e-01 2.4370e+02 … 3.1390e+02 6.0000e+02]
   …
   [2.6444e+03 1.6340e+02 … 1.4750e+02 6.0000e+02]
   [2.6444e+03 1.5580e+02 … 5.5280e+02 6.0000e+02]]
columns = ['t', 'x1', 'y1', 'z1', 'x2', 'y2', 'z2']
attrs = {}
```

The `pept.plots` subpackage provides two PEPT-oriented interfaces to the Plotly visualisation library—`PlotlyGrapher` and `PlotlyGrapher2D`—containing straightforward plotting functions for PEPT data. Let's plot the lines of response from the `lors` array:

```
[7]: # Plot the 3D LoRs that were initialised
from pept.plots import PlotlyGrapher

grapher = PlotlyGrapher(subplot_titles=["The first sample of LoRs"])
grapher.add_lines(lors[0])
grapher.show()
```

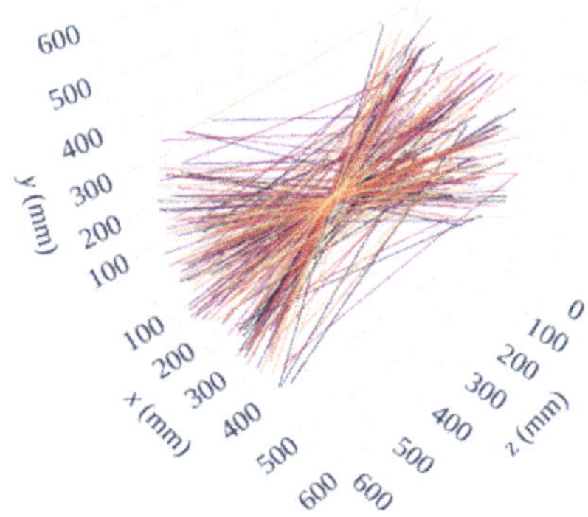

Note the PEPT 3D axes convention, differing from the common mathematical coordinates with Y and Z being switched, such that the Y axis points upwards, while the Z axis is perpendicular to the two detector heads.

Effect of sample size

In the code cell below we track the tracer locations for varying `sample_size` values. For simplicity, the Birmingham method will be used—with which the tracking is done in a single line of Python!

```
[8]: # Use the Birmingham Method to track the tracer with a given number of LoRs
# per sample (i.e. sample size)
from pept.tracking import BirminghamMethod, Stack

sample_sizes = [50, 200, 500, 1500]
titles = []
trajectories = []

# Track the tracer trajectory for the different `sample_sizes`
for sample_size in sample_sizes:
    lors.sample_size = sample_size
    titles.append(f"Sample Size {lors.sample_size} LoRs")

    positions = (BirminghamMethod() + Stack()).fit(lors, verbose=False)
    trajectories.append(positions)

# Plot each trajectory found on a 2x2 subplot grid
grapher = PlotlyGrapher(rows=2, cols=2, subplot_titles=titles)

grapher.add_points(trajectories[0])
grapher.add_points(trajectories[1], col=2)
grapher.add_points(trajectories[2], row=2)
grapher.add_points(trajectories[3], row=2, col=2)

grapher.show()
```

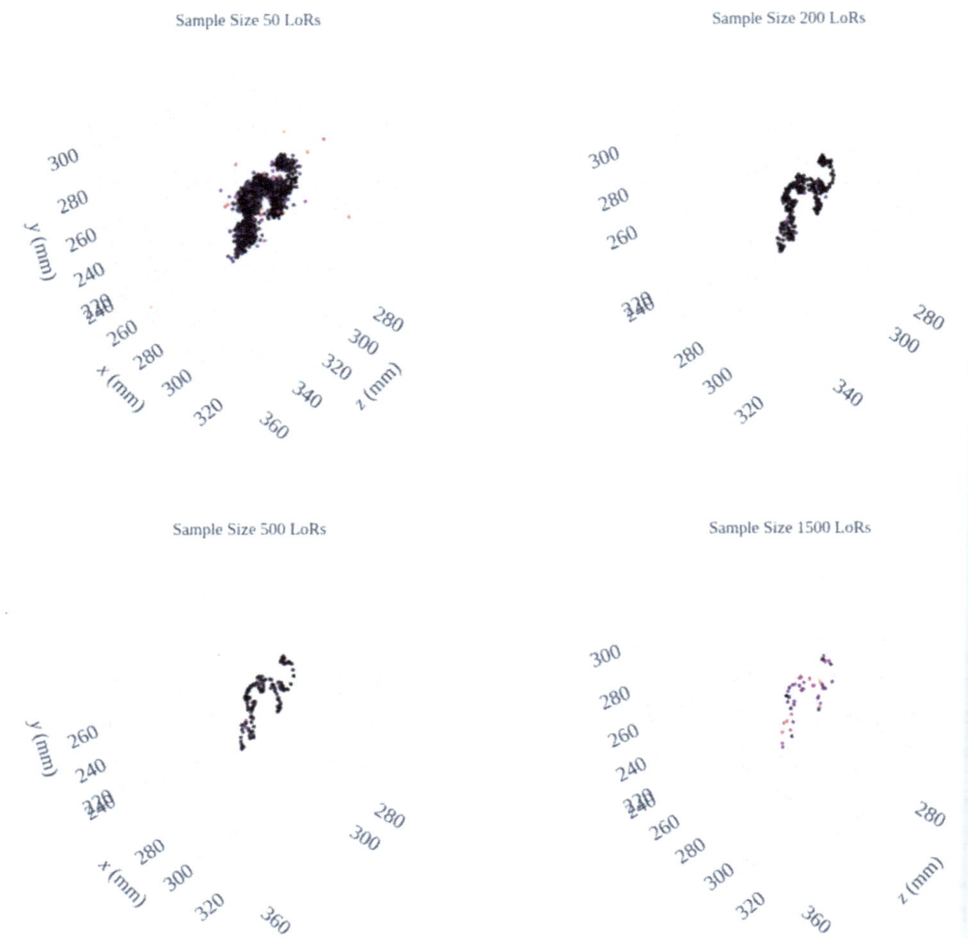

If the sample size is too small, we expect very dense points (i.e. tracking the tracer very often), but with low accuracy; this is seen in the 50 LoRs per sample graph above, as we get many spurious points. If the sample size is too large, the tracer will be tracked fewer times, but the positions found will be more accurate; however, in the extreme, even these positions might suffer from *smoothing effects*, as the tracer could have moved too much during the time a sample's LoRs were intercepted—this is seen in the 1500 LoRs per sample graph above, as the quick turns experienced by the tracer are no longer visible. In the middle, there is always a trade-off between *how often* and *how accurately* we can track the tracer—but what if, without changing the temporal resolution *per se*, we could still have dense, very accurate points? This is the role of the overlap.

Effect of overlap

As mentioned above, the overlap can be used to yield *denser* points on a tracer's trajectory, without affecting the temporal resolution. This allows us to use larger sample sizes, yielding accurate locations, without sacrificing the number of times we can locate the tracer in its tracks—which will prove essential for achieving good statistics when post-processing PEPT data in the later chapters.

[9]:
```
# Use the Birmingham Method to track the tracer with a given number of
# overlapping LoRs between consecutive samples (i.e. overlap)
lors.sample_size = 600    # fix sample size to a relatively large value

overlaps = [0, 150, 300, 450]
titles = []
trajectories = []

# Track the tracer trajectory for the different `sample_sizes`
for overlap in overlaps:
    lors.overlap = overlap
    titles.append(f"Overlap {lors.overlap} LoRs")

    positions = (BirminghamMethod() + Stack()).fit(lors, verbose=False)
    trajectories.append(positions)

# Plot each trajectory found on a 2x2 subplot grid
grapher = PlotlyGrapher(rows=2, cols=2, subplot_titles=titles)

grapher.add_points(trajectories[0])
grapher.add_points(trajectories[1], col=2)
grapher.add_points(trajectories[2], row=2)
grapher.add_points(trajectories[3], row=2, col=2)

grapher.show()
```

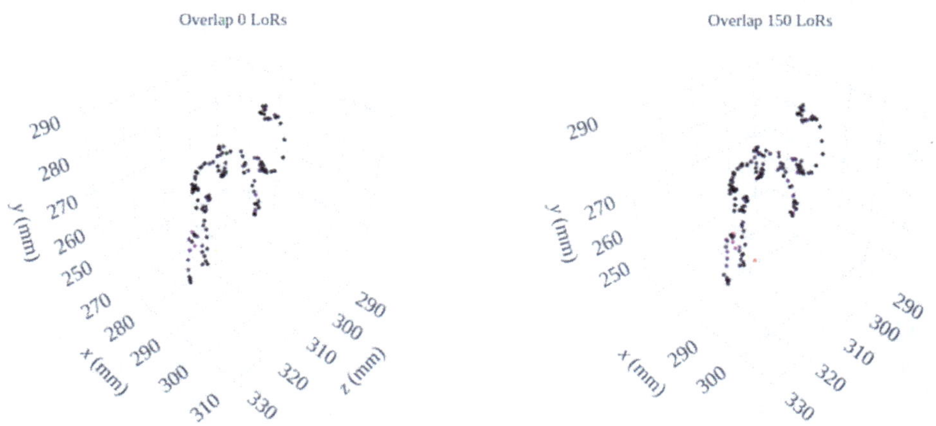

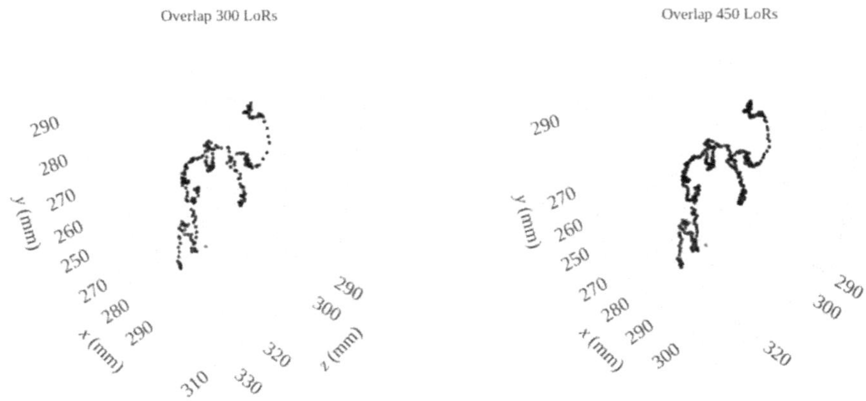

Note how with the help of the overlap, we can have the same high accuracy, as well as dense tracer trajectories! It should be noted, however, that large overlaps of large samples are expected to induce a degree of smoothing.

6.3.2 Trajectory extraction

PEPT typically focuses on tracking particles in long experiments, where they may travel through dense, noisy areas within the imaged system, or may even temporarily leave the scanner's field of view. As we always push for the best temporal resolution—i.e. locating tracers with as few LoRs as possible—it can happen that in specific time intervals, scattered LoR sometimes concentrate within specific regions in space, being indistinguishable from a real tracer. Therefore, virtually all tracer locating algorithms will sometimes yield spurious, erroneous tracer locations depending on the experiment conditions, and so a trajectory filtering step is often employed. The filtration algorithms used in published papers most often fall into one of three categories:

1. Simple removal of all locations with a corresponding spatial error (or proxy thereof) larger than a fixed threshold, for PEPT location algorithms which provide such information (e.g. Gaussian fit standard deviation in the Line Density algorithm, mean distance from the MDP to all LoRs in the Birmingham method).
2. A nearest-neighbour approach, where all locations whose nearest neighbour (or *k*th nearest neighbour) is further than a fixed distance are removed.
3. After some trajectory separation step, all trajectories with fewer points than a fixed threshold are removed.

While research into PEPT tracer location algorithms is a very active field, the same cannot be said about trajectory extraction; at the time of writing, no paper has been published focusing solely on filtering or separating trajectories. The few such algorithms currently developed are most often introduced as subsections in location algorithm developments or 'workflow'-type studies, where they are merely explained, but not systematically studied. Nonetheless, a few types of algorithms can be distinguished:

1. Using trajectory separation methods developed for particle tracking velocimetry (PTV, more details of which can be found in chapter 4.1.1), a quasi-two-dimensional technique used to image transparent systems using video cameras.

As in PTV, multiple tracers are being kept track of across consecutive frames. Two notable trajectory reconstruction algorithms based on PTV methods were mentioned in the development of the G-means clustering and feature point identification (FPI) tracer location algorithms [6, 17].

2. Clustering-based methods, where a minimum distance path—or minimum spanning tree (MST) in graph theory—connecting all tracer locations is constructed, and all paths longer than a certain maximum distance are 'cut'; all remaining clusters of paths are then labelled as individual trajectories.
3. Using system-specific information, e.g. knowing that two tracers being tracked are always at a fixed distance from one another, or that one is significantly more radioactive than the other [3, 52] and using this data to assign trajectory labels as the tracers are being tracked.

As adoption of PEPT becomes wider, it is expected that more sophisticated trajectory extraction methods will be developed, fully utilising PEPT data particularities such as variable temporal and spatial resolutions—approximations of which can be calculated—to aid trajectory filtering steps. Finally, trajectory separation algorithms can be improved beyond the quasi-two-dimensional techniques reminiscent of PTV, or video-based object tracking, to make full use of the three-dimensional data captured using PEPT. As in the previous sections, the interactive examples below illustrate the main trajectory extraction ideas mentioned above.

6.3.3 Interactive example: filtering trajectories

The notebook files accompanying this section are available in the supplementary data online at https://iopscience.iop.org/book/978-0-7503-3071-8.

Filtering trajectories

The most widely-employed trajectory post-processing step is the filtering of tracked PEPT locations to remove spurious positions. Virtually all tracer locating algorithms can sometimes exhibit such artifacts, especially during longer experiments, as sometimes scattered LoRs may happen to concentrate around a region in space while the real tracer is shielded in another dense region of the system, the former being effectively indistinguishable from the latter.

In this notebook we will illustrate the two most widely used methods of filtering trajectories, based on spatial error and nearest-neighbour approaches, respectively.

[1]:
```
# Importing necessary libraries for numerical array manipulation and plotting,
# plus the `pept` library for straightforward PEPT data processing
import numpy as np
import matplotlib.pyplot as plt
import pept

# Set Matplotlib theme (try "darkgrid" too for a more modern look)
import seaborn
seaborn.set_style("whitegrid")

# Set Plotly figure output (try "browser" too for interactive graphs)
import plotly
plotly.io.renderers.default = "png"
```

Initialising lines of response

An example dataset is used from an experiment run at the University of Birmingham Positron Imaging Centre using the ADAC Forté. It consists of a single 1 mm diameter MCC particle activated with fluorine-18 inside a bubbling fluidised bed. The fluidised bed was filled with 90% sand and 10% MCC; air was fed into the bottom of the bed at a rate of 37 litres per minute at 3.5 bar. This dataset was chosen for its high quality captured lines of response, with the tracer still depicting the random particle motion that is inherent to bubbling fluidised beds—and typical in Lagrangian particle tracking.

[2]:
```
# Initialise a set of LoRs from a tracer in a bubbling fluidised bed and plot
# the first sample of LoRs
from pept.plots import PlotlyGrapher

# Initialise the LoRs from the Parallel Screens format into a LineData
lors = pept.scanners.parallel_screens(
    "Data/FluidisedBed1Tracer.csv",
    screen_separation=600,
    skiprows=15,
    sample_size=400,
    overlap=300,
)
# Plot the first sample of LoRs
PlotlyGrapher().add_lines(lors[0]).show()
```

```
Initialised PEPT data in 0.031 s.
```

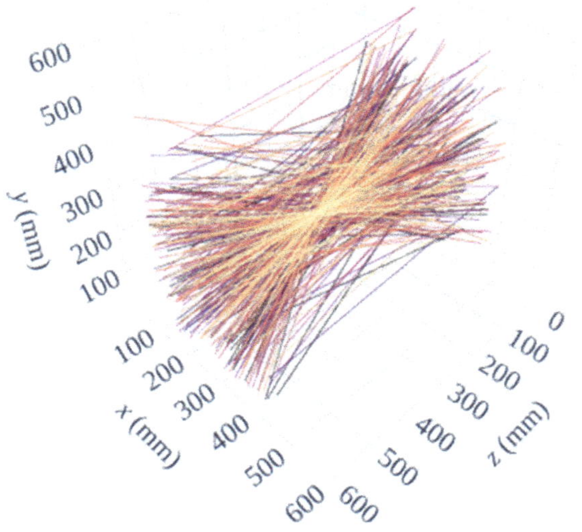

Filtering based on spatial error

A straightforward, but still quite powerful method of filtering trajectories is the removal of locations with corresponding spatial error (or a suitable proxy thereof) larger than a given threshold, *if the tracking algorithm used provides such information.* In the code cell below, we will use the Birmingham method to track a single tracer through a bubbling fluidised bed; the Birmingham method implementation from the `pept` library provides such a measure of the spatial error of a location—given as simply the mean distance from the tracer position to all its constituent LoRs (for the exact details of how the Birmingham method filters scattered LoRs to find the tracer as the LoRs' minimum distance point, see sections 6.2.1 and 6.2.2).

[3]:
```
# Track particle in bubbling fluidised bed using the Birmingham Method, then
# plot the positions found, colour-coding them by their spatial error
from pept.tracking import BirminghamMethod, Stack

# Track tracer locations using the Birmingham Method
positions = (BirminghamMethod() + Stack()).fit(lors)
print(f"Tracer locations found:\n{positions}")

# Plot tracer positions, colour-coding the last column - i.e. the spatial error
PlotlyGrapher().add_points(positions, colorbar_title="Spatial Error").show()
```

```
Batch 1 / 1 :: 100%|                              | 797/797
[00:01<00:00, 420.52it/s]


Processed samples in 1.9627251625061035 s

Tracer locations found:
pept.PointData (samples: 797)
-----------------------------
sample_size = 1
overlap = 0
points =
  (rows: 797, columns: 5)
  [[   7.621      280.41310044 ...  317.09596305    5.3412885 ]
   [  11.864      280.12168546 ...  316.78498205    4.84031552]
   ...
   [2635.313      294.27480764 ...  306.58280854    3.91996215]
   [2638.523      294.46350891 ...  306.47829018    3.76597345]]
columns = ['t', 'x', 'y', 'z', 'error']
attrs = {}
```

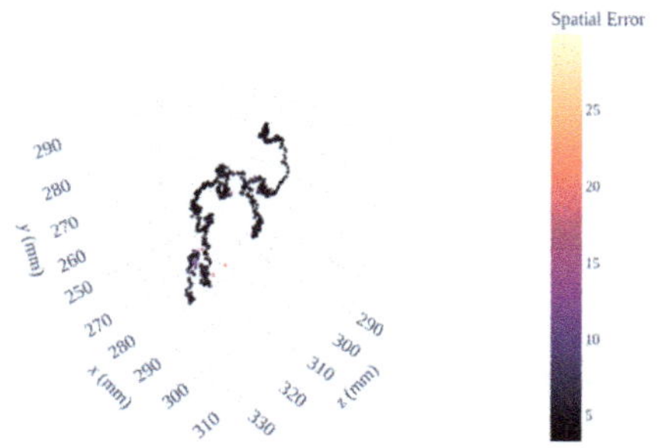

Notice that the tracer locations found have *five* data columns: a timestamp, the 3D point's coordinates, and the *relative spatial error*, or the standard deviation in the distances from the tracer position to its constituent LoRs. Let's plot a histogram of those errors to see how their values are distributed:

[4]:
```
# Plot a histogram of the spatial errors of the Birmingham Method-tracked points
fig, ax = plt.subplots()
ax.hist(positions.points[:, 4], bins=20)
ax.set(xlabel="Spatial Error (mm)", ylabel="Number of Points");
```

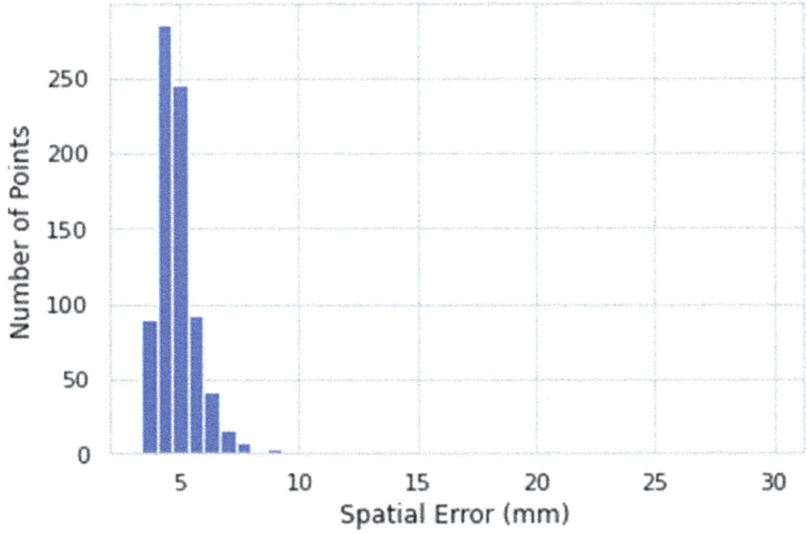

It seems the spatial error tails off, with fewer positions at higher spatial errors. In the code cell below, we filter out tracer locations with spatial errors larger than a given `threshold` and plot both the 'raw' and 'filtered' positions side by side; do change the `threshold` and see its effects!

[5]:
```
# Select only tracer positions with spatial errors smaller than a given
# `threshold` (mm)
threshold = 5
filtered = positions.points[positions.points[:, 4] < threshold]

grapher = PlotlyGrapher(
    cols=2,
    subplot_titles=["Raw Positions", "Filtered Positions"],
)

grapher.add_points(positions)
grapher.add_points(filtered, col=2)

grapher.show()
```

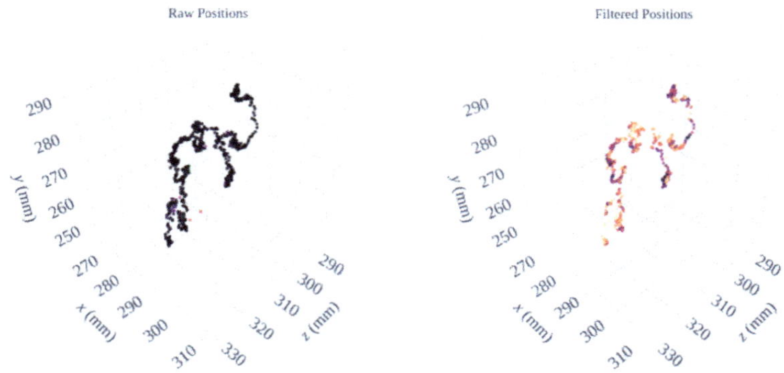

Note that the `PlotlyGrapher.add_points` method automatically colour-codes the plotted points by their last column—which in this case is the Birmingham method-predicted spatial error. Try varying the threshold value to see how it affects you results.

Filtering based on nearest neighbours

Another method of filtering trajectories that does not depend on the tracking algorithm providing some approximation of a spatial error is based on the removal of all positions whose nearest neighbour is further than some maximum distance. On one hand, this technique is applicable to virtually any tracer trajectories; on the other, it does not use any PEPT-specific information to its advantage. A natural extension to this method is the use of the *k*th nearest neighbour for the filtering step, as this will prove more robust in the presence of *dense* noisy points.

In the code cell below we will compute the distance from each tracer location to its *k*-th nearest neighbour, then plot a histogram of these distances; we will use a *k-d tree* to speed up the nearest-neighbour lookup step; feel free to change the value of k and see its effect!

```
[6]: # For each tracer location in `positions`, find the k-th nearest neighbour using
     # a KD-Tree for neighbour lookup acceleration, then plot histogram of the
     # distances found
     from scipy.spatial import KDTree

     # The k-th nearest neighbour
     k = 5

     # Construct a kd tree of the spatial coordinates of the tracer locations to
     # allow fast querying of the nearest-neighbour distances
     tree = KDTree(positions.points[:, 1:4])

     # Find the distance from each point to its k-th nearest neighbour; because we
     # are querying points that are already in the tree, the nearest neighbour will
     # be the point itself - ignore it and use (k + 1)
     distances, indices = tree.query(positions.points[:, 1:4], k=[k + 1])
     distances = distances.flatten()

     # Plot a histogram of the distances found
     fig, ax = plt.subplots()
     ax.hist(distances, bins=40)
     ax.set(xlabel=f"k={k} Nearest Neighbour Distance (mm)",
            ylabel="Number of Points");
```

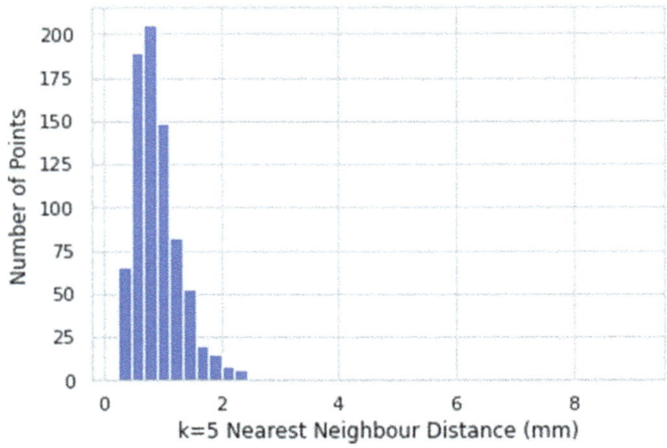

Again, the nearest neighbour distance seems to tail off, with fewer positions at larger distances. In the code cell below, we filter out tracer locations whose kth nearest neighbour is further than `max_distance`, then plot both the 'raw' and 'filtered' positions side by side; do change the `max_distance` and see its effects!

[1]:
```
# Importing necessary libraries for numerical array manipulation and plotting,
# plus the `pept` library for straightforward PEPT data processing
import numpy as np
import matplotlib.pyplot as plt
import pept

# Set Matplotlib theme (try "darkgrid" too for a more modern look)
import seaborn
seaborn.set_style("whitegrid")

# Set Plotly figure output (try "browser" too for interactive graphs)
import plotly
plotly.io.renderers.default = "png"
```

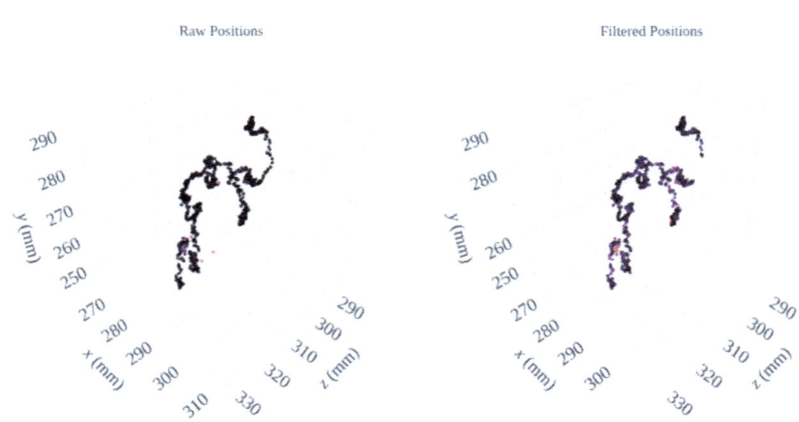

This particular dataset contains relatively clean data, with no points that are too far away; in this case, the nearest-neighbour filtering acts as to remove segments of the trajectory where the tracer is located very rarely—i.e. the temporal resolution is very low. On the other hand, it is also very clear when too high a maximum distance has been used, as important trajectory points are starting to be removed.

Filtering using the PEPT Library

The pept library offers terser filtering of points based on time, location or any other column included in the data using the Condition filter, which transforms a string containing a condition - e.g. "error < 5", "label != -1" - into efficient, parallelised code:

```
[8]: # Select only positions satisfying a condition, e.g. "error < 5", "t > 100"
     from pept.tracking import Condition

     filtered = Condition("error < 5").fit_sample(positions)

     # Plot tracer positions, colour-coding the last column - i.e. the spatial error
     PlotlyGrapher().add_points(filtered, colorbar_title="Spatial Error").show()
```

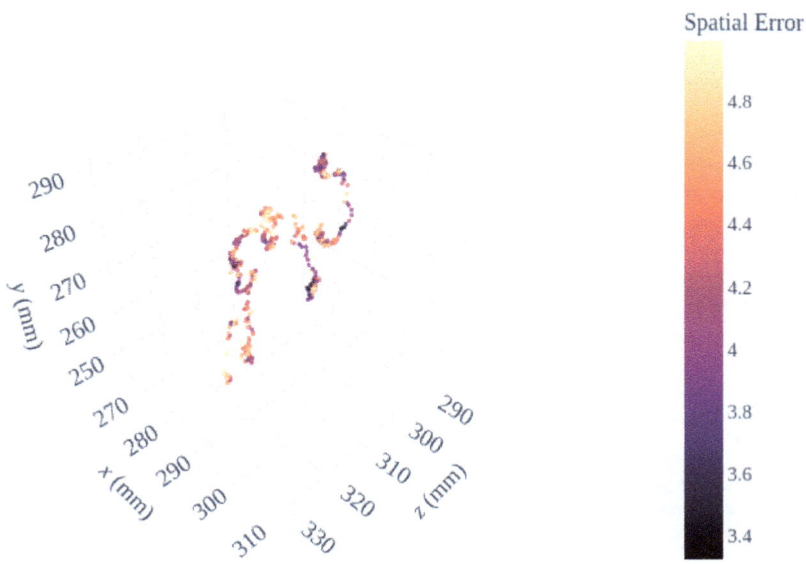

6.3.4 Interactive example: separating trajectories

The notebook files accompanying this section are available in the supplementary data online at https://iopscience.iop.org/book/978-0-7503-3071-8.

Separating trajectories

Separating out individual trajectories from particle positions is essential when performing multiple particle tracking, but can be a highly challenging task, especially when tracers' paths intersect. The tracer *locating* PEPT algorithms described in the previous sections and interactive examples simply find tracer positions within individual *samples* of LoRs, without assigning any trajectory *label* to those positions. In other words, PEPT algorithms typically return time-sorted tracer locations, which in the case of multiple tracers means that their trajectories are intertwined. We must then *separate out the trajectories* to analyse and post-process them individually.

[1]:
```
# Importing necessary libraries for numerical array manipulation and plotting,
# plus the `pept` library for straightforward PEPT data processing
import numpy as np
import matplotlib.pyplot as plt
import pept

# Set Matplotlib theme (try "darkgrid" too for a more modern look)
import seaborn
seaborn.set_style("whitegrid")

# Set Plotly figure output (try "browser" too for interactive graphs)
import plotly
plotly.io.renderers.default = "png"
```

Initialising lines of response

Here we will use a sample of an experiment consisting of two 1 mm diameter alumina particles with fluorine-18 radioactive tracer material. The particles are fixed at the two ends of an arm rotating at 42 rpm, their trajectories describing full circles.

[2]:
```
# Read in two-tracer dataset and plot the first sample of LoRs
from pept.visualisation import PlotlyGrapher

# Initialise data from a Parallel Screens CSV format (i.e. [t, x1, y1, x2, y2])
# into the general "LineData" format ([t, x1, y1, z1, x2, y2, z2]) and set a
# sample size and overlap
lors = pept.scanners.ParallelScreens(
    "Data/RotatingArm2Tracers.csv",
    screen_separation=600,
    skiprows=15,
    sample_size=2 * 200,
    overlap=2 * 120,
)

# Plot the first sample of LoRs
PlotlyGrapher().add_lines(lors[0]).show()
```

```
Initialised PEPT data in 0.032 s.
```

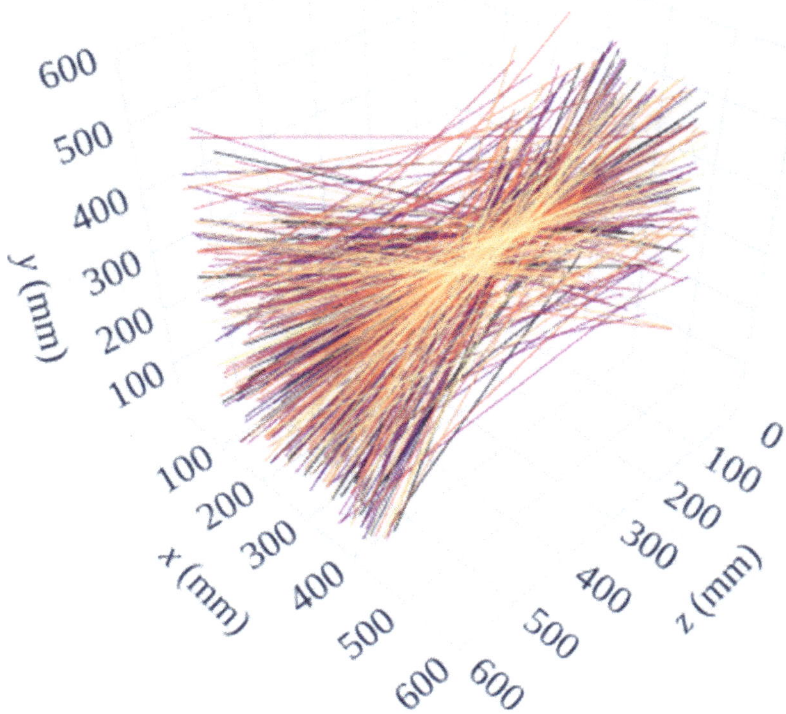

PTV-based trajectory separation

As the particle tracking velocimetry (PTV[17] technique is completely dependent on the ability to reconstruct the trajectories of multiple individual quasi-two-dimensional particles from a video, a significant amount of research has been done in the area of PTV trajectory separation. It is therefore sensible that some of these algorithms are applied to PEPT data. A notable example is the one introduced in [6], where tracer identities are being kept track of on a frame-by-frame basis.

The main idea is that given two consecutive frames (i.e. the tracer locations found in two consecutive samples of LoRs), the tracer pairing that minimises the sum of displacements within the same trajectories is used to assign trajectory labels to new locations. The pairing is found using the well-established minimum weight matching algorithm, which seeks to match a set of 'workers' with a set of 'jobs' such that some cost is minimised; in this case, the cost is simply the sum of distances travelled by tracers on the same trajectories.

An implementation of this trajectory separation method is given below:

[17] For a detailed discussion on PTV see chapter 4)

```
[3]: # Implement the trajectory separation algorithm of Wiggins, Santos and Ruggles,
     # 2019 in which tracer locations from consecutive frames are paired such that
     # the sum of displacements is minimised
     from scipy.spatial.distance import cdist
     from scipy.optimize import linear_sum_assignment

     # If any tracer displacement between two consecutive frames yields a larger
     # velocity than `max_velocity`, assign a new trajectory
     max_velocity = 10


     def match_tracers(frame1, frame2):
         '''Given the tracers' locations found in two consecutive frames (formatted
         as [time, x, y, z, trajectory_label]), set the second frame's tracers'
         trajectory labels such that the frame1-frame2 pairings minimise the sum of
         displacements. If any tracer pairing is found to yield a larger velocity
         than `max_velocity`, a new trajectory label is assigned.
         '''
         positions1 = frame1[:, 1:4]
         positions2 = frame2[:, 1:4]
         distance_matrix = cdist(positions1, positions2)

         # Find tracer pairings which minimise the sum of displacements
         frame1_indices, frame2_indices = linear_sum_assignment(distance_matrix)
         displacements = distance_matrix[frame1_indices, frame2_indices]

         # Assign frame1's matching trajectory labels to frame2; note that the
         # columns are formatted as [t, x, y, z, trajectory_label]
         frame2[frame2_indices, 4] = frame1[frame1_indices, 4]

         # Maximum current label
         max_label = frame1[:, 4].max()

         # If there are more tracers in frame2, assign new trajectory labels to the
         # new tracers
         if len(frame2) > len(frame2_indices):
             new_tracers_indices = set(range(len(frame2))) - set(frame2_indices)
             for index in new_tracers_indices:
                 max_label += 1
                 frame2[index, 4] = max_label

         # If any pairing yields displacements larger than `max_distance`, a new
         # trajectory is considered to have started
         max_distance = max_velocity * (frame2[0, 0] - frame1[0, 0])
         new_trajs_indices = np.argwhere(displacements >= max_distance).flatten()

         # Create new trajectory labels for each pairing with displacement larger
         # than `max_distance`
         for index in new_trajs_indices:
             max_label += 1
             frame2[index, 4] = max_label
```

In the code cell below we will track the tracers from each sample of LoRs and save them in individual frames, before reconstructing the particle trajectories by keeping track of trajectory labels on a frame-by-frame basis using the method outlined in [6].

The PEPT-ML multiple-tracer algorithm is used to find the tracers in each sample of LoRs (for details on how PEPT-ML works, see sections 6.2.14 and 6.2.15).

[4]:
```
# Use PEPT-ML to track the tracers from each sample of LoRs, saving them to
# individual frames, then reconstruct trajectories on a frame-by-frame basis
# using the method outlined in [1]
import pept
from pept.tracking import *
from pept.plots import PlotlyGrapher, PlotlyGrapher2D

# Loading bar
from tqdm import tqdm


# Create a PEPT-ML multi-tracer locating pipeline - using one pass of clustering
pipeline = pept.Pipeline([
    Cutpoints(max_distance = 0.5),
    HDBSCAN(true_fraction = 0.1),
    SplitLabels(remove_labels = False) + Centroids(),
    Stack(),
])

# Use the PEPT-ML pipeline to process each sample of LoRs, saving them in a list
# of frames
frames = pipeline.fit(lors)

# Set the first frame's initial set of (new) trajectory labels
frames[0].points[:, 4] = np.arange(len(frames[0].points))

# Match the trajectory labels from every consecutive pair of frames
for i in range(1, len(frames)):
    match_tracers(frames[i - 1].points, frames[i].points)
```

```
Batch 1 / 1 :: 100%|                    | 498/498
[00:03<00:00, 160.89it/s]


Processed samples in 3.555785655975342 s
```

[5]:
```
# Plot trajectories, colour-coding them by the last column - i.e. the trajectory
# label
num_trajectories = int(frames.points[:, 4].max() + 1)
print(f"Found {num_trajectories} trajectories")

PlotlyGrapher().add_points(frames).show()
```

```
Found 5 trajectories
```

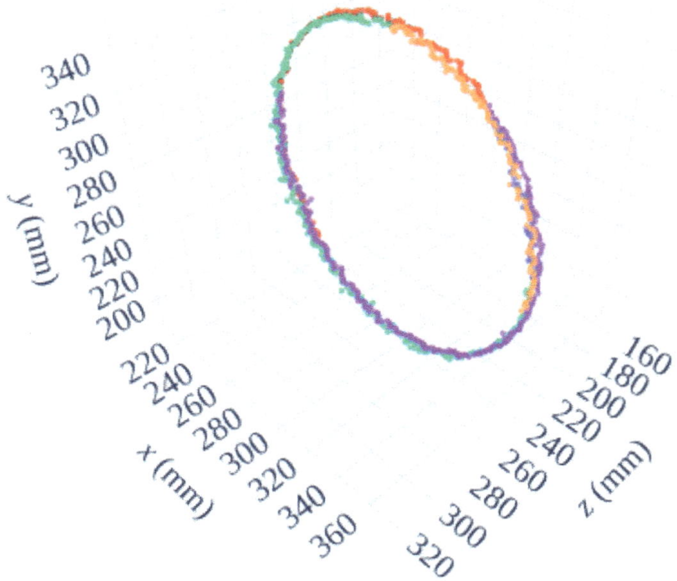


```
[6]: # Create a "timeseries" plot with the tracer positions vs. time
     PlotlyGrapher2D().add_timeseries(frames).show()
```

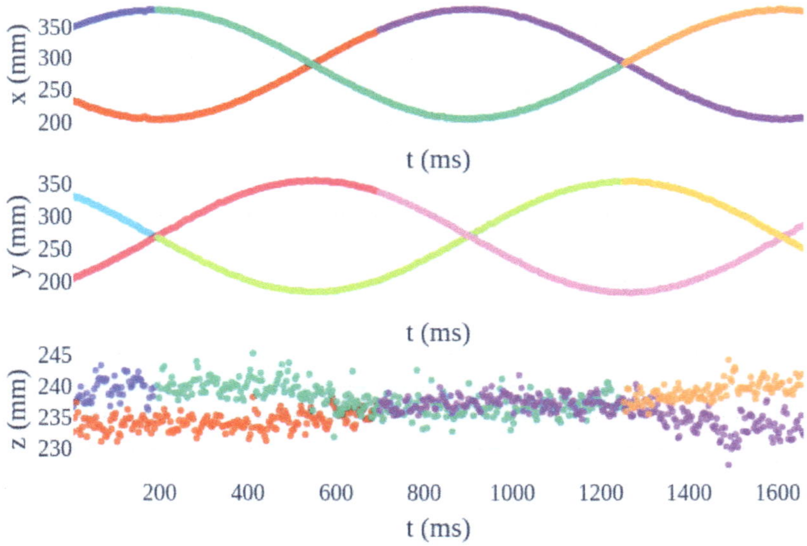

For this trajectory separation algorithm, if one of the two tracers is not found in a frame, it will 'split' its track, starting a new trajectory when it is found again.

Clustering-based trajectory separation

Machine learning is another field from which object tracking methods can be applied to PEPT. The separation of trajectories can be modelled as a clustering problem, where each individual trajectory is considered an individual cluster. The most popular clustering technique, *single linkage clustering*, works by creating a *minimum spanning tree* (MST, i.e. the minimum distance path connecting all points) from a set of points, then 'cutting' all paths connecting a pair of neighbouring points whose distance is larger than a fixed `trajectory_cut_distance`; all remaining connected subtrees from the MST with more than `min_trajectory_size` points are considered new clusters.

However, in its basic form, this algorithm is not applicable to trajectory separation: consider a tracer starting to move freely in space from some initial position, before returning to it after some time; in this case, the minimum spanning tree will connect the first and last points on the tracer trajectory because they are *spatially* close, even though *temporally* they are very far apart. In order to avoid such cases, we need to consider some form of *spatio-temporal* distance, where trajectory points can only be connected if they are both spatially *and temporally* close. This can be achieved by only allowing tracer positions to be connected if they are less than `points_window` indices apart in the sorted vector of timestamps; in other words, we define *reachable* points in a sliding window manner through the time-sorted array of positions.

Because this *point reachability* definition is effectively a custom spatio-temporal metric, it must be implemented separately; an optimised implementation from the `pept` library is used below:

```
[7]: # Implement a clustering-based trajectory separation algorithm which performs
     # single linkage clustering (implementation from the `hdbscan` library) using
     # a custom spatio-temporal metric
     import hdbscan
     from scipy.sparse.csgraph import minimum_spanning_tree
     from pept.tracking.trajectory_separation.distance_matrix_reachable import \
         distance_matrix_reachable

     # Two tracer locations are "reachable" only if they are within `points_window`
     # from one another in the sorted vector of timestamps
     points_window = 20

     # "Cut" all minimum distance paths that are longer than this to form clusters
     trajectory_cut_distance = 10.

     # A cluster / trajectory is only considered valid if it contains more than
     # `min_trajectory_size` points
     min_trajectory_size = 5
```

```
def segregate_trajectories(positions):
    '''Segregate the intertwined points from multiple trajectories into
    individual paths. Adapted from `pept.tracking.segregate_trajectories`.
    '''
    # Calculate the sparse distance matrix between reachable points. This is an
    # optimised Cython function returning a sparse CSR matrix.
    distance_matrix = distance_matrix_reachable(positions, points_window)

    # Construct the minimum spanning tree from the sparse distance matrix. Note
    # that `mst` is also a sparse CSR matrix.
    mst = minimum_spanning_tree(distance_matrix)

    # Get the minimum spanning tree edges into the [vertex 1, vertex 2,
    # edge distance] format, then sort it based on the edge distance.
    mst = mst.tocoo()
    mst_edges = np.vstack((mst.row, mst.col, mst.data)).T
    mst_edges = mst_edges[mst_edges[:, 2].argsort()]

    # Create the single linkage tree from the minimum spanning tree edges using
    # internal hdbscan methods (because they're very fast)
    single_linkage_tree = hdbscan._hdbscan_linkage.label(mst_edges)
    single_linkage_tree = hdbscan.plots.SingleLinkageTree(single_linkage_tree)

    # Cut the single linkage tree at `trajectory_cut_distance` and get the
    # cluster labels, setting clusters smaller than `min_trajectory_size` to
    # -1 (i.e. noise).
    labels = single_linkage_tree.get_clusters(
        trajectory_cut_distance,
        min_trajectory_size,
    )
    # Append the labels to `positions`.
    positions = np.append(positions, labels[:, np.newaxis], axis = 1)
    return positions
```

In the code cell below we will track the two tracers using the PEPT-ML algorithm; for simplicity, the implementation from the `pept` library is used, but the reader is invited to see sections 6.2.14 and 6.2.15 for the method's detailed steps. The tracer locations will then be used to separate out trajectories using the `segregate_trajectories` subroutine.

```
[8]: # Track tracers' locations using PEPT-ML and separate trajectories using the
     # `segregate_trajectories` function defined above
     #
     # Re-use the PEPT-ML pipeline defined above, then stack all frames / positions
     positions = pipeline.fit(lors)
     trajectories = segregate_trajectories(positions.points)
```

```
Batch 1 / 1 :: 100%|                              | 498/498
[00:01<00:00, 341.28it/s]


Processed samples in 1.917762041091919 s
```

Notice that all positions were used at once, avoiding the need for individual PTV-like frames; this allows the reconstruction of trajectories even if a tracer is temporarily not found in a given sample of LoRs, therefore allowing trajectory separation even across multiple frames. The `max_distance` parameter can also have the effect of filtering out spatio-temporally spurious points from the tracer trajectories.

[9]:
```
# Concatenate all positions found across all frames and plot them, colour-coding
# them by the last column - i.e. the trajectory label
num_trajectories = int(trajectories[:, -1].max() + 1)
print(f"Found {num_trajectories} trajectories")

PlotlyGrapher().add_points(trajectories).show()
```

```
Found 2 trajectories
```

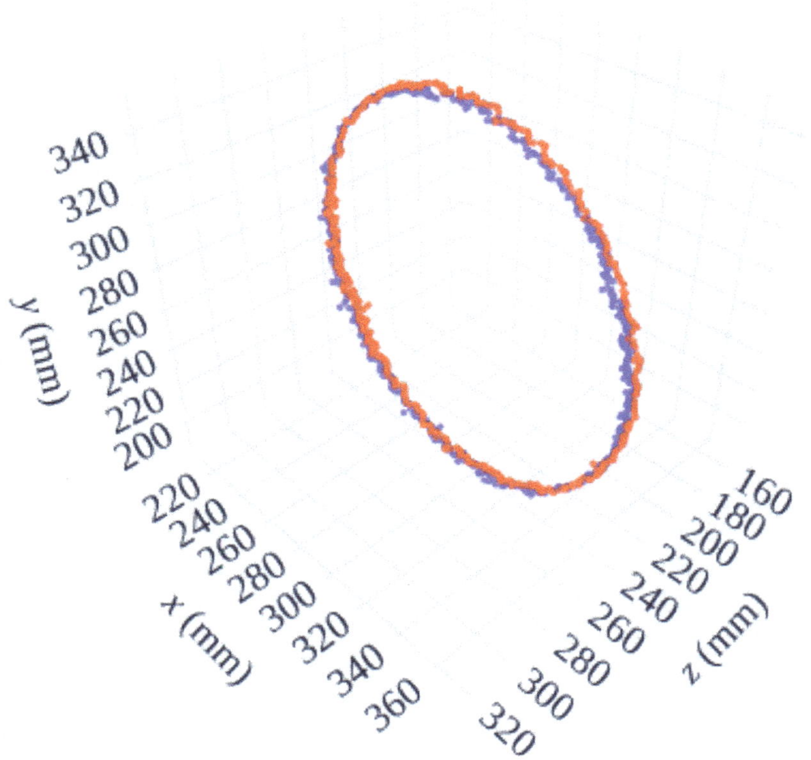

6.4 Horses for courses: comparing algorithm capability for differing tasks

As we have seen in the preceding sections, there exist a number of PEPT pre-processing algorithms, each of which possesses its own unique set of strengths and weaknesses. As such, it is not possible to meaningfully espouse a single 'best' PEPT technique—the methodology most suitable for tracking a single, slow-moving particle is not necessarily the best choice for imaging a more rapidly-circulating particle, nor are either of these guaranteed to be able to optimally locate multiple, distinct tracers, for example. Rather, it is a matter of 'horses for courses'—different algorithms are decidedly more adept than others for different tasks, meaning that to achieve the most with PEPT, it is valuable to be able to wield multiple different algorithms. This was in fact a major motivator behind the creation of this book, and in particular the detailed descriptions and interactive examples of the various available PEPT algorithms provided in this chapter. In addition to these shortened examples, full versions of the Birmingham, FPI and PEPT-ML algorithms, alongside detailed tutorials, can be found in, and freely downloaded from, the Positron Imaging Centre's GitHub[18] repositories.

In this final section, we briefly summarise the main results of a recent collaborative project between the PEPT groups at the Universities of Birmingham, Cape Town, Edinburgh, KwaZulu-Natal and Tennessee, King's College London, Virginia Commonwealth University and Stanford University [54], which directly benchmarked the seven most widely-used PEPT techniques against one another in a series of tests specifically designed to explore various aspects of their functionality.

In terms of the ability to accurately locate a single, stationary source, it is interesting to note that the oldest algorithm—the Birmingham method—was found to generally produce the best results of all algorithms tested, including for the particularly challenging case in which only 10 LoRs (compared to the 'standard' 250) are used to locate a tracer.

In terms of locating a single moving tracer, however, the Birmingham algorithm was found to perform less impressively. In this case, the most recently-developed PEPT-ML and PEPT-EM algorithms were found to produce the strongest performance in terms of raw (spatial) precision, while the PEPT-ML and the SBSR method were found to outperform the other algorithms when considering also the temporal resolution at which data were acquired. In the case of the combined spatio-temporal resolution criterion—defined as the mean location error divided by the root of the detection rate[19]—PEPT-ML was found to perform most strongly for low tracer velocities, but the SBSR method produced significantly better results at high velocities (see figure 6.19).

The ability of the algorithms to cope with high-scatter environments is somewhat less clear-cut: the PEPT-ML and PEPT-EM algorithms were found to be capable of producing the most accurate locations from high-scatter systems in several cases, but their accuracy was observed to fluctuate considerably. Conversely, the FPI, Spherical Density and SBSR methods produced data which were overall slightly less accurate (though still offering sub-millimetre resolution), but were found to do so significantly more reliably.

In terms of their ability to simultaneously detect large numbers of particles, the PEPT-ML and FPI methods were both shown to be able to successfully detect 100% of the tracers present in a system for N up to 80 (the highest value tested) in both planar and ring detector geometries (see section 5.2). The Spherical Density method

[18] https://github.com/uob-positron-imaging-centre

[19] The square root is chosen as the most reliable normalising factor as the location precision of a PEPT measurement is expected to vary as $\frac{1}{\sqrt{N}}$ where N is the number of LoRs used per particle location event [53].

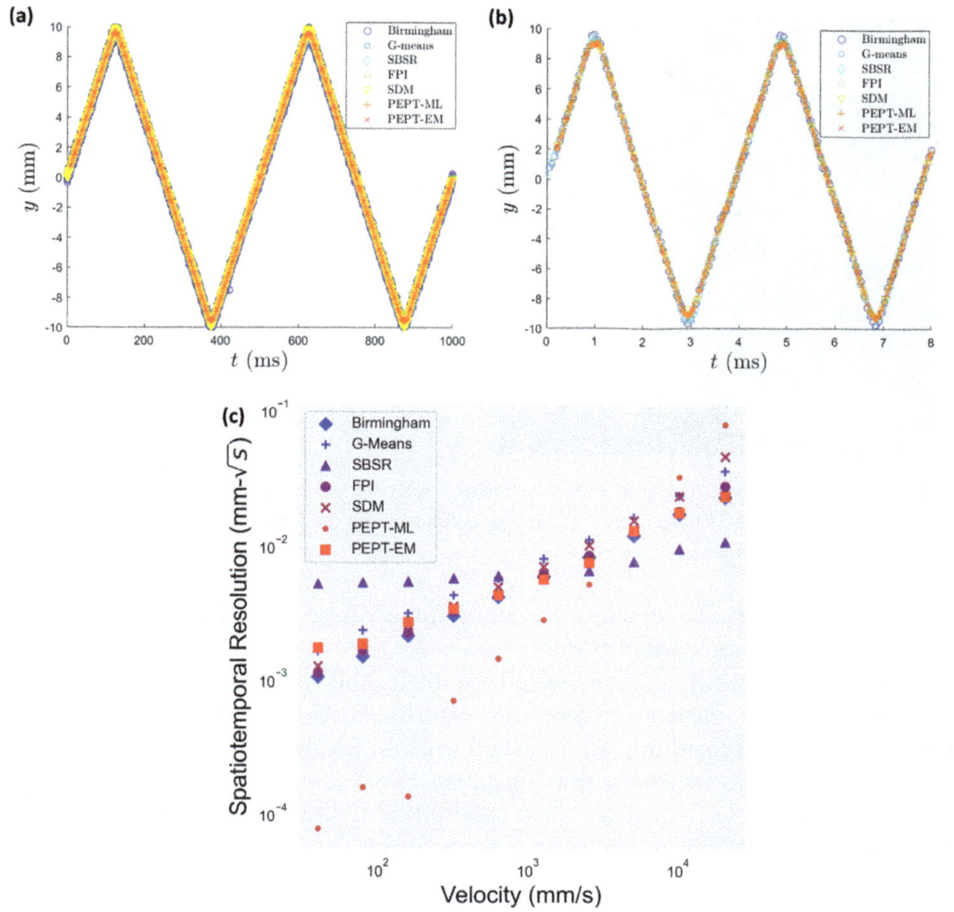

Figure 6.19. Above: Plots showing trajectories, as determined by the seven most widely-used PEPT algorithms, for a tracer moving at (a) 80 mm s^{-1} and (b) 10 240 mm s^{-1} along a sawtooth path. Below: The spatiotemporal resolution (see main text) achieved by each algorithm for different tracer velocities.

was also observed to achieve 100% accuracy for the ring geometry, though begins to 'lose' tracers for $N \gtrsim 40$ in the planar geometry. The G-means algorithm is also observed to lose tracers for $N \gtrsim 35$, but in all cases—even up to $N = 80$—manages to successfully locate at least 80% of the tracers present.

One of the great challenges of multiple-particle tracking is the ability to resolve two or more particles lying in close proximity—often cited as one of the major limitations of PEPT, due to the fact that it makes the reconstruction of inter-particle collisions somewhat complex [3]. It is perhaps surprising, then, that our benchmarking tests showed that a number of algorithms—namely the G-means, FPI, Spherical Density method, PEPT-ML and PEPT-EM algorithms—were able to successfully resolve the locations of two 2 mm tracers in contact. Confirmation of PEPT's ability to successfully locate and delineate multiple particles in such close proximity provides the means to extract a host of additional data from PEPT

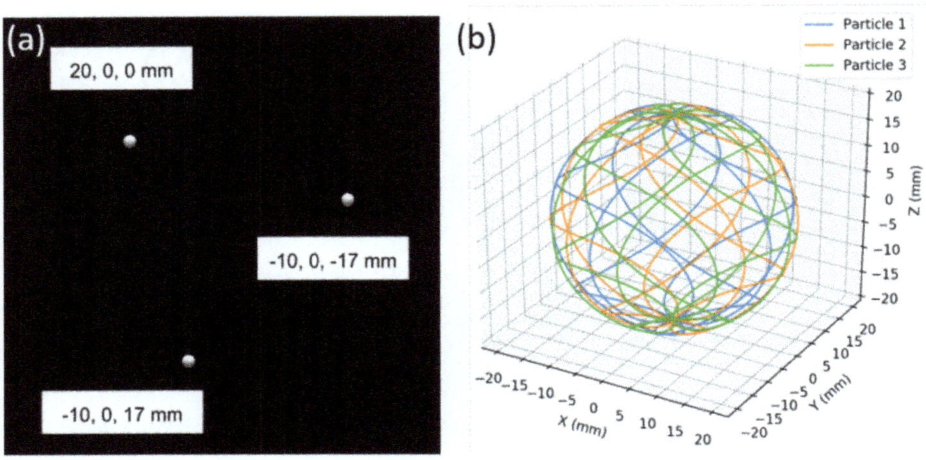

Figure 6.20. (a) Relative placement of tracers used to test the ability of various PEPT algorithms to track and distinguish multiple particle trajectories. (b) The paths followed by the three tracers during the test. Image reused with permission from [54].

experiments, potentially opening the door for a range of new applications where inter-particle collisions are important.

Finally, the algorithms were tested on their ability to successfully track and delineate multiple particle trajectories, as discussed in the preceding section. Specifically, the algorithms were tasked with tracking the trajectories of three particles, each separated by a fixed distance of ~3 cm from one another, as they are translated across the surface of a sphere (see figure 6.20). The G-means, FPI, PEPT-ML and PEPT-EM algorithms were all found to be capable of successfully distinguishing the overlapping trajectories of the different particles and, more impressively still, doing so with sub-millimetre accuracy.

Further details regarding the comparisons of the various PEPT algorithms, including full details of the benchmarking tests upon which these comparisons are based, may be found in reference [54].

References

[1] Parker D J, Broadbent C J, Fowles P, Hawkesworth M R and McNeil P 1993 Positron emission particle tracking–a technique for studying flow within engineering equipment *Nucl. Instrum. Methods Phys. Res.* A **326** 592–607

[2] Parker D J 2017 Positron emission particle tracking and its application to granular media *Rev. Sci. Instrum.* **88** 051803

[3] Yang Z, Parker D J, Fryer P J, Bakalis S and Fan X 2006 Multiple-particle tracking–an improvement for positron particle tracking *Nucl. Instrum. Methods Phys. Res.* **564** 332–8

[4] Bickell M, Buffler A, Govender I and Parker D J 2012 A new line density tracking algorithm for PEPT and its application to multiple tracers *Nucl. Instrum. Methods Phys. Res.* A **682** 36–41

[5] Windows-Yule C R K 2020 Ensuring adequate statistics in particle tracking experiments *Particuology* **59** 43–54

[6] Wiggins C, Santos R and Ruggles A 2016 A novel clustering approach to positron emission particle tracking *Nucl. Instrum. Methods Phys. Res.* A **811** 18–24

[7] Dougherty G 2009 *Digital Image Processing for Medical Applications* (Cambridge: Cambridge University Press)

[8] Hamerly G and Elkan C 2004 Learning the *k* in *k*-means *Advances in Neural Information Processing Systems* pp 281–8

[9] MacQueen J *et al* 1967 Some methods for classification and analysis of multivariate observations *Proc. of the Fifth Berkeley Sym. on Mathematical Statistics and Probability* vol 1 *(Oakland, CA, USA)* pp 281–97

[10] MacKay D J C 2003 *Information Theory, Inference and Learning Algorithms* (Cambridge: Cambridge University Press)

[11] Hartigan J A and Wong M A *et al* 1979 A *k*-means clustering algorithm *Appl. Stat.* **28** 100–8

[12] Anderson T W and Darling D A *et al* 1952 Asymptotic theory of certain "goodness of fit" criteria based on stochastic processes *Ann. Math. Stat.* **23** 193–212

[13] Stephens M A 1974 EDF statistics for goodness of fit and some comparisons *J. Am. Stat. Assoc.* **69** 730–7

[14] Tai Y-C, Chatziioannou A, Siegel S, Young J, Newport D, Goble R N, Nutt R E and Cherry S R 2001 Performance evaluation of the micropet p4: a PET system dedicated to animal imaging *Phys. Med. Biol.* **46** 1845

[15] Wiggins C, Patel N, Bingham Z and Ruggles A 2019 Qualification of multiple-particle positron emission particle tracking (M-PEPT) technique for measurements in turbulent wall-bounded flow *Chem. Eng. Sci.* **204** 246–56

[16] Langford S, Wiggins C, Tenpenny D and Ruggles A 2016 Positron emission particle tracking (PEPT) for fluid flow measurements *Nucl. Eng. Des.* **302** 81–9

[17] Wiggins C, Santos R and Ruggles A 2017 A feature point identification method for positron emission particle tracking with multiple tracers *Nucl. Instrum. Methods Phys. Res.* A **843** 22–8

[18] Crocker J C and Grier D G 1996 Methods of digital video microscopy for colloidal studies *J. Colloid Interface Sci.* **179** 298–310

[19] Sbalzarini I F and Koumoutsakos P 2005 Feature point tracking and trajectory analysis for video imaging in cell biology *J. Struct. Biol.* **151** 182–95

[20] Acharya T and Ray A K 2005 *Image Processing: Principles and Applications* (Wiley: New York)

[21] Wells W A, Rainer R O and Memoli V A 1992 Basic principles of image processing Am *J. Clin. Pathol.* **98** 493–501

[22] Daly E J, Williamson G I, Rusbridge M G and Jacobson A R 1985 The 'boxcar' method for the analysis of non-Gaussian random signals *Plasma Phys. Control. Fusion* **27** 761

[23] Hassan Y A and Canaan R E 1991 Full-field bubbly flow velocity measurements using a multiframe particle tracking technique *Exp. Fluids* **12** 49–60

[24] Patel N, Wiggins C and Ruggles A 2017 Positron emission particle tracking in pulsatile flow *Exp. Fluids* **58** 42

[25] Langford S T, Wiggins C S, Santos R, Hauser M, Becker J M and Ruggles A E 2017 Three-dimensional spatiotemporal tracking of fluorine-18 radiolabeled yeast cells via positron emission particle tracking *PLoS One* **12** e0180503

[26] Lee K S, Kim T J and Pratx G 2014 Single-cell tracking with PET using a novel trajectory reconstruction algorithm *IEEE Trans. Med. Imaging* **34** 994–1003

[27] Knott G D 2000 *Interpolating Cubic Splines* vol 18 (Berlin: Springer)

[28] Analytical Methods Committee *et al* 1989 Robust statistics-how not to reject outliers Part 1. Basic concepts *Analyst* **114** 1693–7
[29] Jung K O, Kim T J, Yu J H, Rhee S, Zhao W, Ha B, Red-Horse K, Gambhir S S and Pratx G 2020 Whole-body tracking of single cells via positron emission tomography *Nat. Biomed. Eng.* **4** 835–44
[30] Blakemore D M, Govender I, McBride A T and Mainza A N 2019 Multiple particle tracking in PEPT using Voronoi tessellations *Chem. Eng. Sci.* **207** 780–9
[31] Aurenhammer F 1991 Voronoi diagrams–a survey of a fundamental geometric data structure *ACM Comput. Surv.* **23** 345–405
[32] Aurenhammer F and Klein R 2000 Voronoi diagrams *Handbook of Computational Geometry* vol 5 (Amsterdam: Elsevier) pp 201–90
[33] Ester M, Kriegel H-P, Sander J and Xu X *et al* 1996 A density-based algorithm for discovering clusters in large spatial databases with noise *KDD Proc.* vol 96 pp 226–31
[34] Schubert E, Sander J, Ester M, Kriegel H P and Xu X 2017 DBSCAN revisited, revisited: why and how you should (still) use DBSCAN *ACM Trans. Database Syst.* **42** 1–21
[35] Breunig M M, Kriegel H-P, Ng R T and Sander J 2000 LOF: identifying density-based local outliers *Proc. of the 2000 ACM SIGMOD Int. Conf. on Management of Data* pp 93–104
[36] Khan K, Rehman S U, Aziz K, Fong S and Sarasvady S 2014 DBSCAN: past, present and future *The Fifth International Conference on the Aplications of Digital Information and Web Technologies (ICADIWT 2014)* (IEEE: Piscataway, NJ) pp 232–8
[37] Odo A E, Govender I, Buffler A and Franzidis J-P 2019 A PEPT algorithm for predefined positions of radioisotopes relative to the tracer particle *Appl. Radiat. Isot.* **151** 299–309
[38] Nicuşan A L and Windows-Yule C R K 2020 Positron emission particle tracking using machine learning *Rev. Sci. Instrum.* **91** 013329
[39] Campello R J G B, Moulavi D and Sander J 2013 Density-based clustering based on hierarchical density estimates *Pacific-Asia Conference on Knowledge Discovery and Data Mining* (Springer: Berlin) pp 160–72
[40] McInnes L and Healy J 2017 Accelerated hierarchical density based clustering *IEEE 2017 Int. Conf. on Data Mining Workshop (ICDMW)* (IEEE: Piscataway, NJ) pp 33–42
[41] McInnes L, Healy J and Astels S 2017 HDBSCAN: hierarchical density based clustering *J. Open Source Softw.* **2** 205
[42] Harris C R *et al* 2020 Array programming with NumPy *Nature* **585** 357–62
[43] Virtanen P *et al* 2020 SciPy 1.0: fundamental algorithms for scientific computing in python *Nat. Methods* **17** 261–72
[44] Manger S, Renaud A and Vanneste J 2021 An expectation-maximization algorithm for positron emission particle tracking *Rev. Sci. Instrum.* **92** 085102
[45] Gundogdu O 2004 Positron emission tomography particle tracking using cluster analysis *Nucl. Instrum. Methods Phys. Res.* A **534** 562–76
[46] Gundogdu O 1999 Multiple particle tracking using the Birmingham positron emission camera *PhD Thesis* University of Birmingham
[47] Kaufman L and Rousseeuw P J 1986 Clustering large data sets *Pattern recognition in practice II* (Amsterdam: Elsevier)
[48] Kaufman L and Rousseeuw P J 2009 *Finding Groups in Data: An Introduction to Cluster Analysis* vol 344 (New York: Wiley)

[49] Gundogdu O and Tarcan E 2004 Location-allocation algorithm for multiple particle tracking using Birmingham MWPC positron camera *Nucl. Instrum. Methods Phys. Res.* A **523** 223–33

[50] Cooper L 1964 Heuristic methods for location-allocation problems *SIAM Rev.* **6** 37–53

[51] Wesolowsky G O and Truscott W G 1975 The multiperiod location-allocation problem with relocation of facilities *Manage. Sci.* **22** 57–65

[52] Yang Z, Fryer P J, Bakalis S, Fan X, Parker D J and Seville J P K 2007 An improved algorithm for tracking multiple, freely moving particles in a positron emission particle tracking system *Nucl. Instrum. Methods Phys. Res.* A **577** 585–94

[53] Parker D J, Forster R N, Fowles P and Takhar P S 2002 Positron emission particle tracking using the new Birmingham positron camera *Nucl. Instrum. Methods Phys. Res.* A **477** 540–5

[54] Windows-Yule C R, Herald M, Nicusan L, Wiggins C and Pratx G *et al* 2022 Recent advances in positron emission particle tracking: a comparative review *Rep. Prog. Phys.* **85** 016101

IOP Publishing

Positron Emission Particle Tracking
A comprehensive guide
Kit Windows-Yule, Leonard Nicuşan, Matthew T Herald, Samuel Manger and David Parker

Chapter 7

Post-processing: extracting physical information from PEPT data

One of the great strengths of the positron emission particle tracking (PEPT) technique is the great depth and diversity of information it can extract from a given scientific or industrial system. This information includes, but is by no means limited to, one-, two- or three-dimensional spatial density distributions, velocity distributions, granular temperature distributions and velocity vector fields, velocity probability distributions, particle dispersion and diffusion, and a variety of measures of segregation, mixing and the rates thereof. In this chapter, we present detailed information regarding these diverse quantities, and how they may be robustly and rigorously extracted from PEPT data, including interactive examples which guide the reader step-by-step through the process, and may even be used directly to process real PEPT data.

In the previous chapter, we have shown how raw line-of-response (LoR) data acquired from a positron camera may be used to reconstruct the pseudo-instantaneous positions of moving particles. In this chapter, we show how these simple, sequential points in three-dimensional space may be further processed so as to yield a diverse range of important physical quantities and fields from a system of interest.

Throughout the chapter, we will provide examples in which we analyse real PEPT data, using as our main example a gas-fluidised binary granular bed (figures 7.1 and 7.2)—a system with a diverse range real-world applications [1–6], and which exhibits a range of interesting phenomena [7–9], thus making it an ideal candidate for PEPT imaging. Using this example, we demonstrate the wealth of valuable information which may be extracted from a single PEPT data set.

7.1 Particle trajectories

7.1.1 Single-particle trajectories and their interpretation

One of the simplest forms of data that can be extracted from PEPT are single-particle position-time plots such as those shown in figure 7.3. Indeed, such plots

Figure 7.1. Video showing the binary sand/plastic fluidised bed system which will be used in the majority of examples in this chapter. Video available at https://iopscience.iop.org/book/978-0-7503-3071-8.

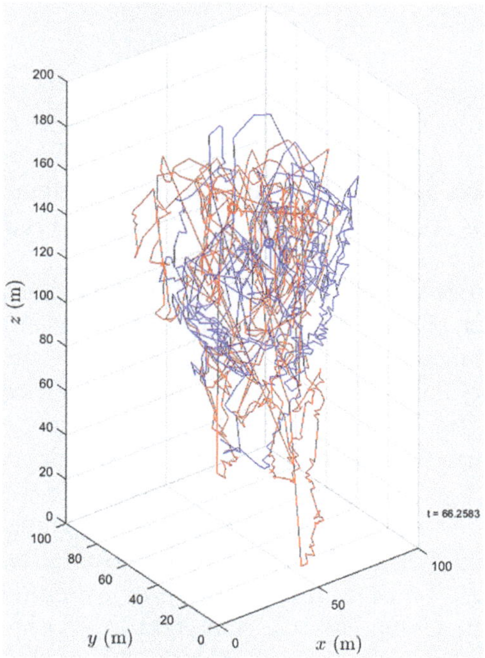

Figure 7.2. Video showing PEPT trajectories extracted for a single sand (red) and plastic (blue) particle within the fluidised bed system from figure 7.1. Video available at https://iopscience.iop.org/book/978-0-7503-3071-8.

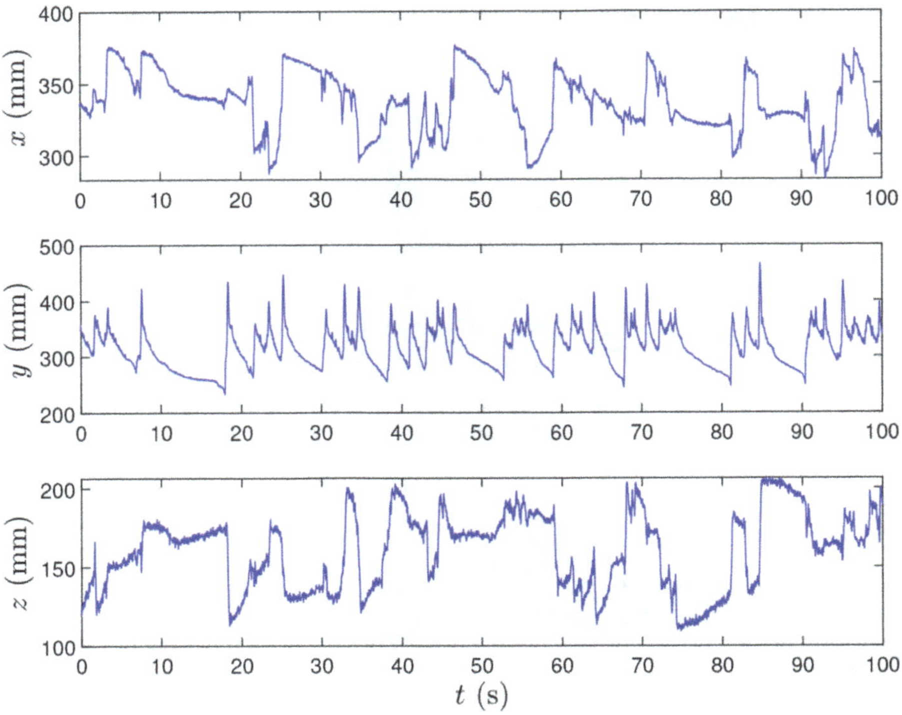

Figure 7.3. Examples of PEPT position versus time data for a single sand particle moving through the fluidised bed system depicted in figure 7.1.

require *no post-processing whatsoever*, as they can be directly produced from the data output by the pre-processing algorithms discussed in chapter 6. Nonetheless, despite their simplicity and the ease with which they may be extracted from pre-processed PEPT data (see section 7.1.2), they may still provide significant insight into the behaviour of a system. To illustrate this point, let us consider the data extracted from our fluidised bed example.

Figure 7.3 shows position versus time plots in the Cartesian x, y, and z coordinates for a single sand tracer particle. Looking, for example, at motion in the vertical y-coordinate, we can see clear evidence of periodic motion of the tracer, suggesting the presence of recirculatory flow within the system, as one would expect for a fluidised bed in the bubbling regime [10]. The plot also clearly shows the upward transit of particles to be significantly faster than their downward motion, again in line with expectations from research (particles are typically transported upwards by fast-moving gas bubbles, but must settle downward through the denser, slower-moving emulsion phase) [10, 11].

If we 'zoom out' and look at the particle motion over a longer timescale, we can extract further information regarding our system. In figure 7.4, we see that for $1400 \lesssim t \lesssim 1700$s the tracer remains relatively static at a position near the base of the system. This observation clearly suggests that, for this particular system, the bed is not adequately fluidised, i.e. there exist 'dead zones' where the gas flow is not adequate to induce motion in the bed's constituent particles. In industry, such an

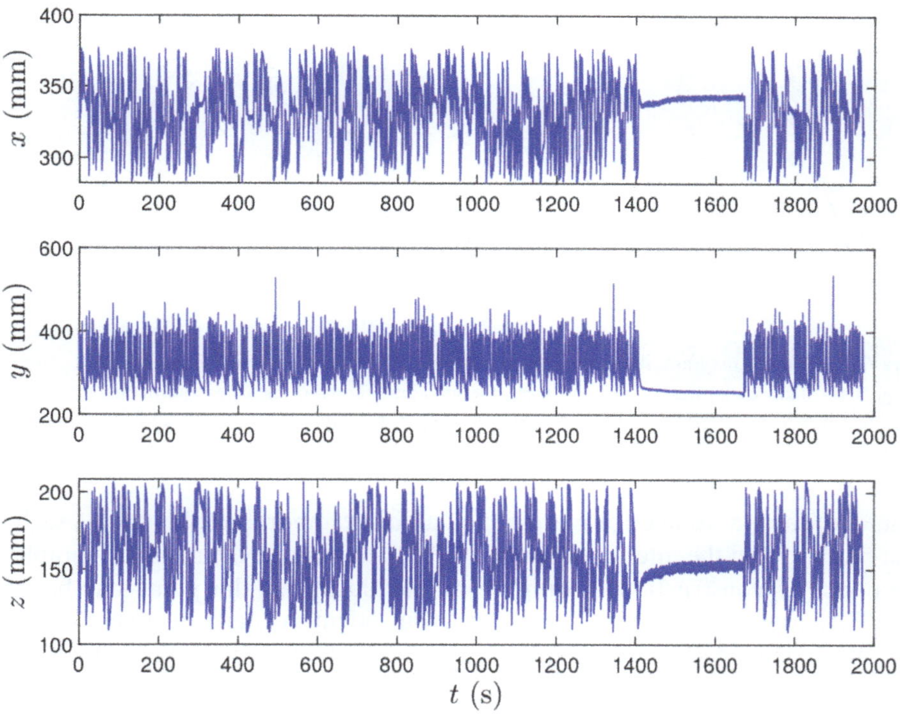

Figure 7.4. PEPT trajectory data for the same system as figure 7.3, but showing a longer time period.

observation may prove crucial in optimising the design or operation of such a particle-handling system, as such dead zones are typically highly undesirable. Observations of such static periods may also be used, for example, to determine the minimum suspension speed in a stirred tank fluid-solids mixing process [23] or infer the presence of caging effects or identify the jamming transition in a sheared or vibrated granular bed [12].

In the above we have illustrated that even the most fundamental form of PEPT output can (with adequate understanding of the system being imaged) be interpreted so as to provide valuable information regarding an experimental system. In the subsequent sections of this chapter, we go on to show the many other ways in which this data can be further manipulated in order to gain still greater and more detailed insight.

As mentioned above, one-dimensional, single-particle trajectories are certainly the simplest form of data discussed in this chapter. That is not to say, however, that their extraction from raw PEPT data and subsequent interpretation is trivial. It is important to note that the algorithms (described in detail in chapter 6) used to extract tracer positions from raw LoR data, though in many cases highly advanced, are not perfect, and require a degree of knowledge and expertise to utilise optimally. In some cases (especially those involving low data rates and/or large amounts of scatter or background noise) they may still produce occasional erroneous data points, and in almost any situation will still (as with any experimental

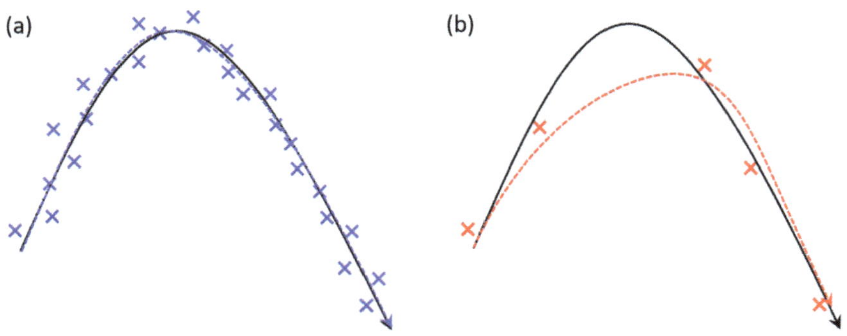

Figure 7.5. Schematic diagrams illustrating the ability of PEPT to reconstruct an identical particle trajectory at (a) high and (b) low data rates. In both instances, the solid lines represent the true (continuous) tracer path and the dotted lines a trajectory reconstructed from (discrete) PEPT data. Each cross represents a single particle location event (i.e. the triangulated centre of a group of LoRs).

technique) carry a finite error margin, and a finite temporal resolution. As such, one must be careful in the interpretation of PEPT data, even in the relatively simple case of trajectory reconstruction. A simple example illustrating this point may be seen in figure 7.5. Figure 7.5 shows two hypothetical, identical particle trajectories (solid lines) imaged at a high data rate (i.e. high particle activity/low attenuation) and a low data rate (i.e. low particle activity/high attenuation). The solid crosses represent particle locations, and the dashed lines the 'best fit' PEPT-measured trajectory connecting these individual locations. Clearly, at a lower data-rate, the PEPT data deviates strongly from the 'true' particle trajectory. While in some cases such a deviation may not be significant, in others it may be highly problematic. For example, if the trajectories in figure 7.5 represent the parabolic flight under gravity of a particle ejected from a fluidised bed, the low-data rate in panel (b) will cause a significant underestimate of the maximum height reached by said particle. This, in turn, may lead to an underestimation of key design parameters such as the transport disengagement height (TDH) of a fluidised bed [13] (see also section 7.2.2 below), which may cause significant issues in an industrial setting.

Issues such as that depicted in figure 7.5 cannot, with presently-available techniques, be fully, reliably resolved through post-processing. Rather, one must ensure that experiments are initially performed with a suitably-active tracer (see section 2), and that they are pre-processed in the optimal manner (see section 6).

7.1.2 Interactive example: plotting single particle trajectories

The notebook files accompanying this section are available in the supplementary data online at https://iopscience.iop.org/book/978-0-7503-3071-8.

Tutorial: Using PEPT data to plot single particle trajectories

In this tutorial, we will show you how to take a pre-processed PEPT data file and use the data contained within to plot a tracer's x-, y- and z-position as a function of time, as discussed in the preceding chapter.

Setting up

We begin, as always, by importing the necessary libraries. In this case, we want to be able to manipulate our data in the form of arrays and matrices, and to be able to plot our data, so we add the following:

```
[1]: # Importing necessary libraries for numerical array manipulation and plotting
     import numpy as np
     import matplotlib.pyplot as plt

     # Set Matplotlib theme (try "darkgrid" too)
     import seaborn
     seaborn.set_style("whitegrid")

     # Set Plotly figure output (try "browser" too)
     import plotly
     plotly.io.renderers.default = "png"
```

Importing data

The next step is to read in the pre-processed PEPT data that we wish to analyse. PEPT data is typically stored as columns of tab-separated ASCII text, with a header explaining the contents of the file.

```
[2]: # Path to the pre-processed PEPT data
     filepath = "Data/FluidisedBedExample.csv"

     # Print the first 20 lines in the file
     with open(filepath) as f:
         for _ in range(20):
             print(f.readline(), end="")
```

```
Example PEPT data of a fluidised bed imaged at the University of Birmingham
Positron Imaging Centre, pre-processed using the Birmingham Method.


Separation=   358

f(opt) :  0.050      Displacement parameters :     300,    400,   1500


Fixed slices:    250 events/slice,     1 locations/slice

 5.4      312.0      266.4      146.7       0.5       0.0       0.00     12
16.6      310.9      265.1      149.4       0.7       0.0       0.00     12
28.8      311.2      265.9      150.0       0.5       0.0       0.00     12
41.2      311.5      265.1      152.6       0.5       0.0       0.00     12
```

We are interested primarily in the first four columns of the file which correspond, in order from left to right, to a timestamp (t) and the calculated x-, y- and z-positions at this timestamp. Each new row of the column corresponds to a new location.

The relevant data from a given PEPT data file can be imported as follows:

[3]:
```
# Loading the file into a matrix named "data", skipping the first 16 rows which
# form the header
data = np.loadtxt(filepath, skiprows=15)

# Extracting the data we need - specifically the time data (first column).
# (Remember that python data indexing starts at 0, so the 1st column is
# labelled 0!)
t = data[:, 0]

# The horizontal x-position data (second column)
x = data[:, 1]

# The vertical y-position data (third column)
y = data[:, 2]

# The horizontal (z) position data (fourth column)
z = data[:, 3]

# For now we will ignore the rest of the columns.

# To check if that has worked properly, we can print the the first 10
# timestamps, and see if they look sensible:
print(t[:10])
```

```
[  5.4  16.6  28.8  41.2  52.1  64.8  76.2  87.5  99.9 109.3]
```

Plotting data

Now that we have our data stored in arrays, we can simply use `matplotlib` to plot our trajectories:

[4]:
```
# Setting up to plot 3 subplots in a single figure
figs, axs = plt.subplots(3, 1)

# Plotting x vs. time in the first subplot
axs[0].plot(t, x)
axs[0].set(xlabel='$t$ (s)', ylabel='$x$ (mm)')

# Plotting y vs. time in the second subplot
axs[1].plot(t, y)
axs[1].set(xlabel='$t$ (s)', ylabel='$y$ (mm)')

# Plotting z vs. time in the third subplot
axs[2].plot(t, z)
axs[2].set(xlabel='$t$ (s)', ylabel='$z$ (mm)')

# Ensuring plots are nicely arranged!
plt.tight_layout()
```

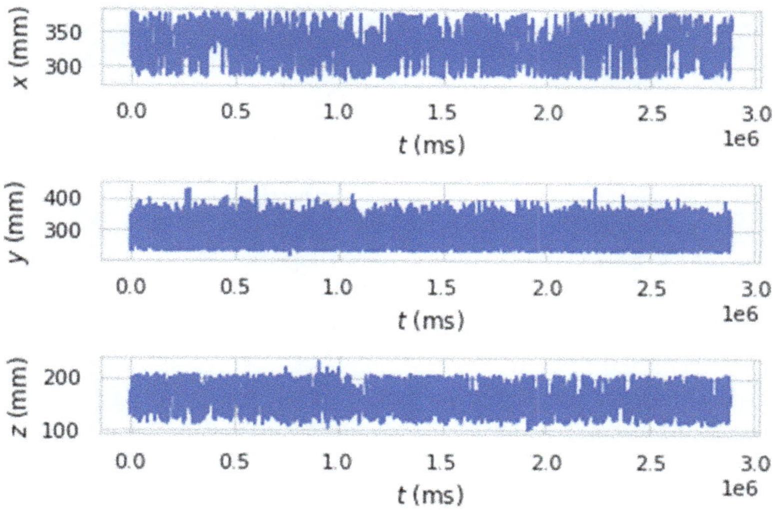

Using the `pept` library

The code presented in the previous sections is explicit and well commented, providing a useful learning tool. In real research, a scientist could use the `pept` Python library (https://github.com/uob-positron-imaging-centre/pept) to automate most tasks and keep logic at a high level. The `pept` library provides a unified interface to PEPT-related algorithms, including tracking, simulation, data analysis and visualisation tools, with high-performance, parallelised implementations of the most computationally-intensive steps.

The code below provides an example of how the `pept` library may be used to plot interactive 3D tracer trajectories, colour-coded by one of the data columns—in this case, the timestamp:

```
[5]: # Import the `pept` library
     import pept

     # Load PEPT data from CSV file
     filepath = "Data/FluidisedBedExample.csv"
     data = pept.utilities.read_csv(filepath, skiprows=16)

     # Create 3D plot of the first 2000 trajectory locations
     grapher = pept.plots.PlotlyGrapher()
     grapher.add_points(data[:2000], colorbar_col=0)
     grapher.show()
```

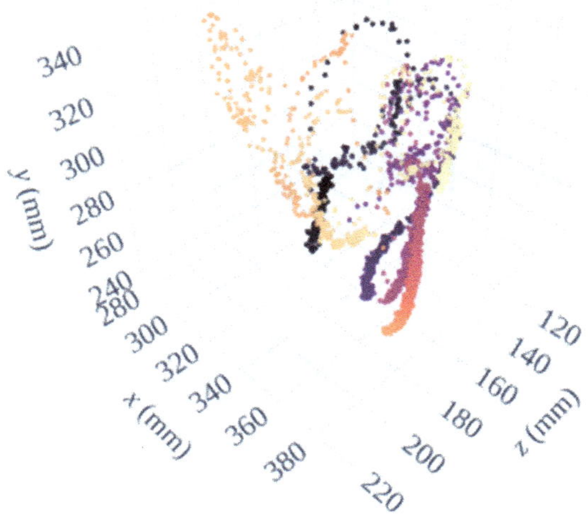


```
[6]: # Create a "timeseries" plot with the tracer positions vs. time
     grapher = pept.plots.PlotlyGrapher2D()
     grapher.add_timeseries(data[:10_000], colorbar_col=0)
     grapher.show()
```

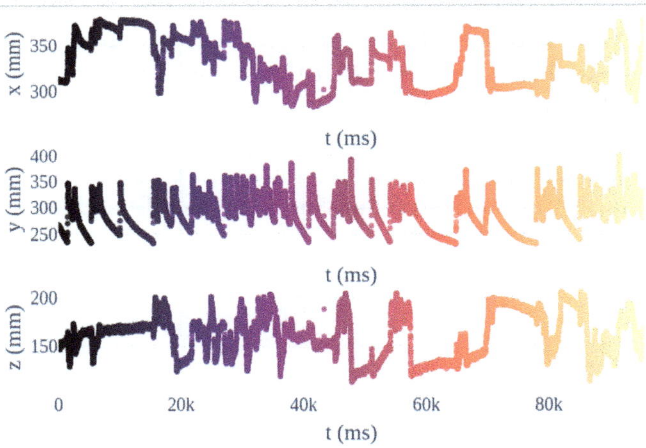

7.1.3 Interactive example: multiple-particle data

The notebook files accompanying this section are available in the supplementary data online at https://iopscience.iop.org/book/978-0-7503-3071-8.

Tutorial: Using PEPT data to plot multiple particle trajectories

In this tutorial, we will show you how to take a pre-processed PEPT data file and use the data contained within to plot the *x*-, *y*- and *z*-position of *multiple tracers* as functions of time. Here we use a sample of an experiment run by Dr Kit Windows-Yule at the University of Birmingham, consisting of two 1 mm diameter alumina particles with fluorine-18 radioactive tracer material. The particles are fixed at the two ends of an arm rotating at 42 rpm, their trajectories describing full circles.

Setting up

We begin, as always, by importing the necessary libraries. In this case, we want to be able to manipulate our data in the form of arrays and matrices, and to be able to plot our data, so we add the following:

[1]:
```
# Importing necessary libraries for numerical array manipulation and plotting
import numpy as np
import matplotlib.pyplot as plt

# Set Matplotlib theme (try "darkgrid" too)
import seaborn
seaborn.set_style("whitegrid")

# Set Plotly figure output (try "browser" too)
import plotly
plotly.io.renderers.default = "png"
```

Importing data

The next step is to read in the pre-processed PEPT data that we wish to process. PEPT data is typically stored as columns of tab-separated ASCII text, with a header explaining the contents of the file.

[2]:
```
# Path to the pre-processed PEPT data
filepath = "Data/RotatingArm2TracersExample.csv"

# Print the first 20 lines in the file
with open(filepath) as f:
    for _ in range(20):
        print(f.readline(), end="")
```

```
University of Birmingham Positron Imaging Centre

This is a sample of an experiment run at the University of Birmingham. It
consists of two 1 mm diameter alumina particles with Fluorine-18 radioactive
tracer material. The particles rotate at 42 rpm, each describing one full
circle. The data has been pre-processed using the PEPT-ML multiple tracer
tracking algorithm.

The columns of data are formatted as: [time (ms), x (mm), y (mm), z (mm),
tracer_signature1, tracer_signature2, trajectory_label]. The tracer signatures
are the mean numbers of cutpoints around each tracer location; it is directly
proportional on the number of LoRs emitted, which is a function of tracer
location within the system. The trajectory label is an index starting from 0.


5.059593714927049035e+01 2.208658715046613565e+02 2.172680609426333262e+02
2.766952363062533777e+02 2.753809523809524080e+02 3.000000000000000000e+01
0.000000000000000000e+00
5.059593714927049035e+01 2.208658715046613565e+02 2.172680609426333262e+02
2.766952363062533777e+02 2.753809523809524080e+02 2.900000000000000000e+01
0.000000000000000000e+00
5.067076559243226797e+01 3.600915547415309561e+02 3.179917231834601807e+02
2.836227456127605819e+02 3.910952380952380736e+02 3.000000000000000000e+01
1.000000000000000000e+00
5.067076559243226797e+01 3.600915547415309561e+02 3.179917231834601807e+02
2.836227456127605819e+02 3.910952380952380736e+02 3.100000000000000000e+01
1.000000000000000000e+00
5.384251723585057903e+01 2.201695493445744773e+02 2.182398549143502748e+02
2.767298995703361584e+02 2.713809523809524080e+02 3.000000000000000000e+01
0.000000000000000000e+00
```

We are interested primarily in the first four columns of the file which correspond, in order from left to right, to a timestamp (t) and the calculated x-, y- and z-positions at this timestamp, plus the last column representing each tracer trajectory's index. Each new row of the column corresponds to a new location.

The relevant data from a given PEPT data file can be imported as follows:

[3]:
```
# Loading the file into a matrix named "data", skipping the first 16 rows which
# form the header
data = np.loadtxt(filepath, skiprows=15)

# Extracting the data we need - specifically the time data (first column).
# (Remember that python data indexing starts at 0, so the 1st column is
# labelled 0!)
t = data[:, 0]

# The horizontal x-position data (second column)
x = data[:, 1]

# The vertical y-position data (third column)
y = data[:, 2]

# The horizontal (z) position data (fourth column)
z = data[:, 3]

# The tracer trajectory label - trajectory 1 has index 0, trajectory 2 has index
# 2, etc.
labels = data[:, -1]

# For now we will ignore the rest of the columns.

# To check if that has worked properly, we can print the the first 10
# timestamps, and see if they look sensible:
print(t[:10])
```

```
[50.59593715 50.59593715 50.67076559 50.67076559 53.84251724 53.84251724
 53.84251724 57.12843755 57.12843755 60.35622976]
```

Plotting data

Now that we have our data stored in arrays, we can simply use `matplotlib` to plot our tracers' positions for each trajectory label:

[4]:
```
# Setting up to plot 3 subplots in a single figure
figs, axs = plt.subplots(3, 1)

# Trajectory labels start from zero and go up to labels.max()
max_label = int(labels.max())

# For each unique trajectory label, plot that tracer's positions
for label in range(max_label + 1):
    # Plotting x vs. time in the first subplot
    axs[0].plot(t[labels == label], x[labels == label])
    axs[0].set(xlabel='$t$ (s)', ylabel='$x$ (mm)')
```

```
    # Plotting y vs. time in the second subplot
    axs[1].plot(t[labels == label], y[labels == label])
    axs[1].set(xlabel='$t$ (s)', ylabel='$y$ (mm)')

    # Plotting z vs. time in the third subplot
    axs[2].plot(t[labels == label], z[labels == label])
    axs[2].set(xlabel='$t$ (s)', ylabel='$z$ (mm)')

# Ensuring plots are nicely arranged!
plt.tight_layout()
```

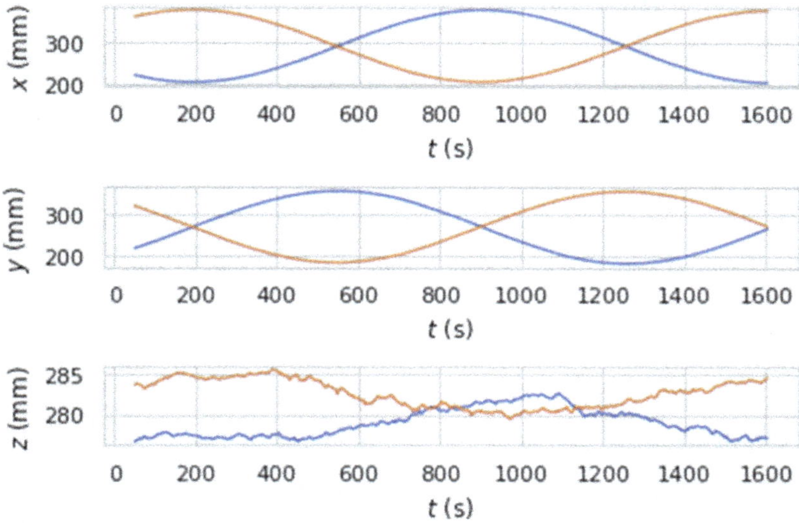

Using the `PEPT` **library**

The code presented in the previous sections is explicit and well commented, providing a useful learning tool. In real research, a scientist could use the `pept` Python library to automate most tasks and keep logic at a high level. The `pept` library provides a unified interface to PEPT-related algorithms, including tracking, simulation, data analysis and visualisation tools, with high-performance, parallelised implementations of the most computationally-intensive steps.

The code below provides an example of how the `pept` library may be used to plot interactive 3D tracer trajectories; by default, the PEPT-oriented `PlotlyGrapher` colour-codes plotted points by their last column, which is the trajectory label for pept-processed multiple tracer data:

```
[5]: # Import the `pept` library
import pept

# Load PEPT data from CSV file
filepath = "Data/RotatingArm2TracersExample.csv"
data = pept.utilities.read_csv(filepath, skiprows=16)

# Create 3D plot of the first 2000 trajectory locations
grapher = pept.visualisation.PlotlyGrapher()
grapher.add_points(data[:2000])
grapher.show()
```

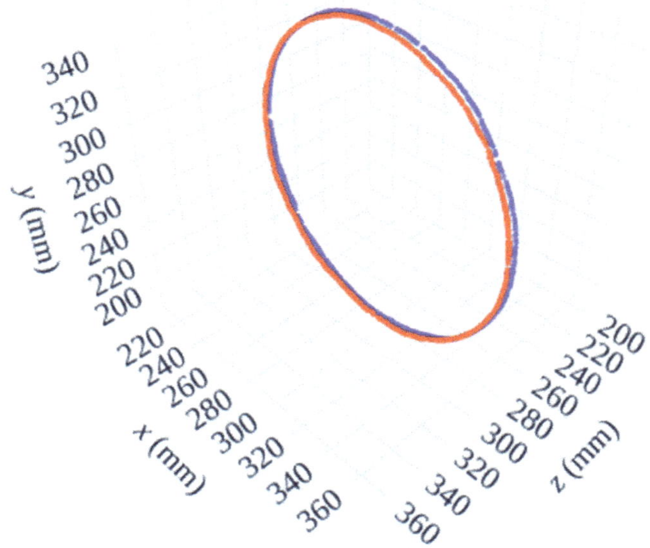

7.2 Occupancy and density

7.2.1 Creating occupancy profiles and distributions

As we have already touched upon in section 2.4, the true power of PEPT lies not in merely plotting the trajectories of a single particle (though, as we have seen in the previous section, this is not without value), but in using this data to determine one-, two- and three-dimensional fields which describe the dynamics of the *whole system* in which our tracer is housed. Perhaps the simplest—or at least most pedagogically useful—example of the manner in which PEPT trajectory data may be converted into 'whole-field' information is the creation of one-, two- or three-dimensional particle density fields.

Let us consider first the simplest case of a one-dimensional vertical density profile. In order to create such a profile, we may begin by subdividing our experimental volume into a series of 'slices' through the vertical y-coordinate, as illustrated in figure 7.6, and simply counting the number of recorded particle locations falling within each slice. Under the assumption of ergodicity [16] (and assuming also that location events occur at regular intervals), the fraction $F_i = N_i^{\text{loc.}}/N_T^{\text{loc.}}$ of the total number of locations ($N_T^{\text{loc.}}$) of our single tracer falling in a given slice, i, can be expected to be equivalent to the average fraction of all the system's particles one may expect to find in said slice at any given time. Given also the known volume, V_s, of our slice, the mean particle number density, n_i, in our slice i may thus be given as:

$$n_i = \frac{NF_i}{V_c} \tag{7.1}$$

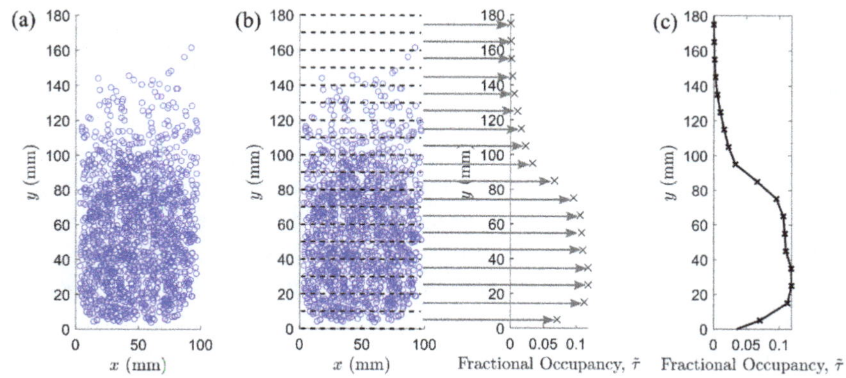

Figure 7.6. (a) 2D projection of a subset of PEPT data points acquired from a monodisperse fluidised-bed system comprising exclusively sand particles. (b) Illustration showing how the experimental volume may be subdivided into a series of 10 mm 'slices' in the vertical direction, and the number fraction of particles falling in each slice used to create a fractional occupancy distribution. (c) The final occupancy distribution produced, whose form broadly matches our expectations for a fluidised-bed system [14, 15].

where N is the total number of particles in the system. If we know also the volume, V_p, of our particles (assuming here, for simplicity, a monodisperse granulate), this equation can then be used to determine the (local) packing density, η_i, as:

$$\eta_i = n_i V_p. \tag{7.2}$$

By performing this calculation for each cell within the system, we may then build up a full one-dimensional density profile for the system.

While for many systems—including the fluidised-bed example used here—the assumption of ergodicity is reasonable (see, for example, [16]), the second assumption is often not valid, for instance in cases where the length of the experiment performed is comparable to the half life of the isotope used to create the tracer, where detector sensitivity is not uniform, and/or where the materials of the system being imaged give rise to significant spatial variations in gamma-ray attenuation. As such, it is more common when calculating occupancy and/or density profiles to consider not the *number* of location events falling within each cell, but rather the total *time spent* by the tracer within each cell. If we take τ_i to be the amount of time spent by the tracer in a given cell, i, during an experiment of total duration T, then we can define our time-based occupancy as:

$$\tilde{\tau}_i = \frac{\tau_i}{T}. \tag{7.3}$$

This occupancy can be used in exactly the same manner as F to produce occupancy, number density and packing density plots.

The principle outlined above can be trivially extended to the case of two- and three-dimensional occupancy and density fields. For a two-dimensional field, instead of subdividing the experimental volume into a series of slices, one would instead divide the 2D projection shown in figure 7.6 into a series of two-dimensional 'pixels' in a given plane of interest, and calculate the occupancy for each of these pixels, in the manner described above. For a three-dimensional distribution, one need simply subdivide the entire three-dimensional experimental volume into a series of 3D voxels and repeat the same process. The choice of the size and geometry of the voxels used is up to the user's discretion, but to achieve optimal results, these parameters should be chosen based on the quality and amount of data, and the geometry of the system. In particular, in order to produce reliable data, each slice, pixel or voxel (which may collectively be termed 'cells') must possess a suitably large number of data points to be able to produce good statistics. Therefore, the lower boundary of the usable cell size is dependent on the amount of data acquired.

It is additionally important to remember that the construction of occupancy/density plots in higher dimensions will, naturally, require a larger amount of data, i.e. a longer experimental run, in order to achieve adequate statistics and thus a suitable degree of accuracy [17]. Indeed, this statement holds true for any and all of the fields discussed in this chapter.

7.2.2 Extracting additional information from density distributions

Just as simple particle trajectories can tell us much more about a system than simply the path followed by a particle (see section 7.1.1), so too can careful analysis of particle density distributions yield significantly more information than simply telling us where particles like to spend most of their time. In the sections below, we detail manners in which these distributions may be more closely analysed and further processed in order to gain valuable additional insight into the dynamics of a system of interest.

Analysing the form of profiles

Even with no further data manipulation, a close inspection of specific features belonging to a given density profile or distribution can provide important new information. For example, the density profiles of dilute or 'gaseous', vertically-vibrated granular systems are known to decay exponentially with height at large y [18, 19]. The decay rate of this exponential decrease can be used to estimate the granular temperature of such a system [19]. It should be noted here, however, that there exist more accurate and reliable methods through which to extract granular temperature values from PEPT data, which we discuss in detail in section 7.6.

Another common feature in a variety of systems is the presence of prominent 'spikes' in a one-dimensional profile (see, as an example, figure 7.7), or small, localised 'hot spots' in two- or three-dimensional, corresponding to a single, comparatively heavily-populated cell or cluster of cells. The presence of such spikes implies that our tracer is liable to spend relatively long periods of time within a given region of the system suggesting, for example, the presence of caging effects or solid-like regions in a dry granular bed [12, 20], or stagnant, defluidised, or non-suspended regions in a fluid or multiphase flow [21–23]. Further distinction between these possibilities can be drawn by analysing other aspects of the PEPT data. For example, one may distinguish between a caged particle and a defluidised particle by studying the

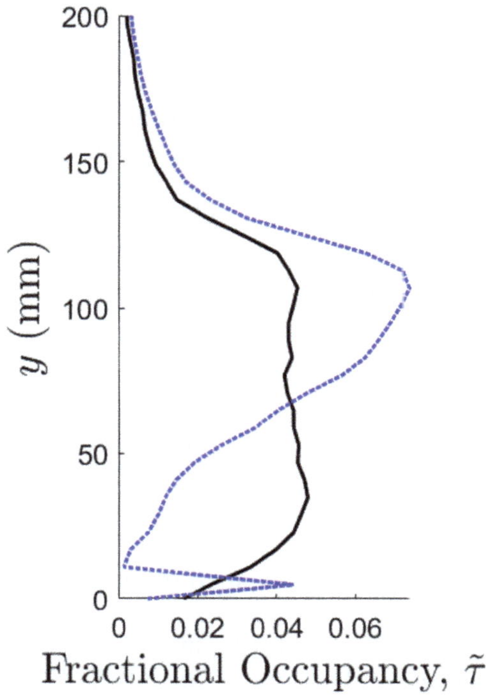

Figure 7.7. Vertical occupancy profiles for both components of a binary gas-fluidised bed comprising small sand particles (solid black line) and large plastic particles (dotted blue line). The spike near the base of the system exhibited by the line representing the plastic component of the bed corresponds to a zone in which a fraction of the plastic particles are partially defluidised. The smaller sand particles represented by the solid black line are more easily fluidised, and therefore show no such spike.

tracer's mean squared displacement behaviour (see section 7.5). One may also, for example, distinguish between a stagnant region and a non-suspended particle by observing the particle's trajectory or velocity—a tracer in the former case will stay exactly still with zero velocity, whereas in the latter it may still maintain a horizontal component of velocity. The appearance of multiple spikes with a characteristic separation comparable to the particle diameter is likely to be evidence of crystallisation [24] within the system.

Quantifying homogeneity

In many industrial processes involving solid–liquid mixing, it is important to be able to accurately quantify the spatial homogeneity of the multiphase system in question [25–27]. One manner in which this may be achieved using PEPT is through the so-called *homogeneity index*, I_H [23]. For the case of a solid-fluid system containing only a single species of solid particle, I_H can be extracted from a known (one-, two- or three-dimensional) occupancy distribution. For clarity, let us once again consider here the simple one-dimensional case. For a truly homogeneous system whose volume can be subdivided into N_c equally-sized cells, the occupancy $\tilde{\tau}_i$ must, by definition, be the same in any given cell, i. In the one-dimensional case, the occupancy profile of a perfectly homogeneous system would therefore correspond to a straight, vertical line with a value equal to the mean occupancy of the system, $\langle\tilde{\tau}\rangle$, as illustrated in figure 7.8. Clearly, the further a given system's occupancy profile deviates from this 'ideal case', the less homogeneous said system must be. As such, we can quantify the homogeneity of a system by defining our homogeneity index as:

$$I_H = \left[\frac{\sum_i^{N_c}(\tilde{\tau}_i - \langle\tilde{\tau}\rangle)^2}{N_c}\right]^{\frac{1}{2}}. \tag{7.4}$$

Using the above-defined metric, a value $I_H = 0$ corresponds to a perfectly homogeneous system, with increasing I_H values corresponding to increasingly inhomogeneous systems. The maximum value of I_H achievable (i.e. that corresponding to a perfectly inhomogeneous system) varies dependent on both the loading of the specific system in question and the cell size used for calculations. As such, one must be careful when comparing I_H values acquired from different systems, or processed with differing mesh sizes.

The homogeneity index, I_H, is very similar in principle to the more widely-known *segregation intensity*, which we shall meet in section 7.3.5.

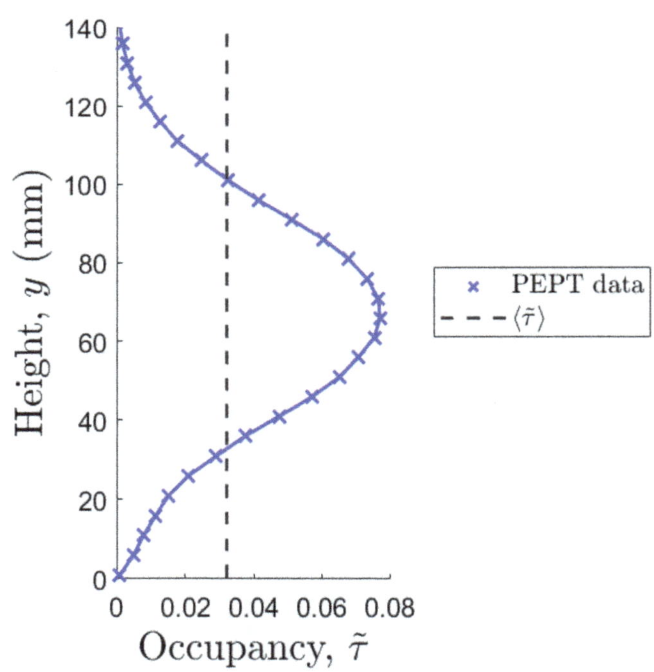

Figure 7.8. Vertical occupancy distribution showing the double-depth-averaged distribution of solid particles along the vertical axis of a stirred tank (symbols and solid blue line). The black dashed line represents the occupancy profile expected for a perfectly-homogeneous system.

Determining the transport disengagement height (TDH)

Though specific to the case of fluidised-bed systems, the transport disengagement height (TDH) is nonetheless an important quantity in a variety of engineering applications, and thus important to characterise correctly.

There exist two main definitions of the TDH, known as the TDH_F (defined as the height at which the change with height of the concentration of [fine] particles being entrained out of the bed becomes negligible [28, 29]) and the TDH_C (the maximum height attained by coarse particles within the system [30, 31]). Clearly, when using a radiation-based technique such as PEPT, it is not desirable for the tracer to be entrained out of a system and into the air of the laboratory! As such, PEPT is typically only used to measure the TDH_C.

The TDH for a fluidised bed can in fact be trivially extracted from a one-dimensional PEPT density profile: for error-free data with a suitably high data rate (see section 7.1.1 and figure 7.5) it corresponds simply to the height above which all cells are equal to 0 (i.e. particles no longer reach)[1].

Measuring the centre of mass and additional inferred parameters

From a one dimensional profile such as those discussed in the main body of section 7.2 above, an estimate of the system's centre of mass may be easily obtained from the known occupancy, $\tilde{\tau}(y)$, at each height, y, within the system as:

$$h_{\mathrm{CM}} = \frac{\sum_i^{N_c} \tilde{\tau}_i(y) \cdot y}{\sum_i^{N_c} \tilde{\tau}_i(y)} \tag{7.5}$$

where N_c represents the total number of cells into which the system is divided. The above represents a general form which may be used for occupancy, number density and/or packing density. In the special case of occupancy, however, we know that by definition $\sum_i^{N_c} \tilde{\tau}_i = 1$, meaning that the above can be reduced to:

$$h_{\mathrm{CM}} = \sum_i^{N_c} \tilde{\tau}_i(y) \cdot y. \tag{7.6}$$

The system's mass centre in the horizontal (x, y) directions may be obtained in an identical manner from the relevant horizontal occupancy distributions.

The knowledge of the position of a system's centre of mass may, in turn, be used to infer other information regarding the system's state and dynamics. For example, work by Luding *et al* on monodisperse one- and two-dimensional vibrated granular beds [32, 33] (later extended to three-dimensional [34] and binary [35] systems) used the vertical mass-centres of the studied beds to infer the gravitational potential energy of the system ($E_g = Mgh_{\mathrm{CM}}$, where $M = \sum m$ is the total mass of all particles in the system), and from this estimate trends relating to the kinetic energy and granular temperature of said system. With the capabilities of PEPT, however, there exist more direct and precise manners in which to measure a system's granular temperature; these are discussed in detail in section 7.6.

[1] It should be noted that in some cases—especially those with high scatter or background radiation—it may be preferable to set a non-zero threshold value to account for the possibility that random events may lead to false locations above the true TDH.

For binary, ternary or polydisperse granular beds, measurements of the relative positions of each individual particle species' center of mass can also be used as a (relatively crude) measure of the degree of segregation exhibited by a given system. Details of how this may be achieved can be found in section 7.3.5. Once again, there exist somewhat more nuanced and sensitive manners in which to quantify segregation strength in granular media from PEPT data, which we also discuss in detail in section 7.3.5.

7.2.3 Interactive example: 1D occupancy profile

The notebook files accompanying this section are available in the supplementary data online at https://iopscience.iop.org/book/978-0-7503-3071-8.

Tutorial: Using PEPT data to produce a one-dimensional vertical occupancy profile

In this tutorial, we will show you how to take a pre-processed PEPT data file and use the data contained within to create and plot a one-dimensional vertical occupancy profile such as those described in the preceding section.

Setting up and importing data

The process for setting up and importing data remains the same as in the preceding examples. This time, however, we are only interested in the time (t) and the vertical (y) position of our tracer, so we don't need to extract the other (x, y) variables as before.

[1]:
```
# Importing necessary libraries for numerical array manipulation and plotting
import numpy as np
import matplotlib.pyplot as plt

# Set Matplotlib theme (try "darkgrid" too)
import seaborn
seaborn.set_style("whitegrid")
```

[2]:
```
# Path to the "pre-processed PEPT data
filepath = "Data/FluidisedBedExample.csv"

# Loading the file into a matrix named "data", skipping the first 16 rows which
# form the header
data = np.loadtxt(filepath, skiprows=16)

# Extracting the data we need - specifically the timestamps...
t = data[:, 0]

# ... and the vertical (y) position data (third column)
y = data[:, 2]
```

Assigning relevant durations to data points

There are many ways to extract occupancy profiles from PEPT data. We present here a method which is comparatively simple to understand yet still relatively accurate.

We begin by calculating the 'midpoints' of each consecutive pair of measured y positions. For each of these midpoints, we can then assign a relevant 'duration',

$dt = t_2 - t_1$, associated with this position, which we assume to be representative of the amount of time spent by the tracer in this approximate position.

```
[3]: # Pre-allocate arrays to store all relevant midpoints of the recorded
     # y-positions and their corresponding durations
     y_mid = np.zeros((len(y) - 1, 1))
     dt = np.zeros((len(y) - 1, 1))

     # Looping over all midpoints, assigning their value based on known particle
     # positions, and calculating the relevant durations for each in turn
     for i in range(0, len(y_mid)):
         # Each value of y_mid is the average of the two y-values on either side
         y_mid[i] = (y[i] + y[i + 1]) / 2

         # The time spent in the vicinity of y_mid can be assumed equal to the
         # difference between the corresponding t values on either side
         dt[i] = t[i + 1] - t[i]
```

Choosing 'slices' and calculating occupancies

Once we know the position and duration associated with each individual data point, we can now—as described in the main text—subdivide our experimental volume into a series of slices in the vertical y-direction. For each slice, we will then determine which data points fall within its vertical bounds, and sum the durations, dt, associated with each of these points, giving us the occupancy for said cell. Plotting the occupancy values acquired against the centre positions of their correspondent cells allows us to visualise the resultant occupancy profile.

We begin by determining the size of the slices (the 'cell size', `dc`) into which our system is divided. Larger values of `dc` provide better statistics—i.e. more accurate data—but coarsen the resultant profile. Smaller `dc` values will hypothetically allow us to see finer details in the data, but only if the statistics remain adequate as the cell size is reduced.

Try playing with the value of `dc` in the code below to see what effect it has on the profile produced!

```
[4]: # Defining our chosen cell size, dc, in mm (the total height of our system is
     # ~200mm)
     dc = 10

     # Finding the necessary number of cells of size dc to span 0 to the maximum
     # y-value detected
     n = int(y_mid.max() / dc)

     # Finding the minimum vertical position detected to know where to start scanning
     y_min = y.min()

     # Pre-allocating a 1D array to store the occupancy profile
     occupancy = np.zeros(n)

     # For each vertical 'slice' within the system, finding all particle locations
     # therein, and summing the durations corresponding to these locations
     for i in range(1, n):
         dt_slice = dt[(y_min + i * dc <= y_mid) & (y_mid < y_min + i * dc + dc)]

         # Adding the sum of the remaining values to occupancy array
         occupancy[i] = np.sum(dt_slice)
```

```
# Normalising
occupancy = occupancy / t.max()

# Defining a suitable axis scale for plotting
scale = np.arange(n) * dc

# Plotting
fig, ax = plt.subplots()
ax.scatter(occupancy, scale)

ax.set(ylabel='$y$ (mm)', xlabel='Occupancy', ylim=[0, 200])
ax.set_aspect(1.0 / ax.get_data_ratio() * 5);
```

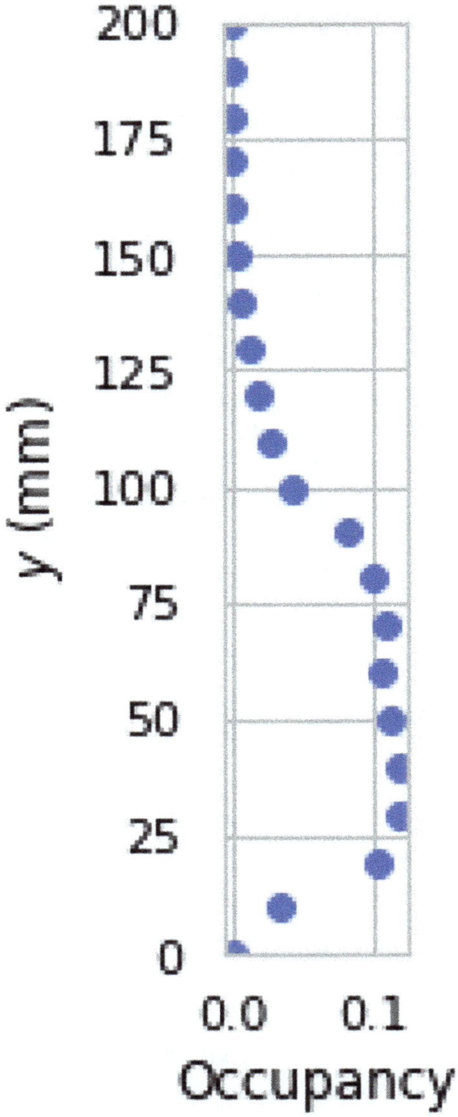

7.3 Binary occupancy distributions and segregation

In the preceding section, our discussion focused predominantly on the case of occupancy and density distributions in *monodisperse* systems, though in the later sections we began to introduce a few examples relating to binary systems. In this section, we explore in greater detail how PEPT may also be used to explore particle distributions within binary, ternary and polydisperse systems, and in doing so gain insight into the important and much-studied phenomenon of *granular segregation* [49]. Throughout this section, for simplicity and clarity, we focus primarily on the case of binary systems. Nonetheless, the methods described can in all cases be trivially extended to systems possessing higher degrees of polydispersity.

7.3.1 One tracer or two?

Clearly, if we wish to perform PEPT in a binary system, we require information from both species present—i.e. we need two tracers (similarly, for ternary systems we would require three, for quaternary four, quinary five, etc). There are two distinct manners in which this may be achieved:

i. On the one hand, we may simply introduce into our system one tracer of each of the N_s species present, and use the trajectory separation techniques described in section 7.1.3 to extract the data pertaining to each particle.
ii. Alternatively, we may repeat the experiment a total of N_s times, for each repeat using a tracer of a different species. Under the assumption of ergodicity, the time-averaged data from all N_s experiments may be combined into a single, complete data set.

Each of the methods given above possesses certain advantages and disadvantages. For example, method (i) requires the least amount of initial experimental time, but involves significantly more effort (both computational and human) in post-processing, due to the complexity of separating particle trajectories (see section 6.3.2). It also risks introducing additional sources of error, as even the most finely-tuned trajectory separation algorithm may under certain circumstances mis-label particles (i.e. mistake one species of tracer for another). The use of multiple particles also limits the maximum data rate, and thus the maximal temporal resolution of the data acquired, as the camera's peak coincidence rate is in essence split between the tracers. For example, if a given positron camera can handle a maximum coincidence rate of 1 MHz, and we require 100 coincidence events to detect a particle, a single particle of suitably high activity may be detected approximately 10,000 times per second. If a second, equally active particle is introduced, then the camera can still only record a maximum of 1,000,000 coincidence events per second, but these will now be shared between coincidences arising from the two detectors. As such, we may expect that each particle will now be detected on average 5,000 times per second—i.e. the overall temporal resolution of the system is halved. Clearly, this problem will be further exacerbated as more tracers are introduced. Further, following from the above, if our particles do not possess equal activities, then their trajectories will be recorded with different effective data rates. For example, if particle A is three times as active as particle B, both particles will still require the same number of LoRs to produce PEPT data points with equal accuracy, but of the LoRs detected by the scanner only $\frac{1}{4}$ will, on average, correspond to particle B, the remaining $\frac{3}{4}$ to particle A. As such, the position of particle A will typically be reconstructed three times as frequently as particle B, leading to a significant disparity in temporal resolution for the two particles. Introducing additional tracers may also begin to 'blind' the camera, as discussed in section 5.2, and will almost certainly increase the probability of false coincidences, even if both particles combined are not active enough to saturate the detectors, due simply to the larger number of radiation sources introduced. On the plus side, this method, unlike (ii), allows greater insight into the interactions between particle species, especially in the case of transient dynamics.

The main disadvantage of method (ii) is simply time: if a measurement of a monodisperse system requires an imaging time T to produce suitable statistics, an equivalent system containing N_S species of particle will require a time period $N_S T$. As a typical PEPT experiment requires an acquisition time $T \sim 1$ h (longer for larger and/or slower-moving systems), it may therefore—for a highly-polydisperse system—take an entire working day to simply acquire one full data set using method (ii). This method also requires that the N_s experimental runs performed are near identical, i.e. that the experiment itself is highly reproducible—though this of course should already be a prerequisite for any good experimentalist!

To date, the majority of PEPT studies of polydisperse systems have utilised method (ii) and, due to the inherent time limitations, have focussed on binary beds, with a small amount of work performed in the ternary case [36]. However, recent advances in the tracking of multiple particles [37–40] mean that in the future, method (i) is likely to become more widely adopted, and thus allow more highly-polydisperse systems to be investigated.

7.3.2 Extracting density from occupancy in polydisperse beds

In section 7.2, we discussed the manner in which one may extract density information from PEPT-measured occupancy values for a monodisperse bed—i.e. one containing only particles of a fixed size (and density, geometry, etc). When the condition of monodispersity is relaxed, though the basic principle for measuring occupancy and converting to density remains the same, we must be slightly more careful in how we calculate both the number density, n_i, and packing density, η_i, in a given cell, i.

For a given tracer species, j, one can calculate both the number-based occupancy, $F_{j,i}$, and the time-based occupancy, $\tilde{\tau}_{j,i}$, of this specific species in a manner identical to the monodisperse case, i.e.:

$$F_{j,i} = \frac{N_{j,\,i}^{\mathrm{loc.}}}{N_{j,\,T}^{\mathrm{loc.}}} \tag{7.7}$$

(where $N^{\text{loc.}}_{j,\,T}$ is the total number of locations recorded for the tracer of species j, and $N^{\text{loc.}}_{j,\,i}$ the number of these locations that fall within the ith cell) and:

$$\tilde{\tau}_{j,i} = \frac{\tau_{j,i}}{T_j} \tag{7.8}$$

where $\tau_{j,i}$ is the amount of time spent by the tracer of species j within cell i, and T_j the total amount of time for which the motion of the j-species tracer was recorded. Note that for the case in which all tracers are used simultaneously, the total time T is by definition equal for all tracer species (i.e. $T_1 = T_2 \cdots = T_{N_S} = T$). For the case of repeated experiments, different T_j values may be used for different tracers. So long as T_j is in all cases suitably large that good statistics may be achieved, the use of differing T is in no way detrimental; indeed, if the tracers used vary significantly in activity, it can be highly beneficial to vary T for differing species.

In equation (7.1), we see that the number density of a tracer may—for a monodisperse system –be determined simply by multiplying the occupancy (F or $\tilde{\tau}$) by a factor $\frac{N}{V_C}$, where N is the total number of particles within the system. To find the local number density, we simply need to replace the total number N with the number, N_j, of particles of our given species within the system, i.e.

$$n_{j,i} = \frac{N_j F_{j,i}}{V_c}\left(= \frac{N_j \tilde{\tau}_{j,i}}{V_c}\right). \tag{7.9}$$

Similarly, the method (equation (7.2)) for calculating the packing density, $\eta_{j,i}$, of a specific species j in cell i can be given as:

$$\eta_{j,i} = n_{j,i} V_p^j, \tag{7.10}$$

where V_p^j is the volume of a single particle of species j.[2]

The *total* number density or solids fraction within a given cell can be determined simply by summing the relevant values for each individual species, i.e.:

$$n_i = \sum\nolimits_j^{N_S} n_{j,i}, \quad \eta_i = \sum\nolimits_j^{N_S} \eta_{j,i}. \tag{7.11}$$

7.3.3 Single-species concentration distributions (local number/volume/occupancy fraction distributions)

In the previous sections, we have seen how PEPT data acquired with multiple tracers may be used to build up a series of individual occupancy and/or density profiles for the different components of a binary or polydisperse particulate system. From these individual profiles, it is possible to build up a picture of the spatial variation in the concentration of one or more given species throughout a system. As the acquisition of homogeneous mixtures is a matter of paramount importance in a variety of industrial processes [41–44], the acquisition of such spatial 'concentration distributions' can be highly valuable.

[2] Similarly to the single-tracer case, equation (7.10) assumes that each species is individually monodisperse.

In order to determine the concentration distribution from PEPT data, one must first acquire an individual occupancy profile for each species present in the system. Importantly, when creating these profiles, the experimental volume must be subdivided into cells of the same size and positioning in all cases, such that the occupancies, O_i (and hence densities n_i, η_i) in all of these cells may be *directly compared.* In this section, we will primarily consider the calculation of the relative volume fraction, ϕ^V, as this is often the most useful, and certainly the most widely-used, parameter through which to give insight into the homogeneity of a system. The calculation of the relative number fraction and the occupancy fraction, however, can be calculated in a manner identical to that demonstrated below, as illustrated in the interactive example following this chapter.

For a given cell, i, within the system, the local volume fraction of a species j is simply the fraction of the total solids volume taken up by species j particles, and can be defined as:

$$\phi_{j,\,i}^{V} = \frac{\eta_{j,i}}{\eta_{\mathrm{Tot.},i}} = \frac{\eta_{j,i}}{\sum_{k=1}^{N_S}\eta_{k,i}} \tag{7.12}$$

where $\sum_{k=1}^{N_S}\eta_{k,i}$ is the sum over all N_S species of their respective individual solids fractions. By determining values of $\phi_{j,\,i}^{V}$ for all cells in the system, one may build up a complete volume fraction profile, such as those illustrated in the interactive example below.

7.3.4 Interactive example: concentration distribution

The notebook files accompanying this section are available in the supplementary data online at https://iopscience.iop.org/book/978-0-7503-3071-8.

Tutorial: Using PEPT data to produce one-dimensional concentration profiles

In this tutorial, we will show you how to use PEPT to determine the spatial variation of the local concentration (mass, volume or number fraction) of a given particle species in a binary bed. The process described here can, however, be trivially extended to systems with higher degrees of polydispersity.

Acquiring individual occupancy distributions

In order to produce a concentration profile, we must first calculate a pair of occupancy profiles—one for each species present in the system. This can be done by following the exact same procedure shown in the preceding tutorial, *Using PEPT data to produce a one-dimensional vertical occupancy profile*, for each of the relevant pre-processed PEPT data files.

The data used in this example correspond to a bidisperse gas-fluidised bed comprising 95 wt% 300 micron sand particles ($\rho_{\mathrm{sand}} = 2800$ kg m^{-3}) and 5 wt% 1 mm plastic particles ($\rho_{\mathrm{plastic}} = 1100$ kg m^{-3}).

[1]:
```
# Importing necessary libraries for numerical array manipulation and plotting
import numpy as np
import matplotlib.pyplot as plt

# Set Matplotlib theme (try "darkgrid" too)
import seaborn
seaborn.set_style("whitegrid")

# Set Plotly figure output (try "browser" too)
import plotly
plotly.io.renderers.default = "png"
```

[2]:
```
# Importing data for the sand component
filepath1 = "Data/ConcentrationSandExample.csv"
data_sand = np.loadtxt(filepath1, skiprows=16)

# Importing data for the plastic component
filepath2 = "Data/ConcentrationPlasticExample.csv"
data_plastic = np.loadtxt(filepath2, skiprows=16)

# Extracting the data we need for each component of the system
t_sand = data_sand[:, 0]
y_sand = data_sand[:, 2]

t_plastic = data_plastic[:, 0]
y_plastic = data_plastic[:, 2]

# Pre-allocate arrays to store all relevant midpoints of the recorded
# y-positions, and their corresponding durations
y_mid_sand = np.zeros(len(y_sand) - 1)
y_mid_plastic = np.zeros(len(y_plastic) - 1)

dt_sand = np.zeros(len(y_sand) - 1)
dt_plastic = np.zeros(len(y_plastic) - 1)

# Calculating midpoints and durations for each species in turn
for i in range(len(y_mid_sand)):
    # Each value of y_mid is the average of the two y-values on either side
    y_mid_sand[i] = (y_sand[i] + y_sand[i + 1]) / 2

    # The time spent in the vicinity of y_mid can be assumed equal to the
    # difference between the correspondent t values on either side
    dt_sand[i] = t_sand[i + 1] - t_sand[i]

for i in range(len(y_mid_plastic)):
    y_mid_plastic[i] = (y_plastic[i] + y_plastic[i + 1]) / 2
    dt_plastic[i] = t_plastic[i + 1] - t_plastic[i]

# Defining our chosen cell size, dc, in mm.
# Note that - if we want to use these data for direct comparison - a single cell
# size must be used for both data sets
dc = 10
```

```
# Finding, for each species, the necessary number of cells of size dc to span 0
# to the maximum y-value detected
n_sand = int(y_mid_sand.max() / dc)
n_plastic = int(y_mid_plastic.max() / dc)

# Setting the final 'N' value to the smaller of the 2 Ns acquired such that
# comparison can be performed
# Note that for the current data sets this works well, but for more strongly-
# segregated systems, where 1 component occupies considerably higher regions
# than another, the shorter array would need to be padded with zeros instead
n = np.min([n_sand, n_plastic])

# Finding the minimum vertical position detected to know where to start scanning
y_min = np.min([y_sand.min(), y_plastic.min()])

# Pre-allocate 1D arrays to store the occupancy profiles corresponding to each
# species
occupancy_sand = np.zeros(n)
occupancy_plastic = np.zeros(n)

# For each vertical 'slice' within each system, finding all particle locations
# therein, and summing the durations corresponding to these locations
for i in range(1, n):
    # Copying all duration values to a new array
    dt_temp_sand = dt_sand.copy()
    dt_temp_plastic = dt_plastic.copy()

    # Removing all values with y-values (i.e. heights) outside the current slice
    dt_temp_sand[y_mid_sand < y_min + i * dc] = 0
    dt_temp_sand[y_mid_sand > y_min + (i + 1) * dc] = 0

    dt_temp_plastic[y_mid_plastic < y_min + i * dc] = 0
    dt_temp_plastic[y_mid_plastic > y_min + (i + 1) * dc] = 0

    # Adding the sum of the remaining values to occupancy array
    occupancy_sand[i] = np.sum(dt_temp_sand)
    occupancy_plastic[i] = np.sum(dt_temp_plastic)

# Normalising
occupancy_sand = occupancy_sand / t_sand.max()
occupancy_plastic = occupancy_plastic / t_plastic.max()

# Defining a suitable axis scale for plotting
scale = np.arange(n) * dc

# Plotting
fig, ax = plt.subplots()
ax.scatter(occupancy_sand, scale, label='Sand')
ax.scatter(occupancy_plastic, scale, label='Plastic')

ax.set(ylabel='$y$ (mm)', xlabel='Occupancy')
ax.set_aspect(1.0 / ax.get_data_ratio() * 5)
ax.set_ylim(0,300)
ax.legend()
```

```
[2]: <matplotlib.legend.Legend at 0x7fab57ea8ed0>
```

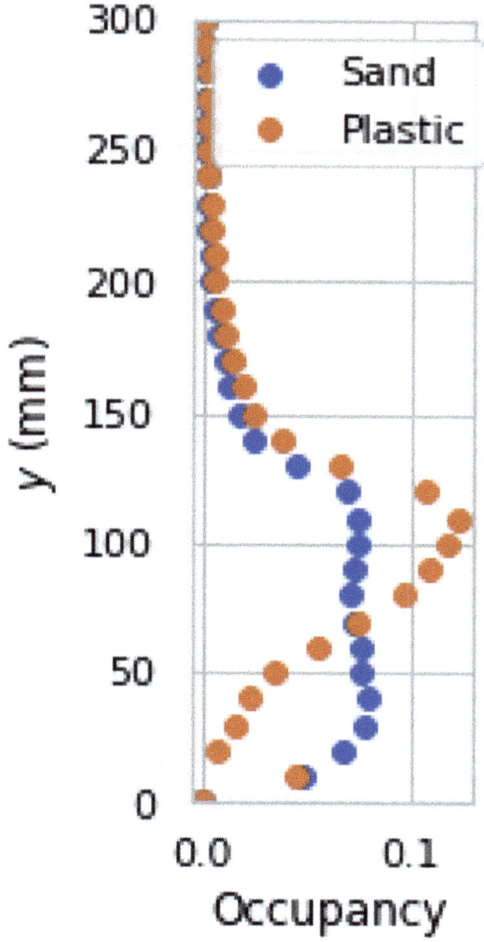

Calculating concentrations

Now that we have occupancy distributions for both of the species present in the system, we can use the data acquired to determine the spatial variation of the *relative concentration* of the two components at each position within the bed.

Various types of relative concentration may be determined from PEPT data, of which we will demonstrate here the three most commonly used. The simplest to produce, yet perhaps most conceptually difficult of the three is the relative occupancy:

$$\phi^{occ} = \frac{\tilde{\tau}_1}{\tilde{\tau}_1 + \tilde{\tau}_2}.$$

The relative occupancy distribution can be extracted from a pair of occupancy profiles such as those calculated above very simply:

```
[3]: # Some values in the division below are zero, let them silently yield 0/0 = NaN
     np.seterr(divide='ignore', invalid='ignore')

     # Determining the relative occupancy of the plastic component of the bed from
     # the 2 occupancies
     # Note that the sand component can be recovered by simply swapping the relevant
     # sand / plastic terms
     phi_occ_plastic = occupancy_plastic / (occupancy_plastic + occupancy_sand)

     # Plotting the resultant profile
     fig, ax = plt.subplots()
     ax.scatter(phi_occ_plastic, scale)
     ax.set(ylabel='$y$ (mm)', xlabel='$\phi^{occ}$')
     ax.set_aspect(1.0 / ax.get_data_ratio() * 5)
     ax.set_ylim(0, 300)
```

```
[3]: (0.0, 300.0)
```

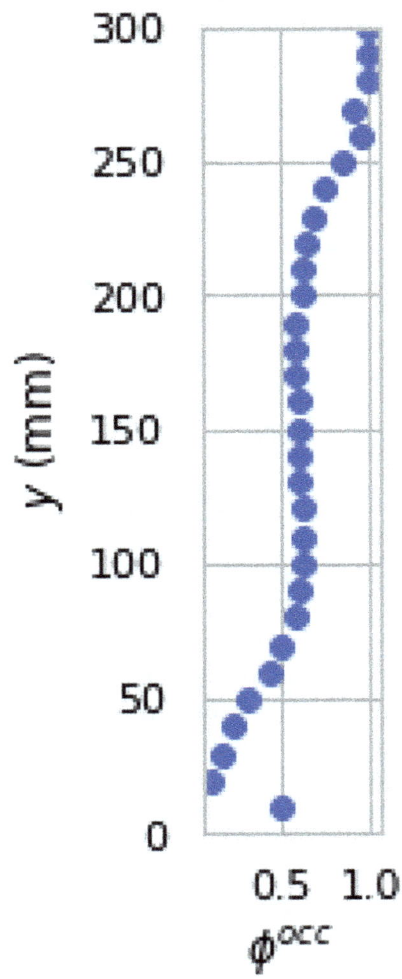

Relative number density

Let us next calculate the distribution of the relative number fraction, ϕ^n, of a our plastic particles throughout the system, which can be defined (for the binary case) as:

$$\phi^n = \frac{n_1}{n_1 + n_2}.$$

Recall from the main text that the number density of a given species, j, can be determined as:

$$n_j = \frac{N_j \tilde{\tau}_j}{V_c},$$

meaning that the equation for the relative number fraction can be re-written as:

$$\phi^n = \frac{\tilde{\tau}_1 N_1}{\tilde{\tau}_1 N_1 + \tilde{\tau}_2 N_2} = \frac{\tilde{\tau}_1}{\tilde{\tau}_1 + \tilde{\tau}_2 \frac{N_2}{N_1}}.$$

In the code below, we produce a relative number fraction plot for the same system as shown above.

```
[4]: # In order to determine the relative number density of our different species,
     # the only additional information we require is the ratio N2 / N1, i.e. the
     # ratio of the total numbers of particles of different species within the system
     # as a whole
     #
     # As we already know the mass fractions, particle sizes and densities of our two
     # different species, we can easily estimate the ratio of particle numbers

     # Calculating the volume of an individual sand particle in mm^3...
     vp_sand = 4 / 3 * np.pi * 0.15 ** 3
     print(f"Single sand particle volume: {vp_sand} mm^3")

     # ...and that of an individual plastic particle (assuming in both cases a simple
     # spherical geometry)
     vp_plastic = 4 / 3 * np.pi * 0.5 ** 3
     print(f"Single plastic particle volume: {vp_plastic} mm^3\n")

     # From their known densities, we can now calculate the masses of a single
     # particle of each species in g
     mp_sand = vp_sand * 2800 / 1e6
     mp_plastic = vp_plastic * 1100 / 1e6

     print(f"Single sand particle mass: {mp_sand} g")
     print(f"Single plastic particle mass: {mp_plastic} g\n")

     # Using the above to determine the number ration n2 / n1 for a known mass ratio
     # 5 wt% : 95 wt%
     number_ratio = (95 / mp_sand) / (5 / mp_plastic)
     print(f"N2 / N1 = {number_ratio}")

     # Determining the relative number density of plastic
     phi_n_plastic = (
         occupancy_plastic /
         (occupancy_plastic + occupancy_sand * number_ratio)
     )
```

```
# Plotting the resultant profile
fig, ax = plt.subplots()
ax.scatter(phi_n_plastic, scale)

ax.set(ylabel='$y$ (mm)', xlabel='$\phi^{n}$')
ax.set_aspect(1.0 / ax.get_data_ratio() * 5)
ax.set_ylim(0, 300)
```

```
Single sand particle volume: 0.014137166941154066 mm^3
Single plastic particle volume: 0.5235987755982988 mm^3

Single sand particle mass: 3.958406743523138e-05 g
Single plastic particle mass: 0.0005759586531581286 g

N2 / N1 = 276.4550264550265
```

```
[4]: (0.0, 300.0)
```

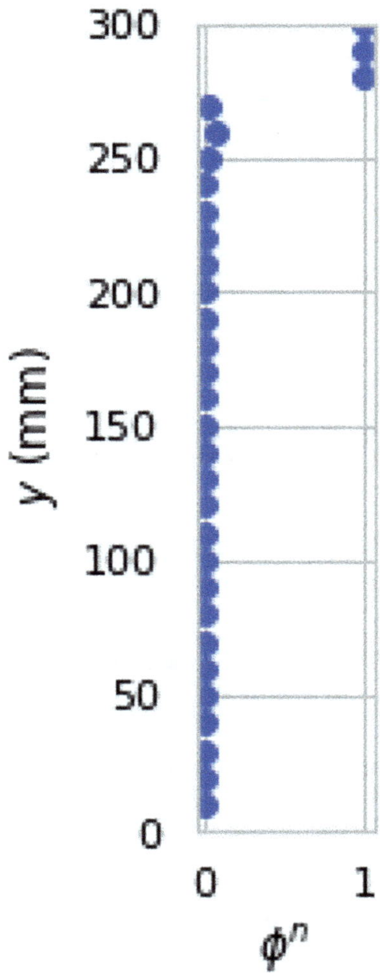

As can be immediately seen above, the plot of the relative number fraction shown above is **strikingly different** from the relative occupancy plot shown earlier. The reason for this is simply that (due both to their larger wt% within the system *and* significantly smaller idividual size, meaning more particles per unit weight) sand particles **significantly outnumber** plastic particles—so much so, in fact, that the number ratio throughout the system is practically zero for all cells except those which contain almost exclusively plastic particles! Clearly, for a system such as that explored here, the relative number density is not a suitable parameter.

Relative solids fraction

Finally, let us calculate the relative volume fraction, i.e. the ratio of individual packing densities or solids fractions, $\eta_{j,i}$ within the various regions of the system. This can be defined, for our simple binary system, as:

$$\phi^{vol} = \frac{\eta_1}{\eta_1 + \eta_2}.$$

From the main text, we know that the solids fraction of a given species, j, can be determined as:

$$\eta_j = n_{j,i} V_p^j = \frac{N_j V_p^j \tilde{\tau}_j}{V_c},$$

meaning that the equation for the relative number fraction can be re-written as:

$$\phi^n = \frac{\tilde{\tau}_1 N_1 V_p^1}{\tilde{\tau}_1 N_1 V_p^1 + \tilde{\tau}_2 N_2 V_p^2} = \frac{\tilde{\tau}_1}{\tilde{\tau}_1 + \tilde{\tau}_2 \dfrac{N_2 V_p^2}{N_1 V_p^1}} = \frac{\tilde{\tau}_1}{\tilde{\tau}_1 + \tilde{\tau}_2 \dfrac{V_T^2}{V_T^1}},$$

where V_T^j is the total volume of particles of a given species j.

In the code below, we calculate and plot the relative volume fraction for the plastic component of our system.

```
[5]: # In order to determine the relative volume fraction of our different species,
     # we start by calculating the relative total volume of the two species present,
     # VT2 / VT1
     #
     # With a known particle (intrinsic) density and a known wt% for each species,
     # this can be easily determined. For our plastic species, it is simply equal to:
     volume_ratio = (95 / 2800) / (5 / 1100)
     print(f"VT2 / VT1 = {volume_ratio}")

     # Determining the relative volume fraction of plastic
     phi_vol_plastic = (
         occupancy_plastic /
         (occupancy_plastic + occupancy_sand * volume_ratio)
     )

     # Plotting the resultant profile
     fig, ax = plt.subplots()
     ax.scatter(phi_vol_plastic, scale)
```

```
ax.set(ylabel='$y$ (mm)', xlabel='$\phi^{vol}$')
ax.set_aspect(1.0 / ax.get_data_ratio() * 5)
ax.set_ylim(0,300)
```

```
VT2 / VT1 = 7.464285714285714
```

```
[5]: (0.0, 300.0)
```

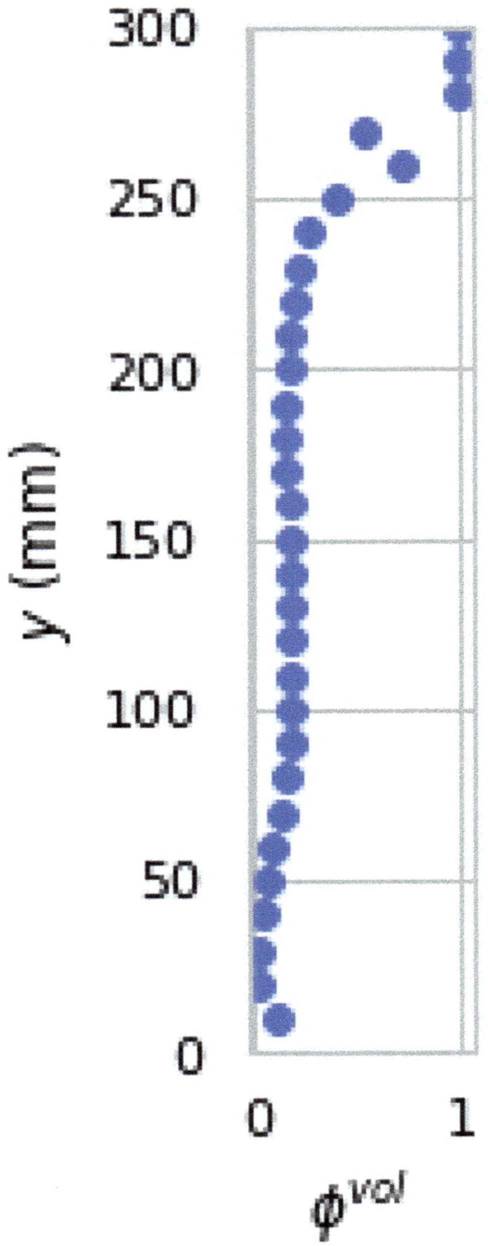

As is clear from the various figures above, both the values and form of the concentration distributions produced **vary significantly** dependent on whether we are dealing with occupancies, number densities or solids fractions. As such, we must be very careful to ensure consistency when working with these data, to ensure clarity when presenting it, and to choose the most suitable distribution for the information we wish to convey.

7.3.5 Segregation/mixing intensity

In the preceding section, we have shown how PEPT may be used to provide detailed distributions showing how the relative concentration of different particle species varies throughout a system. The same principles used above to create 1D profiles may also be trivially extended to produce 2D or even 3D concentration distributions. Though such distributions provide a large amount of information, in many situations—for instance if we wish to directly and easily compare the degree of segregation involved in multiple systems, or indeed one system under differing conditions—it is desirable to instead quantify the degree of mixing or segregation using a single, scalar value. There exist numerous distinct manners in which one may extract such a value from PEPT data though, as is so often the case, each methodology carries its own, distinct strengths and weaknesses. For the sake of simplicity, we will focus our discussion in this section predominantly on the case of binary systems, though the methods here may also be simply extended to the case of polydispersity.

Relative centres of mass

Perhaps the most basic manner in which one can extract a measure of segregation from PEPT data is simply by comparing the relative centres of mass for the species present. In order to do so, it is first necessary to determine the centre of mass position —as described in section 7.2.2—for each species present individually. A simple, dimensionless measure of the segregation strength exhibited by the system can then be determined, for the binary case, as:

$$\alpha = \frac{|h_1 - h_2|}{|h_1^{\text{seg.}} - h_2^{\text{seg.}}|} \tag{7.13}$$

where h_1, h_2 are the steady-state vertical mass centres[3] of the two particle species present, while $h_1^{\text{seg.}}$, $h_2^{\text{seg.}}$ represent the positions of these two species for an equivalent, perfectly-segregated system. For the case of complete, monodirectional segregation (illustrated in figure 7.9(a)), the parameter yields a value of unity, as $h_1 = h_1^{\text{seg.}}$ and $h_2 = h_2^{\text{seg.}}$, $\rightarrow \alpha = \frac{|h_1^{\text{seg.}} - h_2^{\text{seg.}}|}{|h_1^{\text{seg.}} - h_2^{\text{seg.}}|} = 1$. For the case of perfect mixing, meanwhile (figure 7.9(b)), the α parameter will acquire a value of zero, as $h_1 = h_2$, $\rightarrow \alpha = \frac{|h_1 - h_1|}{|h_1^{\text{seg.}} - h_2^{\text{seg.}}|} = 0$.

Despite its pleasing simplicity and ease of implementation, this measure of segregation carries the significant disadvantage that it will only give meaningful

[3] Note that the same calculation can also be performed for the horizontal direction, though horizontal segregation is considerably less common in most granular systems [49].

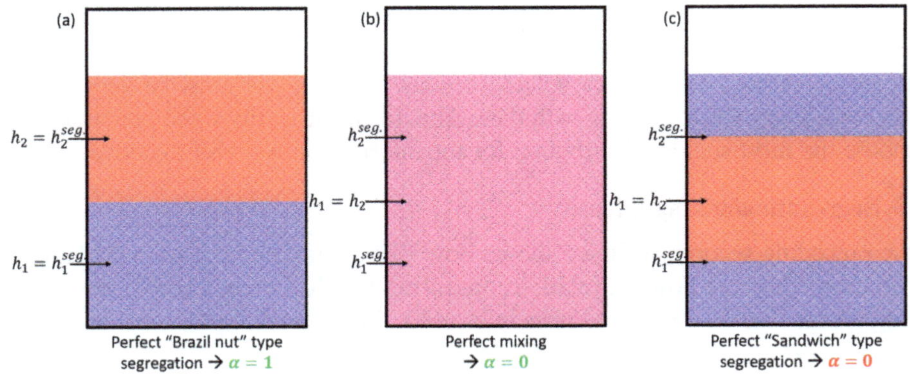

Figure 7.9. Simple schematic illustrations showing some common, broad types of segregation exhibited by granular systems, and the centre of mass positions of the species present. (a) simple 'Brazil nut' type segregation where the system simply separates in the vertical direction, with one species (typically the larger and/or less dense) occupying the upper region of the system and the other the lower [45, 46]. (b) Perfect mixing, in which both species are homogeneously distributed throughout the system (on both the macroscopic and microscopic scales). (c) 'Sandwich' type segregation—more commonly observed in systems comprising finer particles and/or immersed in denser, more viscous fluids [47, 48]—whereby one species occupies both the top and bottom regions of a contain, the other lying in between. Note that there exists a wide range of other segregation patterns beyond those pictured here. The reader may find comprehensive information regarding these in [49].

information if the segregation observed is strictly monodirectional—i.e. all particles of species 1 have a tendency to segregate directly upwards, and all particles of species 2 segregate directly downwards[4]. While this is undoubtedly one of the more common forms of segregation [49], there nonetheless exist certain conditions under which the use of the α parameter may be inaccurate, and indeed seriously misleading. A particularly stark—though nonetheless feasible [47, 48]—example of such a mis-characterisation is illustrated in figure 7.9(c). In this case, despite the two species of particle being clearly spatially separated, the specific, symmetric form of the segregation observed means that the centres of mass of said species align, yielding an α value of zero, thus (falsely) implying a well-mixed system.

Variance-based measures: the Lacey Mixing Index and Segregation Intensity
An alternative—and indeed much more flexible and sensitive—measure of segregation strength can be obtained in a manner first pioneered by Lacey in 1943 [52, 53], and since adopted and adapted by various authors [54, 55]. The general methodology involves—as described in section 7.3.3 for the case of PEPT data—dividing

[4] This simple form of segregation is—for the case in which particles differ predominantly in size—commonly known as the 'Brazil nut effect' (BNE) based on the observation that in a container of mixed nuts, the largest 'species' (typically the Brazil nuts!) are often found at the top of the pile due to segregation [46]. The less-commonly-observed inverse case (larger particles sink to the bottom) is known as the 'reverse Brazil nut effect' (RBNE) [50, 51].

the experimental volume into a series of cells in one, two, or three dimensions and, for each of these cells, determining the local concentration (ϕ) of particles belonging to the chosen reference species. Taking ϕ_i as the solids concentration of our reference species in the ith cell within the system and ϕ_m as the mean concentration of this same species across the system as a whole, the concentration variance, σ^2, for the system may be determined as:

$$\sigma^2 = \sum_{i}^{N_c} \frac{(\phi_i - \phi_m)^2}{N_c - 1} \tag{7.14}$$

where N_c is the number of cells into which the system is divided. The square root of this value (i.e. the standard deviation of the concentration) is often directly used by researchers as a measure of the '*segregation intensity*', I_s, of a system [12, 55–60], often normalised by the maximal value of σ achievable for the system of interest [61, 62], which—for a binary system—can be analytically determined as [63]:

$$\sigma_{\max} = \sqrt{\sigma_m(1 - \sigma_m)}\,. \tag{7.15}$$

Under the condition of such normalisation, a perfectly mixed system ($\phi_i = \phi_m$ for all i) will yield a segregation intensity $I_s = 0$, while a completely segregated system will yield a value $I_s = 1$—just as with the centre-of-mass-based approach described previously.

An alternative normalisation considers also the minimum possible variance within a cell, defined as:

$$\sigma_{\min}^2 = \frac{\sigma_m(1 - \sigma_m)}{n} \tag{7.16}$$

where n represents the number of possible particles within a given cell. Using both of the above parameters, the *Lacey Mixing Index* may be defined as:

$$M_L = \frac{\sigma_{\max}^2 - \sigma^2}{\sigma_{\max}^2 - \sigma_{\min}^2}. \tag{7.17}$$

A similar index, known as the Kramer Mixing Index after its eponymous inventor [64], can be defined as:

$$M_K = \frac{\sigma_{\max} - \sigma}{\sigma_{\max} - \sigma_{\min}}. \tag{7.18}$$

For both the Lacey and Kramer indices, a fully segregated system should yield a value of zero, and a perfectly mixed system a value of unity. It is important to bear in mind when comparing literature using different indices that, somewhat confusingly, this definition is the precise opposite of the previously-described indices!

Unlike the mass-centre method described in the preceding section, all of the indices described above are agnostic to the absolute positioning of particles within the system (see figure 7.10). As such, for a suitably chosen cell size, geometry and

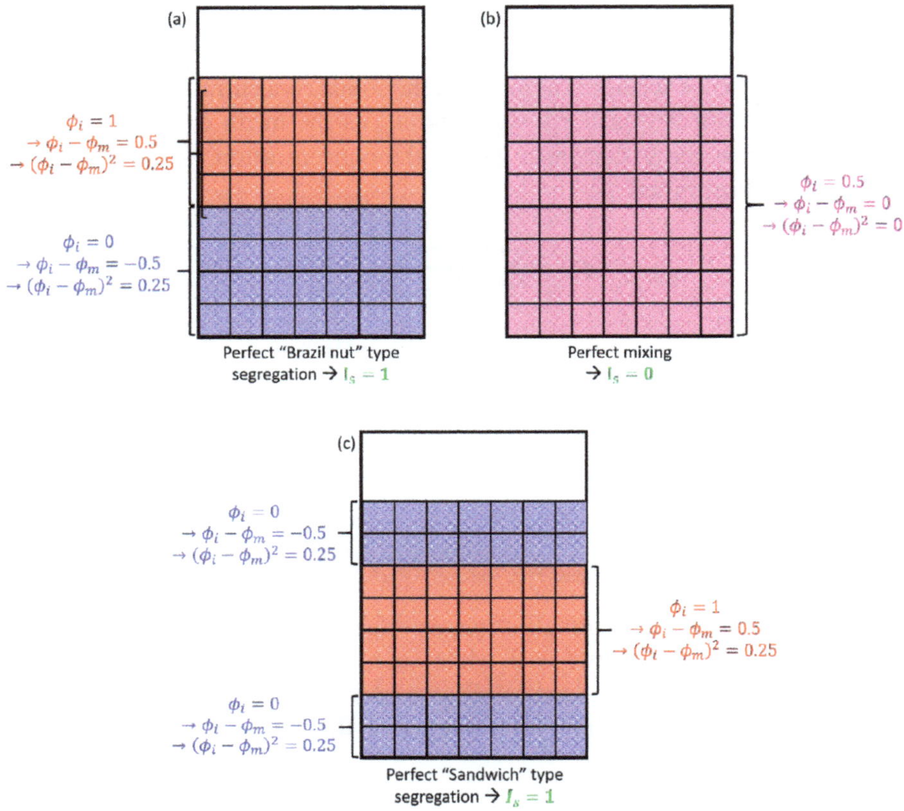

Figure 7.10. Simple schematic depiction of how the systems from figure 7.9 may be subdivided into cells and the local concentration values thereof used to determine the segregation intensity, I_s, as described in the main text.

dimensionality, a fully-segregated system should yield the same I_s, M_L and/or M_K value irrespective of the precise segregation pattern exhibited by the system.

The only major limitation of the above-described methodologies is that (in particular for systems which are neither fully segregated nor fully mixed) the measured indices carry an innate dependence on the size and number of cells used to calculate the concentration variance, making direct comparison of values achieved from studies exploring different systems and/or particle sizes somewhat difficult. This limitation can, however, be overcome to an extent by performing a suitable sensitivity analysis so as to ensure the choice of an optimal cell size—as demonstrated in reference [65].

The entropic mixing parameter

Though, as discussed in chapter 1, granular media do not observe conventional thermodynamics or statistical mechanics, one may nonetheless borrow some of the concepts thereof in order to better characterise and quantify their behaviours [121]. One relevant example of when this may be useful is in the definition of the *entropic mixing parameter*, Φ, developed by Schutyser *et al* [66, 67] and later adapted and popularised in the work of Arntz *et al* [65, 68–71].

As with the previously-discussed parameters, we begin by subdividing the experimental grid into a series of cells and determining the relative concentrations of different species therein (see section 7.3.3). In this case, however, we utilise the relative *number fraction*, ϕ^n, as opposed to solids fraction, of our species of interest. Let $\phi_j^n(i)$ represent the number fraction of particles of species j in the ith cell of a binary system whose other component we denote with the subscript k. The *local mixing entropy* in this particular cell can be defined as:

$$s(i) = \phi_j^n(i)\ln\phi_j^n(i) + \phi_k^n(i)\ln\phi_k^n(i). \tag{7.19}$$

A value for the *global entropy*, S, can then be defined as

$$S = \frac{1}{N_c}\sum_i^{N_c} s(i)n(i), \tag{7.20}$$

where, as previously, N_c represents the total number of cells into which the system is divided. Notably, unlike the indices described in the previous subsection, the calculation in equation (7.20) weights the relevant local quantities by the number of particles, $n(i)$, in the relevant cell. In certain systems—in particular those involving strong density gradients—this weighting may prove very valuable, as it can prevent the value of the calculated segregation or mixing index being skewed by cells containing very few particles of only one species. This issue can be explored directly in the interactive example following this section.

Further information regarding the various manners in which segregation may be characterised, and indeed the segregation phenomenon more generally, can be found in the recent monograph of Rosato and Windows-Yule [49].

7.3.6 Interactive example: segregation strength

The notebook files accompanying this section are available in the supplementary data online at https://iopscience.iop.org/book/978-0-7503-3071-8.

Tutorial: Using PEPT data to quantify segregation

In this tutorial, we will show you how to take the local particle concentration distributions created in the previous tutorial and use them to calculate a concise, scalar parameter which meaningfully quantifies the extent of segregation exhibited by a system. Importantly, the tutorial will also show you some of the complexities and potential pitfalls associated with such calculations!

Acquiring concentration distributions

In this example, we will measure the extent of segregation within a system using the *segregation intensity* parameter described in the main text. In order to do so, we first need to generate a concentration distribution—specifically, the relative solids fraction—to work with. For this example, we will simply work with the same data set explored in the previous tutorial.

[1]:
```
# Importing necessary libraries for numerical array manipulation and plotting
import numpy as np
import matplotlib.pyplot as plt

# Set Matplotlib theme (try "darkgrid" too)
import seaborn
seaborn.set_theme(style = "whitegrid")
```

[2]:
```
# Importing data for the sand component
filepath1 = "Data/ConcentrationSandExample.csv"
data_sand = np.loadtxt(filepath1, skiprows=16)

# Importing data for the plastic component
filepath2 = "Data/ConcentrationPlasticExample.csv"
data_plastic = np.loadtxt(filepath2, skiprows=16)

# Extracting the data we need for each component of the system
t_sand = data_sand[:, 0]
y_sand = data_sand[:, 2]

t_plastic = data_plastic[:, 0]
y_plastic = data_plastic[:, 2]

# Pre-allocating arrays to store all relevant midpoints of the recorded
# y-positions, and their corresponding durations
y_mid_sand = np.zeros(len(y_sand) - 1)
y_mid_plastic = np.zeros(len(y_plastic) - 1)

dt_sand = np.zeros(len(y_sand) - 1)
dt_plastic = np.zeros(len(y_plastic) - 1)

# Calcutaing midpoints and durations for each species in turn
for i in range(len(y_mid_sand)):
    # Each value of y_mid is the average of the two y-values on either side
    y_mid_sand[i] = (y_sand[i] + y_sand[i + 1]) / 2

    # The time spent in the vicinity of y_mid can be assumed equal to the
    # difference between the correspondent t values on either side
    dt_sand[i] = t_sand[i + 1] - t_sand[i]
for i in range(len(y_mid_plastic)):
    y_mid_plastic[i] = (y_plastic[i] + y_plastic[i + 1]) / 2
    dt_plastic[i] = t_plastic[i + 1] - t_plastic[i]

# Defining our chosen cell size, dc, in mm.
dc = 10

# Finding, for each species, the necessary number of cells of size dc to span 0
# to the maximum y-value detected
n_sand = int(y_mid_sand.max() / dc)
n_plastic = int(y_mid_plastic.max() / dc)

# Setting the final 'N' value to the smaller of the 2 Ns acquired such that
# comparison can be performed
n = np.min([n_sand, n_plastic])

# Finding the minimum vertical position detected such that program knows where
# to start scanning
y_min = np.min([y_sand.min(), y_plastic.min()])
```

```
# Pre-allocating 1D arrays to store the occupancy profiles corresponding to each
# species
occupancy_sand = np.zeros(n)
occupancy_plastic = np.zeros(n)

# For each vertical 'slice' within each system, finding all particle locations
# therein, and summing the durations corresponding to these locations
for i in range(1, n):
    # Copying all duration values to a new array
    dt_temp_sand = dt_sand.copy()
    dt_temp_plastic = dt_plastic.copy()

    # Removing all values with y-values (i.e. heights) outside the current slice
    dt_temp_sand[y_mid_sand < y_min + i * dc] = 0
    dt_temp_sand[y_mid_sand > y_min + (i + 1) * dc] = 0
    dt_temp_plastic[y_mid_plastic < y_min + i * dc] = 0
    dt_temp_plastic[y_mid_plastic > y_min + (i + 1) * dc] = 0

    # Adding the sum of the remaining values to occupancy array
    occupancy_sand[i] = np.sum(dt_temp_sand)
    occupancy_plastic[i] = np.sum(dt_temp_plastic)

# Normalising
occupancy_sand = occupancy_sand / t_sand.max()
occupancy_plastic = occupancy_plastic / t_plastic.max()

# Some values in the division below are zero, let them silently yield 0/0 = NaN
np.seterr(divide='ignore', invalid='ignore')

# Calculating the volume fraction distribution
volume_ratio = (95 / 2800) / (5 / 1100)
print(f"V_T^2 / V_T^1 = {volume_ratio}")

phi_vol_plastic = occupancy_plastic / (occupancy_plastic + occupancy_sand *␣
↪volume_ratio)

# Defining a suitable axis scale for plotting
scale = np.arange(n) * dc

# Plotting individual occupancy profiles
fig, ax = plt.subplots(1, 2)
ax[0].scatter(occupancy_sand, scale, label='Sand')
ax[0].scatter(occupancy_plastic, scale, label='Plastic')

ax[0].set(ylabel='$y$ (mm)', xlabel='Occupancy')
ax[0].set_aspect(1.0 / ax[0].get_data_ratio() * 5)
ax[0].set_ylim(0, 320)
ax[0].legend()

ax[1].scatter(phi_vol_plastic, scale)

ax[1].set(ylabel='$y$ (mm)', xlabel='$\phi^{vol}$')
ax[1].set_aspect(1.0 / ax[1].get_data_ratio() * 5)
ax[1].set_ylim(0, 320)
```

```
V_T^2 / V_T^1 = 7.464285714285714
```

```
[2]: (0.0, 320.0)
```

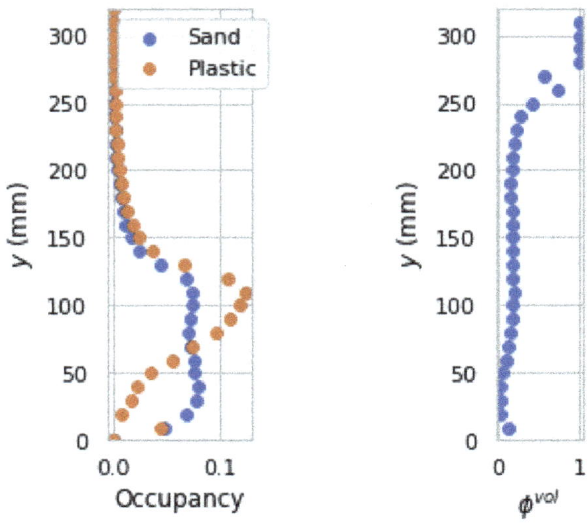

Calculating the segregation intensity

Looking at the above profiles, do you notice anything interesting? One feature which we must consider before calculating the segregation intensity is that even though the first figure shows that the vast majority of solids remain below $y \sim 200$ mm, there exists a significant spike in ϕ^{vol} above this point which—as the I_s parameter does not include any weighting based on the occupancy or number density of a cell—may significantly skew our results. As such, we must decide a) whether to include all of this data in our calculations and, if not, b) how much to remove. In order to ensure scientific rigor, it is vital that any exclusion of data is well-supported by cogent scientific reasoning. A commonly used cut off point—see, for example, [146]—is twice the vertical mass centre of the system, which provides (for a relatively uniform system such as that explored here) a reasonable estimate of the height of the bed's 'bulk' region.

In the example below, we show how the I_s parameter can be calculated from the above data. By altering the value of the *y_cutoff* parameter, you can include more or fewer of the high-y cells in the calculation. Take some time to play with this parameter, and see just how much the chosen cut-off height can influence the measured value of segregation strength (note that you will have to re-run the above section of code if you want to increase your y-cutoff after initially reducing it!). In particular, look at how small perturbations about a given value of *y_cutoff* to see how stable the values achieved are at different points.

```
[3]: # The cutoff point above which data is removed
     y_cutoff = 200;

     # Select only data below the cut-off point
     phi_vol_plastic = phi_vol_plastic[scale < y_cutoff]
     scale = scale[scale < y_cutoff]

     # Re-plotting with only remaining points
     fig, ax = plt.subplots(1, 1)
     ax.scatter(phi_vol_plastic, scale)

     ax.set(ylabel='$y$ (mm)', xlabel='$\phi^{vol}$')
     ax.set_aspect(1.0 / ax.get_data_ratio() * 5)
     plt.ylim(0, 320)
```

```
# Determining the system's mean concentration, phi_m, for the plastic species
phi_m = (0.05 / 1100) / (0.05 / 1100 + 0.95 / 2800)
print("phi_m = ", phi_m)

# Calculating the segregation intensity as defined in the main text
# N_c can simply be found from the length of the remaining array
n_c = len(phi_vol_plastic)
i_s = np.sqrt(np.nansum(np.square(phi_vol_plastic - phi_m)) / (n_c - 1))

# Outputting unnormalised I_s value
print("I_s (unnormalised) = ", i_s)

# Calculating normalisation constant
sigma_max = np.sqrt(phi_m * (1 - phi_m))

# Normalising
i_s = i_s / sigma_max
print("I_s = ", i_s)
```

```
phi_m =  0.11814345991561183
I_s (unnormalised) =  0.05603250987360315
I_s =  0.17359462084202032
```

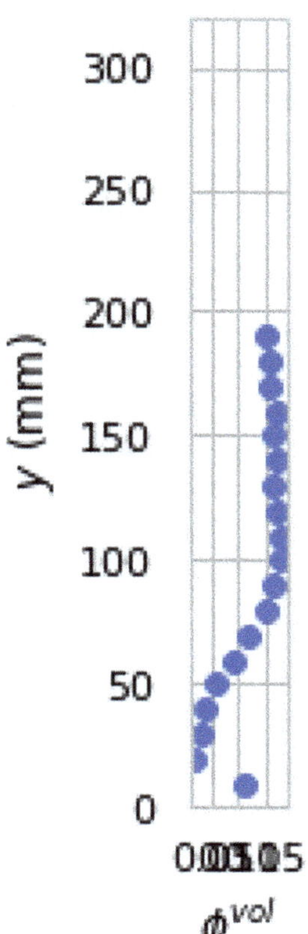

7.4 Velocity

7.4.1 Calculating velocities from PEPT data

In the preceding sections, we have discussed particle position data, and how the averaging of this positional data can be used to determine the spatial distributions of particles throughout a system of interest. If we are to fully understand the behaviours of a dynamic particulate or multiphase system, however, we need to know not only where our tracers are, but also how they are moving. To this end, we discuss in this section how the most elementary dynamic data—the pseudo-instantaneous velocity of a tracer—can be extracted from PEPT data.

As post-processed PEPT data provides us with a list of contiguous times (t_0, t_1, t_2, ...) and spatial coordinates ($\vec{r}_0$, $\vec{r}_1$, $\vec{r}_2$, ...) the most obvious manner in which to determine the tracer velocity is simply calculating $\vec{v} = \frac{\Delta\vec{r}}{\Delta t}$ for each pair of consecutive points, i.e.:

$$\vec{v}_{i+\frac{1}{2}} = \frac{\Delta\vec{r}}{\Delta t} = \frac{\vec{r}_{i+1} - \vec{r}_i}{t_{i+1} - t_i}. \tag{7.21}$$

However, in PEPT as in life, the simplest answer is often not the best. For the case of vanishingly small location errors $\sigma(\vec{r})$ (or, more specifically, for the case in which the spatial error is small compared to the distance moved by the particle between timesteps, i.e. $\frac{\sigma(\vec{r})}{\Delta\vec{r}} \rightarrow 0$) the above approach can be expected to work perfectly well. However, for situations in which this constraint does not hold true (e.g. for the case of slow-moving particles and/or low-activity tracers), the use of only two consecutive data points to calculate a velocity may prove problematic. The issue at hand is perhaps best illustrated visually. Figure 7.11(a) shows a hypothetical example of a single-particle trajectory (black line) alongside a series of PEPT-detected locations (black crosses). The blue arrows shown represent the vector velocities calculated as per equation (7.21). Clearly, in the majority of cases, the vectors predicted in this manner are a poor representation of the particle's true velocity!

An improved estimate of our notional tracer's velocity may potentially be found using a different (though still relatively simple) approach, namely the finite central difference method. Using this approach, the velocity of our particle at its ith location

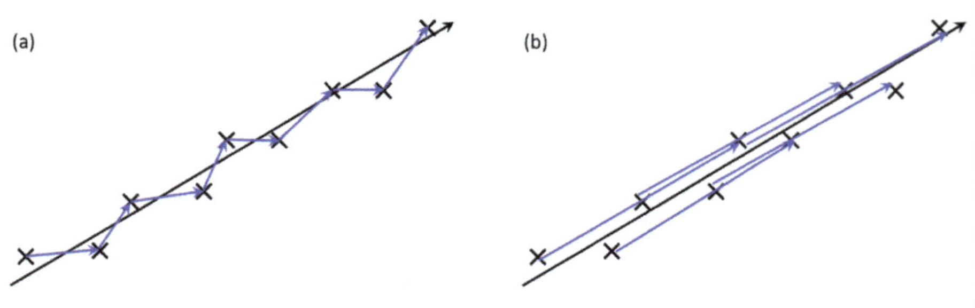

Figure 7.11. Simple schematic illustrating two different manners of calculating particle velocity. In both cases, the black line represents the true particle motion, the black crosses the data points acquired using PEPT, and the blue arrows the velocities determined from said PEPT data. Panel (a) shows the velocity vectors acquired by determining a velocity value for each consecutive pair of points (equation (7.21)). Panel (b) shows the velocity vectors acquired from the same data using a five-point central difference approach (equation (7.22)).

event is estimated based on its measured velocity calculated from two location events a fixed number of 'steps' either side of the location of interest. For example, if we want to estimate the velocity of our particle at our 7th recorded data point, we simply find the average velocity between the 6th and 8th data points (or, alternatively 5th and 9th, 4th and 10th, etc). To put it mathematically,

$$\vec{v}_i = \frac{\vec{r}_{i+j} - \vec{r}_{i-j}}{t_{i+j} - t_{i-j}}. \tag{7.22}$$

As illustrated in figure 7.11, by calculating $\vec{v}$ over a larger number of data points, we are able to average out much of the noise (as $j \to \infty, \Delta\vec{r} \to \infty, \frac{\sigma(\vec{r})}{\Delta\vec{r}} \to 0$), and thus achieve a much closer approximation of our tracer's true velocity. However, the underlying assumption of such a calculation is that our particle maintains an approximately linear trajectory and a constant speed between points $i - j$ and $i + j$. For a particle moving under gravity and/or subject to drag or friction, this assumption can only be expected to hold for a reasonably short space of time. For a particle subject to frequent collisions (and thus sharp changes in trajectory) or immersed in a highly turbulent fluid, this window of opportunity may close still further. As such, when analysing PEPT data using the central difference method, one must make a careful choice as to the number of points to use—too few and the true velocities may be drowned out by noise; too many, and you may smooth out real features of the data. The importance of this careful balancing act is illustrated schematically in figure 7.12. The interactive example at the end of this section provides an opportunity to further explore the concept of determining an optimal j value with which to compute particle velocities.

The central difference approach described above calculates average velocities over a range of data points, but in all cases only considers the two end points ($i - j$ and $i + j$), in essence ignoring all the data in between. A slightly more advanced—though correspondingly more complex—weighted average approach can be taken, wherein the particle's pseudo-instantaneous velocity at point i is determined as:

$$\vec{v}_i = \sum_{j=1}^{N} W_j \frac{\vec{r}_{i+j} - \vec{r}_{i-j}}{t_{i+j} - t_{i-j}} \tag{7.23}$$

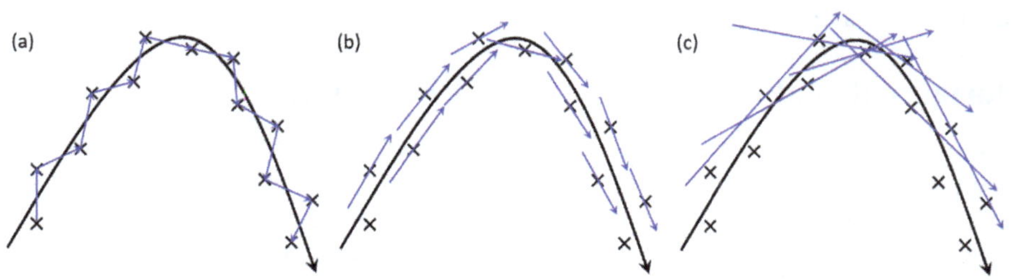

Figure 7.12. Simple schematic illustrating different manners of calculating particle velocity for a nonlinear trajectory. In all cases, the black line represents the true particle motion, the black crosses the data points acquired using PEPT, and the blue arrows the velocities determined from said PEPT data. Panel (a) shows the velocity vectors acquired by determining a velocity value for each consecutive pair of points. Panels (b) and (c) show, respectively, the velocity vectors acquired from the same data using a three-point and seven-point central difference approach.

where $\sum_{j=1}^{N} W_j = 1$, ensuring that the value obtained is correctly normalised. In essence, this approach involves estimating the particle velocity at location i using the finite difference approach for a number (N) of different j values, and assigning each of these calculated velocities a weight, W_j, according to how reliable we expect it to be. The final velocity is then taken as the average of the individual weighted values. Typically, smaller values of j (i.e. velocities calculated using locations closer to the point of interest) are afforded a higher weight; however, there are no 'hard rules' regarding this—the weightings can ultimately be chosen entirely according to the user's discretion.

Though this method may hypothetically achieve more meaningful data than the comparatively simple central difference approach, it introduces a number of free parameters which, if not chosen carefully, may adversely affect the reliability of the data ultimately produced.

The final widely-used methodology for extracting velocity information from PEPT data is to apply least-squares regression analysis [72, 73]. In its simplest form, this can be performed simply by finding the linear best fit to a chosen range of data, the gradient of which can be taken as the pseudo-instantaneous velocity at the central point, though second and third order polynomials may also be used, as may more advanced techniques such as Savitzky–Golay filtering [148]. As with the other methodologies described in this section, the range of data used for fitting must be great enough to minimise the impact of noise, but small enough that the assumption of constant velocity remains reasonable for the majority of the data available.

The least-squares regression approach carries the notable advantage over the central difference method in that it uses all data within the chosen points (as opposed to just the start and end points), and over the weighted average approach in that it involves fewer free parameters. The main disadvantage is that it requires significantly more computational effort than either of the previous approaches. While this is not problematic for small data sets such as those considered in the examples at the end of this section, for real PEPT data sets—which can contain many millions of data points—this computational inefficiency can prove problematic.

7.4.2 Interactive example: finding velocities

The notebook files accompanying this section are available in the supplementary data online at https://iopscience.iop.org/book/978-0-7503-3071-8.

Tutorial: calculating velocities from PEPT data

As we have seen in the main text, there exist a number of distinct manners, varying in complexity, through which we may extract velocity information from PEPT data. In this section, we show you all of these different methods in action, so that you can see first-hand the strengths and limitations of each, and learn how to use them to the best of their indiviudal capabilities.

Before we get started, however, we need some data to work with! In the code below, we import a simulated PEPT dataset for a particle undergoing one-dimensional sinusoidal motion along the x-axis. The sinusoidal motion is chosen for two reasons: firstly, it varies in both direction and speed, making it a challenge for our velocity calculations; secondly, for a known sinuoidal trajectory we have a well known analytical solution for the velocity, meaning that we can easily and directly compare the velocity data calculated using our various PEPT analysis methods to a known true value.

[1]:
```
# Importing necessary libraries for numerical array manipulation and plotting
import numpy as np
import matplotlib.pyplot as plt

# Set Matplotlib theme (try "darkgrid" too)
import seaborn
seaborn.set_style("whitegrid")
```

[2]:
```
# Loading a full 1D particle trajectory
filepath1 = "Data/VelocityDataFull.csv"
data_full = np.loadtxt(filepath1, delimiter=',')
t_full = data_full[:, 0]
x_full = data_full[:, 1]

# Pseudo-pept data
filepath2 = "Data/VelocityDataPEPT.csv"
data_pept = np.loadtxt(filepath2, delimiter=',')
t_pept = data_pept[:, 0]
x_pept = data_pept[:, 1]

# Analytically known velocity data
filepath3 = "Data/VelocityDataVelocity.csv"
data_velocity = np.loadtxt(filepath3, delimiter=',')
v_full = data_velocity

# Plotting
fig, ax = plt.subplots(1, 1)

ax.plot(t_full, x_full)
ax.scatter(t_pept, x_pept)

ax.set(ylabel='$x$ (mm)', xlabel='$t$ (s)')
```

[2]: [Text(0, 0.5, 'x (mm)'), Text(0.5, 0, 't (s)')]

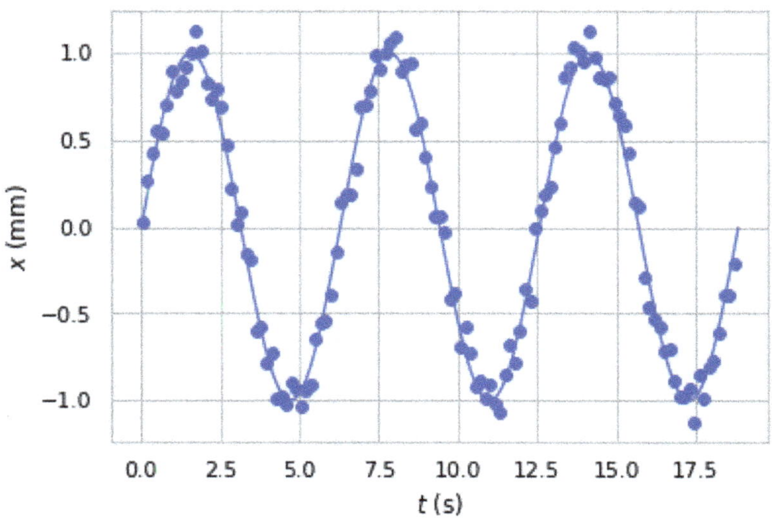

Basic two-point velocity calculation

The image above shows the true path of the hypothetical tracer (solid line) alongside the PEPT data (points). Note that, in order to give ourselves a more realistic challenge, we have made the pseudo-PEPT data deliberately noisy! In the code below, we attempt to extract velocity information from this data using the simple two-point calculation described in the main text.

```
[3]: # Pre-allocating array to hold all velocity values
v_twopoint = np.zeros(len(t_pept) - 1)

# Pre-allocating array to store midpoints of particle location times
t_mid = np.zeros(len(t_pept) - 1)

# Calculating simple two-point velocity of each particle in turn
#
# Note that the velocity calculated technically applies to the midpoint between
# the two points, rather than either individually
for i in range(len(t_pept) - 1):
    # Working out the midpoint between the two points to which the calculated
    # velocity applies
    t_mid[i] = (t_pept[i] + t_pept[i + 1]) / 2

    # Calculating velocity for each midpoint
    v_twopoint[i] = (x_pept[i + 1] - x_pept[i]) / (t_pept[i + 1] - t_pept[i])

# Plotting - comparing calculated velocities to analytically predicted values
fig, ax = plt.subplots(1, 1)

ax.plot(t_full, v_full)
ax.scatter(t_mid, v_twopoint)

ax.set(ylabel='$v_x$ (mm)', xlabel='$t$ (s)')
```

```
[3]: [Text(0, 0.5, '$v_x$ (mm)'), Text(0.5, 0, '$t$ (s)')]
```

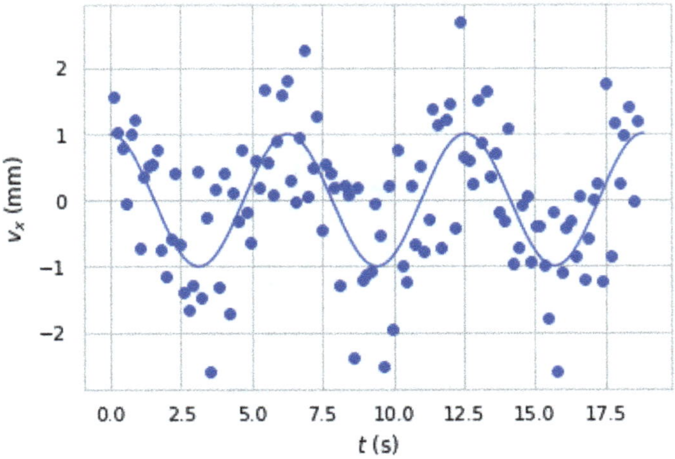

Central difference velocity calculation

Looking at the above figure which compares our expected (solid line) and measured (symbols) velocity values, we can immediately see that the two-point method—for the present dataset—does **not** provide a good representation of the particle's true velocity. The reasons for this are discussed at length in the main text.

For our next attempt, we will try and process the same data using the central difference approach described in section 7.4. In the example below, take some time to explore the effect of using different values of *j*, so that you can see for yourself how important the careful choice of this parameter is!

[4]:
```
# The number of steps before and after the current point over which the velocity
# is calculated - or window half-width
j = 4

# Creating empty array to hold all velocity values
v_cd = np.zeros(len(t_pept))
v_cd[:] = np.NaN

# Calculating velocity for each point using the central difference approach
#
# Note that we skip the first and last j items as the velocity cannot be
# calculated for these without more data!
for i in range(j, len(t_pept) - j):
    # Calculating velocity for each point
    v_cd[i] = (x_pept[i + j] - x_pept[i - j]) / (t_pept[i + j] - t_pept[i - j])

# Plotting - comparing calculated velocities to analytically predicted values
fig, ax = plt.subplots(1, 1)

ax.plot(t_full, v_full)
ax.scatter(t_pept, v_cd)

ax.set(ylabel='$v_x$ (mm)', xlabel='$t$ (s)')
```

[4]:
```
[Text(0, 0.5, '$v_x$ (mm)'), Text(0.5, 0, '$t$ (s)')]
```

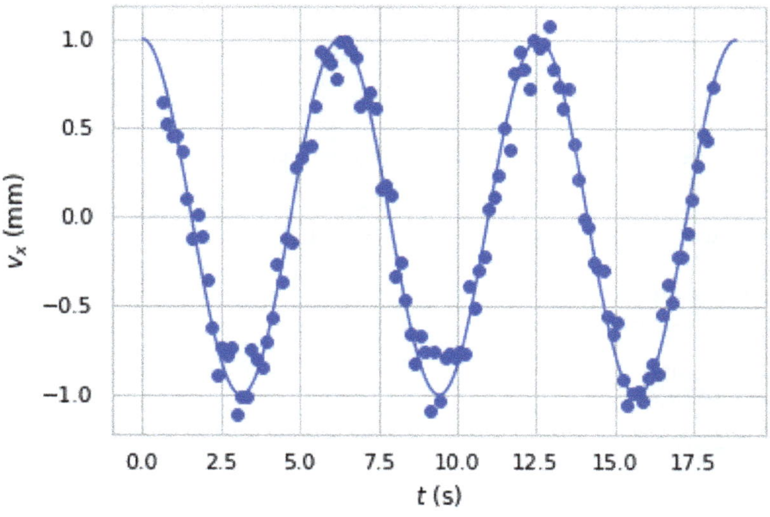

Weighted average velocity calculation

In the above example, we can see that for any reasonable choice of j, our results are much closer to the expected velocity values. Indeed, for a well-chosen j, our results produce a very reasonable approximation of the particle's true velocity.

Let us next look at the weighted average calculation. For the sake of simplicity, for this example we use a fixed range across which to measure our velocities. The choices of the weightings, however, remain free to alter. Try re-running the below code with different weighting combinations to see how this affects the data output. Remember that you can in essence ignore a specific data range by setting the relevant weighting W_i to zero.

[5]:
```
# Choosing values for weighting - note that all values should add up to 1!
# Weighting for velocity calculated with j=1
w1 = 0.25

# Weighting for velocity calculated with j=2
w2 = 0.25

# Weighting for velocity calculated with j=3
w3 = 0.25

# Weighting for velocity calculated with j=4
w4 = 0.25

# Pre-allocating empty array to hold all velocity values
v_wa = np.zeros(len(t_pept))
v_wa[:] = np.NaN

# Calculating velocity for each point in turn using the central difference
# approach with 3 different values of j
for i in range(4, len(t_pept) - 4):
    # Finding value for current velocity with j = 1
    v1 = (x_pept[i + 1] - x_pept[i - 1]) / (t_pept[i + 1] - t_pept[i - 1])

    # Finding value for current velocity with j = 2
    v2 = (x_pept[i + 2] - x_pept[i - 2]) / (t_pept[i + 2] - t_pept[i - 2])

    # Finding value for current velocity with j = 3
    v3 = (x_pept[i + 3] - x_pept[i - 3]) / (t_pept[i + 3] - t_pept[i - 3])

    # Finding value for current velocity with j = 4
    v4 = (x_pept[i + 4] - x_pept[i - 4]) / (t_pept[i + 4] - t_pept[i - 4])

    # Calculating overall average velocity for each point using chosen weighting
    v_wa[i] = w1 * v1 + w2 * v2 + w3 * v3 + w4 * v4

# Plotting - comparing calculated velocities to analytically predicted values
fig, ax = plt.subplots(1, 1)

ax.plot(t_full, v_full)
ax.scatter(t_pept, v_wa)

ax.set(ylabel='$v_x$ (mm)', xlabel='$t$ (s)')
```

[5]:
```
[Text(0, 0.5, '$v_x$ (mm)'), Text(0.5, 0, '$t$ (s)')]
```

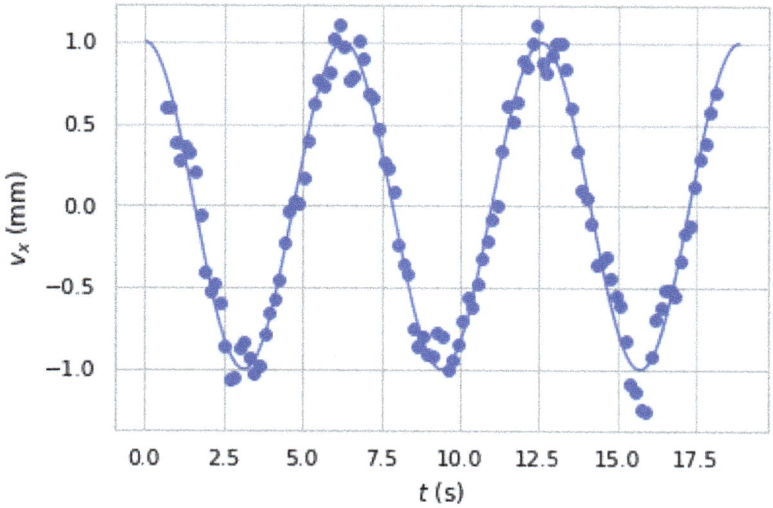

Linear regression

In our final example, we apply a simple linear regression analysis, as discussed in section 7.4 of the main text. As with the central difference calculation, in the code below you can vary the number of points used to calculate each individual velocity, though this method, unlike the former, takes into account **all points** lying between $i - j$ and $i + j$ thus, in the majority of cases, providing more accurate data than the central difference method, which only considers the first and last points.

With a well-chosen j value, you should see that the linear regression method provides a very close reproduction of our particle's true velocity.

[6]:
```
# Choosing the data range over which to perform the regression
# As previously, j is how many points on either side of the point of interest
# we want to use to calculate velocity - or window half-width
j = 6

# Creating empty array to hold all velocity values
v_lr = np.full(len(t_pept), np.NaN)

# Calculating velocity for each point in turn using linear regression
for i in range(j, len(t_pept) - j):
    # Using linear regression to extract the gradient of a linear fit to the
    # relevant range of data - i.e. the velocity!
    p = np.polyfit(t_pept[i - j:i + j], x_pept[i - j:i + j], 1)
    v_lr[i] = p[0]

# Plotting - comparing calculated velocities to analytically predicted values
fig, ax = plt.subplots(1, 1)

ax.plot(t_full, v_full)
ax.scatter(t_pept, v_lr)

ax.set(ylabel='$v_x$ (mm)', xlabel='$t$ (s)')
```

[6]:
```
[Text(0, 0.5, '$v_x$ (mm)'), Text(0.5, 0, '$t$ (s)')]
```

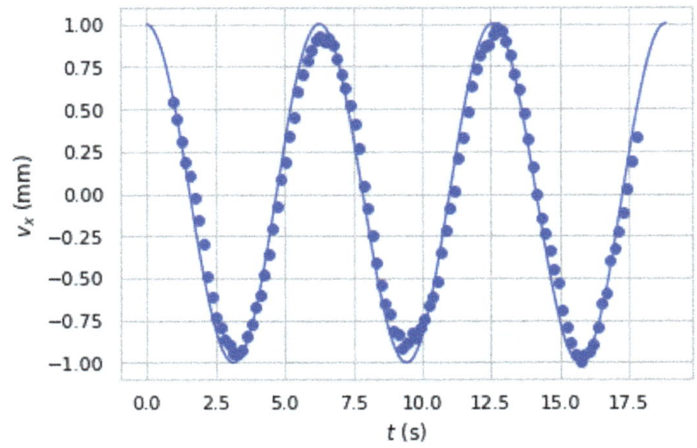

7.4.3 Velocity probability distributions

Now that we know how to calculate particle velocities from PEPT position data, we can build upon this knowledge to extract a number of additional parameters, fields and distributions that may prove highly valuable. A simple, yet still highly informative, quantity which may be extracted from a set of known particle velocities is the velocity probability distribution function (PDF).

In order to calculate a velocity PDF from PEPT data, one must first calculate—as described in the previous section—the tracer's pseudo-instantaneous velocity for each data point recorded. With these values acquired, it is then simply a matter of producing a histogram from the resultant data. As with any histogram, it is necessary to ensure that the width, w, of the velocity 'bins' into which the data is subdivided is suitably large to ensure adequate statistics for the full data range, but also suitably small that the fine detail of the distribution can be accurately interpreted. The histograms produced may be normalised by the total number, N, of recorded data points, to produce a (relative) probability distribution, or by the product $N \cdot w$, to produce an estimate of the probability density function, such as those pictured in figure 7.13.

For a monodisperse system under the ever-important assumption of ergodicity, the distribution of velocities measured for a single tracer particle over a suitably long period can be safely expected to be representative of all particles within the system—i.e. a velocity PDF created using a single particle trace as described above can be taken as the distribution for the system as a whole. For binary, ternary and other polydisperse systems one may, as usual, combine multiple data sets produced using one tracer of each relevant species. The individual velocity distributions for all species present may be acquired simply by following the procedure described above for each tracer individually. The overall distribution may then be simply acquired by taking the average of the acquired distributions, weighted by the relative fraction of each species present.

Though simple to obtain, particle velocity distributions for granular and multiphase systems are very information-rich, potentially providing insight into the particle properties, collision statistics, and driving properties of a system, and much more besides [74–79]—a

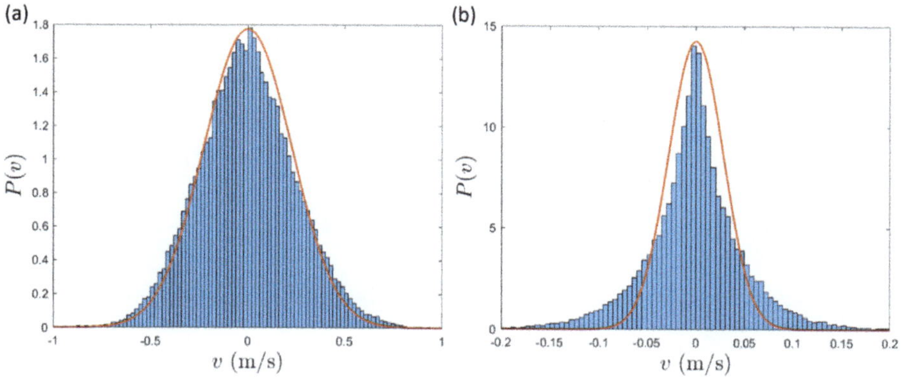

Figure 7.13. Velocity probability density functions for horizontal particle motion in (a) a stirred tank system [23] and (b) a fluidised bed system. In both cases, PEPT data (bars) is compared to a normal distribution (as one may expect to see for the particles comprising a classical, molecular gas). Note the stark differences in the velocity distributions produced by the two systems, and the strongly non-Gaussian nature of the distribution observed for the fluidised bed, where particle–particle interactions dominate system behaviour, as compared to the stirred tank, where particle-fluid interactions are dominant.

matter which we discuss at length in section 8.1. The velocity PDFs of particulate media also differ strongly from our expectations for classical molecular materials with 'granular gases' [80], for example, deviating strongly from the Gaussian velocity distributions expected for ideal (molecular) gases [81–83]—as illustrated in figure 7.13(b).

7.4.4 Interactive example: calculating a velocity probability distribution

The notebook files accompanying this section are available in the supplementary data online at https://iopscience.iop.org/book/978-0-7503-3071-8.

Tutorial: creating velocity probability distributions from PEPT data

In this tutorial, we will apply some of the velocity calculation techniques we learned in the previous tutorial to a full PEPT data set. We will then take the velocity values obtained, and use them to create a velocity probability distribution for this data.

In the examples below, feel free to play with both the j value, which affects the range of data over which velocities are calculated, as well as the histogram bin width, w, both of which will affect the details of the velocity distribution eventually achieved.

Importing data and calculating velocities

We begin in the code below by importing some data—indeed the same fluidised bed data used to produce one of the graphs in the preceding section of the main text—and determining a set of velocity values:

[1]:
```
# Importing necessary libraries for numerical array manipulation and plotting
import numpy as np
import matplotlib.pyplot as plt

# Set Matplotlib theme (try "darkgrid" too)
import seaborn
seaborn.set_style("whitegrid")
```

[2]:
```
# Loading pre-processed fluidised bed PEPT data
filepath = "Data/FluidisedBedExample2.csv"
full_trace = np.loadtxt(filepath, skiprows=16)

# Extracting time and position data in the horizontal (x) direction
t = full_trace[:, 0]
x = full_trace[:, 1]

# The number of steps before and after the current point over which the velocity
# is calculated
#
# Feel free to adjust this parameter to see its influence on the final output!
j = 3

# Pre-allocating empty array to hold all velocity values
v_x = np.full(len(x), np.NaN)

# Calculating velocity for each point in turn using the central difference
# approach (as discussed in detail in the previous tutorial)
for i in range(j, len(x) - j):
    # Calculating velocity for each point
    v_x[i] = (x[i + j] - x[i - j]) / (t[i + j] - t[i - j])
```

Creating a probability distribution

Having determined velocities for our PEPT data points, it is now a relatively simple matter to produce the desired probability distribution using Python's built-in functions (though, of course, the process can be achieved with similar ease in other coding languages).

In the code block below, try adjusting the value of n_{bins}, which will vary the number, and hence width, of the bins used to discretise velocity data, paying particular attention to changes in the distribution when very high and very low values are used. You may also wish to see how the velocity PDF varies when applying different methods for calculating the pseudo-instantaneous velocity—try implementing some of the other methods demonstrated in the previous examples in the code block above.

[3]:
```
# The number of bins into which velocity data is divided
n_bins = 100

# Creating and plotting a suitable histogram showing the velocity probability
# density
fig, ax = plt.subplots()

# Setting 'density = true' gives a PDF; toggle to 'false' for a simple number
# distribution
ax.hist(v_x[j:-j], bins=n_bins, density=True)
ax.set(ylabel='$P(v_x)$', xlabel='$v_x$ (m/s)')

ax.set_xlim(-0.2, 0.2)
```

[3]: (-0.2, 0.2)

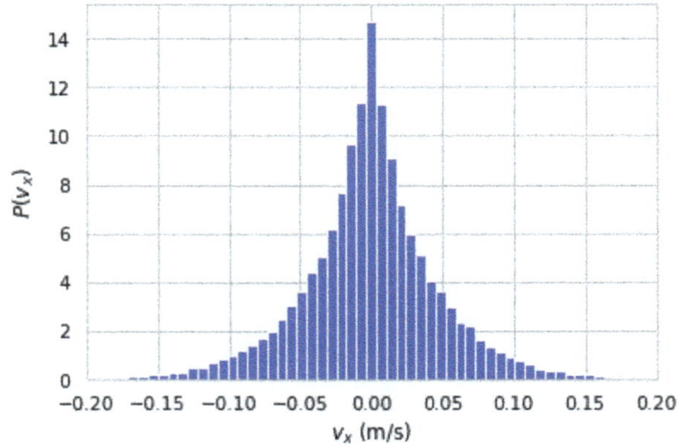

7.4.5 Spatial velocity distributions (velocity vector fields)

One of the most valued applications of PEPT in industry is the visualisation of flow patterns within process equipment. Plots such as those shown in figure 7.14 can provide valuable insight into the functionality of such equipment: for example, in the fluidised bed systems which form our touchstone in this chapter, an axially-asymmetric flow pattern suggests a poorly-designed or obstructed gas distributor [22, 84]. Similarly, for solid–liquid mixing in stirred-tank systems, the presence of multiple distinct vortices may be indicative of an unsuitable impeller design for a given system [85].

The manner in which spatial velocity distributions may be calculated is not dissimilar to the methodology described in section 7.2 for the construction of spatial occupancy or density distributions. Like these, velocity distributions may be computed in 1D, 2D or 3D, though most commonly depth- or azimuthally-averaged two-dimensional fields, such as those depicted in figure 7.14, are used; as such, our main discussion in this section will focus on the creation of such two-dimensional distributions.

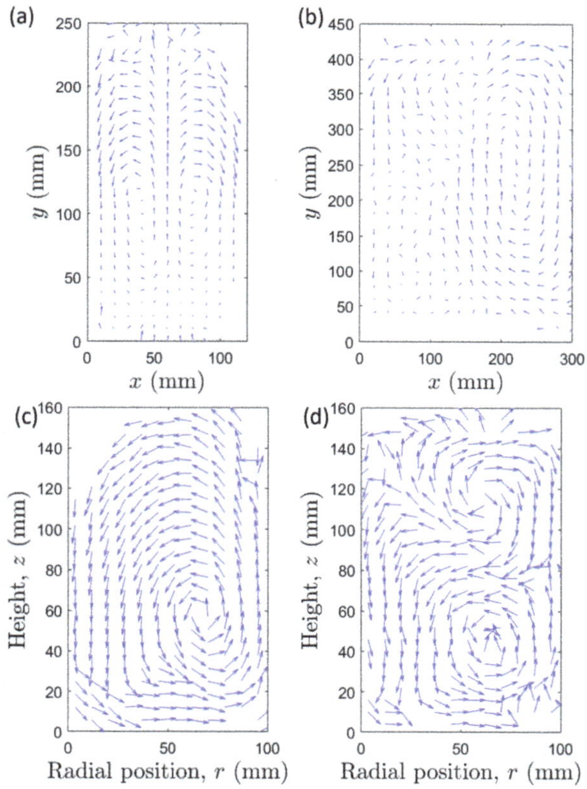

Figure 7.14. Examples of two-dimensional velocity vector fields produced using PEPT data. The top two images represent depth-averaged projections through the Cartesian x–y plane of two different fluidised-bed systems. Panel (a) shows a highly-symmetric flow pattern as would be expected for a well, fluidised bed. Panel (b) shows the case of a poorly-functioning distributor, resulting in asymmetric flow. The bottom two images show azimuthally-averaged flow distributions in the cylindrical polar r–z plane of a stirred tank. Data is shown for (a) a low-viscosity Newtonian fluid and (b) a high-viscosity, non-Newtonian (shear thinning) fluid.

We begin by subdividing the experimental volume into a series of two-dimensional cells in the plane of interest. For each individual cell, i, the velocities corresponding to all N_i data points falling within this cell are recorded, and the average value thereof taken as the mean velocity for that cell, i.e.

$$\bar{v}_i = \frac{1}{N_i}\sum_{j}^{N_i} v_j. \tag{7.24}$$

This analysis may be performed for each spatial velocity component (v_x, v_y, v_z) individually, or for the total velocity ($v_{\text{tot.}} = \sqrt{v_x^2 + v_y^2 + v_z^2}$). The calculation of individual velocities in the x, y and/or z coordinates (or, similarly, the r, θ, and ϕ or r, ϕ and z coordinates if these are more suitable for a given system geometry) allows us to plot velocity vector fields, such as those depicted in figure 7.14, in a chosen plane α–β. In such diagrams, the orientation of the arrow shown represents the direction of flow in the plane of interest, which can be determined as:

$$\varphi = \arctan\left(\frac{v_\beta}{v_\alpha}\right), \tag{7.25}$$

while the length of the arrow represents the magnitude of the velocity in said plane,

$$|v_{\alpha\beta}| = \sqrt{v_\alpha^2 + v_\beta^2}. \tag{7.26}$$

The analysis described above may be trivially extended to the third dimension by considering three-dimensional voxels as opposed to two-dimensional pixels.

As with other previously-discussed quantities, when creating a PEPT velocity (vector) field, one must carefully choose the size of the cells used such that they are large enough to ensure suitable statistics, but small enough that they do not mask the fine detail of the flow fields. For certain systems—e.g. gas-fluidised and vibro-fluidised beds—in which both high and extremely low solids-density regions may coexist [86–91], underpopulated cells may produce noisy data which may adversely affect the interpretation of velocity field data (especially if one's primary interest lies in the more dense regions). In this case, it can be necessary to set a threshold number of data points, or even a threshold number density or solids fraction, for a cell to be considered 'viable' and thus assigned a velocity vector.

In the example exercises following this chapter, we will show the above process in action, and illustrate the effects of varying the cell size used to calculate a flow field.

7.4.6 Interactive example: velocity vector field

The notebook files accompanying this section are available in the supplementary data online at https://iopscience.iop.org/book/978-0-7503-3071-8.

Tutorial: creating velocity vector fields from PEPT data

In this tutorial, we will demonstrate how two-dimensional, depth-averaged velocity vector fields may be extracted from standard PEPT data, exploring in particular the importance of choosing a suitable cell size, dc, to produce both accurate and meaningful data.

Importing data and calculating velocities

As in the previous example, we begin in the code below by importing some data and determining a set of velocity values. Note that, to ensure that we have a suitably large amount of data to ensure good statistics for our two-dimensional plot, the file used here is somewhat larger than those we have used before, and may take a little while to load (and indeed to process in later parts of the exercise).

As with previous tutorials, feel free to play with the parameter *j*, which determines the number of data points over which velocities are calculated, and see what effect it has on your final results.

[1]:
```
# Importing necessary libraries for numerical array manipulation and plotting
import numpy as np
import matplotlib.pyplot as plt

# Set Matplotlib theme (try "darkgrid" too)
import seaborn
seaborn.set_style("whitegrid")
```

[2]:
```
# Loading pre-processed fluidised bed PEPT data
filepath = "Data/FluidisedBedExample3.csv"
full_trace = np.loadtxt(filepath, skiprows=16)

# Extracting time and position data in both the horizontal (x) and vertical (y)
# directions
t = full_trace[:, 0]
x = full_trace[:, 1]
y = full_trace[:, 2]

# For easy plotting, subtracting minima from x and y coordinates so that
# system's left-hand corner lies at (0,0)
x = x - x.min()
y = y - y.min()

# The number of steps before and after the current point over which the velocity
# is calculated
#
# Feel free to adjust this parameter to see its influence on the final output!
j = 3

# Creating empty arrays to hold all velocity values
vx = np.full(len(x), np.NaN)
vy = np.full(len(y), np.NaN)

# Calculating velocity for each point in turn using the central difference
# approach (as discussed in detail in the previous tutorial)
for i in range(j, len(x) - j):
    # Calculating velocity for each point
    vx[i] = (x[i + j] - x[i - j]) / (t[i + j] - t[i - j])
    vy[i] = (y[i + j] - y[i - j]) / (t[i + j] - t[i - j])

print("Total number of points in file: ", len(vx))
```

```
Total number of points in file:  422277
```

Creating a 2D velocity vector field

Now that we have velocities defined for all our data points, our next step is to create a suitable 2D grid, and use the velocity values obtained to determine the mean velocity for each cell thereof, as described in the main text. While there are many possible ways to do this, we have chosen a rather 'old school' method using explicit nested for loops, which—as it does not use too many Python-specific functions—can be easily adapted to almost any programming language.

In the code below, the parameter *dc* defines the size of the cells used for our calculations. Take some time to explore the effect of using different values of `dc` on the velocity vector field finally shown.

[3]:
```
# Some values in the division below are zero, let them silently yield 0/0 = NaN
np.seterr(divide='ignore', invalid='ignore')

# The size, in mm, of the cells into which you want to subdivide the system
#
# Note that, for simplicity, we use square cells of size dc x dc, but in
# reality other cell geometries may also be used.
dc = 10

# Working out the number of cells needed to span the system in both the x and y
# directions
nx = int(np.ceil(x.max() / dc))
ny = int(np.ceil(y.max() / dc))

# Creating a pair of suitably-sized 2D arrays to store resultant velocity field
vx_grid = np.zeros((ny, nx))
vy_grid = np.zeros((ny, nx))

# An additional array to store the number of locations falling in each cell
n_grid = np.zeros((ny, nx))

# Looping through all particle locations in turn
for n in range(j, len(vx) - j):
    # For each location, find the horizontal and vertical indices of the cell in
    # which the particle falls
    ix = int(x[n] / dc)
    iy = int(y[n] / dc)

    # Increment the number of locations found in the current cell...
    n_grid[iy, ix] = n_grid[iy, ix] + 1

    # ...and adding the relevant velocities to the current sum
    # for the current cell
    vx_grid[iy, ix] = vx_grid[iy, ix] + vx[n]
    vy_grid[iy, ix] = vy_grid[iy, ix] + vy[n]

# Finding mean velocities
vx_grid = vx_grid / n_grid
vy_grid = vy_grid / n_grid
```

Plotting the results

From the above code segment, we now have two, two-dimensional grids containing average *x*- and *y*-velocities for all regions of the system's *x*–*y* plane. In the final part of this tutorial, we use these values to plot a velocity vector field, which will allow us to visualise the flow pattern within the system.

In the code below, you can toggle the 'norm_on' parameter true or false to choose whether velocities are plotted as unit length arrows (norm_on = True) or whether they are plotted such that the lengths of the arrows are representative of the local velocity magnitude (norm_on = False). While the latter choice offers more information, if velocity magnitudes vary greatly between different regions of the system, the flow pattern may potentially become difficult to see in low-velocity regions. The former choice does not offer velocity magnitude information, but often allows easier visualisation of a system's flow pattern. Try turning norm_on on and off and see which you think is more appropriate for the current example.

[4]:
```
# Choosing whether to use normalised (equal-length) arrows or arrows whose
# length is representative of |v|
norm_on = False

# Defining a suitable axis scale for plotting
scale_x, scale_y = np.meshgrid(
    np.arange(0, nx * dc, dc),
    np.arange(0, ny * dc, dc),
)

if norm_on == True:
    # Finding relevant normalisation constants for each cell in grid
    norm = np.sqrt(vx_grid ** 2 + vy_grid ** 2)
else:
    norm = 1

fig, ax = plt.subplots(figsize=(9, 8))

ax.quiver(scale_x, scale_y, vx_grid / norm, vy_grid / norm)

ax.set(ylabel='$y$ (mm)', xlabel='$x$ (mm)')
ax.set_aspect(0.66)
```

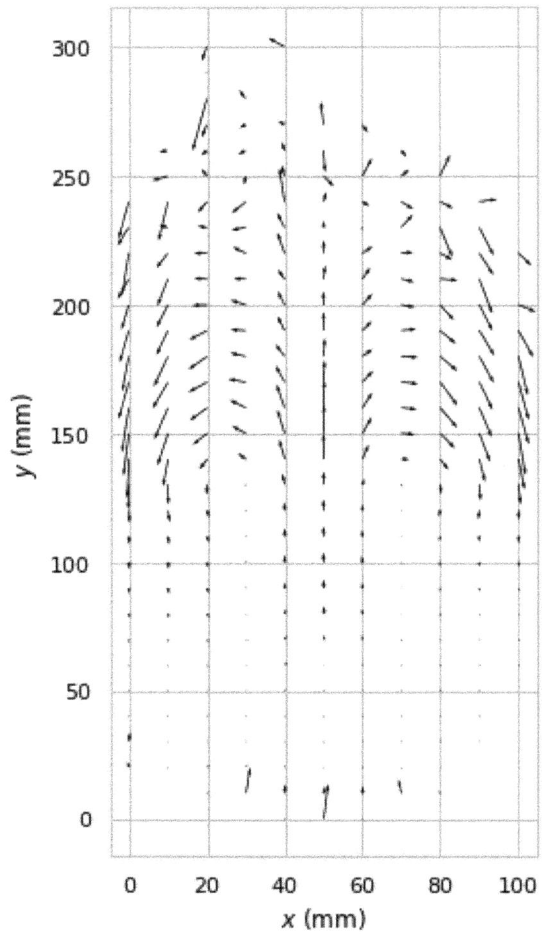

7.4.7 Velocity autocorrelation and cross-correlation

More often used in 'pure science' studies than in industrial applications, a knowledge of the velocity autocorrelation function (VACF) is—due in part to its relation to diffusive motion and thus transport equations [92, 93]—of considerable value to the development of theoretical models of granular and multiphase systems. In simple terms, the VACF is a measure of how quickly a given particle's motion becomes decorrelated from its initial velocity—i.e. how quickly a particle 'forgets' its initial conditions. The specific form of a VACF can give much insight into the dynamics of a system: a classical, 'ideal' gas will exhibit molecular chaos, meaning that one may expect the VACF thereof to decay very rapidly to zero; a dissipative, particulate medium, meanwhile, may retain its 'memory' for much longer [94–97]. The presence of periodic motion on a relatively long timescale may show evidence of vortices or convection rolls within a system [98], while periodicity on a shorter timescale can indicate the presence of 'caging' effects [12, 97].

The velocity autocorrelation function, Z, is typically defined as [99]:

$$\Psi_i(t) = \langle v_i(t + s)v_i(s)\rangle, \tag{7.27}$$

where the index i represents the coordinate along which the velocity is measured ($i = x, y, z$ for the standard Cartesian coordinate system). In the above equation, s represents some arbitrary initial time, and t the 'current' time under consideration; the angled brackets, $\langle\rangle$, represent an ensemble average over all initial conditions. Of course, in PEPT data—since we typically only directly track a small subset of the particles within the system—we cannot directly take an ensemble average. Under the assumption of ergodicity, however, we can, in essence, split a long, single-particle trajectory into a number of individual sections or 'analysis windows', and treat each of these sections as a distinct particle trajectory[5]. If the ergodic assumption holds true, it should not be relevant whether the distinct trajectories belong to different particles, or the same particle at different times.

Once the PEPT velocity data—calculated as described in section 7.4.1—has been divided into a suitable series of N_w distinct time windows, an individual velocity autocorrelation for the nth time window can then be computed as:

$$\psi_i^n = v_i(t + s^n)v_i(s^n), \tag{7.28}$$

where s^n is the start time for the nth trace. The full velocity autocorrelation function can then be calculated as the average over all individual VACFs, i.e.

$$\Psi_i = \frac{1}{N_w}\sum_n^{N_w} \psi_i^n. \tag{7.29}$$

It is important here to recall from chapter 2 that PEPT—due to its reliance on the stochastic process of radioactive decay—does not produce data at regular time-stamps, but rather at random intervals. As such, in order to accurately average the time-series data such as that we are handling here, we must first bin the data points belonging to each individual VACF, ψ^n; the time interval used for binning can be defined by the user, but must remain consistent for all ψ^n.

Examples of velocity autocorrelation functions for both the fluidised-bed and stirred-tank systems explored in the previous section are shown in figure 7.15. The VACF for the fluidised bed system (panel (a)) is observed to decay rapidly to a value ($\Psi \approx 0$), where it remains for the remainder of the trace; this behaviour suggests relatively chaotic, uncoordinated dynamics, implying that the bed is well-fluidised. The stirred tank data (panel (b)), meanwhile, shows clear, regular oscillations in Ψ with time, suggesting the presence of vortices (which, of course, we can confirm from the velocity vector fields shown in figure 7.14!).

It is finally valuable to note that in addition to the *auto*correlation function, one may also calculate *cross*-correlations by using a minor adaptation to equation (7.30). Specifically, an individual cross-correlation $\psi_{\alpha\beta}$ for velocities in two distinct directions α and β ($\alpha, \beta = x, y, z, \alpha \neq \beta$) can be determined as:

---

[5] Note that these sections can be overlapping; indeed, it is common practice to treat each point in a data set as the start of a new trace.

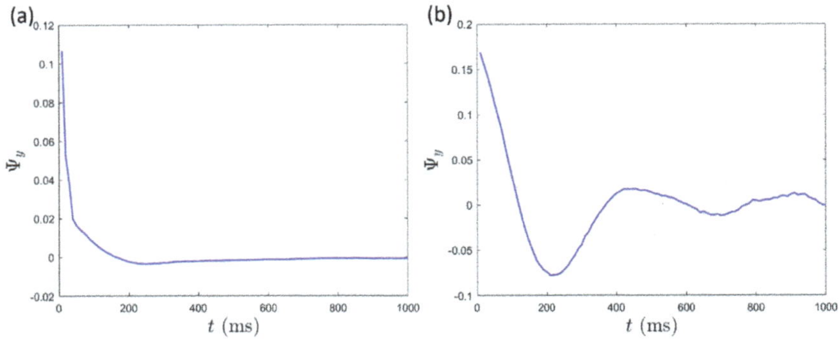

Figure 7.15. Examples of PEPT-measured velocity autocorrelation functions for the vertical (y) coordinate in (a) a gas-fluidised bed and (b) a stirred tank.

$$\psi_{\alpha\beta}^{n} = v_{\alpha}(t + s^{n})v_{\beta}(s^{n}). \tag{7.30}$$

The full, pseudo-ensemble-averaged cross-correlation $\Psi_{\alpha\beta}$ can then be determined by averaging over all individual $\psi_{\alpha\beta}^{n}$, as before.

7.4.8 Interactive example: velocity autocorrelation

The notebook files accompanying this section are available in the supplementary data online at https://iopscience.iop.org/book/978-0-7503-3071-8.

Tutorial: creating velocity autocorrelation functions from PEPT data

In this tutorial, we will demonstrate how to create a velocity autcorrelation function (VACF) from a single PEPT trace. We begin in the code below by importing some data and determining a set of velocity values, in a manner which by now should be very familiar.

[1]:
```
# Importing necessary libraries for numerical array manipulation and plotting
import numpy as np
import matplotlib.pyplot as plt

# Set Matplotlib theme (try "darkgrid" too)
import seaborn
seaborn.set_style("whitegrid")
```

[2]:
```python
# Loading relevant data
filepath = "Data/FluidisedBedExample.csv"
full_trace = np.loadtxt(filepath, skiprows=16)

# Extracting time and position data in both the horizontal (x) and vertical (y)
# directions
t = full_trace[:, 0]
x = full_trace[:, 1]
y = full_trace[:, 2]

# The number of steps before and after the current point over which the velocity
# is calculated
#
# Feel free to adjust this parameter to see its influence on the final output!
j = 3

# Creating empty arrays to hold all velocity values
vx = np.full(len(x), np.NaN)
vy = np.full(len(y), np.NaN)

# Calculating velocity for each point in turn using the central difference
# approach (as discussed in detail in the previous tutorial)
for i in range(j, len(x) - j):
    # Calculating velocity for each point
    vx[i] = (x[i + j] - x[i - j]) / (t[i + j] - t[i - j])
    vy[i] = (y[i + j] - y[i - j]) / (t[i + j] - t[i - j])
```

Splitting data into time windows and calculating the VACF

Now that we have velocity information for a large number of consecutive data points, our next task is to subdivide our experimental data into a series of time windows and, for each window, calculate an individual VACF. For each time window, we must choose both the total length of the window to use, T, as well as the size, dt, of the bins into which our data is divided.

Try adjusting the values of the window length T and bin size dt used in the code segment below, and observe the effect that these parameters have both on the quality of the data produced, as well as the computational time required to produce it!

[3]:
```python
# The duration, in ms, of each individual trace used to calculate an individual
# VACF
window = 1000
# The size, in ms, of the bins into which data will be divided
dt = 10;
# Creating a vector to store VACF values
vacf = np.zeros(int(np.ceil(window / dt)))
```

```
# Creating a second vector of the same length to record the total number of data
# falling in each bin, for averaging purposes
n = np.zeros(int(np.ceil(window / dt)))

# Setting the first relevant timestep as the first value of 's', the arbitrary
# start time of a given trace (see main text)
s = t[j]

# Recording also the index corresponding to this timestamp
s_idx = j

# Cycling through all data points with an assigned velocity
for i in range(1, len(x)):
    # Finding the current time relative to the relevant trace start point
    t_rel = t[i] - s

    # Calculating the current value of the VACF
    vacf_current = vy[i] * vy[s_idx]

    # Putting the value in the relevant time bin
    k = int(t_rel / dt)

    # Ensure relative time falls into the current window and the current time
    # is a real number
    if k < len(vacf) and np.isfinite(vacf_current):
        # If so, adding value to current bin
        vacf[k] = vacf[k] + vacf_current

        #...and also incrementing the number of data points in this bin
        n[k] += 1

    # Checking to see if the current time falls outside the desired trace length
    if t_rel >= window:
        # ...and if so, starting the next trace using the end of the current
        # trace as its starting point...
        s = t[i]

        # ...and saving also the relevant index
        s_idx = i;

# Averaging data
vacf = vacf / n

# Plotting
fig, ax = plt.subplots()

# Creating suitable axis
scale = np.arange(0, window, dt)
ax.plot(scale, vacf)

ax.set(ylabel='$\Psi$', xlabel='$t$ (ms)');
```

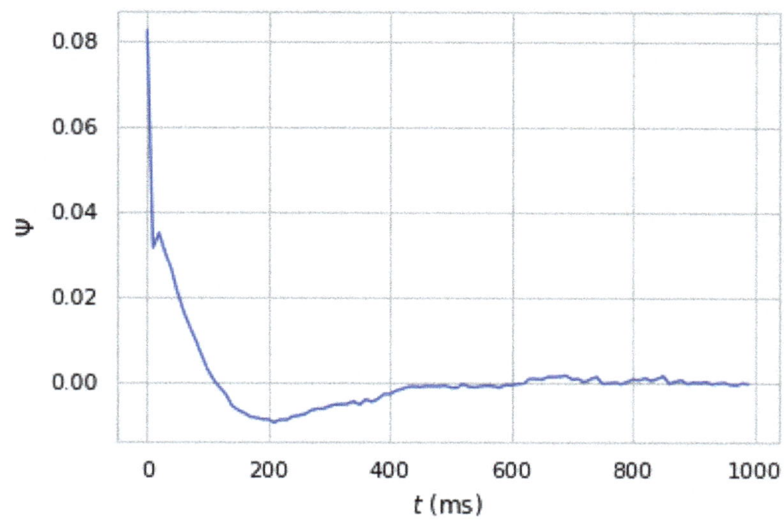

Interpreting the VACF

Based on what we have learned in the main text, what does the form of this VACF tell you about the dynamics of our fluidised bed? Does it produce a reasonable approximation of molecular chaos (and if not, what does this say about the applicability of kinetic-theory-like models to fluidised beds)? Is there evidence of any convective motion on the timescale you have chosen? What if you look at a longer timescale? Finally, what happens if in the code segment above you change the value of `VACF_current` to use x velocity instead of y velocity? Does the form of the VACF change? Why might this be?

7.5 Diffusion, dispersion and mixing

As we have previously mentioned, one of the major industrial applications of PEPT is to monitor and quantify mixing dynamics in process equipment, be it by tracking the dispersion of active ingredients through a pharmaceutical product during secondary manufacture, or exploring the mixing of reactants in the chemical industry[6]. In section 7.3 we explored manners in which the time-averaged, steady-state mixing or segregation of a system may be measured. In the present section, we explore instead parameters from which we may gain an indication of the *rate* of mixing within a system. We do so by using PEPT to extract data relating to the diffusion and/or dispersion of particles through a system, which may in fact be measured in two distinct manners, both of which we describe in detail in the following subsections.

7.5.1 Mean squared displacement and the self-diffusion coefficient

The variation with time of the mean squared displacement (MSD) of a particle from an arbitrary initial position provides valuable insight into the rate of mixing which may be expected for a system of such particles—the more quickly a particle becomes

[6] The diverse industrial applications of PEPT are discussed in detail in chapter 8.

decoupled from its initial position, the more rapidly one may expect mixing to occur. As with the calculation of the VACF described in the preceding section, the mean squared displacement of a particle over a given duration may be calculated by subdividing a single, long PEPT trace into a series of individual, shorter traces, each of the desired length. Under the assumption of ergodicity, each of these individual traces may be considered as a distinct particle trajectory. For each such trajectory, one may thus take the first point in the trace as an initial position $\vec{r}(0)$, and determine the displacement of the particle at an arbitrary subsequent timestamp t as $|\vec{r}(t) - \vec{r}(0)|$. From multiple particle trajectories, the mean squared displacement may then be calculated as [100]:

$$MSD = \langle |\vec{r}(t) - \vec{r}(0)|^2 \rangle \tag{7.31}$$

where, as previously, the ensemble average denoted by the angled brackets, $\langle \rangle$, may—due to the assumption of ergodicity—be approximated by averaging over all individual traces obtained from a given PEPT data set. In the tutorial following this section, we put the above into practice, showing how mean squared displacement (MSD) data can be extracted from a real PEPT data set.

MSD plots produced in the manner described above can provide considerable insight into the dynamics of a given system. An example of an MSD curve extracted from PEPT data may be seen in figure 7.16. The curve shown illustrates ballistic motion [16, 101, 102] at small t, followed by a near-linear increase for large t, indicative of normal, Fickian diffusion [103, 104]—in other words, in this particular instance, our system is seemingly behaving similarly to a classical liquid, suggesting that it is very well fluidised. For still more strongly-excited, highly fluidised systems, one may observe ballistic motion continuing at larger t, indicating that a system is behaving like an ideal, non-interacting gas [79, 105]. Conversely, if our system is in a solid-like, jammed state [106] one will observe the mean squared displacement to rapidly taper off to a horizontal line [12]. In addition to this, one may also observe from plots of the MSD various forms of 'anomalous diffusion', for example, subdiffusive or superdiffusive behaviour [107, 108], or the presence of a plateau

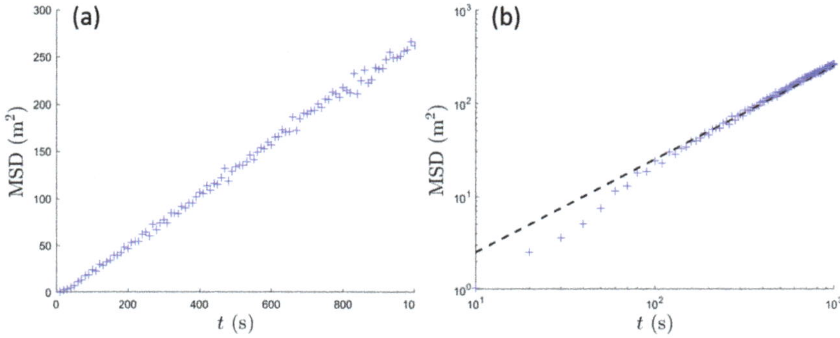

Figure 7.16. Example of PEPT-measured MSD (blue crosses) for a gas-fluidised bed shown on (a) a linear scale and (b) a logarithmic scale. In panel (b), the solid line represents a gradient $\alpha = 1$, corresponding to 'normal' or 'Fickian' diffusion.

region separating the ballistic and normally-diffusive regimes, indicating the presence of caging effects [12, 109].

One of the main uses of MSD measurements—in PEPT studies as well as more generally—is to determine a value of the self-diffusion coefficient, D, a single scalar value which can help us to characterise the diffusivity of a given system or species therein (i.e. the rate at which particles of said species are likely to 'spread out' into their surroundings). This knowledge, in turn, can be very useful as it provides an indication of the rate of mixing one might expect in such a system. With a known MSD, the diffusion coefficient may be calculated using Einstein's relation [99] which, for a three-dimensional system in the limit of large t can be expressed as:

$$D = \lim_{t\to\infty}\frac{1}{6t}\langle|\vec{r}(t) - \vec{r}(0)|^2\rangle. \tag{7.32}$$

In other words, the diffusion coefficient is proportional to the steady gradient achieved by plots such as those in figure 7.16 after a suitably long time.

It is important to note here that the value of D obtained as described above is only strictly valid for systems in which motion is exclusively diffusive—i.e. the displacement of particles is due solely to the random, pseudo-Brownian motion of particles. If additional advective motion (e.g. due to the presence of convective flow, vortices, etc) is present within the system and thus may contribute to the displacement of a particle from its initial position, then equation (7.32) will not yield a meaningful value of the diffusion coefficient. Nonetheless, with a little care, it is possible to extract values of D even for such systems. In order to do so, one must first construct a velocity vector field as described in section 7.4.5. With a known spatial distribution of velocities within the system one may, when calculating the MSD of a particle, subtract off the expected motion due to mean flow within the system, thus leaving only diffusive motion. It should be noted, however, that to achieve this with any degree of accuracy, one requires a spatial velocity distribution with very high spatial resolution (i.e. a very fine mesh size), and thus a very a significant amount of data.

7.5.2 Interactive example: determining the mean square displacement

The notebook files accompanying this section are available in the supplementary data online at https://iopscience.iop.org/book/978-0-7503-3071-8.

Tutorial: measuring mean squared displacement using PEPT data

In this tutorial, we will demonstrate how to use PEPT data to determine the mean squared displacement (MSD) of particles. For this calculation, all we require is time and position data, which we import and extract below:

[1]:
```
# Importing necessary libraries for numerical array manipulation and plotting
import numpy as np
import matplotlib.pyplot as plt

# Set Matplotlib theme (try "darkgrid" too)
import seaborn
seaborn.set_style("whitegrid")
```

[2]:
```
# Loading relevant data
filepath = "Data/FluidisedBedExample.csv"
full_trace = np.loadtxt(filepath, skiprows=16)

# Extracting time and position data in both the horizontal (x) and vertical (y)
# directions
t = full_trace[:, 0]
x = full_trace[:, 1]
y = full_trace[:, 2]
```

Splitting data into time windows and calculating the MSD

The methodology for computing MSD follows a very similar formula to the calculation of the velocity autocorrelation function illustrated in our previous tutorial. As in this prior example, our first task having imported the relevant data is to subdivide this data into a series of individual time windows, each of length T, and bin the data pertaining to each in bins of a suitable width dt.

As in the previous tutorial, try adjusting the values of the window length T and bin size dt used in the code segment below, and observe the effect that these parameters have both on the quality of the data produced, as well as the computational time required to produce it!

[3]:
```
# The duration, in ms, of each individual trace used to calculate an individual
# MSD
window = 1000

# The size, in ms, of the bins into which data will be divided
dt = 10

# Creating a vector to store MSD values
msd = np.zeros(int(np.ceil(window / dt)))

# Creating a second vector of the same length to record the total number of data
# points falling in each bin, for averaging purposes
n = np.zeros(int(window / dt))

# Setting the first relevant timestep as t(0)
t0 = t[0]

# Recording also the index corresponding to this timestamp
t0_idx = 0

# Cycling through all data points with an assigned velocity
for i in range(1, len(x)):
    # Finding the current time relative to the relevant trace start point
    t_rel = t[i] - t0

    # Calculating the current displacement distance from the relevant starting
    # position for this trace
    msd_current = x[i] - x[t0_idx]

    # Putting the value in the relevant time bin
    bin_index = int(t_rel / dt)
```

```
    # Ensure relative time falls into the current window and the current time
    # is a real number
    if bin_index < len(msd) and np.isfinite(msd_current):
        # If so, adding the square of the current displacement value to
        # current bin
        msd[bin_index] = msd[bin_index] + msd_current * msd_current

        #...and also incrementing the number of data points in this bin
        n[bin_index] = n[bin_index] + 1

    # If the current time falls outside the desired trace length...
    if t_rel >= window:
        # ... starting the next trace using the end of the current trace as its
        # starting point...
        t0 = t[i]

        # ...and saving also the relevant index
        t0_idx = i

# Finding the mean for all bins across all traces
msd = msd / n

# Adding an expected (linear) trend for normal diffusion
xx = np.arange(10, 1000, 0.1)
yy = 0.45 * xx

# Plotting
fig, ax = plt.subplots(1,1)

# Creating suitable axis
scale = np.arange(0, window, dt)
ax.scatter(scale, msd)
ax.plot(xx, yy)

ax.set(
    xscale='log',
    yscale='log',
    xlabel='$t$ (ms)',
    ylabel='MSD (mm$^2$)',
    xlim = [100, 1000],
    ylim = [30, 500],
);
```

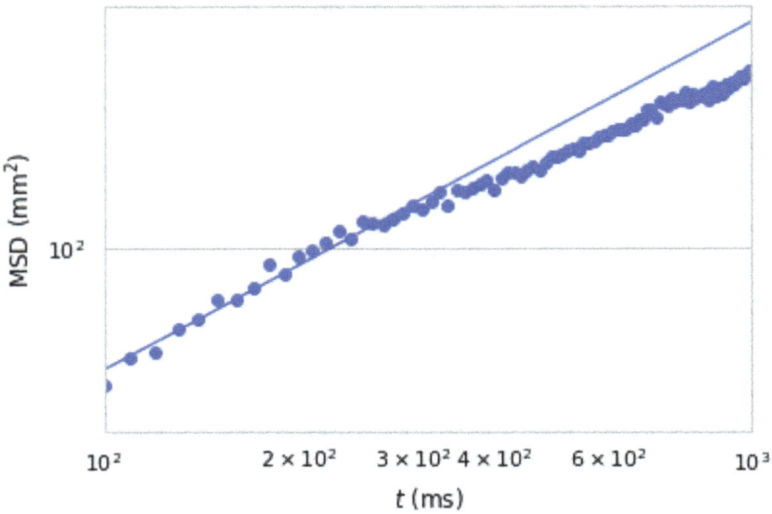

Interpreting the MSD

Based on what we have learned in the main text, what does the form of this VACF tell you about the dynamics of our fluidised bed? Is it normally diffusive, subdiffusive, or superdiffusive? Does this align with your expectations for a fluidised bed? Is there a plateau, or any other evidence of caging effects or convective motion on the timescale of interest?

Try looking at the data for the y direction as well as the x (this can be done by altering the calculation of `msd_current` in the code above). What does the difference between the plots obtained tell you about the dynamics of the system?

Note that in order to fully observe the plots for the y direction you may need to comment out or alter the lines

```
xlim = [100, 1000],
```

and

```
ylim = [30, 500],
```

which restrict the scales in the figure to chosen ranges.

7.5.3 Dispersion and the 'mixer effectiveness'

In the preceding section, we showed how the MSD behaviour of a tracer may be used to provide an estimate of the self-diffusion coefficient, D, for a given particle species, emphasising at the end of the section the complications arising when motion within the system of interest is not strictly *diffusive*, but rather *dispersive*. In this section we demonstrate a method, originally proposed by Martin, Seville and Parker [110], through which the dispersion of particles in such systems may be meaningfully quantified.

Consider a group of tracers in a dynamic system, all possessing similar, known initial positions at some arbitrary time t. If we then follow the motion of these tracers, and then record their positions again at some subsequent time t', we can gauge the extent of dispersion experienced by these particles by how much they have spread out from one another during the interval $t' - t$: if the particles all remain very close to one another, as illustrated in figure 7.17(a), then clearly little dispersion has occurred. Conversely, if the particles have spread out significantly, then one can assume the presence of significant dispersion within the system—or at least the region thereof considered in our measurement (figure 7.17(b)). With known start and end positions for all of the tracers, the dispersion may be quantified by determining the variance on the mean location, i.e.:

$$\sigma^2 = \frac{1}{N}\sum_{i=1}^{N}(x_i - \bar{x})^2 + (y_i - \bar{y})^2 + (z_i - \bar{z})^2. \tag{7.33}$$

In the above, $\bar{x}$ (and similarly $\bar{y}$, $\bar{z}$) represents the mean x (y, z) position of the 'spread out' particles at time t', with x_i, y_i, z_i representing the individual position at t' of the ith particle considered. For the highly idealised case of perfectly uniform flow (particles do not spread out at all, → zero dispersion), $\sigma = 0$, with larger values of σ corresponding to stronger dispersion.

Of course, using PEPT (or indeed any other 3D imaging technique) we cannot (for most systems) know the positions of all particles at all times. Nonetheless, under the assumption of ergodicity, we can consider instead a number of distinct traces starting at different times, t_j, from similar positions. As long as the duration $\Delta t_j = t_j' - t_j$ over which we record the particle motion remains consistent, an average over a number of such traces is (under the ergodic assumption) exactly equivalent to the measurement described above.

An effective manner in which to 'group' particles possessing similar starting points is to subdivide the experimental volume into three-dimensional cells (a process with which we are by now hopefully very familiar!). If the cells are suitably small, then all particles falling within a given cell can be considered to belong to the same initial group, and used to determine the dispersion as per equation (7.33). This subdivision of the experimental volume carries the additional advantage that it allows us to sensitively investigate the spatial

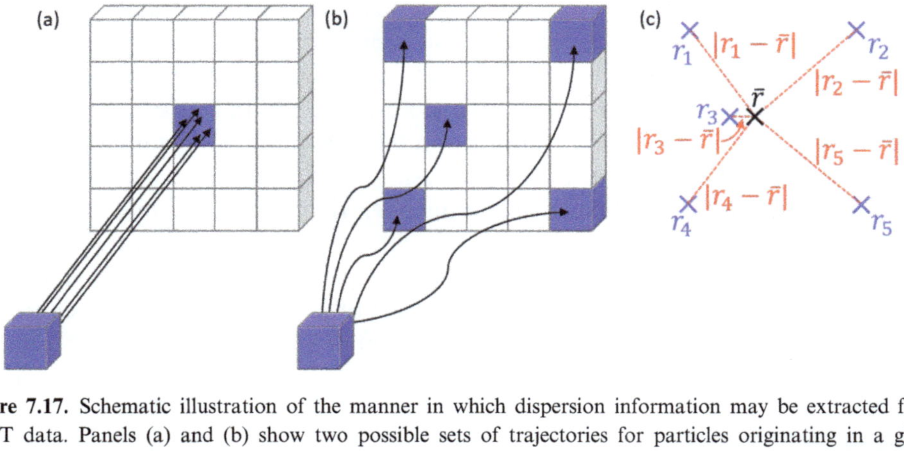

Figure 7.17. Schematic illustration of the manner in which dispersion information may be extracted from PEPT data. Panels (a) and (b) show two possible sets of trajectories for particles originating in a given experimental volume: the former illustrates the case of low dispersion (all particles remain closely grouped), the latter the case of higher dispersion (particles spread out significantly). Panel (c) shows how the end points of the traces from (b) may be used to determine a value of the dispersion as per equation (7.33).

variation of dispersion throughout the system and thus observe, for example, if certain regions of the system are stagnant or otherwise producing undesirably low dispersion. As mentioned previously, in many industrial systems, the acquisition of uniform mixing throughout a system is of paramount importance, meaning that such information may prove highly valuable in modifying the design and/or operation of process equipment.

If an assessment of a system's mixing abilities is indeed a matter of interest, one may use the dispersion parameter described above to define another scalar index, known as the 'mixer effectiveness' (ME), which—as its name implies—characterises the overall mixing performance of the system as a whole. In order to compute ME, in addition to a knowledge of the dispersion σ for each cell within the system, we need to record also the number of passes, n_p, made by the tracer through each cell. This can be simply determined by recording each time the tracer enters and subsequently leaves a given cell, k, incrementing the value of $n_{p,k}$ each time such an event occurs[7].

A knowledge of n_p for each volume element of our system allows us to establish not only the amount of dispersion produced in a given region, but also how much this particular region contributes to the overall mixing process—i.e. if a region both induces strong dispersion *and* is regularly visited by particles, then it will provide a stronger contribution than another region which induces a similar amount of dispersion but through which particles pass more infrequently. The mixer effectiveness is thus defined as:

$$\mathrm{ME} = \frac{1}{N_p} \sum_{k=1}^{N_c} \sigma_k^2 n_{p,k}, \tag{7.34}$$

where N_p is the total number of passes through all N_c cells within the system, and σ_k, $n_{p,k}$ are as described above.

One key consideration to take into account when determining either σ or ME is the length of time, $\Delta t = t' - t$, over which particle motion should be considered—i.e. how long our individual traces should be: too short a time, and the particles will barely move at all, thus giving $\sigma \sim 0$ for any possible case. Too long a duration, and the results may be misleading; for example, in systems in which recirculatory motion is present, if Δt is similar to the characteristic recirculation time, it is possible that a tracer may complete an entire circuit around the system, and thus register a σ value close to zero, despite having moved a considerable distance! In between these extremal cases, however, the user has great flexibility in their choice of Δt. A relatively short Δt will provide information on the micro- or meso-scopic level, and insight into the local forces within the system. A longer Δt, meanwhile, will provide information regarding macroscopic processes. Both of these choices, however, can provide valuable knowledge regarding the system of interest. In the example at the end of this chapter, you will have the chance to directly explore the effect of Δt on dispersion measurements.

It is finally worth noting that the term 'dispersion' is in fact used ambiguously in the PEPT literature. In addition to the definition described above, a number

[7] It is important to note that this value is *not equivalent* to the number occupancy described in section 7.2, which would increment the value of a given cell for each new location within said cell, even if the tracer has not moved from the cell between locations.

of prior works [111–113] use a somewhat simpler measure of dispersion, defined as:

$$D_i = \frac{1}{N-1} \sum_{k=1}^{N-1} \frac{(i_{k+1} - i_k)^2}{t_{k+1} - t_k}, \; i = x, y, z \tag{7.35}$$

where i_k and i_{k+1} are the positions, in the ith coordinate, of the tracer at times t_k and t_{k+1}.

7.5.4 Interactive example: spatial dispersion distribution

The notebook files accompanying this section are available in the supplementary data online at https://iopscience.iop.org/book/978-0-7503-3071-8.

Tutorial: measuring dispersion using PEPT data

In this tutorial, we will demonstrate how to use PEPT data to measure and quantify dispersion. As the determination of dispersion is a rather computationally-intensive process, in the example code below we only consider a single volume element of our system, rather than determining values of σ for all cells, as discussed in the main text. By focussing on a single cell, we not only avoid spending long times waiting for the code below to run, but we are also able to gain a deeper insight into the meaning of the dispersion, σ. The methodologies shown here may be trivially expanded to allow a full system to be analysed—feel free to try this for yourself!

Importing data

As always, our first step is to actually read in a data file to work with, and extract the relevant basic parameters. In this instance, as with our MSD calculations in the previous tutorial, we only need to know values of time and position, and can avoid the necessity of calculating particle velocities.

[1]:
```
# Importing necessary libraries for numerical array manipulation and plotting
import numpy as np
import matplotlib.pyplot as plt

# Set Matplotlib theme (try "darkgrid" too)
import seaborn
seaborn.set_style("whitegrid")
```

[2]:
```
# Loading relevant data
filepath = "Data/FluidisedBedExample.csv"
full_trace = np.loadtxt(filepath, skiprows=16)

# Extracting time and position data for all Cartesian coordinates
t = full_trace[:, 0]
x = full_trace[:, 1]
y = full_trace[:, 2]
z = full_trace[:, 3]

# Re-centering all traces such that they start at 0 - this makes subsequent
# calculations and plots clearer
x = x - x.min()
y = y - y.min()
z = z - z.min()
```

Visualising dispersion

In the section of code below, we choose the region of the system of interest for which we want to measure the dispersion as well as, crucially, the size of this region, and the duration, Δt, over which we wish to let particle travel.

Before we calculate the dispersion 'in anger', it is worth firstly taking the chance to look a little bit more closely at our data—what does the particle motion actually look like in this region? How does the 'real picture' compare to the simplified schematic depicted in the main text? How does this picture change as we adjust the above-mentioned parameters?

In the code below, you can choose to vary all of the above parameters. For each choice, you can then plot the trajectories followed by the particles falling within this range, allowing you to make a more informed decision as to whether these choices seem sensible for calculating the final dispersion value.

[3]:
```
# The duration, in ms, of each individual trace used to calculate an individual
# VACF
dt = 200

# Defining the size and position of our cell of interest in mm
# For simplicity, we choose a cubic cell of side dc
dc = 4

# The x, y and z coordinates corresponding to the centre of the region of
# interest. For simplicity, we start with a position right at the centre of the
# system
xc = x.max() / 2
yc = y.max() / 2
zc = x.max() / 2

# Parameter to record the number of traces found in the chosen range
# Initiated with zero, will be incremented in the for loop below as viable data
# points are found
nt = 0

# Pair of growable lists to append indices of all start and end points
start_ind = []
end_ind = []

# Cycling through all data points
for i in range(len(t) - 1):
    # Seeing if the current point falls within the chosen range
    # Checking first in x direction
    if x[i] >= xc - dc / 2 and x[i] <= xc + dc / 2:
        # Checking next in y-direction
        if y[i] >= yc - dc / 2 and y[i] <= yc + dc / 2:
            # Finally, checking in z-direction
            if z[i] >= zc - dc / 2 and z[i] <= zc + dc / 2:
                # Incrementing the trace number
                nt = nt + 1

                # Finding number of points in trace between t and t'
                n = 0
                while(t[i + n] - t[i] < dt):
                    n = n + 1
```

```
                # Adding indices of start and end points to relevant lists
                start_ind.append(i)
                end_ind.append(i + n)

print(f"{nt} traces found in chosen range.")

# Plotting all traces found in 3D
fig = plt.figure()
ax = fig.add_subplot(projection='3d')

for i in range(len(start_ind)):
    ax.plot(
        x[int(start_ind[i]):int(end_ind[i])],
        y[int(start_ind[i]):int(end_ind[i])],
        z[int(start_ind[i]):int(end_ind[i])],
    )

ax.set(xlabel='$x$ (mm)', ylabel='$y$ (mm)', zlabel='$z$ (mm)')

# For potentially improved clarity, also plotting in 2D!
fig, ax = plt.subplots()
for i in range(len(start_ind)):
    ax.plot(
        x[int(start_ind[i]):int(end_ind[i])],
        y[int(start_ind[i]):int(end_ind[i])],
    )

ax.set(xlabel='$x$ (mm)', ylabel='$y$ (mm)')
```

```
6 traces found in chosen range.
```

```
[3]: [Text(0.5, 0, '$x$ (mm)'), Text(0, 0.5, '$y$ (mm)')]
```

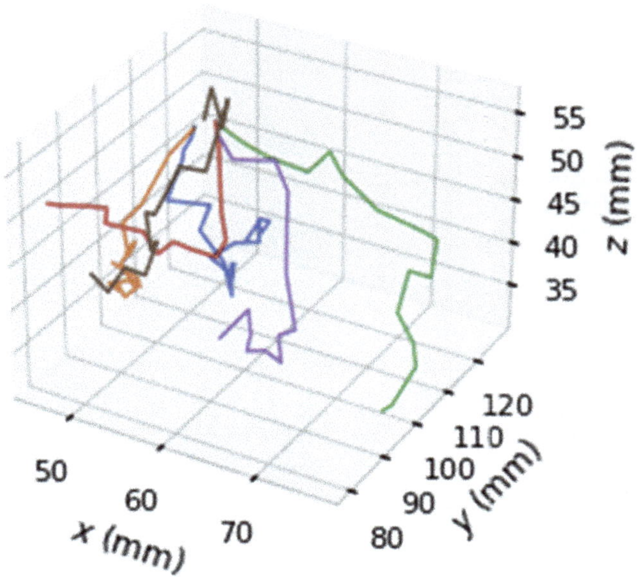

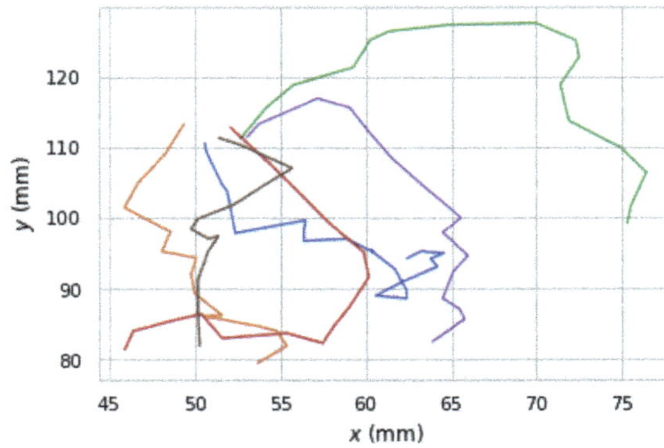

Calculating dispersion

In the above code, you have had the chance to look at the trajectories you will be using for your calculations. Do they look reasonable?

If so, since we already know the indices corresponding to the end points of all our *nt* individual traces, it is fairly simple now to calculate the dispersion as the sum of the variances on these positions for the *x*, *y* and *z* positions (see main text).

See how the dispersion index varies as you vary the trace length, the cell size and various other parameters.

[4]:
```
# Converting endpoint indices into endpoint position values for all 3 Cartesian␣
↪coordinates
# Creating vectors to store values
end_x = np.zeros(len(end_ind))
end_y = np.zeros(len(end_ind))
end_z = np.zeros(len(end_ind))

# Filling vectors with relevant points
for i in range(len(end_ind)):
    end_x[i] = x[int(end_ind[i])]
    end_y[i] = y[int(end_ind[i])]
    end_z[i] = z[int(end_ind[i])]

# Finding mean position values
x_bar = np.mean(end_x)
y_bar = np.mean(end_y)
z_bar = np.mean(end_z)

# Creating a parameter to sum all individual variance values as per equation
running_sum = 0

# Calculating dispersion as per the main text
for i in range(len(end_ind)):
    x_var = (end_x[i] - x_bar) * (end_x[i] - x_bar)
    y_var = (end_y[i] - y_bar) * (end_y[i] - y_bar)
    z_var = (end_z[i] - z_bar) * (end_z[i] - z_bar)
    running_sum = running_sum + x_var + y_var + z_var

dispersion = running_sum / len(end_ind)

print("Dispersion index =", dispersion)
```

```
Dispersion index = 214.5091666666667
```

7.6 Energy and the granular temperature

7.6.1 Defining the granular temperature

As discussed in chapter 1, the dissipative nature of granular and multiphase systems means that their dynamics differ significantly from those of classical, molecular systems [49, 114]. As, at least for the moment, our understanding of classical media far outstrips our knowledge of particulate media, it is highly desirable, where possible, to draw analogy between the two, such that we can attempt to use our existing knowledge of the former to better understand the latter. As such, researchers will often refer to a macroscopic analogue to the thermodynamic temperature, commonly known as the *granular temperature*, T. As things in the world of particle science are seldom simple, there exist several competing definitions of the granular temperature, though the majority of widely-used definitions fall under one of two broad categories.

The first—and in many ways most conceptually simple—definition of the granular temperature, first proposed and popularised by Ogawa [115, 116] considers a direct analogy between the motion of molecules in a classical material and that of macroscopic particles in a granular or multiphase medium. As such, the Ogawa granular temperature is calculated from the fluctuation velocity, $\tilde{v} = |v - \bar{v}|$, of particles about the (local) mean velocity, $\bar{v}$. For a simple, monodisperse system (i.e. $m_1 = m_2 = \cdots = m$) of dimensionality $\mathcal{D}$, the granular temperature can be defined as:

$$T = \frac{1}{\mathcal{D}} m\langle \tilde{v}^2 \rangle. \tag{7.36}$$

As particulate systems can (and very often do [97, 117]) exhibit highly anisotropic temperatures, it can often be useful to define individual temperatures for different spatial coordinates, i.e.

$$T_i = m\langle (v_i - \bar{v}_i)^2 \rangle,\ i = x, y, z. \tag{7.37}$$

The total granular temperature of a system can then be simply obtained from its individual components as [101]:

$$T = \frac{1}{\mathcal{D}} \sum_{i}^{\mathcal{D}} T_i. \tag{7.38}$$

For a three-dimensional system ($\mathcal{D} = 3$), as forms our typical interest with PEPT, the above can be re-written as:

$$T = \frac{1}{3}\left(T_x + T_y + T_z\right). \tag{7.39}$$

The above equations represent the most complete and general form of the granular temperature. However, different researchers often use subtly different constructions, for example not including the dimensionality of the system, or even ignoring the mass term, such that the temperature is defined simply as the square of the fluctuation velocity. It is important to bear in mind these differing definitions when comparing quantitative measurements of T across different research articles.

It is appropriate at this point to also address a common mistake made when discussing and calculating the granular temperature, by clarifying explicitly that the granular temperature of a system or region thereof is not necessarily equivalent to its kinetic energy, E_k: the kinetic energy is calculated using the *absolute* velocity of particles ($E_k \propto \langle v^2 \rangle$), while the granular temperature utilises the fluctuation energy ($T \propto \langle (v - \bar{v}^2) \rangle$). For example, a rapid but perfectly laminar flow of particles immersed in a fluid may hypothetically possess a high kinetic energy ($v \gg 0$) but, if all particles follow the local flow exactly (i.e. $v - \bar{v} = 0$), they will have zero granular temperature.

The above-mentioned confusion arises, perhaps, from the fact that a significant number of studies concerning the granular temperature have explored random 'gases' of particles [80]. As such systems do not possess mean flow, $\bar{v} \sim 0$ for all regions of the system, meaning that $v - \bar{v} \rightarrow v$ and thus that—for this special case—E_k and T can be considered effectively equivalent. For any system in which some form of mean flow exists, however, such equivalence cannot be expected.

For completeness, it is worth discussing also the alternative definition of temperature provided by Edwards *et al* [118–120]. The Edwards temperature, as it is widely known, is predicated on the existence of a granular analogy to free energy, Y, which can be defined as:

$$Y = V - XS \tag{7.40}$$

where V is a system's volume, S its entropy, and X its 'compactivity'[8], which in this formalism represents an analogue to temperature.

For the purposes of working with PEPT, it is typical to adopt the formalism of Ogawa, as opposed to that of Edwards. There are several reasons for this choice, of which we note here the two most compelling: firstly, on a pragmatic level, the Ogawa definition of granular temperature considers the microscopic motion of particles thus making better use of PEPT's particle-level imaging capabilities. Secondly, there exists a great deal of debate in the field as to whether macroscopic systems are subject to entropic effects, and indeed whether statistical mechanics approaches more generally can be meaningfully applied to such systems[9].

Having established what the granular temperature is (or at least what it can be defined as!), in the next sections we detail the manner in which measurements thereof may be obtained from PEPT data.

Due to PEPT's finite temporal resolution and its limitations in terms of explicitly resolving particle collisions, different approaches to the calculation of granular temperature must be taken dependent on the density of the system in question—or more specifically, the characteristic collision rate of said system relative to the characteristic acquisition rate of the imaging apparatus used.

[8] Discussion relating to the compactivity of powders can be found in section 4.4. In simple terms, it is a measure of how 'fluffy' a powder is—a perfectly compacted bed will possess a compactivity $X = 0$, while a bed in its least-dense achievable configuration will yield a value $X = \infty$.

[9] The interested reader may find a detailed review concerning this matter in reference [121].

For relatively dilute systems, where a particle can be successfully located many times before colliding with another particle or any other object, velocity calculations as outlined in section 7.4.1 will typically provide an accurate representation of the particle's instantaneous motion. As such, the values of v obtained thereby may be used to directly calculate the granular temperature as per equation (7.36) or (7.37). This relatively direct method of calculating T is described in detail in section 7.6.2.

For denser systems, the situation is a little more complex: if the mean collision rate of the tracer particle is comparable to or higher than the average rate at which PEPT data points are recorded, velocities measured as per section 7.4.1 no longer represent particles' instantaneous velocities, but rather their net velocities. While these net velocities are suitable for measuring and visualising flow through a system, they 'average out' the finer motion corresponding to particles' fluctuating motion, and thus by definition cannot be used to calculate T. In this case, values of T must instead be obtained through consideration of particles' MSD—the measurement of which is described in section 7.5.1. The methodology for extracting granular temperature measurements from denser systems is described in section 7.6.4.

7.6.2 Dilute systems

For dilute systems, where particle collisions are infrequent compared to PEPT's data acquisition rate, the granular temperature is extremely straightforward to measure. If one simply wishes to calculate the mean granular temperature for the system of interest as a whole, then it is merely a matter of determining the mean velocity of the system:

$$\bar{v}\frac{1}{N}\sum_{i=1}^{N} v_i, \tag{7.41}$$

(where N is the total number of PEPT data points) and then using this mean velocity to calculate the granular temperature as:

$$T = \frac{1}{\mathcal{D}N} m \sum_{i=1}^{N} (v_i - \bar{v})^2, \tag{7.42}$$

where the normalised sum over all individual data points is—under the assumption of ergodicity—equivalent to the ensemble average in equation (7.36). As with the original equation (7.36), the above calculation may also be performed for the individual x, y and z components separately.

In reality, the calculation of a whole-system granular temperature is seldom meaningful, as most dynamic granular systems possess highly spatially-inhomogeneous granular temperatures [122–124]. As such, it is typically of greater value to produce a one, two or three-dimensional granular temperature distribution. In order to do so, we must begin with the now familiar process of subdividing our experimental volume into a series of cells. For any given cell, j, the local granular temperature can then be obtained; firstly, one must calculate the local mean velocity, $\bar{v}_j$, for the cell as per equation (7.41), but as a limit and averaging factor only the number, N_j, of data points falling within the current cell, as opposed to the overall number of data points, N. The local temperature for the jth cell may then be calculated as per equation (7.42), again replacing N with N_j. Repeating this process for all cells within the system yields a discretised approximation of the spatial granular temperature distribution across the system.

For many particulate systems, in particular those—such as vertically-vibrated beds—which are driven by a monodirectional excitation, the temperature distributions thereof can be successfully characterised by a one-dimensional profile [123, 125]. In the tutorial at the end of this section, we show how such a profile may be created from PEPT data.

7.6.3 Interactive example: 1D granular temperature profile

The notebook files accompanying this section are available in the supplementary data online at https://iopscience.iop.org/book/978-0-7503-3071-8.

Tutorial: creating granular temperature profiles from PEPT data

In this tutorial, we will demonstrate how one-dimensional, depth- and width-averaged granular temperature profiles may be extracted from standard single-particle trajectories such as those obtained from PEPT data, exploring in particular the importance of choosing a suitable cell size, *dc*, to produce both accurate and meaningful data.

Importing data and calculating velocities

As discussed in the main text, the granular temperature is mainly relevant for *dilute* systems, where the rate of data acquisition is larger than the rate of particle collisions; for illustrative purposes, this notebook uses a simple discrete element method (DEM, as described in chapter 4) simulation of a monodisperse system of 100 HDPE particles with diameters $d = 5$ mm inside a vibrofluidised bed; the bed is essentially a $100 \times 20 \times 200$ mm box with the bottom plate vibrating with an amplitude of 4 mm and frequency 50 Hz.

As in the previous example, we begin in the code below by importing some data and determining a set of velocity values; feel free to play with the parameter j, which determines the number of data points over which velocities are calculated, and see what effect it has on your final results.

[1]:
```
# Importing necessary libraries for numerical array manipulation and plotting
import numpy as np
import matplotlib.pyplot as plt

# Set Matplotlib theme (try "darkgrid" too)
import seaborn
seaborn.set_style("whitegrid")
```

[2]:
```
# Loading pre-processed fluidised bed PEPT data
filepath = "Data/VibrofluidisedBedTracer.csv"
full_trace = np.loadtxt(filepath, skiprows=15)

# System constants
d = 5e-3                          # Tracer diameter, m
rho = 970                         # HDPE density, kg / m3
mass = rho * np.pi * d**3 / 6     # Tracer mass, kg
```

```
# Extracting time and position data
t = full_trace[:, 0]
x = full_trace[:, 1]
y = full_trace[:, 2]
z = full_trace[:, 3]

# For easy plotting, subtracting minima from the spatial coordinates so that
# system's left-hand corner lies at (0,0)
x = x - x.min()
y = y - y.min()
z = z - z.min()

# The number of steps before and after the current point over which the velocity
# is calculated
#
# Feel free to adjust this parameter to see its influence on the final output!
j = 3

# Creating empty arrays to hold all velocity values
vx = np.full(len(x), np.NaN)
vy = np.full(len(y), np.NaN)
vz = np.full(len(z), np.NaN)

# Calculating velocity for each point in turn using the central difference
# approach (as discussed in detail in the previous tutorials)
for i in range(j, len(x) - j):
    # Calculating velocity for each point
    vx[i] = (x[i + j] - x[i - j]) / (t[i + j] - t[i - j])
    vy[i] = (y[i + j] - y[i - j]) / (t[i + j] - t[i - j])
    vz[i] = (z[i + j] - z[i - j]) / (t[i + j] - t[i - j])

# Calculate absolute velocity
v = np.sqrt(vx * vx + vy * vy + vz * vz)

print("Total number of points in file: ", len(vx))
```

```
Total number of points in file:  500000
```

[3]:

```
# Plot simulated DEM tracer trajectory, colour-coding it by its velocity
fig, ax = plt.subplots()
ax.scatter(x[:3000], y[:3000], c=v[:3000], s=4)
ax.set(xlabel="$x$ ($mm$)", ylabel="$y$ ($mm$)", aspect="equal");
```

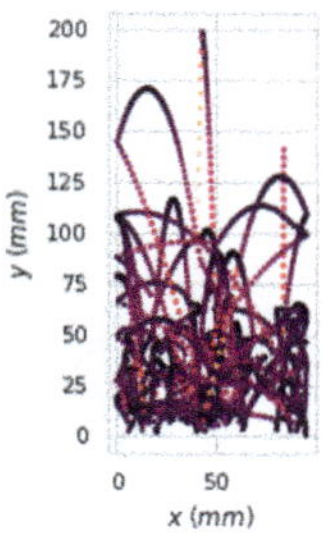

Creating a 1D granular temperature profile

Now that we have velocities defined for all our data points, our next step is to create a suitable grid, and use the velocity values obtained to determine the mean velocity for each cell thereof, as described in the main text. While there are many possible ways to do this, we have chosen a rather 'old school' method using explicit nested for loops, which—as it does not use too many Python-specific functions—can be easily adapted to almost any programming language.

In the code below, the parameter *dc* defines the size of the cells used for our calculations. Take some time to explore the effect of using different values of dc on the velocity vector field finally output.

[4]:
```
# Defining our chosen cell size, dc, in mm (the total height of our system is
# ~200mm)
dc = 10

# Working out the number of cells needed to span the system in the y direction
n = int(y.max() / dc)

# Pre-allocating a suitably-sized 1D array to store granular temperatures in
# each cell
temp_grid = np.zeros(n)

# For each vertical cell, select all particle locations falling inside and
# compute the cell's granular temperature. Note that dimensionality = 3
for i in range(n):
    v_cell = v[(i * dc <= y) & (y < i * dc + dc)]

    # Compute the mean velocity while ignoring NaNs
    vmean_cell = np.nanmean(v_cell)
    temp_grid[i] = 1 / 3 * mass * np.nanmean((v_cell - vmean_cell) ** 2)

# Defining a suitable axis scale for plotting
scale = np.arange(n) * dc

# Plotting
fig, ax = plt.subplots()
ax.scatter(temp_grid, scale)

ax.set(ylabel='$y$ ($mm$)', xlabel='Granular Temperature ($J$)', ylim=[0, 200])
ax.set_aspect(1.0 / ax.get_data_ratio() * 5)
```

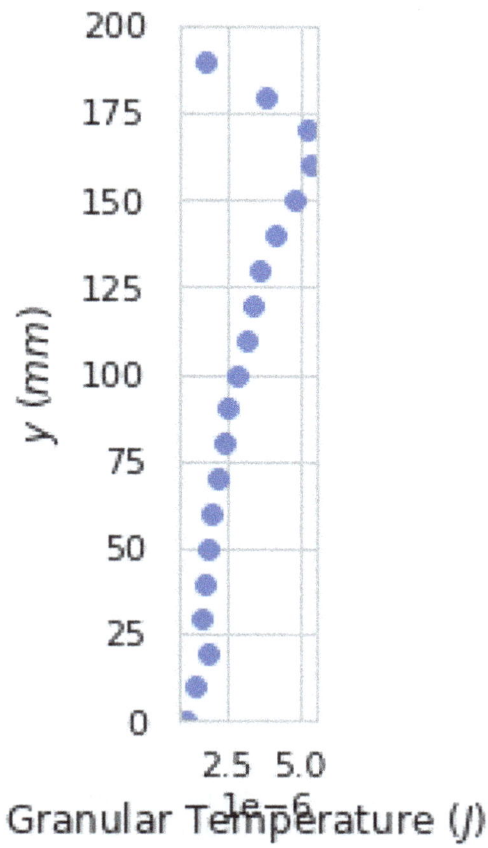

Note that the granular temperature is very low in the dense region at the bottom of the vibrofluidised bed, and increases as the particle movement becomes more 'gas-like' towards the top of the bed.

7.6.4 Denser systems

As we have repeatedly alluded to above, the calculation of T in the manner outlined in the preceding section requires the underlying assumption that the velocities measured by PEPT can be taken as the tracer particle's instantaneous velocity. However, in comparatively dense systems, a tracer particle is likely to collide with other particles and/or its surroundings in between any two consecutive location events (see figure 7.18). If this is the case, then the velocity calculated between two such measured points will not be representative of the particle's true ballistic motion, but rather its net velocity over the time period considered. As such, for such systems, we need another manner to calculate the ballistic motion of our tracer.

One manner in which this may be achieved is through consideration of the MSD of our tracer, which can be determined from PEPT data as outlined in section 7.5.1. In this section, we noted that in the limit $t \to \infty$, the diffusion coefficient D may be extracted from MSD data as per equation (7.32). However, at small t—specifically,

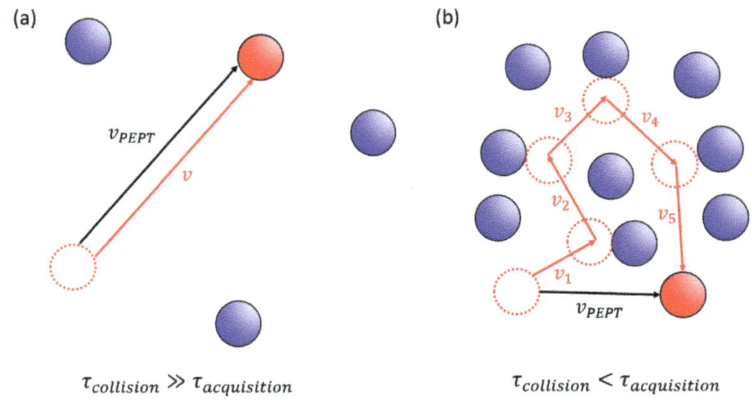

Figure 7.18. Images comparing the true ballistic trajectory(/ies) of a tracer particle (red arrows) to those measured by PEPT (black arrows) for (a) a relatively dilute bed, where the tracer does not undergo any collisions in between two PEPT location events and (b) a relatively dense bed, in which the tracer undergoes multiple collisions in between consecutive locations.

at times smaller than the Enskog mean collision time, τ_E [100, 126, 127]—the MSD is, to a reasonable approximation [101, 127], instead proportional to the square of the particle speed, i.e.

$$\langle |\vec{r}(t) - \vec{r}(0)|^2 \rangle = \langle v^2 \rangle t^2. \tag{7.43}$$

For systems in which there exists no net flow (i.e. $\bar{v} = 0 \rightarrow \tilde{v} = v$) the above equation may be used to directly determine the fluctuation velocity $\tilde{v}$ and thus the granular temperature of a system as per equation (7.36).

In the presence of convective motion or other net flow, the MSD $\langle |\vec{r}(t) - \vec{r}(0)|^2 \rangle$ is a result of both diffusion and advection. In this case, equation (7.43) must be re-written as [128]:

$$\langle |\vec{r}(t) - \vec{r}(0)|^2 \rangle = (\langle \tilde{v}^2 \rangle + \langle v_c^2 \rangle)t^2, \tag{7.44}$$

where v_c is the (local) convective velocity. The above equation tells us that the sum $\langle \tilde{v}^2 \rangle + \langle v_c^2 \rangle$ may be computed simply by determining the gradient of a plot of MSD versus t^2 in the limit of small t. As v_c corresponds to the *net* velocity through a given cell, this quantity can be calculated simply as described in section 7.4.5 and subtracted from the aforementioned product, leaving us with an estimate of the fluctuation velocity from which, in turn, we may calculate T.

7.6.5 Energy and temperature partition

One of the striking features of particulate media is that they can—and indeed commonly do—violate the principle of equipartition observed by classical, molecular fluids [129–133]. That is to say that in a system comprising two or more distinct species of particle, each species present may possess a distinct granular temperature and/or mean kinetic energy. Luckily, the nature of PEPT measurements makes it extremely easy both to accurately measure the individual temperatures possessed by different species, as well as to quantify the degree of non-equipartition within a given system.

As with the extraction of any piece of information from a binary or polydisperse system, we must begin by acquiring a complete and suitably-long PEPT data set using a tracer of each species present within the system (as discussed in greater detail in section 7.3). The granular temperature for each species—or, similarly, a one-, two-, or three-dimensional distribution thereof—may then be determined in either of the manners described in the two preceding sections, dependent on the density of the system in question. If the same grid is used for each species (i.e. the experimental volume is divided into cells of the same size, in the same positions relative to the system of interest) then one may determine the relative temperatures of the different species present in different regions of the system, using a similar approach to that described in section 7.3 for the calculation of the relative species concentration. For a binary system, the temperature ratio, γ, in a given cell i is typically taken simply as:

$$\gamma_i = \frac{T_i^{\alpha}}{T_i^{\beta}} \tag{7.45}$$

where the indices α, β represent different particle species. By calculating the local temperature ratio for each cell in the system of interest, the variation of γ throughout a system may be determined as either a 1D profile or a two- or three-dimensional field. Figure 7.19(a) shows examples of separate,

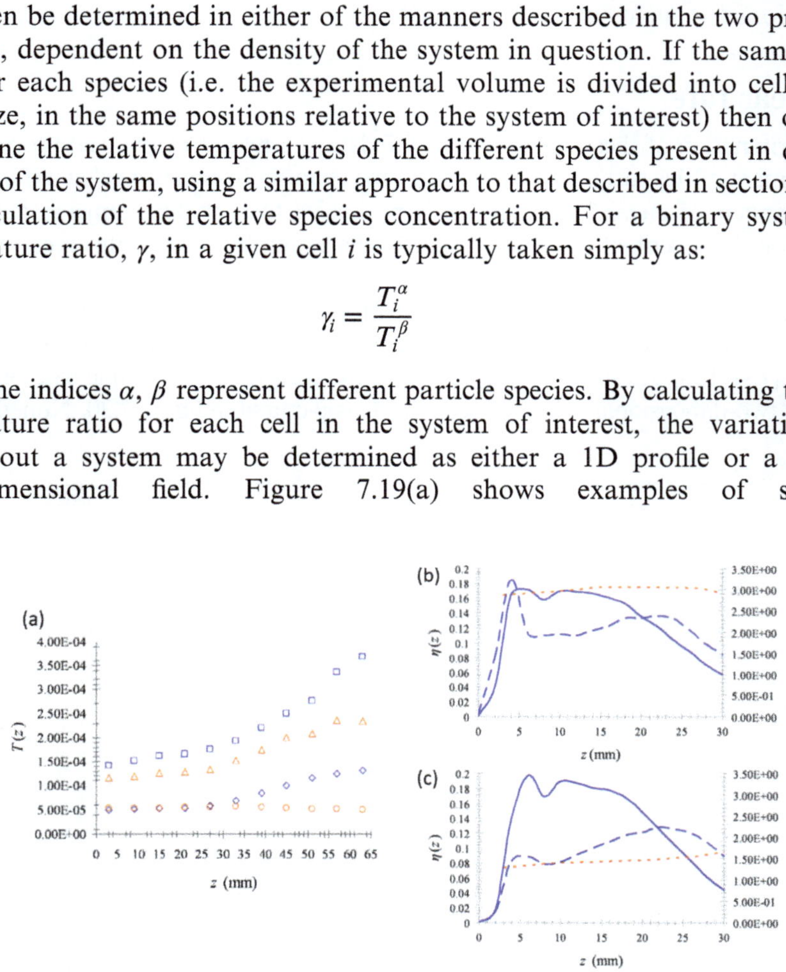

Figure 7.19. (a) Individual temperature profiles for the glass (circles, diamonds) and steel (diamonds, squares) components of a vertically-vibrated binary system of spherical glass and steel particles under different conditions (blue → more elastic bounding walls, orange → more dissipative bounding walls). (b, c) One-dimensional vertical density profiles (blue lines) for both species of the aforementioned systems alongside the 1D distribution of the relative temperature, γ, between the two species. In both panels (b) and (c), solid lines represent the local density of the (heavier) steel component, and dashed lines that of the (lighter) glass component; the two systems show clear though non-complete segregation. Note that despite the presence of segregation, γ remains roughly constant throughout the bulk of the material, even as the relative concentration of the two species varies. Images reprinted by permission from Springer Nature: [133], © Springer Nature 2014.

one-dimensional temperature profiles corresponding to the heavy (steel) and light (glass) components of a binary glass-steel mixture. In both cases shown, the heavier component, as expected [130], possesses a notably higher granular temperature than the lighter in all regions of the system. Interestingly, as shown in panels (b) and (c), the ratio of temperatures, γ, remains roughly constant throughout the bulk of the system, despite significant fluctuations both in the relative particle concentration, as well as the overall packing density—again in line with our expectations for such systems [130, 133].

7.7 Shear rate

The determination of shear rates from PEPT data is a relatively involved process, hence the placement of this section near the end of the chapter! In order to calculate shear rates within a system of interest, we must firstly produce a two- or three-dimensional velocity distribution, as described in section 7.4.5. In the examples used in this section and the following tutorial, we focus on a system—specifically, a horizontal, cylindrical drum rotating about its central axis—which, at least for the case of a comparatively thin drum, can be safely assumed to possess approximately uniform flow dynamics along the axial direction [134–136]. As such, we consider here the comparatively simple case of a depth-averaged two-dimensional velocity vector field, as pictured in figure 7.20. Nonetheless, as

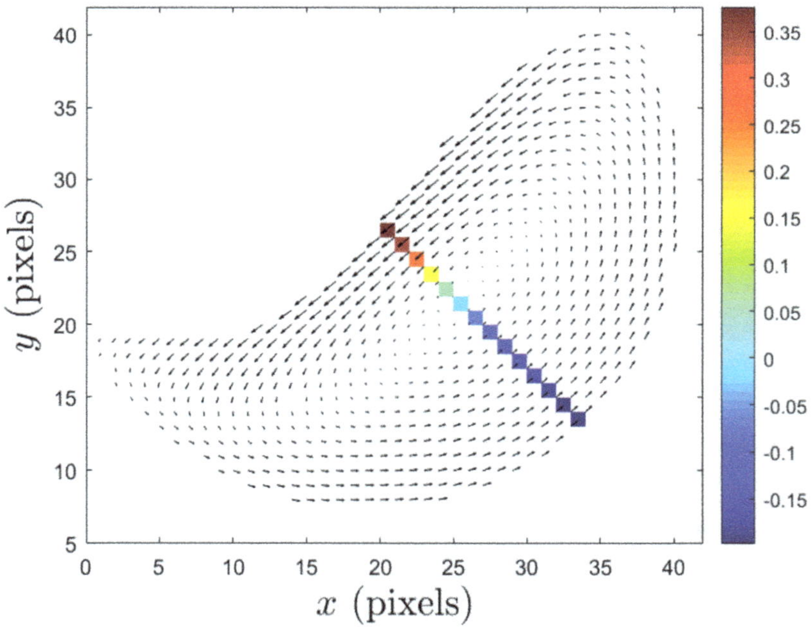

Figure 7.20. Image showing a PEPT-acquired velocity vector field (arrows), determined as described in section 7.4.5, for a rotating-drum system filled with monodisperse MCC particles. The coloured squares represent the tangential velocities corresponding to individual cells lying along a diametrical line passing through the granular bed's approximate centre of circulation [137, 138].

with so many of the quantities discussed in this section, the methodology described may be easily extended to the three-dimensional case and to a wide variety of other systems.

Once a suitable velocity vector field has been determined, we must next choose a suitable line along which to calculate the shear rate. For the current example, we follow the example of Govender *et al* [134] and consider a diametrical line passing through the centre of circulation of the granular bed[10]. From our PEPT data, we can extract the values of the tangential velocities falling along this line—or, more specifically, corresponding to the specific cells in our two-dimensional grid falling closest to said line. The values of these velocities are illustrated using a colour scale in figure 7.20, and plotted as a function of distance, r, along the diametrical line in figure 7.21.

Knowing that $\dot{\gamma} = \frac{dv}{dr}$, the obtained values may be used to obtain the shear rate, $\dot{\gamma}(r, \theta)$ for the current line at angle θ. In order to do so, one may take either of two distinct approaches. The first is to 'manually differentiate' the values obtained—i.e. for each pair of successive points, calculate the change in velocity and divide by the change in distance. While this method is simple, it typically introduces considerable noise in the data obtained, and thus risks obscuring the form of the shear rate profile.

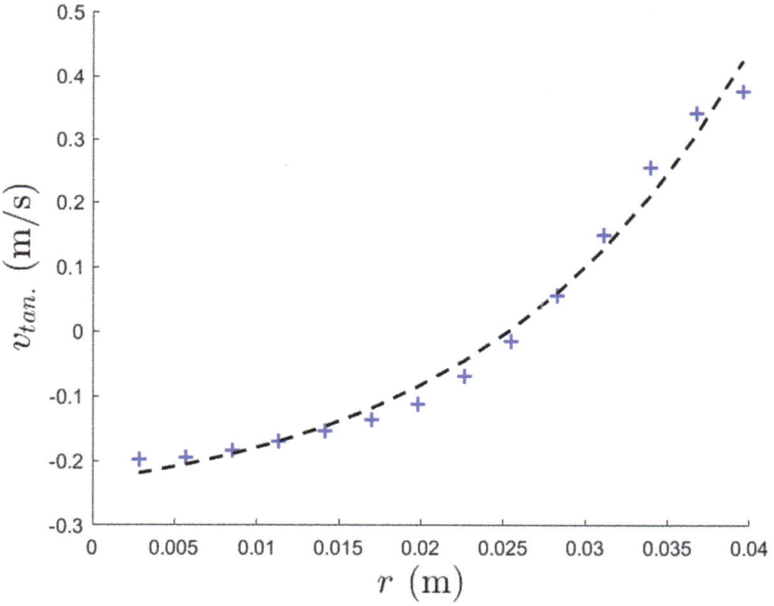

Figure 7.21. Variation of the tangential velocity, $v_{\text{tan.}}$, with distance r along the line of interest pictured in figure 7.20. In the image shown, the discrete points represent PEPT data, while the continuous line represents a least-squares fit to said data to the functional form given in equation (7.46).

[10] The centre of circulation for a system such as that discussed here may be calculated as described in the work of Govender and McBride [137, 138].

An alternative method, adopted by previous researchers both using PEPT and other imaging techniques [134–136] is first to fit an easily-differentiable functional form to the data, and then to determine the shear rate from this function. Our current data set may (as illustrated in figure 7.21) be reasonably well described by a simple exponential of the form:

$$v(r, \theta) = f(r) = a \exp[-b(H - r)] + c \tag{7.46}$$

where H is the total depth of the bed along the line of interest. The relevant coefficients a, b and c can be determined using a nonlinear least squares regression, a process which we illustrate in the tutorial at the end of this section. Once these coefficients have been successfully determined, the shear rate and its variation along the line of interest may be determined simply by differentiating the fitted curve, i.e.

$$\dot{\gamma}(r, \theta) = f'(r) = ab \exp[-b(H - r)]. \tag{7.47}$$

The process described above may be repeated for lines spanning a range of θ values. The results obtained via the simple finite difference method and the model-fitting method are illustrated in figure 7.22.

It is finally important to note here that while the general framework described above can be easily adapted to any number of systems, the specific details presented—in particular the choice of a suitable line along which to measure the tangential

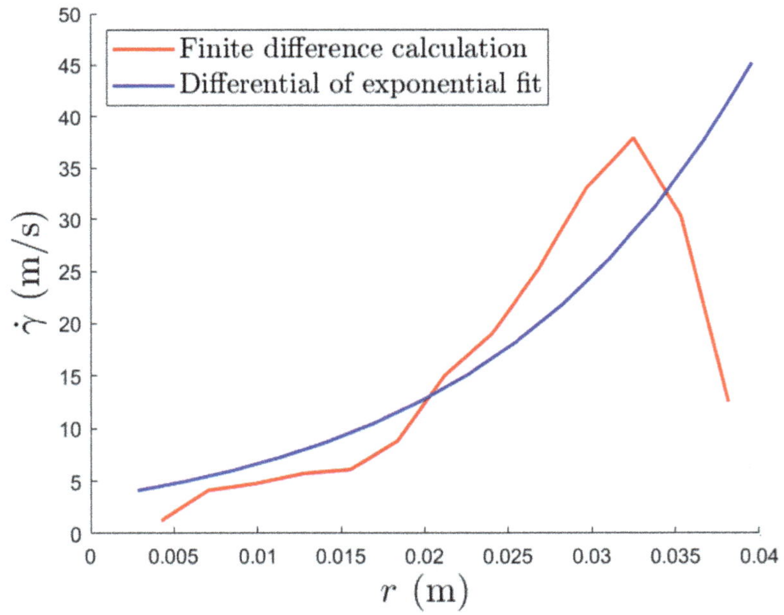

Figure 7.22. Shear rates calculated from the data shown in figure 7.21 from both a central difference calculation using the discrete PEPT data points shown (red line) and by differentiating the exponential curve of best fit (blue line).

velocity and thus shear rate, and the choice of functional form to fit the data acquired—are very much specific to the particular system in question. When performing such analysis in a different system, the choice of these (and other) parameters should be based on a solid understanding of the physics and expected dynamics of said system.

7.7.1 Interactive example: shear rate

The notebook files accompanying this section are available in the supplementary data online at https://iopscience.iop.org/book/978-0-7503-3071-8.

Tutorial: Calculating shear rate from PEPT data

In this tutorial, we will demonstrate how the velocity vector fields we learned to create in the interactive example from section 7.4 can be used to calculate shear rates. In the code below, we go through the (hopefully by now familiar!) process of creating a velocity vector field, though this time for a different system—namely a horizontally-rotating drum filled with granular material.

As usual, feel free to play with the range, *j*, of the central difference velocity calculation, and the size, *dc*, of the cells used in order to produce the best plot possible.

[1]:
```
# Importing necessary libraries for numerical array manipulation and plotting
import numpy as np
import matplotlib.pyplot as plt

# Set Matplotlib theme (try "darkgrid" too)
import seaborn
seaborn.set_style("whitegrid")
```

[2]:
```
# Loading relevant data
filepath = "Data/Drum45RPM.csv"
full_trace = np.loadtxt(filepath, skiprows=16)

# Extracting time and position data for all Cartesian coordinates
t = full_trace[:, 0]
x = full_trace[:, 1]
y = full_trace[:, 2]
z = full_trace[:, 3]

# Re-centering all traces such that they start at 0 - this makes subsequent
# calculations and plots clearer
x = x - x.min()
y = y - y.min()
z = z - z.min()
```

[3]:
```
# Some values in the division below are zero, let them silently yield 0/0 = NaN
np.seterr(divide='ignore', invalid='ignore')

# The number of steps before and after the current point over which the velocity
# is calculated
#
# Feel free to adjust this parameter to see its influence on the final output!
j = 3

# The number of particle locations a cell must possess in order to be considered
# "viable"
thresh = 10

# Pre-allocating empty arrays to hold all velocity values
vx = np.full(len(x), np.NaN)
vy = np.full(len(y), np.NaN)

# Calculating velocity for each point in turn using the central difference
# approach (as discussed in detail in the previous tutorial)
for i in range(j, len(x) - j):
    # Calculating velocity for each point
    vx[i] = (x[i + j] - x[i - j]) / (t[i + j] - t[i - j])
    vy[i] = (y[i + j] - y[i - j]) / (t[i + j] - t[i - j])

# The size, in mm, of the cells into which you want to subdivide the system
# Note that, for simplicity, we use square cells of size dc x dc, but in reality
# other cell geometries may also be used.
dc = 2

# Working out the number of cells needed to span the system in both the x and y
# directions
nx = int(np.ceil(x.max() / dc))
ny = int(np.ceil(y.max() / dc))

# Creating a pair of suitably-sized 2D arrays to store resultant velocity field
vx_grid = np.zeros((ny, nx))
vy_grid = np.zeros((ny, nx))

# An additional array to store the number of locations falling in each cell
n_grid = np.zeros((ny, nx))

# Looping through all particle locations in turn
for n in range(j, len(vx) - j):
    # For each location, find the horizontal and vertical indices of the cell in
    # which the particle falls
    ix = int(x[n] / dc)
    iy = int(y[n] / dc)

    # Once correct cell is located, incrementing the number of locations found
    # in the current cell...
    n_grid[iy, ix] = n_grid[iy, ix] + 1

    # ...and adding the relevant velocities to the current sum
    # for the current cell
    vx_grid[iy, ix] = vx_grid[iy, ix] + vx[n]
    vy_grid[iy, ix] = vy_grid[iy, ix] + vy[n]
```

```
# Finding mean velocities
vx_grid = vx_grid / n_grid
vy_grid = vy_grid / n_grid

# Removing values from cells with too few data points for good statistics
vx_grid[n_grid < thresh] = np.nan
vy_grid[n_grid < thresh] = np.nan

fig, ax = plt.subplots(figsize=(7, 7))

# In this case, just plotting vs. pixel number for clarity in later calculations
ax.quiver(vx_grid, vy_grid)

ax.set(ylabel='$y$ (pixels)', xlabel='$x$ (pixels)');
```

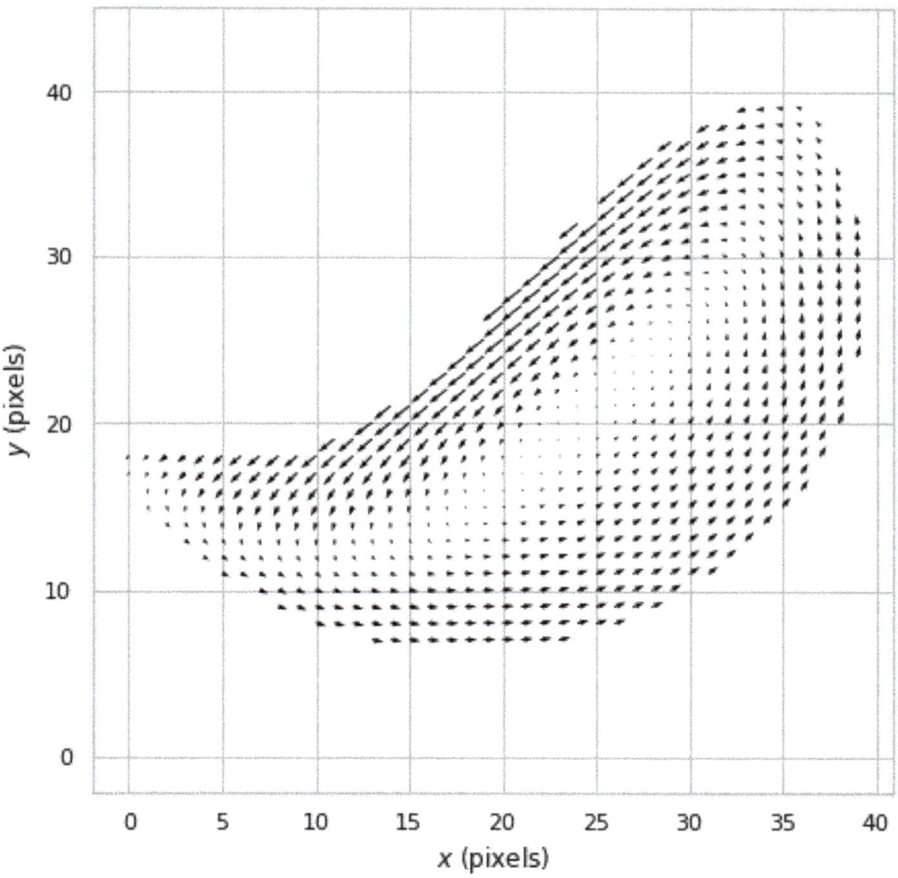

Choosing a suitable line

Having created a velocity vector field for our system, let us now choose a line along which to calculate our shear rate. For the sake of rigour, we will use in this example a diametrical line passing through the centre of circulation of our system—as used in the work of Govender *et al* [147].

[4]:
```
# Creating a line to plot: gradient and offset
m = -1
c = len(vx_grid)

# Feel free to play with the above values to choose a different region through
# which to determine shear rate, though note that choosing a line which does not
# fulfil the criteria outlined above may affect the validity of the shear rate
# calculation in the next step!

# Creating line to plot
x = np.arange(len(vx_grid))
y = m * x + c

# Re-plotting vector field and adding line
fig, ax = plt.subplots(figsize=(8, 8))

ax.quiver(vx_grid, vy_grid)
ax.plot(x,y)

ax.set(
    ylabel='$y$ (pixels)',
    xlabel='$x$ (pixels)',
    xlim = [0, len(vx_grid) - 1],
    ylim = [0, len(vy_grid) - 1],
);
```

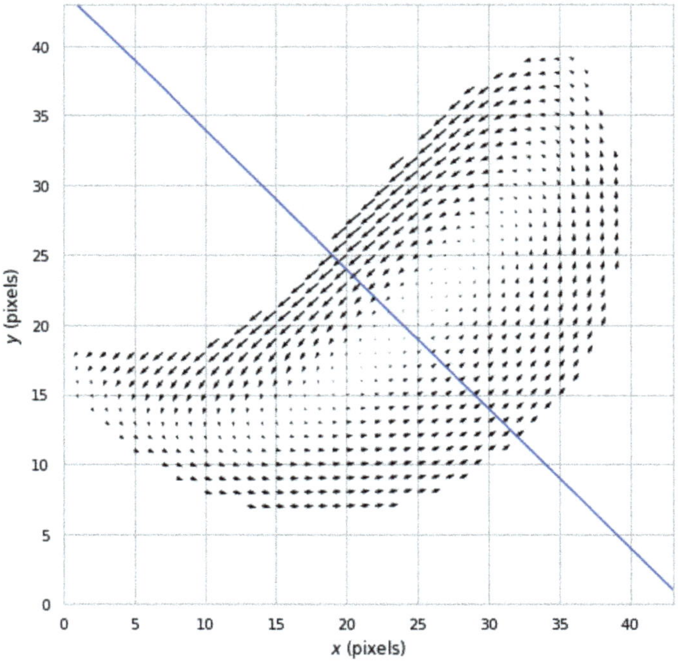

Calculating velocities along the line

Having chosen a line along which to calculate our shear rate, our next job is to find the tangential velocities along this line. If the orientation of our line is well chosen, however (i.e. the line fals through the circulation centre), then these should simply be equal to the net velocities within each cell.

Before we can find the relevant velocity values, however, we must first establish a set of pixels which best approximates our chosen line. This can be achieved in the manner illustrated below, which simply turns the x and y values used to define our line of interest into cell indices.

Once the relevant cells are determined, the velocity in the relevant direction can then be calculated for each cell along the line.

[5]:
```
# Creating an empty array of the same size as our vector field array
line_array = np.full(vx_grid.shape, np.NaN)

# Converting all y values from line to the nearest integer to allow them to be
# used as indices
y = y.astype(int)

# Defining an empty one-dimensional array to store the values acquired in order
v_vec = []

# Looping over all x and filling in relevant velocity values
for i in range(1, len(x) - 1):
    # Ensuring that indices used are within bounds
    if y[i] < len(line_array[0]) - 1:

        # Finding magnitude of current absolute velocity
        v_mag = np.sqrt(vx_grid[i, y[i]] * vx_grid[i, y[i]] +
                        vy_grid[i, y[i]] * vy_grid[i, y[i]])

        # Finding relevant angle
        theta1 = np.arctan(vy_grid[i, y[i]] / vx_grid[i, y[i]])

        # Angle of tangent
        theta2 = np.arctan(-1 / m)

        # Taking dot product with tangential vector to calculate magnitude in
        # direction of interest
        v_rel = v_mag * np.cos(theta2 - theta1)

        # Determining relative direction between 2 vectors
        if vy_grid[i, y[i]] > 0:
            v_rel = -v_rel

        # Adding the velocity to our 2D plot so that we can overlay onto the
        # quiver plot and check the values achieved are a) meaningful and b) in
        # the correct position
        line_array[i, y[i]] = v_rel

        # Adding the velocity to our one dimensional line vector, but only if
        # there is a real value to include
        if ~np.isnan(v_rel):
            v_vec = np.append(v_vec, v_rel)
```

```
# Plotting
fig, ax = plt.subplots(figsize=(8, 8))

ax.quiver(vx_grid, vy_grid)
#ax.contour(line_array)
ax.imshow(line_array, interpolation='nearest')

ax.set(
    ylabel='$y$ (pixels)',
    xlabel='$x$ (pixels)',
    xlim = [0, len(vx_grid) - 1],
    ylim = [0, len(vy_grid) - 1],
);

plt.colorbar(ax.imshow(line_array, interpolation='nearest'));
```

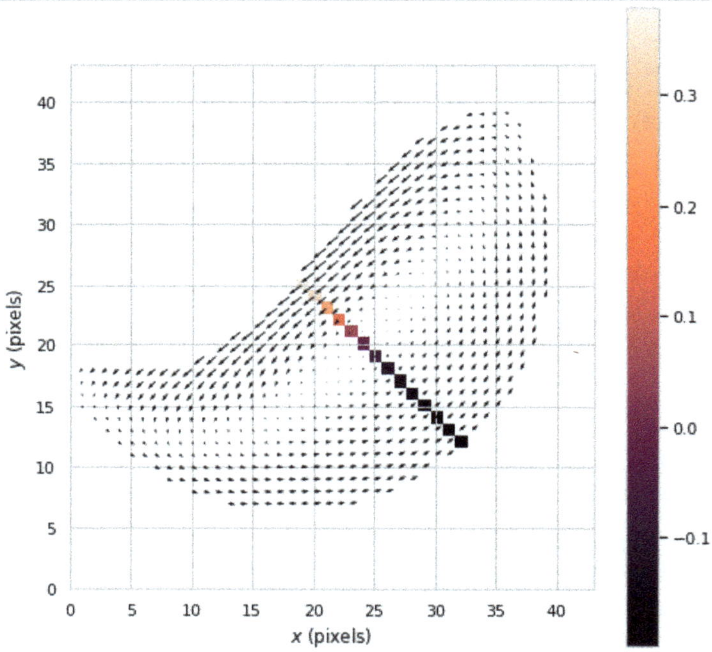

Fitting a functional form

Our next step toward calculating the shear rate from our data is to find a (differentiable) analytical model which fits our data. In the below, we plot the data calculated above as a function of distance along the chosen line, so that its form may be more clearly seen.

```
[6]: # Defining a suitable scale in m
scale = np.arange(len(v_vec)) * (np.sqrt(2) * dc) + (np.sqrt(2) * dc) / 2
scale = scale / 1000

fig, ax = plt.subplots()
ax.scatter(scale, v_vec)
ax.set(
    ylabel='$v_{tan.}$ (m/s)',
    xlabel='Distance along line (mm)',
    xlim=[0, scale.max()],
);
```

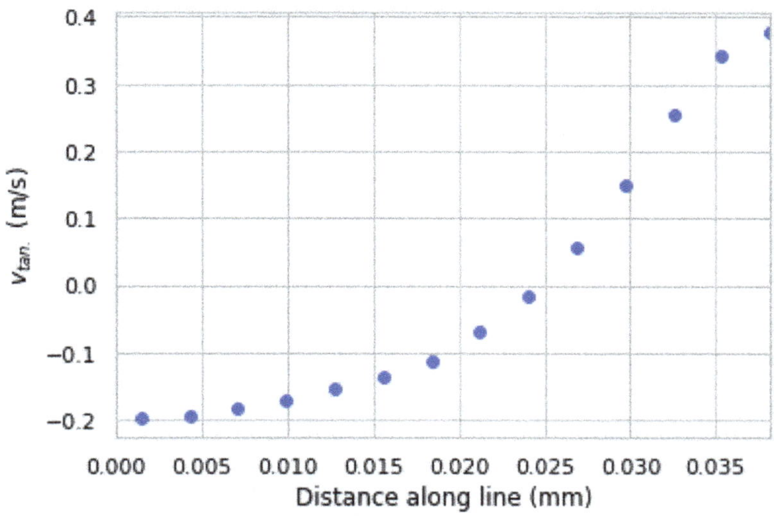

The above form could clearly be reasonably fitted by a range of different functions. In the below, we attempt to fit two—specifically, an exponential fit and a parabolic fit—both of which could sensibly describe the flow profile of a granular free-surface flow.

The fit is achieved using a non-linear least squares fitting algorithm, attempting a fit to a form of our choosing. In the first example, below, we attempt a simple parabolic fit of the form

$$f(x) = ax^2 + b.$$

[7]:
```
# Use the `scipy.optimize.curve_fit` non-linear LSQ fitting subroutine
from scipy.optimize import curve_fit

# Try fitting a parabolic function
def test_func(x, a, b):
    return a * x ** 2 + b

# Initial guess
p0 = [-0.4, 0.1]
params, params_covariance = curve_fit(test_func, scale, v_vec, p0)

print("Fitted function parameters:", params)

# Plot fitted curve
fig, ax = plt.subplots()

ax.scatter(scale,v_vec)
ax.plot(scale, test_func(scale, params[0], params[1]))

ax.set(
    ylabel='$v_{tan.}$ (m/s)',
    xlabel='Distance along line (mm)',
    xlim=[0, scale.max()],
);
```

```
Fitted function parameters: [ 4.21853416e+02 -2.23500200e-01]
```

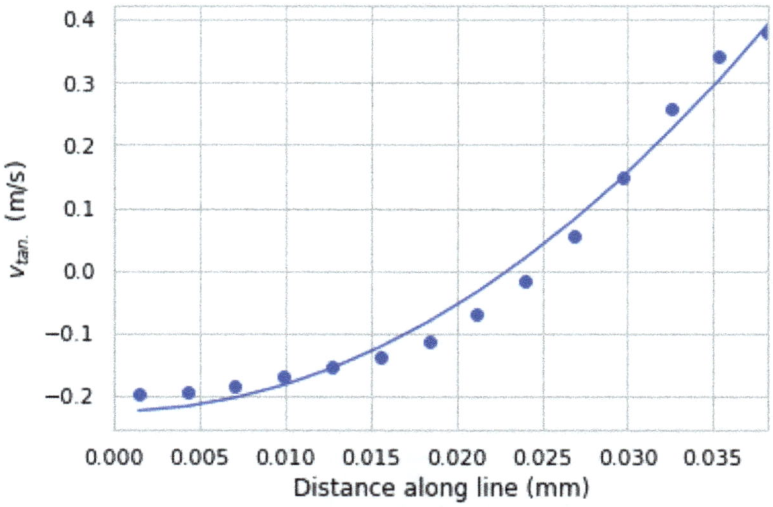

As can be seen in the plot above, the fit produced by our regression analysis provides a reasonable representation of our data. The shear rate can now be calculated by differentiating the above with respect to the distance along the line, yielding

$$\dot{\gamma} = f'(x) = 2ax.$$

[8]:
```
shear1 = 2 * params[0] * scale

fig, ax = plt.subplots()
ax.plot(scale, shear1)
ax.set(
    ylabel='$\dot{\gamma}$ (s$^{-1}$)',
    xlabel='Distance along line (mm)',
    xlim = [0, scale.max()],
);
```

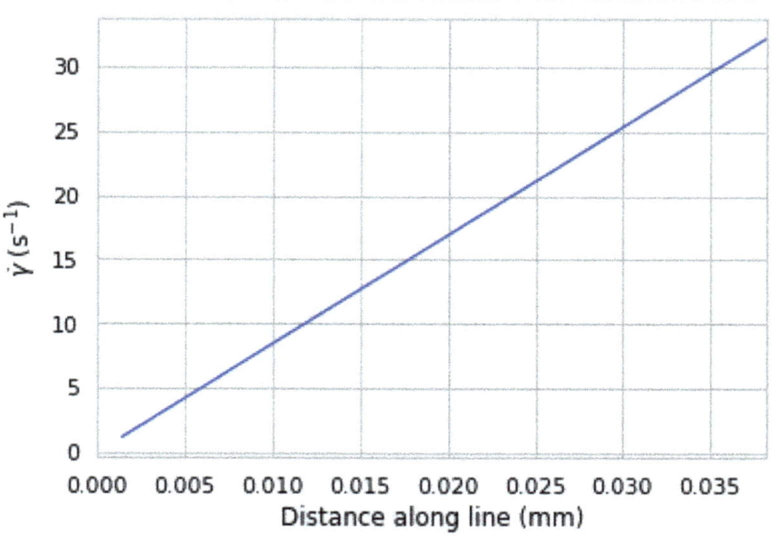

In the second example, we instead attempt a fit of the form

$$g(x) = a \exp[-b(H - x)] + c$$

where H represents the total depth of the bed in the direction of interest.

[9]:
```
# Trying exponential fit
def test_func(x, a, b,c):
    return a * np.exp(-b * (x.max() - x)) + c

# Initial guess
p0 = [0.1, 0.1,0.1]
params2, params_covariance2 = curve_fit(test_func, scale, v_vec, p0)

print("Fitted function parameters:", params)

# Plot fitted function
fig, ax = plt.subplots()

ax.scatter(scale, v_vec, label='PEPT data')
ax.plot(scale, test_func(scale, *params2), label='Fitted function')

ax.set(
    ylabel='$v_{tan.}$ (m/s)',
    xlabel='Distance along line (mm)',
    xlim=[0, scale.max()],
);
```

```
Fitted function parameters: [ 4.21853416e+02 -2.23500200e-01]
```

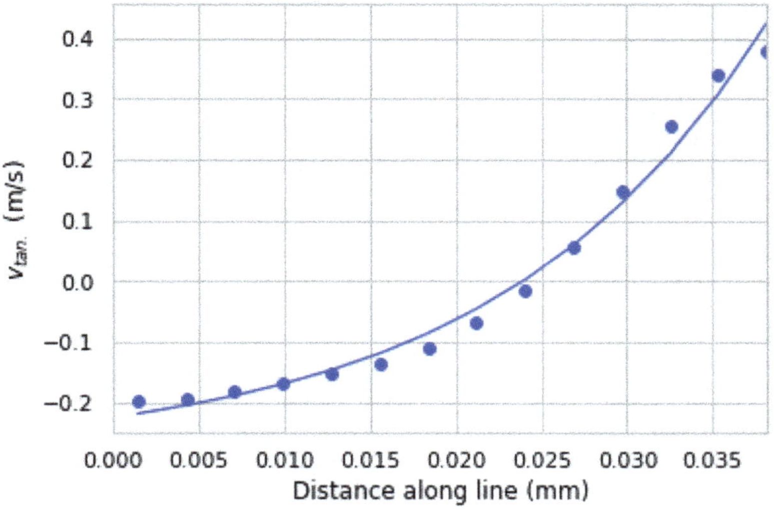

As before, the fit produced by our regression analysis provides a reasonable representation of our data. The shear rate can now be calculated by differentiating the relevant function, this time yielding:

$$\dot{\gamma} = g'(x) = ab \exp[-b(H - x)].$$

The shear rate calculated using this second form can be seen below:

```
[10]: def test_func(x, a, b):
          return a * b * np.exp(-b * (x.max() - x))

      fig, ax = plt.subplots()
      ax.plot(scale, test_func(scale, params2[0], params2[1]))
      ax.set(ylabel='$\dot{\gamma}$ (s$^{-1}$)', xlabel='Distance along line (mm)');
```

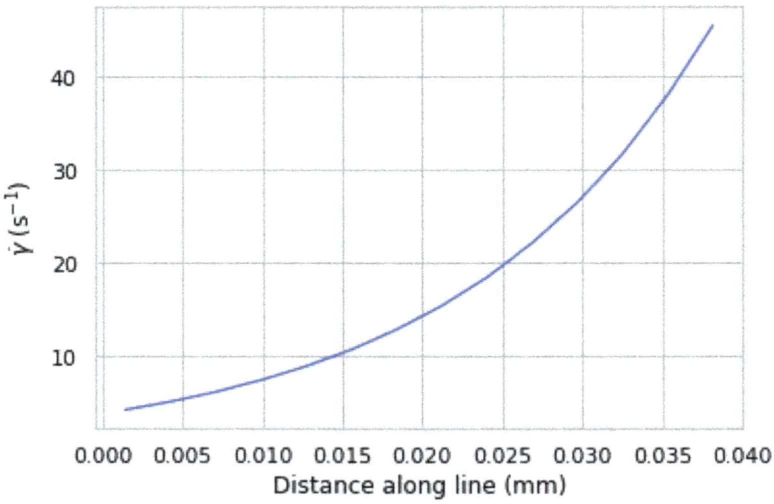

Comparing different analytical functions

It is finally instructive to compare the shear rates measured using both of the analytical forms tested:

```
[11]: fig, ax = plt.subplots()
      ax.plot(scale, shear1)
      ax.plot(scale, test_func(scale, params2[0], params2[1]))
      ax.set(ylabel='$\dot{\gamma}$ (s$^{-1}$)', xlabel='Distance along line (mm)');
```

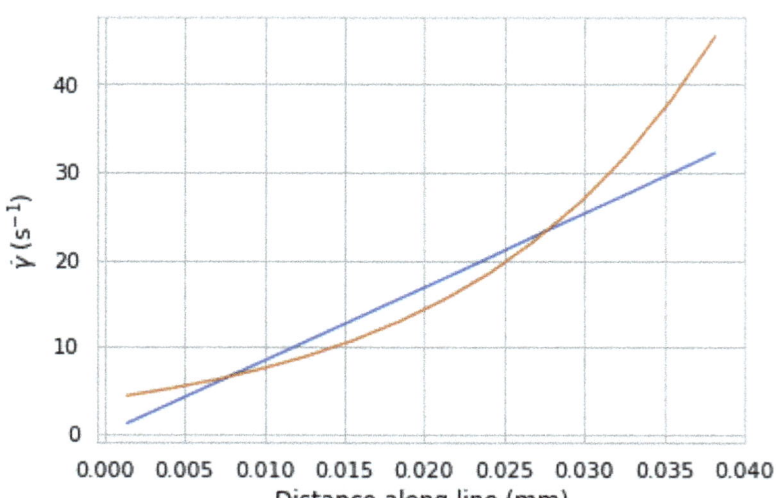

From the graph above, we see that although the shear rates calculated using the different functional forms broadly agree across the majority of the system, there nonetheless exist some notable differences, especially toward the top of the system. As such, it is important, wherever possible, to make an informed choice as to the fitting function used, ideally based on a physical knoweldge of the system in question. For example, in the current example, the parabolic form would make more sense, as the system is relatively fast-flowing, and the exponential form is more typically used to describe creeping flow.

Have a play!

Now that you are familiar with the process of extracting shear rate information from PEPT data, feel free to better familiarise yourself with the process, and see how you can adapt it—for example, what happens if we take a different line through the system—do we still obtain the same approximate functional form for our velocity profile? What about trying to fit a different functional form? Can you improve your data by using smaller or larger grid cells?

7.8 Jump velocities

In certain systems, notably fluidised beds or froth flotation devices (see, respectively, sections 8.3 and 8.7.5), it is valuable to gain insight not only into the motions of particles and/or fluids, but also the dynamics of *bubbles* rising through the system [139–141]. Though it is not possible—or, at least, it is highly difficult—for PEPT to *directly* track the motion of gas bubbles through a system, the technique can nonetheless be used to infer certain key properties of bubble behaviour. One such property is the rise velocity of a bubble (a key indicator of various processes within a fluidised bed), which can be effectively inferred by determining a quantity known as the *jump time* [10, 142, 143].

For simplicity, let us consider for now simply the case of a bubbling fluidised bed, in which bubbles of gas rise through a bulk particulate medium [144]. As bubbles travel upward through the bed, they will entrain particles in their wake [144] (see figure 7.23). As such, if we can identify points in time during which a particle is entrained in the wake of a bubble, and can then measure its velocity whilst entrained, we can measure the bubble rise velocity [142].

We have already discussed, in section 7.4.1, how to measure the velocity of a PEPT tracer. The remaining complexity, therefore, is to distinguish between the motion of a particle as it moves through the bulk phase of the fluidised bed and its motion when entrained by a bubble.

Although Stein *et al* [142] showed that significant, upward vertical particle movement only occurs in the presence of bubbles, a filtering algorithm must be adopted to gather only significant data as, for example, noise or a particle being close to a bubble can distort the mean upwards velocity. In order to perform the necessary filtering, a few conditions must be defined:

i) The particle's upward velocity must exceed a certain threshold.
ii) The particle must travel continuously upward for a minimum duration/distance.

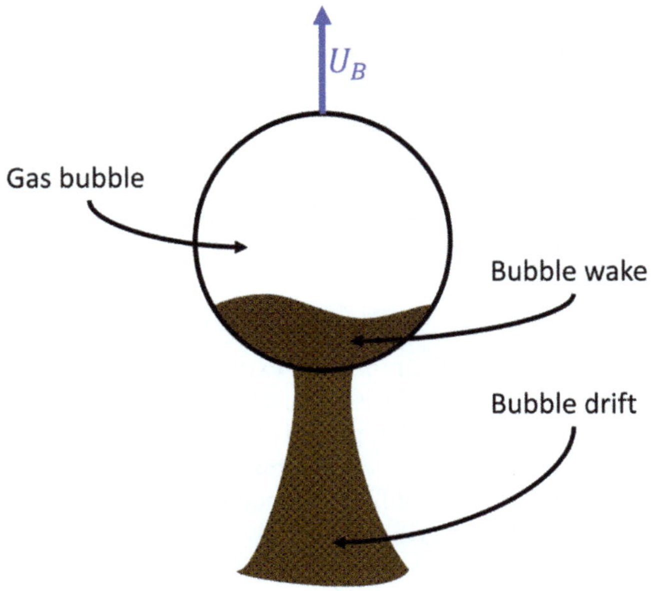

Figure 7.23. Schematic diagram showing the main regions associated with a bubble rising (with velocity U_B) through a gas-fluidised bed.

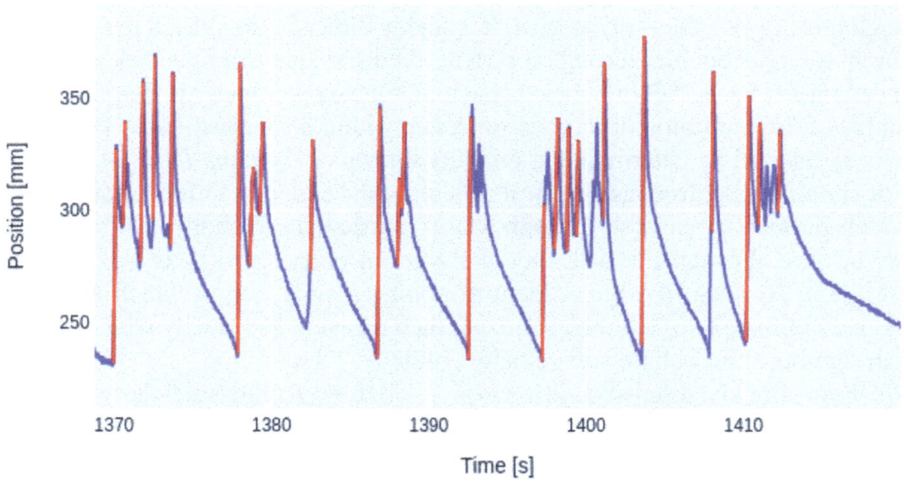

Figure 7.24. Example plot showing the full trajectory of a particle in the vertical direction (blue), with sections identified as 'jumps' (i.e. during which a particle is thought to be entrained in a bubble) by the above-described algorithm coloured in red.

Criterion (i) above allows the algorithm to distinguish between the relatively low upward motion expected due to a particle's pseudo-Brownian motion in a dynamic particulate system in the absence of entrainment [145] and the more significant velocity expected of a rising bubble. Criterion (ii), meanwhile, ensures that noise or erroneous data points are not falsely interpreted as 'jumps' due to particle entrainment. Sadly, there does not exist a single, universally-applicable set of values for the minimum jump velocity and minimum jump duration/distance, though in general the algorithm's output is not overly sensitive to the choice thereof—i.e. there exists a reasonably large range of values which can be chosen without distorting the jump time data. Typically, for relatively clean data, the minimum jump distance need only correspond to a few consecutive data points.

By traversing through a PEPT data-set applying the above criteria to the recorded particle motion, a significant amount of data can be acquired, allowing detailed bubble rise velocity distributions to be determined. An example of an algorithm capable of doing so is shown in the example below, and an example such a method successfully isolating 'jumps' can be seen in figure 7.24.

7.8.1 Interactive example: jump velocities

The notebook files accompanying this section are available in the supplementary data online at https://iopscience.iop.org/book/978-0-7503-3071-8.

Tutorial: computing jump velocities from PEPT data

In this tutorial, we will demonstrate how to use PEPT data to determine a fluidised bed's **jump time**, which can be used to infer the rise velocity of a bubble through the bed. For this calculation, all we require is time and position data, which we import and extract below:

[1]:
```
# Importing necessary libraries for numerical array manipulation and plotting
import numpy as np
import matplotlib.pyplot as plt

# Set Matplotlib theme (try "darkgrid" too)
import seaborn
seaborn.set_style("whitegrid")
```

[2]:
```
# Loading relevant data
filepath = "Data/FluidisedBedExample.csv"
full_trace = np.loadtxt(filepath, skiprows=16)

# Extracting time and position data in both the horizontal (x) and vertical (y)
# directions
t = full_trace[:, 0]
x = full_trace[:, 1]
y = full_trace[:, 2]
```

Initialising variables

The two most important variables to define are the *minimum jump time* and *minimum jump distance* which are used to filter out jumps that were too quick or too short. This can happen due to measurement noise or the particle being entrained in a bubble; however, no universal values exist for which all data will be valid—try changing the values below to see their effects:

```
[3]: # User-defined algorithm parameters:

     # Minimum jump time (ms)
     min_time = 100

     # Minimum jump distance (mm)
     min_distance = 30
```

For this specific implementation, we also need to initialise a few variables to help us keep track of the state of the trajectory:

```
[4]: # Variables to keep track of during computation

     # Boolean flag to save the state of a jump
     #     True  -> Particle is jumping
     #     False -> Particle is not jumping
     flag = False

     # The position at which the current jump started
     start_position = None

     # The time at which the current jump started
     start_time = None

     # Empty lists to solve the end results for
     # The duration of the jump in ms
     jump_time = []

     # The distance of the jump in mm
     jump_distance = []

     # The velocity of the jump in m/s
     jump_velocity = []
```

Traversing through the data to access internal 'jump-state'

While traversing through the tracer locations as the experiment progresses, we are checking whether the particle rises or falls; if the conditions above are met, a jump is recorded in the `jump_time`, `jump_distance` and `jump_velocity` variables.

[5]:
```
# Cycling through all data points with an assigned velocity
i = 0
while i < (len(y) - 1):
    # Incrementing i to move forward to next time step
    i = i + 1

    # Define current time:
    current_time = t[i]

    # Define old and new position
    old_position = y[i - 1]
    current_position = y[i]

    # Find if the particle is moving upwards:
    if current_position > old_position:
        # If the particle is moving upwards and the flag is false
        # A new possible jump started:
        if flag == False:
            # Set the jump flag to True
            flag = True

            # Remember the Current position and time of the particle
            start_position = current_position
            start_time = current_time
        else:
            # Otherwise the particle is already flagged as being in a jump
            # Therefore continuing to the next timestep:
            continue
    else:
        # Particle is steady or moving downwards.
        # If the particle was in a jump, the jump ended now
        if flag == True:
            # Unset the jump flag
            flag = False

            # Check if jump was valid:
            if (current_position - start_position >= min_distance and
                    current_time - start_time >= min_time):
                # Jump is valid and can be counted:
                jump_time.append(current_time - start_time)
                jump_distance.append(current_position - start_position)
                jump_velocity.append(jump_distance[-1] / jump_time[-1])
            else:
                # Jump was invalid and is ignored
                continue
        else:
            # Particle is going down and is already not jumping
            continue
```

After recording particle jumps, we can visualise the data we acquired:

[6]:
```
# Plotting the data as a histogram
fig, ax = plt.subplots()

ax.hist(jump_velocity, density=False, bins=40)
ax.set(ylabel="Counts", xlabel="Jump Velocity ($m / s$)");
```

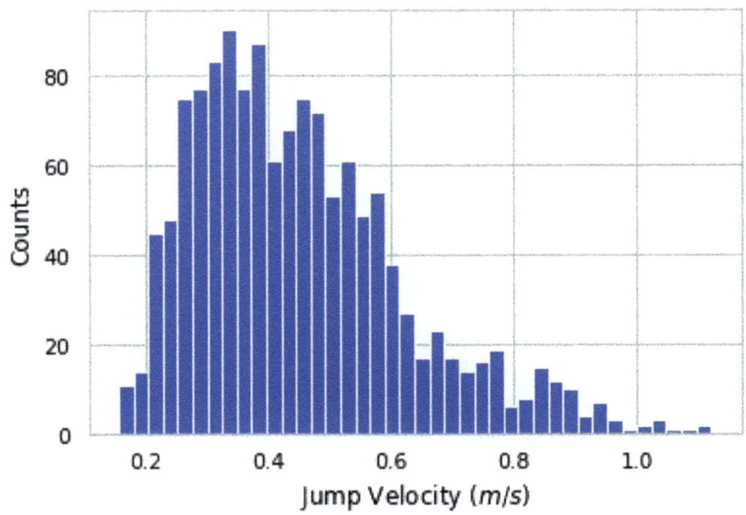

[7]:
```
# Plot the particle velocity recorded relative to the distance travelled
fig, ax = plt.subplots()

xlim = [1, 1000]
ylim = [0.1, 2]

ax.hist2d(jump_distance, jump_velocity, range=[xlim, ylim], bins=[40, 40],
          cmap="Blues", alpha=0.5)
ax.scatter(jump_distance, jump_velocity, s=2, alpha=0.3)

ax.set(
    xscale='log',
    yscale='log',
    xlabel='Distance (m)',
    ylabel='Velocity (m / s)',
    xlim=xlim,
    ylim=ylim,
);
```

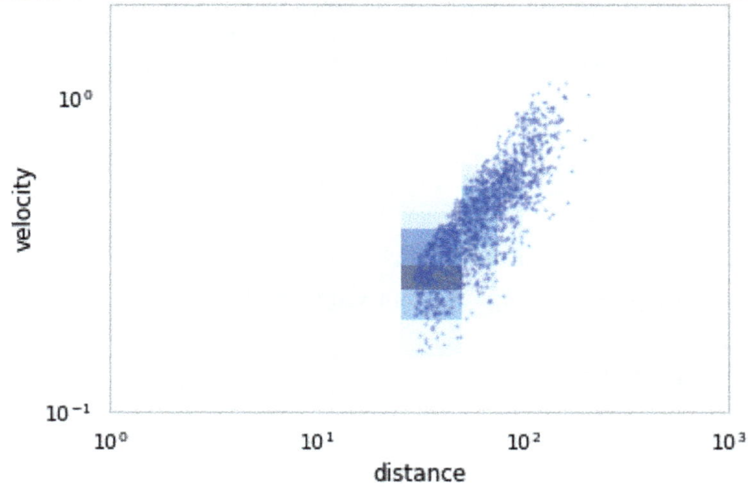

7.9 System-specific quantities

In this chapter, we have introduced some of the most commonly-used and widely-applicable quantities and fields which may be extracted from PEPT data. It is important to emphasise, however, that the above does not represent an exhaustive list; in addition to the above, there exists a wide array of less-commonly-used but nonetheless useful parameters, many of which are (usefully) applicable only to certain systems, processes and/or flow regimes. A number of these more specialised parameters are introduced in chapter 8, where they may be more easily contextualised, and references provided to the papers in which their implementation and use are described.

Finally, we feel it valuable to note that the parameters described in this book—both here and in chapter 8—represent only those that have been utilised by past researchers. There are practically limitless additional manners in which PEPT data may potentially be used to extract meaningful information from diverse systems, and we wholeheartedly recommend that you, as a PEPT user, attempt to develop and utilise new methods of data extraction and analysis above and beyond those described here.

References

[1] Michael B and Michael E A 1991 Fluidised-bed granulation: a chronology *Drug Dev. Ind. Pharm.* **17** 1437–63

[2] Rapagnà S, Jand N, Kiennemann A and Foscolo P U 2000 Steam-gasification of biomass in a fluidised-bed of olivine particles *Biomass Bioenergy* **19** 187–97

[3] Nienow A W 1995 Fluidised bed granulation and coating: applications to materials, agriculture and biotechnology *Chem. Eng. Commun.* **139** 233–53

[4] Hamilton G E, Morton P H, Young T W and Lyddiatt A 1999 Process intensification by direct product sequestration from batch fermentations: application of a fluidised bed, multi-bed external loop contactor *Biotechnol. Bioeng.* **64** 310–21

[5] Bentham A C, Kwan C C, Boerefijn R and Ghadiri M 2004 Fluidised-bed jet milling of pharmaceutical powders *Powder Technol.* **141** 233–8

[6] Di Matteo P, Donsi G and Ferrari G 2003 The role of heat and mass transfer phenomena in atmospheric freeze-drying of foods in a fluidised bed *J. Food Eng.* **59** 267–75

[7] Bokkers G A, van Sint Annaland M and Kuipers J A M 2004 Mixing and segregation in a bidisperse gas-solid fluidised bed: a numerical and experimental study *Powder Technol.* **140** 176–86

[8] Hoomans B P B, Kuipers J A M and van Swaaij W P M 2000 Granular dynamics simulation of segregation phenomena in bubbling gas-fluidised beds *Powder Technol.* **109** 41–8

[9] Porzuczek J 2014 Applications of electrical capacitance tomography for research on phenomena occurring in the fluidised bed reactors *Chem. Process Eng.* **35** 397–408

[10] Stein M, Ding Y L, Seville J P K and Parker D J 2000 Solids motion in bubbling gas fluidised beds *Chem. Eng. Sci.* **55** 5291–300

[11] Stein M, Martin T W, Seville J P K, McNeil P A and Parker D J 1997 Positron emission particle tracking: particle velocities in gas fluidised beds, mixers and other applications *Non-invasive Monitoring of Multiphase Flows* (Amsterdam: Elsevier) pp 309–33

[12] Windows-Yule C R K, Rosato A D, Rivas N and Parker D J 2014 Influence of initial conditions on granular dynamics near the jamming transition *New J. Phys.* **16** 063016
[13] Brems A, Chan C W, Seville J P K, Parker D and Baeyens J 2011 Modelling the transport disengagement height in fluidized beds *Adv. Powder Technol.* **22** 155–61
[14] Wei F, Lin H, Cheng Y, Wang Z and Jin Y 1998 Profiles of particle velocity and solids fraction in a high-density riser *Powder Technol.* **100** 183–9
[15] Duangkhamchan W, Ronsse F, Dewettinck K and Pieters J G 2011 CFD study of solids concentration in a fluidised-bed coater with variation of atomisation air pressure *Powder Technol.* **212** 103–14
[16] Wildman R D, Huntley J M, Hansen J-P, Parker D J and Allen D A 2000 Single-particle motion in three-dimensional vibrofluidized granular beds *Phys. Rev.* E **62** 3826
[17] Windows-Yule C R K 2020 Ensuring adequate statistics in particle tracking experiments *Particuology* **59** 43–54
[18] Sunthar P and Kumaran V 1999 Temperature scaling in a dense vibrofluidized granular material *Phys. Rev.* E **60** 1951
[19] Warr S, Huntley J M and Jacques G T H 1995 Fluidization of a two-dimensional granular system: experimental study and scaling behavior *Phys. Rev.* E **52** 5583
[20] Windows-Yule C R K and Parker D J 2014 Self-diffusion, local clustering and global segregation in binary granular systems: the role of system geometry *Powder Technol.* **261** 133–42
[21] Windows-Yule C R K, Moore A, Wellard C, Werner D, Parker D J and Seville J P K 2020 Particle distributions in binary gas-fluidised beds: shape matters—but not much *Chem. Eng. Sci.* **216** 115440
[22] Windows-Yule C R K, Gibson S, Parker D J, Kokalova T Z and Seville J P K 2020 Effect of distributor design on particle distribution in a binary fluidised bed *Powder Technol.* **367** 1–9
[23] Windows-Yule C R K, Hart-Villamil R, Ridout T, Kokalova T and Nogueira-Filho J C 2020 Positron emission particle tracking for liquid-solid mixing in stirred tanks *Chem. Eng. Technol.* **43** 1939–50
[24] Reimann J, Vicente J, Brun E, Ferrero C, Gan Y and Rack A 2017 x-ray tomography investigations of mono-sized sphere packing structures in cylindrical containers *Powder Technol.* **318** 471–83
[25] Chang C-C, Kuo-Dahab C, Chapman T and Mei Y 2019 Anaerobic digestion, mixing, environmental fate, and transport *Water Environ. Res.* **91** 1210–22
[26] Bowler A L, Bakalis S and Watson N J 2020 A review of in-line and on-line measurement techniques to monitor industrial mixing processes *Chem. Eng. Res. Des.* **153** 463–95
[27] Tagawa A, Dohi N and Kawase Y 2006 Dispersion of floating solid particles in aerated stirred tank reactors: minimum impeller speeds for off-surface and ultimately homogeneous solid suspension and solids concentration profiles *Ind. Eng. Chem. Res.* **45** 818–29
[28] Zenz F A and Weil N A 1958 A theoretical-empirical approach to the mechanism of particle entrainment from fluidized beds *AIChE J.* **4** 472–9
[29] Fournol A B, Bergougnou M A and Baker C G J 1973 Solids entrainment in a large gas fluidized bed *Can. J. Chem. Eng.* **51** 401–4
[30] Sciaskol M, Raczek J and Bandrowski J 1988 Model of gas flow above a bubbling fluidized bed: prediction of splash zone height *Chem. Eng. Process.* **24** 49–55
[31] Sciasko M, Bandrowski J and Raczek J 1991 On the entrainment of solid particles from a fluidized bed *Powder Technol.* **66** 33–9

[32] Luding S, Clément E, Blumen A, Rajchenbach J and Duran J 1994 Studies of columns of beads under external vibrations *Phys. Rev.* E **49** 1634
[33] Luding S, Herrmann H J and Blumen A 1994 Simulations of two-dimensional arrays of beads under external vibrations: scaling behavior *Phys. Rev.* E **50** 3100
[34] Yang X and Candela D 2000 Potential energy in a three-dimensional vibrated granular medium measured by NMR imaging *Phys. Rev. Lett.* **85** 298
[35] Windows-Yule C R K and Parker D J 2014 Center of mass scaling in three-dimensional binary granular systems *Phys. Rev.* E **89** 062206
[36] Windows-Yule K and Parker D 2015 Density-driven segregation in binary and ternary granular systems *KONA Powder Part. J.* **32** 163–75
[37] Nicuşan A L and Windows-Yule C R K 2020 Positron emission particle tracking using machine learning *Rev. Sci. Instrum.* **91** 013329
[38] Odo A E, Govender I, Buffler A and Franzidis J-P 2019 A PEPT algorithm for predefined positions of radioisotopes relative to the tracer particle *Appl. Radiat. Isot.* **151** 299–309
[39] Blakemore D M, Govender I, McBride A T and Mainza A N 2019 Multiple particle tracking in PEPT using Voronoi tessellations *Chem. Eng. Sci.* **207** 780–9
[40] Wiggins C, Santos R and Ruggles A 2016 A novel clustering approach to positron emission particle tracking *Nucl. Instrum. Methods Phys. Res.* A **811** 18–24
[41] Bittorf K J and Kresta S M 2003 Prediction of cloud height for solid suspensions in stirred tanks *Chem. Eng. Res. Des.* **81** 568–77
[42] Yamazaki H, Tojo K and Miyanami K 1986 Concentration profiles of solids suspended in a stirred tank *Powder Technol.* **48** 205–16
[43] Muzzio F J, Robinson P, Wightman C and Brone D 1997 Sampling practices in powder blending *Int. J. Pharm.* **155** 153–78
[44] Cullen P J, Romañach R J, Abatzoglou N and Rielly C D 2015 *Pharmaceutical Blending and Mixing* (New York: Wiley)
[45] Rosato A D, Blackmore D L, Zhang N and Lan Y 2002 A perspective on vibration-induced size segregation of granular materials *Chem. Eng. Sci.* **57** 265–75
[46] Rosato A, Strandburg K J, Prinz F and Swendsen R H 1987 Why the Brazil nuts are on top: size segregation of particulate matter by shaking *Phys. Rev. Lett.* **58** 1038
[47] Burtally N, King P J, Swift M R and Leaper M 2003 Dynamical behaviour of fine granular glass/bronze mixtures under vertical vibration *Granul. Matter* **5** 57–66
[48] Zeilstra C, Van der Hoef M A and Kuipers J A M 2006 Simulation study of air-induced segregation of equal-sized bronze and glass particles *Phys. Rev.* E **74** 010302
[49] Rosato A D and Windows-Yule C 2020 *Segregation in Vibrated Granular Systems* (New York: Academic)
[50] Breu A P J, Ensner H-M, Kruelle C A and Rehberg I 2003 Reversing the Brazil-nut effect: competition between percolation and condensation *Phys. Rev. Lett.* **90** 014302
[51] Hong D C, Quinn P V and Luding S 2001 Reverse Brazil nut problem: competition between percolation and condensation *Phys. Rev. Lett.* **86** 3423
[52] Lacey P M 1943 The mixing of solid particles *Trans. Inst. Chem. Eng.* **21** 53–9
[53] Lacey P M C 1954 Developments in the theory of particle mixing *J. Appl. Chem* **4** 257–68
[54] Yang S C 2006 Density effect on mixing and segregation processes in a vibrated binary granular mixture *Powder Technol.* **164** 65–74
[55] Khakhar D V, McCarthy J J, Shinbrot T and Ottino J M 1997 Transverse flow and mixing of granular materials in a rotating cylinder *Phys. Fluids* **9** 31–43

[56] Yang Z, Parker D J, Fryer P J, Bakalis S and Fan X 2006 Multiple-particle tracking—an improvement for positron particle tracking *Nucl. Instrum. Methods Phys. Res.* A **564** 332–8

[57] Cleary P W, Metcalfe G and Liffman K 1998 How well do discrete element granular flow models capture the essentials of mixing processes? *Appl. Math. Modelling* **22** 995–1008

[58] Porion P, Sommier N, Faugère A-M and Evesque P 2004 Dynamics of size segregation and mixing of granular materials in a 3D-blender by NMR imaging investigation *Powder Technol.* **141** 55–68

[59] Windows-Yule C R K, Weinhart T, Parker D J and Thornton A R 2014 Influence of thermal convection on density segregation in a vibrated binary granular system *Phys. Rev.* E **89** 022202

[60] Windows-Yule C R K 2016 Convection and segregation in fluidised granular systems exposed to two-dimensional vibration *New J. Phys.* **18** 033005

[61] Windows-Yule C R K, Scheper B J, van der Horn A J, Hainsworth N, Saunders J, Parker D J and Thornton A R 2016 Understanding and exploiting competing segregation mechanisms in horizontally rotated granular media *New J. Phys.* **18** 023013

[62] Windows-Yule C R K, Van Der Horn A J, Tunuguntla D R, Parker D J and Thornton A R 2017 Inducing axial banding in bidisperse-by-density granular systems using noncylindrical tumbler geometries *Phys. Rev. Appl* **8** 024010

[63] Chou S-H, Song Y-L and Hsiau S-S 2017 A study of the mixing index in solid particles *KONA Powder Part. J.* **34** 275–81

[64] Kramer H A 1968 *Effect of Grain Velocity and Flow Rate upon the Performance of a Diverter-Type Sampler* (Washington, DC: U.S. Dept. of Agriculture, Agricultural Research Service)

[65] M M H D Arntz 2010 Modeling of particle segregation in a rotating drum *PhD Thesis* University of Wageningen

[66] Schutyser M A I, Padding J T, Weber F J, Briels W J, Rinzema A and Boom R 2001 Discrete particle simulations predicting mixing behavior of solid substrate particles in a rotating drum fermenter *Biotechnol. Bioeng.* **75** 666–75

[67] Schutyser M A I, Weber F J, Briels W J, Boom R M and Rinzema A 2002 Three-dimensional simulation of grain mixing in three different rotating drum designs for solid-state fermentation *Biotechnol. Bioeng.* **79** 284–94

[68] Arntz M M H D, den Otter W K, Briels W J, Bussmann P J T, Beeftink H H and Boom R M 2008 Granular mixing and segregation in a horizontal rotating drum: a simulation study on the impact of rotational speed and fill level *AIChE J.* **54** 3133–46

[69] Arntz M M H D, den Otter W K, Beeftink H H, Boom R M and Briels W J 2011 Repeated segregation and energy dissipation in an axially segregated granular bed *Europhys. Lett.* **92** 54004

[70] Arntz M M H D, den Otter W K, Beeftink H H, Boom R M and Briels W J 2013 The influence of end walls on the segregation pattern in a horizontal rotating drum *Granul. Matter* **15** 25–38

[71] Arntz M M H D, Beeftink H H, den Otter W K, Briels W J and Boom R M 2014 Segregation of granular particles by mass, radius, and density in a horizontal rotating drum *AIChE J.* **60** 50–9

[72] Guida A, Fan X, Parker D J, Nienow A W and Barigou M 2009 Positron emission particle tracking in a mechanically agitated solid-liquid suspension of coarse particles *Chem. Eng. Res. Des.* **87** 421–9

[73] Pérez-Mohedano R, Letzelter N, Amador C, VanderRoest C T and Bakalis S 2015 Positron emission particle tracking (PEPT) for the analysis of water motion in a domestic dishwasher *Chem. Eng. J.* **259** 724–36
[74] Olafsen J S and Urbach J S 1999 Velocity distributions and density fluctuations in a granular gas *Phys. Rev.* E **60** R2468
[75] Brey J J, Cubero D and Ruiz-Montero M 1999 High energy tail in the velocity distribution of a granular gas *Phys. Rev.* E **59** 1256
[76] Rouyer F and Menon N 2000 Velocity fluctuations in a homogeneous 2D granular gas in steady state *Phys. Rev. Lett.* **85** 3676
[77] Brilliantov N V and Pöschel T 2000 Velocity distribution in granular gases of viscoelastic particles *Phys. Rev.* E **61** 5573
[78] Javier Brey J and Ruiz-Montero M J 2003 Velocity distribution of fluidized granular gases in the presence of gravity *Phys. Rev.* E **67** 021307
[79] Scholz C and Pöschel T 2017 Velocity distribution of a homogeneously driven two-dimensional granular gas *Phys. Rev. Lett.* **118** 198003
[80] Pöschel T and Brilliantov N V 2003 *Granular Gas Dynamics* vol 624 (Berlin: Springer)
[81] Puglisi A, Loreto V, Marini Bettolo Marconi U, Petri A and Vulpiani A 1998 Clustering and non-Gaussian behavior in granular matter *Phys. Rev. Lett.* **81** 3848
[82] Kudrolli A and Henry J 2000 Non-Gaussian velocity distributions in excited granular matter in the absence of clustering *Phys. Rev.* E **62** R1489
[83] Kawarada A and Hayakawa H 2004 Non-Gaussian velocity distribution function in a vibrating granular bed *J. Phys. Soc. Japan* **73** 2037–40
[84] Li Y, Fan H and Fan X 2014 Identify of flow patterns in bubbling fluidization *Chem. Eng. Sci.* **117** 455–64
[85] Xu W, Tan Y, Li M, Sun J, Xie D and Liu Z 2020 Effects of surface vortex on the drawdown and dispersion of floating particles in stirred tanks *Particuology* **49** 159–68
[86] Asencio K, Bramer-Escamilla W, Gutiérrez G and Sánchez I 2015 Electrical capacitance sensor array to measure density profiles of a vibrated granular bed *Powder Technol.* **270** 10–9
[87] Eshuis P, van der Weele K, van der Meer D and Lohse D 2005 Granular Leidenfrost effect: experiment and theory of floating particle clusters *Phys. Rev. Lett.* **95** 258001
[88] Lim E W C 2010 Granular Leidenfrost effect in vibrated beds with bumpy surfaces *Eur. Phys. J.* E **32** 365–75
[89] Windows-Yule C R K, Rivas N, Parker D J and Thornton A R 2014 Low-frequency oscillations and convective phenomena in a density-inverted vibrofluidized granular system *Phys. Rev.* E **90** 062205
[90] Menéndez H T, Sack A and Pöschel T Granular Leidenfrost effect in microgravity *Granular Matter* **22** 67
[91] Wang C, Lv Z and Li D 2008 Experimental study on gas-solids flows in a circulating fluidised bed using electrical capacitance tomography *Powder Technol.* **185** 144–51
[92] Lee M H 1983 Can the velocity autocorrelation function decay exponentially? *Phys. Rev. Lett.* **51** 1227
[93] Alder B J and Wainwright T E 1967 Velocity autocorrelations for hard spheres *Phys. Rev. Lett.* **18** 988
[94] Luding S 2000 On the relevance of 'molecular chaos' for granular flows *J. Appl. Math. Mech.* **80** 9–12

[95] Pöschel T, Brilliantov N V and Schwager T 2002 Violation of molecular chaos in dissipative gases *Int. J. Mod. Phys.* C **13** 1263–72
[96] Baxter G W and Olafsen J S 2007 Experimental evidence for molecular chaos in granular gases *Phys. Rev. Lett.* **99** 028001
[97] Windows-Yule C R K and Parker D J 2013 Boltzmann statistics in a three-dimensional vibrofluidized granular bed: idealizing the experimental system *Phys. Rev.* E **87** 022211
[98] Wildman R D, Hansen J-P and Parker D J 2002 Velocity auto-correlation functions in three-dimensional vibro-fluidized granular beds *Phys. Fluids* **14** 232–9
[99] Hansen J-P and McDonald I R 1990 *Theory of Simple Liquids* (Amsterdam: Elsevier)
[100] Wildman R D, Huntley J M and Jean-Pierre H 2001 *Experimental studies of vibro-fluidised granular beds Granular Gases* (Berlin: Springer) pp 215–32
[101] Wildman R D and Huntley J M 2000 Novel method for measurement of granular temperature distributions in two-dimensional vibro-fluidised beds *Powder Technol.* **113** 14–22
[102] Abate A R and Durian D J 2006 Approach to jamming in an air-fluidized granular bed *Phys. Rev.* E **74** 031308
[103] Savage S B 1993 Disorder, diffusion and structure formation in granular flows *Disorder and Granular Media* ed D Bideau and A Hansen (Amsterdam: North-Holland)
[104] Hill K M and Kakalios J 1994 Reversible axial segregation of binary mixtures of granular materials *Phys. Rev.* E **49** R3610
[105] Livorati A L P, Kroetz T, Dettmann C P, Caldas I L and Leonel E D 2018 Transition from normal to ballistic diffusion in a one-dimensional impact system *Phys. Rev.* E **97** 032205
[106] Silbert L E, Ertaş D, Grest G S, Halsey T C and Levine D 2002 Analogies between granular jamming and the liquid-glass transition *Phys. Rev.* E **65** 051307
[107] Bodrova A, Dubey A K, Puri S and Brilliantov N 2012 Intermediate regimes in granular Brownian motion: superdiffusion and subdiffusion *Phys. Rev. Lett.* **109** 178001
[108] Kim E, Martínez A J, Phenisee S E, Kevrekidis P G, Porter M A and Yang J 2018 Direct measurement of superdiffusive energy transport in disordered granular chains *Nat. Commun.* **9** 1–6
[109] Marty G and Dauchot O 2005 Subdiffusion and cage effect in a sheared granular material *Phys. Rev. Lett.* **94** 015701
[110] Martin T W, Seville J P K and Parker D J 2007 A general method for quantifying dispersion in multiscale systems using trajectory analysis *Chem. Eng. Sci.* **62** 3419–28
[111] Kuo H P, Knight P C, Parker D J and Seville J P K 2005 Solids circulation and axial dispersion of cohesionless particles in a v-mixer *Powder Technol.* **152** 133–40
[112] Marigo M, Cairns D L, Davies M, Ingram A and Stitt E H 2011 Developing mechanistic understanding of granular behaviour in complex moving geometry using the discrete element method: Part B: Investigation of flow and mixing in the turbula ® mixer *Powder Technol.* **212** 17–24
[113] Marigo M, Davies M, Leadbeater T, Cairns D L, Ingram A and Stitt E H 2013 Application of positron emission particle tracking (PEPT) to validate a discrete element method (DEM) model of granular flow and mixing in the turbula mixer *Int. J. Pharm.* **446** 46–58
[114] Jaeger H M, Nagel S R and Behringer R P 1996 Granular solids, liquids, and gases *Rev. Mod. Phys.* **68** 1259
[115] Ogawa S 1978 Multitemperature theory of granular materials *Proc. of the US-Japan Seminar on Continuum Mechanical and Statistical Approaches in the Mechanics of Granular Materials, 1978* (Tokyo: Gakajutsu Bunken Fukyu-Kai) pp 208–17

[116] Ogawa S, Umemura A and Oshima N 1980 On the equations of fully fluidized granular materials *Z. Angew. Math. Phys.* **31** 483–93

[117] van der Meer D and Reimann P 2006 Temperature anisotropy in a driven granular gas *Europhys. Lett.* **74** 384

[118] Edwards S F and Oakeshott R B S 1989 Theory of powders *Physica* A **157** 1080–90

[119] Mehta A and Edwards S F 1989 Statistical mechanics of powder mixtures *Physica* A **157** 1091–100

[120] Coniglio A and Nicodemi M 2001 A statistical mechanics approach to the inherent states of granular media *Physica* A **296** 451–9

[121] Windows-Yule C R K 2017 Do granular systems obey statistical mechanics? a review of recent work assessing the applicability of equilibrium theory to vibrationally excited granular media *Int. J. Mod. Phys.* B **31** 1742010

[122] Goldhirsch I 2008 Introduction to granular temperature *Powder Technol.* **182** 130–6

[123] Ramırez R and Soto R 2003 Temperature inversion in granular fluids under gravity *Physica* A **322** 73–80

[124] Paolotti D, Barrat A, Marconi U M B and Puglisi A 2004 Thermal convection in monodisperse and bidisperse granular gases: a simulation study *Phys. Rev.* E **69** 061304

[125] Brey J J, Ruiz-Montero M J and Moreno F 2001 Hydrodynamics of an open vibrated granular system *Phys. Rev.* **63** 061305

[126] van Beijeren H and Ernst M H 1973 The modified Enskog equation *Physica* **68** 437–56

[127] Wildman R D, Huntley J M and Hansen J-P 1999 Self-diffusion of grains in a two-dimensional vibrofluidized bed *Phys. Rev.* E **60** 7066

[128] Wildman R D 2002 Measurement of the first and second moments of the velocity distribution in two-dimensional vibro-fluidised granular beds *Powder Technol.* **127** 203–11

[129] Wildman R D and Parker D J 2002 Coexistence of two granular temperatures in binary vibrofluidized beds *Phys. Rev. Lett.* **88** 064301

[130] Feitosa K and Menon N 2002 Breakdown of energy equipartition in a 2D binary vibrated granular gas *Phys. Rev. Lett.* **88** 198301

[131] Galvin J E, Dahl S R and Hrenya C M 2005 On the role of non-equipartition in the dynamics of rapidly flowing granular mixtures *J. Fluid Mech.* **528** 207–32

[132] Yoon D K and Jenkins J T 2006 The influence of different species' granular temperatures on segregation in a binary mixture of dissipative grains *Phys. Fluids* **18** 073303

[133] Windows-Yule C R K and Parker D J 2014 Energy non-equipartition in strongly convective granular systems Eur *Phys. J.* E **37** 17

[134] Govender I, Mangesana N, Mainza A N and Franzidis J-P 2011 Measurement of shear rates in a laboratory tumbling mill *Miner. Eng.* **24** 225–9

[135] Nakagawa M, Altobelli S A, Caprihan A and Fukushima E 1997 NMR measurement and approximate derivation of the velocity depth-profile of granular flow in a rotating, partially filled, horizontal cylinder *Powders Grains* **97** 447–50

[136] Yamane K, Nakagawa M, Altobelli S A, Tanaka T and Tsuji Y 1998 Steady particulate flows in a horizontal rotating cylinder *Phys. Fluids* **10** 1419–27

[137] McBride A, Govender I, Powell M and Cloete T 2004 Contributions to the experimental validation of the discrete element method applied to tumbling mills *Eng. Comput.* **21** 119–36

[138] Govender I, McBride A T and Powell M S 2004 Improved experimental tracking techniques for validating discrete element method simulations of tumbling mills *Exp. Mech.* **44** 593

[139] Pyle D L and Harrison D 1967 The rising velocity of bubbles in two-dimensional fluidised beds *Chem. Eng. Sci.* **22** 531–5

[140] Fan X, Parker D J, Yang Z, Seville J P K and Baeyens Jan 2008 The effect of bed materials on the solid/bubble motion in a fluidised bed *Chem. Eng. Sci.* **63** 943–50

[141] Burgess J M and Calderbank P H 1975 The measurement of bubble properties in two-phase dispersions–III: Bubble properties in a freely bubbling fluidised-bed *Chem. Eng. Sci.* **30** 1511–8

[142] Stein M G 1999 Particle Motion in Fluidised Beds *PhD Thesis University of Birmingham*

[143] Van Lam Cheun U Y 2010 Solids motion in fluidised beds of fine particles *PhD Thesis* University of Birmingham

[144] Grace J R, Bi X and Ellis N 2020 *Essentials of Fluidization Technology* (New York: Wiley)

[145] Caram H and Hong D C 1991 Random-walk approach to granular flows *Phys. Rev. Lett.* **67** 828

[146] Windows-Yule C R K *et al* 2014 Effects of packing density on the segregative behaviors of granular systems *Phys. Rev. Lett.* **112** 098001

[147] Govender I *et al* 2010 Using positron emission particle tracking (PEPT) to investigate industrial systems *XXV Int. Mineral Processing Congress 2010, IMPC 2010* **vol 4** (no. 7) (Australasian Institute of Mining and Metallurgy)

[148] Press W H and Teukolsky S A 1990 Savitzky–Golay smoothing filters *Computers in Phys.* **4** 669–72

Part III

Applying the PEPT technique

Kit Windows-Yule, Leonard Nicuşan, Matthew T Herald, Samuel Manger and David Parker

Chapter 8

Applications

In this chapter, we discuss the various, diverse applications of PEPT which have been explored by researchers over the past 30 years, providing a review of the extensive literature published during this time. The review covers both the more 'pure-science' investigations using vibrofluidised beds and rotating drums to explore the fundamental physics of particulate systems, as well as the more 'applied' research, focussing on understanding and optimising the dynamics of fluidised beds, stirred-tank reactors, mills, mixers, cyclones, granulators and other widely-used process equipment. We discuss also nascent applications of PEPT in the geosciences and the biomedical field, where its recently-proven ability to track the motion of cells through living organisms may prove highly valuable.

In the preceding chapters, we have discussed extensively *how* PEPT works, and how it may be used. In the present chapter, we shift our focus to *why* we might want to use PEPT, outlining the various scientific and industrial applications of the technique, and providing an overview of the prior work that has been performed in each of these areas, as well as the significant potential for the technique in other, currently unexplored fields. We begin by discussing two experimental geometries (vibrofluidised beds [1] and rotating drums [2]) which have, to date, provided a majority of PEPT's 'pure science' output, before discussing the various industrial systems in which PEPT has been leveraged to perform more applied research.

8.1 Vibrated systems

8.1.1 Industrial and scientific applications

Vibrated systems are the subject of significant academic and industrial interest. Regarding the former, these systems are inherently simple in their design and operation, yet exhibit an extremely rich phenomenology [1], making them a highly valuable canonical system for the study of diverse processes in fields from geophysics [3, 4] to civil engineering [5, 6]. Perhaps most commonly, though, they are used to provide insight into the fundamental physics of particulate systems, exploring important and often scientifically fascinating phenomena such as jamming [7, 8],

doi:10.1088/978-0-7503-3071-8ch8

pattern formation [9, 10] and segregation [1]. The pleasing simplicity of vibrofluidised granular beds also allows the detailed comparison of granular systems and their dynamics to those of equivalent, classical systems, allowing their similarities and, crucially, their differences to be highlighted. To this end, significant work has been performed researching dilute, energetic 'granular gases' [11], whose non-Gaussian velocity distributions [12, 13], heterogeneous density distributions [14], anisotropic temperatures [15] and lack of energy equipartition [16] illustrate striking deviations from their classical counterparts.

In an industrial context, the vibratory excitation of bulk solids is of direct relevance to a number of processes including transport [17], sorting and separation [18–20], vibratory finishing [21, 22], grinding and attrition [23] and compaction [350, 351].

8.1.2 Value of PEPT

Though much work has been done using optical imaging to analyse the dynamics of quasi-two-dimensional vibrated beds [9, 12, 24, 25], such methodologies are of limited use in all but the most dilute three-dimensional systems. The often rapid dynamics of vibrated beds—especially in the study of granular gases, for example—make them similarly unsuitable for conventional tomographic techniques (though these have nonetheless proved of value for the study of less energetic, quasi-static systems [26–28]). Techniques such as MRI have also provided valuable insight into the dynamics of vibrated granular systems [29, 30], however this technique, unlike PEPT, typically carries severe restrictions in terms of system size and the materials used both in the construction of the system and the bed under investigation (see section 4.2.5), making it of limited use for industrial process equipment, where systems are typically larger and/or contain metallic components. As such, PEPT has proven extremely popular as a means of studying highly energetic granular fluids and larger, more industry-relevant systems.

8.1.3 Prior research

In the following subsections, we provide a brief overview of some of the ways in which PEPT has been used to further our understanding of these systems and their diverse properties and phenomena. In each case, we provide a succinct summary of relevant work, alongside references from which the interested reader may gain deeper insight.

8.1.3.1 Granular temperature

In addition to being the first technique to measure the temperature of a three-dimensional granular fluid [31], PEPT was also used to provide first insight into the scaling of a 3D system's granular temperature [32], later extending also to the case of binary 3D systems [33]. PEPT was also—alongside the 2D work of Feitosa and Menon [16]—the first technique to demonstrate directly that binary granular beds violate the equipartition theorem [34], with individual species in the same system possessing different granular temperatures. Later work also showed the extent of nonequipartition between dissimilar species is strongly dependent on the degree of mixing/segregation between coexisting species [35].

In addition to non-equipartition between differing species, granular media are also known to exhibit non-equipartition between different translational degrees of freedom [15, 36]. PEPT imaging of a three-dimensional vibrofluidised system [37] provided the first experimental evidence to support the theory of van der Meer *et al* [15] that this anisotropy is an *inherent property* of a vertically-vibrated bed.

8.1.3.2 Energy transfer

The granular temperature, and in particular its anisotropic and non-equipartitioned nature, are (as touched upon above) intimately linked to the manner in which energy is transferred—be it between different rotational and/or translational modes, different particle species, or between particles and their surroundings. One of the most fundamental processes underpinning the dynamics of vibrated granular systems is the transfer of energy from the driving wall(s) of the system to the granular bed housed within; as such, the determination of a suitable control parameter to characterise this energy input is the focus of much research.

A vibrating wall possesses a characteristic velocity $v = \omega A$ and corresponding dimensionless 'energy input parameter', commonly denoted S, given as [38–40]:

$$S = \frac{\omega^2 A^2}{gd} \tag{8.1}$$

where d is the (mean) particle diameter and g the gravitational acceleration. S and/or v are regularly taken as the key control parameters determining energy transfer to a granular system [41–46]. However, recent PEPT studies have shown that energy transfer to both dilute [47] and dense [48] granular systems depends sensitively on the individual values of the parameters ω and A. PEPT was also used to explore energy transfer in the case of aspherical particles [49], for which interactions with a vibrating surface can impart significant *rotational* energy as well as translational energy. It was shown that dilute systems comprising particles with greater rotational inertia consistently exhibit higher *translational* kinetic energies, even when possessing identical masses and material properties.

8.1.3.3 Convection

In addition to simply visualising convection rolls, as discussed in detail in [50], PEPT may also be used to quantify the strength of granular convection within a system.

The convective flow rate, J [51, 52], may be determined as follows: firstly, a two-dimensional velocity field is determined as described in section 7.4.5. From this field, the centre of convection, across which there is no horizontal component of velocity, may be determined. The flow rate J is then simply determined as the average of the vertical velocity magnitude for all data points falling within a single horizontal slice through the convection centre, i.e.

$$J = \frac{{}^{n}_{i}|v_y^i|}{2n} \tag{8.2}$$

where v_y^i is the vertical velocity corresponding to the ith data point in the central slice and n is the total number of data points in the slice.

PEPT studies have provided great insight into the granular convective phenomenon, providing the first experimental evidence of ‘wall-driven’ pseudo-thermal convection [50], and later showing how the direction, number, and strength of convection rolls may be altered *purely* by changing the properties of a system’s vertical bounding walls, without in any way altering the other properties of the system [53–55] (figure 8.1).

8.1.3.4 Mean squared displacement and self-diffusion

As discussed in section 7.5.1, the mean squared displacement of a PEPT tracer can be used to measure and quantify diffusion within a system, but is also of significant value in its own right.

The variation of mean squared displacement as a function of time can provide significant insight into the dynamics of a system indicating, for example, a shift from fluid-like motion (Fickian diffusion) to sub-diffusive motion as a system approaches the jamming point (see, for example, figure 8.2). Work using PEPT in the middle of the last decade demonstrated for the first time that the jamming point of a vibrated granular system possesses a history dependence [58], and that under suitable conditions a granular medium can in fact be subjected to a granular analogue of supercooling [59], exhibiting fluid-like motion under conditions where it would normally exist in a solid-like, jammed state [57]. Subsequent work showed also that the mean squared displacement of a vibrated granular system is sensitively dependent on the material of the container in which it is housed [60] and the geometry of the particles used [56], demonstrating that aspherical, rod-like particles are more prone to subdiffusive dynamics or jamming than spherical particles, but that the ‘doping’ of a system of such rod-like particles with a small volume of spheres can facilitate faster, more fluid-like dynamics.

The dependence of the self-diffusion coefficient, D, of a (dense) vibrated granular bed on key system parameters was explored by Wong *et al* [61, 62] who showed that dispersion in both the horizontal and vertical directions increases monotonically with

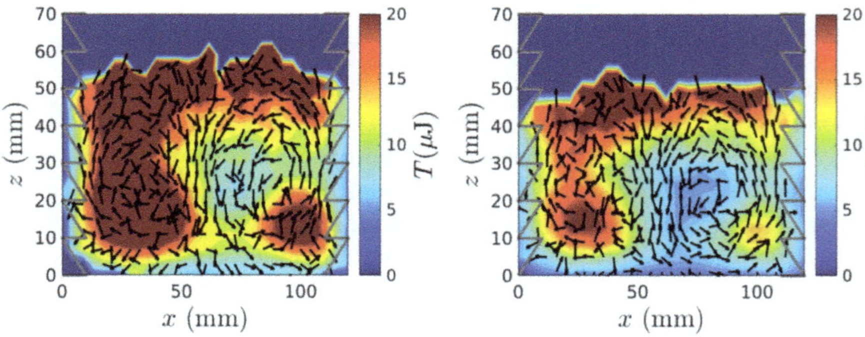

Figure 8.1. Examples of ‘normal’ (left) and ‘inverse’ (right) wall-driven convection imaged using PEPT. The colour map shown represents the variation of the local granular temperature, believed to drive this form of pseudo-thermal convective motion. Image reproduced with permission from Springer (GmbH) [55], Copyright © 2018, The Authors.

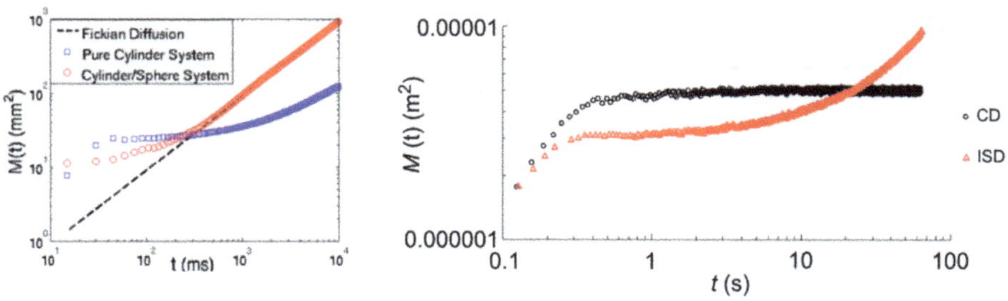

Figure 8.2. Examples of PEPT-measured mean squared displacement plots. The left-hand image (reproduced with permission from reference [56], Copyright 2016 by the American Physical Society) shows a transition from subdiffusive behaviour for a system of vertically-vibrated spherocylinders to 'normal', Fickian diffusion after the system is 'doped' with a small volume of spheres. The right-hand image (reproduced with permission from [57], © 2014 IOP Publishing Ltd and Deutsche Physikalische Gesellschaft. All rights reserved.) shows the two systems, identical other than in terms of their excitation history, existing in two differing dynamical states. Specifically, the system which has been continuously driven (CD) at a constant vibrational amplitude exists in a jammed state, while the system which was subjected to initial strong driving (ISD) maintains fluid-like dynamics.

the driving acceleration $\Gamma = \frac{\omega^2 A}{g}$, though the horizontal dispersion in all cases greatly exceeded that in the vertical direction. Conversely, horizontal dispersion was found to decrease with an increasing layer number, N_l, while, perhaps surprisingly, variations in N_l showed no clear influence on dispersion in the vertical direction. The influence of N_l or, more specifically, the aspect ratio, Æ, on the dispersion coefficient was further explored by Windows-Yule and Parker [63], who found a relation of the form:

$$D \propto \exp(\alpha Æ^{0.3}) \tag{8.3}$$

where α is a coefficient with a dependence on the dissipative properties of the granular bed in question.

8.1.3.5 Segregation

The topic of segregation driven by differences in particle material densities ('density segregation') was explored in detail in a series of PEPT studies performed by Windows-Yule *et al* [33, 57, 58, 63–65]. These experiments showed both dilute and dense, convective [64] and non-convective systems [58] to display a wide variation of segregation rates, steady-state segregation strengths, and segregation patterns, with the basic qualitative forms of these patterns holding for both binary and ternary systems [65] (see figure 8.3).

PEPT has also been used to explore two comparatively under-researched [1] forms of segregation: that due to differences in particles' restitution coefficients ('inelasticity segregation') [58, 66] and that driven by differences in particles' geometric properties ('shape segregation') [56, 67]. The PEPT work performed on inelasticity segregation showed that, for systems of moderate packing density, inelasticity effects significantly influence the segregation dynamics of granular beds, under certain conditions even outweighing those due to differences in density [58]. Conversely, for more extremal values of the packing density, η (i.e. granular gases and quasi-static flows) these effects

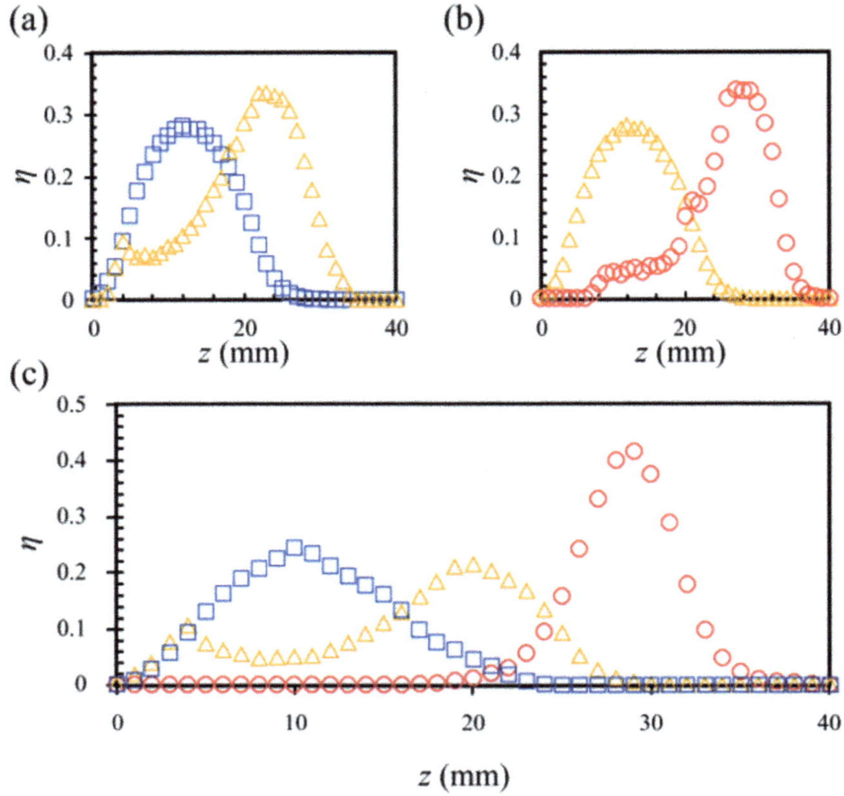

Figure 8.3. PEPT-measured one-dimensional packing density profiles illustrating the density-driven segregation of (a) a binary system of steel (blue squares) and glass (orange triangles) particles, (b) a binary system of glass and polyurethane (red circles) particles and (c) a ternary mixture of steel, glass and polyurethane particles.

become near-negligible [66], perhaps explaining the apparent absence of inelasticity effects in some prior works [16, 68]. Shape segregation, meanwhile, was shown (as expected from prior simulation studies [69, 70]) to represent a relatively weak effect in dilute systems [67]. In dense beds, however, particle geometry was found to be the dominant factor in determining the segregation pattern achieved by a system [56].

8.1.3.6 Comparison with theory and numerics

The detailed kinematic information acquired from PEPT data makes it ideal for the validation of theoretical and numerical models developed to describe the dynamics of vibrated granular systems. PEPT has been used to validate both hydrodynamic [71, 72] and kinetic-theory-based [73] models for vibrated beds through the comparison of theoretically-predicted and experimentally-measured density and temperature distributions [71, 73], and convective flow fields [72]. It has also been used extensively for the calibration and validation of discrete element method (DEM) [74–76] (see section 4.5.1) and event driven [77] simulations of vibrated granular systems [33, 47, 48, 54, 55, 57, 58, 65, 66, 78].

8.2 Rotating drum systems

8.2.1 Industrial and scientific applications

Much like vibrated beds, rotating drum systems also carry both significant industrial relevance [79]—being widely used as driers [80], mills [81], mixers [82], reactors [83, 84], roasters [85] and in various other applications—as well as academic value. In particular, rotating drums represent valuable canonical systems in which to study [86], and develop or validate models regarding [87], granular free-surface flows, thus allowing crucial physical insight into the fundamental physics of granular flows in general [88] as well as important geophysical phenomena such as landslides and avalanches [89, 90].

In addition, rotating-drum systems present a rich and scientifically interesting phenomenology in their own right, exhibiting both radial segregation [91, 92], axial banding [93, 94] and a host of other patterning phenomena [95–97], as well as a number of distinct dynamic regimes [98, 99].

8.2.2 Value of PEPT

The complex, three-dimensional dynamics and segregation patterns exhibited by these systems [100, 101] make them ideal candidates for PEPT imaging. As with vibrated systems, optical imaging techniques are unable to provide insight into crucial phenomena such as the interaction between core formation and axial banding [79]. While techniques such as x-ray CT or RIMS could potentially be used to extract information pertinent to the lower energy (avalanching/rolling [98]) regimes of these systems, their low temporal resolution makes them unsuitable for capturing the dynamics of systems in the more energetic (cascading, cataracting and centrifuging [98]) regimes. While MRI has been successfully applied to the study of rotating drums outside the quasi-static regime [93, 102], the imposition of various constraints—notably the absence of metallic materials—limits its applicability, especially to more industrially-relevant systems which are highly likely to include metal components.

8.2.3 Prior research

8.2.3.1 Characterising bed geometry and flow modes

Rotating-drum systems exhibit a number of distinct flow modes, which greatly affect the mixing, heat transfer, stress distribution and many other properties of particulate beds. As such, the ability to characterise these modes and the transitions between them is an active research area. Ding *et al* [103] used PEPT to study the transition between the low-energy ‘slumping’ and ‘rolling’ modes, showing that said transition can be characterised as the point at which their turnover times are equal. PEPT was later used by Lim *et al* [104] to characterise the avalanching-rolling transition based on a measure of the mean number of avalanches required to convey a particle from the top to the bottom of the granular bed.

The 2016 work of Morrison *et al* [105] explored the more energetic cascading, cataracting and centrifuging modes reached at higher Froude numbers [106], demonstrating how PEPT can be used to extract various features of a rotating-drum system—such as the bulk free surface and centre of circulation—which may be used to characterise, and hence distinguish between, different flow modes. PEPT was also used to analyse the differing (frictional, collisional and intermediate [107]) flow regimes coexisting within these flow modes [108]. Govender *et al* [109] used PEPT to delineate the free-flowing layer of rotated beds in these same regimes, showing that the scaling ($\langle v \rangle \propto h$) of the local depth-averaged velocity, $\langle v \rangle$, with the flowing layer thickness, h, expected for systems in the rolling regime [110] holds also for the higher-Fr cascading and cataracting states.

8.2.3.2 Dispersion and segregation

Dispersion plays a key role in many phenomena—e.g. radial and axial mixing, surface renewal and axial transport—relevant to the effectiveness of rotating-drum systems in industrial processes. Dispersion and diffusion also play key roles in the segregation of rotated beds, which—depending on the application—may be either highly desirable or severely detrimental.

Early PEPT work [111] looked in detail at dispersion in long drums, demonstrated a direct proportionality between the axial dispersion coefficient and the drum's circulation frequency. The coefficient was also found to depend strongly on the particle size, but not the radial width of the drum. Later work [112] investigated both axial and radial dispersion, showing an additional dependence on the drum fill level and polydispersity. Importantly, this work also showed the dispersion behaviour to remain consistent for both batch and continuous processes, meaning that batch data can be reliably used to predict mixing and other relevant properties in continuous processes. PEPT observations of a bidisperse-by-size system [113] again showed an increase in particle mobility, and hence diffusivity, with drum rotation rate, showing also that smaller particles possess a higher diffusivity than larger. This study also showed the presence of radial segregation as expected from previous experimental and theoretical works [91]. Later PEPT work focused on *axial* segregation, showing that such segregation—typically present only in long ($L \gg D$) systems—could be induced in narrow ($L < D$) beds by manipulating the radial geometry of the system [114], and even used to deliberately alter the strength and direction of segregation [78]. PEPT also provided the first experimental evidence of axial banding in bidisperse-by-*density* systems [79] (see figure 8.4).

8.2.3.3 Other quantities

PEPT has been used to provide various other data from rotating drum systems, including dynamic repose angles [115], residence times in continuous systems [116], transaxial and radial velocity fields [115, 116], and shear rates [108]. It has also proved an extremely valuable tool in the validation of both discrete and continuum models of rotating-drum systems [78, 108, 114, 115], as illustrated in figure 8.5.

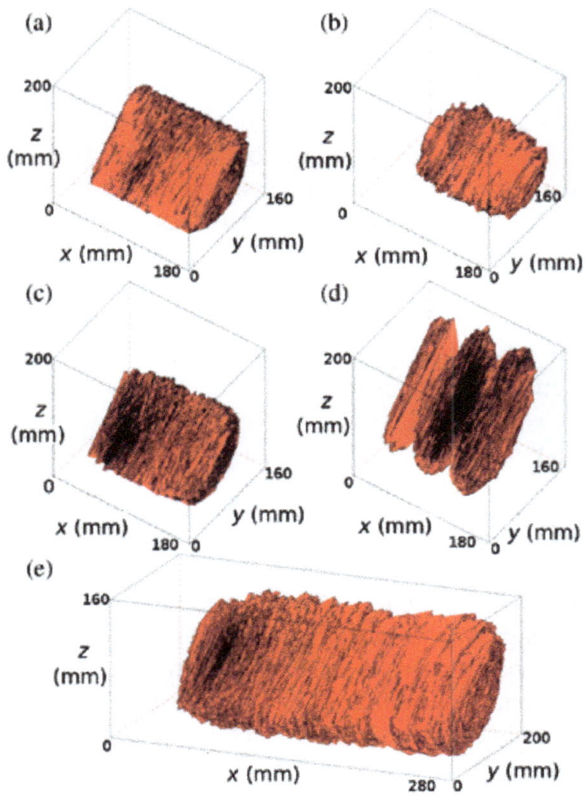

Figure 8.4. Example of PEPT imaging used to visualise the (internal) radial core of an (optically-opaque) rotating-drum system. Note the presence of nascent band formation in panel (c). Image reused with permission from [79], Copyright 2017 by the American Physical Society.

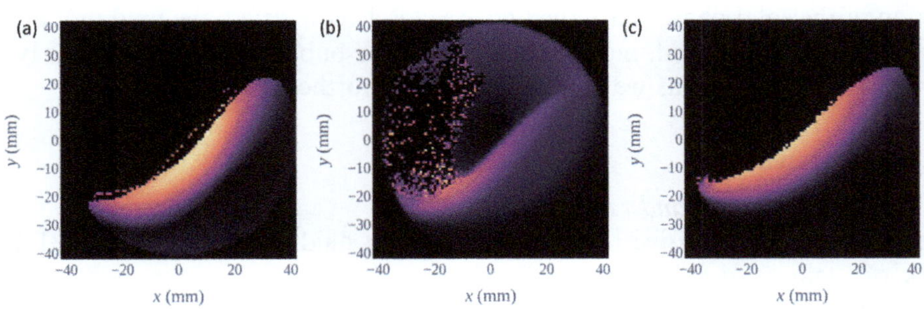

Figure 8.5. Comparison of (a) a PEPT-measured velocity field extracted from an experimental rotating-drum system to (b) a poorly-calibrated DEM simulation and (c) a well-calibrated DEM simulation, illustrating PEPT's value as a validation tool.

8.3 Fluidised beds

8.3.1 Industrial and scientific applications

Fluidised beds are widely utilised in industry, in diverse processes including agglomeration [117, 118], catalysis [119], combustion [120], drying [121, 122], granulation [123], mixing [124], polymerisation [125], heating/roasting [126] and spray coating [127, 128] in commercial activities spanning the agriculture, chemical, food, mining, and pharmaceutical sectors, to name but a few. As such, these systems represent one of the most significant applications of PEPT, where its use has enabled notable advances in our understanding both of the fundamental dynamical behaviours underlying the fluidisation process, and the development and validation of various theoretical and numerical frameworks, as we will discuss in the subsections below.

Though a well-established technology, fluidised beds continue to find significant new applications, for instance atomic layer deposition [129], the recycling of waste plastics [130, 131] or in solar energy applications [132, 133], meaning that their study remains a highly active area of research.

8.3.2 Value of PEPT

Fluidised beds, in particular gas-fluidised beds, typically comprise relatively small particles and exhibit rapid dynamics, meaning that other three-dimensional imaging modalities simply do not possess the necessary combination of high temporal and spatial resolution to successfully record their dynamics. While a significant amount of work has been performed optically imaging quasi-two-dimensional systems [134–136], it is well accepted that the dynamics of such systems diverge strongly from those of (more industrially-relevant) three-dimensional systems. Techniques such as ECT/ERT (see section 4.2.4) are commonly employed to image three-dimensional fluidised beds [137–139]; however, these systems currently only offer relatively coarse, Eulerian information. The finer spatial resolution and Lagrangian tracking offered by PEPT allows a significantly wider range of valuable system properties to be extracted, as we will discuss below.

By its nature, fluidisation arises due to an interaction between a solid particulate phase and a fluid, normally a gas. Although PEPT is only able to directly follow the motion of the solid phase, since the two are closely connected, the fluid motion may in many cases be inferred, most successfully in the bubbling mode (which is typical of many applications), as we will also see in the forthcoming discussion.

8.3.3 Prior research

8.3.3.1 Solids motion and recirculation patterns

The earliest comprehensive research on bubbling fluidised beds using PEPT is the work of Stein *et al* [140, 141]. It has been well-known since the 1970s that the coalescence of bubbles can cause their lateral motion. If the initial gas distribution is uniform, lateral coalescence causes bubbles to move away from the wall so that a more active zone of enhanced bubbling occurs some distance inward from the wall in which the bubbles are larger and more frequent. This tendency is self-reinforcing; at greater heights above the distributor the active bubbling region is found closer to the centre of the bed, so that in deeper beds bubbling is concentrated near the axis (see figure 8.6(a)) [142]. Since the solids travel with the bubbles in their wake and drift regions, the net effect is for strong solids circulation, upwards in the centre of the bed and downwards at the walls (again, see figure 8.6(a)). The details of this motion

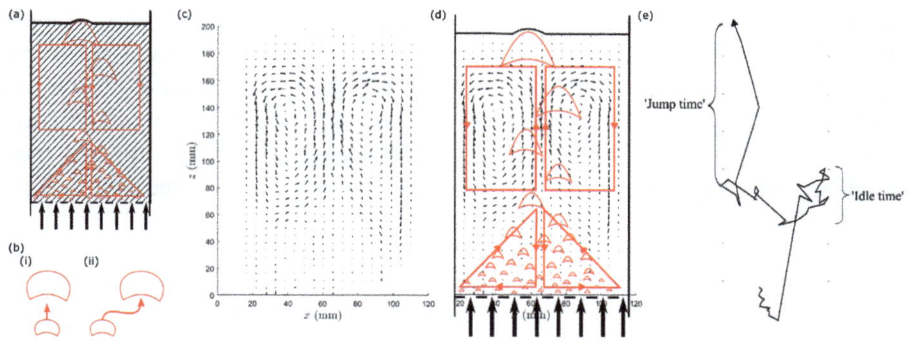

Figure 8.6. (a) Bubble and solids flow pattern and (b) coalescence behaviour—(i) bubble overtaking; (ii) bubble sideways motion (adapted from [142]). (c) PEPT image showing the time- and depth-averaged velocity field and a typical circulation pattern in a fluidised bed. (d) Geldart's schematic overlaid onto the PEPT velocity vector field. (e) Example of real PEPT data showing the trajectory of a particle rising upward through a fluidised bed, demonstrating periods of rapid upward vertical motion ('jumps') interspersed with periods of 'idling'—slow and/or downward motion.

depend on the scale of the equipment as well as the characteristics of the particles and the distributor. PEPT has proved useful in tracking this overall solids motion and showing how it changes with these parameters. Stein *et al* [141], Fan *et al* [143, 144] and Laverman *et al* [145] quantified the solids velocity distribution and related it to the bubble velocity at laboratory scale. Seville *et al* [146] and Ingram *et al* [147] have used PEPT to follow particle motion under semi-industrial scales and conditions, the latter on BP's site at Hull, using a modular PEPT camera as described in section 5.2.4.

8.3.3.2 Bubble motion, jump velocities and idle times

Time averaged circulation patterns such as those shown in figure 8.6(c) can give the impression of ordered motion, but trajectory-following reveals important details of the instantaneous motion at the level of an individual particle (see figure 8.6(e)). Typically, a single particle will be picked up by a rising bubble, rapidly carried some distance upward and then shed into the bulk of the bed where it remains 'idle' for a period before being picked up again. Particles may undergo several such upward 'jumps' in rising through the bed before being ejected into the freeboard with a bursting bubble or otherwise finding a down-flowing region. The jump velocity is close to the bubble velocity [141], while the time duration of a jump is relatively insensitive to gas velocity and comparable with the time between bubble coalescence events. The distribution of idle time shows two peaks, one related to the timescale of down flow at the bed wall and the other to the time between bubbles passing a point and disturbing a stationary particle. It turns out that the idle time is of prime importance in predicting the conditions for particles sintering in fluidised bed processes such as the manufacture of polyethylene; Seville *et al* [148] derived and experimentally verified a predictive equation for onset of sintering based on comparing the idle time with the time required for a sinter neck to grow between two adjacent particles. The manners in which jump velocities and idle times can be calculated from PEPT data are described in detail in section 7.8.

One of the advantages of a fluidised bed as a reactor is the ease with which heat can be added or removed through the walls or through tubes inserted into the bed. The principal mechanism of heat transfer in such cases is particle convection and contact with the heat transfer surface; this has been investigated using PEPT both in the case of in-bed tubes [149] and in the special case of a dense upward moving fluidised bed which is central to the proposed design for a solar receiver [150].

8.3.3.3 Dispersed flows and spouted beds
At higher gas velocities, distinct bubbles disappear and are replaced by more dispersed particle flows. Particles can be collected from the exiting gas and returned to the bed, in which case a circulating fluidised bed (CFB) may be formed, which is a common reactor geometry in which the solids may be a catalyst or a reactant or a combination of both. Again, PEPT has been useful in describing solids flows in such devices [151], the way in which they are influenced by equipment design [152] and the particle motion within ancillary devices such as cyclones [153].

If gas is introduced at a single point on the distributor rather than uniformly across it, a 'spout' may form, which—if it breaks through the bed surface—results in a so-called 'spouted bed' [154], again a common processing geometry, particularly for coating processes [155, 156]. Seiler *et al* [157], for example, show how a comprehensive statistical model of a coating process can be built up from knowledge of the particle cycle time and its trajectory within the region where the spray coat is introduced. A similar approach has been taken to understanding the widely-used pharmaceutical 'Wurster' coater [158]. The detailed structure of a secondary gas jet introduced into a fluidised bed has also been investigated using PEPT [159]. Most recently, PEPT has been used to explore the behaviour of a spouted-bed coffee roaster [160], extracting detailed information regarding density distributions and flow patterns, and their variation with system conditions (see figure 8.7), which can be used to improve existing heat and mass transfer models for the coffee roasting process. The results of this study are currently being used to calibrate and validate a detailed CFD-DEM model of the coffee roasting process.

8.4 Stirred tanks

8.4.1 Industrial and scientific applications

Stirred-tank systems are used extensively in industry for both liquid–solid and liquid–liquid mixing, with three-phase (gas–liquid–solid) systems also regularly utilised. These operations are vital to a number of industrial sectors [161, 162], where they are used in diverse processes including wastewater treatment [163], biofuel production [164, 165], fermentation [166] and in the pursuit of various other chemical reactions [167].

8.4.2 Value of PEPT

As will be discussed in detail in the following section, PEPT offers a number of significant advantages over conventional methodologies when investigating the behaviours of stirred-tank systems. Firstly, optical measurements such as particle image velocimetry—the most commonly used imaging methodology employed to

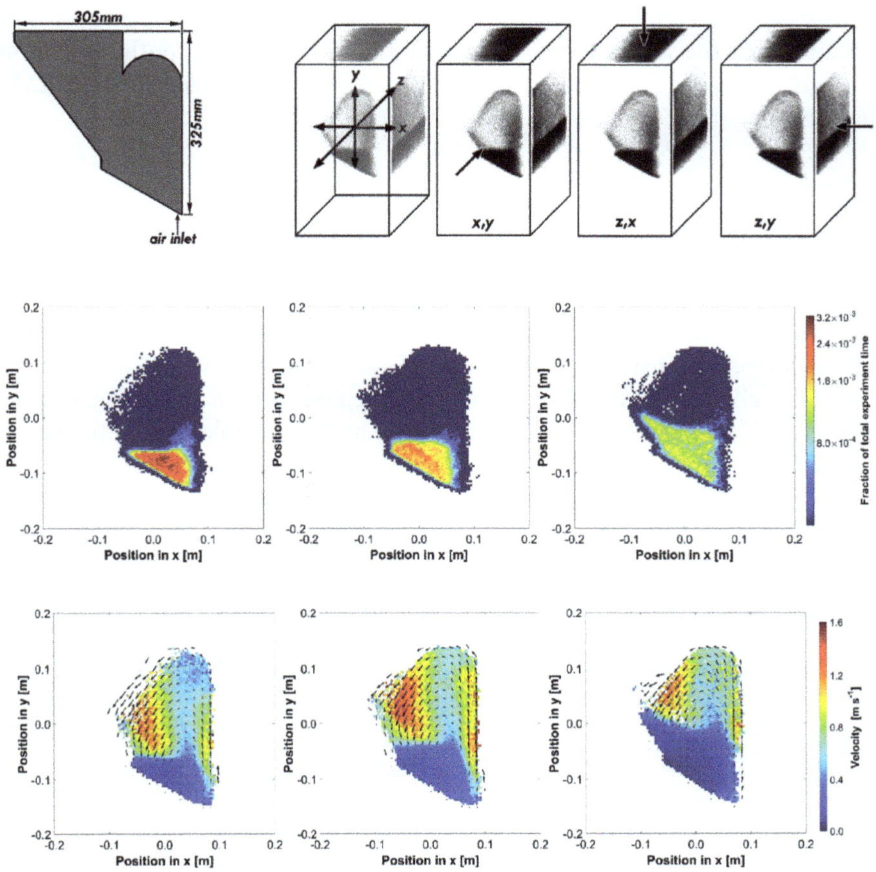

Figure 8.7. Example PEPT images from a spouted bed coffee roaster. *Top row:* Geometry of the roaster and its translation to two-dimensional depth-averaged PEPT projections. *Centre row:* PEPT-measured occupancy distributions showing the existence of two distinct regions within the system. *Bottom row:* PEPT-measured velocity vector fields. Images reused from [160], Copyright 2021, with permission from Elsevier.

study such systems—are only applicable for systems with relatively low solids fractions [168] and fluid phases which are optically transparent. This latter constraint is particularly problematic when trying to study systems—for example biofuel reactors [165] or anaerobic digesters [169]—which utilise complex, optically-opaque, non-Newtonian fluids, as finding a transparent surrogate fluid whose properties precisely match those of the real material is an infamously difficult task. With PEPT, however, it is typically possible to simply use the 'real' fluid, allowing significantly simpler and more real-world-relevant experiments to be conducted. PEPT's ability to perform Lagrangian particle tracking throughout the three-dimensional volume of a system also offers a more rigorous, consistent and automated method [170] for the measurement of key variables such as the minimum speed for just suspension, N_{js} [171], which is conventionally simply estimated by eye [172].

As with other systems described in this chapter, PEPT has also proven a highly valuable tool for the validation of numerical models, including both pure computational fluid dynamics (see section 4.5.2) [173, 174] and coupled CFD-DEM [175] simulations.

8.4.3 Prior research

8.4.3.1 Fundamental studies

The first major study of a stirred-tank system using PEPT was published in 2000 by Fangary *et al* [176], who used the technique to investigate the flow dynamics of non-Newtonian fluids excited by two distinct axial flow impellers. PEPT's ability to image the three-dimensional trajectories of, and (thus) flow fields produced by, particles, as well as to clearly identify stagnant regions within the vessel, allowed the authors to both qualitatively and quantitatively assess the mixing effectiveness of the two impellers studied under a range of conditions. It was shown that, for both impellers, fluid agitation improved with increasing impeller speed, but that as the viscosity of the non-Newtonian fluids was increased, both blades—despite nominally being axial flow impellers—behaved instead as radial flow impellers. Later work by the same authors [177] also used PEPT to measure the minimum speed for just suspension [171], N_{js}, (see figure 8.8) observing significant disagreement between the experimental results obtained and the theoretical predictions of the Zwietering correlation [178] commonly used to predict the N_{js}.

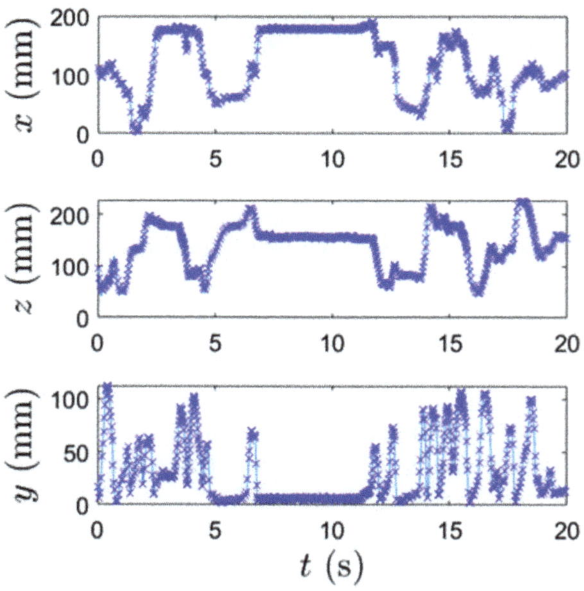

Figure 8.8. As PEPT is able to measure the full, three-dimensional trajectories of particles throughout the volume of a system, it can be highly valuable in determining whether a given system is above or below the N_{js}. In the above, we see that the tracer particle remains effectively static in all three spatial dimensions for several seconds, demonstrating categorically that the impeller speed for the system being imaged is below the N_{js}. Image re-used with permission from [170], John Wiley & Sons. © 2020 The Authors.

8.4.3.2 Three-phase systems

The 2003 work of Fishwick, Winterbottom and Stitt was the first to explore the use of PEPT in *three-phase* (solid-liquid-gas) stirred-tank systems [179, 180]. While mass transfer within stirred tank systems is known to be affected by a variety of parameters [181], the presence of gas dispersion within the system was only observed to alter the influence of the impeller speed—the effects of the system's geometric, physical and hydrodynamic properties were found to be insensitive to the injection of gas. The gassing was, however, observed to markedly influence the slip velocities of particles within the system. Specifically, gas dispersion was observed to decrease slip velocities within the system, thus correspondingly decreasing mass transfer rates. Subsequent work by the same authors [182], again studying three-phase systems, explored the efficiency of upward- and downward-pumping 30° and 45° pitched-blade turbines. The 45°, upward-pumping impeller was found to be the most efficient in all cases, though this efficiency was observed to be notably reduced by the presence of gas dispersion, even when the volume of injected gas remained relatively low. The 30° upward-pumping impeller, though less efficient overall, was nonetheless found to be significantly less susceptible to the effects of aeration. The 45° *downward*-pumping impeller, meanwhile, was shown to be the most strongly influenced, with not only its efficiency but also its flow pattern (see figure 8.9) changing dramatically with the introduction of gassing. The same authors also used PEPT to study a comparatively simple, two-phase system [183], again extracting slip velocities and flow patterns as well as various other data including dispersion and kinetic energy distributions, the latter of which demonstrated the presence of high energy gradients in the vicinity of the impeller and quiescent zones beneath it, expected to negatively impact bulk mixing.

8.4.3.3 Binary and polydisperse systems

The 2009 work of Barigou *et al* [184] used PEPT to achieve a number of firsts in the study of stirred-tank systems, including the first measurement of the spatial distributions of two distinct sizes of dispersed particles, as well as the first three-dimensional visualisation of the flow fields, occupancy maps and Poincaré maps demonstrating the dynamics arising from the mixing of two different non-Newtonian, shear-thinning slurries. The work was later extended to the case of polydisperse solids with up to five distinct fractions [185], showing that the Reynolds number similarity [186] holds for all components of the system, and that the slip velocity of larger particles in the system is greater than that of smaller particles, though the observed differences decrease with increasing total solids fraction. Also in 2009, Guida *et al* [187] used experiments designed around a carefully constructed three-dimensional geometric efficiency map [188] of the PEPT camera used to create high-quality, three-dimensional velocity and concentration fields for both the solid *and* liquid phases of a two-phase stirred tank system driven by a pitched blade turbine.

8.4.3.4 Multimodal studies

Pianko-Oprych, Nienow and Barigou [168] benchmarked the PEPT technique against the more conventional PIV technique using a stirred-tank system agitated

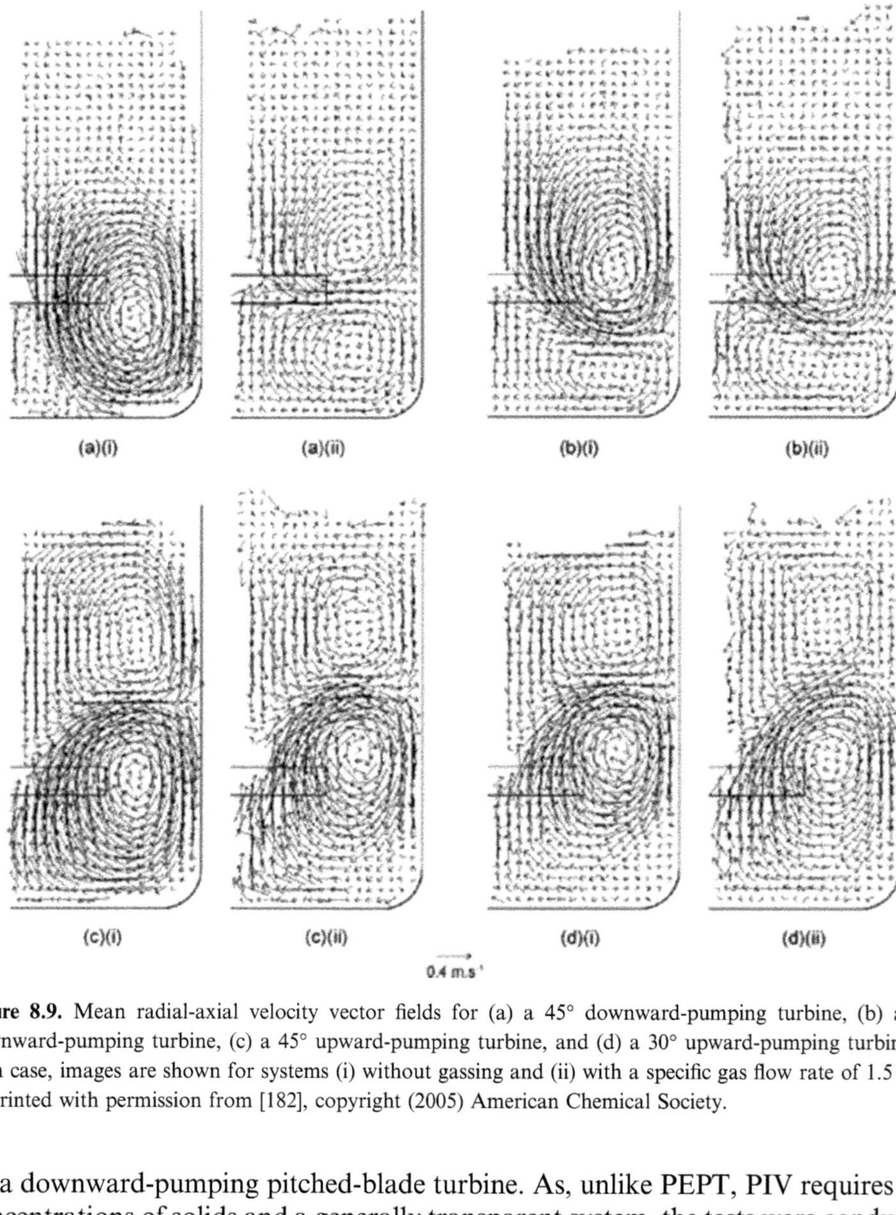

Figure 8.9. Mean radial-axial velocity vector fields for (a) a 45° downward-pumping turbine, (b) a 30° downward-pumping turbine, (c) a 45° upward-pumping turbine, and (d) a 30° upward-pumping turbine. In each case, images are shown for systems (i) without gassing and (ii) with a specific gas flow rate of 1.5 vvm. Reprinted with permission from [182], copyright (2005) American Chemical Society.

by a downward-pumping pitched-blade turbine. As, unlike PEPT, PIV requires low concentrations of solids and a generally transparent system, the tests were conducted using water as a fluid and limiting the solids loading to <5wt%. Generally excellent agreement was found between the two techniques across a range of system conditions. Later work by Chiti *et al* [189] provided a detailed comparison of PEPT data to prior results obtained using laser Doppler anemometry (LDA), similarly finding strong agreement other than minor differences in velocity in the specific region just off the tip of the turbine blade. That the three techniques produce similar results, yet PEPT (unlike PIV, LDA or other optical techniques) has been demonstrated to work successfully in optically-opaque systems and at solids

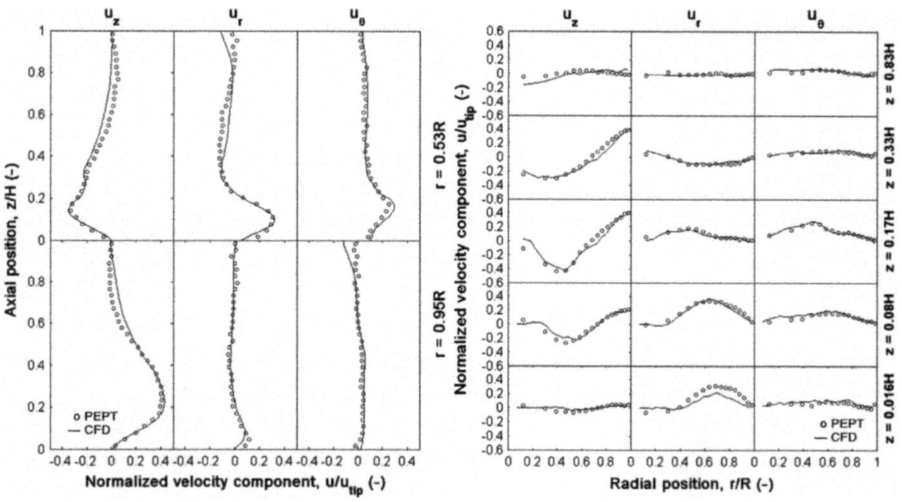

Figure 8.10. Comparison of velocity profiles obtained from PEPT (symbols) and CFD (lines) from a stirred-tank system containing 5.2wt% 1 mm glass spheres, operated at N_{js}. Figure reproduced from [173], Copyright 2013, with permission from Elsevier.

loadings of up to 62.5wt% [170], strongly emphasises the value of this technique to the field.

As with the systems discussed in previous sections, PEPT has also shown significant value in the validation of numerical models. Liu and Barigou used PEPT to validate CFD models of stirred tank systems at both moderate [173] and high [174] solids loadings. In both cases, strong agreement was observed between the CFD-modelled and PEPT-measured flow numbers and velocity profiles for impeller speeds above the N_{js} (see figure 8.10), though the CFD model is found to overestimate the solids density near the base of the system as compared to the PEPT ground-truth data. This issue, however, was found to improve for higher solids loadings. Subsequent work by Boucher *et al* [175] used PEPT to validate a combined CFD-DEM model of a stirred tank, in this case using significantly smaller (~50 μm) particles than explored in the prior works. Though not as detailed as the analysis of Liu and Barigou [173, 174], their results nonetheless suggest good agreement between the numerical and experimental results, and further demonstrate the value of PEPT as a validation tool for diverse simulation methods.

8.4.3.5 High solids loadings and opaque systems

PEPT's ability to operate successfully at high solids loadings was demonstrated in the 2010 work of Guida, Nienow and Barigou [190] who explored concentrations of up to 40wt%, showing that the pumping effectiveness of a pitched-blade turbine remains relatively insensitive to the solids loading for low to moderate values thereof, but that it is significantly hindered at high fractions. The effect is found to be more pronounced for upward-pumping compared to downward-pumping impellers. Their results also show that, in general, downward-pumping turbines are observed to induce more homogeneous mixing within the system. The mixing of stirred tank systems was further explored in a second 2010 paper by the same authors [191], wherein they used PEPT data combined with newly-defined entropic mixing indices to explore the spatial and temporal variation of mixing within the system for a variety of impeller speeds. The data were used to assess the optimal positions in which material inlets and outlets should be placed to ensure efficient mixing.

In 2017, Sindall *et al* [192] used PEPT to perform a more applied study, imaging the interior dynamics of an anaerobic digester. Where conventional studies of such systems require the use of clear-walled systems and optically transparent 'synthetic sludge' [193], the application of PEPT in the current study facilitated the use of real, opaque sewage sludge, thus providing data of more direct relevance to the real problem at hand. In the study, the authors compared data obtained using both 'real' anaerobically-digested sludge and a synthetic equivalent created using carboxymethyl cellulose (CMC)—a common material used to create such surrogates [194]. The study demonstrated clear differences between the dynamics observed in the 'real' and 'synthetic' materials, highlighting the difficulty of choosing a suitable, optically-transparent surrogate for such experiments, and thus once more emphasising the value of PEPT.

Most recently, Windows-Yule *et al* [170] used PEPT to study a simple two-phase system of water and 1 mm glass spheres, exploring a range of solids loadings from 0 to 62.5wt%. The paper provides a detailed methodology for the autonomous measurement of N_{js} from PEPT data, as well as defining a simple, scalar index through which one may meaningfully quantify the homogeneity with which solids are dispersed through such a system.

8.5 Mills

8.5.1 Industrial and scientific applications

Positron emission particle tracking has been widely used to study milling processes in various types of mill—including vertical stirred media mills [195], tumbling mills [196] and oscillating mills [197]. The diverse milling processes explored have significant applications in the agricultural, ceramics, chemical, food, mining and pharmaceutical sectors [198–200]. Milling is used, for example, in the preparation of pharmaceutical powders for granulation and tableting [201], the extraction of cocoa liquor and other key processes for chocolate manufacture [202], the processing of valuable ores in the mining industry [203], and the production of paint [204], to give just a few diverse examples.

8.5.2 Value of PEPT

Unlike many systems such as the previously-discussed rotating drums, vibrated beds and fluidised beds, a two-dimensional model of a stirred media mill is typically unlikely to yield meaningful results! Combined with the fact that such mills are usually densely-packed, techniques such as PIV or PTV are severely limited in the information they can acquire from these systems. Similarly, as a mill cannot be realistically operated in a quasi-static regime, tomographic methods such as CT or

RIMS are also unlikely to be meaningfully applied to these systems, while the previously-mentioned size and material limitations of MRI are again likely to prohibit any realistic experiments. As such, PEPT is one of very few experimental techniques that can obtain detailed, three-dimensional information from an industry-relevant mill system.

8.5.3 Prior research

8.5.3.1 Vertical stirred media mills

The earliest PEPT studies concerning milling processes were performed by Conway-Baker *et al* using a laboratory scale vertically stirred mill [205–207]. The first of these studies [205], conducted in 1999, was in many ways ahead of its time, discussing the potential of PEPT as a means to validate a discrete element method model of a mill. The authors used PEPT to extract various pieces of information—including particle trajectories, velocity profiles and occupancy distributions—from a vertical mill with a pin-bar impeller and discussed the manner in which the data obtained may be used for DEM calibration—though no calibrated DEM data were explicitly shown. Later work by the authors used the same system to provide a somewhat more detailed study imaging, for a variety of impeller speeds, the particle density distributions and flow fields, and identifying three distinct regions thereof—an upper circulating section, a lower circulating section and a central 'impeller band'—mimicking to a degree motion within a complex fluid stirred with an impeller. The authors explored various different impeller designs and rotation speeds, finding both to have a marked impact on the relative sizes and positions of the different regions, which in turn was found to strongly influence the type and extent of breakage observed. Specifically, for low impeller speeds where the regions are not strongly defined, breakage was observed to be abrasion-dominated, producing a larger amount of fine (> 1 μm) particles, while for higher RPM fracture is observed to dominate, producing a coarser final particle size distribution. In a follow-up paper [207], the authors examined also the influence of the volume and size distribution of the media used, the density of the slurry, and the presence of baffles. The authors observe breakage within the system to be fracture-dominated for high media loadings, low slurry densities and/or low baffle settings, and abrasion-dominated for low media loadings, high slurry densities and/or high baffle settings. The authors were also the first to report stratification in polydisperse media for such systems, suggesting that the observed segregation of differently-sized particles may in fact *increase* the energy efficiency of the mill.

In the 2016 work of Riley *et al* [208] PEPT was used to explore the influence of impeller tip speed on both occupancy and kinetic energy distributions within a vertical stirred media mill, in this case a 500 W Cheetah sandgrinder produced by Imerys. The authors found variations in the impeller tip speed to influence the kinetic energy of media in the impeller zone itself, but that it carries comparatively little influence on the distribution of kinetic energy outside this region. Perhaps surprisingly, kinetic energy in the region above the impellers was found to be *lower* at the highest tip speed explored (8.48 ms^{-1}) than for a more moderate velocity of 5.3 ms^{-1}, suggesting that a higher tip speed may in fact inhibit energy transfer through the system between the different 'zones' first identified by Conway-Baker *et al* [205–207].

Work in the same year by Skuse [209], again working with a vertical, stirred-media mill, used PEPT to extract a number of valuable properties including direct measurements of the pseudo-instantaneous forces and thus stresses acting on particles, determined from acceleration data, as well as inferred measurements regarding collision frequencies, calculated using collision theory from PEPT occupancy and velocity data. These two pieces of information were then used to determine the regions of the experimental mill explored most conducive to particle fracture. It was observed that the greatest stresses within the system, as one might expect, were experienced by particles between the impellers, at the impeller tips, with the highest local collision frequencies between particles occurring in the regions above and below the impellers. It was also found that the stresses experienced increased as the size distribution of the material being milled decreased.

8.5.3.2 Tumbling mills

Following the early work of Conway-Baker *et al* [205–207], relatively little PEPT work on milling was performed until the advent of a significant research programme studying horizontal, tumbling mills within the group of Govender at the University of Cape Town, and later the University of KwaZulu-Natal. A series of papers published in 2011 studied a simplified M20 IsaMill [210–213] system (depicted in figure 8.11). The first paper in the series [210] aimed predominantly to demonstrate the ability of PEPT to successfully track the motion of fast-moving media within such a system. Experiments were performed to produce two-dimensional occupancy, velocity and acceleration distributions for two types of media across a range of rotation rates and filling fractions (see, for example, figure 8.13). In addition to demonstrating the applicability of PEPT, the work also demonstrated a direct proportionality between bead velocities and the tip speed of the discs within the system, noting a strong dependence of the relevant proportionality constant on both the fill level of the system and the material of the grinding media used.

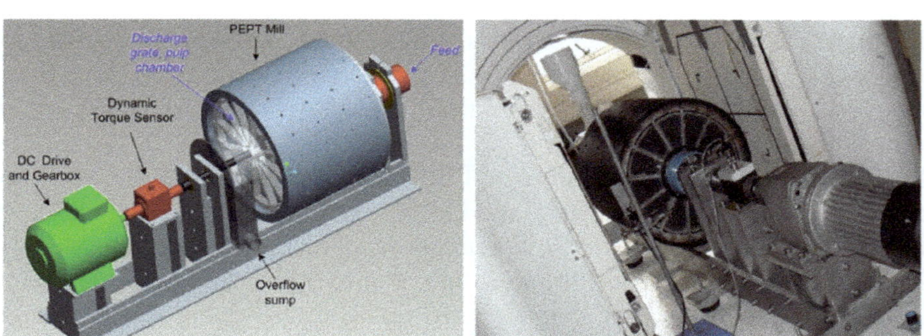

Figure 8.11. Schematic diagram (left) and real image (right) of the Isamill used in the work of Govender *et al* [210–213]. Image reused from [213], Copyright 2011, with permission from Elsevier.

The same batch-mode Isamill system described above was, in a subsequent publication [213], used to provide the first demonstration of PEPT's ability to extract shear rate measurements from opaque systems—a matter discussed in detail in section 7.7. The study used the newly-developed shear-rate measurement method to demonstrate that, for the simple mill and charge used, the shear rate is positively correlated with mill speed and negatively correlated with solids concentration, with the highest shear rates observed at the highest rotation rates and lowest fill levels. The Isamill system was also used [211] to showcase two distinct manners in which PEPT could be used to estimate the power draw of such a system based on the 'torque arm' principle commonly used to calculate power draw in tumbling mills [214–217]. The authors also investigated transitions between differing flow regimes [217, 218], and proposed a new circulation rate model for the system.

The 2017 work of Govender and Pathmathas [109], once again studying tumbling mills, used the detailed Eulerian information which can be extracted from PEPT data to develop an empirical, continuum model describing the granular rheology of the systems explored—a significant open challenge in the field [219–221]. Notably, the authors derived an expression for the (granular) pressure, which is infamously difficult to measure within the interior of a three-dimensional particulate system. In addition to this detailed continuum modelling, the same group also used PEPT to develop a series of scaling laws for tumbling mills [222], finding a linear relation between the depth of the free-flowing layer of material and the velocity at which the mill is rotated, and a constant average shear rate, $\langle\dot{\gamma}\rangle$, across this flowing layer which can be characterised as:

$$\langle\dot{\gamma}\rangle \approx 0.6\sqrt{\frac{g}{d}} \tag{8.4}$$

where g and d represent, respectively, the acceleration due to gravity and the particle diameter. Impressively, the scaling relations observed were found to hold in both the cascading and cataracting [98] flow regimes.

Most recently, de Klerk *et al* [223] developed algorithms allowing PEPT to identify key, bulk geometric features of a tumbling mill, notably its free surface, equilibrium surface, toe and shoulder positions (figure 8.12). The authors then explored the variations of these features with key system parameters including rotation rate and lifter geometry, noting the former to be significantly more influential in determining the system's dynamics and geometric features than the latter.

As with other applications discussed in this section, PEPT has also proven highly valuable in providing detailed calibration and validation data for discrete element method models of both tumbling mills [224] and stirred mills [225]. In the former work, PEPT data is used to create an effective friction model which allows 'dry' DEM simulations (i.e. without coupling to CFD or any other fluid-phase model—see section 4.5.2) to successfully reproduce the dynamics of a system containing fluids or slurries (see, for example, figure 8.13). Such an application further highlights the value of PEPT, as without the detailed, three-dimensional

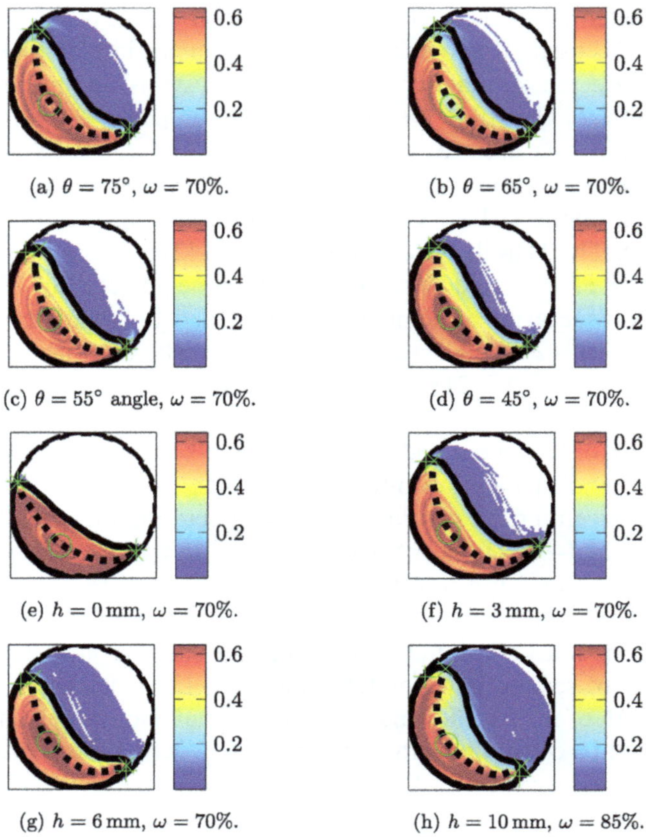

Figure 8.12. PEPT image showing key geometric features of flow within a tumbling mill, including the bed's free surface (solid black line), equilibrium surface (dashed line), centre of circulation (o), re-entry toe and departure shoulder (+) and bulk toe and shoulder (×). The colour map shown represents the local solids concentration. Image reproduced from [223], Copyright 2019, with permission from Elsevier.

experimental data provided thereby, the calibration and validation of such a model would be effectively impossible.

8.5.3.3 Oscillating mills

In 2015, Yu *et al* [226] used PEPT to study still another type of milling process, namely the milling of roller-compacted [227, 228] granulates using an oscillating mill (see figure 8.14)—a process commonly employed in the agricultural, chemical and, in particular, pharmaceutical sectors [229–231] yet which remains comparatively under-investigated in the academic world. The authors investigated the influence of milling speed on the milling process, and the efficiency thereof. Similar to the work of Conway-Baker *et al* exploring vertical, stirred attritor mills [205–207], the authors observed that at low speeds the milling process is abrasion-dominated, while at high speeds it is fracture-dominated, as motion within the system becomes more chaotic and collisional [226]. The latter, collisional regime was found to provide an increased milling efficiency.

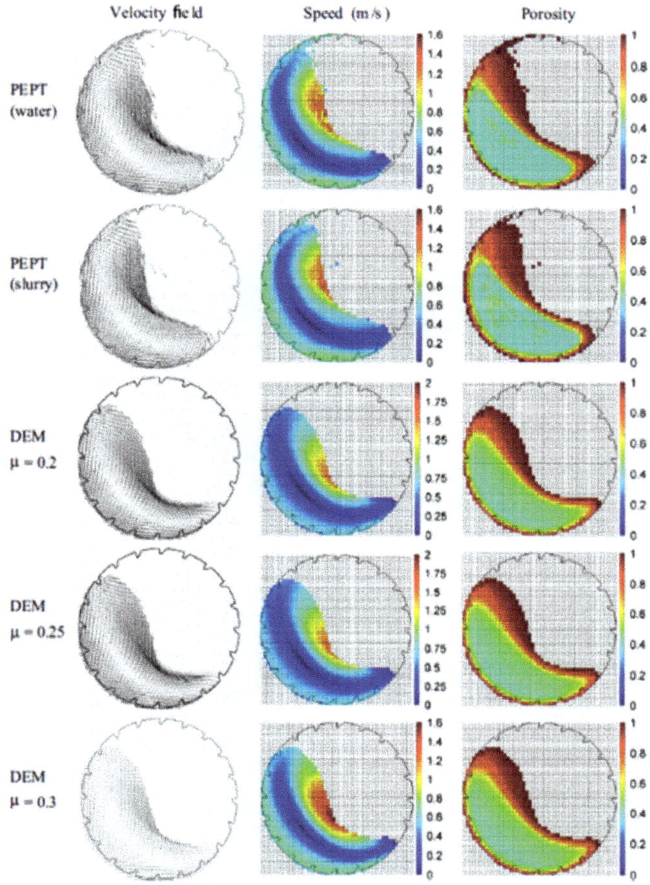

Figure 8.13. Comparison of PEPT data from a (wet) tumbling mill to DEM simulations conducted with varying frictional coefficients, μ, attempting to capture the effective friction of the real, multiphase system using only single-phase simulations. Image reproduced from [224], Copyright 2013, with permission from Elsevier.

8.6 Industrial mixers

8.6.1 Industrial and scientific applications

Mixing is a crucial unit operation in a wide variety of industrial processes spanning almost all industrial sectors. For example, in the pharmaceutical industry, achieving strong, consistent mixing of APIs, excipients, lubricants and other materials is vital not only to the formation of robust and effective tablets, for example, but—more importantly—to ensuring that the end-user receives the correct dose of the desired drug. As such, the correct and reliable operation of industrial mixers can very much be a matter of life and death! Beyond the pharmaceutical world, industrial mixers are used widely in the agriculture, chemical, food, personal care and diverse other sectors [232–234].

It is important to note here that the previous sections concerning vibrated beds (section 8.1) and rotating drums (section 8.2) include significant discussion of mixing

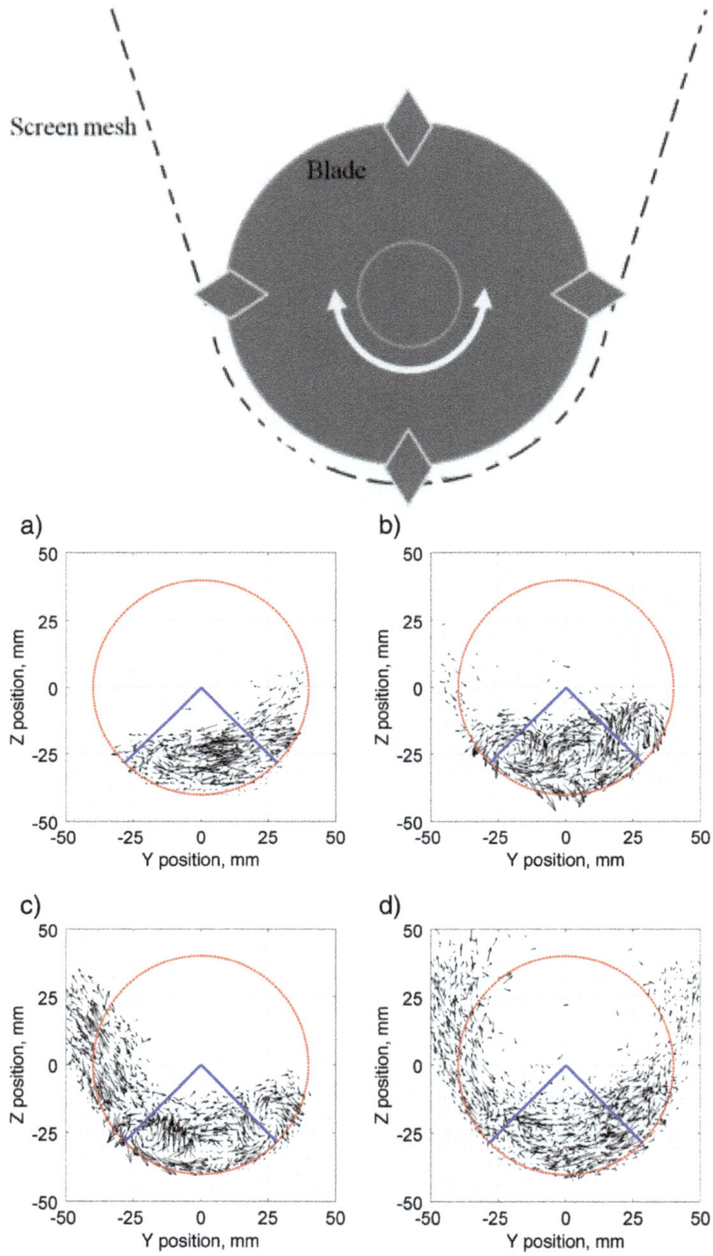

Figure 8.14. Above: Schematic of an oscillating mill. Below: PEPT-measured velocity vector fields extracted from such a mill. Panel (a) shows a low speed case in which relative motion between the mill blades and roller-compacted ribbons being milled is low, resulting predominantly in milling via abrasion. Panel (d) shows the higher-velocity case in which motion is more chaotic and energetic, resulting in a greater degree of fracture. Images reproduced from [226], Copyright 2015, with permission from Elsevier.

and demixing (segregation). In these preceding chapters, however, deliberate emphasis was placed on the study of these processes from an *academic* or 'pure science' standpoint. In the present section, we discuss instead cases in which PEPT has been directly applied in the study of more directly industry-relevant mixing systems.

8.6.2 Value of PEPT

Much like the milling systems discussed previously, industrial mixers, due to their use in many cases of inherently three-dimensional, often asymmetric mixing blades, are extremely difficult to model as quasi-two-dimensional systems, thus limiting the usefulness of optical imaging techniques. Again similarly to milling systems, their geometries and construction materials do not lend themselves to MRI imaging, while their typically fast dynamics preclude the application of techniques such as CT, RIMS and other tomographic methods. Once again, therefore, PEPT represents one of few available options if the detailed imaging of the internal dynamics of such a system is desired.

8.6.3 Prior research

8.6.3.1 Horizontal (ploughshare/paddle) mixers

Industrial mixers, and in particular ploughshare mixers, were among the first systems studied using PEPT. The 1993 work of Broadbent *et al* [235] used the technique to study solids motion within a Lödige mixer for a variety of fill volumes. It was observed that the two central blades of the mixer in essence divided the system into three distinct sections in the axial direction, and that although good intermixing would occur between particles in each of these regions individually, transfer of particles *between* regions was considerably more rare. The dynamics of the system in the axially-central region were also found to differ strongly from those of either 'end' region, suggesting a strong relationship between the performance of such a mixer and its precise geometric design. The authors also observe a transition from an inclined bed at low fill heights to a more homogeneous, fully-stirred system at higher fills.

Subsequent work [236] in a similar but longer system, containing more mixing blades, confirmed the findings of the original study, and additionally showed that the average tangential velocity of particles also increased with fill fraction, with an apparent linear relationship observed between the two parameters, but that the mixing time also increased with increasing fill fraction. The potentially counter-intuitive increase in velocity with fill fraction was suggested to arise due to a reduced degree of slip between the blades and bulk material. Interestingly, varying the tracer size between 2 mm and 8 mm was found to have little effect on the observed velocities (see figure 8.15), though some (secondary) size effect was observed on the system's mixing rate. This longer system was also used [237] to further explore the intermittent transfer of particles between different axial sections first explored in [235]. It was observed that transfer between different regions typically occurs in the bulk of the bed (as opposed to 'through the air') and that most transitions between inter-blade regions are simply repeated crossings across the same blade's plane of rotation. Further study [238] of the

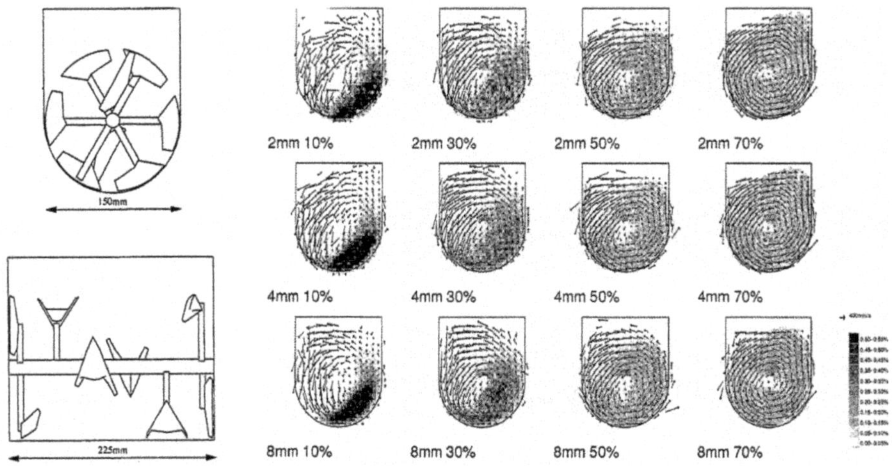

Figure 8.15. Left: Schematic diagram of the ploughshare mixer used by Broadbent *et al* in their early PEPT work [236]. Right: velocity vector fields (arrows) and local solids fractions (shading) for different combinations of fill fraction and tracer size. Images reproduced from [236], Copyright 1995, with permission from Elsevier.

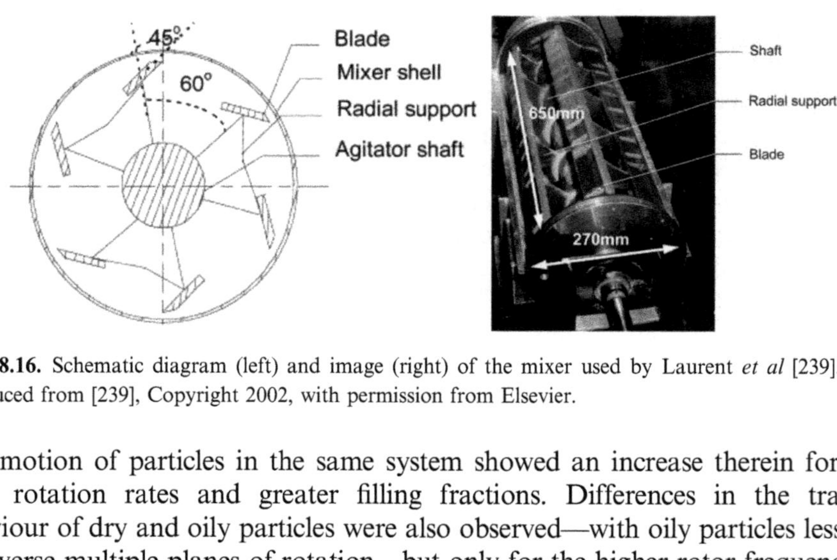

Figure 8.16. Schematic diagram (left) and image (right) of the mixer used by Laurent *et al* [239]. Images reproduced from [239], Copyright 2002, with permission from Elsevier.

axial motion of particles in the same system showed an increase therein for faster blade rotation rates and greater filling fractions. Differences in the transport behaviour of dry and oily particles were also observed—with oily particles less likely to traverse multiple planes of rotation—but only for the higher rotor frequencies.

In the 2002 work of Laurent *et al* [239], another form of bladed mixer—pictured in figure 8.16—was studied. Despite the significant differences in geometry between this system and those studied by Bridgwater *et al* [235–237], the same tendency for particles to remain confined to narrow bands—here defined by the presence of radial blade supports—is observed. By using tracers of different sizes, the authors also inferred valuable information regarding the segregative tendencies of the systems studied, noting in particular that the mean squared displacement of tracers along the axial direction increases more rapidly with time for smaller tracers—i.e. they diffuse more rapidly within the system. Interestingly, however, the size of the tracers used showed little impact on the recorded flow patterns in the cross-sectional plane, despite the aforementioned differences in axial motion. The authors also find a suggestion of size-segregation within the system, though the mechanical origins of this apparent segregation are not concretely determined.

In the subsequent decade, a number of papers combined both PEPT and discrete element method (DEM) modelling to explore the dynamics of these and similar mixing systems in greater detail still. The first such work, performed in 2011 by Hassanpour *et al* [240], compared PEPT and DEM data acquired from a paddle mixer (pictured in figure 8.17), showing good qualitative agreement and reasonable, though imperfect, quantitative agreement. The authors suggest several possibilities—including the lack of precise particle shape modelling in their DEM simulations, or the short duration of the simulations—though the roughness of the PEPT velocity profiles presented suggests that the real issue may in fact simply have been a lack of adequate statistics. Subsequent work involving the same system took a closer look at the equivalence of data acquired from the long-time tracking of one or more individual tracer particles and data acquired from all particles in the system, as is possible to extract from DEM simulations. In particular, the authors compare velocity distributions for (a) single particle PEPT data, (b) single particle DEM data and (c) full DEM data. Interestingly, single-particle measurements for both PEPT and DEM produce a similar peak position in the distribution, though the width of the distribution differs between experiment and simulation. As only 600 s of data are used, however, it is decidedly possible that this disagreement is partially due to less-than-optimal statistics. When comparing against the 'whole-system' data, however, a clear difference in the peak position is observed, likely due to the existence of a non-negligible particle size distribution within the simulation. This is an important observation, as it clearly highlights the importance of using multiple tracers of different sizes if the dynamics of non-monodisperse systems are to be accurately captured by PEPT. The authors also observe that there is no clear advantage to the use of multiple (similar) tracers in a single system over the use of a

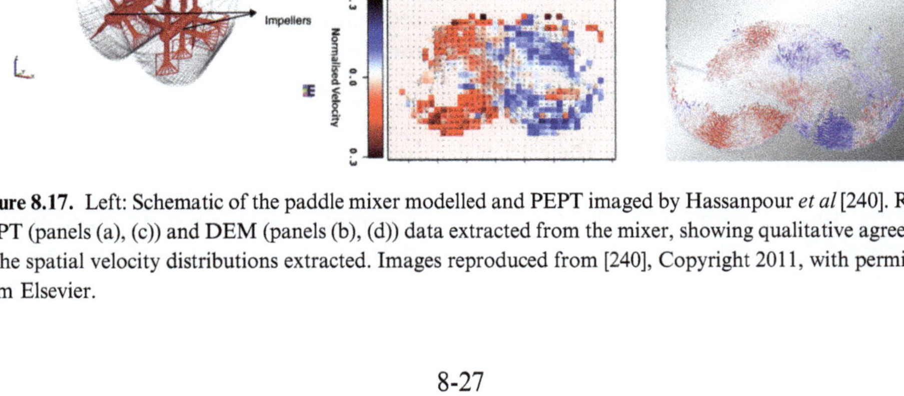

Figure 8.17. Left: Schematic of the paddle mixer modelled and PEPT imaged by Hassanpour *et al* [240]. Right: PEPT (panels (a), (c)) and DEM (panels (b), (d)) data extracted from the mixer, showing qualitative agreement in the spatial velocity distributions extracted. Images reproduced from [240], Copyright 2011, with permission from Elsevier.

single, longer data set, and that the more complicated analysis, and likely loss of data quality, involved with multiple-particle tracking may in fact make the former disadvantageous in practical terms.

The 2012 work of Laurent and Cleary [241] explored a highly simplified system containing only a single rotating plough blade. A number of differences, for example in the repose angle formed as particles flow back into the trench formed by the blade, were observed between experiment and simulation; these differences were thought largely to arise due to the use in the DEM model of spherical particles to represent the highly-aspherical rice grains used in experiment. A second notable limitation, this time in terms of the experimental side, was that the use of only a single plough blade resulted in the existence of significant stagnant regions within the system, bringing into question the validity of the ergodic assumption and thus making this particular system less-than-ideal for study using PEPT. It should be explicitly noted that similar issues were not encountered in the previously-described papers, where the greater density of mixing blades created a significantly more dynamic bed without significant stagnant zones. Nonetheless, despite these issues, strong, quantitative agreement between PEPT and simulation was observed by the authors [241] in terms of the scaling observed between mixing efficiency and plough speed. The PEPT results of Laurent and Cleary [241] were later used by another set of authors [242] to validate a DEM model of a similar system, through comparison of experimental and simulated mean squared displacement measurements. Unlike in the previous work of Cleary and Laurent [241], the authors of [242] explicitly modelled the aspherical geometries of the particles used, thus achieving closer agreement between experiment and simulation (figure 8.18). The validated DEM simulation was then used to gain insight into the mixing dynamics of the system. It was found that in all cases a high, though non-complete, degree of mixing was observed, with the Lacey mixing index increasing rapidly before slowing to an asymptote, and that the final steady-state degree of mixing reached was independent of the system's initial mixing conditions. It was also shown that, as reported in the previously-discussed pure-PEPT studies, the rate of mixing was found to increase with increased blade rotation frequency.

8.6.3.2 Vertical (high shear) mixers

Having focused in the previous section specifically on horizontal ploughshare or paddle mixers, in the current section we shift our focus to another commonly used industrial geometry—vertical high shear mixers [243]. An simple exemplar of such a system, as used in the 2001 work of Stewart *et al* [244] is pictured in figure 8.19, though many geometric variations on this theme are used in industrial practice—as we will see in the discussion below. As with many of the works discussed in the preceding section, the study of Stewart *et al* [244] used PEPT in conjunction with DEM simulations, finding in general a reasonable, though again imperfect, agreement between simulation and experiment. From their PEPT-validated simulations, the authors determined that sliding friction was a dominant factor in determining the behaviour of the systems studied, with rolling friction carrying a smaller, though still non-negligible influence. This work clearly highlights the valuable complementarity between PEPT and DEM, as these parameters are difficult, if not impossible, to isolate and thus

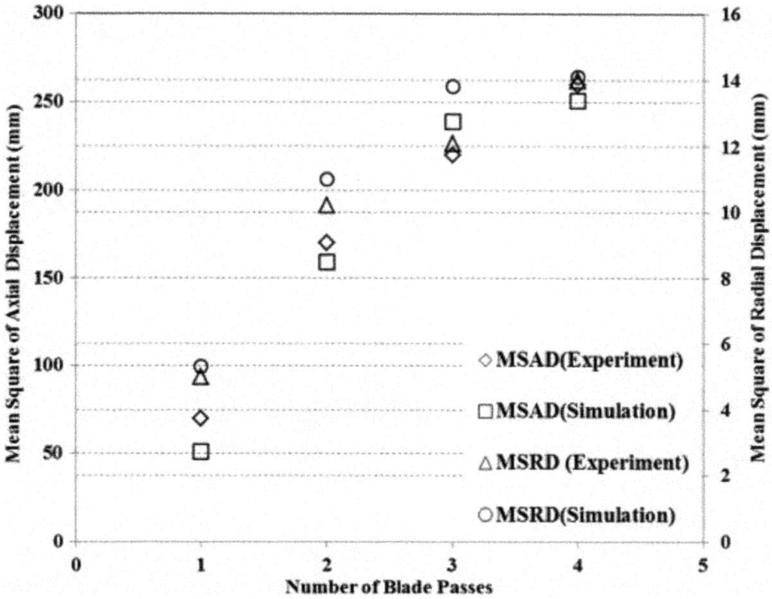

Figure 8.18. Comparison of experimental (PEPT) mean squared displacement measurements to equivalent, simulated (DEM) data, as used to validate the numerical model of Alian *et al* [242]. Both axial (MSDA) and radial (MSDR) displacement are considered in both cases. Images reproduced from [242], Copyright 2015, with permission from Elsevier.

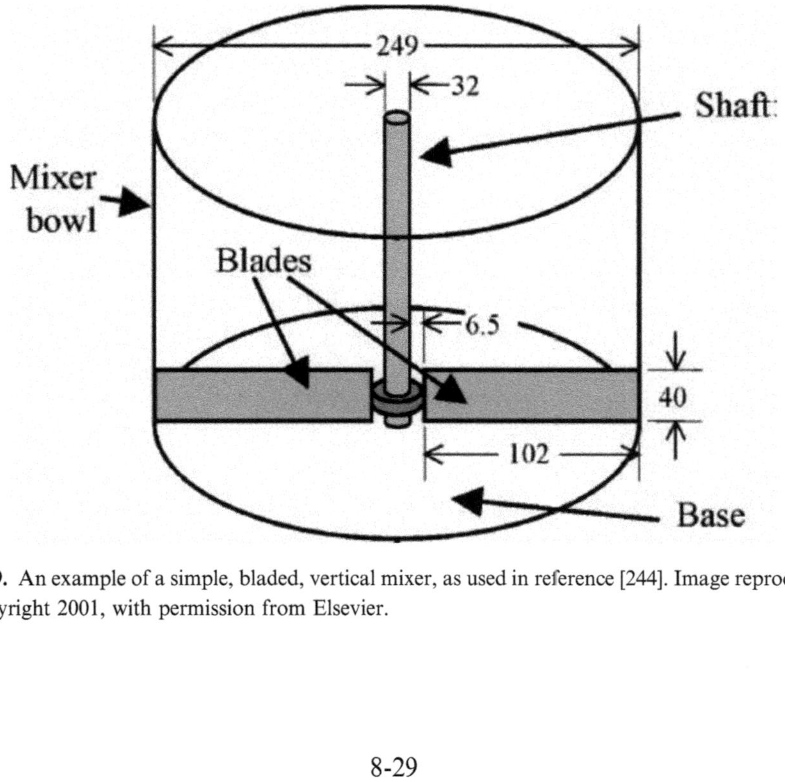

Figure 8.19. An example of a simple, bladed, vertical mixer, as used in reference [244]. Image reproduced from [244], Copyright 2001, with permission from Elsevier.

systematically vary in experiment, yet the DEM simulations performed can only carry scientific weight if validated robustly against detailed experimental data.

The precise influence of the two frictional coefficients on the behaviour of the mixer system studied was explored further by Zhou *et al* [245], who found that increasing either friction coefficient leads to an increase in the bed porosity, while mixing can be promoted by either increasing the sliding friction coefficient or *decreasing* the rolling friction coefficient. The kinetic energy, meanwhile, is observed to be relatively insensitive to the particles' microscopic frictional properties.

PEPT-validated DEM simulations were again employed in the 2003 work of Kuo *et al* [246], this time exploring a high-shear mixer driven by a disk impeller, as well as a simple paddle as employed by Stewart *et al* [244]. For the disc impeller, the DEM simulations were found to qualitatively reproduce the main features observed from the PEPT data, including flow patterns and the general form of the observed velocity distributions, though a quantitative disagreement of about 50% was observed between the numerically predicted and experimentally measured velocities. Agreement was found to slightly improve when using a higher spring coefficient, k (see section 4.5.1), for the DEM simulations, though the improvement was deemed not to be worth the significant increase in simulation time necessitated. For the paddle impeller, meanwhile, significant disagreement was observed between experiment and simulation, with the DEM model failing to reproduce the PEPT-measured flow patterns, and overestimating particle velocities by more than an order of magnitude. The disagreement is thought to stem from an overly-low level of dissipation in the model used for the values of k implemented which, while adequate for the relatively low-energy disc impeller, are unsuitable for modelling higher-energy systems. This paper thus provides a case study clearly underlining the value of using experimental techniques such as PEPT to validate numerical models of particulate systems, as the aforementioned inaccuracies of the DEM models used would likely have gone undetected without the availability of suitable ground-truth data.

That other work [244, 245] using a similar bladed mixer but operating at significantly lower rotation rates (i.e. lower energies) shows markedly better agreement between PEPT and DEM suggests that the previous authors' reasoning regarding the difficulty of the models used to capture the dynamics of highly-energetic systems was indeed correct.

Another PEPT-DEM comparison, this time exploring the dynamics of a helical-blade mixer, reported reasonable agreement between simulation and experiment, though it was noted that experiments involving near-spherical particles (peas) were found to give better agreement than less-spherical wheat grains, most likely due to the simplifying assumption of perfect sphericity employed in the authors' model. The authors found the mixing rate to scale linearly with the blade rotation rate—as observed previously for the case of horizontal mixers—though such a linear scaling was not observed for the mean particle velocity. For the conical, helical-bladed system explored, the authors also identified two distinct mixing mechanisms—convective mixing (macro-mixing) in the recirculating bulk of the system, and dispersion (micro-mixing) nearer the system's upper free surface.

Subsequent work by Ng, Ding and Ghadiri [247] used PEPT to validate not a DEM model, but rather a *computational fluid dynamics* (CFD) model of a particulate system in a high shear mixer. In order to simulate the particles using CFD, the authors employ a continuum description of the (discrete) particulates modelled, based on the kinetic theory of granular flow (KTGF) [248–250]. While the model is able to capture very broad features of the experimental system, such as the expanded bed height and its general flow pattern in the horizontal plane, not only are considerable quantitative differences between predicted and measured velocities observed, but the model fails to predict even significant qualitative features such as the recirculating flow in the vertical-radial plane. The results strongly indicate that the continuum description of granular media requires significant further development before it can successfully describe complex, dense systems such as those explored in this study.

In addition to being used as a valuable DEM validation tool, a number of 'pure PEPT' studies of vertical mixers have also been performed. Ng *et al* performed a series of studies [251–253] investigating the dynamics of a conical frustum-shaped high shear mixer (depicted in figure 8.20). As with the more geometrically simple high shear mixers discussed previously, the authors identify flow both in the horizontal (x–z) plane as well as in the x–y plane, the former being relatively simple and rapid, the latter slower and more complex. An example of the flow patterns generated in the x–y—or, more precisely, the azimuthally-averaged radial–vertical plane—can be seen in figure 8.21.

As can be seen from figure 8.21, the impeller rotation rate strongly affects the flow patterns within the system. The authors also find that while the *absolute* maximum velocity within the system increases with increasing impeller speed, the ratio of maximum particle speed to tip speed reduces, suggesting the existence of an optimal impeller speed for efficient mixing—an observation similar to that of Riley *et al* [208] for milling systems. Subsequent work by the same group [252] extended the above study to explore both dry and wet systems. The authors find a broad similarity between the qualitative dynamics of both dry and wet systems, but note that, in wet

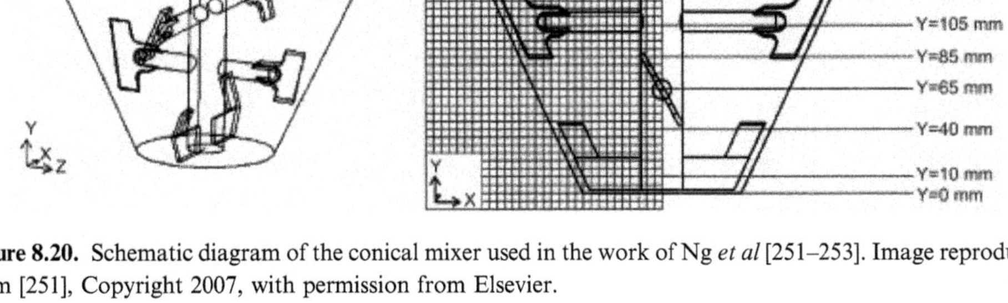

Figure 8.20. Schematic diagram of the conical mixer used in the work of Ng *et al* [251–253]. Image reproduced from [251], Copyright 2007, with permission from Elsevier.

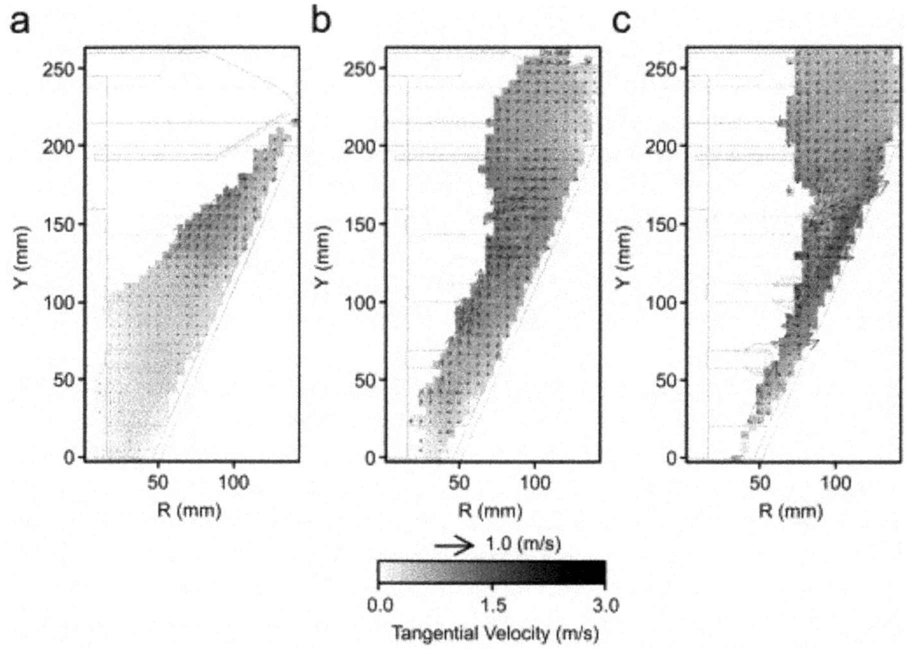

Figure 8.21. Velocity vector fields showing recirculating motion in a conical high shear mixer at varying rotor frequencies: (a) 3 Hz, (b) 6 Hz and (c) 12 Hz. In all cases the orientation of the arrows shown represents the local mean flow direction and their length their velocity magnitude in the plane of interest, while the shading represents the magnitude of the observed velocities in the tangential direction. Image reproduced from [251], Copyright 2007, with permission from Elsevier.

systems, higher concentrations of particles are observed in the upper regions of the systems, and the magnitudes of particle velocities and velocity fluctuations are in general reduced, though the decrease in velocity near the walls of the system observed for dry systems is not observed for wet systems. In a separate study [253], the influence of a number of other factors—namely particle size, density and shape, and system size and fill level—were explored for the case of a dry system. For the ranges explored, the system's dynamics were observed to be relatively insensitive to the size and shape of the particles used. The *density* of the particles, meanwhile, was observed to have a strong influence, with a strong negative correlation between density and tangential velocity. Typical system velocities were also found to be positively correlated to the full level. While larger systems are observed to exhibit more complex flow patterns, as one may inherently expect, the dynamics of the system are nonetheless observed to remain consistent across different scales for constant impeller tip speeds, suggesting the tip speed to be a valuable scaling criterion for such systems.

A pair of pure-PEPT studies were also performed by Saito *et al* [254, 255] on the comparatively simple high shear mixer system first explored by Stewart *et al* [244]. The first of the pair combined PEPT with positron annihilation spectroscopy (PAS) [256] to investigate flow in a system driven by a simple, flat impeller. Their results suggested that different regions within the system would be subject to different characteristic impact velocities, and that mixing within the interior of the system can generally be expected to be strong, though poorer mixing was observed near the boundaries of the system. The second study explored the effect of impeller blade geometry, comparing the dynamics produced by flat and bevelled blades. While the flat blade was found to induce higher velocities within the system, as one may intuitively expect, the use of a bevelled blade was more conducive to turnover, meaning that the latter blade is likely preferable in a mixer system.

8.6.4 Other mixers

While past PEPT research has predominantly focussed on horizontal paddle mixers and vertical high-shear mixers, a variety of other mixer systems have also been explored using the technique. Hiseman *et al* [257], for example, studied the dynamics of granular media within a planetary mixer—that is, a mixer agitated by a 'beater' which rotates around a vertical axis, while this axis in turn rotates around another, central axis, much in the way a planet rotates whilst itself orbiting a central star. The specific system studied by Hiseman *et al* [257] is depicted in figure 8.22. The authors explored the dynamics of the mixer across a range of blade speeds, with a fixed gearing ratio of $\frac{10}{3}$. For all speeds and fill fractions explored, the PEPT-acquired velocity vector fields (see figure 8.23) show generally similar motion, namely a toroidal recirculation cell in the vertical–radial plane. The authors also demonstrated the velocity profiles observed within the system to scale linearly with the applied rotational frequency, as observed in other mixer systems.

In addition to their work on vertical high shear mixers [246], Kuo *et al* [258] also used PEPT to provide a detailed investigation of the dynamics of a v-blender

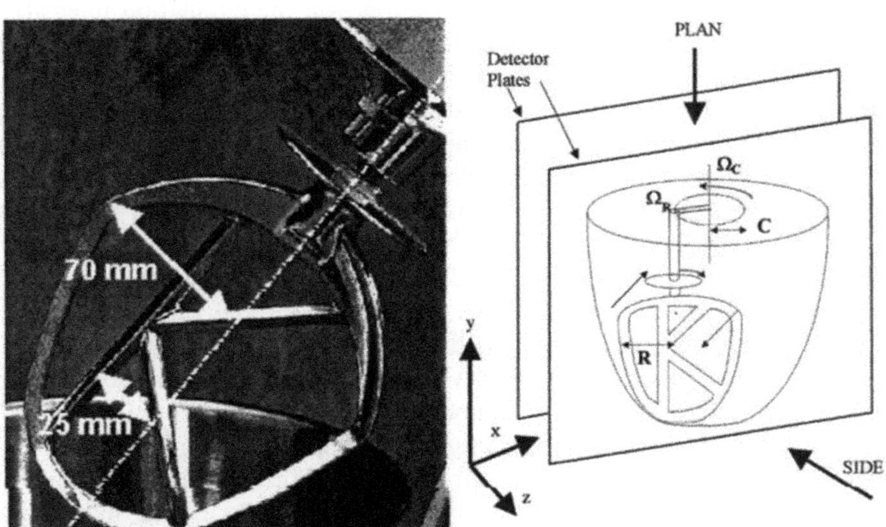

Figure 8.22. Schematic diagram (right) of the planetary mixer used in the work of Hiseman *et al* [257] alongside a close-up (left) of the 'K-beater' agitator used. Image reproduced from [257], Copyright 2002, with permission from Elsevier.

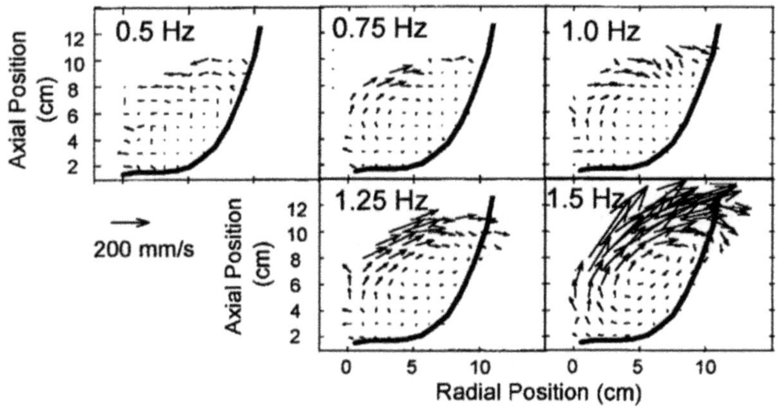

Figure 8.23. Velocity vector fields for the planetary mixer of Hiseman *et al* [257] operated at a number of different agitator speeds. Image reproduced from [257], Copyright 2002, with permission from Elsevier.

(see figure 8.24), another commonly-used type of industrial mixer [259–261]. The authors obtained data for a number of permutations of fill fraction and rotational velocity. In all cases, the broad, fundamental dynamics of the tumbling system were not dissimilar to those of a simple rotating drum as discussed in section 8.2, with the system delineated into a solid body rotation region and a more dilute, fluid-like free surface flow. The v-shaped geometry of the system is also observed to give rise, in the system's rotating frame of reference, to a clearly-defined recirculation in the x–y plane (see figure 8.25). As can be seen in the images shown, both the strength and orientation of the recirculation rolls observed are highly sensitive to the fill fraction of the system. The system's rotation rate, meanwhile, has a strong influence—specifically an approximately quadratic relation—on the circulation intensity, but seemingly no impact on the direction of flow, at least for the range of rotation speeds explored by the authors.

The authors also measured the axial dispersion within the system as a means of directly assessing its effectiveness as a mixer. It was found that the measured axial dispersion coefficient is, just as in the case of a simple rotating drum [111], directly proportional to the rotation frequency of the system. Dispersion was also found to be higher in less-filled systems.

In 2013, Marigo *et al* [262] used PEPT to image the flow dynamics of a *Turbula* mixer and thus validate a discrete element method model thereof. Though the physical geometry of the system studied (a simple cylinder) is notably less complicated than that of the previously-discussed v-blender, the motion used to produce mixing—which comprises a superposition of periodic rotating, translating and inverting movements—is significantly more complex. Unlike for the other mixer systems explored, the PEPT data extracted from the turbula showed a non-monotonic dependence of axial dispersion on the velocity of the mixer, with a minimum at the middle speed of 46 rpm. A similar qualitative trend was observed in the DEM results, though the overall dispersion predicted by DEM typically exceeded that measured by PEPT by approximately 50%. This non-monotonic

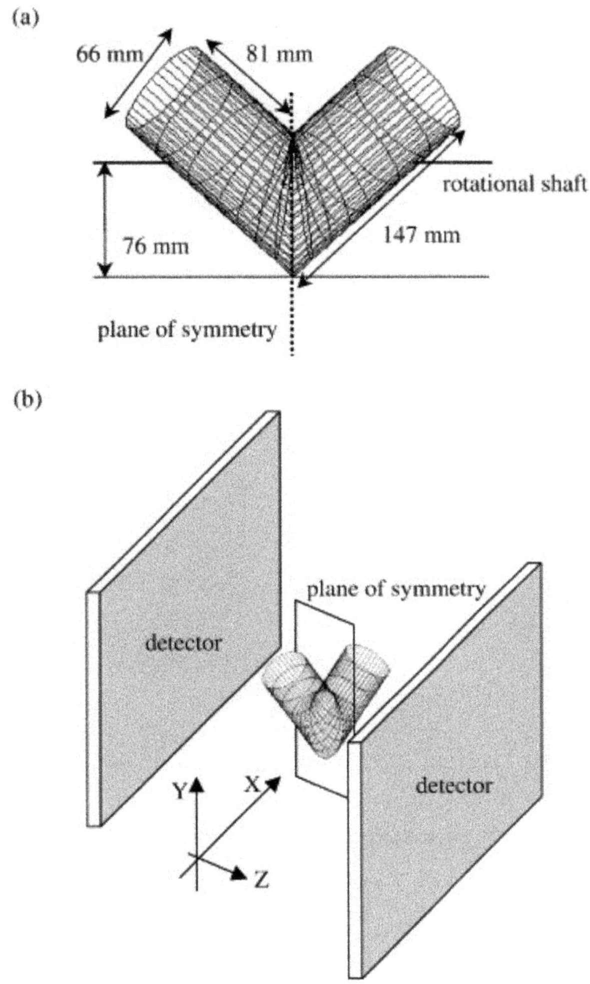

Figure 8.24. (a) Schematic diagram of the simple v-blender geometry used by Kuo *et al* [258]. (b) Schematic showing the system's basal orientation in the field of view of a dual-headed positron camera, showing also the coordinate system used in the main text. Image reproduced from [258], Copyright 2005, with permission from Elsevier.

behaviour is thought to occur due to a transition between two distinct regimes at approximately this rotation rate.

While the majority of PEPT work on mixing has focused on the blending of powders or particulates (as discussed predominantly in the present section) or solid–liquid systems (as discussed in section 8.4) in dynamic mixers, a small number of PEPT studies involving fluid flow in *static* mixers [263] have also been performed. Rafiee *et al* [264] used PEPT to study the motion of highly-viscous Newtonian fluids through a Kenics static mixer (see figure 8.26). These systems, and the manner in which PEPT is applied to them, represent a marked departure from the others previously discussed in this section, as the Kenics mixer is not a batch system, but rather involves a continuous flow of material. As such, PEPT data must be, in

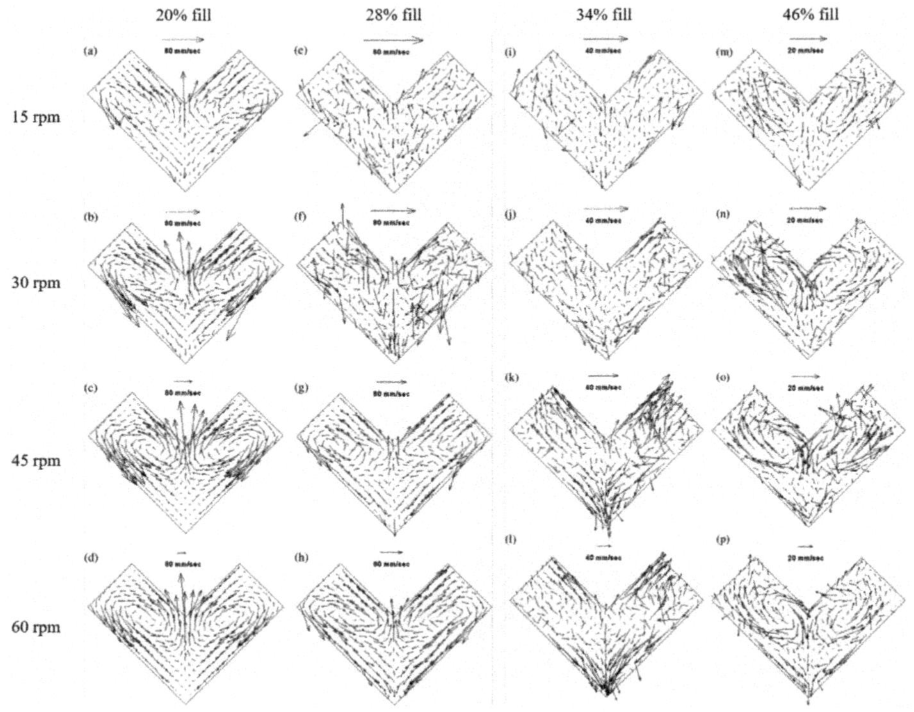

Figure 8.25. PEPT-imaged flow patterns within a v-blender for a range of fill fractions and rotation rates. Image reproduced from [258], Copyright 2005, with permission from Elsevier.

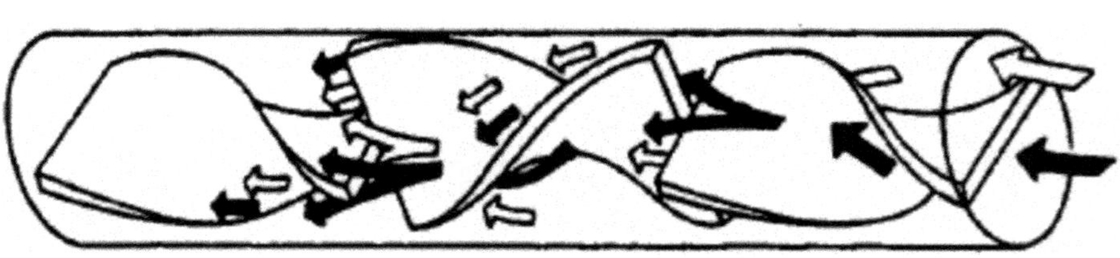

Figure 8.26. Schematic diagram of the Kenics static mixer used in the work of Rafiee *et al* [264]. Image reproduced from [264], Copyright 2011, with permission from Elsevier.

essence, built up 'piece-wise' from a series of distinct passes through the system. Using this technique, the authors found that particles entering the system near the mixer's wall or blades were more likely to have an increased residence time within the mixer as compared to those entering in other regions. In general, the velocities of tracers through the system were found to be relatively constant, though sharp discontinuities in velocity were observed between different segments of the mixer.

Later work by the same group [265] extended the study also to the case of non-Newtonian fluids, as well as introducing some more detailed analysis. The authors again observe relatively constant velocities within elements, with sharp changes typically seen only at the transitions between elements. The distance taken for a tracer to reach a steady-state velocity after such a sharp change is used by the authors as a measure of the flow development; based on this metric, they observe that the length required for flow to develop is shorter for the (shear-thinning) non-Newtonian fluids explored than for their Newtonian counterparts. Their results are also suggestive of a lack of micro-mixing in the directions parallel to the blades of the system. Comparison with prior numerical (CFD) work [266, 267] shows broad, qualitative agreement, even despite dissimilarities in the flow velocities and other details of the systems explored in simulation and experiment.

Comparison of PEPT and CFD data for a static mixer system—here an SMX mixer [268, 269]—was conducted also by Mihailova *et al* [270]. In this case, complementary work using magnetic resonance imaging (MRI—see section 4.2.5) was also performed, providing an extremely comprehensive exploration of the system and its dynamics.

Concentration data showing the mixing of fluids entering the system showed good agreement both between PEPT data and MRI data (see figure 8.27) as well as with prior CFD results [271]. Though the MRI experiments were observed to produce somewhat less noisy data in the image shown, the ability of PEPT to provide both

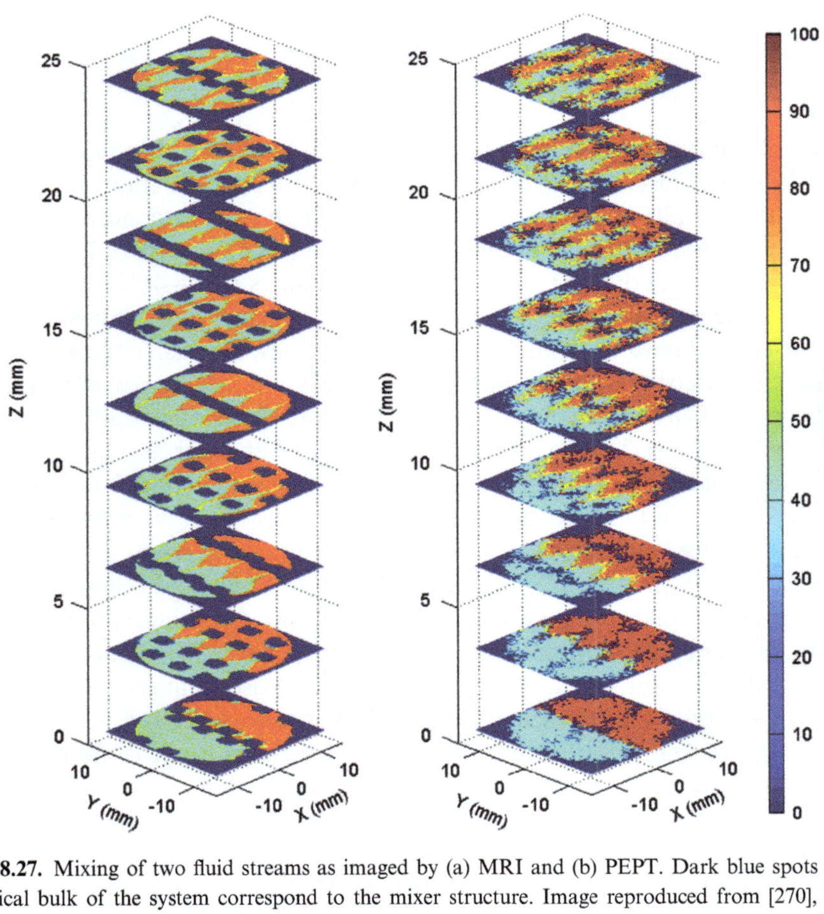

Figure 8.27. Mixing of two fluid streams as imaged by (a) MRI and (b) PEPT. Dark blue spots within the cylindrical bulk of the system correspond to the mixer structure. Image reproduced from [270], Copyright 2015, with permission from Elsevier.

Eulerian *and* Lagrangian data means that it can extract additional information not achievable from MRI, for example local flow patterns, velocity fields and velocity probability distribution functions—which again are found to be consistent with prior CFD work on the same system [271]. The combination of both concentration data and velocity data available from PEPT allowed the authors to also determine the *flux-weighted intensity of segregation*, first proposed by Galaktionov *et al* [272, 273]. The results obtained suggest that mixing is more effective in the first part of the SMX mixer elements, suggesting that shorter elements (and thus shorter overall mixers) could potentially be used without loss of effectiveness.

8.7 Other systems

Though the above represent the systems most intensively studied using PEPT, the technique has also been applied to a significant number of other systems, from simple pipe flows [274] to industrial coffee roasters [160] and even domestic dishwashers [275] and washing machines [276]—a diverse range of applications which clearly emphasises the great versatility and wide-ranging value of the PEPT technique.

8.7.1 Pipe flow

A number of authors have applied PEPT to the study of flow through pipes. As in the case of stirred tanks (and indeed other systems) PEPT, unlike more conventional, optical techniques, allows the exploration of both high particle concentrations and/or opaque systems. This ability was demonstrated in the 2001 work of Fairhurst *et al* [274], who studied the dynamics of high solids loadings (20%–40% v/v) neutrally-buoyant alginate particles. The flow patterns observed in the systems studied were observed to depend sensitively on the size of the particles, the solid concentration and the carrier fluid viscosity. For the more viscous fluids and smaller particles (i.e. lower Reynolds numbers) explored, the flow within the system was observed to exhibit two distinct regions—a fast flowing central core and a slower outer annulus—which the authors note would be decidedly sub-optimal for food flow systems, for example, due to the broad variation in residence times and (hence) processing conditions.

The food science applications of pipe flow were more directly explored in the work of Bakalis *et al* [277, 278] who used PEPT to study the dynamics of both Newtonian and non-Newtonian fluids in an optically-opaque aluminium tube under conditions relevant to real processes in the food industry. PEPT-measured radial velocity distributions for simple Newtonian fluids (sucrose solution), shear-thinning fluids (CMS solution) and Herschel–Bulkley fluids (sucrose + CMC) explored were all found to agree well with theoretical predictions across a range of flow rates. The size of the tracer used was found not to impact greatly on results, other than by influencing the tracer's ability to map out flow near the walls of the system.

Most recently, Wiggins *et al* [279] used a turbulent flow through a pipe to illustrate the power of multiple particle PEPT or 'M-PEPT' (see chapter 6). Using up to four tracers, the authors demonstrated PEPT's ability to accurately extract a range of valuable parameters from the system, including mean velocity and acceleration, Reynolds stress, and turbulent kinetic energy budget.

8.7.2 Cyclones

Cyclones are systems which, via the induction of vortex-like (cyclonic) flow, use centrifugal forces to separate the solid and fluid phases of a multiphase system (see figure 8.28). Due to their relative simplicity, durability and lack of moving parts, these systems are extremely popular in industry for separating particles from both gases (gas cyclones) and liquids (hydrocyclones). The first published PEPT study involving a (gas) cyclone was performed by Chan *et al* [280]. The main goal of the project was to gain a previously-lacking mechanistic understanding of the well-known observation that the pressure drop across a gas cyclone is a strong function of solids loading. To this end, PEPT data concerning tracer velocities, residence times and occupancy distributions were acquired for a variety of different solids loadings. Based on their observations, the authors concluded that the increased pressure drop at high solids loadings arises due to the development of a thick, dense layer of downward-flowing particles forming around the walls of the cyclone.

In 2011, Chang *et al* [282] conducted the first PEPT study of a *hydro*cyclone. The authors use the high temporal-resolution Lagrangian information output by PEPT

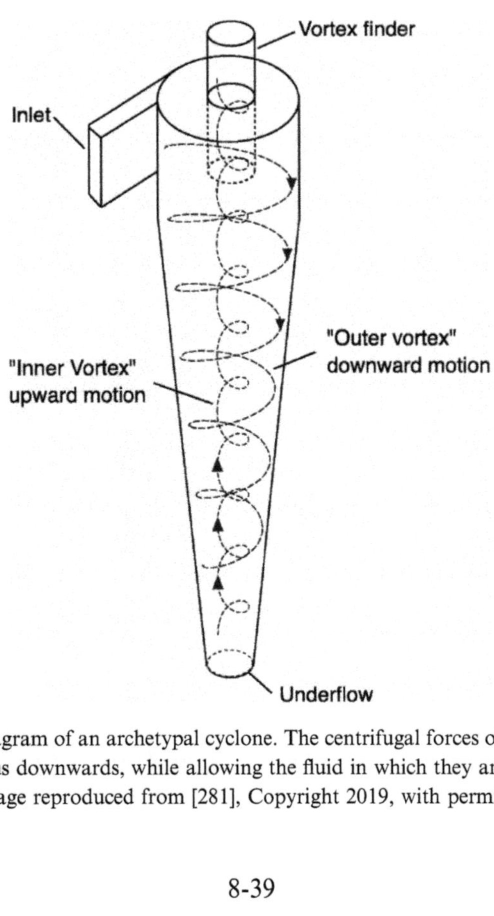

Figure 8.28. Schematic diagram of an archetypal cyclone. The centrifugal forces of the cyclonic gas flow push particles outwards and thus downwards, while allowing the fluid in which they are entrained to flow upward and out of the system. Image reproduced from [281], Copyright 2019, with permission from Elsevier.

(a)

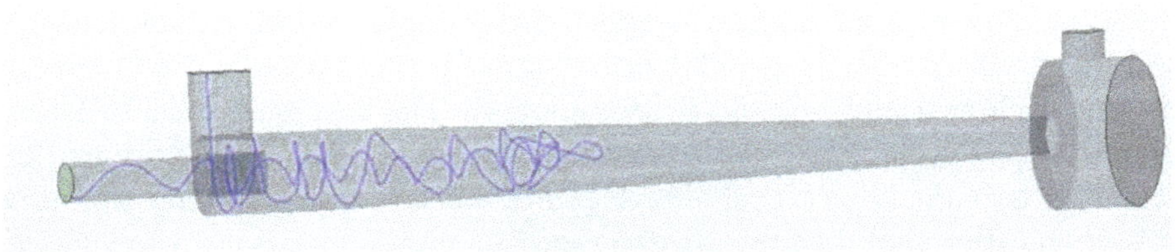

(b)

Figure 8.29. Comparison of PEPT (above) and CFD (below) data showing particle trajectories through a hydrocyclone with a relatively high viscosity (15 cP) fluid. Image reproduced from [281], Copyright 2019, with permission from Elsevier.

to analyse the trajectories followed by particles within the cyclone (see, as an example, figure 8.29). They find the trajectories to be somewhat more erratic than expected, with even relatively heavy particles occasionally making transits between the inner and outer vortices. By analysing the detailed motion of the particle the authors also establish the spatial extend of the vortex region within the system. In subsequent work by the same group [283], experiments were performed in a similar system using neutrally-buoyant tracers, thus mimicking the motion of fluid elements, as opposed to solid-phase particles. The authors use PEPT-measured tangential velocities to validate theoretical expectations, notably those of Alexander [284], describing the radial scaling of tangential velocities within cyclones—suggesting that cyclone motion can be well-described as a near loss-free swirl surrounding a core of near solid-body rotation—and Barth [285, 286] regarding the increase in velocity experienced near the walls of the cyclone as compared to the inlet velocity. The most recent PEPT study of a hydrocyclone [281] provided a first direct insight into the influence of fluid viscosity on the dynamics of such system. It is found that, as one may intuitively expect considering the expected increase in cut size with larger viscosity, excursions of the tracer particle into the upward-flowing inner vortex become more frequent and longer in duration with increasing μ. The study also included CFD simulations, finding good qualitative agreement with PEPT results, and thus yet again emphasising the value of PEPT as a calibration/validation tool suitable for a wide diversity of systems and numerical methods.

8.7.3 Granulators and extruders

Many of the systems discussed previously in the context of mixing (see section 8.6) can, in fact, also be used as granulators. Indeed, the very same vertical, bladed system explored by Saito *et al* [254, 255] predominantly in the context of mixing were, in separate works [287, 288], also used to directly explore the granulation process. In these studies, PEPT was used to explore the dynamics of microcrystalline cellulose powder as it undergoes a wet granulation [289] process using polyvinyl alcohol solution as a binder. The PEPT data obtained was combined with particle (or rather granule) size measurements and torque measurements [290] to provide a deep insight into the process. In their work, the authors used PEPT to measure changes in the bed's expansion, flow patterns, velocity distributions and other parameters during the granulation process—i.e. as binder is added and granules agglomerate and grow in size. Though the qualitative flow pattern was observed to remain self-consistent throughout the granulation process, both the bed height and mean velocity of the system were observed to change markedly during different phases of the granulation processes, these changes correlating well with observed changes in torque.

In addition to high shear mixer-granulators, PEPT has also been used to study granulation processes in other geometries, notably twin screw granulators [291, 292]. The 2012 study of Lee, Ingram and Rowson [291] used PEPT to explore the dynamics of such a system under a variety of different screw speeds and mass feed raters, exploring also various different system configurations. It was found that the residence time of particles within the system decreased with increasing screw speed and/or increasing powder feed rate. The angle of the kneading discs (see figure 8.30) was also found to markedly alter the dynamics of the system, with increasing angles increasing the residence time of particles as well as the fill level of the system (see figure 8.31). Notably, however, the authors find that the fill level never reaches 100% for any configuration of the mixing zone, providing a potential explanation for the comparatively high porosity and low tensile strength observed in granules

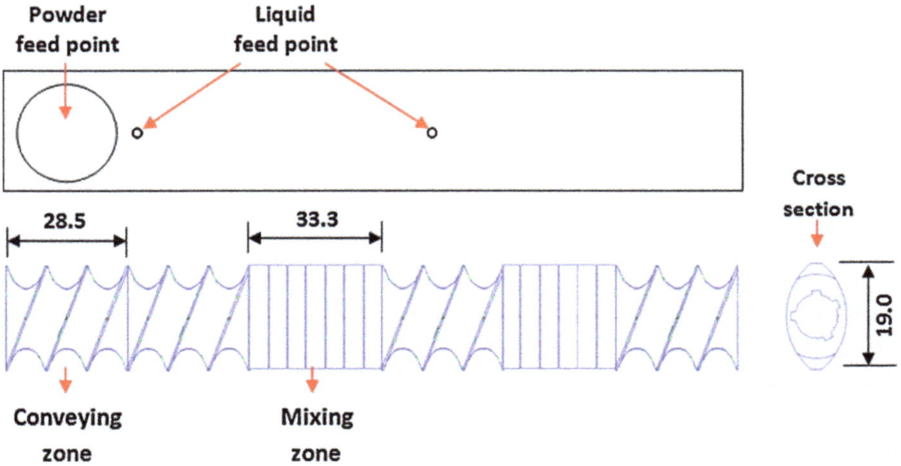

Figure 8.30. Schematic diagram of the twin screw granulator system studied by Lee, Ingram and Rowson [291]. The system consists of screw-like conveying sections for the transport of particles, and two 'mixing zones' each comprising six or seven 'kneading discs' which may be oriented at 30°, 60° or 90°. Image reproduced from [291], Crown copyright © 2012 Published by Elsevier B.V. All rights reserved.

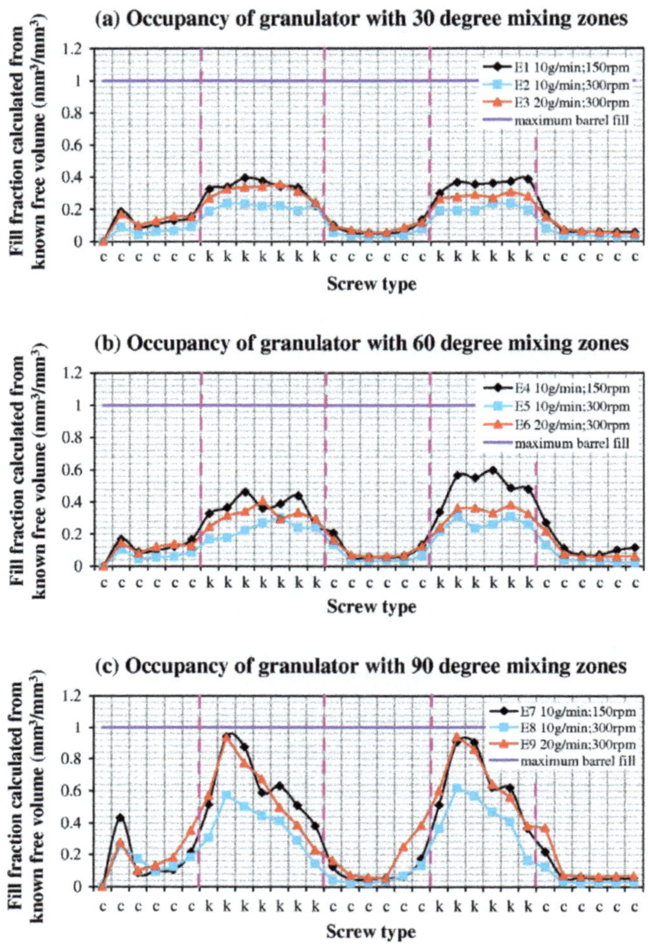

Figure 8.31. PEPT-measured occupancy profiles spanning the axial length of the twin-screw granulator depicted in figure 8.30 under different combinations of kneading disc orientation, feed rate and screw rotation rate. Image reproduced from [291],

produced by twin screw granulators as opposed to those created in high shear mixers [293, 294].

Subsequent work by the same group [292] used PEPT imaging to explore the asymmetric distribution of material often observed in twin screw granulator systems, which is known to negatively affect the uniformity (i.e. quality) of the granules produced thereby. An example of the asymmetry, as observed using PEPT, can be seen in figure 8.32 The study showed the degree of asymmetry to be dependent on both the geometry and fill level of the system, with greater filling fractions reducing the observed heterogeneity. Beside the asymmetry itself, the PEPT data obtained also showed a distinct lack of axial mixing for low fill heights, with material simply moving as two distinct plug flows along each barrel of the twin screw system. At increased fill levels, both mixing and the heterogeneity of wetting are improved.

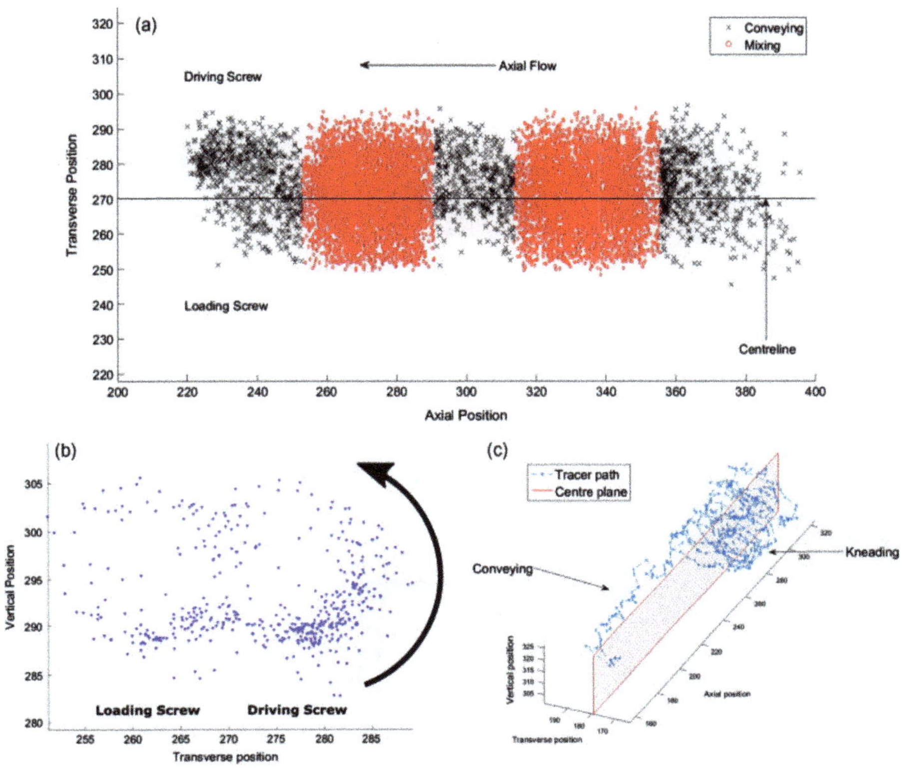

Figure 8.32. Distributions of PEPT data points within the twin screw granulator of Seem *et al* [292] as observed in (a) plan view and (b) end view. From panel (b), one can clearly see the asymmetry in loading in the transverse direction. This asymmetry is further illustrated in panel (c), which shows a single transit of a PEPT tracer through the granulator system. Image reproduced from [292], © 2016 Published by Elsevier B.V.

As well as twin-screw *granulators*, PEPT has also been used to study twin-screw *extruders*. The 2011 work of Diemer *et al* [295] used a specially-designed 'PEPTFlow' modular camera to visualise the extrusion of low-viscosity polymers. As observed for the case of granulation, characteristic residence times in regions containing 90° kneading discs were observed to be longer than those with 60° discs, and an increase in the feed rate resulted in a reduction in residence time in both cases for these (mixing) regions. Interestingly, in the presently-discussed work, the authors note that the decrease in residence time is proportionally larger for the 90° kneading discs than their 60° counterparts. Perhaps surprisingly, for the conveying region of the system, no significant change in residence time distribution is observed in the conveying region as the feed rate is varied, likely due to the generally faster motion in this section and resultant extra capacity. When a reverse element is used to induce back pressure on the flow, thus creating a build up of material and reducing this excess capacity, strong variations in residence time also become evident even in the conveying region.

PEPT has also been used [296] to study flow in more conventional ram extruders [297] (see figure 8.33). PEPT traces, such as those depicted in figure 8.34,

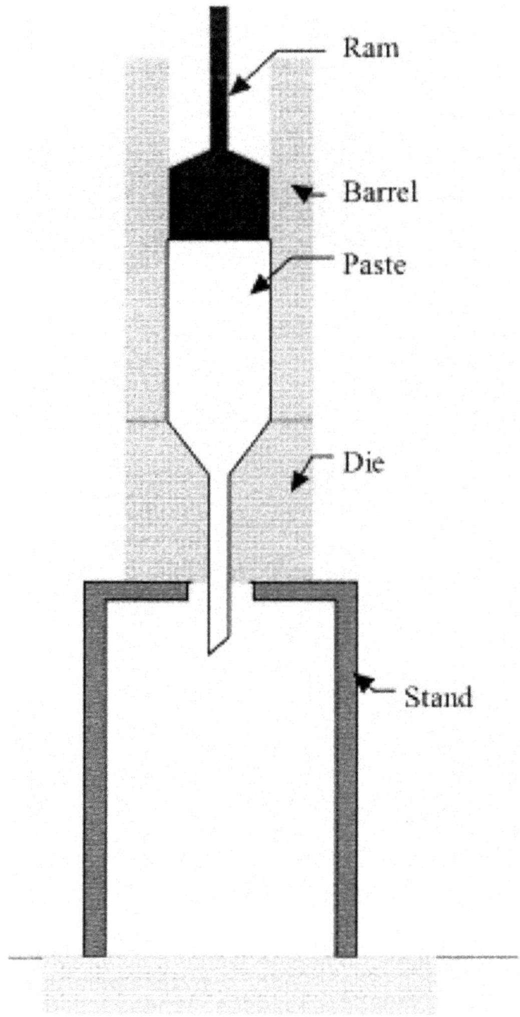

Figure 8.33. Schematic diagram of the ram extruder studied by Wildman *et al* [296]. Image reproduced from [296], Copyright © 1999, reprinted with permission from Elsevier.

demonstrated that forces in the radial direction increase with the radial position of the tracer, with axial and radial forces becoming approximately equal near the radial walls of the system. The extruded material is observed to move with relatively constant and spatially-homogeneous velocity throughout the main bulk of the barrel, experiencing a sharp acceleration when approaching the die entry region, before again obtaining a roughly constant velocity in the die land itself. For both the barrel and die land, the shear strain rate was also measured to be approximately zero, becoming non-zero only in the die entry region. Both shear strain rate and extensional strain rate were observed to be maximal at the radial edges of the system.

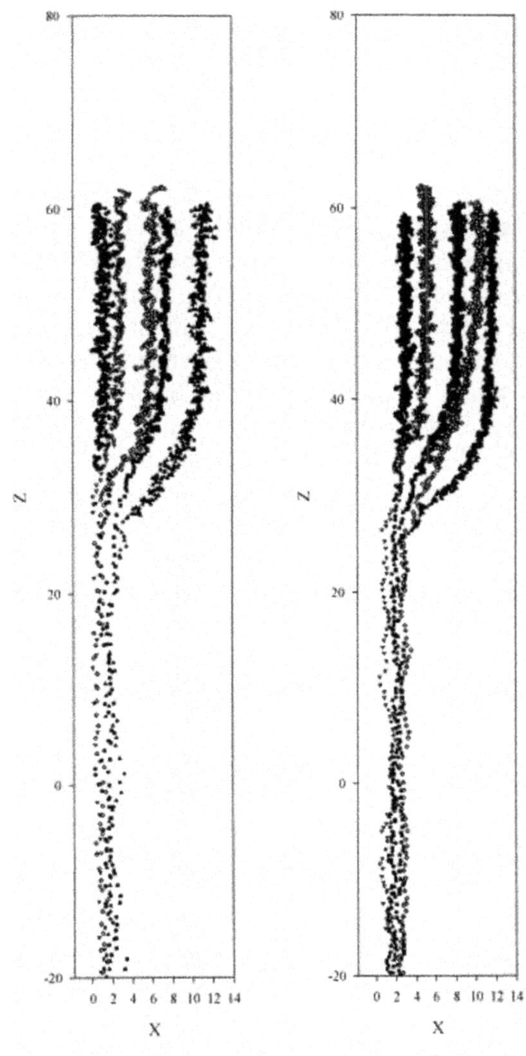

Figure 8.34. Example of PEPT-measured trajectories of particles passing through the ram extruder depicted in figure 8.33 [296]. Image reproduced from [296], Copyright © 1999, reprinted with permission from Elsevier.

8.7.4 Metal casting

One of the more atypical applications of the PEPT technique, which in many ways exemplifies its unique abilities, is in the study of liquid metal casting, requiring the visualisation of the interior dynamics of dense metals, heated in many cases to temperatures in excess of 1000°C. The 2008 work of Griffiths *et al* [298] used a modular PEPT set-up (see section 5.2.4) to directly measure, for the first time, the motion of exogenous inclusions through liquid steel during a casting process. An understanding of the motion of inclusions through a liquid metal is of significant value to various industries, for example in the production of turbine blades [299], as their presence can exacerbate unwanted processes such as fatigue, corrosion and creep [300]. In the experiments of [298], inclusions were represented by alumina and silica particles of sizes in the range 355–710 μm. In each case, a tracer was placed at a known position in a steel mesh over the opening of the mould to be studied. Molten steel heated to 1700°C was then poured into the mould, melting the mesh and thus entraining the irradiated inclusion. PEPT was then used to detect the final position of the particle. Experiments were performed with varying particle sizes and materials, placed in varying initial positions, and in varying numbers. The final positions reached by the particles were found to vary significantly even, in some cases, when all the previously-mentioned control parameters are nominally held constant, demonstrating an inherent variability in the casting process. Placement of a ceramic foam filter in the running system was found to prevent particles from reaching the casting, though the filter itself did not directly trap the particles—rather, it is thought that by slowing the flow of the steel, it allowed the inclusions to float back up through the system.

A set of experiments similar to the above but using instead molten aluminium were performed by the same group in 2010 [302]. Experiments were performed using two different sizes of alumina particles to represent inclusions. In this set of experiments, however, rather than simply measuring the start and end points of the inclusions, their full trajectory through the system during casting was tracked (see, as an example, figure 8.35). The results again indicate a significant variability in the flow dynamics of the system, though several repeat experiments were observed to yield broadly consistent results. The results also suggest that differences in tracer size, at least for the range of sizes (355/710 μm) explored, do not have a strong impact on the motion of an inclusion. Similar observations were also made during subsequent work [301] casting metals—namely Field's metal and Lensalloy-136—with significantly lower melting points than steel or aluminium, though more careful control over the entry of the liquid metal into the inlet, the temperature of the metal and that of the mould were found to improve reproducibility to an extent. Most recently [303], the casting of molten aluminium was once again explored, using particles ranging in size from 200 μm to 600 μm to represent inclusions. Their results again show a relatively limited influence of particle size on the dynamics of the particles. The paper also discusses potential alternative tracers, demonstrating that suitable amounts of activity can potentially be adsorbed onto alumina particles as small as 50 μm—sizes much more representative of typical inclusions in 'real' aluminium casting processes. Though such tracers are not actively included in their experiments, the finding shows promise for future research in the area.

8.7.5 Froth flotation

Froth flotation is a process which uses the bubbling of gases through a liquid–solid slurry to separate hydrophobic and hydrophilic materials [304]. Originally developed for use in the mining sector [305] it is now also widely used in diverse metallurgical and mineral processing applications [306, 307], as well as applications carrying particular contemporary importance such as solid waste management and wastewater treatment in the petroleum industry [308, 309], and waste plastic

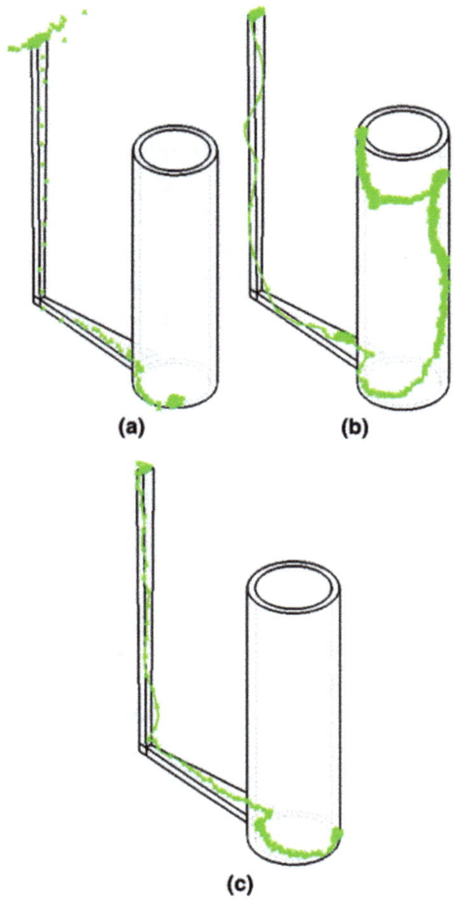

Figure 8.35. Example trajectories of alumina particles in the size range 63–100 μm alumina particles during the casting of Field's metal (panels (a), (b)) and Lensalloy-136 (panel (c)). Image reproduced from [301] Copyright © 2011, THE MINERALS, METALS & MATERIALS SOCIETY and ASM INTERNATIONAL.

recycling [310]. As the froths produced in these flotation devices are typically optically opaque, and the particles within them typically fast-moving, PEPT represents an ideal tool for their study.

The first PEPT experiments to explore the froth flotation process were performed by Waters *et al* in 2008 [312], studying the motion of solid galena particles through the three-phase (solid-liquid-gas) foam created within a Denver cell batch flotation system. The PEPT data show a tendency of particles to rise through the system in a cycle of relatively slow upward motions whilst attached to a bubble, followed by relatively rapid downward 'drops' as the particle detaches, before being collected by another bubble, repeating this process until it eventually spills out over the weir into the concentrate collection tray. Examples of this motion can be seen in figure 8.37.

The initial, comparatively simple, study of Waters *et al* in 2008 [311] was followed up later in the same year by a more detailed analysis, in which PEPT was used not only to analyse single particle trajectories and velocities, but also to build up full pictures of the spatial occupancy and velocity fields of the Denver cell, in this case tracking pyrite particles. The authors compare motion within the impeller-driven system before and after the addition of the froth-inducing airflow through the system. Without airflow, the tracer explores the entire liquid region of the system, exhibiting relatively rapid motion. Once airflow is introduced to the system, the particle rapidly ascends into the upper froth phase, exploring almost exclusively this region, but travelling at much lower characteristic velocities, likely due to the reduced turbulence in the froth phase as compared to the bulk liquid phase. The same authors produced a paper [313] studying both hydrophilic (galena) and hydrophobic (quartz) particles, and comparing the dynamics thereof. The hydrophilic tracer was observed to readily attach to air bubbles, rapidly entering the froth phase and overflowing from the system, as desired. The quartz, meanwhile, was not observed to overflow, remaining predominantly in the bulk liquid phase, and indeed exhibiting dynamics qualitatively similar to those observed without gas flow.

In 2009, Fan *et al* [314] performed further experiments in the same system, this time with ilmenite—a mineral from which titanium is commonly recovered [315], but whose natural properties are not overly conducive to flotation [316, 317]. The authors demonstrated that the floatability of ilmenite could be notably increased by treating the particles with microwaves in such a manner as to enhance the adsorption of surfactants to the ilmenite's surface, thus improving flotation recovery by more than 30%. In the study, PEPT was used to directly demonstrate the observed effect, as well as to elucidate the mechanisms underlying the poor floatability of ilmenite. It was demonstrated the this poor floatability was due not only to the relatively low probability of ilmenite particles being captured by air bubbles, but also their relatively short residence times in the froth phase, with particles observed to frequently make excursions into the lower, bulk-liquid phase. Similar effects were observed in subsequent, more detailed work using ion exchange resins, which can hold higher levels of radioactivity

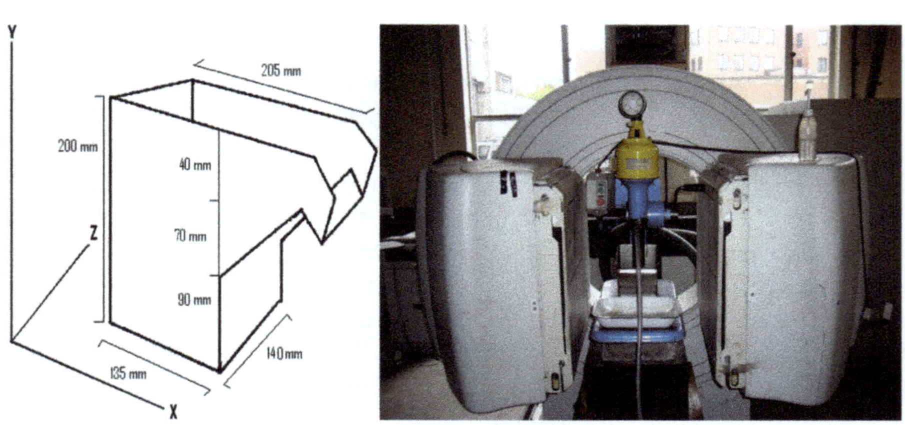

Figure 8.36. Left: Schematic diagram of the Denver cell froth flotation device used by Waters *et al* [311, 312]. Right: The real system, placed in the field of view of Birmingham's ADAC positron camera. Images reproduced from [311], Copyright © 2008, with permission from Elsevier.

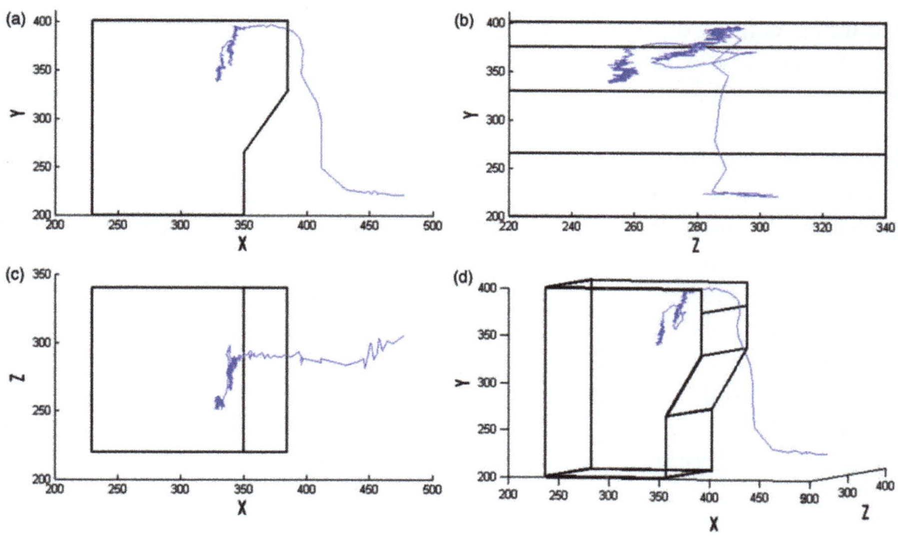

Figure 8.37. Examples of PEPT-measured particle trajectories within the Denver cell froth flotation device (pictured in figure 8.36) used by Waters *et al* [311, 312]. Images reproduced from [312], 1 Sep 2008, reprinted by permission of the publisher (Taylor & Francis Ltd, http://www.tandfonline.com).

[318] and thus produce higher-resolution PEPT data. While ion exchange resins are, in their natural state, hydrophilic, the authors used novel surface treatment methods to create also a hydrophobic variant, allowing them to study the motion of both types of particle with the same, high degree of detail. While the approach of [318], like those preceding it, relied on the use of indirect activation (see section 5.1.3) to radioactively label PEPT tracers, later work by Boucher *et al* [319] was the first to apply *direct activation* (see section 5.1.3) to flotation cell studies. Though requiring significantly larger particles and ultimately being limited to lower activities (and thus lower data rates and resolution) than the indirect methods of [318], for example, the direct activation method carries the significant advantage that the particles used in 'real' flotation procedures can be used directly, with no modification to their physical properties or surface characteristics.

Cole *et al* [320] compared PEPT data with optically-acquired data obtained simultaneously from the same system, as a means of comprehensively validating the application of PEPT to such systems. The two sets of results were observed to agree to a confidence level of 95%, emphatically confirming the accuracy of PEPT in such systems. The results shown here provide valuable reassurance for the use of PEPT in similar systems for cases where tracers are too small, or the froth too optically opaque, to be imaged with conventional optical methods. Further work combining PEPT and optical methods was also performed by the same group [321], though in the case of [321] optical imaging was used synergistically as a complementary technique, as opposed to as a validation tool. Specifically, particle motion data acquired from PEPT was combined with simultaneously-acquired video footage of a rising foam column recording the dynamics and structure of the foam in which the particle is entrained. By simultaneously capturing the dynamics of both phases and thus the interactions between the two, a deeper understanding of the dynamics of the system could be obtained. It was observed that acceleration and deceleration of the tracer could be tied to expansion and contraction of the foam, and that the tracer velocity followed the same streamlines as the bubbles when convective flow was observed within the system, though local coalescence events were observed to be cable of 'shunting' particles between streamlines.

In the most recent addition to the field to date, Sommer *et al* [322] used a device capable of achieving well controlled levels of turbulence to assess the influence thereof on solids motion in flotation cells. Surprisingly, the PEPT data showed no clear impact of the degree of turbulence on either the measured spatial distribution of particles or the displacement of particles for any of the solids loadings explored. It is unclear from the authors' data if this observation is representative of the true dynamics of the system, or whether the tracers used were simply too large to follow the detailed flow profiles of the fluid. The well-controlled system used by the authors nonetheless shows promise for future studies, potentially using more advanced tracers.

8.7.6 Spiral concentrators

Another device widely used in the separation and concentration of minerals and ores in slurry form is the spiral concentrator [323] (see figure 8.38). These devices, like the

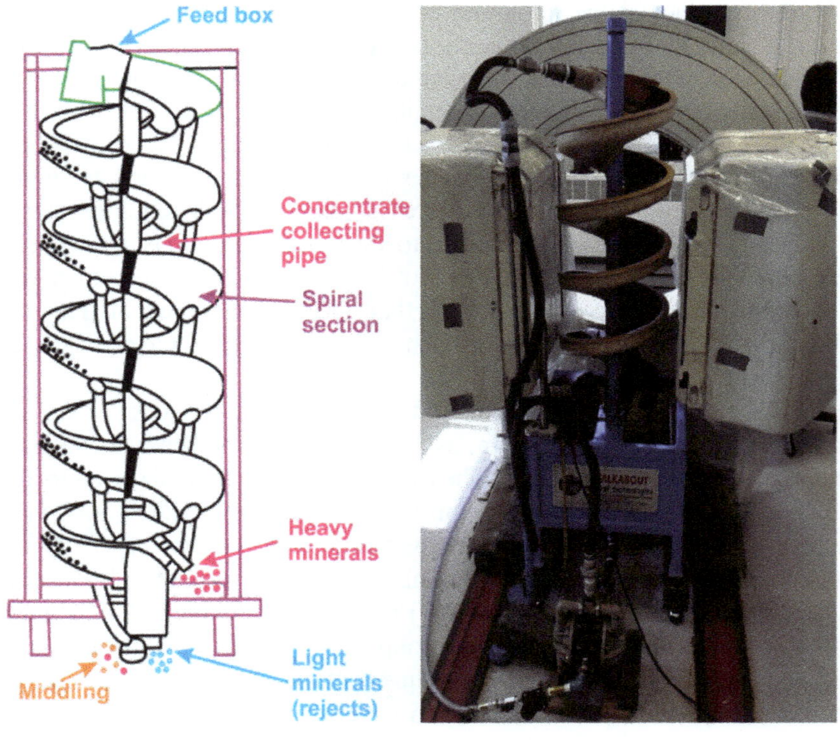

Figure 8.38. Left: Schematic diagram of an archetypal spiral concentrator, reproduced with permission from [323], Copyright © 2018, with permission from Elsevier. Right: Experimental spiral concentrator undergoing PEPT imaging in an ADAC Forté camera (image reproduced with permission from [326], Copyright © 2014, with permission from Elsevier).

previously-discussed froth flotation devices, have significant applications both in the mining industry [323, 324] as well as in the recovery and recycling of valuable materials [325].

The first PEPT study of such a system was performed by Boucher *et al* [326], who used PEPT to analyse the motion of particles possessing various sizes and densities, so as to gain insight into the behaviours of both bulk materials and the valuable materials to be extracted from the bulk. The authors considered the system as being divided radially into three distinct 'zones'—inner, middle, and outer. Unexpectedly, they found that the largest, densest (chromite) particles tended to occupy almost exclusively the outer region of the system—contrary to the widely-held expectation that denser media should in fact report to the inner section of the trough. The authors suggest that this unexpected behaviour arose due to the relatively high flow rates explored and the relatively large size of the tracers [327, 328]; smaller chromite tracers displayed a greater tendency toward the inner zone, more in line with conventional expectations.

Subsequent work by the same authors used a modular PEPT camera to explore the motions of large quartz particles through specific sections of a spiral concentrator system in greater detail, namely the first two turns thereof. The study measured, for the first time, the rate of radial migration of particles within the system, providing valuable insight into the secondary flow within such systems. It is observed that relatively low speeds are required in order to facilitate secondary flow—for the design explicitly tested, particle speeds higher than 5 m s^{-1} were associated with particles becoming in essence 'trapped' in the faster-flowing outer region of the system.

The most recent PEPT study of a spiral concentrator [330] used a modular camera setup similar to that of [329], but this time tracking both large ($d \sim 1.4$ mm) and small ($d \sim 50$ μm) hematite and quartz particles. The authors observe the large, dense hematite particles to split into two distinct bands—one near the centre of the spiral, where denser species would normally be expected, and one near the outer

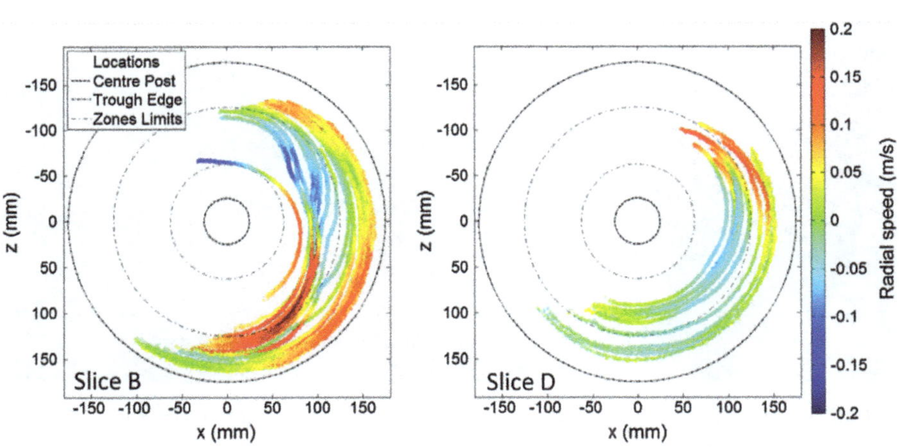

Figure 8.39. Examples of particle trajectories through two distinct segments of a spiral concentrator system. The trajectories are coloured according to their current radial speed. Image reused from [329], Copyright © 2015, with permission from Elsevier.

reaches of the spiral, echoing the observations of [326] (see figure 8.39). The authors found that the path followed by a hematite particle was strongly dependent on its initial feed position—in particular, it was found to be highly unlikely that a large hematite particle entering the system at a large radial position would ever migrate toward the centre of the system—meaning that, in the real process, it would be erroneously discarded. This observation is significant, highlighting the importance of controlling the feed into the system, and potentially highlighting a serious flaw in the design of spiral concentrator systems.

8.7.7 Miscellaneous

Continuing, from the previous subsections, in the vein of minerals processing, PEPT has also—albeit less extensively—been used to investigate another form of separation equipment widely used in the mining industry as well as in plastic recycling efforts [331], namely a batch jig system [332]. The the 2014 study of Roux and Naudé [333] the authors found the settling rates of heavy particles to differ significantly dependent on their placement within the system, with slower downward transits observed near the walls of the system than nearer its centre. A similar effect is not, however, observed for lighter particles. Their results also indicated the presence of a secondary, recirculating flow within the system, as suggested by prior work in the area [334].

In addition to its relatively extensive use in the study of key pharmaceutical processes such as mixing and granulation, as discussed above, PEPT has also, in the 2010 work of Wu *et al* [335], been used to explore the die-filling process [336, 337]. The PEPT-measured motions of tracers during the die filling process are observed to align closely with the macroscopic powder motion within the system. It is also observed that larger particles exhibit higher vertical velocities than smaller particles, suggesting that air effects within the die are significant in determining the dynamics of finer powders.

As well as industrial equipment, PEPT has also been used to image flow within domestic appliances (see, for example, figure 8.40). In a collaboration between the University of Birmingham and Procter and Gamble Ltd, MacNamara *et al* [276] used

Figure 8.40. As well as large industrial equipment, PEPT has also been used to image flow within domestic appliances, such as washing machines and dishwashers. Left-hand image reproduced from [276], Copyright © 2012, with permission from Elsevier. Right-hand image reproduced from [275], Copyright © 1969, with permission from Elsevier.

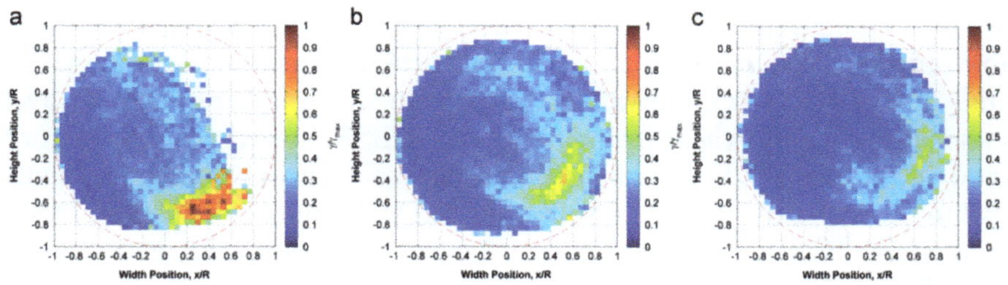

Figure 8.41. Shear rates for different textile loadings in a Miele W1714 domestic washing machine. Data is shown for loadings of (a) 1 kg, (b) 2 kg and (c) 3 kg. Note the decrease in shear—and thus washing effectiveness—with increasing loading. Image reproduced from [276], Copyright © 2012, with permission from Elsevier.

PEPT tracer particles attached to textiles to investigate the dynamics of fabrics in a domestic dishwasher. In particular, as mechanical force is crucial in removing soil and other unwanted materials from clothes during a washing cycle, the authors explored the variation of impact and shear within the system with key system conditions. It was observed that mechanical action was reduced with increasing textile loadings (see figure 8.41), suggesting smaller loads to be more effective at removing stains. In general, motion within the system is observed to be regular, with relatively little axial transport, and an apparent dead zone at the centre of the system, likely to be unfavourable to the washing of any materials existing in this region.

Pérez-Mohedano *et al* [275] worked with both Procter and Gamble and Whirlpool to analyse the fluid dynamics of a household dishwasher. The results show that the flow within the system was most significantly affected by high packing densities of crockery, with tracers exhibiting longer residence times in densely-packed regions. It is thought that this may be beneficial for the cleaning of the crockery, by allowing for prolonged chemical interactions with the detergent used. Interestingly, the inclusion or absence of detergent in the experiments performed was shown to have no clear effect on the dynamics of the system. The authors also used their PEPT results to calibrate a CFD model of the dishwasher system, observing largely good agreement, though the CFD simulations were found to slightly underpredict the downward flow of water after its initial ejection.

Another rather unconventional PEPT experiment was performed in 2017 by Langford *et al* [338], who tracked the motion of yeast cells as they settle through a container of deionised water. Though this simple experiment itself was not overly enlightening, the work demonstrated the ability of PEPT to successfully track tracers carrying very low (~30 Bq) activities, and in doing so demonstrated its potential value for other, more industrially and/or medically-relevant cell-tracking applications.

8.8 Nascent applications

As is abundantly clear from the above sections, PEPT is an established technique in a variety of industrial and scientific applications. However, as illustrated in the above-discussed work of Langford *et al* [338], recent dramatic improvements in terms of the size and number of tracers that can be tracked, and the precision with which this tracking may be performed, have opened the door to numerous other valuable applications, notably in the biomedical and geophysical arenae. The potential of PEPT to track individual cells was in fact first demonstrated by Lee, Kim and Pratx [339] when introducing the 'SBSR' pre-processing algorithm discussed in section 6.2.9, though in this instance using only numerical data, as opposed to the real, experimental results analysed by Langford *et al* [338]. The work of Lee, Kim and Pratx [339], however, was the first to explicitly discuss the potential value of PEPT in the biomedical context, citing its potential value, for example, in the tracking of white blood cells to identify sites of infection or inflammation which cannot be easily identified using other means [340], in the study of cancer metastasis [341], or to assess the progress and effectiveness of stem-cell therapies in treating diverse diseases [342–344]. In 2016, the group of Pratx applied their algorithm to a simple experimental system, achieving imaging with sub-millimetre resolution for tracers with activities as low as 139 Bq, again demonstrating the potential of PEPT to track extremely small and (thus) low-activity particles. Finally, in 2020, the group directly demonstrated [345], for the first time, PEPT's biomedical cell tracking capabilities *in vivo*, tracking a single, gallium-labelled breast cancer cell injected into a mouse (figure 8.42). This direct demonstration of the feasibility of biomedical PEPT has paved the way for significant further work in the field in coming years, representing the dawn of a highly exciting new era in PEPT imaging.

Another area in which PEPT carries great potential which, to date, has yet to be extensively exploited, is that of geophysics/geomechanics and other related areas. Though its value has been hinted at, for example in studies of rotating drum systems (see section 8.2) which provide valuable surrogates for a variety of geophysical phenomena such as landslides, *direct* geomechanical applications of PEPT remain extremely rare. Indeed, the only such application published to date is the 2015 work of Kinugasa *et al* [346] which studies the terramechanics between terrain particles

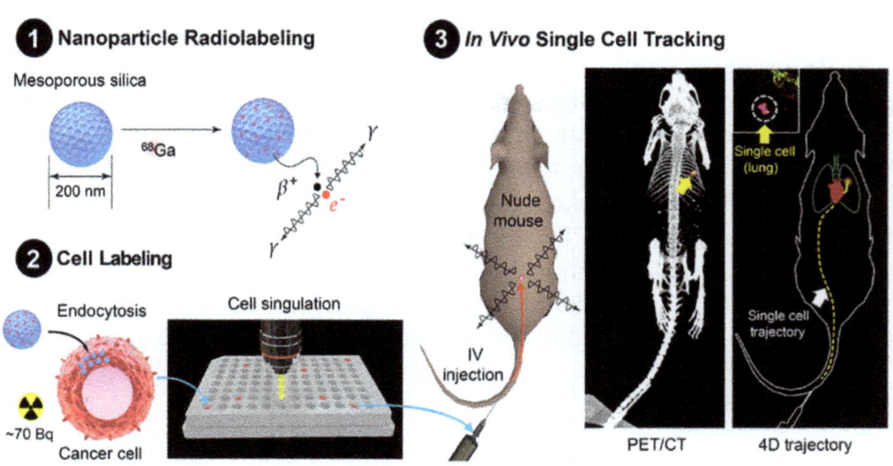

Figure 8.42. Tracking of a single cancer cell in a mouse, showing the labelling process, in which silica nanoparticles (1) are loaded onto a cancer cell (2) before being injected into the mouse and tracked using PEPT (3). Image reproduced from [345] Copyright © 2020, The Author(s), under exclusive licence to Springer Nature Limited.

and the wheels of robotic systems such as those used for planetary exploration, or the study of other environments unsuitable for human exploration, e.g. volcanoes or post-tsunami seashores.

Interestingly, positron emission *tomography* (PET) has, in recent years, found several applications in the geoscientific field, for example, examining the subsurface activity of bacteria or other microbes in soils [347], fracture processes [348] or diffusion in soils and clays [349], suggesting an appetite in the field for new imaging technologies which can help answer questions inaccessible to conventional techniques. It is hoped that PEPT will, in the coming years, provide one such technology and thus fulfil some of its latent potential in this area.

References

[1] Rosato A D and Windows-Yule C 2020 *Segregation in Vibrated Granular Systems* (New York: Academic)

[2] Khakhar D V, Orpe A V and Hajra S K 2003 Segregation of granular materials in rotating cylinders *Physica* A **318** 129–36

[3] Johnson P A, Savage H, Knuth M, Gomberg J and Marone C 2008 Effects of acoustic waves on stick-slip in granular media and implications for earthquakes *Nature* **451** 57–60

[4] Pak H K and Behringer P R 1994 Bubbling in vertically vibrated granular materials *Nature* **371** 231–3

[5] Taslagyan K A, Chan D H and Morgenstern N R 2015 Effect of vibration on the critical state of dry granular soils *Granul. Matter* **17** 687–702

[6] Pak R Y S, Ashlock J C, Kurahashi S and Abedzadeh F 2008 Parametric *G*max sounding of granular soils by vibration methods *Géotechnique* **58** 571–80

[7] D'Anna G and Grémaud G 2001 The jamming route to the glass state in weakly perturbed granular media *Nature* **413** 407–9

[8] Dijksman J A, Wortel G H, van Dellen L T H, Dauchot O and van Hecke M 2011 Jamming, yielding, and rheology of weakly vibrated granular media *Phys. Rev. Lett.* **107** 108303

[9] Umbanhowar P B, Melo F and Swinney H L 1996 Localized excitations in a vertically vibrated granular layer *Nature* **382** 793–6

[10] Ristow G H 2000 *Pattern Formation in Granular Materials* (Berlin: Springer)

[11] Pöschel T and Brilliantov N V 2003 *Granular Gas Dynamics* vol 624 (Berlin: Springer)

[12] Olafsen J S and Urbach J S 1999 Velocity distributions and density fluctuations in a granular gas *Phys. Rev.* E **60** R2468

[13] Rouyer F and Menon N 2000 Velocity fluctuations in a homogeneous 2D granular gas in steady state *Phys. Rev. Lett.* **85** 3676

[14] Puglisi A, Loreto V, Marini Bettolo Marconi U, Petri A and Vulpiani A 1998 Clustering and non-Gaussian behavior in granular matter *Phys. Rev. Lett.* **81** 3848

[15] van der Meer D and Reimann P 2006 Temperature anisotropy in a driven granular gas *Europhys. Lett.* **74** 384

[16] Feitosa K and Menon N 2002 Breakdown of energy equipartition in a 2D binary vibrated granular gas *Phys. Rev. Lett.* **88** 198301

[17] Sloot E M and Kruyt N P 1996 Theoretical and experimental study of the transport of granular materials by inclined vibratory conveyors *Powder Technol.* **87** 203–10

[18] Mohabuth N and Miles N 2005 The recovery of recyclable materials from waste electrical and electronic equipment (WEEE) by using vertical vibration separation *Resour. Conserv. Recycl.* **45** 60–9

[19] Habib M, Miles N J and Hall P 2013 Recovering metallic fractions from waste electrical and electronic equipment by a novel vibration system *Waste Manage.* **33** 722–9

[20] Makinde O A, Ramatsetse B I and Mpofu K 2015 Review of vibrating screen development trends: linking the past and the future in mining machinery industries *Int. J. Miner. Process.* **145** 17–22

[21] Hashimoto F and DeBra D B 1996 Modelling and optimization of vibratory finishing process *CIRP Ann.* **45** 303–6

[22] Wang S, Timsit R S and Spelt J K 2000 Experimental investigation of vibratory finishing of aluminum *Wear* **243** 147–56

[23] Chen W, Huo D, Shi Y and Hale J M 2018 State-of-the-art review on vibration-assisted milling: principle, system design, and application *Int. J. Adv. Manuf. Technol.* **97** 2033–49

[24] Baxter G W and Olafsen J S 2007 Experimental evidence for molecular chaos in granular gases *Phys. Rev. Lett.* **99** 028001

[25] Clément E, Vanel L, Rajchenbach J and Duran J 1996 Pattern formation in a vibrated two-dimensional granular layer *Phys. Rev.* E **53** 2972

[26] Reimann J, Vicente J, Brun E, Ferrero C, Gan Y and Rack A 2017 X-ray tomography investigations of mono-sized sphere packing structures in cylindrical containers *Powder Technol.* **318** 471–83

[27] Weis S and Schröter M 2017 Analyzing x-ray tomographies of granular packings *Rev. Sci. Instrum.* **88** 051809

[28] Gajjar P, Johnson C G, Carr J, Chrispeels K, Gray J M N T and Withers P J 2021 Size segregation of irregular granular materials captured by time-resolved 3D imaging *Sci. Rep.* **11** 1–6

[29] Kawaguchi T 2010 MRI measurement of granular flows and fluid-particle flows *Adv. Powder Technol.* **21** 235–41

[30] Ehrichs E E, Jaeger H M, Karczmar G S, Knight J B, Yu Kuperman V and Nagel S R 1995 Granular convection observed by magnetic resonance imaging *Science* **267** 1632–4

[31] Wildman R D, Huntley J M, Hansen J-P, Parker D J and Allen D A 2000 Single-particle motion in three-dimensional vibrofluidized granular beds *Phys. Rev.* E **62** 3826

[32] Wildman R D, Huntley J M and Parker D J 2001 Granular temperature profiles in three-dimensional vibrofluidized granular beds *Phys. Rev.* E **63** 061311

[33] Windows-Yule C R K and Parker D J 2014 Center of mass scaling in three-dimensional binary granular systems *Phys. Rev.* E **89** 062206

[34] Wildman R D and Parker D J 2002 Coexistence of two granular temperatures in binary vibrofluidized beds *Phys. Rev. Lett.* **88** 064301

[35] Windows-Yule C R K and Parker D J 2014 Energy non-equipartition in strongly convective granular systems *Eur. Phys. J.* E **37** 17

[36] Krouskop P E and Talbot J 2004 Anisotropic energy distribution in three-dimensional vibrofluidized granular systems *Phys. Rev.* E **69** 061308

[37] Windows-Yule C R K and Parker D J 2013 Boltzmann statistics in a three-dimensional vibrofluidized granular bed: idealizing the experimental system *Phys. Rev.* E **87** 022211

[38] Pak H K and Behringer R P 1993 Surface waves in vertically vibrated granular materials *Phys. Rev. Lett.* **71** 1832

[39] Eshuis P, van der Weele K, van der Meer D and Lohse D 2005 Granular Leidenfrost effect: experiment and theory of floating particle clusters *Phys. Rev. Lett.* **95** 258001
[40] Eshuis P, Van Der Weele K, Van Der Meer D, Bos R and Lohse D 2007 Phase diagram of vertically shaken granular matter *Phys. Fluids* **19** 123301
[41] Warr S, Huntley J M and Jacques G T H 1995 Fluidization of a two-dimensional granular system: experimental study and scaling behavior *Phys. Rev.* E **52** 5583
[42] Luding S, Herrmann H J and Blumen A 1994 Simulations of two-dimensional arrays of beads under external vibrations: scaling behavior *Phys. Rev.* E **50** 3100
[43] Kumaran V 1998 Temperature of a granular material "fluidized" by external vibrations *Phys. Rev.* E **57** 5660
[44] Warr S and Huntley J M 1995 Energy input and scaling laws for a single particle vibrating in one dimension *Phys. Rev.* E **52** 5596
[45] McNamara S and Luding S 1998 Energy flows in vibrated granular media *Phys. Rev.* E **58** 813
[46] Lee J 1995 Scaling behavior of granular particles in a vibrating box *Physica* A **219** 305–26
[47] Windows-Yule C R K, Rosato A D, Parker D J and Thornton A R 2015 Maximizing energy transfer in vibrofluidized granular systems *Phys. Rev.* E **91** 052203
[48] Windows-Yule C R K, Rosato A D, Thornton A R and Parker D J 2015 Resonance effects on the dynamics of dense granular beds: achieving optimal energy transfer in vibrated granular systems *New J. Phys.* **17** 023015
[49] Windows-Yule C R K, Maddox B and Parker D J 2014 The role of rotational inertia in the dynamics of vibrofluidised granular gases *Europhys. Lett.* **108** 58006
[50] Wildman R D, Huntley J M and Parker D J 2001 Convection in highly fluidized three-dimensional granular beds *Phys. Rev. Lett.* **86** 3304
[51] Hsiau S-S and Chen C H 2000 Granular convection cells in a vertical shaker *Powder Technol.* **111** 210–7
[52] Hsiau S-S, Wang P-C and Tai C-H 2002 Convection cells and segregation in a vibrated granular bed *AIChE J.* **48** 1430–8
[53] Wildman R D, Martin T W, Krouskop P E, Talbot J, Huntley J M and Parker D J 2005 Convection in vibrated annular granular beds *Phys. Rev.* E **71** 061301
[54] Windows-Yule C R K, Rivas N and Parker D J 2013 Thermal convection and temperature inhomogeneity in a vibrofluidized granular bed: the influence of sidewall dissipation *Phys. Rev. Lett.* **111** 038001
[55] Windows-Yule C R K, Lanchester E, Madkins D and Parker D J 2018 New insight into pseudo-thermal convection in vibrofluidised granular systems *Sci. Rep.* **8** 12859
[56] Windows-Yule C R K, Scheper B J, den Otter W K, Parker D J and Thornton A R 2016 Modifying self-assembly and species separation in three-dimensional systems of shape-anisotropic particles *Phys. Rev.* E **93** 020901
[57] Windows-Yule C R K, Rosato A D, Rivas N and Parker D J 2014 Influence of initial conditions on granular dynamics near the jamming transition *New J. Phys.* **16** 063016
[58] Windows-Yule C R K, Weinhart T, Parker D J and Thornton A R 2014 Effects of packing density on the segregative behaviors of granular systems *Phys. Rev. Lett.* **112** 098001
[59] Frank F C 1952 Supercooling of liquids *Proc. R. Soc.* A **215** 43–6
[60] Windows-Yule C R K, Rivas N, Parker D J and Thornton A R 2014 Low-frequency oscillations and convective phenomena in a density-inverted vibrofluidized granular system *Phys. Rev.* E **90** 062205

[61] Wong Y S, Gan C H and Wang C H 2004 Tracking study on solids motion in vertically vibrated granular beds *AIChE Annual Meeting, Conf. Proc.* pp 3187–95

[62] Wong Y S, Gan C H, Wang C H, Fan X, Parker D J, Ingram A and Seville J P K 2006 Instabilities in vertically vibrated granular beds at the single particle scale *Phys. Fluids* **18** 043302

[63] Windows-Yule C R K and Parker D J 2014 Self-diffusion, local clustering and global segregation in binary granular systems: the role of system geometry *Powder Technol.* **261** 133–42

[64] Windows-Yule C R K, Weinhart T, Parker D J and Thornton A R 2014 Influence of thermal convection on density segregation in a vibrated binary granular system *Phys. Rev.* E **89** 022202

[65] Windows-Yule K and Parker D 2015 Density-driven segregation in binary and ternary granular systems *KONA Powder Part. J.* **32** 163–75

[66] Windows-Yule C R K and Parker D J 2014 Inelasticity-induced segregation: why it matters, when it matters *Europhys. Lett.* **106** 64003

[67] Windows-Yule C R K, Douglas G J M and Parker D J 2015 Competition between geometrically induced and density-driven segregation mechanisms in vibrofluidized granular systems *Phys. Rev.* E **91** 032205

[68] Shi Q, Sun G, Hou M and Lu K 2007 Density-driven segregation in vertically vibrated binary granular mixtures *Phys. Rev.* E **75** 061302

[69] Abreu C R A, Tavares F W and Castier M 2003 Influence of particle shape on the packing and on the segregation of spherocylinders via Monte Carlo simulations *Powder Technol.* **134** 167–80

[70] Roskilly S J, Colbourn E A, Alli O, Williams D, Paul K A, Welfare E H and Trusty P A 2010 Investigating the effect of shape on particle segregation using a Monte Carlo simulation *Powder Technol.* **203** 211–22

[71] Martin T W, Huntley J M and Wildman R D 2005 Hydrodynamic model for a vibrofluidized granular bed *J. Fluid Mech.* **535** 325–45

[72] Viswanathan H, Sheikh N A, Wildman R D and Huntley J M 2011 Convection in three-dimensional vibrofluidized granular beds *J. Fluid Mech.* **682** 185–212

[73] Viswanathan H, Wildman R D, Huntley J M and Martin T W 2006 Comparison of kinetic theory predictions with experimental results for a vibrated three-dimensional granular bed *Phys. Fluids* **18** 113302

[74] Cundall P A and Strack O D L 1979 A discrete numerical model for granular assemblies *Geotechnique* **29** 47–65

[75] Luding S 2008 Introduction to discrete element methods: basic of contact force models and how to perform the micro-macro transition to continuum theory *Eur. J. Environ. Civil Eng.* **12** 785–826

[76] Windows-Yule C R K, Tunuguntla D R and Parker D J 2016 Numerical modelling of granular flows: a reality check *Computat. Part. Mech.* **3** 311–32

[77] Lubachevsky B D 1991 How to simulate billiards and similar systems *J. Comput. Phys.* **94** 255–83

[78] Windows-Yule C R K, Scheper B J, van der Horn A J, Hainsworth N, Saunders J, Parker D J and Thornton A R 2016 Understanding and exploiting competing segregation mechanisms in horizontally rotated granular media *New J. Phys.* **18** 023013

[79] Windows-Yule C R K, Van Der Horn A J, Tunuguntla D R, Parker D J and Thornton A R 2017 Inducing axial banding in bidisperse-by-density granular systems using noncylindrical tumbler geometries *Phys. Rev. Appl.* **8** 024010

[80] Delele M A, Weigler F and Mellmann J 2015 Advances in the application of a rotary dryer for drying of agricultural products: a review *Drying Technol.* **33** 541–58
[81] Powell M S 1993 A study of charge motion in rotary mills, with particular reference to the grinding action *PhD Thesis* University of Cape Town
[82] Jiang M, Zhao Y, Liu G and Zheng J 2011 Enhancing mixing of particles by baffles in a rotating drum mixer *Particuology* **9** 270–8
[83] Fernández F J, Sánchez-Arias V, Rodríguez L and Villaseñor J 2010 Feasibility of composting combinations of sewage sludge, olive mill waste and winery waste in a rotary drum reactor *Waste Manage.* **30** 1948–56
[84] Sherritt R G, Chaouki J, Mehrotra A K and Behie L A 2003 Axial dispersion in the three-dimensional mixing of particles in a rotating drum reactor *Chem. Eng. Sci.* **58** 401–15
[85] Fabbri A, Cevoli C, Alessandrini L and Romani S 2011 Numerical modeling of heat and mass transfer during coffee roasting process *J. Food Eng.* **105** 264–9
[86] Ristow G H 1996 Dynamics of granular materials in a rotating drum *Europhys. Lett.* **34** 263
[87] J M N T Gray 2001 Granular flow in partially filled slowly rotating drums *J. Fluid Mech.* **441** 1–29
[88] Antony S J, Hoyle W and Ding Y 2004 *Granular Materials: Fundamentals and Applications* (Royal Society of Chemistry: Cambridge)
[89] Wu W 2015 *Recent Advances in Modeling Landslides and Debris Flows* (Springer: Berlin)
[90] Schneider D, Kaitna R, Dietrich W E, Hsu L, Huggel C and McArdell B W 2011 Frictional behavior of granular gravel–ice mixtures in vertically rotating drum experiments and implications for rock–ice avalanches *Cold Reg. Sci. & Technol.* **69** 70–90
[91] Khakhar D V, McCarthy J J and Ottino J M 1997 Radial segregation of granular mixtures in rotating cylinders *Phys. Fluids* **9** 3600–14
[92] Dury C M and Ristow G H 1997 Radial segregation in a two-dimensional rotating drum *J. Physique* **I7** 737–45
[93] Hill K M, Caprihan A and Kakalios J 1997 Bulk segregation in rotated granular material measured by magnetic resonance imaging *Phys. Rev. Lett.* **78** 50
[94] Hill K M and Kakalios J 1994 Reversible axial segregation of binary mixtures of granular materials *Phys. Rev.* E **49** R3610
[95] Hill K M, Gioia G and Amaravadi D 2004 Radial segregation patterns in rotating granular mixtures: waviness selection *Phys. Rev. Lett.* **93** 224301
[96] Seiden G and Thomas P J 2011 Complexity, segregation, and pattern formation in rotating-drum flows *Rev. Mod. Phys.* **83** 1323
[97] Zuriguel I, Gray J M N T, Peixinho J and Mullin T 2006 Pattern selection by a granular wave in a rotating drum *Phys. Rev.* E **73** 061302
[98] M M H D Arntz, W K den Otter, W J Briels, P J T Bussmann, H H Beeftink and R M Boom 2008 Granular mixing and segregation in a horizontal rotating drum: a simulation study on the impact of rotational speed and fill level *AIChE J.* **54** 3133–46
[99] Ding Y L, Seville J P K, Forster R and Parker D J 2001 Solids motion in rolling mode rotating drums operated at low to medium rotational speeds *Chem. Eng. Sci.* **56** 1769–80
[100] Choo K, Molteno T C A and Morris S W 1997 Traveling granular segregation patterns in a long drum mixer *Phys. Rev. Lett.* **79** 2975
[101] d'Ortona U, Thomas N and Lueptow R M 2018 Recirculation cells for granular flow in cylindrical rotating tumblers *Phys. Rev.* E **97** 052904
[102] Kawaguchi T, Tsutsumi K and Tsuji Y 2006 MRI measurement of granular motion in a rotating drum *Part. Part. Syst. Charact.* **23** 266–71

[103] Ding Y L, Forster R, Seville J P K and Parker D J 2002 Granular motion in rotating drums: bed turnover time and slumping–rolling transition *Powder Technol.* **124** 18–27
[104] Lim S-Y, Davidson J F, Forster R N, Parker D J, Scott D M and Seville J P K 2003 Avalanching of granular material in a horizontal slowly rotating cylinder: PEPT studies *Powder Technol.* **138** 25–30
[105] Morrison A J, Govender I, Mainza A N and Parker D J 2016 The shape and behaviour of a granular bed in a rotating drum using Eulerian flow fields obtained from PEPT *Chem. Eng. Sci.* **152** 186–98
[106] Juarez G, Chen P and Lueptow R M 2011 Transition to centrifuging granular flow in rotating tumblers: a modified Froude number *New J. Phys.* **13** 053055
[107] Ancey C 2001 Dry granular flows down an inclined channel: experimental investigations on the frictional-collisional regime *Phys. Rev.* E **65** 011304
[108] Govender I and Pathmathas T 2016 A positron emission particle tracking investigation of the flow regimes in tumbling mills *J. Phys. D: Appl. Phys.* **50** 035601
[109] Govender I, Richter M C, Mainza A N and De Klerk D N 2017 A positron emission particle tracking investigation of the scaling law governing free surface flows in tumbling mills *AIChE J.* **63** 903–13
[110] Rajchenbach J 2000 Granular flows *Adv. Phys.* **49** 229–56
[111] Parker D J, Dijkstra A E, Martin T W and Seville J P K 1997 Positron emission particle tracking studies of spherical particle motion in rotating drums *Chem. Eng. Sci.* **52** 2011–22
[112] Ingram A, Seville J P K, Parker D J, Fan X and Forster R G 2005 Axial and radial dispersion in rolling mode rotating drums *Powder Technol.* **158** 76–91
[113] Ding Y L, Forster R, Seville J P K and Parker D J 2002 Segregation of granular flow in the transverse plane of a rolling mode rotating drum *Int. J. Multiph. Flow* **28** 635–63
[114] González S, Windows-Yule C R K, Luding S, Parker D J and Thornton A R 2015 Forced axial segregation in axially inhomogeneous rotating systems *Phys. Rev.* E **92** 022202
[115] Yang R Y, Zou R P and Yu A B 2003 Microdynamic analysis of particle flow in a horizontal rotating drum *Powder Technol.* **130** 138–46
[116] Parker D J, Fan X, Forster R N G, Fowles P, Ding Y and Seville J P K 2005 Positron imaging studies of rotating drums *Can. J. Chem. Eng.* **83** 83–7
[117] Avilés-Avilés C, Dumoulin E and Turchiuli C 2015 Fluidised bed agglomeration of particles with different glass transition temperatures *Powder Technol.* **270** 445–52
[118] Turchiuli C, Eloualia Z, El Mansouri N and Dumoulin E 2005 Fluidised bed agglomeration: agglomerates shape and end-use properties *Powder Technol.* **157** 168–75
[119] Sutton D, Kelleher B and Ross J R H 2001 Review of literature on catalysts for biomass gasification *Fuel Process. Technol.* **73** 155–73
[120] Zhu Q 2013 Developments in circulating fluidised bed combustion *IEA Clean Coal Centre*
[121] Wang H G, Dyakowski T, Senior P, Raghavan R S and Yang W Q 2007 Modelling of batch fluidised bed drying of pharmaceutical granules *Chem. Eng. Sci.* **62** 1524–35
[122] Thomas P P and Varma Y B G 1992 Fluidised bed drying of granular food materials *Powder Technol.* **69** 213–22
[123] Boerefijn R and Hounslow M J 2005 Studies of fluid bed granulation in an industrial R&D context *Chem. Eng. Sci.* **60** 3879–90
[124] Bokkers G A, van Sint Annaland M and Kuipers J A M 2004 Mixing and segregation in a bidisperse gas-solid fluidised bed: a numerical and experimental study *Powder Technol.* **140** 176–86

[125] Bokkers G A 2005 Multi-level modelling of the hydrodynamics in gas phase polymerisation reactors *PhD Thesis* University of Twente
[126] Nagaraju V D, Murthy C T, Ramalakshmi K and Srinivasa Rao P N 1997 Studies on roasting of coffee beans in a spouted bed *J. Food Eng.* **31** 263–70
[127] Dewettinck K and Huyghebaert A 1999 Fluidized bed coating in food technology *Trends Food Sci. Technol.* **10** 163–8
[128] Nienow A W 1995 Fluidised bed granulation and coating: applications to materials, agriculture and biotechnology *Chem. Eng. Commun.* **139** 233–53
[129] Weimer A W 2019 Particle atomic layer deposition *J. Nanopart. Res.* **21** 9
[130] Windows-Yule C R K, Gibson S, Parker D J, Kokalova T Z and Seville J P K 2020 Effect of distributor design on particle distribution in a binary fluidised bed *Powder Technol.* **367** 1–9
[131] Windows-Yule C R K, Moore A, Wellard C, Werner D, Parker D J and Seville J P K 2020 Particle distributions in binary gas-fluidised beds: shape matters—but not much *Chem. Eng. Sci.* **216** 115440
[132] Gomez-Garcia F, Gauthier D and Flamant G 2017 Design and performance of a multistage fluidised bed heat exchanger for particle-receiver solar power plants with storage *Appl. Energy* **190** 510–23
[133] Hamdan M, Sebastia-Saez D, Hamdan M and Arellano-Garcia H 2020 CFD analysis of the use of desert sand as thermal energy storage medium in a solar powered fluidised bed harvesting unit *Computer Aided Chemical Engineering* vol 48 (Amsterdam: Elsevier) pp 349–54
[134] Pyle D L and Harrison D 1967 The rising velocity of bubbles in two-dimensional fluidised beds *Chem. Eng. Sci.* **22** 531–5
[135] Kuipers J A M, Prins W and Van Swaaij W P M 1991 Theoretical and experimental bubble formation at a single orifice in a two-dimensional gas-fluidized bed *Chem. Eng. Sci.* **46** 2881–94
[136] Liu G-Q, Li S-Q, Zhao X-L and Yao Q 2008 Experimental studies of particle flow dynamics in a two-dimensional spouted bed *Chem. Eng. Sci.* **63** 1131–41
[137] Dyakowski T, Jeanmeure L F C and Jaworski A J 2000 Applications of electrical tomography for gas-solids and liquid-solids flows–a review *Powder Technol.* **112** 174–92
[138] Wang H and Yang W 2020 Application of electrical capacitance tomography in circulating fluidised beds–a review *Appl. Therm. Eng.* **176** 115311
[139] Wang H and Yang W 2020 Application of electrical capacitance tomography in pharmaceutical fluidised beds–a review *Chem. Eng. Sci.* **231** 116236
[140] Stein M G 1999 Particle Motion in Fluidised Beds *Thesis* University of Birmingham
[141] Stein M, Ding Y L, Seville J P K and Parker D J 2000 Solids motion in bubbling gas fluidised beds *Chem. Eng. Sci.* **55** 5291–300
[142] Geldart D 1986 *Gas Fluidization Technology* (New York: Wiley)
[143] Fan X, Parker D J, Yang Z, Seville J P K and Baeyens J 2008 The effect of bed materials on the solid/bubble motion in a fluidised bed *Chem. Eng. Sci.* **63** 943–50
[144] Fan X, Yang Z and Parker D J 2011 Impact of solid sizes on flow structure and particle motions in bubbling fluidization *Powder Technol.* **206** 132–8
[145] Laverman J A, Fan X, Ingram A, van Sint Annaland M, Parker D J, Seville J P K and Kuipers J A M 2012 Experimental study on the influence of bed material on the scaling of solids circulation patterns in 3D bubbling gas–solid fluidized beds of glass and polyethylene using positron emission particle tracking *Powder Technol.* **224** 297–305

[146] Seville J P K, Salleh A M, Ingram A, McCormack A, Greenwood R and Reiling V 2004 Particle motion and defluidisation by sintering in the fluidised bed polyethylene process *Fluidization XI* (New York: Engineering Foundation) pp 211–8
[147] Ingram I, Hausard M, Fan X, Parker D J, Seville J P K, Finn N and Evans M 2007 Portable positron emission particle tracking (PEPT) for industrial use *12th Int Conf. on Fluidization—New Horizons in Fluidization Engineering (Engineering Confs. Int. Symp. Series)*
[148] Seville J P K, Silomon-Pflug H and Knight P C 1998 Modelling of sintering in high temperature gas fluidisation *Powder Technol.* **97** 160–9
[149] Wong Y S and Seville J P K 2006 Single-particle motion and heat transfer in fluidized beds *AIChE J.* **52** 4099–109
[150] García-Triñanes P, Seville J P K, Ansart R, Benoit H, Leadbeater T W and Parker D J 2018 Particle motion and heat transfer in an upward-flowing dense particle suspension: application in solar receivers *Chem. Eng. Sci.* **177** 313–22
[151] Chan C W, Seville J P K, Parker D J and Baeyens J 2010 Particle velocities and their residence time distribution in the riser of a CFB *Powder Technol.* **203** 187–97
[152] Chan C W, Brems A, Mahmoudi S, Baeyens J, Seville J, Parker D, Leadbeater T and Gargiuli J 2010 PEPT study of particle motion for different riser exit geometries *Particuology* **8** 623–30
[153] Chan C W, Seville J P K, Fan X and Baeyens J 2008 Particle motion in CFB cyclones as observed by positron emission particle tracking *Ind. Eng. Chem. Res.* **48** 253–61
[154] Epstein N and Grace J R 2010 *Spouted and Spout-Fluid Beds: Fundamentals and Applications* (Cambridge: Cambridge University Press)
[155] Jono K, Ichikawa H, Miyamoto M and Fukumori Y 2000 A review of particulate design for pharmaceutical powders and their production by spouted bed coating *Powder Technol.* **113** 269–77
[156] Conceição Filho R S, Barrozo M A S, Limaverde J R and Ataíde C H 1998 The use of a spouted bed in the fertilizer coating of soybean seeds *Drying Technol.* **16** 2049–64
[157] Seiler C, Fryer P J and Seville J P K 2008 Statistical modelling of the spouted bed coating process using positron emission particle tracking (PEPT) data *Can. J. Chem. Eng.* **86** 571–81
[158] Palmer S E 2007 Understanding and optimisation of pharmaceutical film coating using the Wurster coater *PhD Thesis* University of Birmingham
[159] Hensler T, Tupy M, Strer T, Pöschel T and Wirth K-E 2015 Particle tracking in fluidized beds with secondary gas injection *Proc. Eng.* **102** 850–7
[160] Al-Shemmeri M, Windows-Yule K, Lopez-Quiroga E and Fryer P J 2021 Coffee bean particle motion in a spouted bed measured using positron emission particle tracking (PEPT) *J. Food Eng.* **311** 110709
[161] Myers K J, Janz E E and Fasano J B 2013 Effect of solids loading on agitator just-suspended speed *Can. J. Chem. Eng.* **91** 1508–12
[162] Devarajulu C and Loganathan M 2016 Effect of impeller clearance and liquid level on critical impeller speed in an agitated vessel using different axial and radial impellers *J. Appl. Fluid Mech.* **9** 24824
[163] Chang C-C, Kuo-Dahab C, Chapman T and Mei Y 2019 Anaerobic digestion, mixing, environmental fate, and transport *Water Environ. Res.* **91** 1210–22
[164] Orgill J J, Atiyeh H K, Devarapalli M, Phillips J R, Lewis R S and Huhnke R L 2013 A comparison of mass transfer coefficients between trickle-bed, hollow fiber membrane and stirred tank reactors *Bioresource Technol.* **133** 340–6

[165] Kralova I and Sjöblom J 2010 Biofuels-renewable energy sources: a review *J. Dispersion Sci. Technol.* **31** 409–25

[166] Mohammadi M, Younesi H, Najafpour G and Mohamed A R 2012 Sustainable ethanol fermentation from synthesis gas by *clostridium ljungdahlii* in a continuous stirred tank bioreactor *J. Chem. Technol. Biotechnol.* **87** 837–43

[167] Ni X and Mackley M R 1993 Chemical reaction in batch pulsatile flow and stirred tank reactors *Chem. Eng. J.* **52** 107–14

[168] Pianko-Oprych P, Nienow A W and Barigou M 2009 Positron emission particle tracking (PEPT) compared to particle image velocimetry (PIV) for studying the flow generated by a pitched-blade turbine in single phase and multi-phase systems *Chem. Eng. Sci.* **64** 4955–68

[169] Kariyama I D, Zhai X and Wu B 2018 Influence of mixing on anaerobic digestion efficiency in stirred tank digesters: a review *Water Res.* **143** 503–17

[170] Windows-Yule C R K, Hart-Villamil R, Ridout T, Kokalova T and Nogueira-Filho J C 2020 Positron emission particle tracking for liquid-solid mixing in stirred tanks *Chem. Eng. Technol.* **43** 1939–50

[171] Nienow A W, Edwards M F and Harnby N 1997 *Mixing in the Process Industries* (Oxford: Butterworth-Heinemann)

[172] Brown D, Pip J, Middlenton J, Papadopoulos G and Arik E 2004 *Handbook of Industrial Mixing. Science and Practice* (New York: Wiley)

[173] Liu L and Barigou M 2013 Numerical modelling of velocity field and phase distribution in dense monodisperse solid-liquid suspensions under different regimes of agitation: CFD and PEPT experiments *Chem. Eng. Sci.* **101** 837–50

[174] Liu L and Barigou M 2014 Experimentally validated computational fluid dynamics simulations of multicomponent hydrodynamics and phase distribution in agitated high solid fraction binary suspensions *Ind. Eng. Chem. Res.* **53** 895–908

[175] Boucher D, Gharib N, Deng Z, Leadbeater T, Langlois R and Waters K E 2015 PEPT validation of a CFD-DEM model of a fine quartz particle (60 μm) behaviour in stirred water *Proc. Eng.* **102** 1305–15

[176] Fangary Y S, Barigou M, Seville J P K and Parker D J 2000 Fluid trajectories in a stirred vessel of non-Newtonian liquid using positron emission particle tracking *Chem. Eng. Sci.* **55** 5969–79

[177] Fangary Y S, Barigou M, Seville J P K and Parker D J 2002 A Lagrangian study of solids suspension in a stirred vessel by positron emission particle tracking (PEPT) *Chem. Eng. Technol.* **25** 521–8

[178] Zwietering T N 1958 Suspending of solid particles in liquid by agitators *Chem. Eng. Sci.* **8** 244–53

[179] Fishwick R P, Winterbottom J M and Stitt E H 2003 Effect of gassing rate on solid-liquid mass transfer coefficients and particle slip velocities in stirred tank reactors *Chem. Eng. Sci.* **58** 1087–93

[180] Fishwick R P, Winterbottom J M and Stitt E H 2003 Explaining mass transfer observations in multiphase stirred reactors: particle-liquid slip velocity measurements using PEPT *Catal. Today* **79** 195–202

[181] García-Ochoa F and Gómez E 1998 Mass transfer coefficient in stirred tank reactors for xanthan gum solutions *Biochem. Eng. J.* **1** 1–10

[182] Fishwick R P, Winterbottom J M, Parker D J, Fan X and Stitt E H 2005 Hydrodynamic measurements of up-and down-pumping pitched-blade turbines in gassed, agitated vessels, using positron emission particle tracking *Ind. Eng. Chem. Res.* **44** 6371–80

[183] Fishwick R, Winterbottom M, Parker D, Fan X and Stitt H 2005 The use of positron emission particle tracking in the study of multiphase stirred tank reactor hydrodynamics *Can. J. Chem. Eng.* **83** 97–103

[184] Barigou M, Chiti F, Pianko-Oprych P, Guida A, Adams L, Fan X, Parker D J and Nienow A W 2009 Using positron emission particle tracking (PEPT) to study mixing in stirred vessels: validation and tackling unsolved problems in opaque systems *J. Chem. Eng. Japan* **42** 839–46

[185] Liu L and Barigou M 2015 Lagrangian particle tracking in mechanically agitated polydisperse suspensions: multi-component hydrodynamics and spatial distribution *Int. J. Multiph. Flow* **73** 80–9

[186] Venneker B C H, Derksen J J and Van den Akker H E A 2010 Turbulent flow of shear-thinning liquids in stirred tanks–the effects of Reynolds number and flow index *Chem. Eng. Res. Des.* **88** 827–43

[187] Guida A, Fan X, Parker D J, Nienow A W and Barigou M 2009 Positron emission particle tracking in a mechanically agitated solid-liquid suspension of coarse particles *Chem. Eng. Res. Des.* **87** 421–9

[188] Guida A 2010 Positron emission particle tracking applied to solid-liquid mixing in mechanically agitated vessels *PhD Thesis* University of Birmingham

[189] Chiti F, Bakalis S, Bujalski W, Barigou M, Eaglesham A and Nienow A W 2011 Using positron emission particle tracking (PEPT) to study the turbulent flow in a baffled vessel agitated by a Rushton turbine: improving data treatment and validation *Chem. Eng. Res. Des.* **89** 1947–60

[190] Guida A, Nienow A W and Barigou M 2010 PEPT measurements of solid-liquid flow field and spatial phase distribution in concentrated monodisperse stirred suspensions *Chem. Eng. Sci.* **65** 1905–14

[191] Guida A, Nienow A W and Barigou M 2010 Shannon entropy for local and global description of mixing by Lagrangian particle tracking *Chem. Eng. Sci.* **65** 2865–83

[192] Sindall R C, Dapelo D, Leadbeater T and Bridgeman J 2017 Positron emission particle tracking (PEPT): a novel approach to flow visualisation in lab-scale anaerobic digesters *Flow Meas. Instrum.* **54** 250–64

[193] Hu Y, Wu J, Poncin S, Cao Z, Li Z and Li H Z 2018 Flow field investigation of high solid anaerobic digestion by particle image velocimetry (PIV) *Sci. Total Environ.* **626** 592–602

[194] Carliell-Marquet C 2000 The effect of phosphorus enrichment on fractionation of metals and phosphorus in anaerobically digested sludge *PhD Thesis* Loughborough University

[195] Sinnott M, Cleary P W and Morrison R 2006 Analysis of stirred mill performance using DEM simulation: Part 1–Media motion, energy consumption and collisional environment *Miner. Eng.* **19** 1537–50

[196] Tavares L M 2017 A review of advanced ball mill modelling *KONA Powder Part. J.* **34** 106–24

[197] Lantz R J *et al* 1990 *Size Reduction* vol 2 (New York: Dekker)

[198] Nakach M, Authelin J-R, Chamayou A and Dodds J 2004 Comparison of various milling technologies for grinding pharmaceutical powders *Int. J. Miner. Process.* **74** S173–81

[199] Baláž P and Dutková E 2009 Fine milling in applied mechanochemistry *Miner. Eng.* **22** 681–94

[200] Beckett S T 2001 Milling, mixing and tempering—an engineering view of chocolate *Proc. Inst. Mech. Eng.* E **215** 1–8

[201] Naik S and Chaudhuri B 2015 Quantifying dry milling in pharmaceutical processing: a review on experimental and modeling approaches *J. Pharm. Sci.* **104** 2401–13

[202] Alamprese C, Datei L and Semeraro Q 2007 Optimization of processing parameters of a ball mill refiner for chocolate *J. Food Eng.* **83** 629–36

[203] Rajamani R K, Mishra B K, Venugopal R and Datta A 2000 Discrete element analysis of tumbling mills *Powder Technol.* **109** 105–12

[204] Hong S H and Kim B K 2001 Fabrication of aluminum flake powder from foil scrap by a wet ball milling process *Mater. Lett.* **51** 139–43

[205] Conway-Baker J, Barley R W, Williams R A, Clarke A J, Kostuch J A and Parker D J 1999 Investigation and model validation of media motion in a vertical stirred mill using positron emission particle tracking *Proc. 1st World Congress on Industrial Process Tomography (Buxton, UK)*

[206] Conway-Baker J, Barley R W, Williams R A, Jia X, Kostuch J, McLoughlin B and Parker D J 2002 Measurement of the motion of grinding media in a vertically stirred mill using positron emission particle tracking (PEPT) *Miner. Eng.* **15** 53–9

[207] Barley R W, Conway-Baker J, Pascoe R D, Kostuch J, McLoughlin B and Parker D J 2004 Measurement of the motion of grinding media in a vertically stirred mill using positron emission particle tracking (PEPT) part II *Miner. Eng.* **17** 1179–87

[208] Riley M, Pinkney S, Blackburn S and Rowson N A 2016 Spatial distributions of media kinetic energy as measured by positron emission particle tracking in a vertically stirred media mill *Miner. Eng.* **98** 177–86

[209] Skuse T R 2016 The use of Positron Emission Particle Tracking (PEPT) to determine the grinding mechanisms within a vertically stirred media mill *PhD Thesis* University of Birmingham

[210] Van der Westhuizen A P, Govender I, Mainza A N and Rubenstein J 2011 Tracking the motion of media particles inside an Isamill® using PEPT *Miner. Eng.* **24** 195–204

[211] Bbosa L S, Govender I, Mainza A N and Powell M S 2011 Power draw estimations in experimental tumbling mills using PEPT *Miner. Eng.* **24** 319–24

[212] Kallon D V V, Govender I and Mainza A N 2011 Circulation rate modelling of mill charge using position emission particle tracking *Miner. Eng.* **24** 282–9

[213] Govender I, Mangesana N, Mainza A N and Franzidis J-P 2011 Measurement of shear rates in a laboratory tumbling mill *Miner. Eng.* **24** 225–9

[214] Harris C C, Schnock E M and Arbiter N 1985 Grinding mill power consumption *Miner. Process. Extr. Metall. Rev.* **1** 297–345

[215] Powell M S and Nurick G N 1996 A study of charge motion in rotary mills Part 1–Extension of the theory *Miner. Eng.* **9** 259–68

[216] Powell M S and McBride A T 2006 What is required from DEM simulations to model breakage in mills? *Miner. Eng.* **19** 1013–21

[217] Watanabe H 1999 Critical rotation speed for ball-milling *Powder Technol.* **104** 95–9

[218] Mellmann J 2001 The transverse motion of solids in rotating cylinders–forms of motion and transition behavior *Powder Technol.* **118** 251–70

[219] Jop P, Forterre Y and Pouliquen O 2005 Crucial role of side walls for granular surface flows: consequences for the rheology *J. Fluid Mech.* **541** 167–92

[220] Jop P, Forterre Y and Pouliquen O 2006 A constitutive law for dense granular flows *Nature* **441** 727–30

[221] Forterre Y and Pouliquen O 2008 Flows of dense granular media *Annu. Rev. Fluid Mech.* **40** 1–24

[222] Govender I, Richter M C, Mainza A N and De Klerk D N 2017 A positron emission particle tracking investigation of the scaling law governing free surface flows in tumbling mills *AIChE J.* **63** 903–13
[223] de Klerk D N, Govender I and Mainza A N 2019 Geometric features of tumbling mill flows: a positron emission particle tracking investigation *Chem. Eng. Sci.* **206** 41–9
[224] Govender I, Cleary P W and Mainza A N 2013 Comparisons of PEPT derived charge features in wet milling environments with a friction-adjusted DEM model *Chem. Eng. Sci.* **97** 162–75
[225] Daraio D, Villoria J, Ingram A, Alexiadis A, Stitt E H and Marigo M 2019 Validation of a discrete element method (DEM) model of the grinding media dynamics within an attritor mill using positron emission particle tracking (PEPT) measurements *Appl. Sci.* **9** 4816
[226] Yu S, Wu C-Y, Adams M J, Reynolds G, Gururajan B, Gargiuli J, Leadbeater T, Roberts R and Parker D J 2015 The use of positron emission particle tracking (PEPT) to study milling of roll-compacted microcystalline cellulose ribbons *Powder Technol.* **285** 74–9
[227] Herting M G and Kleinebudde P 2007 Roll compaction/dry granulation: Effect of raw material particle size on granule and tablet properties *Int. J. Pharm.* **338** 110–8
[228] Miller R W 1997 Roller compaction technology *Drugs Pharm. Sci.* **81** 99–150
[229] Kleinebudde P 2004 Roll compaction/dry granulation: pharmaceutical applications *Eur. J. Pharm. Biopharm.* **58** 317–26
[230] Bindhumadhavan G, Seville J P K, Adams M J, Greenwood R W and Fitzpatrick S 2005 Roll compaction of a pharmaceutical excipient: experimental validation of rolling theory for granular solids *Chem. Eng. Sci.* **60** 3891–7
[231] Wu C-Y, Hung W-L, Miguélez-Morán A M, Gururajan B and Seville J P K 2010 Roller compaction of moist pharmaceutical powders *Int. J. Pharm.* **391** 90–7
[232] Fan L T, Chen S J and Watson C A 1970 Annual review solids mixing *Ind. Eng. Chem.* **62** 53–69
[233] Bayer T and Himmler K 2005 Mixing and organic chemistry *Chem. Eng. Technol.* **28** 285–9
[234] Niranjan K, Smith D L O, Rielly C D, Lindley J A and Phillips V R 1994 Mixing processes for agricultural and food materials: Part 5, Review of mixer types *J. Agric. Eng. Res.* **59** 145–61
[235] Broadbent C J, Bridgwater J, Parker D J, Keningley S T and Knight P 1993 A phenomenological study of a batch mixer using a positron camera *Powder Technol.* **76** 317–29
[236] Broadbent C J, Bridgwater J and Parker D J 1995 The effect of fill level on powder mixer performance using a positron camera *Chem. Eng. J. Biochem. Eng. J.* **56** 119–25
[237] Jones J R and Bridgwater J 1998 A case study of particle mixing in a ploughshare mixer using positron emission particle tracking *Int. J. Miner. Process.* **53** 29–38
[238] Jones J R, Parker D J and Bridgwater J 2007 Axial mixing in a ploughshare mixer *Powder Technol.* **178** 73–86
[239] Laurent B F C, Bridgwater J and Parker D J 2002 Convection and segregation in a horizontal mixer *Powder Technol.* **123** 9–18
[240] Hassanpour A, Tan H, Bayly A, Gopalkrishnan P, Ng B and Ghadiri M 2011 Analysis of particle motion in a paddle mixer using discrete element method (DEM) *Powder Technol.* **206** 189–94
[241] Laurent B F C and Cleary P W 2012 Comparative study by PEPT and DEM for flow and mixing in a ploughshare mixer *Powder Technol.* **228** 171–86

[242] Alian M, Ein-Mozaffari F and Upreti S R 2015 Analysis of the mixing of solid particles in a plowshare mixer via discrete element method (DEM) *Powder Technol.* **274** 77–87

[243] Zhang J, Xu S and Li W 2012 High shear mixers: a review of typical applications and studies on power draw, flow pattern, energy dissipation and transfer properties *Chem. Eng. Process.* **57** 25–41

[244] Stewart R L, Bridgwater J, Zhou Y C and Yu A B 2001 Simulated and measured flow of granules in a bladed mixer–a detailed comparison *Chem. Eng. Sci.* **56** 5457–71

[245] Zhou Y C, Yu A B, Stewart R L and Bridgwater J 2004 Microdynamic analysis of the particle flow in a cylindrical bladed mixer *Chem. Eng. Sci.* **59** 1343–64

[246] Kuo H-P, Knight P C, Parker D J, Adams M J and Seville J P K 2004 Discrete element simulations of a high-shear mixer *Adv. Powder Technol.* **15** 297–309

[247] Ng B H, Ding Y L and Ghadiri M 2009 Modelling of dense and complex granular flow in high shear mixer granulator–a CFD approach *Chem. Eng. Sci.* **64** 3622–32

[248] Ding J and Gidaspow D 1990 A bubbling fluidization model using kinetic theory of granular flow *AIChE J.* **36** 523–38

[249] Gidaspow D, Bezburuah R and Ding J 1991 Hydrodynamics of circulating fluidized beds: kinetic theory approach *Technical report* Illinois Inst. of Tech., Chicago, IL (United States). Dept. of Chemical Engineering

[250] Schaeffer D G 1987 Instability in the evolution equations describing incompressible granular flow *J. Differ. Equ.* **66** 19–50

[251] Ng B H, Kwan C C, Ding Y L, Ghadiri M and Fan X F 2007 Solids motion in a conical frustum-shaped high shear mixer granulator *Chem. Eng. Sci.* **62** 756–65

[252] Ng B H, Kwan C C, Ding Y L, Ghadiri M and Fan X F 2007 Solids motion of calcium carbonate particles in a high shear mixer granulator: a comparison between dry and wet conditions *Powder Technol.* **177** 1–11

[253] Ng B H, Kwan C C, Ding Y L, Ghadiri M, Fan X F and Parker D J 2008 Granular flow fields in vertical high shear mixer granulators *AIChE J.* **54** 415–26

[254] Saito Y, Ingram A, Fan X and Seville J P K 2009 Positron emission particle tracking applied to high shear mixer processing *Mater. Trans.* **50** 2341–3

[255] Saito Y, Fan X, Ingram A and Seville J P K 2011 PEPT analysis of impeller blade geometry and averaged velocity in a high shear mixer *J. Fluid Sci. Technol.* **6** 169–76

[256] Grafutin V I and Prokop'ev E P 2002 Positron annihilation spectroscopy in materials structure studies *Phys.-Usp.* **45** 59

[257] Hiseman M J P, Laurent B F C, Bridgwater J, Wilson D I, Parker D J, North N and Merrifield D R 2002 Granular flow in a planetary mixer *Chem. Eng. Res. Des.* **80** 432–40

[258] Kuo H P, Knight P C, Parker D J and Seville J P K 2005 Solids circulation and axial dispersion of cohesionless particles in a v-mixer *Powder Technol.* **152** 133–40

[259] Chowhan Z T and Linn E E 1979 Mixing of pharmaceutical solids. i. effect of particle size on mixing in cylindrical shear and v-shaped tumbling mixers *Powder Technol.* **24** 237–44

[260] Brone D, Wightman C, Connor K, Alexander A, Muzzio F J and Robinson P 1997 Using flow perturbations to enhance mixing of dry powders in v-blenders *Powder Technol.* **91** 165–72

[261] Brone D, Alexander A and Muzzio F J 1998 Quantitative characterization of mixing of dry powders in v-blenders *AIChE J.* **44** 271–8

[262] Marigo M, Davies M, Leadbeater T, Cairns D L, Ingram A and Stitt E H 2013 Application of positron emission particle tracking (PEPT) to validate a discrete element method (DEM) model of granular flow and mixing in the turbula mixer *Int. J. Pharm.* **446** 46–58

[263] Etchells A W III and Meyer C F 2004 Mixing in pipelines *Handbook of Industrial Mixing: Science and Practice* (New York: Wiley) pp 391–478

[264] Rafiee M, Bakalisa S, Fryer P J and Ingram A 2011 Study of laminar mixing in kenics static mixer by using positron emission particle tracking (PEPT) *Proc. Food Sci.* **1** 678–84

[265] Rafiee M, Simmons M J H, Ingram A and Stitt E H 2013 Development of positron emission particle tracking for studying laminar mixing in kenics static mixer *Chem. Eng. Res. Des.* **91** 2106–13

[266] Hobbs D M and Muzzio F J 1998 Reynolds number effects on laminar mixing in the kenics static mixer *Chem. Eng. J.* **70** 93–104

[267] Rahmani R K, Keith T G and Ayasoufi A 2006 Numerical simulation and mixing study of pseudoplastic fluids in an industrial helical static mixer *J. Fluid Eng.* **128** 467–80

[268] Thakur R K, Vial C, Nigam K D P, Nauman E B and Djelveh G 2003 Static mixers in the process industries–a review *Chem. Eng. Res. Des.* **81** 787–826

[269] Singh M K, Anderson P D and Meijer H E H 2009 Understanding and optimizing the SMX static mixer *Macromol. Rapid Commun.* **30** 362–76

[270] Mihailova O, Lim V, McCarthy M J, McCarthy K L and Bakalis S 2015 Laminar mixing in a SMX static mixer evaluated by positron emission particle tracking (PEPT) and magnetic resonance imaging (MRI) *Chem. Eng. Sci.* **137** 1014–23

[271] Zalc J M, Szalai E S, Muzzio F J and Jaffer S 2002 Characterization of flow and mixing in an SMX static mixer *AIChE J.* **48** 427–36

[272] Galaktionov O S, Anderson P D, Peters G W M and Meijer H E H 2002 Morphology development in kenics static mixers (application of the extended mapping method) *Can. J. Chem. Eng.* **80** 604–13

[273] Meijer H E H, Singh M K and Anderson P D 2012 On the performance of static mixers: a quantitative comparison *Prog. Polym. Sci.* **37** 1333–49

[274] Fairhurst P G, Barigou M, Fryer P J, Pain J P and Parker D J 2001 Using positron emission particle tracking (PEPT) to study nearly neutrally buoyant particles in high solid fraction pipe flow *Int. J. Multiph. Flow* **27** 1881–901

[275] Pérez-Mohedano R, Letzelter N, Amador C, VanderRoest C T and Bakalis S 2015 Positron emission particle tracking (PEPT) for the analysis of water motion in a domestic dishwasher *Chem. Eng. J.* **259** 724–36

[276] Mac Namara C, Gabriele A, Amador C and Bakalis S 2012 Dynamics of textile motion in a front-loading domestic washing machine *Chem. Eng. Sci.* **75** 14–27

[277] Bakalis S, Cox P W, Wang-Nolan W, Parker D and RYER P J F 2003 Use of positron-emission particle tracking (PEPT) technique for velocity measurements in model food fluids *J. Food Sci.* **68** 2684–92

[278] Bakalis S, Fryer P J and Parker D J 2004 Measuring velocity distributions of viscous fluids using positron emission particle tracking (PEPT) *AIChE J.* **50** 1606–13

[279] Wiggins C, Patel N, Bingham Z and Ruggles A 2019 Qualification of multiple-particle positron emission particle tracking (M-PEPT) technique for measurements in turbulent wall-bounded flow *Chem. Eng. Sci.* **204** 246–56

[280] Chan C W, Seville J P K, Fan X and Baeyens J 2009 Particle motion in CFB cyclones as observed by positron emission particle tracking *Ind. Eng. Chem. Res.* **48** 253–61

[281] Hoffmann A C, Skorpen and Chang Y-F 2019 Positron emission particle tracking and CFD investigation of hydrocyclones acting on liquids of varying viscosity *Chem. Eng. Sci.* **200** 310–9

[282] Chang Y-F, Ilea C G, Aasen Ø L and Hoffmann A C 2011 Particle flow in a hydrocyclone investigated by positron emission particle tracking *Chem. Eng. Sci.* **66** 4203–11
[283] Chang Y-F and Hoffmann A C 2015 A Lagrangian study of liquid flow in a reverse-flow hydrocyclone using positron emission particle tracking *Exp. Fluids* **56** 1–14
[284] Alexander R 1949 Fundamentals of cyclone design and operation *Proc. Aust. Inst. Mining Met.* **152** 203
[285] Barth W 1956 Berechnung und auslegung von zyklonabscheidern auf grund neuer untersuchungen *Brennst.-Waerme.-Kraft* **8** 1–9
[286] Hoffmann A C, Dechsiri C, Van de Wiel F and Dehling H G 2005 Pet investigation of a fluidized particle: spatial and temporal resolution and short term motion *Meas. Sci. Technol.* **16** 851
[287] Saito Y, Fan X, Ingram A and Seville J P K 2011 A new approach to high-shear mixer granulation using positron emission particle tracking *Chem. Eng. Sci.* **66** 563–9
[288] Saito Y, Ingram A, Fan X and Seville J P K 2011 Effects of blade design in wet granulation in a high shear mixer determined by positron emission particle tracking *J. Environ. Eng.* **6** 233–41
[289] Kristensen H G and Schaefer T 1987 Granulation: a review on pharmaceutical wet-granulation *Drug Dev. Ind. Pharm.* **13** 803–72
[290] Bouwman A M, Henstra M J, Hegge J J M E, Zhang Z, Ingram A, Seville J P K and Frijlink H W 2005 The relation between granule size, granule stickiness, and torque in the high-shear granulation process *Pharm. Res.* **22** 270–5
[291] Lee K T, Ingram A and Rowson N A 2012 Twin screw wet granulation: the study of a continuous twin screw granulator using positron emission particle tracking (PEPT) technique *Eur. J. Pharm. Biopharm.* **81** 666–73
[292] Seem T C, Rowson N A, Ingram A, Huang Z, Yu S, De Matas M, Gabbott I and Reynolds G K 2015 Twin screw granulation–a literature review *Powder Technol.* **276** 89–102
[293] Lee K T, Ingram A and Rowson N A 2010 Continuous granulation of pharmaceutical powder using a twin screw extruder *World Congress on Particle Technology* vol 6
[294] Lee K T 2013 Continuous granulation of pharmaceutical powder using a twin screw granulator *PhD Thesis* University of Birmingham
[295] Diemer J, Chilles C, Colbert J, Miri T, Ingram A, David P, Fard A S and Anderson P D 2011 Flow visualisation in co-rotating twin screw extruders: positron emission particle tracking and numerical particle trajectories *Int. Polym. Process.* **26** 540–50
[296] Wildman R D, Blackburn S, Benton D M, McNeil P A and Parker D J 1999 Investigation of paste flow using positron emission particle tracking *Powder Technol.* **103** 220–9
[297] Crowley M M, Zhang F, Repka M A, Thumma S, Upadhye S B, Battu S K, McGinity J W and Martin C 2007 Pharmaceutical applications of hot-melt extrusion: Part I *Drug Dev. Ind. Pharm.* **33** 909–26
[298] Griffiths W D, Beshay Y, Parker D J and Fan X 2008 The determination of inclusion movement in steel castings by positron emission particle tracking (PEPT) *J. Mater. Sci.* **43** 6853–6
[299] Beshay Y K 2010 The application of positron emission particle tracking to study non-metallic inclusions in metal castings *PhD Thesis* University of Birmingham
[300] Zerbst U, Madia M, Klinger C, Bettge D and Murakami Y 2019 Defects as a root cause of fatigue failure of metallic components. III: Cavities, dents, corrosion pits, scratches *Eng. Fail. Anal.* **97** 759–76

[301] Griffiths W D, Beshay Y, Caden A J, Fan X, Gargiuli J, Leadbeater T W and Parker D J 2012 The use of positron emission particle tracking (PEPT) to study the movement of inclusions in low-melting-point alloy castings *Metall. Mater. Trans.* B **43** 370–8

[302] Griffiths W D, Parker D J, Fan X and Hausard M 2010 Tracking inclusions in aluminium alloy castings using positron emission particle tracking (PEPT) *Mater. Sci. Technol.* **26** 528–33

[303] Dybalska A, Caden A J, Parker D J, Wedderburn J and Griffiths W D 2020 Liquid metal flow studied by positron emission tracking *Metall. Mater. Trans.* B **51** 1912–7

[304] Pomykala J A Jr, Jody B J and Daniels E J 2007 Thermoplastic separation and recovery from various mixed scrap plastics by the Argonne developed froth flotation technology *Presentation at the 13th Annual Global Plastics Environmental Conf., Orlando, FL* pp 6–7

[305] Trass O 2011 Characterization and preparation of biomass, oil shale and coal-based feedstocks *Advances in Clean Hydrocarbon Fuel Processing* (Amsterdam: Elsevier) pp 3–53

[306] Rubio J and Tessele F 1997 Removal of heavy metal ions by adsorptive particulate flotation *Miner. Eng.* **10** 671–9

[307] Ramaswamy B, Kar D D and De S 2007 A study on recovery of oil from sludge containing oil using froth flotation *J. Environ. Manage.* **85** 150–4

[308] Al-Shamrani A A, James A and Xiao H 2002 Separation of oil from water by dissolved air flotation *Colloids Surf.* A **209** 15–26

[309] Jafarinejad S 2017 Solid-waste management in the petroleum industry *Petroleum Waste Treatment and Pollution Control* (Amsterdam: Elsevier) ch 7 pp 269–345

[310] Merrington A 2017 Recycling of plastics *Applied Plastics Engineering Handbook Plastics Design Library* ed M Kutz 2nd edn (Norwich, NY: William Andrew Publishing) pp 167–899

[311] Waters K E, Rowson N A, Fan X, Parker D J and Cilliers J J 2008 Positron emission particle tracking as a method to map the movement of particles in the pulp and froth phases *Miner. Eng.* **21** 877–82

[312] Waters K E, Rowson N A, Fan X F, Parker D J and Cilliers J J 2008 The use of positron emission particle tracking as a method for tracking particles in flotation froths *Phil. Mag. Lett.* **88** 735–9

[313] Waters K E, Rowson N A, Fan X and Cilliers J J 2009 Following the path of hydrophobic and hydrophilic particles in a Denver cell using positron emission particle tracking *Asia-Pac. J. Chem. Eng.* **4** 218–25

[314] Fan X, Waters K E, Rowson N A and Parker D J 2009 Modification of ilmenite surface chemistry for enhancing surfactants adsorption and bubble attachment *J. Colloid Interface Sci.* **329** 167–72

[315] Bulatovic S and Wyslouzil D M 1999 Process development for treatment of complex perovskite, ilmenite and rutile ores *Miner. Eng.* **12** 1407–17

[316] Drzymala J, Luszczkiewicz A and Simiczyjew P 1983 Flotation study on high-hercynite ilmenite ores *Int. J. Miner. Process.* **10** 289–96

[317] Fan X and Rowson N A 2000 The effect of Pb $(NO_3)_2$ on ilmenite flotation *Miner. Eng.* **13** 205–15

[318] Cole K, Buffler A, Cilliers J J, Govender I, Heng J Y Y, Liu C, Parker D J, Shah U V, van Heerden M and Fan X 2014 A surface coating method to modify tracers for positron emission particle tracking (PEPT) measurements of froth flotation *Powder Technol.* **263** 26–30

[319] Boucher D, Jordens A, Sovechles J, Langlois R, Leadbeater T W, Rowson N A, Cilliers J J and Waters K E 2017 Direct mineral tracer activation in positron emission particle tracking of a flotation cell *Miner. Eng.* **100** 155–65

[320] Cole K E, Waters K E, Parker D J, Neethling S J and Cilliers J J 2010 PEPT combined with high speed digital imaging for particle tracking in dynamic foams *Chem. Eng. Sci.* **65** 1887–90

[321] Cole K E, Waters K E, Fan X, Neethling S J and Cilliers J J 2010 Combining positron emission particle tracking and image analysis to interpret particle motion in froths *Miner. Eng.* **23** 1036–44

[322] Sommer A-E, Ortmann K, Van Heerden M, Richter T, Leadbeater T, Cole K, Heitkam S, Brito-Parada P R and Eckert K 2020 Application of positron emission particle tracking (PEPT) to measure the bubble-particle interaction in a turbulent and dense flow *Miner. Eng.* **156** 106410

[323] Haldar S K 2018 Mineral processing *Mineral Exploration* ed S K Haldar 2nd edn (Amsterdam: Elsevier) ch 13 pp 259–90

[324] Gupta A and Yan D S 2016 *Mineral Processing Design and Operations: An Introduction* (Amsterdam: Elsevier)

[325] Ramachandra Rao S R 2011 *Resource Recovery and Recycling from Metallurgical Wastes* (Amsterdam: Elsevier)

[326] Boucher D, Deng Z, Leadbeater T, Langlois R, Renaud M and Waters K E 2014 PEPT studies of heavy particle flow within a spiral concentrator *Miner. Eng.* **62** 120–8

[327] Holtham P N 1992 Particle transport in gravity concentrators and the Bagnold effect *Miner. Eng.* **5** 205–21

[328] Loveday G K and Cilliers J J 1994 Fluid flow modelling on spiral concentrators *Miner. Eng.* **7** 223–37

[329] Boucher D, Deng Z, Leadbeater T W, Langlois R and Waters K E 2016 Speed analysis of quartz and hematite particles in a spiral concentrator by PEPT *Miner. Eng.* **91** 86–91

[330] Boucher D, Deng Z, Leadbeater T, Langlois R and Waters K E 2016 Observation of iron ore beneficiation within a spiral concentrator by positron emission particle tracking of large (ø = 1440 μm) and small (ø = 58 μm) hematite and quartz tracers *Chem. Eng. Sci.* **140** 217–32

[331] Pita F and Castilho A 2016 Influence of shape and size of the particles on jigging separation of plastics mixture *Waste Manage.* **48** 89–94

[332] Cui J and Forssberg E 2003 Mechanical recycling of waste electric and electronic equipment: a review *J. Hazard. Mater.* **99** 243–63

[333] Roux W P and Naudé N 2014 Positron emission particle tracking inside a laboratory batch jig *J. South Afr. Inst. Min. Metall.* **114** 525–8

[334] Williams R A, Jia X, Clarke A J and Parker D J 1998 Tomographic visualisation of particle motion during jigging *Innovation in Physical Separation Technologies: Richard Mozley Symp. Volume* pp 139–54

[335] Wu C Y, Fan X F, Motazedian F, Seville J P K, Parker D J and Cocks A C F 2010 Quantitative investigation of powder flow during die filling using positron emission particle tracking *Proc. Inst. Mech. Eng.* E **224** 169–75

[336] Wu C-Y and Cocks A C F 2004 Flow behaviour of powders during die filling *Powder Metall.* **47** 127–36

[337] Wu C-Y 2008 DEM simulations of die filling during pharmaceutical tableting *Particuology* **6** 412–8

[338] Langford S T, Wiggins C S, Santos R, Hauser M, Becker J M and Ruggles A E 2017 Three-dimensional spatiotemporal tracking of fluorine-18 radiolabeled yeast cells via positron emission particle tracking *PLoS One* **12** e0180503

[339] Lee K S, Kim T J and Pratx G 2014 Single-cell tracking with PET using a novel trajectory reconstruction algorithm *IEEE Trans. Med. Imaging* **34** 994–1003

[340] Doherty P W, Bushberg J T, Lipton M J, Meares C F and Goodwin D A 1978 The use of indium-111-labeled leukocytes for abscess detection *Clin. Nucl. Med.* **3** 108–10

[341] Hong H, Yang Y, Zhang Y and Cai W 2010 Non-invasive cell tracking in cancer and cancer therapy *Curr. Top. Med. Chem.* **10** 1237–48

[342] Soria B, Skoudy A and Martin F 2001 From stem cells to beta cells: new strategies in cell therapy of diabetes mellitus *Diabetologia* **44** 407–15

[343] Garbern J C and Lee R T 2013 Cardiac stem cell therapy and the promise of heart regeneration *Cell Stem Cell* **12** 689–98

[344] Goldman S 2005 Stem and progenitor cell-based therapy of the human central nervous system *Nat. Biotechnol.* **23** 862–71

[345] Jung K O, Kim T J, Yu J H, Rhee S, Zhao W, Ha B, Red-Horse K, Gambhir S S and Pratx G 2020 Whole-body tracking of single cells via positron emission tomography *Nat. Biomed. Eng.* **4** 835–44

[346] Kinugasa T, Kuwagi K, Leadbeater T W, Gargiuli J, Parker D J, Seville J P K, Yoshida K and Amano H 2015 Three-dimensional dynamic imaging of sand particles under wheel via gamma-ray camera system *J. Terramech.* **62** 5–17

[347] Kinsella K, Schlyer D J, Fowler J S, Martinez R J and Sobecky P A 2012 Evaluation of positron emission tomography as a method to visualize subsurface microbial processes *J. Hazard. Mater.* **213** 498–501

[348] Kulenkampff J, Gründig M, Richter M and Enzmann F 2008 Evaluation of positron-emission-tomography for visualisation of migration processes in geomaterials *Phys. Chem. Earth* **33** 937–42

[349] Kulenkampff J, Gründig M, Zakhnini A and Lippmann-Pipke J 2016 Geoscientific process monitoring with positron emission tomography (GEOPET) *Solid Earth* **7** 1217–31

[350] De Richter S K *et al* 2015 Vibration-induced compaction of granular suspensions *European Phys. J. E* **38** 74

[351] Bement R A P and Selby A R 1997 Compaction of granular soils by uniform vibration equivalent to vibrodriving of piles *Geotechnical & Geological Eng.* **15** 121–43

IOP Publishing

Positron Emission Particle Tracking
A comprehensive guide

Kit Windows-Yule, Leonard Nicuşan, Matthew T Herald, Samuel Manger and David Parker

Chapter 9

Future outlook

In this chapter, we look forward to what the future may hold for the PEPT technique, including the detectors, tracers and algorithms used, and the new applications made possible thereby. We discuss the manners in which ongoing improvements in detector physics may facilitate still greater temporal and spatial resolution in future PEPT systems, while the further exploration of modular, portable positron cameras may facilitate the study of a still wider range of systems, in particular through the use of 'on site' PEPT to explore those too large, dangerous or otherwise unsuitable for laboratory studies. We also discuss the manners in which—in tandem with new algorithms and the aforementioned developments in camera resolution—improvements in our abilities to effectively manipulate, radiolabel and thus track micron-scale particles may open the door to significant new applications in the biomedical field. In addition to advancements in PEPT's current abilities, we consider also how novel PEPT algorithms may facilitate entirely new functionality, facilitating, for example, the study of phase-changing media.

9.1 Future detector systems

9.1.1 Improved components

As is evident from the detailed discussion in section 5.2, the development of new PEPT detectors facilitating faster and more precise imaging is currently a highly active field. This is expected to remain the case for the foreseeable future, as even the most recently-developed PEPT detectors (Birmingham's 'SuperPEPT' and Cape Town's HR++) have been constructed using components which are several years of development behind the current cutting-edge, which itself continues to improve at a pace [1, 2].

In particular, as the coincidence windows achievable by detectors become increasingly small, the value of incorporating time-of-flight information [1, 2] into PEPT algorithms increases. Such time-of-flight algorithms carry the potential to provide a genuine step-change in both the temporal and spatial resolution achievable with PEPT, as well as further ameliorating current limitations on the numbers of LoRs required to detect a particle, and thus the sizes of tracers that can be imaged.

9.1.2 Improved geometries

Even without improving the properties of the detector components themselves, there exists significant scope to advance PEPT through improvements to the *geometries* of detectors, and the adaptability thereof. The advent of modular positron camera systems (see section 5.2.4) has provided the potential to develop specialised camera geometries tailored to specific system geometries. In the simplest sense, this can mean merely imaging systems that are taller or wider than a standard positron camera can accommodate. However, having theoretically unlimited flexibility in the placement, orientation and number of detectors used to image a given system, there exists the potential to create a *truly optimised* camera for any given system. While the optimisation of such a geometry would previously have required a somewhat laborious, ad hoc process, with numerical modelling tools such as GATE (see section 5.3) increasingly being applied to PEPT problems, this process may be expedited considerably. The use of such computational models also opens the possibility of directly applying numerical optimisers to further increase the efficiency of—and decrease the human effort requisite for—the development of such optimised geometries. Current research at the University of Birmingham is aiming to realise this possibility using recently developed artificial evolution algorithms [3].

Moreover, even for relatively simple, standard geometries such as the planar ADAC cameras described in section 5.2, there exists significant scope for improvement in the quality of the data acquired simply by improving the manner in which said detectors are used. At present, the standard practice for a PEPT experiment in a planar geometry is to perform said experiment at a fixed detector separation. As we have seen in section 5.2, however, there exists, for a given camera geometry (and thus similarly a given detector head separation for a planar camera) an optimal tracer activity for the acquisition of clean, high-resolution PEPT data. From this, we can naturally intuit that as a tracer's activity decreases, the optimal detector separation will decrease also (i.e. we compensate for a loss of activity with an increase in geometric efficiency to maintain an optimal count rate for the camera in question). As such, it is hypothetically possible that we can program a planar camera to continually adjust its head separation such that even as a tracer decays, an optimal acquisition rate, and thus a stable acquisition rate and optimal data quality, are maintained. Similarly, with suitable mechatronics, it would be possible to create a mobile camera capable of 'following' a tracer so as to maintain optimum efficiency, or even to image systems larger than the camera's (static) field of view. Ongoing work to develop detailed models for the prediction of positron camera count rates as a function of tracer activity and head separation will prove highly valuable in this pursuit.

9.1.3 Multimodality

Another interesting but currently under-explored possibility which may bring new depth and detail to the study of particulate media is the application of multimodal imaging—that is, the synergistic use of PEPT with one or more complementary imaging modality. As discussed extensively in chapter 4 (and illustrated graphically in figure 4.1 which, for convenience, we re-plot here in figure 9.1), all experimental techniques—PEPT included—carry distinct strengths and weaknesses. As such, by simultaneously applying more than one technique to a system, we stand to gain more complete and more detailed information. For instance, combining PEPT with fast x-ray radiography or CT may allow us to simultaneously obtain information regarding the structure of a system *and* the motion of particles through it, or to simultaneously record the motion of gas bubbles *and* particles through a fluidised bed, allowing us to gain direct insight into how these two phases interact. Combining PEPT and ECT would allow the user to simultaneously gain three-dimensional, transient, Eulerian information as afforded by the latter technique alongside the more precise, Lagrangian information provided by the former. Researchers at the University of Birmingham are currently exploring the simultaneous application of PEPT and MPT, marrying PEPT's ability to record translational motion with high spatial and temporal resolution and MPT's unique ability to precisely track particle rotation with high temporal resolution.

Figure 9.1. Table summarising the main strengths and limitations of the various experimental imaging modalities explored in this chapter. Green colouring indicates a technique is strong in a given area. Properties coloured in orange and red, respectively, indicate relatively minor and comparatively significant weaknesses—for example orange colouring in 'temporal resolution' indicates an inability to track fast-moving flows, whereas red indicates that only static or quasi-static systems can be imaged. Similarly, for 'orientation information', orange would indicate that particle orientation *can* be tracked, but doing so is comparatively difficult and/or only achievable under specific circumstances (e.g. with PEPT, where such tracking is possible, but only with relatively large particles). The specific reasoning underlying the rating of each feature for each technique is discussed in detail in chapter 4.

While, as can be seen from the literature reviewed in chapter 8, some attempts at multimodal imaging with PEPT have already been made, they are few and far between, and typically involve sequential studies with the techniques involved, as opposed to the *simultaneous* multimodal imaging of a system, from which one may potentially gain more insight.

9.2 Future tracers

As has been discussed repeatedly in the preceding chapters, the miniaturisation of tracers has been, and continues to be, a major goal in the field, as has the creation of biocompatible tracers. As we push below the 10 μm barrier, more and more new applications involving the tracking of cells become possible. The tracking of cancerous cells *in vivo* has in fact already been demonstrated [4] in mice and, as tracer development continues, it is likely we will also be able to track, for example, blood cells, T-cells and various other biological agents, the study of whose motions and functions will prove of significant value to our understanding of diseases and their potential cures. The translation to PEPT imaging directly in humans is also a major goal for the coming years.

To achieve the above, advances must be made in two distinct areas. Firstly, we must find ways to increase the uptake of positron-emitting material by our tracers of choice. One currently under-explored possibility is to use covalent bonding—i.e. directly reacting fluorine-18 or other suitable β^+ emitters with a material, as opposed to relying simply on diffusion or ionic bonding. Secondly, though it may sound mundane, we must develop reliable methods for the labelling, manipulation and separation of micron-scale particles, such that they may be effectively labelled and inserted into the systems of interest. At present, the field of microfluidics [5] seems to offer some of the most compelling potential solutions in this area [6–9].

Continuing in the vein of possible medical applications of PEPT, there exists significant interest (discussed further below) in the labelling of pharmaceutical powders, such that their dissolution—either *in vitro* (e.g. to test tablet quality as a means of assessing the effectiveness of a given manufacturing process) or *in vivo* (e.g. to ensure their dissolution occurs at an appropriate rate, and in an appropriate part of the digestive system)—may be directly and rigorously monitored. The significant open challenge in this application is determining a manner to ensure that the positron emitter of interest preferentially bonds to the material of interest (i.e. an active pharmaceutical ingredient) as opposed to, for example, excipients, lubricants or any other molecules in the surrounding medium.

9.3 Future algorithms

9.3.1 Harnessing time-of-flight

The introduction of time-of-flight data to PEPT imaging, as discussed above, would require not only advanced detector systems, but also the development of new algorithms to meaningfully incorporate and utilise this information. In the simplest case, this additional information could simply be used in conjunction with existing algorithms as a way of 'pre-filtering' the data—i.e. removing any LoRs which the time-of-flight data show cannot belong to the tracer of interest. Greater gains, however, stand to be made by building entirely novel algorithms around this new ability which would allow, for example, the weighting of cutpoints, or grid cells on a line-density map, based on their distance from the expectation value of the position of a given positron–electron annihilation event as predicted by time-of-flight measurements.

9.3.2 PEPT in the information age

Another area in which other fields have recently demonstrated significant expansion and development which could be easily adapted for use in PEPT imaging is that of computer vision. Machine-learning and AI-based computer vision algorithms allowing the tracking of objects in videos—i.e. sequential pixelated frames, much as we acquire from PEPT data—are now commonplace, but have yet to be applied to PEPT imaging. The availability in PEPT data of additional information, for example indirect tracer signatures, as discussed at various points in chapter 6, means that these algorithms could, potentially, not only be adapted for use with PEPT, but also improved upon and expanded if applied to positron imaging problems.

As well as simply aiding in trajectory separation, computer vision and AI methods also stand to form the basis of other significant advancements in the PEPT technique. A promising current development which may benefit from such methods is '*volumetric PEPT*', which aims to track a moving 'cloud' of radioactivity, as opposed to a single point, and could potentially provide information regarding the dispersity, distribution, shape and orientation thereof. Volumetric PEPT can be thought of, in a way, as a 'half-way house' between PEPT and PET—offering less detail than PET regarding the *specific* distribution of the tracer medium, but with significantly higher temporal resolution. A particular advantage of this approach would be in the tracking of particles' rotational motion, currently only possible by tracking three distinct points on a relatively large tracer [10], and thus somewhat limited in terms of 'realistic' applications. As PEPT tracers with distributed activity are known to possess distinct 'signatures' in the LoR distributions and thus point clouds (see section 6.2.14) produced thereby, using volumetric PEPT one could hypothetically apply AI algorithms to track the orientation and rotation of significantly smaller particles. This approach has already proven successful in simple, quasi-two-dimensional experiments (see figure 9.2), and shows significant promise for the full, 3D case.

As well as developing PEPT algorithms with new capabilities, significant work is also underway improving the efficiency and accessibility of existing algorithms. Efforts have already begun to allow PEPT algorithms to work in distributed environments, and to facilitate GPU-based processing. These capabilities would be of particular value for complex techniques such as the above-discussed volumetric PEPT, and for methods (such as the Voronoi multiple particle tracking method described in section 6.2.10) which are currently too computationally intensive to be used on real data sets. It is also hoped that future PEPT algorithms will be able to make use of all cloud-computing resources, taking advantage of the recent explosion of Infrastructure as a Service (IaaS) and Platform as a Service (PaaS) web platforms, such as Google Colab, Amazon AWS and Microsoft Azure.

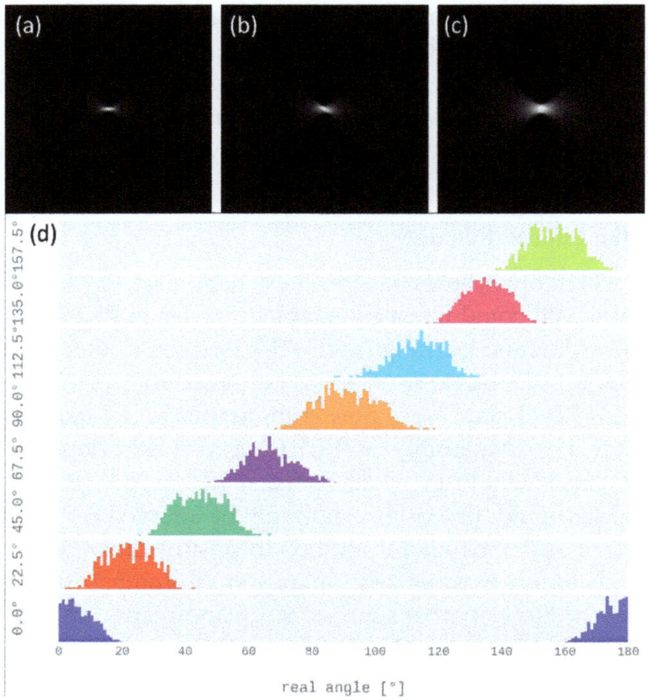

Figure 9.2. Examples of characteristic PEPT line density patterns for a cylindrical particle oriented at (a) 0, (b) 45 and (c) 90 degrees relative to the horizontal. A convolutional neural network (CNN) [11] was trained to recognise the patterns associated with particles oriented at eight distinct angles, and then tasked with classifying a series of images showing line density distributions for particles oriented at arbitrary angles. The results are shown in panel (d), with the x-axis showing the 'real' orientation of the particle, and the y-axis the CNN's 'best guess' of its angle. The results show that, even with a limited, highly discretised training set, the algorithm was able to successfully determine the angle of orientation to within a reasonable margin of error, even for angles it had not been directly trained with.

Beyond solid tracers

Another currently-active research area is the imaging of phase-changing materials—for example the dissolution of tablets in liquids, with PEPT being used to determine the dissolution rate of a specific active pharmaceutical ingredient. Such a measurement, if found to be viable, stands to provide a faster and more rigorous alternative to current methods for testing the dissolution rates of tablets [12–14] and, crucially, one that can be easily performed in optically opaque and/or closed systems—including the human body. The fundamental concept is to use PEPT in combination with positron emission projection imaging (PEPI) [15, 16] and/or conventional PET (see section 4.2.3). The three reconstruction methods used in conjunction potentially allow for a powerful synergy: PET offers full, three-dimensional, volumetric information but a low frame rate; PEPI offers increased temporal resolution, but only a two-dimensional projection of the tracer material as it spreads through a fluid; PEPT offers the highest frame rate, but will be able to track only the solid body dissolving (i.e. our tablet in the preceding example). The PEPT component, combined with a suitable sensitivity map [17, 18] of the detector system used, will allow the measurement of the loss of activity of the dissolving, melting or volatilising object, which can be compared to the expected loss due to radioactive decay, and thus used to provide information regarding its dissolution rate with high temporal frequency—thus allowing insight into processes considerably faster than can be measured using conventional sampling or imaging methods, in opaque fluids and/or systems. By tracking the particle, we are also able to perform 'subtractive PEPT'—removing LoRs belonging to the dissolving medium so as to leave only those corresponding to the dissolved/melted material, which can then be used to provide cleaner, more accurate PEPI or PET images. While these represent just two of a variety of possibilities, we hope it provides an indication of the potential value of PEPT in this area.

9.4 New applications for PEPT

9.4.1 Beyond science and engineering

As is clear from chapter 8, the vast majority of PEPT studies to date have focussed on scientific and engineering applications, in particular in the fields of physics, and chemical and mechanical engineering. The improved scanners, tracers and algorithms discussed above, however, have the chance to facilitate the translation into other fields, including those beyond the boundaries of the physical and engineering sciences, such as medicine and even the humanities.

As discussed in detail in chapter 8, recent developments in the tracking of small tracers [19], including cells [4, 20], have highlighted the considerable promise offered by PEPT in the biomedical field. Notable potential applications include the tracking of bloodflow (for which current diagnostic measurements are relatively crude), the study of cancer metastasis [21], and research into T-cell therapy [22–24] and autologous chondrocyte implantation [25, 26], to provide just a few possible avenues.

The advent of smaller tracers also opens the door for studies concerning the behaviour of clays, soils and other micron-scale materials. Such an ability would carry clear applications in currently under-explored fields including civil engineering, soil science and geomechanics, but is also of potential interest even outside the sciences. A notable example is the field of archaeology, where better understanding of the dynamics of soils and artefacts buried in said soils during various geophysical and human-induced events may help us better understand the formation processes of archaeological sites [27, 28] and thus, in turn, help us to better understand our cultural past, as well as our scientific present.

9.4.2 If the mountain will not come to Muhammad, then Muhammad must go to the mountain

Researchers at the University of Birmingham have recently been awarded a Royal Academy of Engineering proof of concept grant to further develop PEPT's capabilities for *in situ* imaging on real industrial sites, and more broadly to perform research assessing the appetite for such imaging in the industrial community.

Though a very limited number of *in situ* PEPT studies (indeed only one [29]) have been performed previously, this project aims to provide a catalyst for the expansion of such studies. In an age where continuous [30] (as opposed to batch [31]) processing is becoming increasingly popular in industry (meaning that, increasingly, imaging a single piece of equipment in isolation from its upstream and downstream processes is not ideal) the ability to perform PEPT imaging on site, potentially on multiple individual systems forming a continuous process, is clearly of significant value.

In addition to continuous processes, *in situ* PEPT imaging is of significant value in cases where systems are simply too big to be transported for imaging in a laboratory or, similarly, are too dangerous, or require too much additional infrastructure. One application, currently the subject of active PEPT research, which ticks all three of these boxes is the recycling of plastic waste using fluidised-bed pyrolysis reactors, which are not only larger than can be accommodated in typical laboratory scanners, but also operate at high (>600 °C) temperatures, and produce combustible, often noxious, gases which require significant additional equipment to safely treat [32]. *In situ* PEPT imaging using the modular camera systems described in section 5.2.4 represents a rare manner in which detailed information regarding the hydrodynamics of these systems can be obtained. This hydrodynamic information, coupled with more conventional measurements and chemical analysis, is vital to the development and optimisation of these systems, which may ultimately play a significant role in our fight against plastic waste and the larger climate crisis. While this represents but one of myriad industrial processes to which *in situ* PEPT may be valuably applied, it provides a particularly clear example of the significant positive impact that *in situ* PEPT, and PEPT in general, may have in the world.

References

[1] Ota R, Kwon S I, Berg E, Hashimoto F, Nakajima K, Ogawa I, Tamagawa Y, Omura T, Hasegawa T and Cherry S R 2021 Direct positron emission imaging: ultra-fast timing enables reconstruction-free imaging arXiv:2105.05805

[2] Schaart D R 2021 Physics and technology of time-of-flight PET detectors *Phys. Med. Biol.* **66** 09TR01

[3] Igel C, Hansen N and Roth S 2007 Covariance matrix adaptation for multi-objective optimization *Evol. Comput.* **15** 1–28

[4] Jung K O, Kim T J, Yu J H, Rhee S, Zhao W, Ha B, Red-Horse K, Gambhir S S and Pratx G 2020 Whole-body tracking of single cells via positron emission tomography *Nat. Biomed. Eng.* **4** 835–44

[5] Tabeling P 2005 *Introduction to Microfluidics* (Oxford: Oxford University Press)

[6] Sajeesh P and Sen Ashis Kumar 2014 Particle separation and sorting in microfluidic devices: a review *Microfluid. Nanofluidics* **17** 1–52

[7] Inglis D W 2009 Efficient microfluidic particle separation arrays *Appl. Phys. Lett.* **94** 013510

[8] Bayareh M 2020 An updated review on particle separation in passive microfluidic devices *Chem. Eng. Process.* **153** 107984

[9] Zhang S, Wang Y, Onck P and den Toonder J 2020 A concise review of microfluidic particle manipulation methods *Microfluid. Nanofluidics* **24** 1–20

[10] Yang Z, Parker D J, Fryer P J, Bakalis S and Fan X 2006 Multiple-particle tracking—an improvement for positron particle tracking *Nucl. Instrum. Methods Phys. Res.* A **564** 332–8
[11] Rawat W and Wang Z 2017 Deep convolutional neural networks for image classification: a comprehensive review *Neural Comput.* **29** 2352–449
[12] United States Pharmacopeia 2000 United States Pharmacopeial Convention *Summary of Information Submitted to MedMARX in the Year*
[13] Stamatopoulos K, Batchelor H K, Alberini F, Ramsay J and Simmons M J H 2015 Understanding the impact of media viscosity on dissolution of a highly water soluble drug within a USP 2 mini vessel dissolution apparatus using an optical planar induced fluorescence (PLIF) method *Int. J. Pharm.* **495** 362–73
[14] Mehta A C 1994 Review of analytical methods used in the dissolution testing of pharmaceuticals *Anal. Proc.* **31** 245–8
[15] Loggia D, Gouze P, Greswell R and Parker D J 2004 Investigation of the geometrical dispersion regime in a single fracture using positron emission projection imaging *Transp. Porous Media* **55** 1–20
[16] Gouze P, Tenchine S, Loggia D, Bruderer C, Fedorov A, Adler P M, Thovert J F, Greswell R, Riley M and Parker D J 2003 A study of the dispersion in a single thin fracture using positron emission projected imaging *Groundwater in Fractured Rocks* pp 407–8
[17] Dey T and Rodrigue P 2015 Evaluation of the Intel Xeon phi co-processor to accelerate the sensitivity map calculation for PET imaging *J. Instrum.* **10** C07011
[18] Vandenberghe S, D'Asseler Y, Kolthammer J, Van de Walle R, Lemahieu I and Dierckx R A 2002 Influence of the angle of incidence on the sensitivity of gamma camera based PET *Phys. Med. Biol.* **47** 289
[19] Lee K S, Kim T J and Pratx G 2014 Single-cell tracking with PET using a novel trajectory reconstruction algorithm *IEEE Trans. Med. Imaging* **34** 994–1003
[20] Langford S T, Wiggins C S, Santos R, Hauser M, Becker J M and Ruggles A E 2017 Three-dimensional spatiotemporal tracking of fluorine-18 radiolabeled yeast cells via positron emission particle tracking *PLoS One* **12** e0180503
[21] Hong H, Yang Y, Zhang Y and Cai W 2010 Non-invasive cell tracking in cancer and cancer therapy *Curr. Top. Med. Chem.* **10** 1237–48
[22] Soria B, Skoudy A and Martin F 2001 From stem cells to beta cells: new strategies in cell therapy of diabetes mellitus *Diabetologia* **44** 407–15
[23] Garbern J C and Lee R T 2013 Cardiac stem cell therapy and the promise of heart regeneration *Cell Stem Cell* **12** 689–98
[24] Goldman S 2005 Stem and progenitor cell-based therapy of the human central nervous system *Nat. Biotechnol.* **23** 862–71
[25] Minas T and Chiu R 2000 Autologous chondrocyte implantation *Am. J. Knee Surg.* **13** 41–50
[26] Harris J D, Siston R A, Pan X and Flanigan D C 2010 Autologous chondrocyte implantation: a systematic review *J. Bone Joint Surg. Am* **92** 2220
[27] Luria D, Fantalkin A, Zilberman E and Ben-Dor E 2020 Identifying the Brazil nut effect in archaeological site formation processes *Mediterr. Geosci. Rev.* **2** 267–81
[28] Sapir Y, Sarah P, Sapir Y, Katz H and Faust A 2021 Topsoil formation processes as indicated from geoarchaeological investigations at Tel'Eton, Israel, and its environment *Mediterr. Archaeol. Archaeom.* **21** 85–107
[29] Ingram I, Hausard M, Fan X, Parker D J, Seville J P K, Finn N and Evans M 2007 Portable positron emission particle tracking (PEPT) for industrial use *12th Int Conf. on*

Fluidization—New Horizons in Fluidization Engineering (Engineering Confs. Int. Symp. Series)

[30] Rathore A S, Agarwal H, Sharma A K and Pathak M mar 2015 Continuous processing for production of biopharmaceuticals *Prep. Biochem. Biotechnol.* **45** 836–49

[31] Rippin D W T 1993 Batch process systems engineering: a retrospective and prospective review *Comput. Chem. Eng.* **17** S1–13

[32] Yuan G, Chen D, Yin L, Wang Z, Zhao L and Wang J Y 2014 High efficiency chlorine removal from polyvinyl chloride (PVC) pyrolysis with a gas–liquid fluidized bed reactor *Waste Manage.* **34** 1045–50

Chapter 10

Accessing a PEPT facility

Having, over the past several hundred pages, provided a detailed account of the PEPT technique and how it is used, including multiple interactive examples in which illustrative PEPT data sets can be explored and manipulated, it seems only sensible that we should also inform you, the reader, as to how you might take the next logical step and go about performing PEPT 'for real'. In this very brief chapter we do just that, providing details of currently-active PEPT facilities offering collaboration and/or 'PEPT as a service'. There are currently two main, established PEPT facilities regularly performing collaborative work and commercial PEPT imaging.

The University of Birmingham's Positron Imaging Centre (https://www.birmingham.ac.uk/research/activity/physics/particle-nuclear/positron-imaging-centre/index.aspx), located in the United Kingdom, is the original home of PEPT (see chapter 3), and the (academic) home of the authors of this book. The PIC is the only PEPT facility to possess its own, dedicated cyclotron, facilitating efficient and flexible access to a diverse range of radioisotopes for use in the creation of radiotracers (see chapter 5). As well as laboratory studies, it also offers 'on site' PEPT, for the imaging of systems which cannot easily be transported to Birmingham. Those interested in working with the Positron Imaging Centre can contact its director, Professor Tzany Kokalova Wheldon (t.wheldon@bham.ac.uk)—or indeed the authors of this book (c.r.windows-yule@bham.ac.uk, d.j.parker@bham.ac.uk). We wholeheartedly welcome enquiries from those in both Science and Industry with ideas for interesting potential applications of PEPT.

The second longest-running PEPT facility, established in 2009, belongs to South Africa's University of Cape Town (http://www.peptcapetown.uct.ac.za/), and is located at the iThemba Laboratory for Accelerator-Based Sciences national laboratory, Faure, South Africa. The centre possesses particular expertise in mining, mineral processing and related fields. Those interested in working with PEPT Cape Town can contact peptcapetown@uct.ac.za.

doi:10.1088/978-0-7503-3071-8ch10

At the time of writing, a new facility is being established at Virginia Commonwealth University by the group of Lane Carasik (https://fastresearchgroup.weebly.com/), and active PEPT research is also being conducted at Stanford University in the group of Guillem Pratx (https://med.stanford.edu/pratxlab.html).

Kit Windows-Yule, Leonard Nicuşan, Matthew T Herald, Samuel Manger and David Parker

Appendix A

Python tutorial

Python tutorial

For engineers, scientists and other people who just want a computer to do the work.

The aim of this tutorial is to get you up to speed with Python programming. Besides the basics, it should teach you how to use it effectively for usual programming tasks, like creating dynamic arrays, iterating them, saving them as CSV, etc.

It assumes a very basic understanding of programming languages—e.g. knowing what a variables and conditionals are. If you've written more than 100 lines of code in, say, MATLAB, you should be fine.

A.1 The Python programming language

Python is a general-purpose, interpreted programming language that aims to be as simple and clear as possible. Its main design philosophy dictates: >There should be one—and preferably only one—obvious way to do it.

Indeed, the Python community and developers put a high premium on simplicity, sometimes even at the expense of computational efficiency. However, for this exact reason Python code is perhaps the easiest to read and understand, even for people without any specific knowledge of the language. This ease of expressing ideas as succintly as possible is also responsible for Python's adoption as the *de facto* programming language for data science. There is a very active—and welcoming!—community developing an ecosystem of *libraries* that can help with almost any task at hand.

Even in the field of scientific computing, which traditionally prefers more explicit lower-level languages such as Fortran and C++, almost all numerical code developed nowadays also includes at least a Python interface. As such, Python also found use as a powerful *orchestrating* language, gluing together external high-performance subroutines into cohesive scripts that are easy to develop, understand and extend. For example, the ubiquitous library for matrices and N-dimensional arrays, NumPy (short for **Num**erical **Py**thon), contains subroutines written in heavily-optimised, low-level C code, making array computing in a dynamic language like Python oftentimes faster than straightforward implementations in much more complex languages.

Because it is an interpreted programming language, Python code can exist and run in a variety of environments:

- As a script.
- As functions or classes grouped into packages.
- As a command-line (or REPL—Read-Eval-Print-Loop).
- As mixed text and code cells in a Jupyter Notebook (such as this one!).

A.1.1 Some terminology

Perhaps it would be useful to clear up some of the terms that are usually passed around when talking about programming in Python, and programming languages in general:

- **Python** is the programming *language* itself, containing a set of syntactic rules for expressing code (much like the grammar of a human language!)
- **CPython** is the most popular Python *interpreter*, the actual program that takes code written following Python's syntax rules, understands it, and executes whatever commands the program outlines. Indeed, you can have different interpreters for the same language!
- **IDE**, or **I**ntegrated **D**evelopment **E**nvironment is, like the name suggests, a programming environment; at the minimum, it includes a text editor with programming syntax highlighting and the ability to execute the code written in it. *Spyder* is an example IDE specially tailored to Python programming.
- **Library** in the context of programming, is a collection of pre-written code that often automates or simplifies a task. *NumPy* is such a library allowing simple expression of array operations without worrying about low-level computer details (e.g. adding two matrices without needing to think about requesting memory from the operating system, loading each value from the matrices, telling the computer processor to add them, then saving them in the requested memory—sounds pretty tedious put this way, doesn't it?)
- **Jupyter Notebook**—essentially what you are seeing here!—is a programming environment containing *text cells* and interactive *code cells* that can be executed. If a *Jupyter Notebook* is the file itself (normally ending in *".ipynb"*), *JupyterLab* is the environment that opens and displays notebooks such as this one.
- **Anaconda** is a 'batteries included' programming environment bringing together CPython, the most important libraries for data science, Spyder, an IDE for Python development and JupyterLab, which can host mixed text and

interactive code cells. If you would like to code in Python, downloading Anaconda is the easiest way to get started!

By the way, *Python* comes from the famous British comedy troupe Monty Python, not the snake!

A.1.2 The 'pip' package manager

It is a very good idea to encapsulate code into packages, modules, and the like. Thankfully, python has a pretty smart package manager called 'pip' that takes care of installing and updating different packages.

Let's install the Numerical Python (NumPy) package. It provides classic MATLAB-like array functionality to Python, while being a bit more extensible. It is simply the most important library in Python for numerical computing.

Run the next cell. You will see the output right under it.

[1]:
```
# Python comments start with a hashtag
# pip is a system command. To run shell-like commands
# in a Notebook environment, put an exclamation mark (!)
# in front of it
!pip install numpy
```

```
Requirement already satisfied: numpy in
/home/andreinicusan/anaconda3/envs/peptbook/lib/python3.7/site-packages (1.21.0)
```

A.1.3 Using modules

Once you have a module installed, you can import it inside your Python code (be it script, function, sandwich toaster, etc.):

[2]:
```
# A simple import:
import numpy
print( numpy.array([1, 2, 3]) )


# An import that creates a shorthand:
import numpy as np
print( np.array(['a', 'simple', 'array', 5.13]) )


# Or you can import a specific module from inside the package:
from numpy import array
print( array([1., 3.14, -2.718]) )
```

```
[1 2 3]
['a' 'simple' 'array' '5.13']
[ 1.     3.14  -2.718]
```

A.2 Basic constructs

Now we'll start actually programming in Python!

A.2.1 Basic data structures

The simplest (i.e. most usual) data types that you will use are:

- `int` : normal integers. Python integers have a variable size: they are as big as your memory permits!
- `float` : real numbers. For the C++ nerds: it is equivalent to a double-precision floating-point number.
- `str` : character strings.

```
[3]: # Integers:
     print(1 + 2 + 13213241342431241324132431 2 // 5)  # Modulo division is //, not %

     # Real numbers:
     print(3.14 *  2. ** 2)  # Raise a number to a power using **

     # Strings:
     print("I am a" + ' "concatenated" string.')
```

```
2642648268486248264826 4865
12.56
I am a "concatenated" string.
```

A.2.1.1 Lists

The `list` is a built-in data structure. Unlike a MATLAB array, it is a variable-length container. This makes it very handy when we are dynamically allocating/reading in data:

```
[4]: # Define a list with []. Only use semicolons when writing
     # multiple commands on the same line.
     x = [1, 'hey', 2.718] ; print(x)

     # Dynamically add elements to it:
     x.append(42) ; print(x)

     # It is a list, not a mathematical array!
     print( [1, 2, 3] + [4, 5, 6] )
     print( [-1, 1] * 3 )
```

```
[1, 'hey', 2.718]
[1, 'hey', 2.718, 42]
[1, 2, 3, 4, 5, 6]
[-1, 1, -1, 1, -1, 1]
```

Conditionals and loops:

In Python we have the usual `if` / `elif` / `else` statements and the `for` / `while` loops.

Python is a whitespace-delimited language. This means that blocks of code are delimited by the indentation level, rather than curly braces or start/end statements.

```
[5]: # Conditionals:
x = 3.14

# The start of a block is usually denoted with a colon:
if type(x) == int:
    print("x is an integer!")
elif type(x) == float:
    print("x is a float!")
else:
    print("well who knows?")
```

x is a float!

```
[6]: # Loops:

# The while loop is exactly like in other programming languages
i = 0
while i < 4:                # Note the colon!
    print("while")
    i += 1

print() # Print a newline

# The for loop, on the other hand, might be a bit different.
# It iterates through the elements of an iterable, like a list:
data = [1, 2, 3]
for i in data:
    print(i)
print()

# range() is an iterable
for i in range(5):
    print(i)
```

```
while
while
while
while

1
2
3

0
1
2
3
4
```

A.2.2 Functions

Functions in Python have a few extra features: - Positional arguments. - Keyword arguments. - Default values.

[7]:
```
# A simple function with one positional argument.
def double(x):        # Don't forget the colon!
    x = x * 2         # There are no braces or start/end, only tabs/spaces
    return x

print(double(3.14))
```

```
6.28
```

[8]:
```
# A function with one positional argument and one optional keyword argument.
def add(x, value = 10):            # value is 10 by default
    return x + value

print(add(2, value = 40))          # keyword argument
print(add('Hello', ' world!'))     # positional arguments
print(add(6))                      # default value
```

```
42
Hello world!
16
```

A.2.3 Classes

A `class` is a powerful way to bundle data (called 'properties' and/or 'attributes') and functions that operate on it (called 'methods'):

[9]:
```python
# Define a simple class:

class Employee:

    # This is the constructor, which is called when instantiating the class.
    # The `self` argument is supplied as the first parameter to every method.
    def __init__(self, name, surname, age):
        # We 'attach' the parameters to the class instance via self.
        self.name = name
        self.surname = surname
        self.age = age

        # Create new class attributes:
        self.wage = 6.15
        self.position = "intern"

    # Define a method, supplying the default `self`
    def show(self):
        print('Employee data:')
        print('- name: ' + self.name + ' ' + self.surname)
        print('- age: ' + str(self.age) )
        print('- wage: ' + str(self.wage) )
        print('- position: ' + self.position)
        print()


# Instantiate the class, providing the 3 positional arguments of __init__
# Note that `self` is omitted
worker = Employee('John', 'Doe', 20)

# We can access the class properties with dot-syntax:
print( worker.name + ' ' + worker.surname )

# We can call a method, omitting `self` again:
worker.show()

# We can change the class attributes just as we access them:
worker.age += 1
worker.wage = 7.7
worker.position = "engineer"
worker.show()
```

```
John Doe
Employee data:
- name: John Doe
- age: 20
- wage: 6.15
- position: intern

Employee data:
- name: John Doe
- age: 21
- wage: 7.7
- position: engineer
```

It's important to note that there is nothing that classes can do that couldn't also be achieved with functions! Don't worry if you do not *deeply* grasp what a `class` is—it is a more advanced concept that only becomes useful once your project grows in size. The most important things to understand are: - A class is a container for properties and methods. - You 'instantiate' a class and it will remember the properties you set. - You access the properties and methods of instances of classes using a dot.

So a class helps you 'bundle' together related properties and behaviours, which you don't really need to do for smaller projects (say < 1000 lines of code). Using classes and objects that encapsulate data and functions is commonly called *Object-Oriented Programming*, or *OOP*. Teaching OOP would take an entire tutorial on its own, and its usefulness would only become apparent for large codebases.

If you do, however, find yourself writing more than 1000 lines of code for a single project, then you're probably ready to learn and understand OOP. In that case we recommend the following tutorial—it is quite comprehensive but doesn't bore you with all the theoretical aspects of OOP: https://realpython.com/python3-object-oriented-programming/.

A.3 NumPy basics

Now let's extend our Python arsenal with some NumPy-fuelled mathematical power! The easiest way to explain it is by example:

A.3.1 NumPy array creation and basic properties

Starting from lists:

```
[10]: import numpy as np

# From a list:
x = np.array([1, 2, 3])

print(x)            # The whole array: [1, 2, 3]
print(type(x))      # Variable type: numpy.ndarray
print(x.dtype)      # Type of the elements of the array: int32
print(x.shape)      # Shape of array: (3)
print(x[0])         # Element at index 0. Yes, Python is 0-indexed, like a proper␣
 ↪language.

print('----------------')

y = np.array( [ [1.1, 2.2] , [3.3, 4.4] ] )
print(y)
print(y.dtype)
print(y.shape)
print(y[0][1])
print(y[0, 1])
```

```
[1 2 3]
<class 'numpy.ndarray'>
int64
(3,)
1
----------------
[[1.1 2.2]
 [3.3 4.4]]
float64
(2, 2)
2.2
2.2
```

Using NumPy functions:

[11]:
```
import numpy as np

# Using numpy functions:
x = np.zeros(5) ;          print(x)
y = np.ones( (2, 3) ) ;    print(y)  # (2, 3) is a tuple
z = np.eye(2) ;            print(z)
```

```
[0. 0. 0. 0. 0.]
[[1. 1. 1.]
 [1. 1. 1.]]
[[1. 0.]
 [0. 1.]]
```

A.3.2 NumPy array indexing

This part is very similar to MATLAB's arrays.

[12]:
```
import numpy as np

x = np.arange(10)
print(" Original array: ")
print(x)

print("\n Selecting slices: ")
print( x[3:] )    # Everything after index 3
print( x[3:6] )  # Indexes 3, 4, 5. It excludes 6.

print("\n Logical indexing: ")
print( x > 5 )
print( x[ x > 5 ] )
```

```
 Original array:
[0 1 2 3 4 5 6 7 8 9]

 Selecting slices:
[3 4 5 6 7 8 9]
[3 4 5]

 Logical indexing:
[False False False False False False  True  True  True  True]
[6 7 8 9]
```

[13]:
```
import numpy as np

x = np.arange(18).reshape(3, 6)
print(" Original array: ")
print(x)

print("\n Slicing rows: ")
print( x[0] )
print( x[0, :] )

print("\n Slicing columns: ")
print( x[:, 3] )

print("\n x[:, 2] is not equivalent to x[:][2] !")
print(x[:, 2])     # Selects all rows at column 2
print(x[:][2])     # Selects the whole array, THEN selects row 2
```

```
 Original array:
[[ 0  1  2  3  4  5]
 [ 6  7  8  9 10 11]
 [12 13 14 15 16 17]]

 Slicing rows:
[0 1 2 3 4 5]
[0 1 2 3 4 5]

 Slicing columns:
[ 3  9 15]

 x[:, 2] is not equivalent to x[:][2] !
[ 2  8 14]
[12 13 14 15 16 17]
```

A.3.3 Operations on NumPy arrays

Inspired by MATLAB, extended by NumPy (TM). Here are some basic operations:

[14]:
```
import numpy as np

x = np.ones( (3, 3) )
y = np.eye(3)

# Element-wise operators:
print(" Element-wise addition: ")
print(x + y)

print("\n Element-wise multiplication: ")
print(x * y)

# Scalar operations:
print("\n Scalar multiplication: ")
print(x * 2)

print("\n Add a scalar to the whole array: ")
print(x + 41)
```

```
 Element-wise addition:
[[2. 1. 1.]
 [1. 2. 1.]
 [1. 1. 2.]]

 Element-wise multiplication:
[[1. 0. 0.]
 [0. 1. 0.]
 [0. 0. 1.]]

 Scalar multiplication:
[[2. 2. 2.]
 [2. 2. 2.]
 [2. 2. 2.]]

 Add a scalar to the whole array:
[[42. 42. 42.]
 [42. 42. 42.]
 [42. 42. 42.]]
```

[15]:
```
import numpy as np

# Create a 3x3 array with random values between (0, 1)
x = np.random.random( (3, 3) ) * 10

# ... or filled with 1.
y = np.ones( (3, 3) )

# Matrix-wise operations:
print(" Matrix-wise multiplication:")
print( np.matmul(x, y) )

print("\n Matrix inverse:")
print( np.linalg.inv(x) )
```

```
 Matrix-wise multiplication:
[[24.47349246 24.47349246 24.47349246]
 [19.92173465 19.92173465 19.92173465]
 [10.99102813 10.99102813 10.99102813]]

 Matrix inverse:
[[ 0.39529359 -0.37074506 -0.1172165 ]
 [-0.0157042  -0.03341652  0.18652046]
 [-0.15817011  0.30337897 -0.1067107 ]]
```

[16]:
```
import numpy as np

x = np.arange(3)
y = np.arange(9).reshape(3, 3)

print(" Dot product of vectors: ")
print(np.dot(x, x))
print(x @ x)

print("\n Dot product of vector and matrix: ")
print(x @ y)
```

```
 Dot product of vectors:
5
5

 Dot product of vector and matrix:
[15 18 21]
```

[17]:
```
import numpy as np

x = np.arange(9).reshape(3, 3)

print(" Functions are applied element-wise by default: ")
print(np.sin(x))
```

```
 Functions are applied element-wise by default:
[[ 0.          0.84147098  0.90929743]
 [ 0.14112001 -0.7568025  -0.95892427]
 [-0.2794155   0.6569866   0.98935825]]
```

A.4 Matplotlib basics

Now let's create some simple plots using Matplotlib. The quickest way is using MATLAB-style 'stateful' plotting (stateful in the sense that Matplotlib saves a global figure for all plotting commands you run):

```
[18]: import numpy as np
      import matplotlib.pyplot as plt

      x = np.linspace(0, 2, 100)

      plt.plot(x, x, label='linear')
      plt.plot(x, x**2, label='quadratic')
      plt.plot(x, x**3, label='cubic')

      plt.xlabel('x label')
      plt.ylabel('y label')

      plt.title("Simple Plot")
      plt.grid()
      plt.legend()

      plt.show()
```

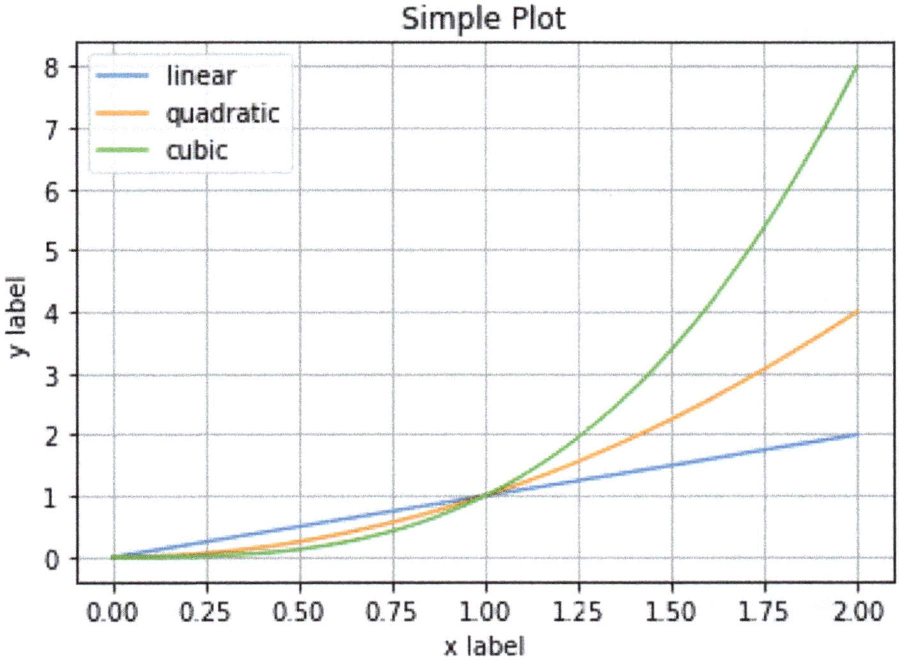

A more programmatic way to plot data would be to use individual figures and axes in the object-oriented style:

```
[19]: import numpy as np
      import matplotlib.pyplot as plt


      x = np.arange(0, 10, 0.1)
      y = np.sin(x)


      fig, ax = plt.subplots() # Creates a subplot with one figure by default


      ax.plot(x, y)
      plt.show()
```

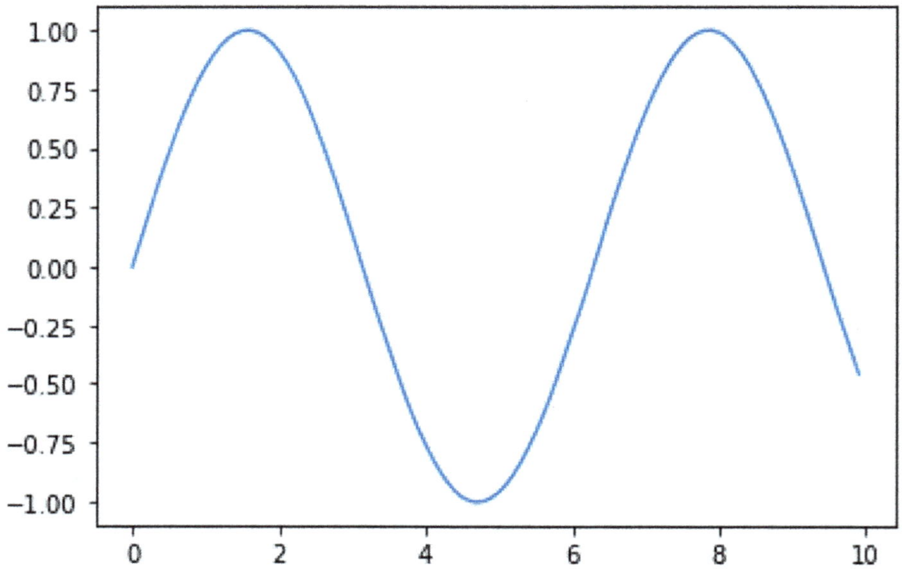

When using subplots, the object-oriented style becomes more useful:

```
[20]: import numpy as np
      import matplotlib.pyplot as plt

      x = np.arange(0, 10, 0.5)
      y = np.sin(x)

      # Create a figure and an *array of axes*
      fig, axes = plt.subplots(1, 2)

      print("Created a list of axes:")
      print(axes)

      axes[0].plot(x, y, 'go')
      axes[1].plot(x, y, marker = 'x', color = 'r')
      plt.show()
```

```
Created a list of axes:
[<AxesSubplot:> <AxesSubplot:>]
```

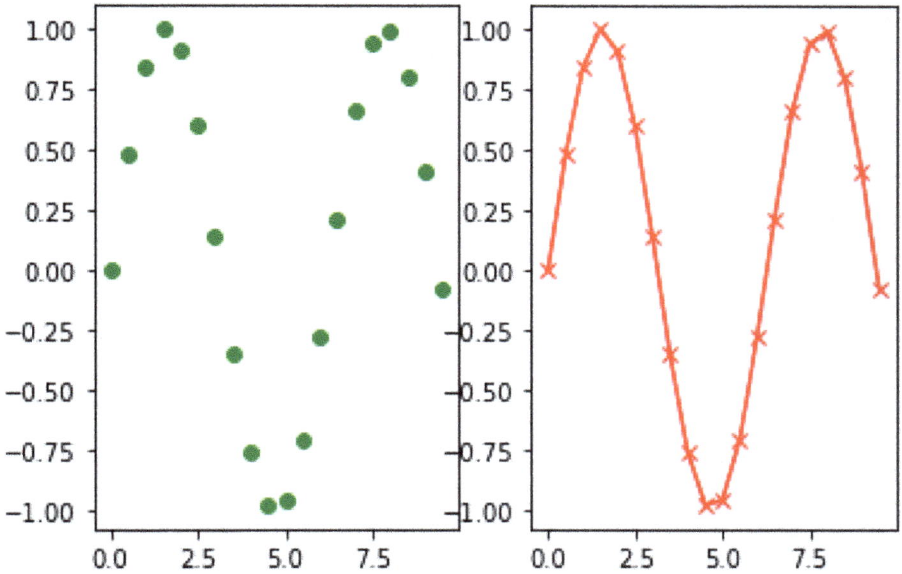

A.5 Plotly basics

Plotly is JavaScript-based plotting engine that creates beautiful figures. They can be exported as interactive webpages and can be more efficient than Matplotlib in handling big datasets.

Plotly is not a 'core' package like NumPy or Matplotlib. Let's install it using `pip`:

```
[21]: # Again, use an exclamation mark in front of system commands:

      !pip install plotly
```

```
Requirement already satisfied: plotly in
/home/andreinicusan/anaconda3/envs/peptbook/lib/python3.7/site-packages (4.14.3)

Requirement already satisfied: six in
/home/andreinicusan/anaconda3/envs/peptbook/lib/python3.7/site-packages (from
plotly) (1.16.0)

Requirement already satisfied: retrying>=1.3.3 in
/home/andreinicusan/anaconda3/envs/peptbook/lib/python3.7/site-packages (from
plotly) (1.3.3)
```

Plotly is a high-level, declarative charting library. This means that every setting of the plot can be accessed and set as a property of the instance.

```
[22]: # Set Plotly figure output (try "browser" too for interactive graphs)
import plotly
plotly.io.renderers.default = "png"


import plotly.graph_objects as go
import numpy as np

data_x = np.random.random(100) * 10
data_y = np.random.random(100) * 10
data_z = np.random.random(100) * 10

fig = go.Figure()

# A trace is an individual graph. Multiple traces can be
# displayed on a single figure.
trace = go.Scatter3d(
    x = data_x,
    y = data_y,
    z = data_z,
    mode = 'markers',
    marker = dict(
        color = data_z,    # Encode z as a colorbar
    )
)

fig.add_trace(trace)
fig.show()
```

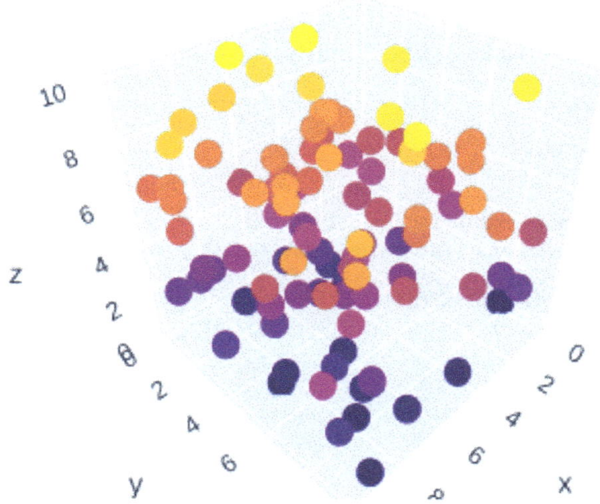

A.6 Extra

This section is entirely optional. It shows some more advanced, but oftentimes useful features of Python. We recommend you skim through it, such that you are at least familiar with the concepts. If a need for them arises, you'll know where to find them:

A.6.1 Other data structures

Some other data structures are: - `tuple` : an ordered, immutable collection. - `dict` : a dictionary, mapping keys to values. - `set` : a set, like the mathematical definition—i.e. no repeating values.

Also, `None` is a special value that signifies an undefined or unset variable. It can be used as a default function parameter.

Each of them defines some special operations:

[23]:
```
# Tuple:
x = ("Sam", 3)
print('Tuple x: ', x)

# Dictionaries can be defined in two ways.
y = dict(sam = 5, john = 18)
z = {'sam': 5, 'john': 18}
print('\nDict y: ', y)
print('Dict z: ', z)

# Dictionaries have a special operation to check the existence of a key:
print("Key 'sam' in dict y: ", ('sam' in y))
print("Key 'peter' in dict y: ", ('peter' in y))

# Sets
t = {1, 2, 3, 1, 1}     # repeated entries are ignored
print("\nSet 't': ", t)

# The existence of an element in a set is checked with 'in', again:
print("Element 1 in set t: ", (1 in t))
print("Element 5 in set t: ", (5 in t))
```

```
Tuple x:  ('Sam', 3)

Dict y:  {'sam': 5, 'john': 18}
Dict z:  {'sam': 5, 'john': 18}
Key 'sam' in dict y:  True
Key 'peter' in dict y:  False

Set 't':  {1, 2, 3}
Element 1 in set t:  True
Element 5 in set t:  False
```

[24]:
```
# Using None as default function parameter:
def database_query(query, database = None):
    # Use 'is' for checking specifically for None,
    # rather than False
    if database is None:
        print("Database is unset!")
        return
    if database is not None:
        print("Database is ", (database))
        query

database_query("select")
database_query("select", database = "test")
```

```
Database is unset!
Database is  test
```

A.6.2 Documentation

The documentation of a function/class/module can be accessed in a couple of ways:

[25]:
```
import numpy as np

# Using the `help` method
help(np.linalg.inv)

# Short documentation using a question mark in front of the method:
#? np.linalg.inv

# Long documentation:
#?? np.linalg.inv


# Using the __doc__ method
#print(np.linalg.inv.__doc__)
```

```
Help on function inv in module numpy.linalg:

inv(a)
    Compute the (multiplicative) inverse of a matrix.

    Given a square matrix `a`, return the matrix `ainv` satisfying
    ``dot(a, ainv) = dot(ainv, a) = eye(a.shape[0])``.


    Parameters
    ----------
    a : (..., M, M) array_like
        Matrix to be inverted.


    Returns
    -------
    ainv : (..., M, M) ndarray or matrix
        (Multiplicative) inverse of the matrix `a`.

    Raises
    ------
    LinAlgError
        If `a` is not square or inversion fails.

    See Also
    --------
    scipy.linalg.inv : Similar function in SciPy.

    Notes
    -----

    .. versionadded:: 1.8.0


    Broadcasting rules apply, see the `numpy.linalg` documentation for
    details.

    Examples
    --------
```

```
>>> from numpy.linalg import inv
>>> a = np.array([[1., 2.], [3., 4.]])
>>> ainv = inv(a)
>>> np.allclose(np.dot(a, ainv), np.eye(2))
True
>>> np.allclose(np.dot(ainv, a), np.eye(2))
True

If a is a matrix object, then the return value is a matrix as well:

>>> ainv = inv(np.matrix(a))
>>> ainv
matrix([[-2. ,  1. ],
        [ 1.5, -0.5]])

Inverses of several matrices can be computed at once:

>>> a = np.array([[[1., 2.], [3., 4.]], [[1, 3], [3, 5]]])
>>> inv(a)
array([[[-2.  ,  1.  ],
        [ 1.5 , -0.5 ]],
       [[-1.25,  0.75],
        [ 0.75, -0.25]]])
```

You can easily document your own code, and Python will automatically generate the documentation strings:

```
[26]: # Show how documentation works. Triple quotes are comments that
      # span multiple lines.

      def doc_test(x):
          '''The first line is the short documentation. This is a quick description of
          the function.

          This string is only shown when using the single question mark in front of
          the function name. The implementation below is only shown when using two
          question marks.
          '''

          print("This is the implementation of the function, only shown with ??.")

      # Only check one at a time:
      ? doc_test
      #?? doc_test
```

```
Signature: doc_test(x)
Docstring:
The first line is the short documentation. This is a quick description of
the function.

This string is only shown when using the single question mark in front of
the function name. The implementation below is only shown when using two
question marks.
File:      ~/Prog/PEPT/PEPT-Book/Appendix/<ipython-input-26-013b43a50ccb>
Type:      function
```

A.6.3 Python virtual environments

Once you start building larger applications, it is important to have a reliable programming environment. What happens if: - Your app only works on one version of the Python interpreter. - Your app only works with older versions of some modules.

The set of Python and modules' versions forms a 'Virtual Environment'. You can 'freeze' the state of your Virtual Environment and know that what you are developing surely works on those versions.

This is the main idea. There are quite a few ways of managing multiple virtual environments (but conda is probably the nicest *virtualenv* manager). I will let you read about them once the need for them arises.

A.7 Common *how-tos*

This is the final chapter of this tutorial, congratulations!

We will show you the 'easy' (or 'right', following Python's philosophy) way of doing a few usual tasks. This should save you a few trips to *StackOverflow* and NumPy's documentation.

A.7.1 Formatting strings

We frequently need to output data in a pretty format. Thankfully, Python has some nice ways of printing stuff:

[27]:
```
# Format strings to output data nicely:
import numpy as np

num = 12
lst = [1, 2, 3]
vec = np.array(lst)

# Supply multiple arguments to print:
print("Number: ", num, "\nList: ", lst, "\nVector: ", vec, "\n")

# Use the format identifier. Add curly braces {} inside a string,
# then supply the variables to format:
print("Number: {}\nList: {}\nVector: {}\n".format(num, lst, vec))

# Use f-strings. These are the shortest:
print(f"Number: {num}\nList: {lst}\nVector: {vec}\n")
```

```
Number:  12
List:  [1, 2, 3]
Vector:  [1 2 3]

Number: 12
List: [1, 2, 3]
Vector: [1 2 3]

Number: 12
List: [1, 2, 3]
Vector: [1 2 3]
```

A.7.2 Iteration + list comprehensions

If you start using C++-style for/while loops, stop and think if there isn't a more concise (i.e. better) way of doing what you want. Python has some nice features for different types of iteration:

[28]:
```
# Iterate using an index range:
x = 0
for i in range(5):
    x += 1

print("Looped using range: x = {}\n".format(x))

# Iterate through the elements of a list, while also
# having the index:
data = ["apple", "pears", "oranges"]
for index, elem in enumerate(data):
    print(f"At index {index}, we have {elem}")
print() # For the newline
```

```
Looped using range: x = 5


At index 0, we have apple
At index 1, we have pears
At index 2, we have oranges
```

List comprehensions provide a concise and powerful way of generating or accessing iterable data.

The main syntax is:

```
[ expression for element in iterable if condition ]
```

[29]:
```
# For-comprehensions are powerful ways of accessing data
data = "The world is definitely not deterministic".split()
print(data)

# Collect words whose length is > 4
new_data = [elem for elem in data if len(elem) > 4]
print(new_data)

# You can also use list comprehensions to generate data
squares = [x ** 2 for x in range(4)]
print(squares)
```

```
['The', 'world', 'is', 'definitely', 'not', 'deterministic']
['world', 'definitely', 'deterministic']
[0, 1, 4, 9]
```

[30]:
```
# List comprehensions are great for concise data extraction:
# Generate a list of arrays:
import numpy as np

data = [np.arange(10).reshape(2, 5)] * 4
print(data)

# Extract the second row of every array
# and create a new NumPy array
rows = np.array( [arr[1] for arr in data] )
print(rows)
```

```
[array([[0, 1, 2, 3, 4],
       [5, 6, 7, 8, 9]]), array([[0, 1, 2, 3, 4],
       [5, 6, 7, 8, 9]]), array([[0, 1, 2, 3, 4],
       [5, 6, 7, 8, 9]]), array([[0, 1, 2, 3, 4],
       [5, 6, 7, 8, 9]])]
[[5 6 7 8 9]
 [5 6 7 8 9]
 [5 6 7 8 9]
 [5 6 7 8 9]]
```

A.7.3 Dynamic array creation

We frequently need to create arrays dynamically, without knowing how much data there will be (e.g. reading lines from a file). Python `lists` have variable size, but are not great for maths, while NumPy `ndarrays` are very powerful but are not resizable. How can we achieve the best of both worlds? Combining them, of course:

[31]:
```
# First append stuff to a dynamic list
import numpy as np


data = []
# Random number of iterations:
for i in range(int(2 + 10 * np.random.random())):
    data.append(np.ones(5) * i)


print(data) # list of 1D arrays
# After reading stuff into a list, transform it into a numpy array
data = np.array(data)

print(data) # 2D array
```

```
[array([0., 0., 0., 0., 0.]), array([1., 1., 1., 1., 1.]), array([2., 2., 2.,
2., 2.])]
[[0. 0. 0. 0. 0.]
 [1. 1. 1. 1. 1.]
 [2. 2. 2. 2. 2.]]
```

A.7.4 Saving and loading data

We frequently need to load data from a file, do something to it, then save it somewhere else. In engineering, we most often handle big piles of numerical data that is usually saved as in a CSV (comma-separated-values) format. Again, Python has a few tricks up its sleeve.

Reading or writing files line by line is laborious and oftentimes slow. NumPy has some optimised functions for this:

[32]:
```
import numpy as np

# Generate some random data
data = np.random.random( (10, 5) ) * 10
print("Generated data: \n", data)


# Save the data as .csv using savetxt
filepath = "pretty_data.csv"
np.savetxt(filepath, data)


# We can afterwards load the data, with a few options
loaded_data = np.loadtxt(filepath, skiprows = 2, max_rows = 3)
print("\nLoaded data: \n", loaded_data)
```

```
Generated data:
 [[9.44642986 9.40445387 9.80144088 4.88482918 7.69583183]
 [4.79478469 2.91000579 9.06200523 8.44680124 6.90280663]
 [8.51620386 9.67796492 4.44184629 4.0658947  6.84146845]
 [2.74897723 9.61950009 5.0128295  9.21609805 2.82352351]
 [1.03235896 7.39804024 0.78871263 1.58392272 7.53382301]
 [3.76552982 8.09121346 4.96635772 0.14574747 8.68161159]
 [9.56584111 6.86981845 1.68892208 8.41088842 5.75440539]
 [1.72628578 8.55631365 5.02452022 1.1191789  2.16620918]
 [5.29169923 6.58650424 3.76816671 5.23018613 1.50814383]
 [9.66246611 7.42195788 2.74245101 1.89319542 4.20191717]]
Loaded data:
 [[8.51620386 9.67796492 4.44184629 4.0658947  6.84146845]
 [2.74897723 9.61950009 5.0128295  9.21609805 2.82352351]
 [1.03235896 7.39804024 0.78871263 1.58392272 7.53382301]]
```

A.7.5 Sorting

Sorting data is terribly common. NumPy has one of the best sorting functions available, called TimSort (read about it, it's great). For example, we often want to sort the rows of an array based on one of the columns (e.g. time from a simulation):

[33]:
```
# Some typical sorting
import numpy as np

vec = np.random.random(10) * 10
print("Vector of random values between 0 and 10: \n", vec)
print("Sorted: \n", np.sort(vec))

arr = np.random.random((7, 5)) * 10
print("\nArray of random values between 0 and 10: \n", data)

print("\nNumPy arrays have the .argsort function. We can use it on, say, column␣
↪3: \n", data[:, 2].argsort())
print("Those are the indices that the elements should have such that the vector␣
↪would be sorted.")

print("\nUsing argsort and logical indexing, we can sort an array based on, say,␣
↪column 3: \n", data[ data[:, 2].argsort() ])
```

```
Vector of random values between 0 and 10:
 [5.00738119 7.13273267 1.50111289 1.79344121 8.11077912 7.23466646
 8.97428667 8.71313566 3.13670247 7.83439306]
Sorted:
 [1.50111289 1.79344121 3.13670247 5.00738119 7.13273267 7.23466646
 7.83439306 8.11077912 8.71313566 8.97428667]
Array of random values between 0 and 10:
 [[9.44642986 9.40445387 9.80144088 4.88482918 7.69583183]
 [4.79478469 2.91000579 9.06200523 8.44680124 6.90280663]
 [8.51620386 9.67796492 4.44184629 4.0658947  6.84146845]
 [2.74897723 9.61950009 5.0128295  9.21609805 2.82352351]
 [1.03235896 7.39804024 0.78871263 1.58392272 7.53382301]
 [3.76552982 8.09121346 4.96635772 0.14574747 8.68161159]
 [9.56584111 6.86981845 1.68892208 8.41088842 5.75440539]
 [1.72628578 8.55631365 5.02452022 1.1191789  2.16620918]
 [5.29169923 6.58650424 3.76816671 5.23018613 1.50814383]
 [9.66246611 7.42195788 2.74245101 1.89319542 4.20191717]]


NumPy arrays have the .argsort function. We can use it on, say, column 3:
 [4 6 9 8 2 5 3 7 1 0]
Those are the indices that the elements should have such that the vector would
be sorted.

Using argsort and logical indexing, we can sort an array based on, say, column
3:
 [[1.03235896 7.39804024 0.78871263 1.58392272 7.53382301]
 [9.56584111 6.86981845 1.68892208 8.41088842 5.75440539]
 [9.66246611 7.42195788 2.74245101 1.89319542 4.20191717]
 [5.29169923 6.58650424 3.76816671 5.23018613 1.50814383]
 [8.51620386 9.67796492 4.44184629 4.0658947  6.84146845]
 [3.76552982 8.09121346 4.96635772 0.14574747 8.68161159]
 [2.74897723 9.61950009 5.0128295  9.21609805 2.82352351]
 [1.72628578 8.55631365 5.02452022 1.1191789  2.16620918]
 [4.79478469 2.91000579 9.06200523 8.44680124 6.90280663]
 [9.44642986 9.40445387 9.80144088 4.88482918 7.69583183]]
```

A.8 The end

If you reached this point, congratulations! You're officially a Python programmer, ready to take on the beautiful, vast world of data science. We covered a lot of ground here, from introducing the `Python` interpreter, the `pip` package manager and basic programming constructs, all the way to using NumPy arrays and list comprehensions to slice and dice our data, be it nicely-formatted or not. Because that's what data science is all about, right?

Don't forget that it's perfectly fine to not remember everything in this tutorial by heart—all that matters is knowing where to get the information you need, using `help` or *StackOverflow* or your neighbour. From here on, the best way to learn is by practice: start coding a few projects you might be interested in (there are lots of examples on Google) and continue to develop your programming skills; the hardest, but most rewarding aspect isn't knowing a programming language, but **making the computer do the work** using the right tools at your disposal—be they programming languages or libraries.

Glossary

Activity:	The level of radioactivity possessed by a tracer. In the specific context of PEPT, this typically refers to the number of positrons and thus gamma photon *pairs*, as opposed to individual gamma photons, emitted per second
ASIC:	Application-specific integrated circuit—i.e. a chip that was specially designed for a given purpose, such as identifying single events in PET / PEPT systems.
Attenuation:	The decrease in the intensity of (i.e. number of photons in) a beam of electromagnetic radiation (e.g. x-rays or gamma-rays) as it passes through a given medium.
Becquerel (Bq):	The SI derived unit of measure characterising the number of nuclei undergoing radioactive decay in a quantity of radioactive material. For example, a PEPT tracer particle with an activity of 20 MBq undergoes 20 million radioactive disintegrations per second. Indeed, Becquerels represent frequency (being equivalent to s^{-1} or Hz), but this special name was introduced to avoid potentially dangerous measurement mistakes.
Bidispersity:	A bidisperse or 'binary' granular system is one containing exactly two distinct 'species' of particle (see also Polydispersity).
Binary system:	See 'Bidispersity'
Body force:	A force, such as that occurring due to a gravitational or electromagnetic field, that acts throughout the volume of a system (or a body, if you will).
Coincidence event:	The recording, by a gamma-camera system, of two photons within a time window short enough that said photons are likely to have been produced simultaneously, i.e. due to a positron annihilation.
Coincidence rate:	The number of coincidence events (see above) registered by a gamma camera per unit time.
Collimation:	The use of a collimator (see below) to ensure that a detector receives only photons travelling in or close to a specific direction (typically orthogonal to the detector face).
Collimator:	A device which filters a beam of x-rays or gamma rays such that all photon trajectories are approximately parallel, travelling in a

single chosen direction (in the case of nuclear imaging, normally perpendicular to the face of a detector).

Cyclone: A device which uses the centrifugal forces induced by cyclonic fluid flows for the separation of particulate solids from fluids.

Detector sensitivity: The number of events detected per unit time per unit of activity (i.e. the fraction of photons emitted by a source that are successfully detected by a given camera).

Direct activation: The introduction of radioactivity into a material via the direct conversion of stable nuclei in said material into radioactive nuclei.

Eddy (current): In the context of fluid mechanics, an eddy current—often referred to simply as an eddy—is a current whose flow direction diverges from that of the general (i.e. 'bulk' or 'net') flow within a system.

Electrical Permittivity: See 'Permittivity'.

Ergodicity: An ergodic system is one whose time-averaged statistical behaviour is equivalent to its ensemble-averaged behaviour. In our present context, it is more clear to explain it in terms of a system for which the average behaviour of a single particle over a suitably long time is representative of the behaviour of an entire system of identical particles.

False coincidence: A coincidence event (see definition above) which does not correspond to a positron-electron annihilation pair, or otherwise does not yield a line of response passing through the point of annihilation of such a pair.

FoV: Field of view, the experimental volume which is actively being imaged by a detector system.

FPGA: Field-programmable gate array, a circuit that can be configured by end-users 'in the field'.

FWHM: Full-width at half-maximum, the width of a bell-like distribution at half the maximum probability.

Half-life: The time taken for a sample of material to reduce to half of its initial quantity. In nuclear science, it represents the time needed for half of the radioactive nuclei in a sample to undergo radioactive decay.

Hysteretic: A hysteretic material—granular media being prime examples—is one whose behaviour depends upon its history.

Indirect activation: The application of radioactivity to a particle through the absorption (or otherwise application) of an already radioactively-labelled material.

Granular material: A system composed of multiple, discrete, macroscopic solids.

Inverse square law: The amount of x-ray or gamma radiation received from a point-like radiation source (e.g. a tracer particle) decreases as the square of the distance from the source to the receiving object.

Labelling: Marking (either literally or figuratively) a particle or fluid such that its motion may be recorded. In the context of PEPT, this is usually achieved via the attachment of a radioactive substance (see also 'Radiotracer').

Line of Response (LoR): The line, or vector, representing the path taken by a pair of back-to-back gamma photons emitted from a positron-electron annihilation event.

Mass absorption coefficient:	The rate at which a given form of radiation with a given initial energy is absorbed by a given material.
Monodispersity:	A monodisperse granular system is one which contains only one single 'species' of particle—i.e. a system of particles all of whose sizes, densities, geometries and other physical properties can be considered effectively identical.
Non-invasive:	In the context of an experimental measurement, a non-invasive reading is one which acquires data from a system without disrupting its usual behaviour. As an example, inserting a physical sensor into a flowstream is an example of an invasive measurement, while performing PEPT using a tracer that is physically identical to others in the system would be non-invasive.
Occlusion:	The 'loss' of one or more tracer particles during PEPT (or indeed other) imaging, for example when the count rate registered by a particle is too low for it to be successfully detected.
Permittivity:	The (electrical) permittivity of a material is a measure of how strongly said material is polarised by an applied electric field. It can, in essence, be thought of as a measure of how much energy can be stored by a material in an electric field.
Polydispersity:	Any granular system which is not monodisperse, i.e. which contains more than one distinct 'species' of particle, can be considered polydisperse. However, the term is typically only used for systems comprising more than two or three distinct species, the terms binary/bidisperse or ternary/tridisperse being more commonly used to describe systems with lower orders of polydispersity.
Positron camera:	A detector system designed to detect the interactions of gamma rays emitted from positron-electron annihilation events, and thus reconstruct their trajectories as required for PEPT (and indeed PET) imaging.
Powder characterisation:	The use of experimental measurements to extract information regarding the physical properties—typically *bulk* properties—of a powder or granulate. Examples of powder characterisation devices are provided in section 4.4.
Quasi-two-dimensional:	A system whose constituents themselves are three-dimensional (e.g. spheres as opposed to discs) but whose dimensions are constrained such that all, or at least a majority, of motion occurs in only two of three spatial dimensions.
Radioisotope:	A form (isotope) of a given element which undergoes radioactive decay.
Radiolabelling:	Attaching radioactivity to a particle or fluid such that it may be tracked using a suitable imaging technique. See also 'labelling'.
Radiotracer:	A particle (or, in the case of PET, fluid) 'labelled' with a radioactive material such that its trajectory may be followed using a suitable imaging technique.
Secondary radiation:	Radiation emitted by a tracer or a system being imaged which does not arise from β^+ annihilation, and thus cannot be used for the location of tracers. Note that this term can also be defined more broadly—the specific definition provided here is applicable only to positron imaging techniques such as PEPT.

Segregation:	The spontaneous separation of the individual components of a dynamic binary or polydisperse system based on differences in particles' sizes, shapes, densities and/or other physical properties.
Sensitivity:	See *detector sensitivity*.
Slurry:	A semi-solid mixture of particles immersed/suspended in a fluid.
Steady state:	A state in which a system's behaviour does not vary significantly with time—for example a stream of fluid circulating at a constant rate and in a consistent pattern within a stirred container.
Ternary system:	A ternary granular system is one containing exactly three distinct 'species' of particle.
Tomography:	From the Greek *tomos* meaning slice or section; the process of reconstructing a full three-dimensional image of an object from a series of two-dimensional projections of said object.
Tracer:	See 'Radiotracer'.
Tracer activity:	In the context of PEPT, the level of β^+ radioactivity possessed by a tracer particle—i.e. the number of positrons emitted per unit time.
Z number:	The number of protons in the nucleus of an atom.

www.ingramcontent.com/pod-product-compliance
Ingram Content Group UK Ltd.
Pitfield, Milton Keynes, MK11 3LW, UK
UKHW051911100926
8556IPUK00005B/848

9 780750 330695